Instrumental Analysis

Instrumental Analysis

Robert M. Granger II
Sweet Briar College

Hank M. Yochum
Sweet Briar College

Jill N. Granger
Western Carolina University

Karl D. Sienerth
Elon University

New York   Oxford

OXFORD UNIVERSITY PRESS

Oxford University Press is a department of the University of Oxford.
It furthers the University's objective of excellence in research,
scholarship, and education by publishing worldwide.

Oxford is a registered trade mark of Oxford University Press
in the UK and certain other countries.

Published in the United States of America by Oxford University Press
198 Madison Avenue, New York, NY 10016, United States of America.

© 2017 by Oxford University Press

CIP data is on file at the Library of Congress
ISBN 978–0–19–994231–2

Printing number : 9 8 7 6 5 4 3 2 1

Printed by Edwards Brothers Malloy, United States of America

Brief Contents

Contents

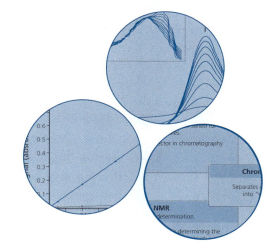

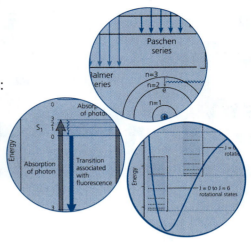

Chapter 3   An Introduction to Optics   49

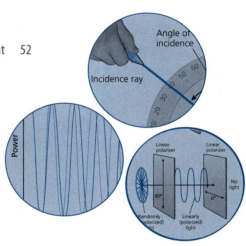

Chapter 4   An Introduction to Instrumental Electronics   89

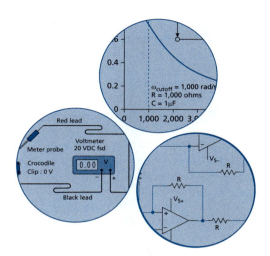

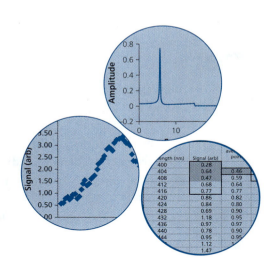

SECTION B: SPECTROSCOPY & SPECTROMETRY

Ultraviolet and Visible Absorption Techniques

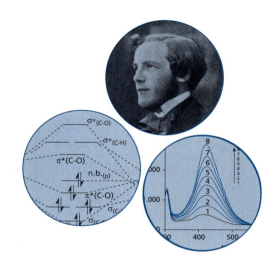

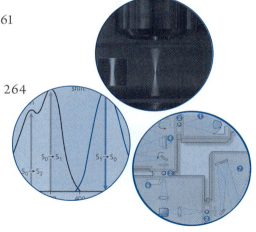

Chapter 9   Atomic Emission Spectroscopy   277

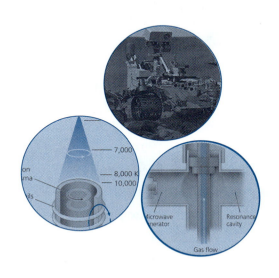

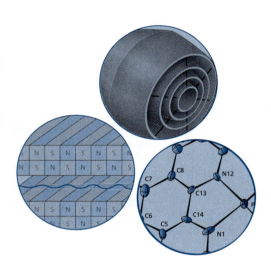

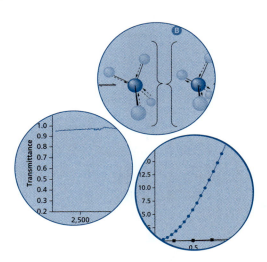

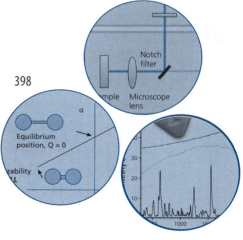

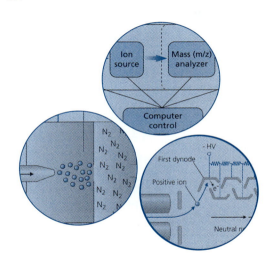

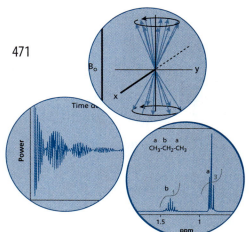

SECTION C: SEPARATION SCIENCE

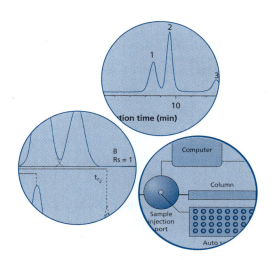

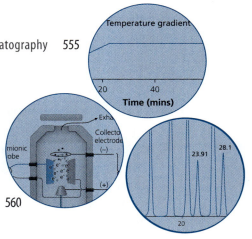

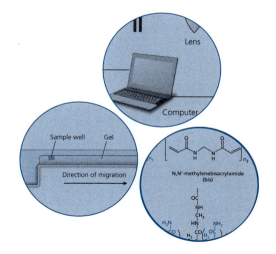

SECTION D: ELECTROANALYTICAL TECHNIQUES

Chapter 18   Potentiometry and Probes   633

Chapter 19   Analytical Voltammetry   663

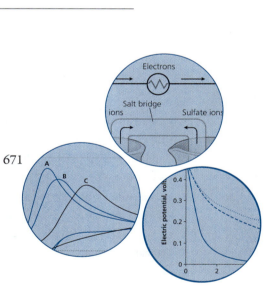

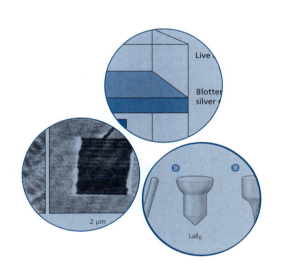

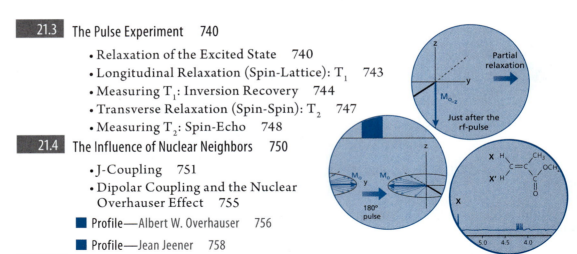

Chapter 22   Statistical Data Analysis   779

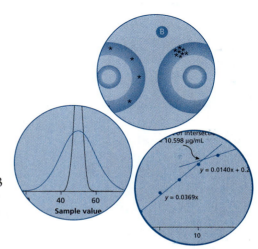

Preface

Instrumental Analysis provides a rigorous, modern, and engaging coverage of chemical instrumentation, written with the undergraduate student in mind. At its core, this book considers the underlying theory, instrumental design, applications, and operation of *spectroscopic, electroanalytical, chromatographic,* and *mass spectral* instrumentation. Also included is a discussion of *surface imaging, thermal analysis,* modern *statistical tools,* and *signal processing.*

In order to give the student the requisite skill in identifying the comparative advantages and disadvantages in choosing one analytical technique over another, this textbook includes direct comparisons between the differing analytical techniques with a discussion of how these choices affect the interpretation of the data in its final form. In addition, today's analyst must understand the trade-offs resulting from the advancements in the miniaturization of electronics. Many manufacturers of chemical instrumentation now provide "labs-in-a-box," field kits, and hand-held sensors whose predecessors would have filled entire laboratories in the past. These portable instruments provide fast, on-site analysis, often with minimal sample preparation and waste—although sometimes at the expense of data quality. When appropriate, we make side-by-side comparisons between the bench-top and portable versions of modern analytical instrumentation.

Organization

Instrumental Analysis is organized into five sections.

- **Section A: Foundations.** The textbook starts with an introductory chapter titled "The Analyst's Toolbox," which establishes a perspective of considering instrumental techniques as tools used by the analyst to answer questions. The competent analyst must master the use of multiple tools in order to solve complex chemical problems. This theme is reinforced regularly throughout the textbook. The remainder of this section is devoted to a review of foundational knowledge and includes the underlying quantum mechanics of spectroscopy, optics, basic electronics, and signal processing. Although many undergraduate chemistry programs will have covered these topics in a prerequisite course, a review of these topics is helpful prior to delving into the individual instrument types.

- **Section B: Spectroscopy & Spectrometry.** This nine-chapter section covers the core topics of instrumental analysis, beginning with a discussion of electronic absorption spectroscopy (molecular and atomic) followed by electronic emission spectroscopy (molecular and atomic). This is followed by an expanded discussion of X-ray spectroscopic techniques, and we have been deliberate in discussing the role of X-ray spectroscopy as an elemental analysis technique compared against atomic absorption spectroscopy and atomic emission spectroscopy. We follow our discussion of X-ray spectroscopy with two separate chapters on vibrational spectroscopy: infrared (IR) and Raman. We introduce the Fourier transform in our discussion of IR spectroscopy with several activities that allow students to perform a Fourier transform using Microsoft Excel. With the advent of miniaturized lasers, the utility of Raman spectroscopy has greatly improved in the past few decades; therefore, we have devoted an entire chapter to the discussion of Raman spectroscopy. Next, we discuss mass spectrometry (MS) using a modular approach that outlines ion generation, mass selection, and detection as separate events. We present the most common MS applications as "plug and play" models, thus allowing the students to "synthesize" a great many different MS instrument types. We conclude this section with a discussion of nuclear magnetic spectroscopy. Although nuclear magnetic resonance (NMR) has most likely been discussed in courses on organic chemistry, our textbook does not focus on data interpretation but rather on instrument design, with an emphasis on how the signal is generated, detected, and processed. (An advanced NMR chapter on pulse sequences and two-dimensional signal processing is also included in Section E.)

- **Section C: Separation Science.** Before a chemical analysis can take place, the analyte must be isolated from the matrix. This section discusses the three most common instrumental separation techniques: liquid chromatography, gas chromatography, and electrophoresis. Each chapter begins with the fundamental principles governing the technique, then considers the most common applications, and finally discusses the "state of the art" for each technique.

- **Section D: Electroanalytical Techniques.** We begin this section with a chapter covering potentiometry and probes followed by a separate chapter on analytical voltammetry. Because the analytical application of potentiometry has benefited greatly from the evolution of miniaturized electronics, we have emphasized the design and application of *probes* in our discussion of potentiometry. Today, probes find regular use in medicine, environmental monitoring, and quality-control systems, as well as in law enforcement, forensics, and security systems. Analytical voltammetry provides a quick, inexpensive, and reliable means of obtaining both quantitative and qualitative information about an electroactive system. Voltammetry can provide important evidence regarding the mechanism by which systems undergo chemical and electrochemical reactions. Importantly, voltammetric experiments can be scaled down to the intracellular level. In the chapter on analytical voltammetry, we discuss the basics of voltammetry, include some common waveforms used in research laboratories, and demonstrate the utility of and limits imposed by microscale and nanoscale electrodes.

- **Section E: Additional Topics**. We recognize that most teachers will not have sufficient time to cover all of the chapters in this textbook. However, we end *Instrumental Analysis* with three chapters that represent less common, yet important, topics. At the very least, more aggressive students can read these chapters on their own time or the text may provide useful background for a laboratory project. The first chapter in this section considers material and surface analysis, in which we discuss important microscopic, thermoanalytical, and mechanical stress techniques. The next chapter specifically details the application of pulse sequences as they apply to advanced NMR techniques. Students are taught the significance of T_1 and T_2 relaxation, and the basics of programming two-dimensional experiments such as *correlation spectroscopy,* or COSY; *total correlation spectroscopy,* or *TOCSY;* and *nuclear Overhauser effect spectroscopy,* or NOESY. Finally, we end this section with a review of statistical data analysis. We placed this chapter last in recognition of the fact that many schools teach this material in a separate course, while others choose to include it in their instrumental analysis course.

Teaching and Learning Features

- **Compare and Contrast Boxes:** As the book develops, strategically placed *Compare and Contrast boxes* allow for a direct comparison of several different instrumental techniques. These boxes have been specifically designed to help the student understand the different trade-offs one makes when choosing one technique over another. When appropriate, Compare and Contrast features are presented as tables that allow for direct comparison of such things such as resolution, speed of analysis, detection limits, purchase cost, operation cost, maintenance cost, sample preparation time, and destructive versus nondestructive analysis. The feature may be presented as a Venn diagram that highlights the similarities and differences in an instrument's design and/or utility. Students are also prompted to think critically about the similarities and differences as they move from one section to another.

- **Profile Boxes:** To capture the student's imagination, each chapter begins with a real-world application and additional *Profile boxes* are sprinkled throughout each chapter. These boxes are designed to answer the "so what?" and "who cares?" questions often asked by students.

- **Activity Boxes:** When feasible, the text includes *Activity boxes* that allow a student to recreate a physical phenomenon or to build a working instrument component using easily obtainable materials. These activities have been designed so that the student can perform the activity outside of the classroom; however, they are also suitable as classroom demonstrations or small group activities.

- **Worked Examples:** Each chapter contains several worked examples that walk the student through key calculations and/or exercises. Additional exercises are also included at the end of each chapter.

- **Sample Problems:** Several practice Problems are place in close proximity to each worked Example feature, allowing the student to acquire their own expertise.

- **Further Reading:** Each chapter ends with a bibliography of influential texts and/or journal articles pertaining to that topic.

- **Exercises:** Exercises that reinforce key concepts are strategically placed at the end of each chapter and offer an opportunity for further reinforcement of key ideas.

Support Package

Oxford University Press offers a comprehensive ancillary package for instructors who adopt *Instrumental Analysis*. The Ancillary Resource Center (ARC), located at **www.oup-arc.com/granger**, contains the following teaching tools:

- **Digital image library.** The image library includes electronic files in PowerPoint format of every illustration, graph, photograph, figure caption, and table from the text—in both labeled and unlabeled versions. Images have been enhanced for clear projection in large lecture halls.

- **Lecture notes for each chapter.** Editable lecture notes are in PowerPoint format. Each chapter's presentation includes a succinct outline of key concepts and incorporates the graphics from the chapter.

- **Instructor's Solutions Manual.** The solutions manual provides step-by-step answers to the chapter exercises.

Acknowledgments

From Robert Granger: I would like to thank my coauthors for all of their encouragement, technical expertise, and editorial assistance, as well for their collaboration. I would also like to acknowledge Sweet Briar College, particularly for a semester-long sabbatical in support of this textbook. In addition, I am grateful to the librarians and staff at Sweet Briar College, especially Joe Malloy, for their assistance in acquiring the numerous manuscripts needed during the research phase of this project. I and my coauthors are also indebted to Shelbie Filson for her clerical assistance. Lastly, my family has been immensely patient, encouraging, and enduring through the many years in which I have spent weekends, evenings, and vacations hunched over my laptop.

From Jill Granger: I am grateful for the opportunity of being included as part of the writing team for this book project, and I would like to thank my colleagues for their perseverance, collaboration, patience, and good humor. This project afforded me the chance to draw heavily on the excellent education I received as a graduate student at Purdue University, and I am thankful to the professors who instilled in me a curiosity and enthusiasm for the ways in which chemical problems can be solved analytically. Likewise, I would like to thank my husband, Rob, whom I also found at Purdue, for our two-plus decades of partnership, as well as our four children and extended family for their encouragement, understanding, and support during this six-year adventure in which "the book" became another member of our family. Also, I would like to thank the Sweet Briar College community for the flexibility and encouragement to engage in this work alongside my other responsibilities. Finally, I would like to acknowledge the support of Western Carolina University, which has provided me with a new academic home as we go into final production.

From Karl Sienerth: I would like to express sincere appreciation of the strong support for this project by the Elon University Department of Chemistry, particularly in the form of release time and creative class scheduling to provide me with much-needed time to work on the book. Furthermore, the office of the Dean of the Elon College of Arts and Sciences has been very supportive throughout the multiyear effort, most recently in the form of funding for technical editorial support. Finally, I deeply appreciate that in all I attempt, I am encouraged and inspired by my wife, Sonya.

From Henry (Hank) Yochum: I would like to acknowledge Sweet Briar College for supporting this project. In particular, I would like to also acknowledge the support of faculty colleagues from the Department of Engineering and Physics. I am also exceptionally thankful to my wife, Marcia, and son, Max, for their patience and support.

From the entire team: We would all like to thank our editorial team at Oxford University Press (OUP), especially the support of Jason Noe, Senior Editor, who has been with us from the project's inception and who provided the research and encouragement that got us started, as well as valuable team meetings that helped us to incorporate reviewers' feedback into making this a much better book. Also, we are grateful for the support of Caitlin Kleinschmidt, the OUP editorial assistant who got us off to a good start, and Andrew Heaton, the OUP assistant editor for Chemistry, who has managed our project through completion and provided the organizational structure to move us through the writing and review stages and into production; for the support of Barbara Mathieu, the production editor who has helped turn our manuscript into a book; and for the support of Rob Duckwall, Caitlin Duckwall, Craig Durant, and their colleagues at Dragonfly Media Group, who have helped us to make the world of instrumental analysis more visually appealing and educationally valuable for the students who will use the book. We would also like to thank Patrick Lynch, editorial director; John Challice, vice president and publisher; Frank Mortimer, director of marketing; David Jurman, marketing manager; Ileana Paules-Bronet, marketing assistant; Jolene Howard, director of market development; Meghan Daris, market development associate; Bill Marting, national sales manager; Michele Laseau, art director; Lisa Grzan, production manager; and Theresa Stockton, production team leader.

Manuscript Reviewers

We have greatly benefited from the perceptive comments and suggestions of more than 50 talented scholars and instructors who reviewed the manuscript of *Instrumental Analysis*. Their insight and suggestions contributed immensely to the published work. We are especially grateful to the following reviewers for their insight and support on the project:

Tarab Ahmad	Western Illinois University
Silvana Andreescu	Clarkson University
Craig Aspinwall	University of Arizona
Dean Atkinson	Portland State University
Ryan Bailey	University of Illinois Urbana-Champaign
Rebecca Barlag	Ohio University
Steven Brown	University of Delaware
Clifton Calloway	Winthrop University
Andres Campiglia	University of Central Florida
Yunwei Charles Cao	University of Florida
Tami Lasseter Clare	Portland State University
David Dobberpuhl	Creighton University
Jeremiah Duncan	Plymouth State University
Joe Emily	South Carolina State University
Richard Foust	James Madison University
Mahdi Garelnabi	University of Massachusetts, Lowell
Teresa Golden	University of North Texas
C. Michael Greenlief	University of Missouri
Erin Gross	Creighton University
Justin Harris	Ohio State University
Joseph Hornak	Rochester Institute of Technology
Jinmo Huang	The College of New Jersey
Takashi Ito	Kansas State University
Jiri Janata	Georgia Institute of Technology
Frank Keutsch	University of Wisconsin Madison
Chul Kim	California State University, East Bay
Jun Li	Kansas State University
Xiujun Li	The University of Texas at El Paso
Bryce Marquis	University of Central Arkansas
Nelly Mateeva	Florida A&M University
Peter Palmer	San Francisco State University
Dmitiri Pappas	Texas Tech University
David Patrick	Western Washington University
Daniel Pharr	Virginia Military Institute
Robert Richter	Chicago State University
David Ryan	University of Massachusetts, Lowell
Omowunmi Sadik	Binghamton University
Jonathan Scaffidi	Miami University

Shahab A. Shamsi   Georgia State University
Jianguo Shao   Midwestern State University
Greg M. Swain   Michigan State University
Dan Sykes   Pennsylvania State University
Steven Symes   University of Tennessee at Chattanooga
Brian Tissue   Virginia Polytechnic Institute
Grant Wangila   University of Arkansas
Robert White   University of Oklahoma
Kathryn Williams   University of Florida
Yi Xiao   Florida International University
Zhibo Yang   University of Oklahoma
Wei Zhan   Auburn University

Chapter 1

The Analyst's Toolbox

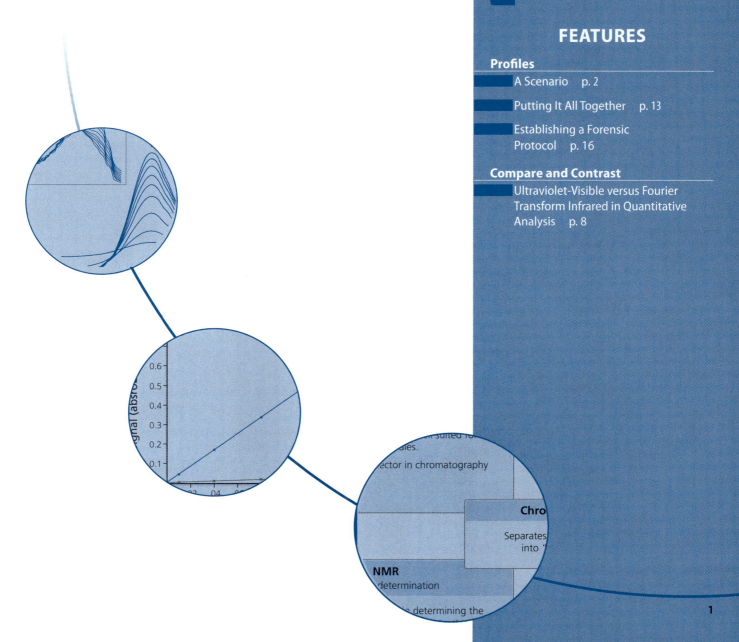

1

PROFILE

A Scenario

A man is found lying dead on his living room floor with white powder on his upper lip. The man's roommate has a prior arrest for cocaine possession. However, there is also a box of powdered doughnuts on the kitchen counter. It is your job to determine (1) whether the powder on the man's lip is cocaine; (2) if it is cocaine, whether the cocaine is in the HCl form or freebase form (the penalties for possession of the two forms are different); and (3) if it is cocaine, whether the man died of an overdose (i.e., what is the concentration of cocaine in his blood?).

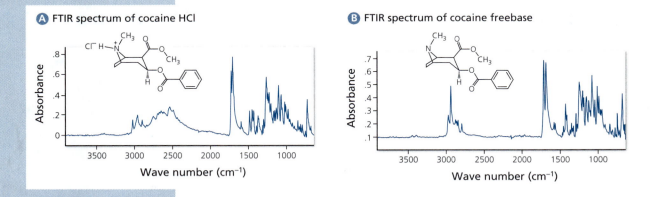

Figure 1.1 FTIR of cocaine·HCl (A) and the freebase (B).

To answer these questions, you will need to use several different tools. For instance, a simple Fourier transform infrared (FTIR) spectrum of the white powder could identify it as cocaine based on identifying structural components of that molecule. FTIR would also allow you to determine the form (freebase or HCl salt) of the compound. Figure 1.1 shows the FTIR of cocaine·HCl and freebase cocaine. These two molecules are structurally almost identical, and yet they are easily differentiated by an FTIR spectral analysis. If you determine that the powder is cocaine, you would next measure the concentration of cocaine in the man's blood. High-performance liquid chromatography (HPLC) could be used to isolate the cocaine, and the HPLC would most likely be coupled with an ultraviolet-visible (UV-vis) spectrometer to detect and quantify the cocaine. Another option would be to extract the cocaine into an organic solvent and use gas chromatography (GC) to isolate the cocaine. The GC could be coupled with a mass spectrometer as the detector, which would also detect and quantify the cocaine.

What this scenario demonstrates is the importance of knowing and appreciating the various strengths and limitations of each technique at your disposal. For instance, as demonstrated by Figure 1.1, FTIR spectroscopy is extremely efficient at qualitative analysis; however, it is vastly inferior to UV-vis spectroscopy for quantitative analysis. On the other hand, by itself, the UV-vis spectrometer is almost useless for conducting qualitative analysis, and that is why we isolated the cocaine using HPLC prior to conducting quantitative UV-vis analysis of the sample. As you move through the chapters of this text, you will gain an appreciation for the abilities and limitations of a great number of analytical techniques. ∎

1.1   Introduction

Instrumental methods stem essentially from endeavors to extend the human senses. In this text we will be describing, in significant detail, a fairly large range of instruments and the methods associated with each. As we introduce new techniques, it is important that you incorporate each new technique in the context of the techniques with which you are already familiar. The intention of this chapter is to profile five common instrumental techniques. Although much of your laboratory experience to date has likely been focused on one technique at a time, in reality an analyst must draw on a breadth of knowledge in order to achieve his or her goals in the laboratory. Each of the five techniques presented in this chapter, as well as many others, will be covered in greater detail in later chapters.

Five important *types* of instruments are applied to the majority of problems or applications in chemistry. These include *ultraviolet-visible (UV-vis) spectroscopy, Fourier transform infrared spectroscopy*[1] *(FTIR), nuclear magnetic resonance spectrometry (NMR), mass spectrometry (MS)*, and *chromatography* (liquid chromatography [LC] and gas chromatography [GC]). Visit any practicing chemistry facility (academic or industrial), and you will undoubtedly find a version of each of these instruments.

It is interesting to note that, although four of the instrument types discussed here are spectroscopic (UV-vis, FTIR, NMR, and MS), only two of them (UV-vis and FTIR) are considered *optical methods*.[2] UV-vis spectroscopy measures electronic transitions within a molecule,[3] whereas FTIR spectroscopy measures the vibrational frequencies associated with each bond type.[4] Both of these are absorption methods, meaning that energy is put into the system and we monitor a change in that energy. In UV-vis and FTIR, the source of that energy is within and just outside the visible range (ultraviolet through infrared). NMR spectroscopy also involves absorption; however, energy is put into the system by an externally applied magnetic field. These spectrometric methods extend our ability to "see" molecules and detect molecular interactions that are outside our normal human senses.

The terms *spectrum* and *spectrometry* refer, historically, to the observance of light over a range of wavelengths. However, MS does not inherently deal with light or photons at all—instead, MS measures a "mass spectrum," or a tally of the number of ions of specific mass-to-charge ratios over a range of values. MS extends, down to the atomic level, our ability to weigh objects. Although MS can be used to determine the molar mass of specific isotopes with a high degree of accuracy, the vast majority of mass spectrometers are used for molecular analysis to determine structure and composition.

Of the five techniques introduced, the odd one out is chromatography, which deals with the efficient separation of components of a mixture. Chromatography is not a spectrometric method, and it is not viable as an analytical method alone without some *other* type of instrument being used as a detector to observe the outcome of the experiment. But chromatography allows us to isolate discrete substances from complex mixtures and it often precedes other analytical techniques. Chromatography is absolutely essential to the analyst, and every analytical chemist should acquire substantial experience in chromatographic separation methods.

---

[1] Raman spectroscopy will likely surpass FTIR in the coming decades.

[2] By *optical* we mean that the method involves measuring the power of light from a source (P_0), followed by measuring the power of light after it has passed through an absorbing material (P).

[3] Although many atomic spectroscopic methods (Chapters 7 and 9) involve the absorption and/or emission of ultraviolet/visible light, the term *UV-vis* is understood to mean molecular spectroscopy, usually in solution.

[4] Infrared spectroscopy is exclusively a molecular technique. A single atom cannot undergo vibrational excitation.

1.2 Ultraviolet-Visible Spectroscopy

UV-vis spectroscopy involves the absorption of photons in the ultraviolet region from 195 to 400 nm and the visible region from 400 to 900 nm of the *electromagnetic spectrum* (EMS). To be precise, the region from 700 to 900 nm is the near-infrared region of the EMS; however, we simply refer to the technique as UV-vis spectroscopy. Specific details regarding UV-vis instrumentation are covered in Chapter 6. Here we introduce UV-vis spectroscopy in the context of one of many tools an analyst can use to solve a problem. Absorption of photons in the UV-vis range of the EMS is associated with the transition of electrons in bonding molecular orbitals to higher (usually antibonding) orbitals.[5] We expect to see absorption of light when the energy of the incident photon equals the energy required to elevate the electron from the *ground state* (E_0) to the *excited state* (E_1). If that absorption were the only event taking place, we would expect to see a line spectrum for all UV-vis absorptions. However, you know from experience in your prior chemistry courses that UV-vis absorption peaks are typically broad. Chapter 6 provides the theoretical explanation for that empirical observation, which we summarize here as:

1. There are multiple vibrational and rotational transitions associated with each electronic transition.

2. Interactions between absorbing species and the solvent cause changes in the ground and excited states for all transitions.

These conditions create a relatively wide range of very closely spaced energy levels for each nominal molecular orbital energy level. The net result is the typically broad UV-vis absorption peaks which exhibit baseline widths ranging from 20% to 50% of the nominal peak wavelength. It is the wide nature of a UV-vis peak that makes UV-vis spectroscopy a poor choice for doing qualitative analysis, and yet it is this same feature that makes UV-vis a good choice for *quantitative* analysis (we will explain why in the Compare and Contrast segment after Section 1.3).

As a rule, we plot UV-vis spectra as absorbance (ordinate or *y* axis) versus wavelength in nanometers (abscissa or *x* axis). In order to measure absorbance, it is necessary to first measure the strength of the unabsorbed signal (the blank) and follow this with the measurement of the sample. The computer will calculate and render the absorbance spectrum for us, but fundamentally the instrument is measuring light throughput at each wavelength and determining the transmittance ratio. (Chapter 6 provides more details.)

One of the reasons that UV-vis spectroscopy lends itself to quantitative analysis is the fact that the photons involved in UV-vis spectroscopy are sufficiently energetic that they can overcome the binding energy for electrons in certain photoemissive materials,[6] thus making it possible to take advantage of some of the most sensitive photon detectors available such as photomultiplier tubes and cascade photodiodes. We will examine the construction of some of these detectors in Chapter 6.

A typical UV-vis spectrum exhibits very few (and often as few as one) peaks, and the natural width of those peaks often makes it impossible to resolve them. This poor resolution makes UV-vis spectroscopy an inadequate tool for conducting qualitative analysis but as we will soon see, the poor resolution can be exploited. Oddly, the broad peak widths found in UV-vis spectra are well suited to high-precision analysis. Figure 1.2 shows the structure (inset) and UV-vis spectrum of *di-2-pyridyl ketone (DPK)* in acetonitrile.[7] In Figure 1.2(A), we see two strong, poorly resolved peaks at 240 and 272 nm and a third very weak peak at 356 nm. Without additional information, this spectrum would tell us virtually nothing qualitative about the analyte other

---

[5] UV-vis spectroscopy of discrete atoms is a technique called atomic absorption spectroscopy (see Chapter 7).

[6] This is an application of the photoelectric effect.

[7] Sienerth, K. D.; Granger, R. M.; et al. *Inorg. Chem.* **2004**, *43*, 72–78.

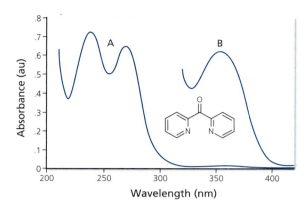

Figure 1.2 Ultraviolet-visible spectrum of DPK (structure inset) in dry acetonitrile. The concentration of DPK is 0.0828 mM (A) and 4.17 mM (B). The cell path length was 1.0 cm.

than that it *probably* exhibits pi-bonds and/or lone pair electrons. However, UV-vis peaks often exhibit high *molar absorptivities,*[8] which significantly enhance the sensitivity of methods based on UV-vis spectroscopy.

Although straightforward quantitative analysis using *Beer's law*[9] is the most common application of UV-vis, we must not discount qualitative applications of UV-vis entirely. The high sensitivity allows experimenters to distinguish very small changes in concentration, and in a curious twist, that ability lends itself to a qualitative application: kinetic studies. Kinetic measurements are used to understand the fundamental molecular interactions involved in more complex chemical reactions. In order to successfully make such measurements, it is often necessary to be able to note small changes in molar concentration in a very short time frame, a factor supported by the rapid response and recovery of most detectors used in UV-vis spectroscopy. In a recent report describing investigation of diazirines,[10] researchers were able to observe the UV-vis spectrum of a transient species having a lifetime of less than a nanosecond, which required measuring micromolar changes on the picosecond time scale.[11] UV-vis spectroscopy also provides valuable qualitative results in many biochemical studies. Because biochemical reactions often require aqueous media and involve large and complex molecules (both of which present significant challenges for FTIR and NMR methods), UV-vis spectroscopy is used to monitor changes in absorption peak wavelengths resulting from the interaction of large biomolecules with other species. Recently, Matera and Clark used UV-vis spectroscopy to study the oxidation of fatty acids (as analogs to low-density lipoproteins) by the enzyme *myeloperoxidase,* which has been implicated as a possible culprit in the initiation of atherosclerosis (Figure 1.3).

A good-quality, research-grade spectrophotometer will typically be able to achieve a resolution of 1 to 2 nm over the range of 190 to 1,100 nm, and in 2016 it cost less than $10,000. A basic-level spectrophotometer that is suitable for many laboratory experiments can be acquired for less than $3,000, but these instruments typically require the user to select a fixed wavelength for any measurement, rather than providing a full spectrum across a range of wavelengths.

---

[8] The molar absorptivity is the proportionality coefficient that relates absorption to molar concentration and path length. In Beer's law (further discussed in Chapter 6), we see that A = εbc, where ε is the molar absorptivity coefficient.

[9] See the Compare and Contrast feature later in this chapter and Beer's law in Chapter 6.

[10] Important precursors in biochemical and organometallic studies.

[11] Platz, M. S.; et al. *J. Amer. Chem. Soc.,* **2009**, *13 (38)*, 13784–13790.

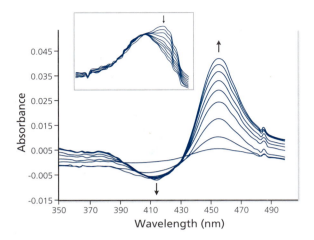

Figure 1.3 Difference spectra of the decay of a reaction product of the enzyme myeloperoxidase (MPO). Each difference spectra was generated by subtracting the spectrum in 20-second intervals from the initial MPO spectrum. Inset shows the actual spectral changes. Figure used with permission of the author: Matera, K. M.; Clark, A. L. *Bioorg. Med. Chem. Lett.*, **2010**, *20(19)*, 5643–5648.

Problem 1.1: Absorption of UV-vis photons usually is associated with electronic transitions—that is, such absorptions cause electrons to move from one energy level to another. In your course on general chemistry, you learned that energy levels are quantized. Based only on that information, what types of peaks (broad or narrow) would you expect to see in a UV-vis spectrum? Explain your answer. Propose some ideas for why we usually observe smooth, broad peaks in UV-vis spectroscopy?

Problem 1.2: Beer's law states that absorbance = εbc. In this equation, ε is the molar absorptivity (in units of $cm^{-1} \cdot M^{-1}$), b is the absorbent path length in cm, and c is the concentration in molarity. Estimate ε for the DPK peak seen at 272 nm in Figure 1.2.

Problem 1.3: Using the molar absorptivity calculated in Problem 1.2, what would be the concentration of DPK required to produce an absorbance of 0.433 at 272 nm?

Problem 1.4: A 13.50-mg sample of impure DPK was dissolved in 50.00 mL acetonitrile and placed in a sample cell with a path length of 0.05 cm. The absorbance was found to be 0.513 at a wavelength of 272 nm. What was the percent purity of the DPK sample? *Hint*: Use the molar absorptivity calculated in Problem 1.2.

Problem 1.5: What is the underlying basis for the relationship between the absorptivity (strength or height) of the peak and quantitative sensitivity?

1.3   Infrared Spectroscopy

When an analyte molecule absorbs infrared (IR) photons, the molecule transitions from the ground vibrational state to an excited vibrational state. Each vibrational excited state has a specific energy (frequency or wavelength) associated with it. Because the vibrations are a result of the bonds that exist in the molecules being studied, observing the absorbance of infrared energy at specific frequencies makes it possible to deduce what types of bonds are present in the molecule. The frequency of vibration also allows us to make judgments about what types of atoms or functional groups are involved in the bonds. For instance the molecule *DPK* (structure shown in Figure 1.2) exhibits an absorption peak at 1,690 cm^{-1} (λ = 5.92 μm, or ν = 5.36 × 10^{13} Hz),[12] which is characteristic of the stretching of the bond between carbon and oxygen in the carbonyl (C=O) group. In this chapter we will discuss infrared spectroscopy—both traditional infrared and the more modern FTIR techniques—in the context of its role as one of the tools in an analyst's toolbox. Specific details about infrared spectroscopy are covered in Chapter 11.

In contrast to UV-vis, infrared peaks are generally quite narrow, exhibiting a baseline width of 1% to 10% of the nominal peak wavelength. An infrared spectrum typically shows many such peaks, each of which adds to what we can learn about the analyte. Because of structural information provided, infrared spectroscopy is an excellent *qualitative* tool. For instance, consider the FTIR spectrum of DPK dissolved in carbon tetrachloride (Figure 1.4). The peak width (0.17 μm) of the carbonyl peak in the FTIR spectrum of DPK is only 2.9% of the nominal peak absorbance wavelength (5.92 μm).[13]

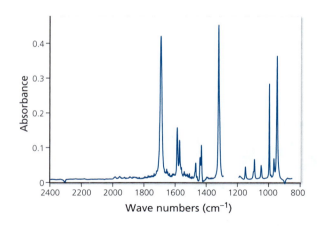

Figure 1.4 Infrared absorbance spectrum of 0.058 M DPK in CCl$_4$ plotted as a function of wave numbers. The cell path length was 0.2 mm.

In contrast to the UV-vis spectrum, the FTIR spectrum has many peaks, each of which can be attributed to a stretching or bending motion associated with specific bonds in the molecule. The strong peak at 1,690 cm^{-1} indicates a carbonyl, the two peaks around 1,560 cm^{-1} indicate a pyridyl ring, and the peaks around 1,430 cm^{-1} indicates an aromatic ring. The multiple absorption bands provide a wealth of qualitative information that we can use to achieve a good estimate of the molecular structure; indeed, the region from about 1,500 to 500 cm^{-1} is considered the "fingerprint region" because it exhibits many peaks that, collectively, can be used to identify a substance. However, compared to UV-vis peaks the narrowness and low absorptivities of FTIR peaks makes them less useful for *quantitative* analysis. We will expand on this idea in the Compare and Contrast feature following this section.

We use the term *infrared spectroscopy* in a chemistry laboratory in a broad sense. By far, the most common type of instrument that you would find in a working laboratory is the FTIR spectrometer. The FTIR spectrometer uses a Michelson interferometer instead of a monochromator or other wavelength selection system. There are many advantages inherent in using the FTIR spectrometer in comparison to a scanning spectrometer, and we will come back to those details in Chapter 11. *Infrared spectroscopy* also has to do with the actual region of the electromagnetic spectrum being probed. The infrared region includes wavelengths from about 800 to almost 1,000,000 nm (or 13,000–10 cm^{-1}). This region is subdivided into the *near-IR*,[14] from 800 to 2,500 nm (or 12,500–4,000 cm^{-1}); the *mid-IR*, from 2,500 to 50,000 nm (or 4,000–200 cm^{-1}); and the *far-IR*, from 50,000 to 1,000,000 nm (or 200–10 cm^{-1}). The mid-IR provides the most useful information about organic and organometallic substances, so when we say *infrared spectroscopy*, it is assumed that we are referring to the *mid-IR region* unless specifically stated otherwise.

A good research-grade transmission FTIR instrument can be purchased (in 2016) for around $30,000. The addition of reflectance capabilities can add $5,000 to $25,000 to that price, depending on the type and quality desired.

Problem 1.6: Estimate ε for the peak seen at 1,690 cm^{-1} in Figure 1.4. (See Problem 1.2.)

---

[12] The unit cm^{-1} is referred to as a wave number (it is the reciprocal of the wavelength when the wavelength is reported in centimeters, and it is commonly used in infrared spectroscopy).

[13] In order to provide direct comparison to the UV-vis peak width given above, we use units of wavelength here instead of the units of *wave numbers* (cm^{-1}) given in the spectrum.

[14] This region is called "near" because it is near the visible region. Our anthropocentric perspective of nature puts the visible region at the center of the electromagnetic world.

Problem 1.7: This is an open-ended question, so you will need to defend your answer. You are setting up a new analytical laboratory. Which do you purchase first: a UV-vis or an FTIR instrument? In every such instance, a chemist must take into account cost and benefit and also must consider what is needed in his or her analyses. Be sure to explain your answer.

COMPARE AND CONTRAST

Ultraviolet-Visible versus Fourier Transform Infrared in Quantitative Analysis

The rule of thumb among chemists is that UV-vis spectroscopy is very poor for qualitative analysis but excellent for quantitative analysis, and the opposite is true for infrared spectroscopy. We have already made the case comparing UV-vis and FTIR with respect to qualitative analysis; UV-vis spectra exhibit very few, broad peaks, and FTIR spectra exhibit many narrow peaks that provide a wealth of information about the chemical nature of the analyte.

So let us take a moment to discuss what makes UV-vis superior to FTIR for quantitative analysis. There are two contributing factors: the broad peaks of UV-vis tend to lead to lower errors in measurements, and in general, the sensitivity of UV-vis detectors provide significantly lower detection limits. In this section, we explore these two phenomena in more detail. Figure 1.5 shows an FTIR absorption band centered at 1,800 cm⁻¹ (5,556 nm) with a typical baseline breadth of about 4% of the nominal peak position. If our instrumental error was 0.5% in the wave number reading, the experimental error in the signal is almost 6%, and a 1.0% error in the wave number reading imparts a whopping 20% error in the measured signal. The sharp peaks that make FTIR an excellent qualitative tool work against its use for quantitative analysis.

Figure 1.6 shows a simulated UV-vis peak centered at 400 nm having a typical breadth of 35% of the nominal peak position

(140 nm). If our instrument has an error of 0.5% in measuring the wavelength, the error in our signal measurement is only 0.08%. A 1% error in wavelength measurement imparts a 0.33% error in the signal, and a 4% error in the wavelength measurement imparts a 5% error in signal. Figure 1.7 is a comparison of errors for narrow FTIR peaks relative to broad UV-vis peaks.

The other reason UV-vis excels over FTIR for quantitative analysis is due to the relative peak intensities. In Problem 1.2, we learned the fundamental relationship:

$$A = \varepsilon b c$$

where A is the measured absorbance (no units), ε is the molar absorptivity (cm⁻¹·M⁻¹), b is the path length (cm) of incident light through the cell, and c is the concentration in molarity.* In a quantitative analysis experiment, we would establish a calibration plot by measuring the absorbance of a series of solutions of known concentration and then plot absorbance versus concentration. From that point forward, we could simply measure the absorbance of any solution of unknown concentration and use the plot (or the linear regression) to determine the concentration. In this scenario, the slope of the A versus c plot is referred to as the *sensitivity,* and a high sensitivity

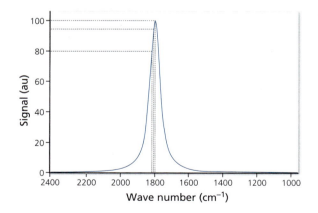

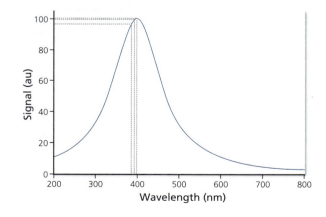

Figure 1.5 Simulated infrared absorbance peak at 1,800 cm⁻¹ (5,556 nm) with a breadth of 4% (72 cm⁻¹). The vertical line immediately to the left of the line centered at 1,800 cm⁻¹ represents the position consistent with a 0.5% error in wave number (1,809 cm⁻¹). The horizontal lines associated with those two verticals represent the error in signal. The line farthest from the 1800-centered one represents the position and signal consistent with a 1.0% error in wave number (1,818 cm⁻¹).

Figure 1.6 UV-vis peak with peak width of 35% (140 nm). The vertical line immediately to the left of the line centered at 400 nm represents the position consistent with a 0.5% error in wave number (398 nm). The horizontal lines associated with those two verticals represent the error in signal. The line farthest from the 400-centered one represents the position and signal consistent with a 1.0% error in wave number (384 nm).

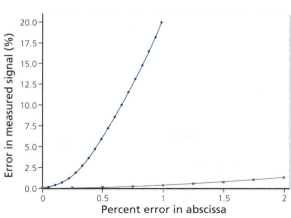

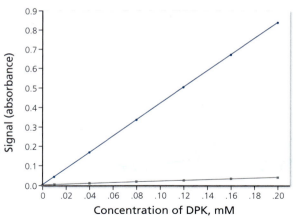

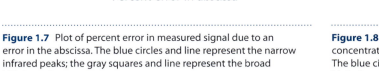

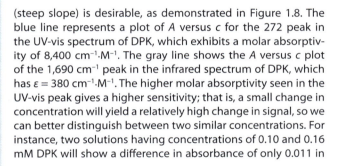

Figure 1.7 Plot of percent error in measured signal due to an error in the abscissa. The blue circles and line represent the narrow infrared peaks; the gray squares and line represent the broad UV-vis peaks.

Figure 1.8 Plot of calculated signal (absorbance) vs. analyte concentration for methods exhibiting different molar absorptivities. The blue circles and line represent UV-vis peaks which typically have high absorptivities; the gray squares and line represent infrared peaks which typically have low absorptivities.

(steep slope) is desirable, as demonstrated in Figure 1.8. The blue line represents a plot of A versus c for the 272 peak in the UV-vis spectrum of DPK, which exhibits a molar absorptivity of 8,400 $cm^{-1} \cdot M^{-1}$. The gray line shows the A versus c plot of the 1,690 cm^{-1} peak in the infrared spectrum of DPK, which has $\varepsilon = 380$ $cm^{-1} \cdot M^{-1}$. The higher molar absorptivity seen in the UV-vis peak gives a higher sensitivity; that is, a small change in concentration will yield a relatively high change in signal, so we can better distinguish between two similar concentrations. For instance, two solutions having concentrations of 0.10 and 0.16 mM DPK will show a difference in absorbance of only 0.011 in

the FTIR spectrum but a difference of 0.25 (2,000% larger!) in the UV-vis spectrum.

For a 0.011 AU change in signal in the UV-vis spectrum, we could distinguish between a 0.10 M solution and a 0.1025 M solution, a difference of 2.5 µM. DPK is not unique; it is a general rule that molar absorptivities associated with UV-vis peaks range from 10^2 to 10^6, whereas those for FTIR peaks are much smaller, typically ranging from 10 to 10^3.

The relationship $A = \varepsilon bc$ is a form of Beer's law. It is discussed further in Chapter 6.

Nuclear Magnetic Resonance Spectrometry

Although the specific instrumental details of NMR spectroscopy are covered in Chapters 14 and 21, we introduce it here to put the technique in context with UV-vis and FTIR spectroscopy and to build on the theme of filling your analytical toolbox with a collection of complementary analytical tools. Although one can perform quantitative analysis with NMR, it is by most accounts a *qualitative* tool, giving the analyst specific information about molecular structure, connectivity, and orientation. In comparison, FTIR gives the analyst information about the *types of bonds* present in a molecule (such as the presence of a carbonyl or cyano group), whereas the NMR gives the analyst information about the *connectivity* of the atoms in the molecule. For example, in an NMR spectrum, you can differentiate the protons in a methyl ($-CH_3$) from those of a methylene ($-CH_2$). Furthermore, the NMR spectrum can tell you if the methyl group is directly adjacent to the methylene group or further away from it. NMR methods are used to probe the details of molecular structure.

Although NMR is a spectroscopic method (such as UV-vis and FTIR), it is not considered optical[15] because the source of energy is a low-energy radiofrequency wave generator rather than a "light" source. Under the influence of an externally applied magnetic field,

---

[15] Both UV-vis and FTIR are considered optical methods because they involve a source of light energy that is measured directly before and after it interacts with a sample.

which provides the necessary environment for the NMR phenomenon, the nuclei in the sample are stimulated (into resonance) by a continuous wave or pulse of radiofrequency energy. As the nuclei transition (resonate) from one state to another, the signal is measured. Whereas UV-vis spectroscopy helps us "see" absorbances in electrons stimulated using radiation in the ultraviolet and visible regions of the EMS, NMR helps us "see" absorbances in nuclei stimulated using an oscillating magnetic field generated in the radio frequency range.

Two important phenomena provide NMR spectrometry with unique value as a qualitative tool: chemical shift and spin-spin coupling. The basic idea behind chemical shift is that the chemical environment near each nucleus is unique and produces a unique magnetic environment. For instance, the hydrogens on the carbon and the hydrogen on the oxygen in methanol (CH_3–OH) have very different environments. These different environments help spread the signals along the x axis in an NMR spectrum. The spectrum is oriented such that those nuclei experiencing greater magnetic *shielding* are on the right (nearer zero) and those experiencing less magnetic shielding by nearby electrons are on the left (with increasingly positive numbers on the abscissa, δ, or ppm scale).

The proton (^1H) NMR spectrum of diethylmalonate is given in Figure 1.9. The positions (chemical shifts) of each set of peaks tells us something about the compound. The methyl group is observed at about 1.25 ppm. Those ^1H nuclei are experiencing the greatest level of shielding. On the other hand, the methylene protons adjacent to the methyl group are significantly less shielded (at 4.25 ppm) due to the adjacent oxygen in the ester group. The position of the resonance peaks tells us a great deal about the types of elements that are near each nucleus in the molecular environment.

Figure 1.9 also demonstrates the other important phenomenon that provides valuable qualitative information: spin-spin coupling. Nearby nuclei also create small magnetic variations in the field experienced by the analyte nucleus. Thus, protons on adjacent carbons can "split" the signal of the analyte proton. You likely have encountered at least a brief description of this phenomenon in previous classes, and we will explore it in greater detail in Chapters 14 and 21. For now it is sufficient to remember that (n) protons on adjacent carbons will split a proton absorption peak into (n + 1) peaks, and we can use that information to learn about the *specific* chemical structure of a molecule. For instance, in Figure 1.9, the triplet at 1.25 ppm tells us that a well-shielded aliphatic proton is attached to a carbon (a methyl) that is adjacent to a CH_2 group. The two protons on the adjacent carbon split the methyl group signal into (2 + 1 = 3), a triplet.

Note that the total area under each peak is proportional to the number of nuclei involved in the signal; the peak heights of the triplet and quartet are much lower than each singlet would have been in the absence of splitting, but the total area under the

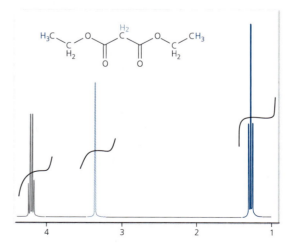

Figure 1.9 ^1H NMR spectrum of diethylmalonate. The x-axis is the chemical shift, in units of ppm, not as concentration but as ppm of a fundamental reference resonance frequency. This will be explained in Chapter 14.

peak is proportional to the number of nuclei of that type. The horizontal S-curved line in Figure 1.9 provides a measure of the *integration* for each peak group, which tells us the relative number of protons involved in each signal.

The chemical shift, splitting patterns, and integration obtained from a simple ^{1}H NMR experiment provide a remarkable supply of information about an analyte. Unfortunately, one of the main drawbacks of NMR spectrometry is the cost of the instrumentation and the cost of maintenance. Certainly the most expensive part of the system is the magnet. Although some solid-core magnets and electromagnet-based systems still exist, most commercially available systems use a magnetic field generated by current flowing through coils of superconducting wires. To achieve a superconducting state, those wires must be cooled to about 6K, which requires the continuous presence of liquid helium in a large Dewar flask. To prevent the very costly liquid helium from boiling off too rapidly, a secondary Dewar flask filled with liquid nitrogen surrounds the central Dewar flask. The overall initial cost of a routine use superconducting NMR instrument will be around $250,000, with maintenance costs of $10,000 to $20,000 per year. Furthermore, if you accidently allow the magnet to warm above the superconducting temperature, the magnet will "quench" and the cost to recharge the magnet can be many thousands (> $40,000) of dollars.

If you are like most students of instrumental analysis, you have likely studied NMR spectral interpretation in previous classes in your chemistry curriculum. You may have even used an NMR for sample analysis in your course on organic chemistry. We are not going to spend much time in this text on spectral analysis; however, in the later chapters in this book, we will present you with an instrumental perspective on how the signals are generated, measured, and processed. As an important part of your analytical toolbox, we also provide you with an overview of modern NMR experiments in the form of pulsed FT techniques. You will see that the signal processing of the FT NMR is quite analogous to the FTIR.

Problem 1.8: In a standard ^{1}H NMR spectrum, what type of splitting pattern would you expect for each unique proton in 1-chloropropane? In 2-chloropropane?

Problem 1.9: The spectra of 2-butanone, methylethylether, and 1-butyne will each exhibit a triplet, a quartet, and a singlet similar to the spectrum seen for diethylmalonate (Figure 1.9). For each of those compounds, explain how you would distinguish it from diethylmalonate using the NMR spectra. *Hint*: Consider the shielding effects of nearby nuclei and the integration ratios you might expect.

1.5 Mass Spectrometry

MS is not, strictly speaking, a *spectrometric* technique, because it does not involve electromagnetic radiation in any way. So in this regard, the technique is fundamentally different than UV-vis, FTIR, or NMR spectroscopy. In MS, the term *spectrometry* relates to the fact that the output from the instrument is in the form of a "mass spectrum," or rather a plot of relative ion abundance on the ordinate over a range of mass-to-charge ratio (m/z) values on the abscissa. MS is covered in detail in Chapter 13, but the basics are introduced here as another important analytical tool in your toolbox. Although stand-alone mass spectrometers have their place in the analytical laboratory, MS is also frequently found in tandem with chromatographic instruments, GC-MS and LC-MS as prime examples, where the MS serves as the detector.

Like NMR, the results of an MS experiment can provide very specific information that can help users elucidate the chemical structure of an analyte molecule. Indeed, the GC-MS and LC-MS[16] rival NMR for general utility in providing substantive qualitative information, and they are also quite excellent for quantitative purposes as well.

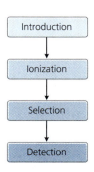

Figure 1.10 Basic steps involved in an MS experiment.

---

[16] GC-MS stands for gas chromatography with mass spectrometric detection. LC-MS stands for liquid chromatography (usually high-performance liquid chromatography, or HPLC) with MS detection.

The function of a typical MS instrument in the analyst's toolbox is relatively straightforward (see Figure 1.10). Unlike the spectroscopic methods we have so far discussed, the sample is not subjected to an impinging energy source; it is not acted on while staying in a fixed location. Instead, in MS, the sample is *introduced* into the system in a dynamic fashion and the analysis is governed essentially by mobility of the sample through the system. For now, we will describe a system where the sample is introduced as a gas into the MS system.

Next, the analytes within the sample must be *ionized*. Minimally, we must impart a single charge to the analyte molecule, either by removing an electron or by adding a charged species such as a proton, H^+. In many MS systems, the analyte precursor is not only ionized but also fragmented to produce *product ions* that represent different pieces of the molecule. The fragmentation pattern can provide additional information that allows the analyst to determine the various parts of the overall molecule.

Once we have ionized our precursor molecule, and possibly generated a subset of product ions, we must determine which species are present. To do this, we accelerate all ions by sending them through a voltage gradient and then passing the cluster of ions through a mass analyzer, which under computer control can *select* ions of a specific m/z ratio. In a given fraction of a second, only ions of a single m/z get through the mass analyzer in order to be detected and counted. By varying certain parameters (see Chapter 13 for more details), we can select a series of m/z values in the period of about 1 second, resulting in an output of counts (number of ions) at each m/z—that is, a mass spectrum. In the majority of cases, the charge imparted (in a standard instrument) is unity (+1), and so we can presume that the abscissa essentially gives us the mass of each ion. However, we must always remain cognizant of the possibility of ions with higher charges. For instance, if we see a peak in the mass spectrum at m/z = 57 amu, it is very likely the result of the *tert*-butyl ion $(CH_3)_3C^+$ but *might* be due to the dication of octane, $C_8H_{18}^{2+}$.

Figure 1.11 shows the mass spectrum of diethyl oxalate. We observe the *precursor ion* peak (i.e., singly charged diethyl oxalate) at around 146 amu. The largest peak in the spectrum is called the *base peak*, and here it occurs at 74 amu, corresponding to the fragment created by the breakage of the oxalate symmetrically in half resulting in two $\cdot COOCH_2CH_3^+$ product ion fragments.

We see a notable peak at 59 amu from the loss of a methyl from the $\cdot COOCH_2CH_3^+$ product ion, resulting in a $\cdot COOCH_2^+\cdot$ fragment and another notable peak at 56 amu after loss of both $-OCH_2CH_3$ substituent from the precursor ion.

A great deal of information about the chemical structure of the analyte molecule can be garnered from the mass spectrum, but it is often difficult to determine exactly how all the fragments were put together. However, under consistent experimental

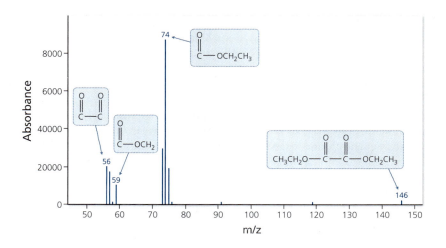

Figure 1.11 Mass spectrum of diethyl oxalate.

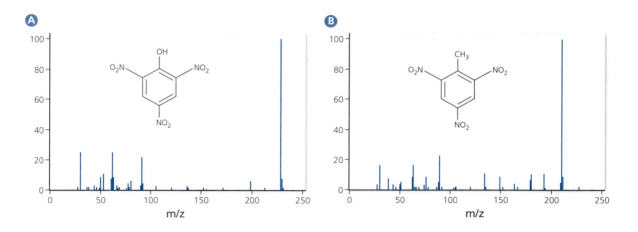

Figure 1.12 Mass spectra of two similar explosives: (A) picric acid, or trinitrophenol, and (B) trinitrotoluene, or TNT.

conditions, any given substance will generate the same pattern of peak positions and peak height ratios: the mass spectrum. Therefore, the mass spectrum serves as a spectrometric *fingerprint* of that substance. If we collect and store the mass spectrum of a certain compound, we can later compare the stored spectrum with that of an unknown substance and know with high certainty if they are the same. In fact, it is common for chemistry laboratories to purchase and install libraries (databases) of several thousand mass spectra in order to facilitate identification of unknown species.[17] This ability to search and compare spectra has become a mainstay in many fields such as forensics, biochemistry, environmental chemistry, and organometallic chemistry. Figure 1.12 shows the mass spectra of two explosives that vary chemically by only one atom in the structure.[18] It is easy to discern differences in the two mass spectra and if these spectra were obtained on an instrument that contained a database, the spectra would be displayed with a percent probability match to the spectra stored in the database.

Even if you lack access to an MS database, using the other tools in your toolbox will allow you to put the pieces back together. For instance, NMR is a very useful technique for determining connectivity in a molecule, so comparing the MS and NMR spectra usually leads to a full understanding of how each fragment in the MS is connected. Similarly, both $C=O^+$ and $CH_2=CH_2^+$ have an m/z = 28 amu, so observation of a fragment with m/z = 28 amu in a MS spectrum might be ambiguous. But as we have shown, the FTIR is a very useful tool for determining bond types in a molecule so addition of FTIR data to your analysis will discern between the existence of a C=O bond and $CH_2=CH_2$ bond in your molecule. Once you know what you are dealing with, a simple UV-vis experiment will allow you to very precisely determine the quantity of analyte in your sample. ▬

PROFILE

Putting It All Together

---

[17] See http://www.nist.gov/pml/data/asd.cfm.

[18] These spectra were obtained from the National Institute of Standards and Technology (NIST) Chemistry Webbook, http://webbook.nist.gov/chemistry/ (accessed August 13, 2015).

Stand-alone MS instruments are not common in the standard chemistry laboratory setting. The cost of a GC-MS system will range, depending on various user-selected options, from around $60,000 to $120,000 (in 2016). A tandem LC-MS system is usually more expensive, ranging from $100,000 to $250,000.

Problem 1.10: Consider the spectrum shown in Figure 1.11. Looking at the structure of the compound given in the figure, what gives rise to the specific peaks we see? That is, why do you think we observe certain fragments but not others?

Problem 1.11: Both spectra shown in Figure 1.12 exhibit a notable peak at m/z = 30 amu. Postulate at least one fragment that could give rise to that peak in both spectra.

1.6   Chromatography

Before we conclude our introduction to instrumental analysis, we need to add one more tool to the analyst's toolbox. Rarely in the real world do we have pure samples. Most often real samples are mixtures—and often quite complex ones. Typically, the first action taken by the analyst is the isolation of the analyte from the sample mixture, and chromatography is the workhorse technique that allows us to efficiently separate the components of complex mixtures, allowing for the analysis of the components individually. We must note that the term *chromatography* refers to a *very* broad range of methods based on a wide variety of instruments (and noninstrumental techniques as well). If you have studied organic chemistry, then you most likely have conducted at least two different types of chromatography in the laboratory, thin-layer chromatography (TLC) and GC. The two most common chromatographic instrument types found in the chemical instrument laboratory are HPLC and GC. The principles governing these techniques will be covered in detail in Chapters 15 and 16.

As might be inferred from the name, GC uses an inert gas, such as helium or nitrogen, as a vehicle for the sample to move through the instrument. We call this carrier gas the *mobile phase*. In some ways, this is analogous to MS, which also has a dynamic sample profile. These two techniques are readily coupled into a tandem GC-MS instrument, in which the separated sample components are directly delivered into the mass spectrophotometer for analysis.

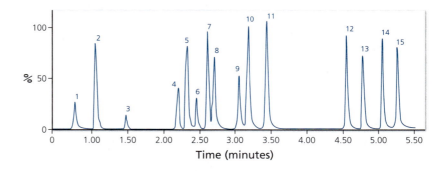

Figure 1.13 UPLC chromatogram demonstrating the separation of 15 drugs. The mobile phase was $CHOONH_4$ (30 mM, pH 3.5) at 0.5 mL/min. The drugs are 1, nicotine; 2, cotinine; 3, paraxanthine; 4, caffeine; 5, amphetamine; 6, 3,4-methylenedioxyamphetamine (MDA, or "sass"); 7, methamphetamine; 8, 3,4-methylenedioxymethamphetamine (MDMA, or "ecstasy"); 9, benzoylecgonine; 10, 3,4-methylenedioxyethamphetamine (MDEA, or "Eve"); 11, ketamine; 12, cocaine; 13, lysergic acid diethylamide (LSD); 14, phencyclidine (PCP, or "angel dust"); and 15, fentanyl.

Although the HPLC system is generally more complex and expensive than a GC one, the HPLC system has a significant advantage in that it is not limited to volatile and thermally stable samples. If the analyte can be dissolved in a solvent, it can be introduced into the HPLC system. Improved technology in instrument design (in particular, development in liquid pumping technology) has led to the development of ultrahigh performance liquid chromatography (UPLC) systems, which can achieve much higher performance.[19] The output from a chromatography system is called a *chromatogram*. Figure 1.13 shows a UPLC chromatogram for the separation of 15 drugs in just over 5 minutes using MS detection.

Chromatographic instrumentation allows for customization of analysis via an assortment of detectors. The choice of detector type provides the user with valuable spectroscopic or other physical data at the time of separation. UV-vis spectrophotometers and fluorescence spectrophotometers (in HPLC) and the thermal conductivity detector (in GC) are among the most common. Mass spectrometers are becoming more common as HPLC detectors, providing a very powerful combination. However, the addition of an MS detector adds significant cost.

In a simple chromatographic separation, the analytes are separated based on differential interactions between the mobile phase (the carrier gas in a GC technique or a pumped solvent in an HPLC technique) and the *stationary phase*. The stationary phase, for example, in a TLC experiment, would be the thin glass or plastic plate that is coated with a stationary (unmoving) material. Analytes in the mixture that do not interact with the stationary phase will move at the same rate as the mobile phase, and will *elute* with the mobile phase (or travel the furthest on the TLC plate). At the other extreme, analytes that interact strongly with the stationary phase will be retained for relatively long periods and will elute much later (not travel very far if at all on the TLC plate). Compounds that interact moderately with the stationary phase will be retained for an intermediate period of time and thus will elute between those described in the first and second cases. In instrumental techniques, the stationary phase material is loaded into a column; thus, the stationary phase is often referred to as the column. Column technology and column design will be discussed in the later chapters on these techniques. The most expensive part of the HPLC system is the solvent pump (or pumps) used to push the mobile phase through the column. These pumps must achieve pressures as high as 300+ atm in order to obtain high performance in term of resolution criteria for the separation. These pumps add significant cost to the instrument.

Problem 1.12: Consider what you know about MS and HPLC. Why is it only relatively recently that MS has been used as a detector for HPLC?

Problem 1.13: Consider the chromatogram given in Figure 1.13, and note that peak 7, methamphetamine, and peak 8, 3,4-methylenedioxymethamphetamine, are not fully separated. Why do you think the experimenters had difficulty in that separation?

In this chapter, we have discussed some relative advantages and disadvantages of five common instruments: UV-vis is great for *quantitative analysis*; FTIR, NMR, and MS are great *qualitative tools*; and chromatography often precedes the other methods and/or uses them as detectors (Figure 1.14). But we mentioned in the introduction to this chapter that a typical analytical chemistry laboratory would have *each* of these instruments. That is, an analytical facility that has only FTIR instruments or only NMR instruments would more likely be a specialty center, involved in research or consulting, and not a general analytical laboratory. Keeping to the analogy of a toolbox, we would

---

[19] UPLC pressures are much higher and thus take advantage of the efficiency gains associated with particle sizes smaller than 2 μ.

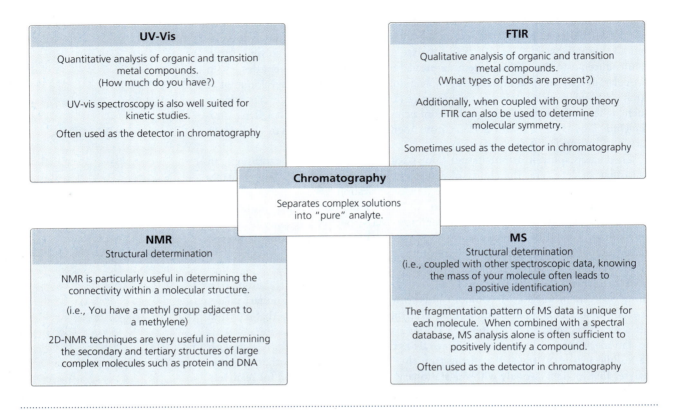

Figure 1.14 Summary diagram comparing five instrumental "tools" found in many analytical laboratories.

not expect a carpenter to have only a hammer or only a saw in his or her toolbox. The name of this chapter plays on this analogy. On occasion, the carpenter only requires the use of one tool from the toolbox, but to be a truly versatile carpenter, you need to gain proficiency with a wide range of tools.

Likewise in the laboratory, you will occasionally encounter an analysis that only requires the use of one instrument, or perhaps two. But truly complex chemical problems require several to be used in concert to arrive at a satisfactory determination. As you work your way through this textbook, be mindful to put each new technique into the *context* of your overall analytical toolbox. Ask yourself these questions:

- What does this new technique do (better or worse) when compared to other techniques I have already studied?

- How might I use this new technique in conjunction with the techniques I have already studied?

Most people have heard of γ-hydroxybutyric acid (GHB) because of its illegal use as a recreational street drug and from its more nefarious moniker, the "date rape drug." It is a depressant and an intoxicant, and it is known to cause euphoria, disinhibition, amnesia, and unconsciousness in users. Over the past decade, researchers have sought to overcome a core difficulty in the forensic analysis of aqueous GHB-containing samples. The drug easily converts to the free acid and other byproducts through reaction with water. Additionally, the street "quality" drug usually has impurities left over from clandestine synthetic techniques.* Furthermore, the free acid form of GHB is produced by hydrolysis of another common street drug, γ-hydroxybutyrolactone (GBL) in aqueous samples. The forensic analysis of these samples was hindered by the inability of FTIR to exclusively distinguish among the various forms of the drugs, and by the lack of availability of standard reference samples for the free acid, because GHB is usually produced in the carboxylate salt form. Chemists at the US Drug Enforcement Administration and US Food and Drug Administration solved the problem, but it took a combination of many instrumental methods to achieve that accomplishment.[†,‡]

First, the researchers synthesized a pure reference sample of the free acid form of GHB. They tested a series of synthesis protocols and used HPLC with UV detection to measure the yields of each method and to monitor the unreacted salt form of GHB and any GBL formed in the process. Second, they tested a series of real-world samples, such as GHB added to a beverage (where the pH is often low). In such a case, the GHB might exist in equilibrium between the salt and free acid forms, and some portion of the GHB might have been converted to GBL. Established GHB forensic protocols were not attuned to looking for GHB in all of its forms in such a situation. At a pH of less than 2.7, GHB exists almost entirely as the free acid form, but above that pH (as would be the case in the beverage of a would-be date rape victim) the interconversion reactions make the other forms viable. Throughout the study, researchers used a combination of HPLC-UV, HPLC-MS, GC-MS, ^{1}H-NMR, and FTIR. In the end, they proposed that forensic samples suspected of containing GHB or GBL be analyzed with a combination of FTIR and ^{1}H-NMR: FTIR would distinguish between free acid and carboxylate GHB, and ^{1}H-NMR would distinguish between GHB and GBL. See Figures 1.15 and 1.16. ■

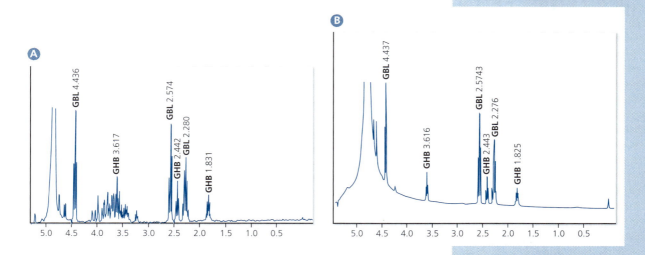

Figure 1.15 Proton nuclear magnetic resonance spectra of aqueous-based GBL forensic exhibits containing γ-hydroxybutyric acid (GHB) free acid: (A) a commercially formulated γ-butyrolactone (GBL) "beverage," (B) residue from a soft drink bottle used for storage of GBL and rinsed with water.

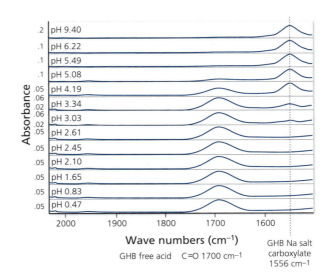

Figure 1.16 Stacked plot of expanded region infrared spectra of NaGHB/HCl reaction mixtures in D$_2$O as a function of solution pH after reaction.

---

* *Ciolino, L. A.; et. al. J. Forensic Sci., 2001, 46, 1315–1323.*

† *DeFrancesco, J. V.; et al. J. Forensic Sci., 2006, 51(2), 321–329.*

‡ *Witkowski, M. R.; et al. J. Forensic Sci., 2006, 51(2), 330–339.*

1.7 Further Reading

BOOKS

Bell, S. *Forensic Chemistry*, 2nd ed.; Prentice Hall: Upper Saddle River, NJ, **2012**.

de Hoffmann, E.; Stroobant, V. *Mass Spectrometry: Principles and Applications*, 3rd ed.; Wiley-Interscience: Chichester, England, **2007**.

Griffiths, P.; De Haseth, J. A. *Fourier Transform Infrared Spectrometry (Chemical Analysis: A Series of Monographs on Analytical Chemistry and Its Applications)*, 2nd ed.; Wiley-Interscience: Chichester, England, **2007**.

Jacobsen, N. E. *NMR Spectroscopy Explained: Simplified Theory, Applications and Examples for Organic Chemistry and Structural Biology*; Wiley-Interscience: Chichester, England, **2007**.

Snyder, L. R.; Kirkland, J. J.; Dolan, J. W. *Introduction to Modern Liquid Chromatography*, 3rd ed; John Wiley & Sons: Hoboken: NJ, **2009**.

Sparkman, O. D.; Penton, Z.; Kitson, F. G. *Gas Chromatography and Mass Spectrometry: A Practical Guide*, 2nd ed.; Academic Press: Amsterdam, **2011**.

JOURNALS

Brettell, T. A.; Butler, J. M.; Almirall, J. R. Forensic Science, *Anal. Chem.* **2011**, *83(12)*, 4539–4556.

Leclercq, L.; Cuyckens, F.; Mannens, R.; de Vries, P. T.; Evans, D. C. Which Human Metabolites Have We MIST? Retrospective Analysis, Practical Aspects, and Perspectives For Metabolite Identification and Quantification in Pharmaceutical Development. *Chem. Res. Toxicol.*, **2009**, *22(2)*, 280–293.

1.8 Exercises

EXERCISE 1.1: Given the relationship absorbance = εbc, where ε is the molar absorptivity (in units of $cm^{-1} \cdot M^{-1}$) and c is the concentration in molarity, estimate ε for the peak seen at 356 nm in Figure 1.2.

EXERCISE 1.2: The molar absorptivities for all three peaks exhibited by DPK in Figure 1.2 have been determined in this chapter. What conjectures can you make about the transitions that give rise to those peaks?

EXERCISE 1.3: Consider the peak at about 1,000 cm^{-1} in Figure 1.4.

(a) Convert the wave number (cm^{-1}) to wavelength in micrometers.

(b) Estimate the absorbance of that peak and convert it to percent transmittance.

EXERCISE 1.4: The Fourier transform is mentioned with respect to two instruments in this chapter. Find a reliable website that describes the Fourier transform and summarize your findings in 250 words or less.

EXERCISE 1.5: Rationalize the positions of the base peaks of the two spectra shown in Figure 1.12.

EXERCISE 1.6: What region of the EMS (in Hertz) is utilized by each of the following spectroscopies?

(a) UV-vis   (b) FTIR

Chapter 2

Quantum Mechanics and Spectroscopy

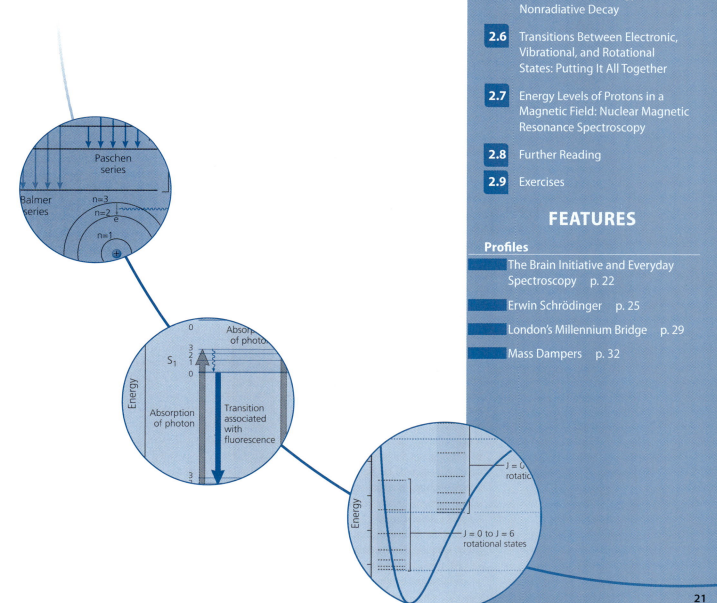

PROFILE

The Brain Initiative and Everyday Spectroscopy

In April of 2013, President Obama unveiled the White House Brain Initiative with $100 million in seed money and pledges of nearly the same from private sources to map the human brain.* The Brain Initiative promises to revolutionize our understanding of the human mind and uncover new ways to treat, prevent, and cure brain disorders such as Alzheimer's disease, schizophrenia, autism, epilepsy, and traumatic brain injury. One of the key tools that will be used in the White House Brain Initiative is *functional magnetic resonance imaging,* which is a specialized form of *nuclear magnetic resonance (NMR)* spectroscopy. The underlying principles of NMR spectroscopy and magnetic resonance imaging (MRI) are essentially the same, and we will examine the underlying quantum mechanics of NMR in this chapter. A more complete coverage of NMR occurs in Chapters 14 and 21. NMR spectroscopy is also a powerful technique used by chemists to characterize molecules. NMR is a beautiful example of the abstract world of quantum chemistry playing an important role in helping society; in addition to medical imaging (MRI), NMR is an important tool used today in drug design and discovery.† All forms of spectroscopy are based on the measurement of energy as an atom or molecule undergoes a change in quantum states. This chapter reviews the basic quantum mechanics of spectroscopy. ■

* *http://www.whitehouse.gov/infographics/brain-initiative.*

† *Klages J.; Coles M.; Kessler H. NMR-Based Screening: A Powerful Tool in Fragment-Based Drug Discovery. Analyst,* **2007** *Jul;132(7):693–705.*

2.1   Introduction

Quantum mechanics exists at the interface of chemistry and physics. This chapter reviews aspects of quantum mechanics necessary to understand instrumental techniques and analyses involving the interaction of light and matter. The material in this chapter will provide a foundation for many of the instrumental techniques covered in this textbook. Fundamentally, spectroscopy explores particular energy level differences between quantum states in molecules and atoms.

Quantum mechanics is used to accurately model systems on the atomic or molecular level. Without an understanding of quantum mechanics, we could not have developed the spectroscopic techniques we use to investigate atomic composition and chemical structures or the advanced medical imaging devices such as magnetic resonance imaging (MRI). For example, energy levels associated with the vibrational states of molecular bonds are found in the infrared region of the electromagnetic spectrum (FTIR), and the energy levels associated with the valence electron energy levels of atoms and molecules occur in the ultraviolet-visible and near-infrared regions of the electromagnetic spectrum (UV-vis). Details associated with the appropriate instruments will come in later chapters, but suffice it to say that infrared absorption spectroscopy, Raman spectroscopy, UV-vis absorption spectroscopy, fluorescence spectroscopy, X-ray fluorescence spectroscopy, and nuclear magnetic resonance (NMR) spectroscopy can only be fully understood in terms of the underlying quantum mechanics. You are encouraged to review sections of this chapter as you study instrument specific techniques later in the text.

At the core of many experimental approaches in spectroscopy is the notion that atoms and molecules exist in well-defined energy states. For example, the cover art of this

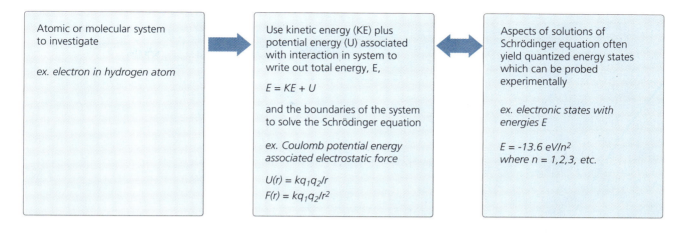

Figure 2.1 Quantized energy states in a nutshell. From left to right, this is the process to determine energy states from the Schrödinger equation. From far right to left, one can think of this as how one can come to understand fundamental interactions from experimentation.

chapter is a glimpse of Figure 2.4 which outlines the quantum mechanic events leading to the line spectrum of hydrogen in a gas in a discharge tube. Figure 2.4 represents transitions between well-defined electronic states for hydrogen. These well-defined, or quantized, states can be calculated by solving the *Schrödinger equation* for a given potential energy of an atom or molecule. Potential energy is related to force (i.e., attraction and repulsion). The Schrödinger equation, in general, is a partial differential equation,[1] and learning to solve it for different systems is a skill developed in a physical chemistry course. A schematic of the process involving the Schrödinger equation and the associated inputs and useful outputs is shown in Figure 2.1. An important part of the process of quantum mechanics is formulating the mathematical sum of the kinetic energy and potential energy of the atom or molecule being investigated. Aspects of the solution of the Schrödinger equation can yield the form of the energy states for the system. Depending on the nature of the system, the energy level differences can take on a wide range of values.

The focus of this textbook is not to perform quantum mechanical calculations of this type. However, it is important to realize that spectroscopic analysis is a way of probing the fundamental quantum states of the atom or molecules under investigation and we use conclusions of quantum mechanics to understand the results. The scientist matches spectroscopic data to changes in quantum states and uses his or her understanding of quantum mechanics to draw conclusions. Fundamentally, it is the uniqueness of the properties of an atom's or molecule's quantized energy states that allows us to make determinations of the identity, energy, structure, and properties of unknown compounds. We are reviewing quantum mechanics at this time so that we can better understand the spectroscopic techniques outlined later in the text.

The hydrogen atom, having only one proton and one electron, is a key case study for demonstrating the principles of quantized energy states and the connection to spectroscopy. In this case, the electron in the hydrogen atom can occupy only one of a set of very well defined states, as depicted in Figure 2.2. These energy states, described mathematically in Equation 2.1, are found by solving the Schrödinger equation for the forces acting on an electron (the Coulombic force that exists between the charged electron and nucleus). For hydrogen, the energy of a given state is written as:

$$E_n = \frac{-13.6 \text{ eV}}{n^2} = \frac{-2.179 \times 10^{-18} \text{ J}}{n^2}$$ **Eq. 2.1**

---

[1] We show the Schrödinger equation (although a less general version that is time independent and one dimensional) in the profile on Schrödinger later in this section.

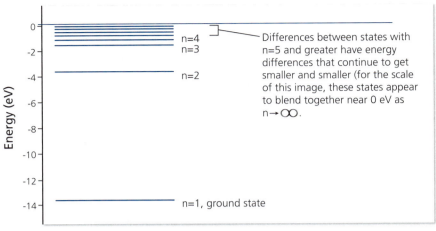

Figure 2.2 Hydrogen atom electron energy levels.

where n is the primary quantum number and can take on positive integer values from 1 to infinity.[2] Solving the Schrödinger equation for multielectron atoms is challenging and generally involves approximations and computer-based techniques. In other words, no tidy equation exists for energy levels of these atoms, but this approach leads to an increased understanding of multielectron atoms.

Figure 2.2 and Equation 2.1 show only the principal quantum number *n*. As we discuss spectroscopy, in this and subsequent chapters, we will introduce additional quantum numbers: most notably, vibrational quantum numbers and rotational quantum numbers for molecules and spin quantum numbers for electrons within an atom or molecule. In general, spectroscopy is the measure of electromagnetic radiation (either emitted or absorbed) as atoms or molecules undergo changes in quantum states.

2.2 The Interaction Between Electromagnetic Radiation and Matter: Absorption and Emission of Light

Quantized energy states in atomic or molecular systems can be probed experimentally with electromagnetic radiation (light). The electromagnetic spectrum, introduced in Chapter 1, is worth reviewing. The interaction of light with matter can take several forms.

A photon can be absorbed or emitted only when its energy, $E_{photon,}$ is the same as the difference in energy between two quantum states of an atom or molecule, ΔE. Mathematically, this means that absorption or emission can occur if:

$$\Delta E = E_{photon}$$ **Eq. 2.2**

where

$$\Delta E = E_f - E_i$$ **Eq. 2.3**

and

$$E_{photon} = \frac{hc}{\lambda} = \frac{1.99 \times 10^{-25} \text{ J} \cdot \text{m}}{\lambda} = h\nu$$ **Eq. 2.4**

---

[2] The value of 13.6 eV (2.187×10^{-18}J) is known as the Rydberg constant.

Erwin Schrödinger was a brilliant theoretical physicist who is credited with developing what we now call *quantum mechanics*. In 1926 Schrödinger published a paper whose title translates as "quantization as a problem of proper values."* In this paper, Schrödinger solved his new wave equation, shown below, for the electron in a hydrogen atom. The energy values from Schrödinger's calculations were essentially the same as those determined experimentally by spectroscopists for hydrogenic atoms.† The one-dimensional, time-independent version of the Schrödinger equation is written as:

$$\frac{-h^2}{8\pi^2 m}\frac{d^2\,\psi(x)}{dx^2}+U\,\psi(x)=E\,\psi(x)$$

where h is Planck's constant, m is the mass of the particle described by the wave function ψ, U is the potential energy of the system, and E is the total energy.

Born in 1887, Schrödinger was the only child of a botanist. His maternal grandmother was a professor of chemistry. Home schooled to the age of 11, he entered formal education in the classics in 1898. In 1910 he received a PhD in theoretical physics from the University of Vienna (at age 23!). After he completed his schooling, he served as an artillery officer during World War I. In Zurich, between World Wars I and II, he invented quantum mechanics! In 1927 he moved to Berlin to be closer to Einstein, but in 1933 he fled to Oxford, England, to escape working for Hitler. He received the Nobel Prize shortly thereafter. He returned to Austria and remained there until Hitler invaded Austria in 1938. After a short stay at Princeton, he took a position as Director of the School of Theoretical Physics in Dublin, Ireland, and then returned to Austria, where he remained until his death in 1961. He was also an accomplished poet and sculptor. ■■

PROFILE
Erwin Schrödinger

---

* Schrödinger, E., *Quantisierung als Eigenwertproblem. Annalen der Physik,* **1926**, *(384), 361–377.*

† *Atoms with only one electron.*

E_f and E_i are the energies of the final and initial states of the atom or molecule, λ is the wavelength of the absorbed photon, h is Planck's constant, c is the speed of light, and ν is the frequency of the absorbed photon. Various textbooks may use different notation; for example, Equation 2.3 may sometimes be written in terms of E_2 (instead of E_f) and E_1 (instead of E_i). It is the *difference* in the energies of the states in the atom or molecule that corresponds to the energy of the absorbed photon. It is important to see that energy conservation must be upheld in this relationship.

A generic energy level sketch showing photons being absorbed at three different energies is shown in Figure 2.3. Although many wavelengths of light are incident on the sample, the only photons absorbed are those photons whose energy matches the energy gap between two quantum states within the sample. In this example, there are only three wavelengths of light that match this criterion. The corresponding absorption and transmission spectra are also sketched. In this sketch, the incident light excites electrons from the ground state (the lowest energy state of the system) and the electrons

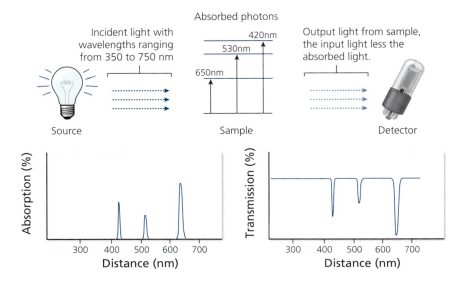

Figure 2.3 Incident light on a molecule is absorbed at three wavelengths. The energy of the absorbed photons corresponded to energy differences between quantum states within the sample. The corresponding absorption spectrum and transmission spectrum are also shown.

are promoted to three different excited states. A system can also absorb forms of energy other than electromagnetic. For example, thermal energy or electron beams can promote the energy state of a system.

Substitution of Equation 2.3 into Equation 2.4 yields Equation 2.5. This is known as the Bohr equation. The Bohr equation correctly predicts the energy levels for the quantum states in the hydrogen atom. If one includes the square of an ion's atomic number (Z^2) in the Bohr equation, one will produce an equation that correctly predicts the energy levels for hydrogenic ions[3] (ions that contain only one electron). See Equation 2.6.

$$\Delta E = E_f - E_i = h\nu = \frac{hc}{\lambda} = -2.179 \times 10^{-18}\,\text{J}\left(\frac{1}{n_f^2} - \frac{1}{n_i^2}\right) = E_{photon} \qquad \textbf{Eq. 2.5}$$

$$\Delta E = E_f - E_i = h\nu = \frac{hc}{\lambda} = -2.179 \times 10^{-18}\,\text{J}\left(Z^2\right)\left(\frac{1}{n_f^2} - \frac{1}{n_i^2}\right) = E_{photon} \qquad \textbf{Eq. 2.6}$$

We can see that the energy states for an electron in a hydrogen atom (or hydrogenic ion) are governed by the quantum states (n) in which the electron resides.

Problem 2.1: Calculate the wavelength and energy of photons necessary to promote an electron from the n = 3 state to the n = 4, 5, 6, and 7 states in hydrogen. Label the region of the electromagnetic spectrum for each of the emission lines.

---

[3]One must account for electron–electron repulsion if one wishes to predict the energy states for atoms or ions that contain more than one electron.

Example 2.1

(a) Suppose the electron in the hydrogen atom is in the n = 2 state. Use Equations 2.3 and 2.4 to determine what wavelength photon could excite this electron to the n = 3 state.

$$E_3 - E_2 = \frac{hc}{\lambda}$$

$$\lambda = \frac{hc}{E_3 - E_2} = \frac{1.99 \times 10^{-25}\,\text{Jm}}{-2.42 \times 10^{-19}\,\text{J} - (-5.45 \times 10^{-19}\,\text{J})} 6.56 \times 10^{-7}\,\text{m} = 656\,\text{nm}$$

(b) What would happen if we shine 800 nm light on the atom? Would the electron move to a different energy level? The electron would be excited to another energy level only if there was a state, $E_?$, such that

$$E_? - E_2 = \frac{hc}{800\ \text{nm}}$$

Solving for $E_?$, we find

$$E_? = -5.45 \times 10^{-19}\,\text{J} + \frac{1.99 \times 10^{-25}\,\text{J m}}{8.00 \times 10^{-7}\,\text{m}} = -2.96 \times 10^{-19}\,\text{J}$$

But is there a state with this energy? Check the quantized energy level equation for hydrogen and one will find that the answer is no. Therefore, there is no absorption of 800 nm photons.

A system can also *emit light* as the system decays from a higher quantum state to a lower one. This is seen in Figure 2.4, where we show several transitions that may result in emission from the hydrogen atom. There are several types of emission that we will encounter; our example here is atomic fluorescence. Figure 2.4 demonstrates *electronic transitions* in an *atomic system*. Hydrogen is the simplest atomic system and when we study atomic spectroscopy later in the text (Chapters 7 and 9), we will move beyond the hydrogen model and explore larger atoms.

Problem 2.2: Can an electron in a hydrogen atom moving from one energy level to a lower energy level cause a transition in the microwave region of the spectrum? Investigate only transitions between adjacent states of principal quantum number n.

In Figure 2.3, we represented the quantum mechanical changes that occur as a sample absorbs light energy. In Figure 2.5, we represent the same sample as electrons in an excited quantum state relax to the ground state. The sample emits at three wavelengths as the sample returns to the ground state. In Figure 2.5, there are three radiative transitions from three excited states to the ground state. A corresponding emission spectrum is shown, indicating the three emission lines associated with the relaxation events.

Quick Connection Between Quantum Mechanics and Spectroscopy

It is important to point out that historically, spectroscopy is what first allowed us to understand the quantum mechanical nature of the universe; in fact, much of our understanding of quantum mechanics came from early spectroscopic experiments. For example, the observed emission of radiation from excited gases played a pivotal role in the development of quantum mechanics.

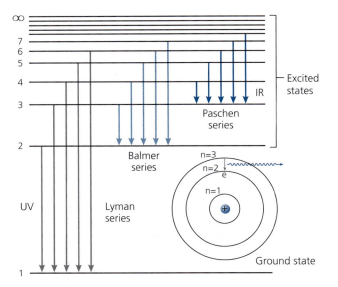

Figure 2.4 Emission from various states in hydrogen.

Example 2.2

A large voltage is used to excite hydrogen gas in a transparent tube. You measure many emission wavelengths, including light at 397.4 nm, 410.9 nm, and 433.9 nm. By trial and error, find the initial and final states for each of these emission lines.

STRATEGY –

Solve Equation 2.5 for λ and input different n-values until you find the corresponding wavelengths. Use a spreadsheet program so that you only need to input the n-values.

SOLUTION –

We might start by calculating the wavelengths for transitions from various values of n to the ground state, n = 1. We can calculate these wavelengths following the above example and will see that the n = 2 to n = 1 transition results in a 121.6 nm photon and the n = 3 to n = 1 in a 102.6 nm photon. As we increase the initial state, the wavelength decreases, showing that our emission lines do not come via emission to the ground state. We should then calculate the wavelengths associated with transitions down to n = 2.

The following table shows the results for more transitions.

Answer	
n = 3 to n = 2	656.6 nm
n = 4 to n = 2	486.6 nm
n = 5 to n = 2	433.9 nm
n = 6 to n = 2	410.9 nm
n = 7 to n = 2	397.4 nm

When London's Millennium Bridge opened to pedestrian traffic in 2000, the bridge began to experience large sideways oscillations. The oscillations were the result of a resonance between the stepping frequency of the pedestrians and one of the bridge's natural vibrational modes. Apparently, an inherent low frequency oscillation of the bridge was close enough to the natural frequency of a human pedestrian that it caused the pedestrians to subconsciously synchronize their steps to the bridge's natural frequency. The result being that the human traffic reinforced the oscillation and drove the bridge's oscillation to a higher amplitude.* ■

* Abdulrehem M. M. and Ott E.; *Low Dimensional Description of Pedestrian-Induced Oscillation of the Millennium Bridge, Chaos, 2009, (1)*

PROFILE

London's Millennium Bridge

Problem 2.3: This problem makes use of the four-level system and associated emission lines from Figure 2.5. Assume there is an additional energy level E_5 that has an associated emission to E_4 with a wavelength of 710 nm. Sketch the energy level diagram for this five-level system, clearly labeling the energies of all $E_{excited}$ to E_1 transitions.

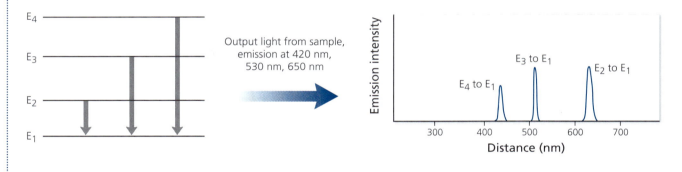

Figure 2.5 Emission of three different energy photons from a generic molecule as excited state electrons relax to the ground state.

2.3 Molecular Vibrations Lead to Quantized Energy Levels

So far, we have discussed electronic states for the hydrogen atom. We will now describe the quantum states that are unique to molecules. Any molecule will certainly have electronic energy levels, analogous to the atomic system, but it will also have other energy states. The goal in this section is to provide a foundation in quantum mechanics for these other states, in particular, states associated with *vibrating molecules*. We will start by describing diatomic molecules (examples include H_2, N_2, CO, HCl, KBr) because they lay the groundwork for understanding infrared absorption spectroscopy. Molecular vibrations have characteristic absorption values that correspond to the infrared region of the electromagnetic spectrum.

We will use the harmonic oscillator model—at its simplest representing a system as a mass on a spring—as a starting point to model molecular vibrations. The harmonic oscillator model and modifications with various levels of sophistication are used to

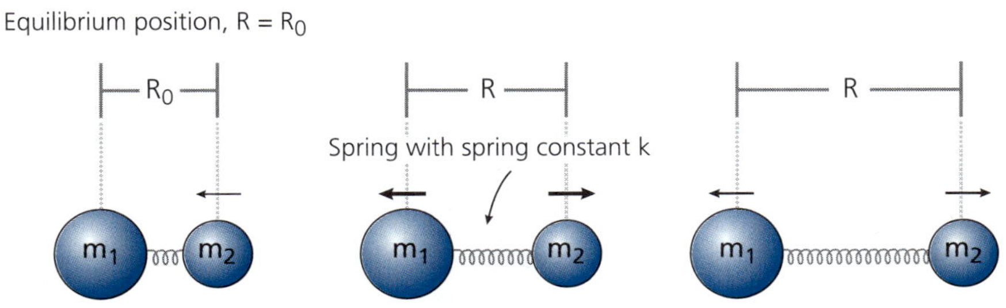

Equilibrium position, R = R$_0$

Figure 2.6 Molecular arrangement of diatomic molecule and model with connecting spring representing molecular bond.

describe physical systems throughout nature.[4] In particular, we will use the simple harmonic oscillator to model the two-atom system by treating the atoms as two objects with mass connected by a spring (the chemical bond). Our diatomic system is shown in Figure 2.6, where atoms 1 and 2 (of mass m$_1$ and m$_2$, respectively) of the molecule are separated by distance R. Attraction between atoms is the result of molecular bonding. We will model the molecular bonding as a spring that holds the two atoms together, with the atoms allowed to move along the axis as shown in Figure 2.6. We now need a mathematical form of the force between the two atoms so that it is possible to calculate the energy differences between the different quantum levels. Again, we turn to the Schrödinger equation, this time with a parabolic potential energy function:[5]

$$U(R) = U(R_0) + \frac{1}{2}k(R - R_0)^2 \qquad \textbf{Eq. 2.7}$$

where R is the distance between the atoms in the diatomic molecule, R_0 is the equilibrium distance (when the spring is in its relaxed state) between the atoms where the energy is a minimum, and k is an *effective spring constant*. $U(R_0)$ is the value of the potential energy at the lowest energy value. The stiffer the molecular bond, the larger the spring constant.[6] This potential energy is associated with a Hooke's law–type restoring force (Equation 2.8).

$$F = -k(R - R_0) \qquad \textbf{Eq. 2.8}$$

The plot of energy versus atomic separation, R, is shown in Figure 2.7. The plot shows the actual molecular potential (solid line) and an approximation for the potential using a parabolic simple harmonic potential to approximate the full molecular potential (dashed line). This approximation allows one to solve the Schrödinger equation for this potential. This simplification is valid as long as the amplitude of the atomic oscillations does not become too large. One can see how the parabola matches the molecular potential for values of R around the equilibrium position, R_0.

Using the simple harmonic oscillator model, the vibrational energy for a diatomic molecule has quantized energy levels that take the form

$$E = \left(n + \frac{1}{2}\right)\hbar\omega_{oscillator} \qquad \textbf{Eq. 2.9}$$

[4] Abdulrehem, M. M.; Ott E. Low Dimensional Description of Pedestrian-Induced Oscillation of the Millennium Bridge, *Chaos*, **2009**, (*1*), 013129.

[5] See Levin, I. N. *Physical Chemistry*, 6th ed. (Chapter 20); McGraw-Hill: New York, **2009**.

[6] See Robinett, R., *Classical Results, Modern Systems, and Visualized Examples*, 2nd ed.; Oxford University Press: New York, **2006**.

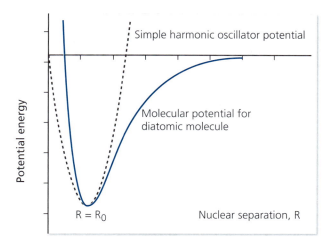

Figure 2.7 Molecular potential and simple harmonic approximation for diatomic molecule.

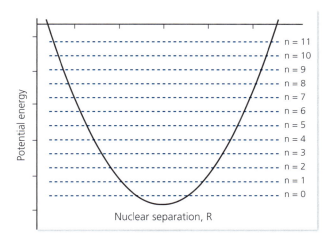

Figure 2.8 Quantized energy levels associated with molecular vibration of a diatomic molecule.

where n must be an integer $(n = 0, 1, 2 \ldots)$, and ω is the angular frequency (see Equation 2.10) of the oscillation of the molecule. The quantized energy levels are shown in Figure 2.8 along with the simple harmonic oscillator potential.

In this context, we are using n for the quantum number associated with the harmonic oscillator model; other sources may use a different letter for this quantum number, such as the Greek letter ν, which can easily be confused with frequency.

Unlike the hydrogen atom energy states, n *can be zero* for the diatomic molecule oscillations. This means that the lowest energy state is just above the minimum of the potential energy, as can be seen in Figure 2.8. Figure 2.8 also shows us that ΔE between adjacent vibrational states is *constant*, unlike ΔE for electronic states where the distance between adjacent states decreases as $n \to \infty$. When we consider a more sophisticated potential (see Problem 2.6 for an example using the Morse potential[7]), which takes the asymmetry of

---

[7] For a review of Morse potential, see Atkins, P. E.; de Paula, J. *Physical Chemistry*, 8th ed. Oxford University Press: Oxford, UK, **2006**.

PROFILE
Mass Dampers

In addition to molecular vibrations, the concept of matching the natural frequency of an object with a driving frequency (resonance) has many applications. Just like molecules, buildings (as well as all structures) have particular natural frequencies that depend on a structure's size and rigidity. When an earthquake or winds drive the building at these special natural frequencies, extremely large amplitude motion is possible. One technology to deal with this resonance issue in skyscrapers is the tuned mass damper. For instance, the Taipei 101 building has a tuned mass damper that weighs 660 tons suspended from the 87th to the 92nd floors of the building. When the building is driven by an earthquake or winds near the natural frequency of the building, the tuned mass damper is set to move such that it exactly opposes the motion of the building, effectively canceling out the effects of resonance and keeping the building from serious damage. The building is designed to withstand winds in excess of 134 mph and the estimated strongest possible earthquakes in the region. ■

the potential into account, we see that the energy levels are fairly equally spaced for low energies but that they do get closer together at higher energies. For most chemical systems, it is reasonable to assume the vibrational energy states are equally spaced.

The *angular frequency of oscillation* is

$$\omega_{oscillator} = \sqrt{\frac{k}{\mu}}$$

Eq. 2.10

where k is again *the effective spring constant* (related to the rigidity of the chemical bond between the two atoms in the molecule) and μ is *the reduced mass of the system*. The reduced mass can be found by

$$\mu = \frac{m_1 m_2}{m_1 + m_2}$$

Eq. 2.11

where m_1 and m_2 are the masses of each atom in the diatomic molecule.

We will at times need to relate *angular frequency* $\omega_{oscillator}$ (units of rad/s) to the frequency $v_{oscillator}$ (units of 1/s or Hz) by

$$\omega_{oscillator} = 2\pi v_{oscillator} \qquad \textbf{Eq. 2.12}$$

The quantized states discussed here are associated with vibrations of the molecule and are not due to electronic transitions. Quantum mechanics dictates that the energy of the vibration can only take on certain values. We probe these energy states using absorption or emission of radiation just as we discussed above for the electronic states of atomic systems. As we have already mentioned, transitions associated with molecular *vibration* correlate with energies in the infrared region of the electromagnetic spectrum. In other words, the photon's frequency and the molecular vibration must be in resonance in order for the molecule to absorb the photon's energy.

As we have seen when we discussed electronic transitions, we often use wavelength to describe the energy of a photon. Another commonly used quantity to describe photon energy, especially when investigating vibrational absorption, is the *wave number*, $\tilde{v}$. To calculate the wave number, convert the wavelength to cm and take the reciprocal.

$$\tilde{v} = \frac{1}{\lambda \, \text{cm}} \qquad \textbf{Eq. 2.13}$$

We have just seen that the force that binds two atoms to form a diatomic molecule can be modeled as a system of two masses (the atoms) connected by a spring (the bond) and that this system exhibits quantized energy levels that are equally spaced. This model works well as long as the molecular separation is reasonably close to the equilibrium separation.

Until now, we have ignored an important aspect of transitions between energy levels for our model systems as we have assumed that *transitions can occur between any two states*. In fact, transitions between some states occur with much higher probability than others; these are deemed *"allowed"* transitions (compared to *"forbidden"* ones). Allowed transitions follow *selection rules*, which are derived from quantum mechanics. For the vibrational energy states, modeled by the simple harmonic oscillator: allowed vibrational transitions have $\Delta n = \pm 1$.

Example 2.3

The spring constant between the oxygen atoms in $^{16}O_2$ is around 1177 N/m. Calculate the angular frequency and the vibrational frequency of O_2 molecule.

For the angular frequency:

$$\omega_{oscillator} = \sqrt{\frac{k}{\mu}} = \sqrt{\frac{k}{\dfrac{m_1 m_2}{m_1 + m_2}}} = \sqrt{\frac{1177 \, \text{N/m}}{\dfrac{(16\,\text{u})(16\,\text{u})}{32\,\text{u}}\left(\dfrac{1.661 \times 10^{-27}\,\text{kg}}{1\,\text{u}}\right)}} = 2.97 \times 10^{14} \, \text{rad/sec}$$

For the frequency:

$$v_{oscillator} = \frac{\omega}{2\pi} = \frac{2.97 \times 10^{14}}{2\pi} = 4.73 \times 10^{13} \, \text{s}^{-1}$$

Example 2.4

The spring constant between the atoms in HBr is around 412 N/m. Determine the energy difference between the n = 0 and n = 1 states for this molecule in joules. What photon wavelength, in units of microns, and wave number, would be associated with absorption from n = 0 to n = 1? Label the region of the electromagnetic spectrum.

For the angular frequency:

$$\omega_{oscillator} = \sqrt{\frac{k}{\mu}} = \sqrt{\frac{k}{\frac{m_1 m_2}{m_1 + m_2}}} = \sqrt{\frac{412\,\text{N/m}}{\frac{(1\,\text{u})(80\,\text{u})}{81\,\text{u}}\left(\frac{1.661 \times 10^{-27}\,\text{kg}}{1\,\text{u}}\right)}} = 5.01 \times 10^{14}\,\text{rad/sec}$$

For the frequency:

$$V_{oscillator} = \frac{\omega_{oscillator}}{2\pi} = \frac{5.01 \times 10^{14}}{2\pi} = 7.98 \times 10^{13}\,\text{s}^{-1}$$

Calculating the energy level difference:

$$E_1 - E_0 = \left(1 + \frac{1}{2}\right)\hbar\omega - \left(0 + \frac{1}{2}\right)\hbar\omega = \hbar\omega = (1.05 \times 10^{-34}\,\text{J})(5.01 \times 10^{14}\,\text{rad/sec})$$
$$= 5.26 \times 10^{-20}\,\text{J}$$

Calculating the wavelength associated with a photon of this energy:

$$E_1 - E_0 = 5.26 \times 10^{-20}\,\text{J} = \frac{hc}{\lambda} \text{ and } \lambda = 3.78 \times 10^{-6}\,\text{m} = 3.78\,\mu\text{m}$$

This is in the infrared.

To calculate the wave numbers, convert the wavelength to cm and take the reciprocal:

$$\tilde{v} = \frac{1}{\lambda\,\text{cm}} = \frac{1}{3.78 \times 10^{-4}\,\text{cm}} = 2645\,\text{cm}^{-1}$$

Quantum mechanics also dictates that the net dipole moment of the molecule must change during the vibration for transitions between vibrational states. This means a diatomic molecule composed of two of the same atom (H_2, N_2, O_2) will not have *allowed* transitions associated with vibration. Nondiatomic molecules will have vibrations that may not have a changing dipole moment, so some of these oscillations will also not result in absorption. This critical aspect of vibrational spectroscopy will be discussed in more depth in Chapter 11 when we explore infrared spectroscopy and in Chapter 12 when we discuss Raman spectroscopy.

Problem 2.4:

(a) Calculate the reduced mass for NO and HCl.

(b) Would the reduced mass of H–^{35}Cl be different than H–^{37}Cl? If so, might this lead to a difference in infrared absorption? Support your response with a calculation.

(c) For Example 2.3, would you expect to measure infrared absorption for O_2?

Problem 2.5: The spring constant between atoms in HCl is around 516 N/m. Calculate the angular frequency and vibrational frequency of HCl molecule. Determine the energy difference between the n = 0 and n = 1 states for this molecule in joules. Is this an allowed transition? What photon wavelength, in units of microns and wave numbers, would be associated with absorption from n = 0 to n = 1? Label the region of the electromagnetic spectrum.

Problem 2.6: The absorption from n = 0 to n = 1 occurs at 2,170 cm^{-1} in ^{12}C–^{16}O.

(a) Calculate the spring constant between atoms in this molecule.

(b) State why this is an allowed transition.

2.4 Molecular Rotation Leads to Quantized Energy Levels

We just explored the diatomic molecule and assumed the atoms could vibrate along one direction. In addition, a molecule can rotate. We will assume that the rotation of the molecule does not actually affect the vibration (it does, but not much, so we can ignore it for now). This model is referred to as the rigid rotator. The rotating diatomic molecule has a kinetic energy that leads to quantized energy levels. Figure 2.9 has a sketch of the rigid rotator system.

The kinetic energy of the rotating molecule is:

$$KE = \frac{1}{2}I\omega^2 = \frac{1}{2}\frac{L^2}{I}$$

Eq. 2.14

where I is the rotational inertia of the molecule and ω is the angular velocity of the rotation. The rotational inertia is the rotational analog of mass and tells us, roughly, how difficult it is to change the angular velocity of a system. This is the same equation we would use to calculate, for example, the kinetic energy of a rotating merry-go-round or a spinning ice skater. For example, when a spinning ice skater holds her arms away from her body, her rotational frequency decreases, and when she drops her arms, her rotational frequency increases. There is a smaller rotational inertia when her arms are close to her body because the mass of the system is closer to the rotational axis. It is also possible to write the kinetic energy in terms of the angular momentum, L, of the system. Quantum states associated with the square of angular momentum, L^2, arise elsewhere, in particular, when a molecule is placed in an external magnetic field. These states are critical for understanding NMR spectroscopy and MRI, where an external magnetic field is used to split[8] energy levels. The square of angular momentum has associated quantized values, which when rewritten in terms of the kinetic energy yield, for the rigid rotator,

$$KE = J(J+1)\frac{\hbar^2}{2I} = E$$

Eq. 2.15

where the quantum number J must be zero or a positive integer. J is the *quantum number for total rotational angular momentum*. The moment of inertia is written as:

$$I = \mu R_0^2$$

Eq. 2.16

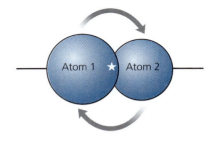

Figure 2.9 Sketch of rotating diatomic molecule; the star represents the center of mass of the molecule.

---

[8] In this context, we use the word *split* to indicate a situation in which we induce a difference in energy between two energy states originally at the same energy (degenerate).

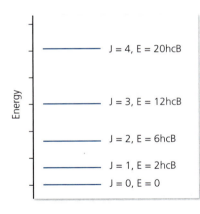

Energy

J = 4, E = 20hcB

J = 3, E = 12hcB

J = 2, E = 6hcB

J = 1, E = 2hcB

J = 0, E = 0

Figure 2.10 Quantized energy levels due to rotation in diatomic molecule.

where μ is *the reduced mass* of the molecule and R_0 is the equilibrium separation of the atoms. We see that the larger the distance between the atoms, the larger the rotational inertia. Combining Equations 2.15 and 2.16 yields the energy levels associated with rotation.

$$E = J(J+1)\frac{\hbar^2}{2\mu R_0^2} = hcBJ(J+1)$$

Eq. 2.17

where B is defined by:

$$B = \frac{\hbar}{4\pi cI}$$

and c is the speed of light. This energy relationship is quite useful because it contains information concerning the rotational inertia and hence information about the equilibrium atomic distance, also known as the bond length. Unlike the energy spacing for vibrational transitions, the spacing between adjacent energy levels *increases* with increasing J (see Figure 2.10). Check for yourself! Absorption will occur at energies 2hcB, 4hcB, 6hcB, and 8hcB.

Absorption due to transitions between rotational states tends to be in the microwave region of the spectrum. For the rotational energy states, modeled as rigid rotator, the selection rule is "allowed rotational transitions have $\Delta J = \pm 1$."

We have now calculated energy level differences for electronic, vibrational, and rotational states. Table 2.1 summarizes roughly the energy ranges and associated wavelengths for electronic, vibrational, and rotational states.

Example 2.5

The bond length for HCl is 1.3 Å. The molecule is rotating. What is the difference in energy of the J = 0 and J = 1 states? If electromagnetic radiation were to be absorbed with this energy, what would be the corresponding wavelength and wave number? In what region of the electromagnetic spectrum would this lie?

$$E_1 - E_0 = 1(1+1)\frac{\hbar^2}{2\mu R_0^2} = \frac{\hbar^2}{\dfrac{m_1 m_2}{m_1 + m_2}R_0^2}$$

$$= \frac{(1.05\times10^{-34}\text{ J s})^2}{\dfrac{(1\text{ u})(35\text{ u})}{36\text{ u}}\left(\dfrac{1.661\times10^{-27}\text{ kg}}{1\text{ u}}\right)(1.3\times10^{-10}\text{ m})^2} = 4.04\times10^{-22}\text{ J or } 0.0025\text{ eV}$$

$$\lambda = \frac{hc}{E} = \frac{(6.63\times10^{-34}\text{ J s})(3\times10^8\text{ m/s})}{4.04\times10^{-22}\text{ J}} = 4.92\times10^{-4}\text{ m or } 0.0492\text{ cm}$$

$$\tilde{v} = \frac{1}{\lambda\text{ cm}} = \frac{1}{0.0492\text{ cm}} = 20.3\text{ cm}^{-1}$$

This falls at the border between microwave and infrared.

TABLE 2.1: Summary of Electronic, Vibrational, and Rotational States

	ELECTRONIC STATES	VIBRATIONAL STATES	ROTATIONAL STATES
	Ultraviolet-visible	Infrared/microwave	Microwave
Rough energy range (J)	1×10^{-18}–2.9×10^{-19} J	2.9×10^{-19}–2×10^{-22} J	2×10^{-22}–2×10^{-25} J
Wavelength (m)	1.8×10^{-7}–7×10^{-7} m	7×10^{-7}–1×10^{-3} m	$> 1 \times 10^{-3}$ m

Problem 2.7: The frequency spacing between rotational levels is 511 MHz in HBr. Calculate the difference in energy of the $J = 0$ and $J = 1$ states. Calculate the bond length for HBr.

Problem 2.8: Rotational absorption is observed at 0.13 cm for the CO molecule. The transition is associated with the $J=1$ to $J=2$ energy levels. Use this information to estimate the rotational inertia for CO.

In Table 2.1, we ignored what transitions might occur when X-rays are directed onto an atom. X-rays have wavelengths between around 0.01 and 10 nm and are more energetic than ultraviolet light. The transitions associated with X-rays are quite different from those associated with visible light. An X-ray directed onto an atom can cause an inner orbital electron to be ejected from the atom, leaving behind a vacancy (a place where there *used to be* an electron). Another electron from an outer orbital will move to this vacancy, emitting an X-ray in the process. A secondary electron will fill that new vacancy, also potentially emitting an x-ray in order to obey conservation of energy. This emission of X-rays is called X-ray fluorescence (XRF) and the wavelength of the emitted X-ray can yield important information about the atomic composition of a sample. XRF is discussed in Chapter 10. One can also learn a lot about a sample by studying the ejected electrons via a technique called X-ray photoelectron spectroscopy. This is also covered in Chapter 10.

2.5 Transitions Between Vibrational and Rotational States: The Role of Thermal Energy and Nonradiative Decay

We have not yet discussed the role of thermal energy in electronic, vibrational, or rotational transitions. So far we have assumed that our system was in the ground state before excitation but thermal energy can populate excited states if the energy difference between quantum states is thermally accessible. For example, the thermal energy associated with room temperature (300K) is around 0.04 eV or 6.4×10^{-21} J. Example 2.5 showed us a situation where ΔE is on the order of 10^{-22} J, so it is very reasonable to assume that excited state rotational levels are occupied for molecules and atoms at room temperature.

A calculation using the Boltzmann distribution provides a more complete picture of populated energy levels due to thermal excitation. The *Boltzmann distribution* is

$$\frac{N_2}{N_1} = \frac{g_2}{g_1} e^{\frac{-\Delta E}{k_B T}}$$ **Eq. 2.18**

where N_1 and N_2 are the populations of two different states 1 and 2, k_B is the Boltzmann's constant (1.38×10^{-23} J /K), T is absolute temperature (Kelvin), g_1 and g_2 are the degeneracies of these states, E_1 and E_2 are the energies of the states (with $E_2 > E_1$), and ΔE is $E_2 - E_1$. The degeneracy of an energy level is the number of quantum states that produce *this particular energy*. For example, there might be a case where two states

The Boltzmann Distribution

The Boltzmann distribution is an important idea that will be used in several chapters in this text. In Chapter 9 on atomic emission spectroscopy, we use the Boltzmann distribution to calculate the percentage of copper atoms that will be in an excited state as a function of temperature. In Chapter 14 on NMR spectroscopy, we use the Boltzmann distribution to calculate the relative population of states for a rotating hydrogen atom as a function of an applied magnetic field. Compare Examples 2.6 and 2.7 to Example 9.1 and also see the Compare and Contrast feature in Section 14.3.

Example 2.6

Suppose a system (with no degeneracy so that $g_1 = g_2 = 1$) has energies $E_2 = 1 \times 10^{-22}$ J and $E_1 = 0.1 \times 10^{-22}$ J and is at T = 300K. What is the ratio of the populations? Interpret this ratio.

$$\frac{N_2}{N_1} = \frac{g_2}{g_1} e^{\frac{E_1 - E_2}{k_B T}} = e^{\frac{0.1 \times 10^{-22} \text{ J} - 1 \times 10^{-22} \text{ J}}{(1.38 \times 10^{-23} \text{ J/K})(300 \text{ K})}} = e^{-0.021} = 0.979$$

The interpretation is that the population N_1 is only very slightly more than that in N_2. This is because the thermal energy is close to the value of the difference in the energies of the states.

Example 2.7

Sodium has two closely spaced emission lines. One at 589.0 nm and the second at 589.6 nm. These are associated with transitions from two closely spaced 3p orbitals to the ground state 3s orbital. The degeneracy of the lower state is two and the degeneracy of the upper state associated with the 589.0 nm state is four. Produce a plot of the ratio of the number of electrons in the upper state to the lower state as a function of temperature from 2,000K to 5,000K. Use your plot to describe the population difference between the 3p and 3s orbitals at 3,500K and at 4,000K respectively, understanding that the larger the ratio, the more emission one is likely to detect.

We can write out the Boltzmann distribution equation with the relevant quantities as:

$$\frac{N_2}{N_1} = \frac{g_2}{g_1} e^{\frac{-\Delta E}{k_B T}} = 2 e^{\frac{-3.371 \times 10^{-19} \text{ J}}{(1.38 \times 10^{-23} \text{ J/K})(T)}}$$

This can be plotted in a spreadsheet program with T varying from 2,000K to 5,000K to yield the following plot.

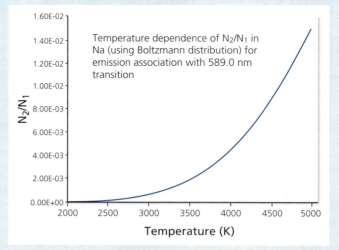

By comparing the populations at 3,500K and at 4,000K, we see that the N_2/N_1 ratio is more than twice as large at 4,000K ($\sim 5 \times 10^{-3}$) compared to 3,500K ($\sim 2 \times 10^{-3}$). We would expect more than twice the emission at 4,000K compared to 3,500K. We will revisit this method of determining relative populations via the Boltzmann distribution later in this chapter as we discuss NMR spectroscopy and again in Chapter 7 when we discuss flame temperatures in atomic absorption spectroscopy.

described by two different quantum numbers would have the same energy. The degeneracy of this energy level would be two. The degeneracy of a particular atomic state and orbital degeneracies for molecules is typically discussed in a physical chemistry or inorganic chemistry course.

It is important to have a sense of scale for these population values. Note that the distribution ratio (N_2/N_1) becomes smaller as ΔE becomes larger. Many transitions between atomic energy levels fall into the ultraviolet and visible region of the electromagnetic spectrum, and so the distribution ratio is on the order of 10^{-5} at room temperature, indicating that the overwhelming majority of the atoms are in the ground state. In NMR, the transitions fall in the radiofrequency range and the distribution ratio is very close to 0.5, placing almost half of the nuclei in the excited state from ambient thermal energy alone.

Problem 2.9: Suppose two different states (with no degeneracy so that $g_1 = g_2 = 1$) have energies $E_2 = 2 \times 10^{-22}$ J and $E_1 = 0.5 \times 10^{-22}$ J. At what temperature will N_2 have a population so that the ratio of $N_2/N_1 = 100$? Suppose now that the degeneracies are $g_1 = 3$, $g_2 = 2$. Is it possible to find a temperature where the population in N_2 is larger than N_1?

So far, we have seen that absorption of light (or in some cases thermal energy) can excite a system to a higher energy state and that emission of light results in the system transitioning to a lower energy state. A third class of transitions allows for a system to move to a lower energy state through nonradiative decay. A nonradiative process must conserve energy, so if a system moves from some E_i to E_f, the energy must have gone somewhere. For radiative decay, that energy is in the form of an emitted photon. For nonradiative decay, the energy of the transition is transferred to thermal energy.

Nonradiative decay is shown in Figure 2.11. Two electronic states and associated vibrational and rotational states are shown. The molecule has an electronic transition from the ground state to the first excited electronic state and then decays *nonradiatively* (without emitting light) to the ground *vibrational* state of the first excited electronic state. Note the light blue shading between the dashed lines. The light blue shading represents many closely spaced rotational energy levels. On the scale of electronic energy level differences, the rotational energy states are essentially "smeared" together about the vibrational levels. Due to nonradiative decay, the photon that is emitted as the molecule subsequently relaxes back to the ground electronic state has less energy

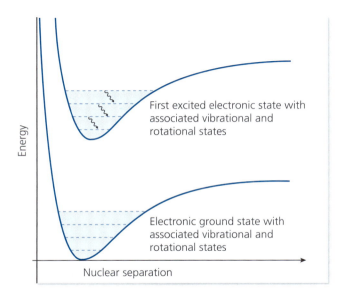

First excited electronic state with associated vibrational and rotational states

Electronic ground state with associated vibrational and rotational states

Energy

Nuclear separation

Figure 2.11 Electronic states for a diatomic molecule. Dashed lines correspond to vibrational states, and solid lines correspond to rotational states.

(longer wavelength) than the photon that was absorbed during excitation (total conservation of energy). Spectroscopically, this *shift to longer wavelength* for emitted photons is referred to as a *Stokes shift*. We will examine Stokes shifts in greater detail in Chapter 8 when we introduce luminescence spectroscopy and again when we study Raman spectroscopy in Chapter 12.

2.6   Transitions Between Electronic, Vibrational, and Rotational States: Putting It All Together

We introduced selection rules for vibrational and rotational states previously. These selection rules reduce the number of transitions that one might actually observe. There are also selection rules associated with electronic transitions, for example, a rule associated with spin. These will be presented in more detail in Chapter 6.

In molecules, each electronic state has associated vibrational states and rotational states. In Figure 2.12, we sketch energy levels of a diatomic molecule, including the vibrational and rotational states. You see that the vibrational states are equally spaced in energy and that the difference in energy between adjacent rotational states increases. You also see that the rotational states start at the same energy as the vibrational state but reach beyond to the next vibrational state. One can imagine sketching the vibrational and rotational states onto Figure 2.12 and getting quite an array of energy levels. Detangling these transitions via absorption and emission is the job of the spectroscopist.

Although the selection rule for vibrational transitions stipulate that all transitions occur with $\Delta n = \pm 1$, a key distinction must be made when investigating transitions *between* two different *electronic* states. A molecule can undergo an electronic transition with no change in vibrational quantum number so long as it moves into a new electronic state. For samples at room temperature, we can expect a system to be in the ground electronic and vibrational state because the thermal energy is less than the energy differences between states. We cannot make such an assumption about the rotational states because they will be populated above $J = 0$ at room temperature. In practice, these rotational states will be unresolved (smeared together) for most optical spectrometers; the difference in rotational energy levels is very small compared to vibrational and electronic energy levels, as shown in Figure 2.11. When measuring UV-vis and infrared absorption spectra of molecules, the "smeared together" rotational states result in somewhat broader absorption spectra.

In order to understand transitions with electrons, we must deal with spin and the Pauli Exclusion principle. Figure 2.13 shows a simplified energy level diagram showing electrons as either spin up or spin down. The *ground electronic state* is labeled with S_0. The capital "S" indicates that the ground state is *a singlet*,[9] and the subscript "0" indicates that it is the ground state. The right side of Figure 2.13 also shows the first excited state singlet (designated S_1) and the first excited state *triplet* (designated T_1).[10] It is important to appreciate that when a paired electron is excited to a higher energy orbital, the two electrons are no longer occupying the same orbital, and so are no longer bound by Pauli's Exclusion principle. It is now possible for the electron in the excited state to flip spin states, leaving the two electrons with parallel spins. This produces an excited state triplet; see Figure 2.13B).

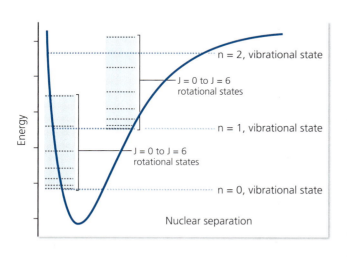

Figure 2.12 Close-up image of vibrational and rotational states associated with an electronic level.

n = 2, vibrational state

J = 0 to J = 6 rotational states

n = 1, vibrational state

J = 0 to J = 6 rotational states

n = 0, vibrational state

Energy

Nuclear separation

---

[9] For a review of spectroscopic term symbols, see Douglas B.; McDaniel D.; Alexander J. *Concepts and Models of Inorganic Chemistry*, pp 32–41, 442–443. John Wiley & Sons: New York **1994**.

[10] A full mathematical description of singlet and triplet states can be found in Griffiths, D. *Introduction to Quantum Mechanics*, 2nd ed., Pearson Education: Essex, UK, **2005**.

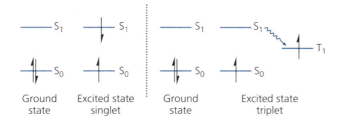

Figure 2.13 A simplified energy level diagram showing the ground state and first excited state singlet (A) and the ground state and first excited state triplet (B).

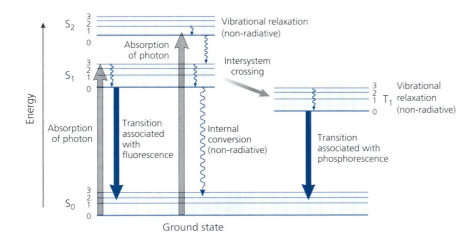

Figure 2.14 Jablonski diagram for molecular spectroscopy.

We do not observe excited state triplets for every molecule. Within each state, there are also several vibrational states that were left out of Figure 2.13 for clarity.

The Jablonski Diagram

As we learn about the specifics of molecular spectroscopy, we will use expanded energy level diagrams, called Jablonski diagrams, to understand the possible processes associated with transitions (nonradiative decay, emission, and absorption). A *Jablonski diagram* is shown in Figure 2.14. We will now discuss the various aspects of this important figure.

Due to Pauli's Exclusion principle, a ground state T_0 is forbidden, but *Hund's rule* allows us to surmise that the energy of T_1 should be lower than that of S_1 for a given system and so for some systems we see excited triplet states with designations of T_1, T_2, T_3, and so on. We have also denoted the triplet states in our Jablonski diagram (see Figure 2.14) along with the corresponding vibrational states. The triplet state is typically placed to the right of the singlet states, and (per Hund's rule) it is lower in energy compared to the corresponding singlet excited state.

Fluorescence and Phosphorescence

Absorption occurs when an appropriate wavelength photon causes a transition from the singlet ground state to an excited singlet state: for example, from S_0 to S_1 or from S_0 to S_2, as shown in Figure 2.14. Absorption events occur very rapidly—on the order of 1 fs. The excited electron can relax through several vibrational levels within the same electronic level, imparting vibrational energy to the molecule in order to conserve energy. These transitions are called vibrational relaxation and occur in the fs to ps time scale. Such vibrational relaxations are sometimes called *nonradiative decay.*

When an electron relaxes from an excited electronic singlet state to a lower electronic state without emitting light, this is referred to as *internal conversion*: for example, from S_2 to S_1. This is sometimes called *radiationless* relaxation and results in the energy being transformed into heat and subsequent excitation of vibrational quantum states. These occur over a range of times—from 10 fs to 10 ns.

When an electron transitions nonradiatively from a singlet state to a lower triplet state, this is called an *intersystem crossing*. Transitions from singlet to triplet (and triplet to singlet) are quantum mechanically forbidden because they disobey the spin selection rule that states that the total spin cannot change ($\Delta s = 0$). Singlet-to-triplet transitions involve a spin flip for the excited state electron. A transition from ground singlet state to excited singlet state is allowed because the total spin does not change. Strictly speaking, forbidden transitions do occur, but they have a much lower probability compared to other transitions.

The transition from the triplet excited state T_1 to the singlet ground state S_0 is forbidden because Δs is not 0. The relaxation from an excited triplet state to a ground singlet state is referred to as *phosphorescence*. This occurs with a very low but nonzero probability. Relaxation for forbidden transitions is very slow compared to allowed transitions. Because of this low probability, it takes much longer for phosphorescence to occur; lifetimes can be on the millisecond timescale. The allowed transition that occurs between an excited singlet state and the singlet ground state is referred to as *fluorescence*. Both fluorescence and phosphorescence are important spectroscopic techniques and will be the topic of Chapter 8, on luminescence spectroscopy, and Chapter 9, on atomic emission spectroscopy.

Internal conversion and intersystem crossing are similar in that they both involve a radiationless relaxation from a higher excited state to a lower excited state. However, for internal conversion, the excited state electron's spin remains unchanged and for intersystem crossing, the excited state electron flips spin states.

Problem 2.10: Suppose an electron is trapped in one dimension to a length of 2 nm, the length of some polymer.

(a) What is the wavelength absorbed by such a system from the ground state to the first excited state? *Hint*: This can be modeled as a particle in a one-dimensional box.[11]

(b) How does the length of the polymer (the length of the box) affect the absorption wavelength from the ground state to the first excited state? In order to answer this question, use a spreadsheet to calculate the wavelengths associated with this transition for one-dimensional boxes that are from 1 to 200 nm in size at intervals of 10 nm. Make a plot of wavelength of absorption versus polymer length.

2.7 Energy Levels of Protons in a Magnetic Field—Nuclear Magnetic Resonance Spectroscopy

NMR spectroscopy, to be covered in Chapters 14 and 21, is an extremely powerful and sophisticated tool used to determine the structure of compounds. In this section, we will discuss the elements of quantum mechanics that help make NMR such a powerful technique. For now, we will restrict our discussion to the hydrogen atom and focus on what happens to the energy of the proton when it is placed in an external magnetic field. This particular type of NMR is called proton NMR or ^{1}H NMR.

---

[11] Note that h = Planck's constant, n = quantum number, m = mass, and a = length of the "box." $E_n = n^2 h^2 / 8 m a^2$

As is common in quantum mechanics, we will start with the potential energy function that describes the system of interest. The potential energy (U) of any magnetic dipole moment ($\bar{\mu}$) when placed in an external magnetic field ($\bar{B}$), is:

$$U = -\bar{\mu} \cdot \bar{B} \qquad \textbf{Eq. 2.19}$$

The magnetic dipole moment, ($\bar{\mu}$), is an important quantity to consider because it helps determine how an object would behave when placed in a magnetic field. For an object, such as a current-carrying loop, the magnetic dipole moment is the product of the current in the loop and the loop's area. Because a nucleus is a charged particle, we can use the concept of a spinning nucleus to draw analogies to a current-carrying loop. The dipole moment of a proton with spin can be written as:

$$\mu = \gamma \hbar I \qquad \textbf{Eq. 2.20}$$

where I is the nuclear spin quantum number[12] for the proton, γ is the *gyromagnetic ratio* for the proton (267.5 MHz/T or 2.675×10^8 1/[s T]), and $\hbar$ ("h-bar") is the reduced Planck constant.

Solving the Schrödinger equation for the potential energy function yields:

$$E = -m_z \hbar \gamma B = -\left(\pm \frac{1}{2}\right)\hbar \gamma B \qquad \textbf{Eq. 2.21}$$

The proton magnetic moment (m_z) then has two orientations in the external magnetic field and the two orientations have different energies. The difference in energy between these two proton spin states in the magnetic field is:

$$\Delta E = \frac{1}{2}\hbar \gamma B - \left(-\frac{1}{2}\right)\hbar \gamma B = \hbar \gamma B \qquad \textbf{Eq. 2.22}$$

This equation can be written in an alternate form by setting $\Delta E = \hbar \omega = \hbar \gamma B$ and solving for the angular frequency to obtain:

$$\omega = \gamma B \qquad \textbf{Eq. 2.23}$$

The angular frequency here is sometimes referred to in a classical description of NMR as the *Larmor frequency*. In this interpretation, the dipole moment of the proton precesses around the applied magnetic field at this frequency, see Figure 2.15. It is important to note that NMR is a fundamentally quantum mechanical technique, and the analogy to spinning particles is only a model used to help us visualize the system.

It is this magnetic field dependence of the energy levels (shown in Equation 2.21 and Figure 2.16) for a nucleus that gives NMR its power to probe the environment of the proton. The splitting of levels due to a magnetic field is called the *Zeeman effect* (we will revisit the Zeeman effect in Chapter 7, on atomic absorption spectroscopy). Note that if B were zero, Equation 2.22 would show no energy level difference and hence no possible transitions between levels. It is the application of an *external magnetic field* that produces the energy difference between the two states and the energy difference increases as the external magnetic field strength increases (see Figure 2.16).

As you will see in the next example, the wavelength range that causes transitions between the two spin states is in the radiowave part of the electromagnetic spectrum.

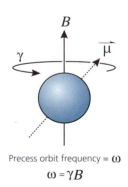

Precess orbit frequency = ω

$$\omega = \gamma B$$

Figure 2.15 In the classical analogy, you can visualize the dipole moment as precessing about the direction of the magnetic field.

---

[12] Unfortunately, I is also used for rotational inertia, introduced earlier in this chapter.

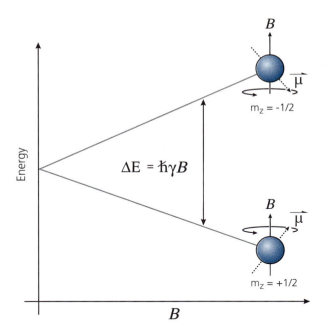

Figure 2.16 Energy level difference versus applied magnetic field for protons in a magnetic field.

Example 2.8

Suppose you could put a bare proton in a magnetic field of 3 T. What is the energy difference between nuclear spin states? Give your answer in joules. What is the frequency of this photon in MHz? What wavelength photon would promote the proton from the lower state to the upper state?

$$\Delta E = \hbar \gamma B = (1.05 \times 10^{-34}\,\text{Js})(2.675 \times 10^{8}\,\text{s}^{-1}\text{T}^{-1})(3\,\text{T}) = 8.43 \times 10^{-26}\,\text{J}$$

Using

$$E = h\nu,$$

$$E = h\nu, \; \nu = \frac{E}{h} = \frac{8.43 \times 10^{-26}\,\text{J}}{6.6 \times 10^{-34}\,\text{Js}} = 127\,\text{MHz}$$

$$\lambda = \frac{hc}{\Delta E} = \frac{(6.63 \times 10^{-34}\,\text{Js})(3 \times 10^{8}\,\text{m/s})}{8.43 \times 10^{-26}\,\text{J}} = 2.36\,\text{m}$$

This wavelength falls in the range of radio waves. In fact, radiofrequency waves are used in all NMR spectrometers as the excitation source.

Problem 2.11:   Carbon-13 (^{13}C) has a gyromagnetic ratio of 6.73×10^{7} T^{-1}, sec^{-1}. Repeat Example 2.8 for a carbon-13 nucleus.

If electromagnetic radiation of the same energy as the energy level difference is sent through the sample, absorption will occur as the proton undergoes a spin flip. The particular wavelength of the radiation that causes the proton to spin flip depends on the

Example 2.9

You are studying a molecule with ^{1}H NMR. The external applied magnetic field is 1.259 T. There are two protons in two different chemical environments. One "sees" a total field of 1.258 T, and the other "sees" a total field of 1.257 T. Calculate the two different wavelengths that would cause absorption.

$$\Delta E = \hbar\gamma B_{total} = (1.05\times10^{-34}\,\text{Js})(2.675\times10^{8}\,\text{s}^{-1}\text{T}^{-1})(1.258\,\text{T}) = 3.533\times10^{-26}\,\text{J}$$

$$\lambda = \frac{hc}{\Delta E} = \frac{(6.63\times10^{-34}\,\text{Js})(3\times10^{8}\,\text{m/s})}{3.533\times10^{-26}\,\text{J}} = 5.6297\,\text{m}$$

$$\Delta E = \hbar\gamma B_{total} = (1.05\times10^{-34}\,\text{Js})(2.675\times10^{8}\,\text{s}^{-1}\text{T}^{-1})(1.257\,\text{T}) = 3.531\times10^{-26}\,\text{J}$$

$$\lambda = \frac{hc}{\Delta E} = \frac{(6.63\times10^{-34}\,\text{Js})(3\times10^{8}\,\text{m/s})}{3.531\times10^{-26}\,\text{J}} = 5.6329\,\text{m}$$

We see that the two different magnetic field environments result in two different radiofrequency absorption energies or wavelengths.

magnitude of the external magnetic field (B) and the local magnetic environment produced by the molecule to which the proton is attached.

Note that the magnetic field, B_{total}, that the proton "sees" is due to both the external magnetic field (B_0) and the magnetic fields generated by its magnetic neighbors $(B_{internal})$ within the molecule. This can be written as:

$$\vec{B}_{total} = \vec{B}_0 + \vec{B}_{internal} \qquad\qquad \textbf{Eq. 2.24}$$

This means that the energy of radiation that causes absorption depends *on the local chemical environment of the proton.* In a simplified sense, NMR is a type of radiofrequency absorption spectroscopy where the energy level splitting dependency on magnetic field is used as a fingerprint for different proton environments. In practice, the total field is written in terms of the screening constant, σ, because the internal field is a small percentage of the external field:

$$\vec{B}_{total} = \vec{B}_{external} - \sigma\vec{B}_{external} \qquad\qquad \textbf{Eq. 2.25}$$

We can combine our relationship for the energy level differences with our equation for the total B field:

$$\Delta E = \hbar\gamma(B_{external} + B_{internal}) = \hbar\gamma B_{external}(1 - \sigma) \qquad\qquad \textbf{Eq. 2.26}$$

We see that the energy difference between states depends on the screening constant that is sensitive to the chemical environment of the proton.

Problem 2.12: Carbon-13 (^{13}C) has a gyromagnetic ratio of 6.73×10^7 T^{-1}, s^{-1}. Repeat Example 2.9 for a carbon-13 nucleus.

We saw earlier in Example 2.6 that thermal energy can play an important role in the relative populations between two states. What is the relative population of two *magnetic states* in NMR? If the populations N_1 and N_2 are exactly identical, the absorption of electromagnetic radiation caused by a transition from N_1 to N_2 would be exactly canceled out by emission from N_2 to N_1. We need more protons in the lower energy state N_1 than N_2 if we want to measure absorption. We turn again to the Boltzmann distribution discussed earlier in this chapter to address this issue. Calculating the ratio of the populations N_2 to N_1 at 300K using Example 2.9, we find:

$$\frac{N_2}{N_1} = e^{\frac{E_1 - E_2}{k_B T}} = e^{\frac{-\hbar \gamma B}{k_B T}} = e^{\frac{-3.531 \times 10^{-26} \text{ J}}{(1.38 \times 10^{-23} \text{ J/K})(300\text{K})}} = 0.99999$$

Put another way, $N_1 = 1.00001\ N_2$. We see that the two different states N_2 and N_1 have *very nearly* the same population. Although there are more protons in the lower energy state N_1 compared to the higher energy state N_2, there are very few excess protons in the N_1 state compared to the total number of protons. This means that only an extremely small percentage of the protons are being probed by NMR absorption, and hence there is a small signal for the experimenter to use. One way to increase the number of protons in N_1 relative to N_2, as we see in the Boltzmann equation, is to increase B_0, an important parameter when one chooses an NMR instrument. For example, if one uses a field five times as large in the above calculation, $N_1 = 1.00004\ N_2$ and the signal is now four times greater.

> **Problem 2.13:** You are considering purchasing a new NMR instrument so that you can do ^{1}H NMR. Suppose one instrument would provide a magnetic field of 10 T and another would provide a field of 3 T. For a given molecule, would these two instruments have two different absorption wavelengths between spin states? If so, calculate the two different wavelengths that would cause absorption, ignoring the internal magnetic field of the molecule.

2.8 Further Reading

BOOKS

Atkins, P. E.; de Paula J. *Physical Chemistry*, 9th ed.; Oxford University Press: Oxford UK, **2010**.

Engle, T. *Quantum Chemistry and Spectroscopy*, 3rd ed.; Prentice Hall: Upper Saddle River, NJ, **2012**.

Hollas, J. M. *Modern Spectroscopy*. John Wiley & Sons: Chichester, UK, **2005**.

McQuarrie, D. A.; Simon, J. D. *Physical Chemistry: A Molecular Approach*. University Science Books, **1997**.

Reed, B. C. *Quantum Mechanics*. Jones and Bartlett: Sudbury, MA, **2008**.

Schatz, G. C.; Ratner M. A. *Quantum Mechanics in Chemistry*. Dover: Mineola, NY, **2002**.

Turro, N. J.; Scaiano, J. C.; Ramamurthy V., *Modern Molecular Photochemistry of Organic Molecules*. University Science Books, **2010**.

2.9 Exercises

EXERCISE 2.1: The spring constant associated with NO is 1530 N/m. Calculate the frequency of oscillation for this molecule.

EXERCISE 2.2: The spring constant of the diatomic molecule NO is 1530 N/m.

  (a) Calculate the energy level difference from n = 0 to n = 1

  (b) Calculate the energy level difference from n = 1 to n = 2.

EXERCISE 2.3: The spring constant associated with CO is 1860 N/m. Calculate the frequency of oscillation for this molecule.

EXERCISE 2.4: The molecular vibration of HF is well described with a spring constant of 970 N/m. What is the wavelength of photon absorbed from the n = 1 to n = 2 states?

EXERCISE 2.5: Using spreadsheet software, calculate the energies for vibrational states from n=0 to n=10 for HI. HI has a spring constant of 320 N/m. Using your spreadsheet, calculate the energy level differences between adjacent levels up to n=10.

EXERCISE 2.6: It can be quite useful to move beyond the simple harmonic oscillator model of diatomic molecules. Using the Morse potential, a more accurate model for the molecular potential that takes into account the asymmetric molecular potential, the energy levels are:

$$E = \left(n + \frac{1}{2}\right)\hbar\omega - \frac{(h\upsilon)^2}{4D_e}\left(n + \frac{1}{2}\right)^2$$

where D_e is the bond energy (the energy from the minimum of the potential the energy at which the bond is broken). Note that the first term in the equation is what we found using the simple harmonic oscillator potential. The second term is often referred to as the anharmonic correction.

The dissociation energy for CO is 11.2 eV (1.79×10^{-18} J) the force constant is 1860 N/m. Calculate the energy difference between the ground state and the first excited state two ways, with the energy equation for the simply harmonic oscillator and with the anharmonic correction. How large is the correction associated with the anharmonic potential?

EXERCISE 2.7: For two cases, using the energy levels from the harmonic oscillator and then with the full Morse potential energy state, use spreadsheet software to produce a plot of energy level difference for vibration as a function of quantum number n. Describe the relationship and discuss the role of the anharmonic correction. Use values for HCl for your calculation (k = 80 N/m, disassociation energy = 7.0×10^{-19} J).

EXERCISE 2.8: The disassociation energy for NO is 7.0 eV (1.12×10^{-18} J) and the effective spring constant is 1530 N/m. Calculate the energy level difference between the n = 5 and n = 6 states and the associated absorption wavelength using the harmonic oscillator model and the Morse potential model (see Exercise 2.7).

EXERCISE 2.9: A gas sample is thought to be either LiBr or LiI. For both molecules calculate the difference in energy due to rotation alone between J = 0 and J = 1. Then compare your answer to the energy due to rotation alone between J = 1 and J = 2 states. Are these energies above or below the thermal energy of 0.04 eV?

EXERCISE 2.10: We have now plotted energy level diagrams for several systems (hydrogen atom, vibrating diatomic molecules, and rotating diatomic molecules) in this chapter. Make such an energy level sketch of the particle in a box system described in Problem 2.10.

EXERCISE 2.11: At what temperature might you expect to have populated an excited state E_2 to about 10% of N_1 from a ground state E_1 when the energy level difference is 5×10^{-20} J? What wavelength photon would be associated with the absorption between these two states? Degeneracies can be assumed to be one.

EXERCISE 2.12: Produce a spreadsheet tool that can be used to investigate the relationship between temperature, energy level difference, and populations in N_1 and N_2. Your tool should allow you to input the temperature and then plot the ratio of the populations for a range of energy level differences. What other relationships can you explore with your tool?

EXERCISE 2.13: You are doing proton NMR and would like to investigate the role of the external magnetic field on the ratio of populations N_2/N_1 between two spin states in proton NMR at 300K. Ignore any particular local magnetic field contributions from the molecule being investigated. Use a spreadsheet tool to produce a plot of N_1/N_2 as a function of external magnetic field. Use your tool to determine the B field at which $N_1 = 1.02 N_2$. The largest NMR magnets can achieve a B field of around 23 T. Comment on the viability of getting to your B field.

An Introduction to Optics

Angle of incidence

Incidence ray

Power

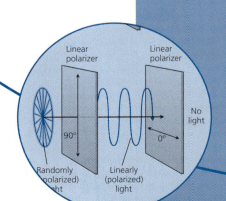

Linear polarizer

Linear polarizer

No light

90°

0°

Randomly (polarized) light

Linearly (polarized) light

One could make the claim that the diffraction grating is the most important optical component used in optical spectrophotometers. This component gives the chemist the ability to discriminate among the different energies of light, and because the energy of light is associated with different energy levels within atoms and molecules, it would be fair to say an optical grating is the "heart" of the spectrophotometer. For example, one can find a diffraction grating in an ultraviolet-visible spectrometer, an atomic absorption and atomic emission spectrometer, a fluorometer, and a Raman spectrometer.

At its most basic level, the *diffraction grating* is a surface with uniform sized, evenly spaced, parallel grooves that cover the surface of the grating. Panel (A) of Figure 3.1 is a close-up image of a commercial grating showing the parallel grooves. From left to right, the scale is around 6 mm. The distance between grooves is one aspect of the grating that determines how light of different wavelengths will diffract on the surface of the grating. One can observe diffraction on the surface of a compact audio disk (CD) or digital video disk (DVD), as shown in panel (B). The regular structures (~500 nm wide) etched onto its surface are essentially a diffraction grating, as shown in the second image from the top. Panel (C) shows white light (composed of a broad range of wavelengths in the visible spectrum) incident on a reflection diffraction grating. The diffracted light leaves the surface of the grating at an angle that depends on the wavelength of the light.

In an instrument, the incident light on the grating may be the signal from a spectrophotometer, and the diffraction grating selects a specific energy of light to send onto a detector. The diffraction grating will be discussed in more detail in this chapter. ■

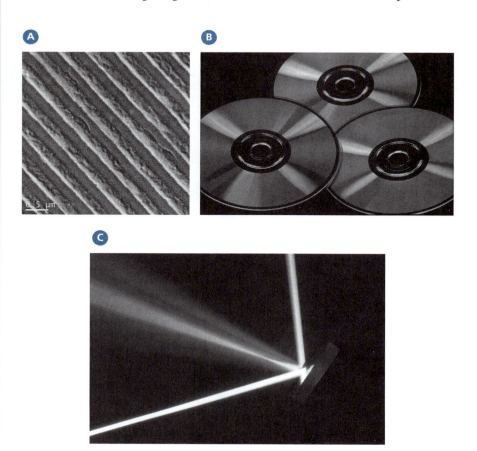

Figure 3.1 (A) Image of grating showing small scale features (scale set on bottom); (B) CDs acting as diffraction gratings; (C) incident white light reflected off diffraction grating, with diffracted light separated into different wavelengths.

3.1 An Introduction to the Properties of Light

Several of the most important "workhorse" instrumental techniques (ultraviolet-visible [UV-vis] absorption, Fourier transform infrared [FTIR], fluorescence, and Raman spectroscopy) as well as more specialized techniques involve the interaction of matter and electromagnetic radiation (light). Optical components are used to control the direction, power, and wavelength of that electromagnetic radiation. As we will discuss in Chapter 5 on signal processing, the ratio of signal to noise can hinder our ability to make conclusions based on our experiments. One perspective to keep in mind as we discuss optics is how different optical processes may lead to a decrease or an increase in the level of an optical signal. The goal of this chapter is to give the reader a streamlined and applied introduction to topics in optics that are relevant to chemical instrumentation.

Light can be described using either a wave model or a particle model. Despite these two approaches, light is, in some sense, both a wave and a particle, and a model based on one approach will at some level be incomplete. A pure wave model of light is referred to as *physical optics*. Physical optics deals with the behavior of light where one must consider the wave nature of light in order to understand the behavior. The physical optics view of light as an oscillating electric and magnetic field is represented by Figure 3.2.

In the wave model, the energy of the light wave is proportional to the square of the electric field amplitude. This notion of energy being related to the amplitude of the light wave played an important role in debates over the wave and particle model of light, as we'll see when we discuss the photoelectric effect. When one discusses *interference* (superposition or "adding up" of waves), diffraction (wave deviation from linear behavior), and polarization, one is applying ideas adopted from physical optics (wave mechanics). For example, the colorful fringes one sees on an oil slick and the ability to use a diffraction grating to select a band of wavelengths are the result of interference due to the wave nature of light.

To obscure the description of light further, we sometimes describe a light wave as a ray (*geometric optics*). In geometric optics, we often talk about the "ray" being bent when it passes through media of different density (*refraction*) or split by a beam splitter, for example. When we focus light with a simple magnifying glass, we understand this effect using a geometric optics (ray) model. Geometric optics deals with behavior that is not explicitly associated with the wave nature of light.

Also, light is sometimes described using a *particle model* (the *photon*). The particle model plays an important role in spectroscopy. For example, we think of an absorption event for an atom (or molecule) as the promotion of an electron from one quantum state to another, and this event occurs when the energy gap between the two quantum states equals the energy of the impinging photon (particle). We presented an overview of the key ideas about these quantum states for atoms and molecules in the discussion of quantum mechanics and spectroscopy in Chapter 2.

Wavelength, Energy, and Frequency

The model of light we choose to use depends on the application, so it is important to move between these complementary models easily. For example, the energy of a photon is related to its frequency and wavelength by:

$$E = h\nu = \frac{hc}{\lambda} \qquad \text{Eq. 3.1}$$

where E is the energy in units of Joules, ν is the frequency in units of Hertz (s^{-1}), h is Planck's constant $(6.626 \times 10^{-34} \text{ J·s})$, and c is the speed of light in a vacuum $(2.99 \times 10^{8} \text{ m/s})$.

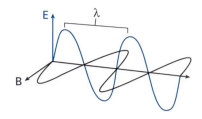

Figure 3.2 The wave model of light existing as mutually perpendicular time-varying electric and magnetic fields. The waves travels to the right.

TABLE 3.1: Visible Light and Associated Wavelength, Frequency, and Energy

VIOLET	RED
400 nm	700 nm
7.48×10^{14} Hz	4.27×10^{14} Hz
3.1 eV/photon	1.7 eV/photon
4.96×10^{-19} J/photon	2.83×10^{-19} J/photon

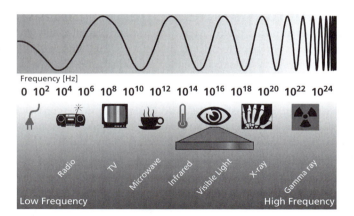

Figure 3.3 The electromagnetic spectrum.

Although we usually report energy in Joules (J), one can find photon energy in units of electron volts (eV). The conversion between Joules and eV is 1 eV = 1.6022 $\times$ 10^{-19} J. In order to calculate directly between eV and nm, one can use hc = 1240 eV nm in Equation 3.1. For a sense of scale, the energy of photons in the visible region is shown in Table 3.1. Figure 3.3 compares the visible spectrum to other regions of the electromagnetic spectrum.

Chemists also use a unit called the *wave number* ($\overline{v}$) to describe light. This unit is defined as the reciprocal of the wavelength when the wavelength is defined in cm. The wave number, $\overline{v}$, is:

$$\overline{v} = \frac{1}{\lambda(cm)}$$

Eq. 3.2

This unit is the preferred unit of measure in vibrational spectroscopy (infrared and Raman). Because wavelength and frequency have a reciprocal relationship, the wave number is essentially a frequency unit. And frequency and energy are directly proportional. So when one plots an infrared spectrum using wave numbers, features of a higher wave number represent molecular vibrations of a higher frequency (or higher energy).

PROFILE

The Photoelectric Effect Shows the Particle Nature of Light

By the late 1800s, the debate on the nature of light seemed to be largely over. It was thought that light was indeed a wave. However, in 1887 Heinrich Hertz observed that sparks could be made to jump off of a charged electrode at lower voltages when subjected to ultraviolet light. Further investigation of this photoelectric current showed that when light is incident on a metal cathode, electrons can be ejected from the cathode if the light is of a sufficiently high frequency. Below a threshold frequency, no electrons are ejected regardless of the light's intensity. These ejected electrons, called photoelectrons, can be drawn to a neighboring metal plate held at positive voltage and can then be measured as current. The maximum kinetic energy of the ejected electrons can be determined by setting the applied voltage between the two plates such that photoelectrons stop reaching the anode (i.e., no current). At this stopping voltage, the electric potential energy equals the initial kinetic energy of the photoelectrons.

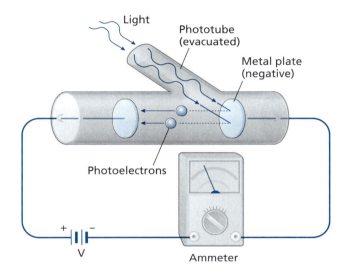

Experimental setup for the photoelectric effect: ammeter measures photoelectron current and voltage V can be set to determine stopping voltage.

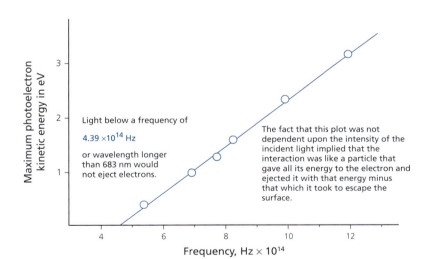

Light below a frequency of

4.39×10^{14} Hz

or wavelength longer than 683 nm would not eject electrons.

The fact that this plot was not dependent upon the intensity of the incident light implied that the interaction was like a particle that gave all its energy to the electron and ejected it with that energy minus that which it took to escape the surface.

Plot of maximum kinetic energy of electrons versus frequency of incident light.

Let us think through the implications of these data for a particular metal. Imagine a 750-nm light source incident on metal cathode. We find that *no matter* how intense we make the light, no photoelectrons are emitted. However, when we change the wavelength to 650 nm, we measure a photoelectron current. As we increase the intensity of light at this wavelength, we observe a proportional photo-electron current.

However, when we measure the stopping voltage, we see that no matter how intense the light, the maximum kinetic energy of the photoelectron is the same for a given wavelength. This is shown in the data here. These were two very curious observations.

A wave model of light suggests that the energy of light is dependent on the amplitude of the electric and magnetic fields. In other words, one increases the energy of the light simply by increasing the intensity. But the photoelectric observations previously described contradict the wave model. In order to eject electrons, it seems some sort of "light packet" must have a minimum energy. The experiment clearly shows that the energy of the packet is solely determined by the wavelength. In 1905, Albert Einstein presented a theory of light that correctly described the experimental data from the photoelectric effect experiment. In the paper, he described light using a particle-like model. These "light packets" are now called photons. ■■

Problem 3.1: An argon ion laser has a power of 0.5 W at 415 nm.

(a) What is the energy of a 415 nm photon?

(b) What is the wave number of a 415 nm photon?

(c) For this power, how many photons strike a sample in 1 second? Recall that a W is a J/s.

(d) If the wavelength was tuned to 418 nm and we assume the same power at 0.5 W, do any results for parts (a) to (c) change?

Coherence

A light source is said to be *monochromatic* if it is composed of "one" wavelength (of course, one cannot have "one" wavelength; even monochromatic sources are composed of a small bandwidth or range of wavelengths). When it comes to "real" light sources, the image of the single electromagnetic wave composed of one sine wave is an oversimplification. For instance, the light that is emitted from a standard light bulb or from the sun contains light of a great many different wavelengths. Light sources composed of a wide range of wavelengths are referred to as a *continuum* light source. In contrast, if the monochromatic light source is a laser, then the light also has a property called coherence. When light is *coherent,* all of the photons have the same wavelength[1] and the waves are in phase. In other words, the peaks and troughs line up with each other. The result can be a very intense light source.

In Figure 3.4, we see three examples of light waves. Sunlight is neither monochromatic nor coherent; light from a light-emitting diode, or an LED, is monochromatic but not coherent; and light from a laser is monochromatic and coherent.

---

[1] Notice the use of both particle and wave mechanics to describe the properties of laser light.

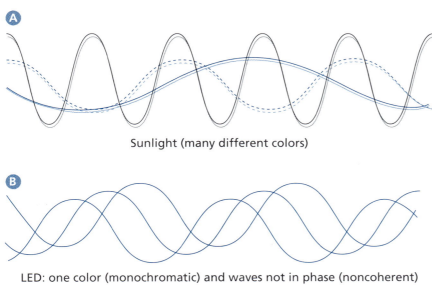

Sunlight (many different colors)

LED: one color (monochromatic) and waves not in phase (noncoherent)

LASER: one color (monochromatic) and waves in phase (coherent)

Figure 3.4 (A) A continuum light source, (B) a monochromatic noncoherent light source, and (C) a monochromatic coherent light source.

Polarization

The *polarization* describes the direction of the time-varying electric field component of light. The polarization of light can play a critical role in particular spectroscopies (e.g., polarized Raman spectroscopy) where the orientation of the sample and the orientation of the electric vector of the light wave (the polarization) can be used to gain information about the material.

Certain light sources are referred to as *unpolarized* because the beam has equal components of the electric field vector in each direction (vertical and horizontal).

Figure 3.5 shows a representation of both a vertically polarized light source in panel (A), a horizontally polarized light source in panel (B), and a representation of an unpolarized light source in panel (C). Note that instead of drawing a large number of arrows, each denoting the direction of the electric field, the unpolarized light source is sketched as equal parts vertical and horizontal polarization. Light from the sun and from conventional light bulbs are examples of unpolarized light. More complicated polarizations are also possible, including circularly polarized light, which is created by two sinusoidally varying electric fields with a particular relationship between the peaks and troughs of the light waves.[2]

---

[2] More precisely, there is a particular phase relationship between the light waves.

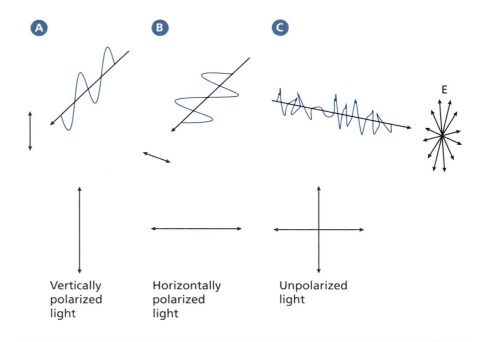

Vertically
polarized
light

Horizontally
polarized
light

Unpolarized
light

Figure 3.5 Sketches of electric field oscillations for vertical, horizontal, and unpolarized light.

Interference

When two light waves (or any type of wave) exist in the same location and at the same time, the resulting wave can be thought of as the mathematical sum of both waves. When the peaks and troughs of both waves line up, we refer to the two waves as being *in phase* and say that they *constructively* interfere. Conversely, when the peaks of one wave line up with the trough of the other wave, we refer to the two waves as being exactly *out of phase* and say that they *destructively interfere*. Depending on how the peaks and troughs line up, one can have a light wave that may be the result of completely constructive interference, completely destructive interference, or anything in between. To simplify matters, we will consider two waves with the same amplitude and the same wavelength.

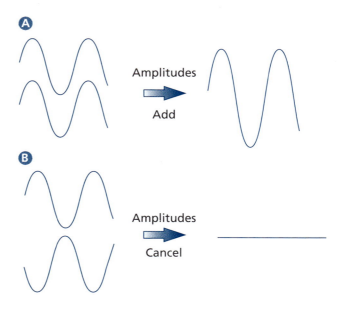

Amplitudes
Add

Amplitudes
Cancel

Figure 3.6 Extreme cases of interference:
(A) fully constructive, and (B) fully destructive.

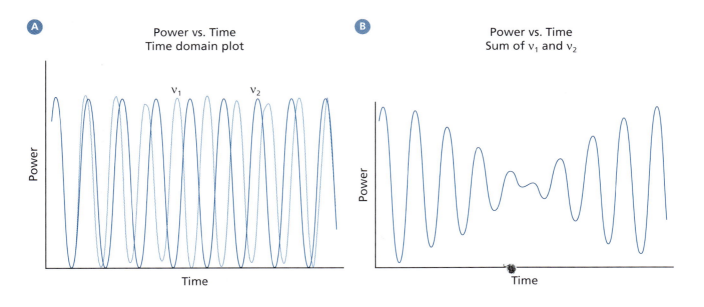

Figure 3.7 Panel (A) shows two sine waves with slightly different frequencies. Panel (B) is the sum of those two waves and represents a beat pattern derived from the two waves. The two waves come in and out of phase with time.

In Figure 3.6, we see two waves of the same wavelength that add constructively to yield a wave with twice the original amplitude and two waves that add destructively to yield a wave with zero amplitude (no wave!).

Interference can also occur with waves of different wavelengths. A case of interference for two waves with slightly different wavelengths is shown in Figure 3.7, and the resulting sum of the two waves' amplitude as a function of time results in what is called a beat pattern. We will examine the significance of the beat pattern in detail when we explore FTIR in Chapter 11. The *Fourier transform* is a mathematical algorithm that allows one to convert a time domain signal into the frequency domain. In other words, the Fourier transform allows us to find the individual wave frequencies that contributed to the net beat pattern. We will more fully introduce the Fourier transform in Chapter 5, which discusses signal processing. The Fourier transform will also be used in Chapter 14 when we discuss Fourier transform nuclear magnetic resonance spectroscopy.

Demystifying Interference

The interference of light (as well as water waves and sound waves) can be explored through the simulation "wave interference" from http://phet.colorado.edu.

Example 3.1

$I_1(t)$ and $I_2(t)$ describe the intensity of two light waves.

$$I_1(t) = 10 \sin\left(30t + \frac{\pi}{4}\right)$$

$$I_2(t) = 10 \sin\left(30.4 + \pi\right)$$

Assume that both waves occur at the same spatial location. How would the resulting wave amplitude vary as a function of time?

STRATEGY –

The voltage registered at the detector represents the resulting intensity of the combined light waves and is the sum of the individual intensities at each point in time.

SOLUTION –

Program each sine wave into a spreadsheet and plot the sum as a function of time. The result is shown below.

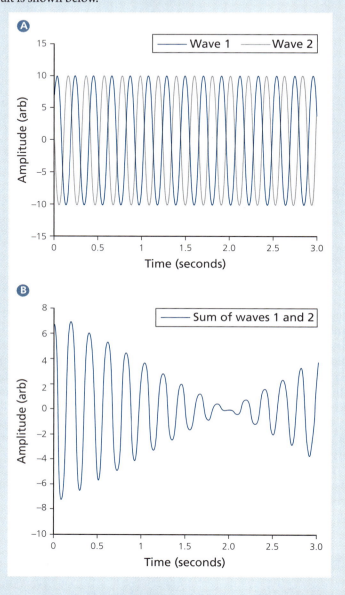

Note that each wave has the same amplitude (10 and 10) and that the frequency of the waves is similar, but not exactly the same (30 and 30.4). The waves are not in phase and the resulting interference is a beat pattern.

Problem 3.2: Two waves are described by the following two equations for amplitude I as a function of time, t:

$$I_1(t) = 5\sin(2t + \pi)$$
$$I_2(t) = 5\sin(2t + 3\pi)$$

Assume that both waves occur at the same spatial location. Use a calculator to calculate the sum of I_1 and I_2 for times 0, 1, 2, and 3. Decide if the sum of these waves is fully constructive, fully destructive, or likely to result in a beat pattern.

Problem 3.3: The equations for the intensities of three waves are presented below.

$$I_1(t) = 12\sin\left(500t + \frac{\pi}{4}\right)$$

$$I_2(t) = 5\sin\left(500t + \frac{\pi}{2}\right)$$

$$I_3(t) = 6\sin\left(500t + \frac{3\pi}{4}\right)$$

Assume that all three waves occur at the same spatial location. Plot the sum of these waves and comment on the interference.

Any device that makes use of interference of light requires significant control over the path length differences of the optical beams. The required control is on the order of the wavelength of the associated light. Similarly, an observed effect from a material involving interference will involve distances on the order of the wavelength of the light used.

Diffraction

When a water wave encounters a barrier with an opening, the water spreads out from the opening, as shown in panel (A) of Figure 3.8. Notice the circular wave pattern after the opening. The effect of a wave spreading out from a barrier is known as *diffraction*. When you walk through a door, you continue on your merry linear path forward because you are essentially a particle. Waves do not behave this way.

When the water wave is directed onto two small openings, as shown in panel (B) of Figure 3.8, we see the water waves spread after the openings. At any given point after the openings, the water waves that reach that point would travel different distances, causing interference there. From the image, the water waves form regions where the wave amplitude is very small and regions where the wave amplitude is large because of the interference of the water waves. This is known as double slit diffraction. Perhaps

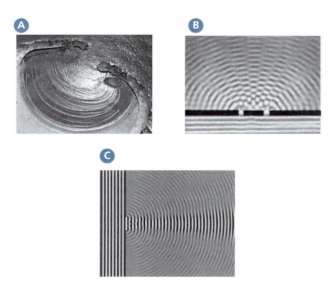

Figure 3.8 (A) Water wave diffraction through an opening, (B) water waves and double-slit diffraction, and (C) water wave and single-slit diffraction.

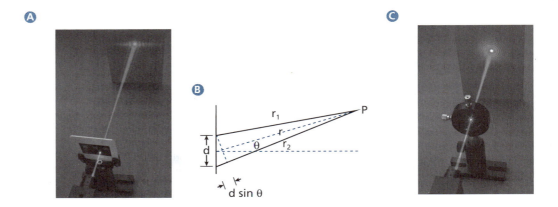

Figure 3.9 (A) Light moves through double slits producing a diffraction pattern of bright and dark spots. (B) The difference in path length that light travels from the two slits to P causes bright and dark spots. (C) Diffraction through a circular aperture—Airy disk.

surprisingly, water waves can also form a diffraction pattern after traveling *through only one opening*. This is shown in panel (C) of Figure 3.8. Because the slit has a finite width and because each point of the wave can be thought of as a secondary spherical point source according to *Huygens' principle*, different point sources in the one slit travel different distances for any given position, causing *interference*. This is known as single slit diffraction.

If light is incident on a single slit or double slit (or more complicated arrangement), the transmitted light will produce a diffraction pattern in an analogous manner to the diffraction observed for water waves. In the case of light, the interference is associated with the electric field of the light as opposed to the water wave height. As with the water wave, each aperture can be considered a new point source and so, diffraction is fundamentally caused by the interference of a great many point sources. The resulting *diffraction pattern* is composed of bright (constructive interference of the point sources) and dark spots (destructive interference of the point sources).

In panel (A) of Figure 3.9, we show the diffraction pattern when light moves through a double slit and the resulting bright and dark aspects of the diffraction pattern. In panel (B), we show the path length difference between light that moves from each slit to a screen at point P. When the path length difference, $d\sin\theta$, is equal to an integer value of the wavelength of the light, constructive interference occurs and a bright spot will be seen at P. When the path length difference is a multiple of the wavelength/2, then destructive interference occurs and a dark spot will be seen at P. When light moves through two-dimensional openings, such as a circular aperture or square aperture, the patterns from diffraction can be quite stunning. In panel (C), we show laser light incident on a circular aperture. For the circular aperture, note the bright and dark concentric circles and the bright central maximum.

Scattering

Scattering occurs when light interacts with particles and some percentage of the incident light leaves these particles in many different directions. An example of light scattering is shown in Figure 3.10(A) where a laser beam is incident on a cuvette filled with water with a few drops of milk added. The light scatters off the fat particles from the milk and makes the beam visible. Notice that the intensity of the scattered light decreases across the cuvette from left to right. Chemists refer to the scattering which occurs in a colloidal suspension as the *Tyndall effect*. This effect can be used to determine if a sample is a solution or if it is a colloidal suspension. Only the colloidal suspension will scatter light, as shown in Figure 3.10(B), where one cuvette (left,

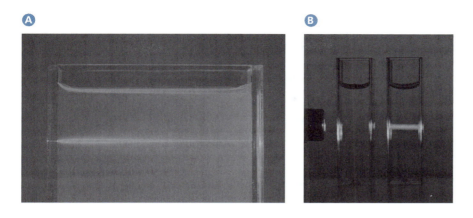

Figure 3.10 (A) Scattering of fat particles from milk in water, light attenuated. (B) Demonstration of scattering off colloids in right cuvette.

no scatter) holds a solution and the other (right, scattering visible) holds a colloidal suspension.

Scattering behavior differs depending on the wavelength of light and the size of the scattering particles. Other factors that influence light scattering can include the shape of the molecule, the polarization of incident light, the orientation of the molecule, and the density of molecules. The following rules, while they hold for spherical scatterers, provide an overview of the role of particle size and wavelength. *Rayleigh scattering* (roughly defined as when the particle size is less than $\lambda/15$), has a scattering probability that varies as $1/\lambda$.[4] This means that for Rayleigh scattering, short wavelengths are more likely to scatter than longer wavelengths. Rayleigh scattering is often used to describe the preferential scattering of sunlight that we see as the blue sky. *Mie scattering* refers to scattering by particles whose size ranges from roughly the wavelength of the incident light to roughly 100λ. The wavelength dependence on scattering probability, $P_{scattering}$, can be written to include Rayleigh and Mie scattering as:

$$P_{scattering} \sim \frac{1}{\lambda^n}$$

Eq. 3.3

where $n = 4$ for Rayleigh scattering and as the particle size increases, n changes from 4 to 0, meaning there is essentially no wavelength dependence for particles that are much larger than the wavelength.

Because particle size and particle geometry are factors in the intensity of scattered light, techniques have been developed to analyze the microscopic topography of surfaces and particle geometries in solution. These include instruments such as the dynamic light scattering system that can provide the analyst information on particle sizes.

Both Mie and Rayleigh scattering are forms of elastic scattering. This means the scattering process does not result in energy transfer to or from the scatterer to the scattered light. When energy is transferred to or from the scatterer to the scattered light this is an inelastic process. Raman scattering is an example of inelastic scattering. Raman scattering is a powerful spectroscopic tool and is the focus of Chapter 12.

Problem 3.4: Consider visible light incident on particles in a sample. For each of these particle size dimensions, roughly compare the wavelength dependence of the scattered photons:

(a) O_2 molecule (size ~0.29 nm)

(b) Small aerosol (size ~0.1 μm)

(c) Very large aerosol (size ~2 μm)

3.2 Controlling Optical Beams

The designer of an optical instrument needs to know how to steer light to a given location (e.g., a sample holder in a UV-vis absorption spectrometer). Also, he or she may need to control the beam size at that location, collect light after it has interacted with a sample, select (and/or discriminate against) a narrow bandwidth of light, and focus light onto a detector.

Mirrors and Reflection

A ray of light incident on the interface between two materials obeys the law of reflection, as seen by the light rays incident on the mirror in Figure 3.11.

The law of reflection states that the angle of incidence equals the angle of reflection. These angles are measured from the normal, which is a line drawn perpendicular to the surface, as shown in this figure. The *reflectance (or reflectivity)* of a mirror is the ratio of power of the reflected light to the power of the incident light.

Depending on the needs of the experiment, there are several different mirror types one can choose from: dielectric mirrors, metallic mirrors, and other specialized mirrors, including those with a narrow set of reflected wavelengths for lasers. *Dielectric mirrors* have a shorter high reflectance wavelength range but a very high reflectance (99%) in that range. Metal mirrors tend to have broad reflectance ranges but somewhat lower reflectance (~90%) than dielectric mirrors. Because all mirrors have a reflectivity less than 100%, there will be some signal loss when mirrors are used to direct a light source.

Reflection generally occurs at any interface between two different optical materials. In other words, the interface does not need to be between air and a metallic-coated mirror surface for there to be reflection. Some reflection will occur when a light ray is incident on a glass surface, for example, from a lens. Although the relative amount of light reflected may be very small, roughly 3%, the total power in the reflected light wave may be significant if using a high power laser source. These reflected beams can be a hazard.

Concave mirrors (or sometimes called *focusing mirrors*) are often used in grating monochromators, a topic discussed later in this chapter. As shown in Figure 3.12(A), a concave mirror focuses parallel light rays to a single *focal point* after reflection.

A concave mirror can also be used to collect light from a point source and after reflection, direct those rays as a parallel output. This can be visualized by imagining a light source at the focal point F in Figure 3.12(B) and reversing the directions of the arrows. The focal length, f, is an important parameter for a concave mirror. The focal length is the distance, along the principal axis (along its center) of the mirror, from the mirror to where parallel light is brought to a point. This is shown in Figure 3.12(B). The radius of curvature, R, is the radius of the mirror should it be drawn as a complete circle.

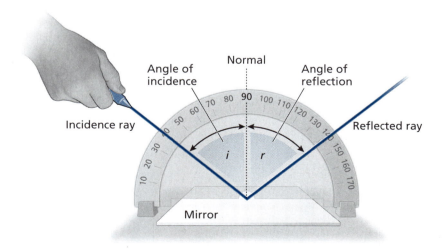

Figure 3.11 Light ray reflection, demonstrating law of reflection.

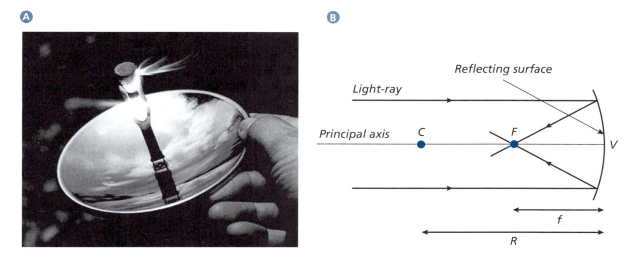

Figure 3.12 (A) Photograph of a curved mirror focusing sunlight onto a combustible material (from sundancesolar.com). (B) Concave mirror showing focal length and radius of curvature.

Lenses and Refraction

Lenses redirect light because of refraction. *Refraction* is the change in direction associated with a ray of light when it travels between two different materials. Figure 3.13 shows a light ray refracting as it moves from air to glass and then from glass to air. Refraction is described by Snell's law:

$$n_1 \sin \theta_1 = n_2 \sin \theta_2 \qquad \textbf{Eq. 3.4}$$

where n_1 is the refractive index for material 1 and n_2 is the refractive index for the material through which the beam refracts.

In Figure 3.13, the incident material is air (n = 1) and the material with the refracted beam is glass (n = 1.5). The angles θ_1 and θ_2 are measured with respect to the normal, as shown in Figure 3.14. When a light ray moves from air to a larger index material, the light ray bends toward the normal.

The focusing property of a converging lens (one in which incident parallel rays of light are brought roughly to a point) can be understood through Snell's law. A photograph of a converging lens with multiple parallel light rays is shown in Figure 3.15. Also included in the figure is a sketch of a lens with two light rays incident on the material, showing the focal length of the lens. Light is moving from air to glass then to air again. At each interface, the light beam is redirected because of Snell's law. The resulting behavior of many of these light rays is that the beam is focused. Like the concave mirror, we can define the focal length, *f*, as the distance from the center of the lens to the position where the lens takes parallel light rays and focuses them at the focal point.

The *index of refraction* for a given material is wavelength dependent. The index of refraction of BK7, a common glass used in optical components,[3] is shown in Figure 3.16. This general shape in the visible to infrared is found in most materials and is the so-called *normal dispersion* region. It is important to be aware of this normal dispersion region. For example, when using a prism to disperse a continuum light source, the nonlinear relationship between the refractive index and wavelength results in a nonlinear relationship between spectral resolution and wavelength. Similarly, when using a lens to focus light composed of many wavelengths, one must be cognizant that the true focal point is wavelength dependent.

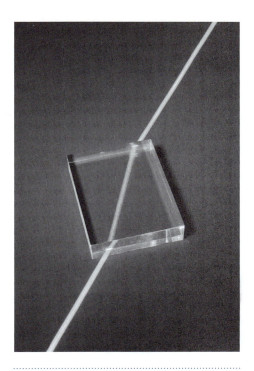

Figure 3.13 Refraction of light as it moves from air to glass and then glass to air.

---

[3] Borosilicate/potassium oxide optical glass.

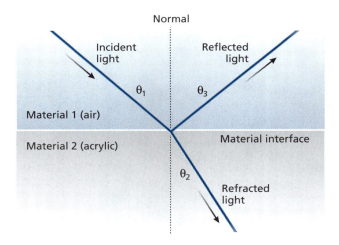

Figure 3.14 Reflection and refraction of light between air and acrylic showing angle of incidence, reflection, and refraction.

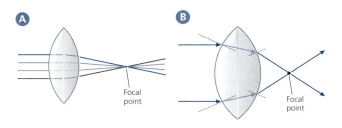

Figure 3.15 (A) Lens focuses light because of Snell's law. (B) Refraction of two light rays with a converging lens.

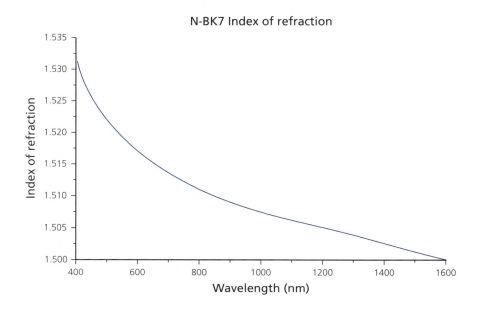

Figure 3.16 Index of refraction versus wavelength for N-BK7 glass.

Example 3.2

A broadband light source is composed of wavelengths from 400 nm to 800 nm. The beam is incident on a block of BK7 glass at an angle of 30 degrees from the normal. What happens to the beam as it refracts through the glass? Answer this quantitatively by describing the refraction of the 400 nm and 800 nm extremes in wavelength.

We can rearrange Snell's law to solve for θ_2 to obtain:

$$\theta_2 = \sin^{-1}\left[\frac{n_1}{n_2}\sin\theta_1\right]$$

From the plot of index of refraction versus wavelength, the index at 400 nm is approximately 1.53 and the index at 800 nm is 1.51. The 400-nm light will refract at:

$$\theta_2(400\,\text{nm}) = \sin^{-1}\left[\frac{n_1}{n_2}\sin\theta_1\right] = \sin^{-1}\left[\frac{1}{1.53}\sin 30°\right] = 19.07°$$

and the 800-nm light will refract at:

$$\theta_2(800\,\text{nm}) = \sin^{-1}\left[\frac{n_1}{n_2}\sin\theta_1\right] = \sin^{-1}\left[\frac{1}{1.51}\sin 30°\right] = 19.34°$$

Using your calculator, either set the default for angle to degrees or convert the angle in radians to degrees by multiplying by $2\pi/360$. The smaller angle for the 400-nm light compared to the 800-nm light means that the 400-nm light bends more toward the normal than the 800-nm light. Note that because the angle is measured with respect to the normal, the smaller angles mean closer to the normal.

Problem 3.5:   A light source is composed of wavelengths from 550 nm to 1.2 μm. The beam is incident on a block of BK7 glass at an angle of 45 degrees from the normal. What happens to the beam as it refracts through the glass? Answer this quantitatively by describing the refraction of the 550 nm and 1.2 μm extremes in wavelength.

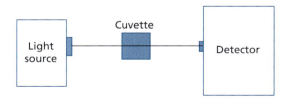

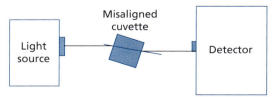

Figure 3.17 Misaligned cuvette and Snell's law.

A very practical case of Snell's law in action can occur with a misaligned cuvette. Figure 3.17 shows an instrument with a light source, cuvette, and detector. When the cuvette is oriented as shown in the top of the figure, the light enters the entrance slit of the detector. However, if the cuvette is misaligned, the light will refract, potentially causing the light to miss the entrance slit of the detector. In the figure, we are sketching the refraction caused by the fluid in the cuvette, not making the distinction between refraction from the glass cuvette and the fluid.

Collecting and Collimating Light

There are many times when light needs to be collected from a source, collimated, and then focused onto a detector. This basic configuration will be revisited when we discuss the UV-vis spectrophotometer in Chapter 6, the fluorimeter in Chapter 8, and the Raman spectrometer in Chapter 12. In Figure 3.18, we see how to collect, collimate, and focus a source with two lenses. Light is *collimated* when it propagates with no (or at least very little) change in beam size.

In Figure 3.18, the source is on the left hand side and has height of h. This may be the size of the lamp source. The source lens is placed at a distance from the source equal to the source lens's focal length, f_s. This lens collects the light and collimates it. The second lens is placed at a distance from the detector equal to the second lens's focal length, f_d. The second lens takes the collimated light and focuses it onto the detector. For spectroscopy experiments where the source is a lamp, the sample would reside between the two lenses. You would like to collect as much of the light as possible and then focus that light onto a detector in order to maximize the intensity of the signal. The height of the source spot, h, and the focused spot, H, are related by:

$$H = -\frac{f_d}{f_s} h$$

Eq. 3.5

The minus sign in the equation means that the image at the detector is inverted when compared to the source.

Problem 3.6: In your laboratory you have a converging lens with a focal length of 6 cm that you plan to use to collect light from a sample. If your source aperture is 0.5 cm in height, what focal length lens would you need to acquire if you wished to focus the collimated beam onto a 1 mm diameter detector aperture? Discuss the feasibility of focusing the beam to such a small diameter.

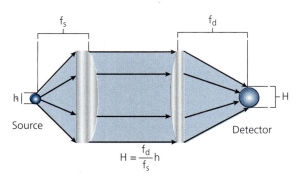

Figure 3.18 Collecting, collimating, and focusing light.

Usually the goal in an optical experiment is to collect as much light as possible because this will generally lead to the largest signal. One measure of the light collecting ability of a lens is the numerical aperture (NA). The NA is approximately:

$$N.A. \approx \frac{D}{2f}$$

Eq. 3.6

where D is the diameter of the lens and *f is the focal length*. In order to collect as much light as possible, this equation says you should use a physically large lens with a short focal length. This relationship does have its limits, however. One might hope to use, for example, a 10 meter diameter lens with a 1 mm focal length in order to create an extremely large NA. This lens would be impossible to actually construct. Imagine the radius of curvature needed for a lens to have such a short focal length at this size.

A related quantity, the *f-number* (or f/# as it is sometimes written), also describes the light gathering ability of an optical component. The f/# is defined as

$$f/\# \approx \frac{f}{D}$$

Eq. 3.7

The smaller the f/#, the larger the light gathering ability of the optical component. We will revisit f/# when we discuss the monochromator.

Problem 3.7: Calculate NA and the f/# for each of the lenses here. Rank the lenses in terms of their light collecting ability.

LENS	FOCAL LENGTH (CM)	DIAMETER (CM)
1	5	3
2	6	3
3	7	3
4	7	5
5	9	5

Focusing a Collimated Laser Beam

Laser sources are routinely used by chemists to excite and/or probe chemical systems. One may need to use a laser source to generate high intensity light, for example, in Raman spectroscopy. Or one may need a light beam focused as small as possible because the beam is used to scan over small scale structures. In order to completely understand the pros and cons of different focusing options, it is important to know the relationship between spot size, focal length of the lens, wavelength, beam intensity, and other beam parameters. For Gaussian laser beams (a valid model for many lasers), the following relationship can be used to determine the *beam waist* or the diameter of the focused beam $(2w_0)$:

$$2w_0 = \frac{4\lambda f}{\pi D}$$

Eq. 3.8

where λ is the wavelength of the light, *f* is the focal length of the lens, and D is the diameter of the beam at the lens face. This relationship assumes that the beam fills the lens. Figure 3.19 shows a Gaussian beam incident on a lens as well as the focused laser beam and the associated beam waist $(2w_0)$. For applications where the instrumental designer wants to maximize the intensity of a laser source, the sample will be placed at or near the focal length where the beam size is $2w_0$. Also of interest is the focus depth, shown in Figure 3.19, known as the Rayleigh range. The size of the focus depth would provide a range for the level of position control needed between your lens and the sample. If the

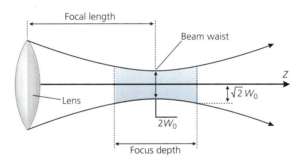

Figure 3.19 Focusing a Gaussian laser beam.

focus depth is only 1 μm, this means that the sample must be at this particular position to get maximum intensity.

The relationship in Equation 3.8 conveys an extremely important idea. If you want to use a Gaussian beam to illuminate a small region of space and hence image a very small area, the lower the wavelength of the light used, the smaller the possible laser spot. At a basic level, it is diffraction, discussed previously here, that creates fundamental limits for the focusing power of lenses. For this reason, X-ray laser sources hold significant promise for providing extremely small scale images (essentially images of atoms). We will revisit X-ray spectroscopy in Chapter 10.

Example 3.3

You need to focus a 10 mW, 632.8 nm Gaussian laser beam that is 5 mm in diameter into a sample. You have access to a lens with a focal length of 6 cm and focal length of 12 cm. For both lenses, the light fills the size of the lens. Using the Gaussian beam equations, what is the smallest diameter of the beam (known as the beam waist) for each lens?

STRATEGY –

In order to find the beam waist size, use Equation 3.8.

SOLUTION –

For the 6 cm focal length lens:

$$2w_0 = \frac{4\lambda f}{\pi D} = \frac{4(632.8\times10^{-9}\,\text{m})(0.06n)}{(3.14)(5\times10^{-3}\,\text{m})} = 9.6\times10^{-6}\,\text{m} = 9.6\,\mu\text{m}$$

For the 12 cm focal length lens:

$$2w_0 = \frac{4\lambda f}{\pi D} = \frac{4(632.8\times10^{-9}\,\text{m})(0.12n)}{(3.14)(5\times10^{-3}\,\text{m})} = 1.9\times10^{-5}\,\text{m} = 19\,\mu\text{m}$$

Problem 3.8: You are comparing the beam waist for two different situations with the goal of using the smallest beam waist possible. A Nd-YAG laser system emits light at 532 nm and the beam is 8 mm in diameter. You also have a Ti-sapphire laser that emits at 855 nm and has a beam diameter of 6 mm. Compare the beam waist for both laser systems using a focusing lens with a focal length of 10 mm. Assume the light fills the lenses in each case.

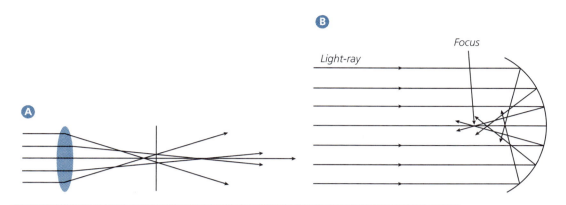

Figure 3.20 (A) Light rays entering the top and bottom parts of the lens cause spherical aberration and a blurring of the focal point. (B) Concave mirror showing spherical aberration/blurred focus.

As we saw in our discussion of focusing laser light, in reality, lenses do not produce a point of focused light or a perfectly replicated image. There are limitations to what a system of lenses and mirror can do to form an image. These limitations are called *aberrations*. Chromatic aberration occurs when multiple wavelengths move through an optical system. Because the index of refraction is wavelength dependent, different wavelengths will have slightly different focal lengths for the same lens. Several other types of aberrations, including spherical aberration, occur for monochromatic light. For most of geometric optics, we assume that light rays enter optical components close to the center of the component, which is referred to as the optical axis. Figure 3.20 shows an example of spherical aberration, where light enters the lens far from the optical axis. Spherical aberration is also shown for a concave mirror. One benefit of mirrors over lenses is that mirrors do not suffer from chromatic aberration.

Polarizers

Several spectroscopic techniques can be modified so that information about molecular orientation or geometry can be further enhanced by studying the effect of polarized incident light and/or polarized emitted light from the sample. For example, UV-vis, fluorescence, atomic absorption, and Raman spectroscopy each benefit from controlling and measuring the polarized output as a function of different polarized inputs. An important component used to control the polarization of a light source is the *polarizer*. Many sunglasses are polarizers. These are effective because light that is reflected off a surface (e.g., water) is preferentially polarized. The sunglasses can be made so that the polarizers reduce the amount of transmitted light from reflections. This reduces the glare.

Polarizers are generally made of long stretched polymers that lie parallel to each other imbedded in or coating an optical component such as a lens or a plate of glass. The parallel orientation of the polymer strands only allows light whose electric field vector is similarly oriented to transmit through the optical component. A linear polarizer is essentially a filter that transmits the component of the light containing a particular polarization (e.g., vertical). Figure 3.21 shows unpolarized light as it moves through a linear polarizer. The output is linearly polarized (vertical) light. If a second polarizer is added perpendicular to the first polarizer, none of the vertically polarized light would move through the horizontal polarizer.

Beyond the techniques where polarization plays a role, the properties of some optical components actually depend on the polarization of light. We have learned that the degree of refraction a ray experiences is a function of the wavelength of the ray. Similarly, for certain optical components the performance of the component can be influenced by the *polarization* of the light.

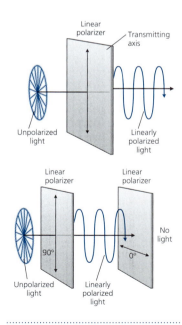

Figure 3.21 Unpolarized light incident on linear polarizer and unpolarized light with two polarizers 90 degrees apart.

3.3 Wavelength Selection

Selecting for a particular range of wavelengths of light is a common task for optical instruments. This may be to control the wavelength of light that is being absorbed or emitted by a chemical sample or to control the wavelength output from a broadband source used to excite a chemical sample. This task is typically accomplished in one of three ways: (1) grating monochromators, (2) optical filters, and (3) interferometers.

Introduction to Prism and Grating Monochromators

A *monochromator* selects a narrow range of wavelengths of light and allows only that narrow range to pass through the device. Of course, the wavelength resolution required is dependent on the type of experiment the analyst is performing. We will discuss the wavelength resolutions of monochromators shortly. The word *monochromator* comes from two Greek roots (mono = 1 and chroma = color) and a Latin suffix (-ator = something that acts in a specific manner). So the name *monochromator* literally means *to pick a single color*. Be careful not to spell it monochro*meter*. Monochromators are not meters of light intensity; rather, they select a specific color (or more precisely a narrow band of colors) of light. Monochromators come in one of two types, the prism monochromator or the grating monochromator. We will describe the basic design of each type and variations on the grating monochromator will appear in subsequent chapters as particular instruments are discussed.

Prism monochromators make use of the wavelength dependence of index of refraction discussed earlier this chapter. The light source is focused onto an entrance slit that serves to decrease the amount of stray light from the room that enters the monochromator. The light that passes through the entrance slit travels to a collimating lens placed a distance (f) equal to the *focal length* of the lens. The collimated light is then directed to a prism where it is refracted based on wavelength. The dispersed light is directed to an exit slit and a detector is placed at the exit slit. By rotating the prism, you can select the bandwidth of light that reaches the exit slit (see Figure 3.22A). Most monochromators have adjustable slit widths, so one can adjust the spectral resolution by adjusting the slit widths. A discussion of slit widths follows the introduction of the grating monochromator. Because the angle of refraction is wavelength dependent, the distance between two wavelengths on the slit varies over the spectral range of the spectrophotometer. In other words, the spectral resolution is wavelength dependent. For this reason, prism monochromators are seldom found in modern instruments.

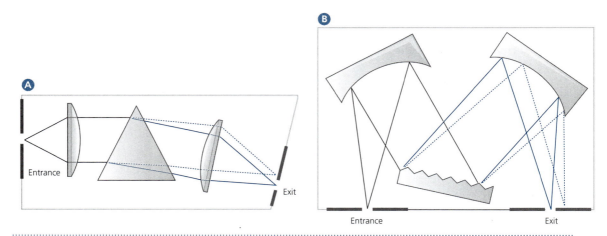

Figure 3.22 (A) Schematic of a basic prism monochromator. (B) Schematic of a basic grating monochromator.

As we opened this chapter, we noted that the diffraction grating[4] may be the most important optical component of a spectrophotometer. Like the prism monochromator, the *grating monochromator* is also used to disperse light as a function of wavelength. However, the dispersion of light by a grating is the result of *diffraction* rather than *refraction*. A grating is an optical structure with regular lines, grooves, or rulings across the length. Transmission or reflection gratings can be purchased with a range of groove densities and sizes made for specific wavelength ranges. A grating monochromator is shown in Figure 3.22B. The instrument has an entrance slit, two concave mirrors, a reflective grating, and an exit slit. The instrument has an associated focal length, which is the focal length of the curved mirrors (assuming they are identical).

The Diffraction Grating

Our focus will be on the *reflective diffraction grating* because it is the form most commonly used in monochromators. In particular, we will emphasize the use of blazed reflective gratings. Blazing refers to the sawtooth shape of the grating surface shown in Figure 3.23. The manufacturer of the grating uses the angle of the sawtooth to control the wavelength range of the grating. When choosing a grating, one can choose between different wavelength blazes. For example, a 300 nm blaze is optimized for wavelengths near 300 nm.

In Figure 3.23, white light is incident on a blazed reflective diffraction grating. In the figure, two adjacent incoming light waves each strike a different face on the diffraction grating and are reflected away. The incident ray on the right side of the figure traveled further than the incident ray on the left side. If the additional distance traveled by the ray on the right was an integer value (d) of the wavelength, then the two rays will be in phase and constructively interfere as they leave the grating surface. Note that this is very similar to the example of double slit diffraction discussed in Figure 3.8 panel (b) earlier. Specifically, we see that the left side beam travels a distance $d\sin\beta$ (compared to the adjacent wave) and the right incident beam travels an "extra" distance $d\sin\alpha$. When the difference in path length for two nearby diffracted waves, is an integer number of wavelengths, then constructive interference occurs

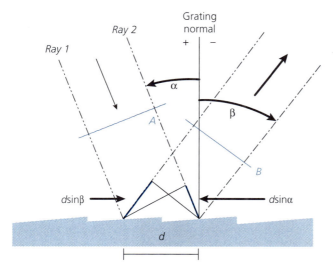

Figure 3.23 Two light rays incident on diffraction grating and the path length difference for these rays. Sign convention for A and B using the grating equation.

---

[4] An excellent resource for further study on gratings is Palmer, C.; Loewen, E. *The Diffraction Grating Handbook*, 7th ed.; Richardson Gratings: Rochester, NY, 2014.

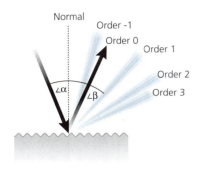

Figure 3.24 Diffraction of light as a function of wavelength using a blazed reflective grating showing different orders of diffraction.

and a bright spot or line is seen. One will only measure bright lines off a grating at particular angles as a function of wavelength. In other words, because the incident angle (α) and the reflected angle (β) are different, the angle β (or α) at which constructive interference will occur is wavelength dependent. This condition can be written mathematically as the *grating equation*:

$$\Delta s = m\lambda = d(\sin\alpha + \sin\beta) \qquad \text{Eq. 3.9}$$

where Δs is the difference in path length, m is the order of the reflection ($m = 0, +/-1, +/-2$, and so on), λ is the wavelength of the incident ray on the grating (generally composed of a wide range of wavelengths), d is the grating spacing, α is the incident angle of light, measured from the normal (just as in Snell's law), and β is the reflected light angle measured from the normal where angles to the right of normal are negative. Make sure to notice this statement about negative β values. The signs of the orders should also be noted. The grating spacing is often determined by measuring the length of the grating and counting the number of lines or grooves on a grating and dividing these to determine d, the groove distance.

When $m = 0$ (order 0), this corresponds to simple reflection and $\alpha = \beta$. The grating equation can be used to determine the *diffraction angle, β,* for a given wavelength and order. The first-order diffraction is what is typically used in a monochromator. (The first order $m = -1$ and $m = +1$ diffracted beams are shown in Figure 3.24 as well as higher order beams.)

Example 3.4

You have a laser (a monochromatic light source) and want to estimate the wavelength. You decide to use a reflective diffraction grating to do this. The light is incident on the grating at 20 degrees with respect to normal. Using a grating with a groove distance of 5 mm, the first-order diffraction, is found at an angle of −12.5 degrees. Determine the wavelength of the source. You might find a review of Figure 3.23 useful.

STRATEGY –

Solve the grating equation for λ.

SOLUTION –

For the 6 cm focal length lens:

$$\lambda = \frac{d}{m}(\sin\alpha + \sin\beta) = \left(\frac{5 \times 10^{-6}\,\text{m}}{1}\right)(\sin(20) + \sin(-12.5)) = 6.28 \times 10^{-7}\,\text{m} = 628\,\text{nm}$$

Problem 3.9: Suppose that you found out that the actual wavelength of the laser from Example 3.4 was 632.8 nm.

(a) Calculate the angle of the first-order diffracted light. Compare this angle to what was measured in the example.

(b) Show how you could use the known wavelength of the laser light to determine the line spacing of the grating.

Problem 3.10: Given a reflective diffraction grating with a groove distance of 3 mm, calculate the diffraction angle for a first-order $(m = 1)$ diffraction of a 526.5 nm beam with an incident angle α of 20 degrees. At what angle would you expect to find the second order $(m = 2)$ diffracted beam?

The ability to separate wavelengths effectively is called the *resolving power, R*, of the grating. This is the ability to separate a wavelength λ and $\lambda + \Delta\lambda$. R is defined as:

$$R = \frac{\lambda}{\Delta\lambda} \qquad \textbf{Eq. 3.10}$$

A grating that can separate a very small spread of wavelengths of a source would have a large R. A low resolving power grating may have an R of 1,000 and a large resolving power up to 100,000 or larger. It is important to note that the resolving power describes the grating and not the complete ability of the monochromator to separate different wavelengths at the exit slit. After all, the monochromator is constructed of mirrors, a grating, and slits. Each component contributes to the overall resolving power of the monochromator. The relevant quantity to describe the monochromator's ability to separate wavelengths at the exit slit is the linear dispersion, D. Linear dispersion will be introduced after our discussion of gratings.

The resolving power of a grating is related to the order, m, and the number, N, of grooves or lines illuminated by the source radiation.

$$R = mN \qquad \textbf{Eq. 3.11}$$

Gratings are often characterized by the groove density, the number of grooves per millimeter. It is common to find gratings with groove densities ranging from the few hundreds to a few thousand grooves/mm. The resolving power increases for higher orders of the diffracted beam. However, the intensity of diffracted light at these higher orders generally decreases. This means that if the $m = 2$ diffracted light was used instead of the $m = 1$, the resolution would improve but the optical signal would be reduced, ultimately affecting the signal-to-noise ratio.

Example 3.5

A grating is 25 mm long and has 600 lines/mm. Calculate (a) distance between grooves; (b) the total number of grooves for the grating; (c) the resolving power assuming first-order diffraction; and (d) assuming a center wavelength of 550 nm, the minimum wavelength separation associated with this resolving power.

SOLUTION –

(a) The distance between grooves is: $d = \dfrac{1}{600\,\text{lines/mm}} = 0.00166\,\text{mm} = 1.66\,\mu m$

(b) The number of lines is: $N = (600\,\text{lines/mm})(25\,\text{mm}) = 15{,}000\,\text{lines}$

(c) The resolving power is: $R = mN = (2)(15{,}000) = 30{,}000$

(d) The minimum wavelength separation is: $\Delta\lambda = \dfrac{\lambda}{R} = \dfrac{550\,\text{nm}}{30{,}000} = 0.018\,\text{nm}$

Problem 3.11: You need to be able to resolve a wavelength difference of 0.1 nm around 500 nm. You have a reflective diffraction grating that has 600 lines/mm and is 12.7 mm long. You plan to use the first-order diffracted light. Will this grating have the resolving power needed? Support your response with a calculation.

Example 3.6

An optical beam is composed of two wavelengths, 589.00 nm and 589.59 nm strikes a diffraction grating at an incident angle of $\alpha = 45°$. These are the wavelengths associated with the sodium doublet. What resolving power do you need in order to separate the two beams? You have a 5.0 cm grating with 12,000 lines. For the first order $(m = 1)$, at what angles do these two wavelengths diffract? If a screen is put 0.6 m from the grating, what is the distance between the two beams?

SOLUTION –

The resolving power R is:

$$R = \frac{\lambda}{\Delta\lambda} = \frac{589\,\text{nm}}{589\,\text{nm} - 589.59\,\text{nm}} = 999$$

The groove distance is:

$$d = \frac{5.0 \times 10^{-2}\,\text{m}}{12,000} = 4.2\,\mu\text{m}$$

Solve the grating equation for the angle β:

$$\beta = \sin^{-1}\left(\frac{m\lambda}{d} - \sin\alpha\right)$$

Solve for β for the two cases, $\lambda = 589$ nm and $\lambda = 589.59$:

$$\beta_{589.00} = \sin^{-1}\left(\frac{589 \times 10^{-9}}{4.2 \times 10^{-6}} - \sin 45°\right) = -34.454°$$

$$\beta_{589.59} = \sin^{-1}\left(\frac{589.59 \times 10^{-9}}{4.2 \times 10^{-6}} - \sin 45°\right) = -34.444°$$

The difference in these angles may seem quite small at 0.01 degrees. However, if these two wavelengths were projected a distance away from the grating, one can easily attain reasonable separations. With the screen a distance L from the grating, the two wavelengths are separated by a distance (Δx) apart

$$\Delta x_{\text{between wavelengths}} = L\left(\tan\beta_{\lambda_1} - \tan\beta_{\lambda_2}\right)$$

$$\Delta x_{\text{between wavelengths}} = L\left(\tan\beta_{\lambda_1} - \tan\beta_{\lambda_2}\right) = 0.6\text{m}\left(\tan(-34.444) - \tan(-34.454)\right)$$
$$= 0.006\text{m} = 0.6\text{cm}$$

We see that the farther away we put the screen from the grating, the farther apart the diffraction maxima. With L=0.6 cm we can easily spatially resolve the sodium doublet.

Problem 3.12:   You have purchased a 25 mm wide reflection grating with 1,200 lines/mm. What is the resolving power and what minimum wavelength can be resolved around 750 nm?

Problem 3.13: White light contains wavelengths that range from 400 to 700 nm. Given a reflective diffraction grating with a groove distance of 5 μm,

(a) Calculate the diffraction angle for first-order diffraction of a 400 nm, a 550 nm, and a 700 nm beam with an incident angle of $\alpha = 20$ degrees.

(b) If the exit slit for this monochromator is 5 cm from the grating, what will the dispersion distance (Δx) be between these three beams at the exit slit?

(c) If the exit slit is 20 μm wide, what is the bandwidth of radiation that passes through the exit slit for the 550 nm beam?

(d) Does the bandwidth change for the other two beams?

Grating Efficiency   Of course, gratings cannot have an *efficiency* of 100% in the diffracted wavelengths. The absolute efficiency of the grating gives the percentage of diffracted light at a given wavelength in a particular order as a function of the incident light intensity at that wavelength. A plot of absolute efficiency as a function of wavelength is shown in Figure 3.25 for a 500 nm blazed grating. Note that the absolute efficiency also depends on the polarization of the incident light. Light polarized perpendicular to the grating has a somewhat higher efficiency above the blaze wavelength. The efficiency of the grating is one of several instrumental realities that will reduce the signal. Many instruments will have several gratings mounted on a wheel so that the analyst can choose between several gratings with different blaze wavelengths.

The shape of the efficiency plot may also affect the way one perceives the wavelength dependence of the signal. This is part of the wavelength-dependent response of the instrument itself. In other words, because the grating does a better job at diffracting 500 nm light compared to 650 nm light (70% absolute efficiency compared to 40%), the measured intensities will scale accordingly. For this reason, we typically run a "blank" before we scan an actual sample. This topic will be discussed further in later chapters.

The structure of the grating itself, the inclusion of imperfections and irregularities in the lines, can lead to spectral lines where they should not otherwise be. This problem is referred to as *ghosting*. Another type of grating, the holographic grating, can be made by writing the grating with laser light holographically. Holographic gratings do not suffer from these imperfections and therefore have very low stray light reflectance, although typically lower efficiency.

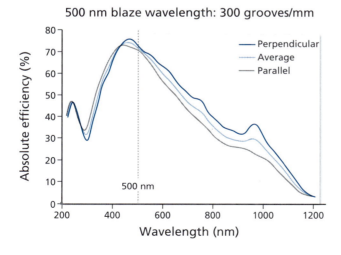

Figure 3.25 Efficiency as a function of wavelength for a blazed reflective grating.

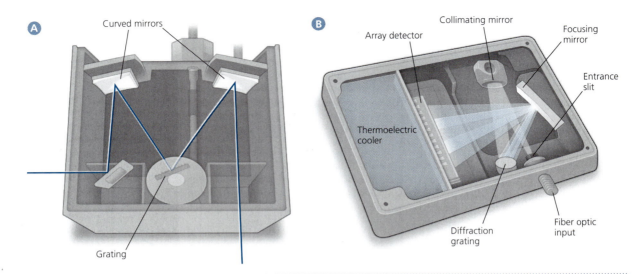

Figure 3.26 (A) Czerny-Turner Monochromator, including entrance and exit slits; (B) monochromator where light is dispersed across an array detector.

> **Problem 3.14:** An optical signal has an intensity of 1.1 (arbitrary units) at 500 nm and 1.0 at 700 nm after diffracting off the 500 nm blazed grating (see Figure 3.24). By taking the absolute efficiency of the grating into account, what was the intensity of the 500 nm and the 700 nm signal before interacting with the grating?

Putting It All Together: Details on the Grating Monochromator

We can now return to discussing the monochromator. A schematic of a grating monochromator was shown initially in Figure 3.22. In Figure 3.26 we show a photograph of a monochromator (in this case a *Czerny-Turner* type) composed of two slits, two convex mirrors, and a diffraction grating. Light from the entrance slit enters the monochromator and is reflected off a concave mirror. The concave mirror takes the light from the point source entrance slit and collimates it onto the reflective grating. Light is diffracted off the grating and is incident on another concave mirror. This mirror takes the light from the grating and focuses it on the exit slit. From the figure, one can see how light diffracts off the grating and is "selected" for measurement outside the instrument at the exit slit. In the grating based monochromator, the grating is rotated in order to select the wavelength that reaches the *exit slit*.

We also show a fiber optic spectrometer, which takes light into the device through a fiber optic coupler. Light from the entrance slit is also collimated with a mirror and directed onto the gratings. In this device, the diffracted light is directed onto an array detector.

We saw that refraction is what makes lenses focus light. One can also use diffraction to focus light. This is used for X-ray optics where diffraction must be used because optical materials are not transparent to X-rays. We saw examples of diffraction earlier in this chapter, where light incident on a circular aperture made a concentric circle pattern of light. This pattern is known as the Airy disk. We can make an optical component out of concentric circles alternating between transparent and opaque materials. This is called a Fresnel zone plate. It is not obvious, but a collimated light source incident on the zone plate can be focused. The focusing action is the byproduct of the sum of all the light waves interacting with each part of the zone plate. This technique is used to control X-ray beams, for example, in X-ray photoelectron spectroscopy or X-ray fluorescence, as described in Chapter 10. ▬

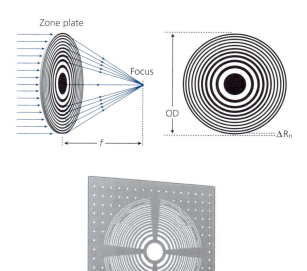

Overlapping Diffraction Orders   When working with grating monochromators, it is important to understand the challenge of *overlapping diffraction orders*. The image from Figure 3.24 does not tell the whole story about the diffracted light. Sometimes we experience overlap between adjacent diffraction orders. For example, with white light incident on a reflective grating, we might find a first-order $(m = 1)$ diffracted beam for 700 nm and a second-order $(m = 2)$ diffracted beam for 350 nm at the same diffraction angle β:

$$(1)(700\text{nm}) = d(\sin\alpha + \sin\beta) = (2)(350\text{nm})$$

This means that at this angle, where we would like to be measuring the power at the detector for one wavelength (the whole point of the monochromator), we would actually be susceptible to measuring power at the detector from both the 700 nm first-order diffraction, as well as the 350 nm second-order diffraction!

As an example of where this might cause signal challenges, if one is using 350 nm light as an excitation source and measures emitted light from a sample at 700 nm through a monochromator, the overlapping orders would contaminate the signal. If the detector responds to both 350 nm and 700 nm light, this would potentially suggest that there is much more signal at 700 nm than there actually is. This problem would be further exacerbated because the 350 nm excitation is much higher in intensity than the 700 nm emission from the sample. The issue of overlapping orders is shown for a general case of a range of input wavelengths in Figure 3.27.

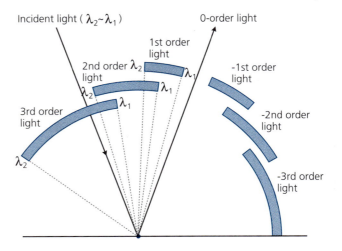

Figure 3.27 Diffraction off grating showing overlapping wavelength ranges for diffracted light.

One way to reduce the issue of overlapping orders is to use optical filters to block particular wavelengths. Filters will be discussed later in this chapter.

Problem 3.15: A signal composed of wavelengths from 375 to 780 nm is being measured. The light is sent through a grating monochromator. Should you be concerned with diffraction order overlap? Justify your response with a calculation.

The dispersing capability of the instrument can be quantified through two related quantities, the angular dispersion and the linear dispersion. The *angular dispersion*, D_β, describes how the angle of diffraction varies with wavelength. The *linear dispersion*, D, is related to the distance between adjacent wavelengths diffracted by the monochromator at the focal length.

The larger the angular dispersion the larger the angle will be between two diffracted beams with wavelengths of λ and $\lambda + d\lambda$. The angular dispersion, D_β, can be defined as the derivative of the diffraction angle β with respect to wavelength:

$$D_\beta = \frac{d\beta}{d\lambda}$$

Eq. 3.12[5]

We can take the derivative of the grating equation to find that the angular dispersion:

$$D_\beta = \frac{m}{d \cos\beta}$$

Eq. 3.13

The units for the angular dispersion are radians/length. This quantity is frequently reported in degrees/nm. As the groove spacing, d, decreases, the angular dispersion increases. For many applications in spectrometers, the diffraction angle β will be small (< 15 degrees). Mathematically, this means that the cosine of a small angle is approximately 1. The angular dispersion then becomes:

$$D_\beta \approx \frac{m}{d}$$

Eq. 3.14

---

[5]Be careful not to confuse the symbols used to describe the derivative D and the d that represents the groove spacing in the grating equation.

where the d is the groove spacing and m is the *diffraction order*. In this case, the angular dispersion depends only on the diffraction order and the groove spacing. Neither of these quantities are wavelength dependent. Many monochromators are designed so that the diffracted angle is small.

The larger the linear dispersion the larger the distance will be between the two diffracted beams of wavelength λ and of $\lambda + d\lambda$. The linear dispersion, D, can be related to the angular dispersion D_β by:

$$D = fD_\beta \frac{mf}{d\cos\beta} \qquad \text{Eq. 3.15}$$

where f is the focal length of the monochromator. When the two mirrors in the monochromator have the same focal length, the focal length of the mirror is the focal length of the monochromator. Linear dispersion is often quoted in mm/nm, where the mm refers to the distance between wavelengths on a detector or monochromator slit and the nm refers to the wavelengths. The reciprocal linear dispersion, D^{-1}, is also a quantity used to describe grating systems. For cases where b is small, the linear dispersion reduces to:

$$D = fD_\beta \approx \frac{mf}{d} \qquad \text{Eq. 3.16}$$

Increasing the focal length increases the linear dispersion. This corresponds to a longer path length for the diffracted beam to spread through. This larger linear dispersion corresponds to a larger distance between adjacent wavelengths on the exit slit.

Example 3.7

You have access to two diffraction gratings, one with 600 grooves/mm and a length of 50 mm and one with 1,200 grooves/mm and a length of 25 mm. Calculate the linear dispersion and resolving power for both. You also plan to use the first-order diffracted beam and use a configuration where β is around 10 degrees and the focal length of the system is 0.3 m.

SOLUTION –

The groove spacing, d, can be found as the reciprocal of the grooves/mm.

For the grating with 600 groove/mm that is 50 mm long:

The linear dispersion is:

$$D \approx \frac{mf}{d} = \frac{(1)(300\,\text{nm})}{(1/600)\text{mm}\left(\dfrac{1\times10^{-3}\,\text{m}}{1\,\text{mm}}\right)\left(\dfrac{1\,\text{nm}}{1\times10^{-9}\,\text{m}}\right)} = 0.180\,\text{mm/nm}$$

Note that the groove spacing was converted to nm in the dominator.

The resolving power is: $R = mN = (1)(600\,\text{grooves/nm})(50\,\text{mm}) = 30{,}000$

For the grating with 1,200 groove/mm that is 25 mm long:

The linear dispersion is:

$$D \approx \frac{mf}{d} = \frac{(1)(300\,\text{mm})}{(1/1{,}200)\text{mm}\left(\dfrac{1\times10^{-3}\,\text{m}}{1\,\text{mm}}\right)\left(\dfrac{1\,\text{nm}}{1\times10^{-9}\,\text{m}}\right)} = 0.360\,\text{mm/nm}$$

The resolving power is: $R = mN = (1)(1{,}200\,\text{grooves/mm})(25\,\text{mm}) = 30{,}000$

We see that both gratings have the same resolving power but different linear dispersions.

Monochromator Slits   Several variables influence the spectral resolution of both prism and grating monochromators (e.g., prism material, grating rulings, distance from prism/grating to *exit slit*, uncertainty of angular position of prism or grating). In practice, the instrument user only has access to the *entrance slit width* and the *exit slit width*. These slit widths are set to be identical for most applications. The slit width plays a critical role in spectral resolution or bandwidth resolution and power/intensity measured. As you reduce the exit slit width, you will decrease the "number" of wavelengths that can enter the slit—hence increasing the resolution. As you reduce the slit width, you will also reduce to total amount of light that will get through the exit slit, thus reducing the total light output of the monochromator. When using a monochromator, the user must balance their need for resolution with total signal power needed. By maximizing resolution by narrowing the slit width, the trade-off made is to reduce the power reaching the detector, and one may very well compromise the ability to measurement the signal at all as you approach the limits of your detector. The other extreme is to set the exit slit at their widest setting.[6] This would pass the maximum amount of radiant power (larger signal) at the expense of spectral purity (resolution). The operator must balance the trade-off of radiant power and spectral resolution in the performance of a particular analysis.

One can determine the theoretical slit width, *wslit*, required to completely separate two wavelengths, $\Delta\lambda$, with a given diffraction grating by:

$$\Delta\lambda = \frac{w_{slit}}{D} \qquad\qquad \textbf{Eq. 3.17}$$

where D is the linear dispersion of the grating. We have assumed that the exit slit and entrance slits are at the focal points of the instrument. The quantity $\Delta\lambda$ is also referred to as the *bandpass* or *bandwidth* of the monochromator.

Example 3.8

Calculate the slit widths necessary for a 0.1-nm bandpass for the two grating cases in Example 3.7 and comment on their relative utility in a spectrometer.

SOLUTION –

We will rearrange Equation 3.17 and solve for the slit width.
For the 600-groove/mm grating:

$$w = \Delta\lambda D = (0.1 \text{nm})(0.180 \text{mm/nm}) = 1.8 \times 10^{-2} \text{mm} = 18 \mu\text{m}$$

For the 1,200-groove/mm grating:

$$w = \Delta\lambda D = (0.1 \text{nm})(0.36 \text{mm/nm}) = 3.6 \times 10^{-2} \text{mm} = 36 \mu\text{m}$$

The 1,200-groove/mm grating has a slit width twice that for the 600-groove/mm grating. Because a larger slit width will allow for more signal intensity, we would choose the 1,200-groove/mm grating.

Problem 3.16:   You are trying to separate wavelengths that are 505.1 nm and 505.18 nm. The grating is 25 mm long and has 1,800 grooves/mm. Use the $m = 1$ order diffraction to determine the dispersion angle between the two wavelengths. Repeat the calculation for $m = 2$. Comment on the difference.

---

[6] In practice, the exit slit width maximum is usually the same as the width of the entrance slit.

The Michelson Interferometer

The modern Fourier Transform Infrared Spectrophotometer (FTIR) uses neither a prism nor a grating to select a wavelength of light. In fact, it passes all incident radiation from the source through the sample simultaneously and uses a device called a *Michelson interferometer* to discern absorption as a function of wavelength. In some respects, the interferometer is similar to a diffraction grating monochromator in that it makes use of constructive and destructive interferences between different wavelengths of light. However, it operates in a completely different manner.

In the FTIR, the source light enters an interferometer, as shown in Figure 3.28. Radiation from the source (A) travels to a semitransparent beam splitter (B). A portion of the incident radiation passes through the beam splitter to a stationary mirror (C), and a portion is directed to a mirror whose position moves linearly (back and forth) about a set position (D). Both beams return to the beam splitter and pass through to the sample and detector (E). For a moment, consider a single wavelength of light; as the translating mirror moves, it brings the first portion of the split beam in and out of phase with the second portion of the split beam. When the two beams recombine at the semitransparent beam splitter, a *beat pattern* results (see Figure 3.7). This beat pattern is a function of the time-dependent location of the moving mirror. When all wavelengths of light are simultaneously passed throughout the Michelson interferometer, a net beat pattern is created as the individual wavelengths of light come in and out of phase. By knowing the exact position of the moving mirror as a function of time, a Fourier transform analysis can determine the frequencies of light that contributed to create the net beat pattern.[7] The detector in the FTIR measures the intensity of all of the wavelengths at the same time as the translating mirror moves back and forth. Surprisingly, the FTIR is capable of very high resolution (measuring differences in wavelength of tiny fractions of 1 nm). We will cover this in more detail in Chapter 11 when we discuss infrared spectroscopy.

One of the great advantages of the Michelson interferometer is speed. Older infrared spectrometers would scan through each wavelength using a scanning monochromator. A typical infrared spectrum could take up to 5 minutes to measure. The modern FTIR can

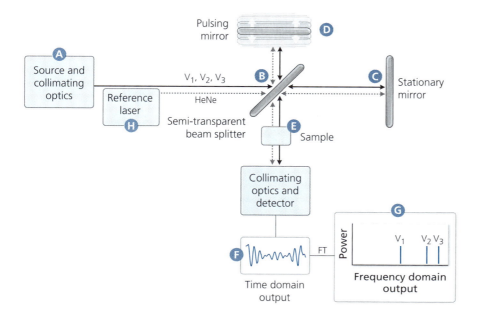

Figure 3.28 Michelson interferometer in an FTIR.

---

[7]See Chapter 5 on signal processing for a discussion of the Fourier transform and Chapter 11 on infrared spectroscopy for an Excel activity utilizing the Fourier transform.

take a "scan" in less than one second because all of the source radiation passes through the sample at the same time. The rapid scanning, combined with computerized data collection, allows for signal averaging of FTIR spectra, which greatly increases the signal-to-noise ratio.

Optical Filters and Power Reduction

Many times, one needs to control the power as a function of wavelength of a light source or sample. This can be done in one of several ways. The most straightforward way is to use *neutral density (ND) filters*. ND filters have fairly flat absorption spectra over the visible wavelength region of the electromagnetic spectrum and can be purchased with a large range in transmission values (from ~80% to less than 0.01%). The absorption spectrum will vary according to the material used (e.g., quartz, BK7). There are also ND filters that work by reflection as opposed to absorption. One can control a beam's optical power quite effectively with ND filters so long as the light source in use is in the appropriate range of the filter. The transmission properties of ND filters are described by the equation:

$$T = 10^{-OD}$$ **Eq. 3.18**

where T is the transmission (0 to 1) of the filter, and OD *is the optical density* of the filter. ND filters can be easily found with OD of 0.1 ($T = 0.79$ or 79%) to 4 ($T = 0.0001$ or 0.01%) or more (see Figure 3.29).

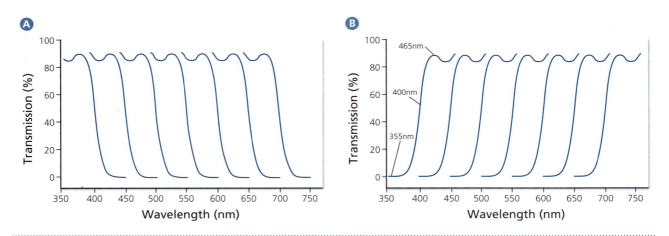

Figure 3.29 (A) Short-pass and (B) long-pass filters. The graph shows the percent transmittance as a function of wavelength.

Example 3.9

You are using an Argon Ion laser in a spectroscopy experiment and need to vary the power as a part of your investigation. The laser operates at 514.5 nm and has a maximum power of 50 mW. You have access to ND filters with ODs of 0.3 and 0.5. What powers can you obtain?

SOLUTION –

For the 0.3-OD filter: $T = 10^{-OD} = 10^{-0.3} = 0.501$

   50.1% transmitted, so P = (0.501)(50 mW) = 25.1 mW

For the 0.5-OD filter: $T = 10^{-OD} = 10^{-0.5} = 0.316$

   31.6% transmitted, so P = (0.316)(50 mW) = 15.8 mW

We can combine the 0.3-OD and 0.5-OD filters to obtain:

$$T = 10^{-OD} = 10^{-0.8} = 0.158$$

   15.8% transmitted so P = 0(.158)(50 mW) = 7.92 mW

Problem 3.17: You have access to five ND filters with optical densities of 0.1, 0.3, 0.5, 1.0, and 2.0. You have a helium-neon laser at 632.8 nm at 13 mW. What optical powers can you attain with these filters, making sure to use all possible combinations of the filters? Create a spreadsheet to answer this question.

A chemist may need to selectively control the wavelength of a broadband (continuum) light source with filters. For example, if you are exciting a sample with a wavelength of 355 nm and measuring emission from 400 to 700 nm with a sensitive photodetector, you may need to use a filter to remove the scattered 355 nm excitation light from your emission signal. *Color filters* are one possible solution to this problem. Color filters come in a variety of absorption (or reflection) profiles. If looking for a special color filter, it is best to investigate the commercial options on vendor sites. Another option for wavelength control is the long-pass or short-pass filter. These are available in a selection of wavelength ranges (see Figure 3.29 for one set of examples). Each curve in the figure represents a different type of filter that could be chosen to meet given needs.

An additional option, the *interference filter,* is particularly useful when a selected narrow wavelength range should be passed while rejecting other wavelengths. The interference filter has two glass substrates with carefully chosen reflecting layers and a spacer. The spacer size is chosen so that light of a very narrow wavelength range can be transmitted while other wavelengths destructively interfere as they reflect across the spacer width. Because the filter properties can be tuned with the spacer size, a wide range of commercial wavelength filtering options are available. An example of a transmission spectrum for a 532 nm interference filter is shown in Figure 3.30.

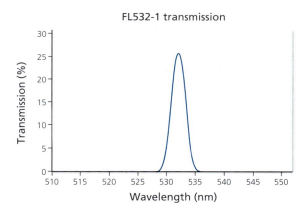

Figure 3.30 Transmission spectrum for an interference filter (band-pass) from Thorlabs; the full width at half maximum is ~1 nm.

Example 3.10

In an optical experiment, you need to dramatically reduce the amount of scattered 355 nm light from a laser system that reaches the detector while simultaneously transmitting as much of your signal as possible. Your signal is expected to be in the range of 400 to 500 nm. You have access to the short-pass and long-pass filters shown in Figure 3.29. Choose a filter and explain your choice.

The first transmission curve from the left for the long-pass filters is the best choice. The curve on the far left shows a filter with nearly zero transmittance at wavelengths below 375 nm but the same filter has 60% transmittance at 400 nm and 90% transmittance at 465 nm. The other long-pass filters would reduce the signal intensity. For example, the transmission curve second from the left would transmit 60% of the signal at 450 nm and 0% of the light at 400 nm.

3.4 Common Optical Materials

There are many different types of material commonly used in optical instrumentation and it is important to have a sense of the optical properties of several of the very common materials used, especially the absorption spectrum of various materials. For example, can you use a BK7 glass lens to focus a 266 nm ultraviolet laser? Will potassium bromide (KBr) absorb at 1.2 mm? Which type of plastic cuvette (acrylic or polystyrene) can be used at 325 nm? Table 3.2 shows the optically transparent range for several common optical materials. From this table, it is clear that KBr is the material of choice for optics associated with the infrared region of the spectrum as the material is 80% transparent out to 2.5 mm. KBr windows are routinely found in FTIR spectrometers.

One challenge associated with KBr is that it absorbs moisture readily, which degrades the optical quality of the material. Calcium fluoride optics are excellent for use with excimer lasers (194 nm) because of the high transmission at this wavelength and because the damage threshold is high. Crystal quartz and ultraviolet-fused silica are both transparent in the ultraviolet and are common materials found in UV-vis spectrometers.

Many times, a key resource for information about a particular material's optical properties can be found in the optical component catalogs and websites of the optical manufacturers.

TABLE 3.2   Transparent Range of Some Common Optical Materials

	μm	nm	$\bar{v}$ (cm^{-1})
Crystal quartz	0.150–2.5	150–2,500	6,667–400
UV fused silica	0.175–2.0	175–2,000	5,714–500
Borosilicate crown glass, N-BK7	0.375–2.0	375–2,000	2,667–500
Polystyrene (type of plastic)	0.340–0.90	340–900	2,491–1,111
Acrylic/PMMA (type of plastic)	0.300–0.90	300–900	3,333–1,111
CaF$_2$	0.13–7.0	130–7,000	7,700–143
KBr	0.250–2.50	250–2,500	4,00–400

3.5 Beyond Linear Optics

More than likely, all of your experiences with light have been in the linear optics regime. It may seem obvious, but if you send a low-powered light source (even a low-powered laser) through a cube of glass, the light that comes out of the glass has the same wavelength that was sent in. However, you may be surprised to learn that if a suitably powerful laser is sent through a particular type of optical material, the light that exits the optical material will be both the wavelength of the incoming light and a component that is *half* the wavelength (twice the frequency) of the incident light. What has occurred is the generation of the second harmonic of the fundamental wavelength. This is an example of a classic *nonlinear optical phenomenon*. In Figure 3.31, infrared laser light at 1064 nm drives (or makes oscillate) atoms in a nonlinear optical material such that light at half the wavelength (twice the frequency) emerges. There are several other nonlinear optical effects that have a wide variety of uses (e.g., X-ray imaging of ultrafast chemical processes). What is important to know is that the higher the peak power of the laser source, the more likely nonlinear optical effects are to occur.

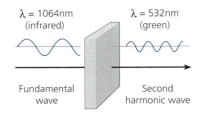

Figure 3.31 Schematic of a nonlinear optical process, second harmonic generation.

Metamaterials have gained significant attention in the field of optics because of their potential for very unique optical properties. These include the ability to focus a light beam beyond the limitations caused by diffraction (this is referred to as the *super lens**) and the ability to have light bend around an object suggesting the object is "invisible"[†] to observers. These applications make use of the extreme diffracting behavior of metamaterials.

The optical properties of a metamaterial are derived not from the particular atoms of the structure so much as the geometric arrangement of very small structures composed of a given material. Much of the progress being made in this field is due to the latest developments in controlling structures on the nanoscale. The size of the structures is determined by the wavelength of interest, so that an optical cloak for ~1 micron light might have structures that are on the order of this size. Metamaterials can have exotic optical properties; for example, they may have negative refractive indices (you will not find that in your introductory physics book!). The example shown in Figure 3.32 uses microwave radiation, and hence the metamaterial structure must be constructed with sizes on the order of the wavelength of microwaves (centimeters).

The super lens holds great promise for microscopy applications where the diffraction limit sets the limit on the smallest structures a microscope can resolve. Beating the diffraction limit with a super lens would mean being able to image smaller structures than can currently be investigated or perhaps allow us to image small systems more inexpensively.

Although you may think of invisibility cloaks as something magical from a science fiction, story physicists, chemists, and engineers are actually making significant progress toward making material objects "invisible" by designing metamaterials. In order to make an object "invisible," light must bend around the object so that an observer cannot tell that an object is in the way of any light source. Figure 3.31 shows an example of a cloaked object in the microwave region and how microwaves bend around a cloaked object. Recent papers from the literature demonstrate the progress being made in this exciting and rapidly developing field of optics. For example, the authors in the cited *Science* article here discuss cloaks that may operate in the near-infrared, an important step toward cloaks that might operate at visible wavelengths in the future. ■

PROFILE

Innovation and Discovery in Optics: Metamaterials Hold Promise for the Perfect Lens, Invisibility Cloaks, and More

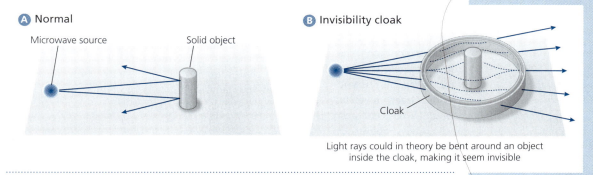

Figure 3.32 Schematic of microwaves being bent around a cloaked item.

---

*Zhang, X.; Liu, Z. Superlenses to Overcome the Diffraction Limit. *Nature Mater.* 2008, 7, 435–441.

[†]Ergin, T.; Stenger, N.; Brenner, P.; Pendry, J. B.; Wegener, M. *Three-dimensional Invisibility Cloak at Optical Wavelengths. Science* 2010, 328, 337–339.

3.6 Further Reading

BOOKS

Born, M.; Wolf, E. *Principles of Optics: Electromagnetic Theory of Propagation, Interference and Diffraction of Light*, 7th ed.; Cambridge University Press: New York, **1999**.

Haus, J. *Optical Sensors: Basics and Applications*, 1st ed.; Wiley-VCH: Weinheim, Germany, **2010**.

Hecht, E. *Optics*, 5th ed.; Addison Wesley: Boston, 2016.

Pedrotti, F.; Pedrotti, L. S.; Pedrotti, L. M. *Introduction to Optics*, 3rd ed.; Pearson Prentice Hall, Upper Saddle River, NJ, **2007**.

Taylor, J. R.; Zafirator, C. D.; Dubson, M. A. *Modern Physics for Scientists and Engineers*, 2nd ed.; Pearson Prentice Hall, Upper Saddle River, NJ, **2004**.

ONLINE RESOURCES

CVI Laser Optics; catalog and technical library. http://cvilaseroptics.com (accessed Aug, 2015).

Melles Griot; catalog and technical library. www.cvimelles-griot.com (accessed Aug, 2015).

Newport Corporation; resources such as tutorials, applications notes, and catalog. www.newport.com (accessed Aug 13, 2015).

O'Haver, T. Interactive Computer Models for Analytical Chemistry Instruction; Professor O'Haver has created an excellent set of simulation tools to understand various topics in instrumental analysis, including several tools for understand gratings. http://terpconnect.umd.edu/~toh/models/index.html (accessed Aug, 2015).

Thorlabs Products, home. https://www.thorlabs.com/navigation.cfm (accessed Aug, 2015).

3.7 Exercises

EXERCISE 3.1: You are using an Nd:YAG laser as an excitation source for a time-resolved fluorescence experiment. The Nd:YAG laser emits at 1064 nm. You are using a nonlinear optical material (KTP) to generate the second harmonic of this beam at 532 nm. The 532 nm output still contains a percentage of the laser fundamental at 1064 nm and you want to only use the 532 nm photons. Search online or in an optics catalog for a filter that will block 1064 nm and pass 532 nm. Characterize the loss at both 532 nm and 1064 nm for the filter.

EXERCISE 3.2: When light of wavelength λ moves from air (n = 1) to a material with index of refraction n(λ), the fraction of light reflected, R, is:

$$R = \frac{(1-n)^2}{(1+n)^2}$$

This relationship holds when the light is incident at 90 degrees. You are to use BK7 glass in your experimental setup. What percentage of light is reflected off the BK7 if you use λ = 355 nm (recall that the index of refraction versus wavelength was provided in this chapter)? Would the percentage of reflected light increase or decrease if you use λ = 785 nm?

EXERCISE 3.3: A light source contains wavelengths at 501 nm and 502 nm. A grating with a groove spacing of 2.5 μm is used and the light is incident on the grating at 45 degrees. For the m = 1 order, what is the angle of reflection for the 501 nm and 502 nm components of the beam?

EXERCISE 3.4: The intensity, I, of a light source is:

$$I = \frac{P}{A}$$

where P is the power of the source and A is the area of the beam. A diode laser with wavelength of 488 nm and power of 10 mW is to be used in a Raman spectroscopy experiment, but you are not sure you will get a large enough intensity to study your sample (you have been told that you need at least 3 W/cm^2). In order to decide the minimum radius of beam to attain this intensity, produce a plot of the intensity of the beam versus the beam radius, from r = 0.001 to 0.5 cm.

EXERCISE 3.5: One of your laboratory partners is doing experiments measuring the index of refraction of a new polymer film. In the experiment, two laser beams, wavelength 632.8 nm and 594 nm, enter a polymer surface from air at an angle of 35 degrees from the normal and refract. He reports that the 594 nm beam refracts at an angle of 27 degrees from the normal and the 632.8 nm beam refracts at 25 degrees. Calculate the index of refraction for the polymer film at these two wavelengths. There is something strange about the data he has proposed. What is it?

EXERCISE 3.6: Your boss has asked you to buy a new grating for your grating monochromator. She has told you that you need to measure peak intensities at wavelengths of 501.1 nm and 501.3 nm. Assume slit widths of 0.3 nm and a grating that must be 5 cm long. What groove spacing should you choose? Does this grating spacing seem physically possible? Search commercial gratings online to see if such a grating is commercially available.

EXERCISE 3.7: Load the PhET simulation "Bending Light" from http://phet.colorado.edu. Set the top materials to Mystery A and the bottom material to water. Measure the angle of refractions with the protractor in the simulation and then use Snell's law to calculate the index of refraction of Mystery A material. Change the top material to Mystery Material B and the bottom material to air and determine the index of refraction for Mystery Material B.

EXERCISE 3.8: It turns out it is possible to construct an object, a metamaterial, that has a negative index of refraction. A 855 nm beam is incident at an angle of 30 degrees from the normal onto a negative index object. Assume an index of refraction of –1 for the object. Calculate the angle of refraction and comment on what this means.

EXERCISE 3.9: Load the PhET simulations "Wave Interference" from http://phet.colorado.edu. Go to the light tab. Set up the two-slit experiment. Qualitatively answer the following question: What is the relationship between the slit width and wavelength and the extent to which interference occurs?

EXERCISE 3.10: Load the PhET simulation "Bending Light" from http://phet.colorado.edu. Set the top material to glass and the bottom material to air. Is it possible for the laser light to completely reflect off the interface? If so, what is the maximum angle that allows for this? Is this angle dependent on the index of refraction of the top material?

EXERCISE 3.11: The third and fourth harmonic of a Nd:YAG laser produces radiation at 355 and 266 nm, respectively. Calculate the beam waist size $(2w_0)$ for each harmonic of the Nd:YAG laser assuming a beam that is 8 mm in diameter for a lens with a focal length of 6 cm.

EXERCISE 3.12: Choose two focal length lenses to collect, collimate, and focus light from a xenon lamp that is 8 mm in size. Sketch your optical system and determine the size of the focused spot. Calculate the NA of your collection lens.

EXERCISE 3.13: In the photograph in the figure, three light rays are incident on a diverging lens. In the image is a sketch of a diverging lens and two incident light rays. Use Snell's law to sketch the light rays as they refract from air to glass and then from glass to air.

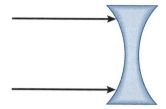

EXERCISE 3.14: Two waves are described by the following two equations for their intensities:

$$I_1(t) = 100\sin\left(5t + \frac{\pi}{9}\right)$$

$$I_2(t) = 100\sin\left(5t + \frac{\pi}{3}\right)$$

Assume that both intensities occur at the same spatial location. Plot each of these two waves in spreadsheet software. Construct your spreadsheet so that you can easily change any of the parameters (amplitude, 100 for both; frequency, 5 for both; and phase constant, $\pi/9$ and $\pi/3$) and see the corresponding plot of I_1, I_2, and their sum. Change the phase constant for the first wave to 0 and the second to $\pi/2$. What happens to the sum of I_1 and I_2?

EXERCISE 3.15: When focusing a beam of light, what advantages are gained by using a focusing mirror over a focusing lens?

EXERCISE 3.16: Why are grating monochromators more commonly found in modern spectrophotometers than prism monochromators?

Chapter 4

An Introduction to Instrumental Electronics

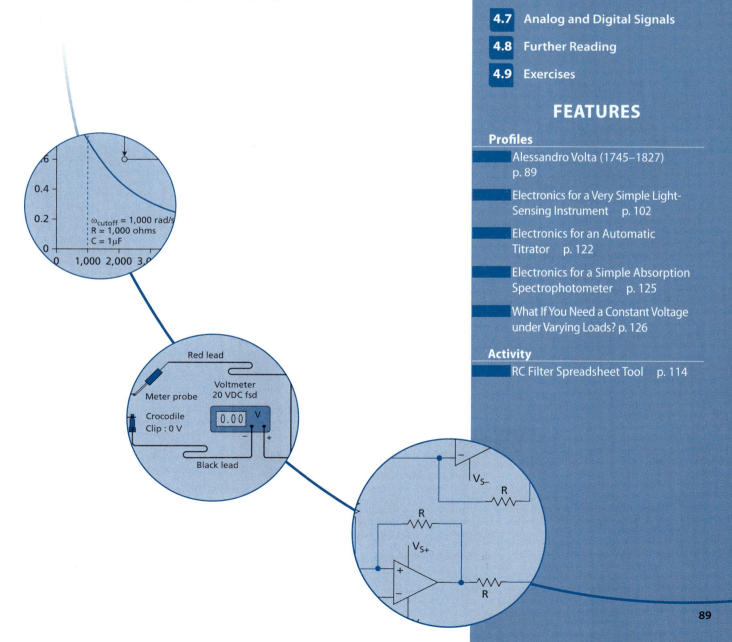

Early studies of electricity mostly involved static electricity. For example, in 1791 Luigi Galvani demonstrated that a dissected frog's leg could be made to twitch if the nerve ending was put in contact with a conducting metal.* He coined the term *animal electricity*. On the basis of these and other early studies, it was believed that electricity was produced by living things. However, in 1799 Alessandro Volta produced the first chemical battery, thus demonstrating that electricity can be produced by a nonbiological reaction. Volta's initial battery was latter termed a *voltaic pile* and consisted of copper and zinc wafers separated by felt or paper that had been soaked in salt water. The basic design of the voltaic pile is similar to how AA and AAA batteries are constructed to this day.

Napoleon Bonaparte very much admired Volta's work, and Napoleon and Volta became friends. You can see Napoleon's influence on Volta in this portrait of Volta, where Volta has imitated Napoleon's classic pose, with one hand inserted into his jacket.

The SI unit of electric potential (the volt) is named in honor of Alessandro Volta. ■

* See the Profile "Behind Frankenstein" in Chapter 19 for more about Luigi Galvani.

4.1   Introduction

At the heart of every instrumental technique is the measurement of either voltage or current (the *signal*). The signal is then processed and the results are displayed and/or archived for later analysis. So it is safe to say that all modern instruments rely on electronics and computer control. The goal of this chapter is to introduce students to the basic electronic components typically found in modern scientific instrumentation and to serve as a launch point for students who wish to learn more about instrumental electronics.

An understanding of electronics helps us diagnose instrument problems and allows us to push the state of the art in instrument design. Examples of electronics associated with chemical instrumentation are shown in profiles throughout this chapter.

The basic passive electrical components are the resistor, capacitor, and inductor. Passive components do not add energy to the electronic system. In general, instrumentation will also include other components such as diodes, transistors, operational amplifiers, as well as other integrated circuits (ICs). Two types of ICs, the 555 timer and the op amp, are shown in Figure 4.1 as well as examples of resistors, capacitors, and inductors. Although these components might look similar, the microcircuitry inside each chip causes them to function very differently.

In a chemical instrument, often the flow of electricity associated with a detector and circuit used to generate a signal can be used to understand a chemical system. In general, this flow of electricity is the *signal*[1] of interest. Signals in circuits can be classified into two groups: direct current (DC) and alternating current (AC). The distinction largely depends on whether the power source varies in time. DC circuits have current that is constant in time and can be powered by batteries, power supplies, or even solar cells. AC circuits have currents that are changing in time. The wall socket is an AC power source; however, many devices that are plugged into an AC output wall socket convert the power to a fixed DC voltage by use of an AC to DC transformer.

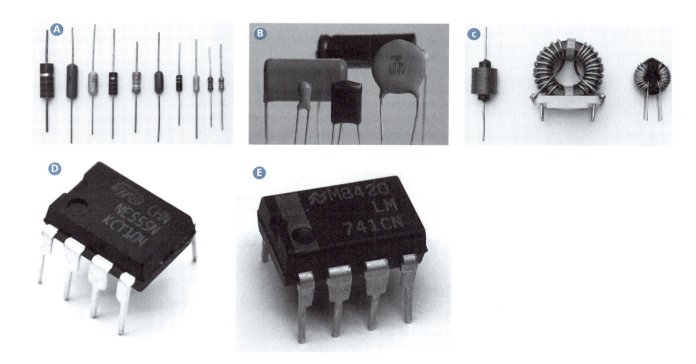

Figure 4.1 Various circuit components: (A) resistors, (B) capacitors, (C) inductors, (D) integrated circuit—555 timer chip, (E) integrated circuit–operational amplifier.

---

[1] To be clear, there are cases when it is not the actual flow of electricity but the potential difference or voltage (discussed in short order) that is the signal—for example, in potentiometry or voltometry.

Circuit Symbols

Figure 4.2 names many common circuit components and provides a photograph and the symbol used to denote each component in a circuit diagram. We will discuss more details of several of these components later in the chapter. Some components are known by several different symbols, and some components also come in a variety of "packages."

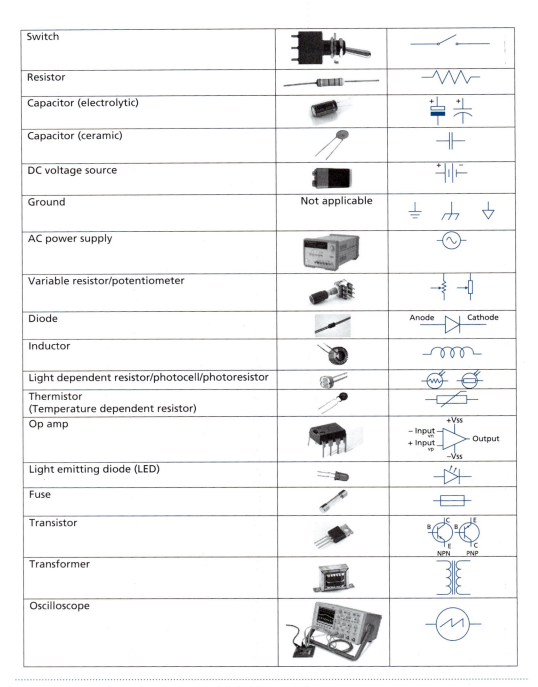

Figure 4.2 Various electrical components and their circuit symbols.

4.2 Direct Current (DC) Circuits

Current, Voltage, and Multimeter Basics

When describing the electrical behavior of circuits, we talk about the current that flows *through* a given circuit element and the potential difference or voltage *across* a given circuit element. Current, measured in amperes, is the amount of charge that flows past a point in the circuit in a given amount of time. As a loose analogy, if current was water in a river or water flowing in a pipe, the current can be thought of roughly as the amount of water that flows past a point in a certain amount of time. Voltage is the potential energy per charge, measured in J/C or volts. Voltage, using the same water analogy, can be roughly thought of as the applied force or the pressure that "pushes" on the free charges to flow in a conducting material due to an external energy source. A common source of this voltage is a battery. Comparing an electric circuit to the water pipe system, the voltage source or battery takes on the same role as a water pump. The width of the river or size of the pipe can be thought of loosely as the resistance where a wider river or pipe allows for more water flow. For resistors, there is a direct relationship, described by Equation 4.1, between the "push" applied by a battery, V (in volts or V for short), the amount of current flowing in the circuit element, I (in amps, A, or coulombs/second), and the resistance of the resistor, R, (measured in ohms or Ω):

$$V = IR \qquad \textbf{Eq. 4.1}$$

For a given resistor, Equation 4.1 says that the larger the applied voltage V, the larger the current I flowing in the circuit. This equation is referred to as *Ohm's law*,[2] which states that for a given resistor, the ratio of voltage to current is constant, and is ideally independent of the applied voltage. Most components obey Ohm's law. However, Ohm's law is limited at larger applied voltages, where the ratio of applied voltage to current will not be constant—hence deviating from Ohm's law. Furthermore, some components are intrinsically non-Ohmic in their behavior (e.g., the diode). They do not obey Ohm's law; there is not a linear relationship between voltage and current with increasing applied voltage.

The power dissipated, or energy lost per time in a resistor, is measured in watts. Power is the product of the voltage across the component and the current and is described in Equation 4.2.

$$P = IV = I^2 R \qquad \textbf{Eq. 4.2}$$

Resistors are rated for the maximum power they can withstand. A common resistor power rating is 1/4 W, although one can find lower power ratings (e.g., 1/8 W and lower) and higher power ratings (1/2 W up to many W).

Example 4.1

The current flowing in a resistor when 9 V is applied is 50 mA.

(a) Calculate the resistance of the circuit element.

Rearranging Ohm's law gives

$$R = \frac{V}{I} = \frac{9 \text{ V}}{50 \times 10^{-3} \text{ A}} = 180 \ \Omega$$

---

[2] We will see Ohm's law again in Chapters 17 and 19. However in Chapter 19 it appears as E =I R. In the area of *electrochemistry*, it is more common to represent voltage using an "E" instead of a "V" where the symbol "E" is referred to as the potential and the symbol "V" represents the unit volts instead of the voltage.

(b) What current will flow in the element when 6 V is applied?

The resistance is a property of the resistor, so it does not change.

$$I = \frac{V}{R} = \frac{6\ V}{180\ \Omega} = 33.3\ mA$$

We see that the lower the applied voltage, the lower the current.

(c) What is the power dissipated in the resistor with the 9 V supply? Would a 1/4 W power rated resistor be appropriate to use?

$$P = IV = \left(50 \times 10^{-3}\ A\right)\left(9\ V\right) = 0.45\ W$$

No, you should not use a 1/4 W-rated resistor. This power is above the maximum power rating.

Problem 4.1: A voltage of 12 V is applied to a 780 Ω resistor. The power rating for the resistor is 1/8 W. What current flows in the resistor? Is this 1/8 W power rating an appropriate choice?

Problem 4.2: The current through a resistor is 0.8 mA and the applied voltage is 9 V. What is resistance of the resistor? What is the power dissipated in the resistor?

Problem 4.3: A variable voltage supply is used to characterize the current versus voltage behavior in an electrical component. An experiment is done as follows—a voltage is chosen on the supply and then the current through the component is measured. The table here shows the data. Does this electrical component obey Ohm's law?

APPLIED VOLTAGE (V)	MEASURED CURRENT (A)
0.50	0.001
0.60	0.003
0.65	0.005
0.70	0.013
0.75	0.039
0.80	0.200

Series Circuit Elements and the Voltage Divider

The electrical behavior of resistors, as well as all electrical components, depends on how they are arranged. The different arrangements of electrical components in a circuit can produce very different circuit behaviors. Two electrical components are in *series* when they are connected such that the current that flows through the first component also passes through the second. In Figure 4.3, we show two resistors in series. Going back to our water analogy, in a series circuit there are no "forks in the river" so the all the current that flows past R_1 also flows past R_2. Hence, the current is the same in each of the resistors, *even when the resistances are not the same.*

A common misconception is that current gets "used up" and decreases from one resistor to the next in series. The current in a series circuit is the same at every point in the circuit. This is a statement of conservation of charge: no charge was lost in the resistor or wire. However, the same is not true of the voltage. Starting from a point in the circuit just after the battery or power supply the voltage (pressure) drops as it passes through each resistor.

In order to use Ohm's law to determine the current in a series circuit, it is convenient to calculate a net resistance in the circuit. A circuit with multiple resistors in series can be reduced to a circuit with a single equivalent resistance, R_{eq}, according to:

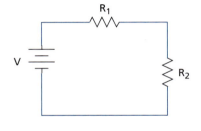

Figure 4.3 Resistors arranged in series with a voltage source.

$$R_{eq} = R_1 + R_2 + \ldots \qquad \textbf{Eq. 4.3}$$

and Ohm's law solved for current leads to:

$$I = \frac{V}{R_{eq}}$$

Eq. 4.4

Example 4.2

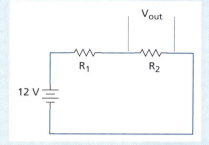
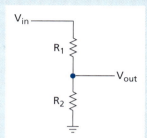

You have put two resistors, $R_1 = 100\ \Omega$ and $R_2 = 200\ \Omega$, in series with a 12 V source (shown here). An alternate way to draw this same circuit is also shown. Note that V_{in} is 12 V and the bottom symbol represents ground (0 V). V_{out} is shown in both figures. The current in R_1 is 40 mA.

(a) What is the current in R_2?

The current in R_2 is the same as R_1 because they are in series.

(b) What is the voltage across R_1?

Be careful, the voltage across R_1 is not 12 V. Let us use Ohm's law. The voltage across a resistor is the current through the component times the resistance.

$$V_1 = IR_1 = (40 \times 10^{-3}\ A)(100\ \Omega) = 4\ V$$

(c) What is the voltage across R_2?

We will use Ohm's law again. Note that the current in R_2 is the same as in R_1.

$$V_2 = IR_2 = (40 \times 10^{-3}\ A)(200\ \Omega) = 8\ V$$

The voltage across R_2 is also V_{out} as drawn here.

(d) What is V_{out}?

V_{out} is the same as the voltage across R_2.

Notice that the total voltage drop in Example 4.2 was, 4 V + 8 V = 12 V. This is the same as the total voltage supplied. This is no coincidence! The total voltage does get "used up" in some sense. If you imagine "walking" around *any* closed loop in this circuit (and any circuit), the sum of voltages from batteries *must* equal the sum of voltages across each component you have "walked" through. This is known as *Kirchhoff's voltage law*.

The circuit from Example 4.2 and Figure 4.3 shows a simple but important circuit: the *voltage divider*. By choosing the values of R_1 and R_2 in the voltage divider, you can vary the output voltage to any value you choose up to the supply voltage, V. The voltage divider equations help make this idea clearer. The voltage, V_1, across R_1, and the voltage V_2, across R_2 can be found via:

$$V_1 = \left(\frac{R_1}{R_1 + R_2}\right)V \qquad V_2 = \left(\frac{R_2}{R_1 + R_2}\right)V$$

Eq. 4.5

We will revisit this circuit in a more useful example in the Profile—Electronics for a Very Simple Light Sensing Instrument: Voltage Divider Photoresistor Circuit.

Example 4.3

You would like the voltage across a resistor to be 5.5 V in the instrument you are designing. Unfortunately, you only have access to a 9 V battery. You do have access to any combination of resistors you would like. Design a voltage divider circuit to meet this requirement.

Using one voltage divider equation (Equation 4.5, we know that $V = 9$ V and $V_1 = 5.5$ V. We can choose R_1 and then find the R_2 that fulfills the equation. We'll choose $R_1 = 330\ \Omega$.

$$\frac{V_1}{V} = \left(\frac{R_1}{R_1 + R_2}\right) = \left(\frac{330\ \Omega}{330\ \Omega + R_2}\right) = \frac{5.5\ V}{9\ V}$$

$$\frac{330\ \Omega}{0.611} = (330\ \Omega + R_2)\ \text{and}\ R_2 = 210\ \Omega$$

Problem 4.4:   Design a voltage divider circuit that provides 3.3 V across one resistor. The voltage source is an array of solar cells that provide 4.6 V. What two resistor values will you choose? The following resistors are available: 330 Ω, 390 Ω, 450 Ω, 510 Ω, and 1,000 Ω.

Parallel Circuit Elements and the Current Divider

In Figure 4.4, we show two resistors in parallel. Unlike the resistors in series, each of these resistors will not necessarily get the same current. Using our water flow analogy, in the parallel circuit, there is a "fork in the river" where some of the current from the battery can flow into R_1 and some will flow into R_2. From our water flow analogy, whatever flows into the fork must flow out of the fork.

We need to make an important distinction about the direction of the current and the sign of the moving charges. In Figure 4.4, we have drawn the direction of the current associated with moving positive charges. This is known as the *conventional current*, the direction that positive charges would move in a circuit. The electrons, which are flowing in the circuit, are flowing in the opposite direction to the conventional current. Although we now know that it is the electrons that flow, it has become common to use the direction of the conventional current. We will use the conventional current direction throughout this chapter.

In a manner similar to a series circuit, we can also calculate an equivalent resistance for a parallel circuit. The equivalent resistance in a parallel circuit is determined by:

$$\frac{1}{R_{equivalent}} = \frac{1}{R_1} + \frac{1}{R_2} + \dots \qquad \text{Eq. 4.6}$$

With only two resistors, we can reduce this equation to make this easier to use:

$$R_{equivalent} = \frac{R_1 R_2}{R_1 + R_2} \qquad \text{Eq. 4.7}$$

Using Kirchhoff's voltage law, our rule for "walking" around any closed loop in a circuit and accounting for voltage changes, we can discover an important rule for voltage

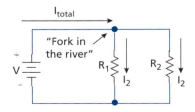

Figure 4.4 Resistors arranged in parallel.

across parallel components. From Figure 4.4, one loop is to go through the battery (e.g., a supply of 9 V) and then go through the R_1 resistor and then "walk" back to the battery. Kirchhoff's voltage law states that the voltage across the resistor is the same as across the battery, namely 9 V. But you could also choose the loop through the battery and the R_2 resistor. This means R_2 also gets the 9 V from the battery (said another way, the 9 V are applied across R_2), even though R_1 and R_2 do not have the same value. This is a key conclusion: electrical components are in *parallel* when they are connected such that each component gets the same voltage, independent of the values of resistance.

Because of the current-splitting behavior of the parallel resistor circuit, it is sometimes referred to as the *current divider* circuit. The splitting of the current is shown in Figure 4.5. The current I_{total} can be written as:

$$I_{total} = I_1 + I_2 \qquad \text{Eq. 4.8}$$

so that the sum of currents I_1 and I_2 equals the total current I_{total} in the circuit. Put another way, whatever current enters a junction must leave the junction. This concept is known as *Kirchhoff's current law*.

The current in each resistor can be found using the current divider equations:

$$I_1 = \left(\frac{R_2}{R_1 + R_2} \right) I_{total}$$

$$\text{Eq. 4.9}$$

$$I_2 = \left(\frac{R_1}{R_1 + R_2} \right) I_{total}$$

where I_{total} is the current that flows in the battery.

> In a series circuit, the current is the same through all components.
> In a parallel circuit, the voltage drop is the same across each parallel branch of components.

Example 4.4

For the circuit shown here, calculate the current in each resistor. What is the voltage across each resistor?

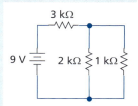

We need to find the equivalent resistance of the three-resistor network. The 2 kΩ and 1 kΩ are arranged in parallel. Their equivalent resistance is:

$$R_{equivalent} = \frac{R_1 R_2}{R_1 + R_2} = \frac{(1{,}000\ \Omega)(2{,}000\ \Omega)}{1{,}000\ \Omega + 2{,}000\ \Omega} = 0.667\ \text{k}\Omega$$

The 0.667 kΩ equivalent resistor is in series with the 3 kΩ resistor. We can easily add these together to find that the equivalent resistance of the network is 3.667 kΩ. We have reduced the circuit to a single equivalent resistor with a voltage supply. The current in the equivalent resistor is:

$$I = \frac{V}{R_{eq}} = \frac{9\ \text{V}}{3{,}667\ \Omega} = 2.454\ \text{mA}$$

This is the current that flows in the battery and hence also the current in the 3 kΩ resistor. We can use the current divider equation to find the current in the 2 kΩ resistor:

$$I_1 = \left(\frac{R_2}{R_1+R_2}\right)I_{total} = \left(\frac{1,000\Omega}{2,000\Omega+1,000\Omega}\right)2.454 \text{ mA}$$

$$= 0.818 \text{ mA}$$

The current in the 1 kΩ resistor is:

$$I_2 = I_{total} - I_1 = 2.454 \text{ mA} - 0.818 \text{ mA} = 1.636 \text{ mA}$$

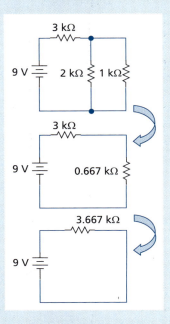

We can calculate the voltage across each resistor using the current and Ohm's law:

$$V_{3k\Omega} = I_{3k\Omega}R_{3k\Omega} = (2.454\times10^{-3} \text{ A})(3,000 \text{ }\Omega) = 7.362 \text{ V}$$
$$V_{2k\Omega} = I_{2k\Omega}R_{2k\Omega} = (0.818\times10^{-3} \text{ A})(2,000 \text{ }\Omega) = 1.636 \text{ V}$$
$$V_{1k\Omega} = I_{1k\Omega}R_{1k\Omega} = (1.636\times10^{-3} \text{ A})(1,000 \text{ }\Omega) = 1.636 \text{ V}$$

Note that the voltages across the two parallel components (2 kΩ and 1 kΩ resistors) are the same. Also, note that the voltage drop across the 3 kΩ and 1 kΩ resistors adds up to 9 V and the voltage drop across the 3 kW and 2 kΩ resistors adds up to 9 V.

Problem 4.5: Two resistors are in parallel with a 5 V voltage supply. The resistors are 1,000 Ω and 2,000 Ω. Determine the current and voltage drop for each resistor.

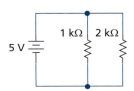

Example 4.5

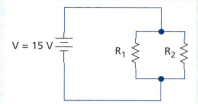

You have been given a new assignment in your laboratory to investigate what is happening with the circuit shown here. The voltage supply is 15 V and the specifications for the circuit show that R_1 should be 1,000 Ω and R_2 should be 3,000 Ω. The current in R_1 is required to be less than 20 mA.

You have measured the resistances and found that R_1 is 1,000 Ω but that R_2 is actually 180 Ω. What are the effects of this mistake? Compare the current in the battery with the 180 Ω and the 3,000 Ω resistors. Also, calculate the current in R_2 and voltage across R_2 for each resistor value.

The equivalent resistance for the 1,000 Ω/180 Ω circuit is:

$$R_{equivalent} = \frac{R_1 R_2}{R_1 + R_2} = \frac{(1,000)(180)}{1,000 + 180} = 152\ \Omega$$

The current for this equivalent circuit, and hence the current in the battery, is:

$$I = \frac{V}{R} = \frac{15\ V}{152\ \Omega} = 99\ mA$$

The equivalent resistance for the 1,000 Ω/3,000 Ω circuit is:

$$R_{equivalent} = \frac{R_1 R_2}{R_1 + R_2} = \frac{(1,000)(3,000)}{1,000 + 3,000} = 750\ \Omega$$

The current for this equivalent circuit—and hence the current in the battery—is:

$$I = \frac{V}{R} = \frac{15\ V}{750\ \Omega} = 20\ mA$$

The current in the battery is much larger (99 mA compared to 20 mA) with the incorrect, low value of R_2. We know that the voltage across either value of R_2 would be the same because the component is in parallel with R_1. We can calculate the current in R_1 to determine if we have exceeded the current requirement.

$$I_{1,000\Omega} = \frac{V}{R} = \frac{15\ V}{1,000\ \Omega} = 15\ mA$$

With the incorrect low value of resistor, the current in R_1 has not exceeded the stated maximum.

Problem 4.6:

For the circuit shown here, answer the following questions as true or false:

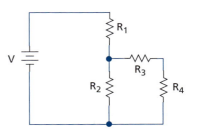

(a) R_1 is in series with R_2
(b) R_1 is in series with R_3
(c) R_3 is in series with R_4
(d) R_2 is in parallel with R_3
(e) R_2 is in parallel with the R_3 and R_4 combination
(f) R_2 is in parallel with R_4

Problem 4.7: Three resistors are in parallel with a 9 V voltage supply. The resistors are 1,000 Ω, 12,000 Ω, and 33,000 Ω. What is the equivalent resistance of the circuit? Determine the current and voltage across each resistor. If you added another resistor (5,000 Ω) in parallel, would the equivalent resistance increase, decrease, or stay the same?

Example 4.6

You have been asked to build a circuit with a 9 V supply and three resistors (200 Ω, 400 Ω, 100 Ω) in series. You would expect the current in the 200 Ω resistor to be:

$$I = \frac{V}{R_{eq}} = \frac{9\ V}{200\ \Omega + 400\ \Omega + 100\ \Omega} = 13\ mA$$

Your laboratory and circuit are a bit messy and you have several stray wires lying around your bench. You realize there is a stray wire across the 400 Ω resistor, as shown in the figure. What is the effect of this wire connection across the 400 Ω resistor on the current in the 200 Ω resistor?

The electrical connection across the 400 Ω resistor is known as a short circuit, a conductive pathway that has essentially no resistance and hence makes all the current flow through it instead of an important component. The effect of the wire is to make the parallel combination of the wire and resistor to have essentially no resistance. If we remove the 400 Ω resistance that has been shorted we obtain:

$$I = \frac{V}{R} = \frac{9\ V}{200\ \Omega + 100\ \Omega} = 30\ mA$$

The effect of the short circuit is that more than twice the current flows in the 200 Ω resistor with the short in place.

The Multimeter

The multimeter is a device used to measure current, voltage, and resistance (some also measure capacitance). The multimeter is a two-terminal measuring device with a variety of settings. The terms *voltmeter* and *ammeter* are used to describe instruments

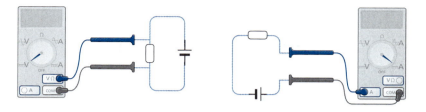

Figure 4.5 Using a multimeter to measure voltage (left) and current (right).

that measure voltage and current, respectively. The multimeter does both of these functions (and often more). In order to measure voltage, the two multimeter terminals are placed across the component of interest, as shown in Figure 4.5 (and the meter is switched to voltage mode). In order to measure current, one "breaks into" the circuit so that the current in the circuit element of interest must flow through the multimeter (and the meter is switched to current mode). The multimeter reading accuracy is limited by the internal resistance of the device and the introduction of loading error.

Voltage and Current Loading Error

From Figure 4.5 we see that measuring a voltage across a circuit component requires us to place the voltmeter in parallel with the component being measured. The existence of the voltmeter's circuit parallel to the circuit being measured provides an alternate path for the current and therefore decrease the current flowing through the component being measured. In other words, the act of measuring the circuit alters the circuit being measured. This deviation from the actual voltage across a resistor due to the existence of the voltmeter is called *loading error*. To eliminate loading error we would need an infinite resistance in the parallel circuit. This would force all of the current to flow through the component being measured. An infinite resistance is not possible since some current must flow through the meter in order for it to function. For the circuit shown in Figure 4.6, the loading error is:

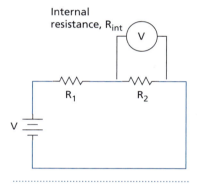

Figure 4.6 Voltage measurement and voltage loading error.

$$\% \text{ loading error} = 100\% \times \left(\frac{V_{w/meter} - V_{w/o\,meter}}{V_{w/o\,meter}} \right) = \frac{-1}{1 + \dfrac{R_{int}\left(R_1 + R_2\right)}{R_1 R_2}} \times 100\% \qquad \textbf{Eq. 4.10}$$

The voltmeter circuit symbol is shown in Figure 4.6 as a circle with a V inside. To minimize the loading error, multimeters are designed with a very high input resistance. We can see in our loading error equation that as R_{int} becomes larger, the percentage loading error becomes smaller. This idea is illustrated in Figure 4.6, where we have two resistors in series with a voltage supply and a multimeter set to measure voltage across R_2. In practice, the internal resistance of the multimeter depends on the quality of the meter and the particular voltage range in use. The internal resistance could be as low as $M\Omega$ (M is for mega, 10^6) for voltages more than 100 V to as high as $G\Omega$ (G is for giga, 10^9) for voltages less than around 10 V.

In current mode, a standard resistor in the meter causes a deviation in the actual current flowing in the component being measured, which causes a loading error for current. Ideally, the standard resistor would have much less resistance compared to the rest of the circuit. Because of this additional resistance introduced, the meter-measured current will be less than the current that would flow without the meter.

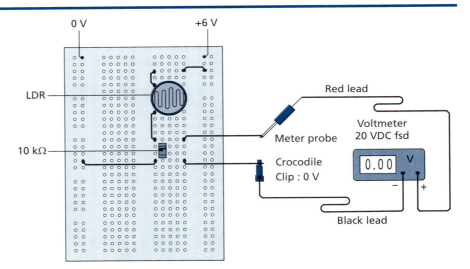

PROFILE

Electronics for a Very Simple Light-Sensing Instrument: Voltage Divider Photoresistor Circuit

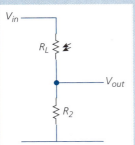

This profile describes a simple system to measure light intensity using inexpensive and readily available components. The system could be used to measure fluorescence or adapted to measure absorption. The voltage divider was presented as two resistors in series with an applied voltage. The voltage divider is perhaps most useful when one resistor is a sensor that has a varying resistance as a function of some change in the environment. The light-dependent resistor (LDR) or photocell has a resistance that decreases with the amount of light on the sensor. We can build a voltage divider with an LDR and a resistor in series with a voltage source, as shown here. These components have been placed on a *solderless breadboard*. Breadboards are often used when prototyping a circuit because it is easier to move components around without the permanence of solder. Working with solderless breadboard can be nonintuitive at first. It is conventional to supply power to your breadboard at the top left and right, shown as 0 V and +6 V here. The 0 V and +6 V column are electrically connected under the board that provide 0 V and +6 V to the entire column. There is no electrical connection between each parallel row of five connections unless a component is placed across them (e.g., the LDR or resistor in the image above).

The drawing here shows a breadboard. Note the trough that runs vertically down the center of the board. That trough separates the +6V side of the board from the ground (0V) side of the board. In this image the line across the trough represents your lead to ground and the circuit of interest is constructed on the +6V side of the board.

We also show an equivalent circuit using circuit symbols instead of the image on the solderless breadboard.

We can use a simple laser pointer to excite fluorescence in a sample of interest and direct the fluorescence onto the LDR. We can measure the voltage drop across the LDR or the resistor. In the circuit shown here, we have chosen the voltage across the resistor, although there is no reason to choose this over the LDR.

The voltmeter is set to measure voltage V_2. Using the voltage divider equation, V_2 depends on the LDR resistance via:

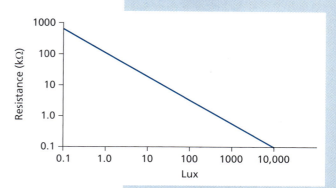

$$V_2 = \left(\frac{R_2}{R_1 + R_2} \right) V = \left(\frac{10{,}000\ \Omega}{R_{LDR} + 10{,}000\ \Omega} \right) 6\ V$$

A plot of R_{LDR} versus input intensity for a particular photoresistor is shown here. When more light (or lux—the SI unit of illuminance, also defined as 1 lumen/square meter) strikes the LDR the resistance through the LDR decreases.

What type of voltages will we measure for low light intensities, using the light dependent resistor shown in the plot? We will use the voltage divider equation and estimate a low light intensity of ~ 0.1 lux causes a resistance of 700 kΩ.

$$V_2 = \left(\frac{R_2}{R_1 + R_2} \right) V = \left(\frac{10{,}000\ \Omega}{700{,}000\ \Omega + 10{,}000\ \Omega} \right) 6\ V = 0.08\ V$$

The next step would be to calibrate the system so that the voltage measurement can be scaled to an optical power measurement. There are other interesting ways* to make a simple spectrophotometer—we will discuss another example in our next profile. ■

---

* Thal, M. A.; Samide, M. J. Applied Electronics: Construction of a Simple Spectrophotometer, J. Chem. Ed. **2001**, 78, 1510–1512.

Example 4.7

A circuit has two resistors (both 2.5 MΩ) in series with a 240 V voltage source. Calculate the current and voltage across one of the 2.5 MΩ resistors for two cases, assuming the multimeter has infinite resistance and assuming a finite (100 MΩ) resistance multimeter.

Infinite resistance multimeter:

$$I = \frac{V}{R} = \frac{240\ V}{2.5 \times 10^6\ \Omega + 2.5 \times 10^6\ \Omega} = 48\,\mu A$$

$$V = IR = (48 \times 10^{-6}\ A)(2.5 \times 10^6\ \Omega) = 120\ V$$

The voltage is evenly split between the two equal value resistors as we would expect.

Finite resistance multimeter: The equivalent resistance of the circuit is that of the 2.5 MΩ resistor plus the parallel group of the 2.5 MΩ and the 100 MΩ multimeter. The equivalent resistance is found via

$$R_{equivalent} = 2.5 \times 10^6\ \Omega + \frac{(2.5 \times 10^6\ \Omega)(100 \times 10^6\ \Omega)}{2.5 \times 10^6\ \Omega + 100 \times 10^6\ \Omega} = 4.93 \times 10^6\ \Omega$$

The equivalent resistance with the finite resistance multimeter is less than the theoretical multimeter with infinite input resistance (5 MΩ). The additional pathway for current has lowered the resistance. The current in the battery with the finite resistance meter is

$$I = \frac{V}{R} = \frac{24 \text{ V}}{260 \times 10^6 \text{ }\Omega} = 48.6\,\mu\text{A}$$

The voltage across the 2.5MΩ resistor (not the resistor with the meter attached) is

$$V = IR = (48.6 \times 10^{-6}\text{ A})(2.5 \times 10^6 \text{ }\Omega) = 121.5 \text{ V}$$

This calculation shows that we are left with 118.5 V (less than 120 V) for the second resistor because of the loading error.

Problem 4.8:

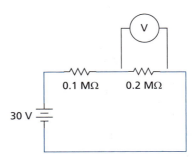

You have a circuit with two resistors (0.1 MΩ and 0.2 MΩ) in series with a 30 V DC supply.

(a) Using Ohm's law, calculate the theoretical voltage drop (without loading error) across the 0.2 MΩ resistor.

(b) What would the measured voltage across the 0.2 MΩ resistor be using a voltage meter with an internal resistance of 250 MΩ.

(c) What would the measured voltage across the 0.2MΩ resistor be using a voltage meter with an internal resistance of 1,000 MΩ.

(d) Calculate the percentage loading error for parts c and d.

Problem 4.9:   In the profile (Electronics for a Very Simple Light-Sensing Instrument: Voltage Divider Photoresistor Circuit), we mentioned that the simple voltage divider approach with the photoresistor would be prone to voltage loading error. Using the values in the profile, calculate the voltage drop across the LDR assuming an infinite resistance multimeter and assuming an internal resistance of 1 MΩ for the multimeter.

4.3 Capacitors and Resistor–Capacitor (RC) Circuits

Capacitors are two-terminal circuit elements and are essentially two conductive plates separated by an air gap or insulator. The symbol for a capacitor is two parallel lines (⊣⊢) and represents the two parallel plates (also see Figure 4.2). Capacitance is the ability of an electrical component to store charge and depends on the geometry of the

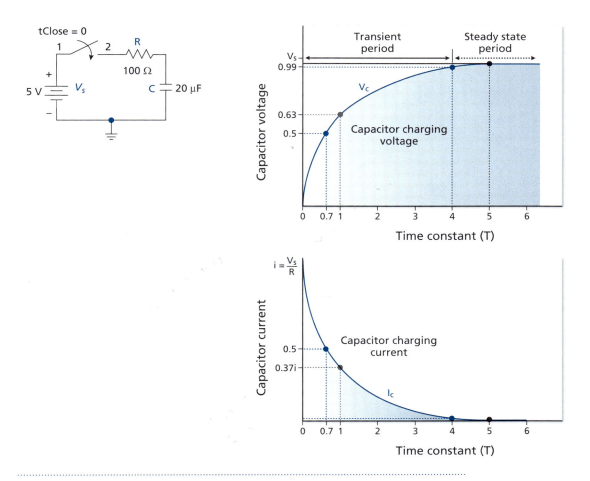

Figure 4.7 RC circuit (top-left), voltage changing in time after switch is closed (top-right), and current changing in time after switch is closed (bottom-right).

parallel plates. The unit used to quantify a capacitor is the farad (F). A farad is defined as the capacitance that will produce a one-volt potential across the capacitor when it has been charged by one coulomb of charge.

The circuit shown in Figure 4.7 shows a resistor, capacitor, and voltage supply connected in series with a switch. This is called an *RC circuit*. The RC circuit is a useful model for instrumental electronics when trying to understand the time response of an instrument. It would be important to know that particular instrumental electronics can respond quickly enough for the timescale over which your signal may be changing. This application of the RC circuit model is described in Chapter 5 on signal processing. When the switch is closed, current (by convention, positive charge) moves to one end of the capacitor plate and the positive charge builds up. Because of the positive charge on this one plate, the other plate becomes negatively charged. After a time, the plates become fully charged and the current stops flowing. The capacitor stores charge and voltage. During the charging process, assuming an initially uncharged capacitor, the current and voltage for the capacitor vary in time (plotted in Figure 4.7) according to:

$$I(t) = I_0 e^{-\frac{t}{RC}}$$
                                            Eq. 4.11

$$V(t) = V_0 \left(1 - e^{-\frac{t}{RC}}\right)$$
                              Eq. 4.12

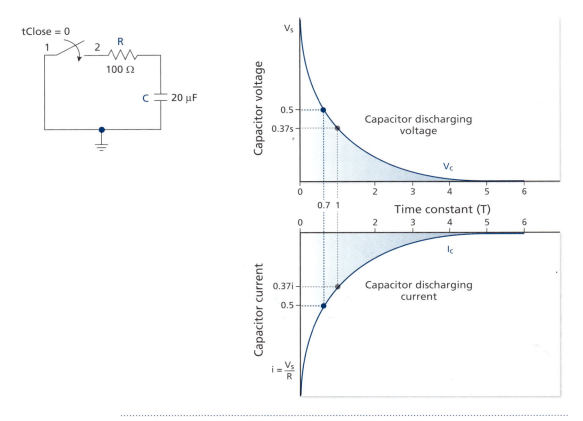

Figure 4.8 RC circuit with capacitor initially charged (top left), voltage changing in time during discharging (top right), and current changing in time during discharging (bottom).

where V_0 is the voltage of the power supply and I_0 is the maximum (initial) current, which is equal to V_0/R. The product of R (resistance) and C (capacitance) is called the RC time constant.

$$\tau = RC \qquad\qquad \textbf{Eq. 4.13}$$

The time constant is an important part of the description of the charging process (as well as the discharging process, as we will see shortly). In terms of charging a capacitor, the RC time constant is the time it takes for a capacitor, as shown in Figure 4.7, to go from zero voltage to 63% of the voltage supply. The larger the time constant, the longer the time it will take the capacitor to get to the full voltage of the supply. Note that the top-right panel in Figure 4.7 has a label for the voltage of the capacitor showing when the capacitor's voltage is $0.63V_s$. The time it takes for a capacitor to reach $0.63V_s$ is the time constant.

An initially charged capacitor can be discharged, as shown in Figure 4.8. During this process the charges on the plates of the capacitor are free to move off the plates. The charge leaves the plates, creating a current that decreases in time. The initial voltage across the charged capacitor also decreases in time once the switch is closed. For discharging a capacitor in an RC circuit, the current and voltage across the capacitor are:

$$I(t) = -I_0 e^{-\frac{t}{RC}} \qquad\qquad \textbf{Eq. 4.14}$$

$$V(t) = V_0 e^{-\frac{t}{RC}} \qquad\qquad \textbf{Eq. 4.15}$$

In contrast to the charging of the capacitor, both the current and the voltage approach zero during the discharging. In general, only time varying voltage or current signals (AC) can flow for capacitors. Because of this, capacitors are sometimes used to block DC signals.

Example 4.8 – Time Behavior of the RC Circuits and the Oscilloscope

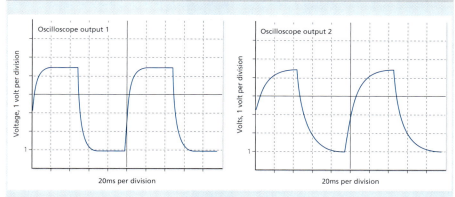

We wish to observe the effect on the time dependence of the voltage in an RC circuit using either a 3 µF or 8.2 µF capacitor in series with the resistor (1.3 kΩ). The source has a voltage of 4.5 V for a short time and then 0 V for a short time so that the capacitor charges and discharges. The voltage across the capacitor is monitored by an *oscilloscope* and is displayed on the oscilloscope's screen as a plot of voltage versus time. An oscilloscope is a common tool used to measure time-varying voltages. Many oscilloscopes can measure fairly fast voltage signals down to the nanosecond time scale.

The two oscilloscope outputs are shown here. Note that the volts/division (the conversion factor between "divisions" and volts) are both set to 1 V. The time per division for each are set at 20 ms.

(a) Which oscilloscope output (1 or 2) corresponds to which capacitor (3 µF and 8.2 µF)?

The larger the capacitance, the longer it takes for the capacitor to be charged to the 4.5 V of the source (and the longer the time to discharge). Output 2 takes a longer amount of time to get to the 4.5 V when compared to output 1. The RC time constant is larger for output 2.

(b) We have changed the seconds/division to 2 ms. Use this output to estimate the time constant for output 1. Compare to the calculated value.

The time constant can be estimated from the output by determining the time at which the voltage is at 63% of its maximum, 4.5 V.

$$(0.63)(4.5 \text{ V}) = 2.84 \text{ V}$$

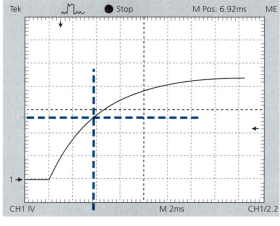

Oscilloscope output 1 zoomed in

The time constant is the time it takes for the voltage to get to 2.8 V from being initially uncharged. Using the lines we have sketched on the output, the time is approximately 4 ms. This is in close agreement with:

$$\tau = RC = (1{,}300 \ \Omega)(3 \times 10^{-6} \ \mu F) = 3.9 \ ms$$

Oscilloscopes also have a cursor capability that makes these types of measurements easy to accomplish.

The key role of the RC circuit in chemical instrumentation is related to how quickly an instrument can respond to a change in voltage. This change in voltage is often the signal for the analyst, so the time response can be important. Because the voltage across a capacitor is time dependent, the presence of capacitance in a detector component of an instrument limits the speed at which it can respond to a voltage change.

Problem 4.10: You would like to have a series RC circuit that will discharge in as long a time as possible. You have access to the following components (see table). Which components would you choose?

RESISTORS	CAPACITORS
200 Ω	0.1 μF
330 Ω	1 μF
500 Ω	10 μF
780 Ω	100 μF
1,000 Ω	1,000 μF
	10,000 μF

Problem 4.11: You are working with a short pulsed electrical voltage signal and need this voltage to discharge as fast as possible. You have access to the components (see table). Which components would you choose?

RESISTORS	CAPACITORS
1,200 Ω	0.1 pF
1,800 Ω	0.9 pF
2,400 Ω	1.8 pF
3,000 Ω	2.2 pF

Problem 4.12: You have a resistor (10,000 Ω), a capacitor (100 μF), a 9 V battery, and a switch (currently open) all connected in series. You close the switch.

(a) What is the voltage across the capacitor 0.01 seconds later?

(b) What is the current 0.01 seconds later?

(c) What is the voltage across the capacitor 5 seconds later?

(d) What is the RC time constant for this circuit?

4.4   Alternating Current (AC) Circuits

AC sources cause charges to move back and forth because of a sinusoidal applied voltage or current. In Figure 4.9, we show a DC voltage source with a resistor and an AC voltage source with a resistor to highlight the difference in charge movement. Note the two different symbols used for the DC and AC sources. In the DC circuit, the flow of positive charges moves clockwise as shown by the arrows. In the AC circuit, charges move clockwise (solid arrows) and then counterclockwise (dashed arrows), and then clockwise, and so on. The key elements of an AC signal are frequency (f) and amplitude (V_m). A sinusoidal voltage can be written mathematically as:

$$V = V_m \sin(\omega t + \phi) = V_m \sin(2\pi f t + \phi) \qquad \textbf{Eq. 4.16}$$

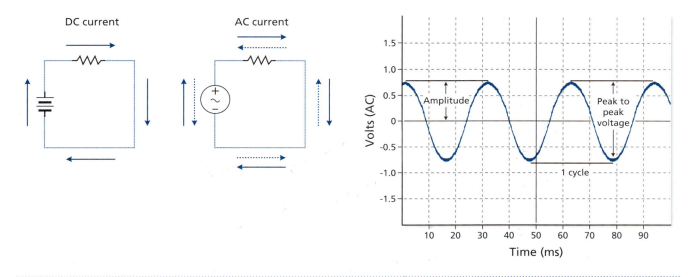

Figure 4.9 Comparison of DC and AC current and AC voltage waveform.

where V_m is the voltage amplitude, ω is the signal angular frequency (measured in rad/s), ϕ is the phase constant, and f is the frequency (measured in Hz or cycles/s). Be careful to keep track of whether you are using frequency or angular frequency because they differ by a factor of 2π. In addition to the voltage amplitude, an AC voltage can also be characterized by the peak to peak voltage ($V_{pk\text{-}pk}$) and the root mean square (RMS) voltage (V_{RMS}). V_{RMS} is related to V_m via:

$$V_{RMS} = \frac{\sqrt{2}}{2} V_m$$

Eq. 4.17

The time to complete one cycle, the period, is related to the frequency through:

$$T = \frac{1}{f}$$

Eq. 4.18

The standard wall outlet is our most familiar AC voltage source. In the United States, this output has a frequency of 60 Hz and a voltage amplitude (peak voltage) of 110 to 120 V. Many instruments make use of sinusoidal AC for power, and some also use signals that vary sinusoidally.

Example 4.9

For the AC voltage shown in Figure 4.9, determine the (a) peak voltage, (b) period, (c) frequency, (d) angular frequency, and (e) V_{RMS}.

The peak voltage can be read off the plot to be ~ 0.7 V.

The period is the time to complete one cycle and is ~ 30 ms.

The frequency is $f = \dfrac{1}{T} = \dfrac{1}{30 \times 10^{-3} \text{ s}} = 33.3 \text{ Hz}$

The angular frequency is $\omega = 2\pi f = 2\pi(33.3 \text{ Hz}) = 209.4 \text{ rad/sec}$

The RMS voltage is $V_{RMS} = \dfrac{\sqrt{2}}{2} V_m = \dfrac{\sqrt{2}}{2}(0.7 \text{ V}) = 0.49 \text{ V}$

Problem 4.13: For the AC voltage shown, determine the (a) amplitude, (b) peak-to-peak voltage, (c) period, (d) frequency, (e) angular frequency, and (f) V_{RMS}.

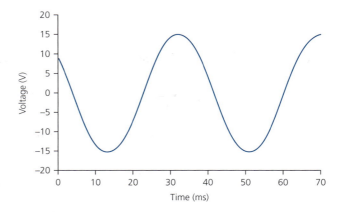

Ohm's Law for Alternating Current (AC) Circuits

We can use an AC version of Ohm's law, but we need to replace the use of resistance with a new quantity, the impedance, Z. The impedance is related to the resistance. The AC version of Ohm's law is:

$$V = IZ \qquad\qquad \textbf{Eq. 4.19}$$

where Z is the impedance, measured in ohms. The impedance can be thought of as the opposition to electric conduction as a result of all of the electronic components in a circuit (e.g., resistors, capacitors, inductors). The general formula for impedance is:

$$Z = R + jX \qquad\qquad \textbf{Eq. 4.20}$$

where R is resistance, X is the reactance, and j is the square root of -1. We show the resistance and reactance for resistors, capacitors, and inductors in Table 4.1.

For the capacitor, the reactance depends on the capacitance and the frequency of the AC signal. We have introduced a new quantity, inductance, L, measured in henrys. An inductor (see Figure 4.1), at its most basic, is a coil of wire. As with capacitance. which depends on the size and distance between the plates, inductance of a coil depends on the area of the coil, the number of turns in the coil, and the length of the coil. The reactance of the inductor depends on the inductance and the frequency of the AC signal.

Note from Table 4.1 that the reactance for both the capacitor and inductor depends on ω, the angular frequency of the signal. This frequency dependence has the very practical effect that circuits can be designed (combinations of resistors, capacitors, and inductors) to filter sinusoidal signals. For example, we can build a circuit that allows low-frequency voltages to pass while blocking high-frequency voltages. This is known

TABLE 4.1: Resistance and Reactance for Resistor, Capacitor, and Inductor

COMPONENT	RESISTANCE	REACTANCE
Resistor	R	0
Capacitor	0	$X_C = \dfrac{1}{\omega C}$
Inductor	0	$X_L = \omega L$

as a *low-pass filter*. Conversely, we can build a circuit that blocks low-frequency voltages while allowing higher frequency voltages to pass. This is known as a *high-pass filter*.

Low-Pass, High-Pass, Band-Pass, and Band-Stop Filters

The classic *low-pass filter* is simply a resistor and capacitor in series with a sinusoidal input voltage, shown in Figure 4.10. The voltage across the capacitor, V_{out}, is the filtered output. Choices of resistance (R) and capacitance (C) will lead to different filtering behavior. In particular, the output of the RC low-pass filter can be written as:

$$\frac{V_{out}}{V_{in}} = \frac{1}{\sqrt{1 + \omega^2 R^2 C^2}}$$ **Eq. 4.21**

and the cutoff or corner frequency (the frequency where power of the filtered signal has dropped by half) is:

$$\omega_{cutoff} = \frac{1}{RC}$$ **Eq. 4.22**

Equation 4.22 is found using Ohm's law for AC circuits and making use of the impedance of the relevant components. The derivation of equations 4.21 and 4.22, as well as associated equations for other passive filters later in this section, are beyond the scope of the chapter. [3]

The cutoff angular frequency is sometimes referred to as W_0. A plot of V_{out}/V_{in} as a function of frequency is shown in Figure 4.10. For this low-pass filter, voltages with a frequency below 1,000 rad/s, are passed with less loss than voltages with a frequency larger than 1,000 rad/s. Note that the largest V_{out} possible is simply V_{in}. The RC circuit is a passive filter as opposed to an active filter. Active filters both filter signals and provide amplification (usually with operational amplifiers, see later section) so that the output can be larger than the input.

The RC series circuit is a high-pass filter when the output is taken across the resistor instead of the capacitor. The high-pass RC filter is shown in Figure 4.11 along with a plot of V_{out}/V_{in} as a function of frequency. For this high-pass filter, voltages with a

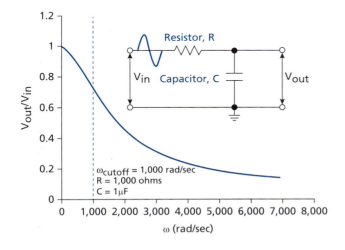
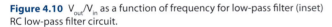
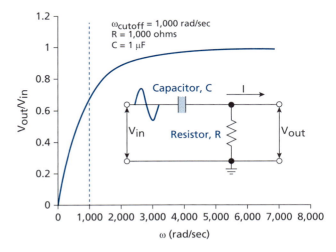

Figure 4.10 V_{out}/V_{in} as a function of frequency for low-pass filter (inset) RC low-pass filter circuit.

Figure 4.11 V_{out}/V_{in} as a function of frequency for high-pass filter (inset) RC high-pass filter circuit.

---

[3] The interested readers should see, for example, Alexander, C.; Sadiku, M. *Fundamentals of Electric Circuits*, 5th ed.; McGraw Hill: New York, **2012**.

frequency above 1,000 rad/s, are passed with less loss than voltages with a frequency lower than 1,000 rad/s. The output of the RC high-pass filter is:

$$\frac{V_{out}}{V_{in}} = \frac{\omega RC}{\sqrt{1 + \omega^2 R^2 C^2}}$$

Eq. 4.23

and the cutoff or corner frequency is:

$$\omega_{cutoff} = \frac{1}{RC}$$

Eq. 4.24

Equation 4.23 can be found using Ohm's law for AC circuits. The different frequency response between the RC low pass filter and the RC high pass filter comes from the different frequency dependence of the resistor and capacitor. V_{out} for the low pass is $IZ_{capacitor}$ where V_{out} for the high pass filter is $IZ_{resistor}$. As we see in Table 4.1, $Z_{capacitor}$ and $Z_{resistor}$ do not share the same frequency dependence. V_{in} is the same for both arrangements ($V_{in} = IZ_{total}$).

We can check the high-pass behavior of the filter described by Eq. 4.23 by looking at the limiting cases:

$$\omega \text{ goes to zero: } \frac{V_{out}}{V_{in}} \rightarrow \frac{0}{\sqrt{1+0}} \rightarrow 0$$

$$\omega \text{ goes to infinity: } \frac{V_{out}}{V_{in}} \rightarrow \frac{\infty}{\sqrt{1+\infty}} \rightarrow 1$$

In addition to these RC passive filters, band-pass filters and band-stop filters, which pass a band or block a band of frequencies defined by the circuit parameters, can be made with basic components such as resistors, capacitors, and inductors. One arrangement for the band-pass filter, shown in Figure 4.12, combines an inductor, capacitor, and resistor in series and the output voltage is taken off the resistor. Figure 4.12 also shows V_{out}/V_{in} for the band-pass filter. The same circuit becomes a band-stop filter if the output voltage is taken across the capacitor and inductor combination. The frequency

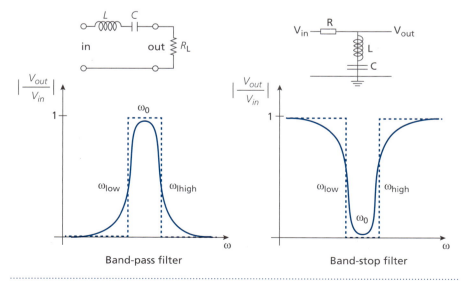

Figure 4.12 Band-pass filter (RLC version) (left) and band-stop filter (RLC version) (right).

at the center of the band-pass ω_0 (and band-stop) is related to resistance (R), inductance (L), and capacitance (C) with:

$$\omega_0 = \frac{1}{\sqrt{LC}}$$ **Eq. 4.25**

and the low and high cutoff frequencies can be found using:

$$\omega_{low,\,cutoff} = \frac{-R}{2L} + \sqrt{\left(\frac{R}{2L}\right)^2 + \frac{1}{LC}} \quad \omega_{high,\,cutoff} = \frac{R}{2L} + \sqrt{\left(\frac{R}{2L}\right)^2 + \frac{1}{LC}}$$ **Eq. 4.26**

The filtering bandwidth for both the band-pass and band-stop filters is set by:

$$\Delta\omega = \omega_{high} - \omega_{low} = \frac{R}{L}$$ **Eq. 4.27**

Derivations of these equations are beyond the scope of the chapter.

Filtering AC signals can be important to the analyst, as we will see in Chapter 5 on signal processing. In particular, many voltage measurements are the result of several signals, in addition to the one that the analyst is investigating. These additional signals mask the key signal for the analyst. An AC filter may remove some of these unwanted signals. For example, the low-pass filter can reduce high-frequency thermal noise. These topics are discussed in Chapter 5. The high-pass filter can reduce low-frequency signals such as 1/f or flicker noise.

Problem 4.14: Given the following resistors, what capacitance should you choose in order to design a low-pass filter with a cutoff angular frequency at 5,000 rad/s?

(a) 1 kΩ

(b) 10 kΩ

(c) 1 MΩ?

Sketch your design and label where you would measure V_{out}.

Problem 4.15: Given the following capacitors, what resistance should you choose in order to design a high-pass filter with a cutoff angular frequency at 15,000 rad/s?

(a) 0.1 μF

(b) 1 μF

(c) 10 μF

Sketch your design, and label where you would measure V_{out}.

Problem 4.16: You have put a 1,000 μF capacitor and 2,000 Ω resistor in series with an AC voltage source with amplitude of 45 V and frequency of 4,000 Hz. What is V_{out} across the capacitor? What is V_{out} across the resistor?

We have now discussed the charging and discharging behavior of capacitors as well as the filtering properties of the RC circuit. We can find evidence in electronics for an effective capacitance when we have two conducting elements at different potentials physically near each other. Such stray capacitance can lead to unwanted effects in a circuit, essentially adding a capacitor that we did not want. We can see from this section on RC filters that an unintended capacitor added to a circuit, depending on how it is arranged, could produce filtering that was not wanted.

ACTIVITY

RC Filter Spreadsheet Calculation

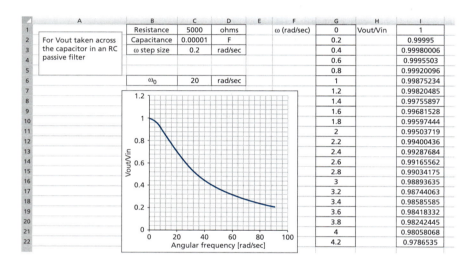

In this activity, we create a spreadsheet that helps describe RC, low-pass filters by allowing the user to input values for resistance and capacitance and display the cutoff frequency as a plot of V_{out}/V_{in} versus angular frequency.

Set up the spreadsheet, as shown here, to allow the user to input resistance, capacitance, and a value for step size for angular frequency. The step size is the amount we add to each frequency cell and will help us produce a plot that is most useful in terms showing the behavior of V_{out}/V_{in}. In the first cell

for V_{out}/V_{in}, input the equation as shown here and fill down. We now plot V_{out}/V_{in} versus angular frequency for the values of resistance, capacitance, and step size. We have also added a cell that calculates the cutoff frequency, Ω_0, for this type of filter.

Use the spreadsheet tool to investigate the behavior of varying the resistor value by producing plots using 5,000 W; 7,000 Ω; 9,000 Ω; and 11,000 Ω. Comment on how V_{out}/V_{in} varies at angular frequency at 40 rad/s. Produce a plot of V_{out}/V_{in} for these different resistor values.

4.5   Operational Amplifiers

The *operational amplifier* (op amp) is a very powerful and commonly used circuit component. It is useful because it can provide an output voltage that is the result of a mathematical operation on input voltages. Before we discuss this further, we should point out that the op amp is an important and common example of an *integrated circuit* (IC). ICs are an arrangement of very small resistors, capacitors, diodes, and transistors (we will discuss IC components later in this chapter), generally built on a silicon substrate. ICs may contain a very large number of components. There are a great many ICs that can provide a wide range of functionalities.

Although we will treat the op amp as a black box,[4] inside the op amp chip (in particular the 741, a common op amp) are about 20 transistors and about 11 resistors. The op amp symbol is the triangle shown in Figure 4.13. The circuit symbol is superimposed on a sketch of the actual eight pins of the op amp. Each pin has a particular function, and different op amps may have different pin allocations. For the 741, pins 2 and 3 are voltage inputs (2 is known as the inverting input and is drawn on the circuit symbol as a minus sign, and 3 is known as the noninverting input and is drawn on the circuit symbol as a plus sign), 6 is the voltage output, and pins 4 and 7 are where the op amp obtains its power. It is important to remember that op amps must be powered; this is not always drawn on a circuit diagram, and the maximum gain an op amp can provide cannot exceed the voltage of the op amp's power supply. The other pins are used less frequently and are beyond the scope of this introduction to the op amp.

The various options for how one can set the relationship between the input voltage(s) and the output is what makes the op amp so useful. The op amp can, for example, multiply an input voltage by a constant (voltage amplifier), add multiple inputs (summing), subtract inputs (differential), take derivatives (differentiating) or integrals (integrating) of input, and act as an active filter. We will describe several of these op amp circuits later in this section. The input voltages for the op amps described in this section might be voltages from some transducer (e.g., photoresistor, photodiode, thermistor, pressure sensor)

The op amp is one of the key circuit components that can be used to drive other components, such as a readout device, without drawing current from the input components, such as a detector. The very large input impedance of the op amp is why the component does not draw much current. This means that essentially no current flows into the op amp inputs. This key feature of the op amp minimizes loading error. The input can be DC or AC and a variety of op amp ICs are available, including single ICs with multiple op amps.

Inverting and Noninverting Operational Amplifiers

The inverting amplifier is a common op amp circuit and is shown in Figure 4.14. The input signal voltage, V_{in}, is connected to a resistor, R_i, and this combination is sent into the inverting input of the op amp. A feedback resistor is connected from the output pin to the input side (in this case, the inverting input). The V_{s+} and V_{s-} correspond to the voltage supplies that power the op amp. V_{out} can be found via:

$$V_{out} = -\frac{R_f}{R_i}V_{in}$$ **Eq. 4.28**

where $\dfrac{R_f}{R_i}$ = the gain. We can choose R_f and R_i to get a larger output voltage than input voltage. This configuration is known as an inverting op amp because, due to the negative sign in the equation, it will convert a positive voltage into a negative voltage (or vice versa).

The noninverting amplifier is shown in Figure 4.15. The input voltage is connected to the noninverting input of the op amp. A resistor, R_i, is connected to ground and to the noninverting input and a feedback resistor, R_f, connects back to the inverting input. The output voltage for the noninverting op amp configuration is:

$$V_{out} = \left(1 + \frac{R_f}{R_1}\right)V_{in}$$ **Eq. 4.29**

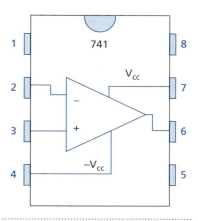

Figure 4.13 Op amp (741) with eight-pin layout and inset circuit symbol (triangle), with pins 4 and 7 supplying power, + and $-V_{cc}$.

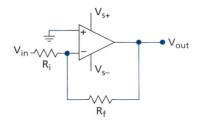

Figure 4.14 Inverting op amp circuit.

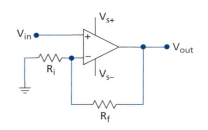

Figure 4.15 Noninverting op amp circuit.

---

[4] This means that we are not working to understand, from first principles, how these components each contribute to how the circuit component works.

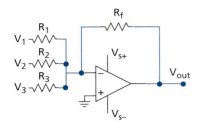

Figure 4.16 Summing op amp circuit.

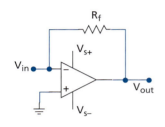

Figure 4.17 Current-to-voltage amplifier.

Summing Operational Amplifier

The summing op amp circuit, shown in Figure 4.16, takes several input voltages and produces an output voltage based on the feedback and input resistors according to:

$$V_{out} = -R_f \left(\frac{V_1}{R_1} + \frac{V_2}{R_2} + \frac{V_3}{R_3} \right)$$

Eq. 4.30

This op amp is useful when it is necessary to add voltage signals.

Problem 4.17: You have access to resistors with values 1 kΩ, 1.5 kΩ, and 5.0 kΩ. You have a voltage signal from a transducer with a maximum voltage of 0.2 V. Using a noninverting op amp configuration, what maximum output voltages could you provide with the combinations of these resistors? Sketch your op amp circuit options.

Current-to-Voltage Amplifier

When working with small currents (less than 1 mA), it may be necessary to convert this small current (which a multimeter may not have the range to measure) to a voltage that is easier to measure. The current-to-voltage amplifier (also called the transimpedance or transresistance amplifier) takes an input current I_{in} to the inverting voltage input and a feedback resistor back to the inverting input. The current-to-voltage amplifier is shown in Figure 4.17. The noninverting voltage input is connected to ground. The output voltage V_{out} is:

$$V_{out} = -I_{in} R_f$$

Eq. 4.31

Example 4.10

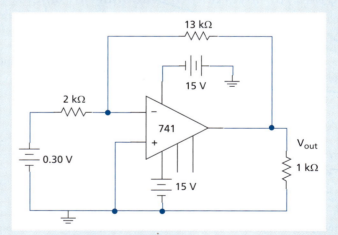

What is the voltage across the 1 kΩ resistor for the op amp circuit shown here? The two lines beneath the op amp correspond to pins not being used.

Because the input voltage is connected to the inverting input and the feedback resistor is connected to the inverting side, this is an *inverting op amp*. The output voltage of the op amp is across the 1 kΩ resistor so we can use the inverting op amp equation to calculate the voltage across this resistor.

$$V_{out} = -\frac{R_f}{R_1} V_{in} = -\frac{13,000 \ \Omega}{2,000 \ \Omega} (0.3 \ \text{V}) = -1.95 \ \text{V}$$

Example 4.11

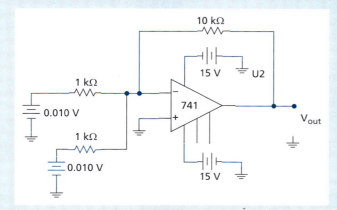

In the op amp circuit shown here, the input voltages are from two pressure transducers that have been placed in two different industrial containers. A pressure sensor voltage of 10 mV for each sensor corresponds to the maximum safe pressure. What is the output voltage for this circuit when the pressure has just reached the unsafe level for both vessels?

This is a *two-input summing op amp* circuit.

$$V_{out} = -R_f \left(\frac{V_1}{R_1} + \frac{V_2}{R_2} \right) = -10,000 \ \Omega \left(\frac{10 \times 10^{-3}}{1,000 \ \Omega} + \frac{10 \times 10^{-3}}{2,000 \ \Omega} \right) = -0.2 \ V$$

The current-to-voltage amplifier is often used in conjunction with photodiodes and photomultiplier tubes (PMTs; these will be introduced in Chapter 6). A small current, for example 50 mA, can be converted to a voltage of, say, 1 V using a feedback resistor of 20,000 Ω.

The Voltage Follower

The voltage follower or voltage buffer is an important op amp circuit used in applications where one wants to measure a voltage and does not want the measuring process to cause loading errors. The very large input resistance (or impedance) of the op amp is important here. From the discussion of loading error, we know that adding a meter in voltage mode to an electronic component affects the voltage measured because of the current that would flow through the meter. The *voltage follower* allows for extremely large resistances, potentially 10^{12} ohms, and hence reduces loading errors. This is the main reason op amps are so ubiquitous in chemical instruments. The small currents generated by a detector (e.g., a PMT) need to drive an output device such as a voltmeter. A voltmeter typically requires several milliamps to operate properly. Drawing this much current off of the PMT circuitry would produce large loading errors. The output current (or voltage) of an op amp is proportional to the input current (or voltage) but is driven by the op amp's power supply and not by the instrument's detector output.

The voltage follower is shown in Figure 4.18. The voltage of interest, V_{in}, is connected to the noninverting input, and a feedback connection is made from the output of the op amp into the inverting input. The relationship between input voltage and output voltage is quite simple, $V_{out} = V_{in}$. This circuit has a feedback connection from output to input, but no resistor is placed between the two. The voltage follower is essentially a noninverting amplifier with a gain of 1 where the output current is driven by the

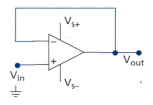

Figure 4.18 Voltage follower.

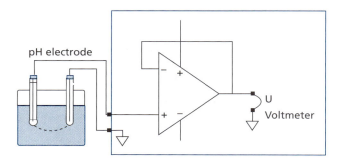

Figure 4.19 pH electrode and op amp voltage follower.

op amp's power supply. For example, the voltage follower is used in a pH sensor where a voltage is measured across two electrodes. The internal resistance in this circuit can be 100 MΩ or more. For the pH meter, the voltage V_2 from the figure is the electrode voltage. A sketch showing this basic idea for the pH meter is shown in Figure 4.19. The current generated at V_{out} is used to drive the pH meter's readout device.

Problem 4.18: A pH meter has electrodes that produce a small voltage that depends on the pH of the analyte. We will assume the circuit can be modeled as shown in the circuit here, where we have included the large resistance of the electrodes and the load associated with a meter. If $V_{electrode} = 0.15$ V and $R_{electrode} = 150$ MΩ, what voltage is measured across R_L in the following situations:

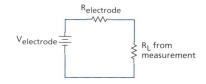

(a) R_L is associated with an inexpensive multimeter with an input impedance of 100 MΩ.

(b) R_L is associated with a op amp voltage follower with multimeter and is 10×10^{12} Ω.

Problem 4.19: Assume the voltage follower in Example 4.12 has an input impedance of 10^{12} Ω. Calculate the parallel equivalent resistance for the 3 kΩ/10^{12} Ω circuit. Report the loading error as a ppb.

Operational Amplifier Comparator

The op amp comparator compares two voltages: the noninverting voltage input and the inverting voltage input. Depending on which voltage is larger, the op amp outputs the positive power supply voltage (V_{S+}) or the negative power supply voltage (V_{S-}). The op amp comparator is shown in Figure 4.20. The output voltage options (or in other words, the logic states for the comparator) are:

$$\text{If } V_{inverting} > V_{noninverting}, \text{ then } V_{output} \text{ is } V_{S-}$$

$$\text{If } V_{noninverting} > V_{inverting}, \text{ then } V_{output} \text{ is } V_{S+}$$

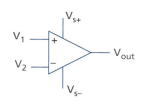

Figure 4.20 Op amp comparator.

The comparator provides a sort of digital output, either V_{S+} or V_{S-}. If the op amp negative supply is instead connected to ground, V_{S-} is replaced with a voltage of 0 V and the ouput is either V_{S+} or 0 V. We can attach resisistors, light-emitting diodes (LEDs), motors, and so on, to be turned on or off with the output voltage.

Example 4.12 – Voltage Divider, Loading Error, and Op Amps

A laboratory needs a 3 V supply for a 1,000 Ω resistor but the only available supply is 5 V. With the voltage divider consisting of a 2 kΩ resistor in series with a 3 kΩ shown here, the voltage across the 3 kΩ resistor is 3 V.

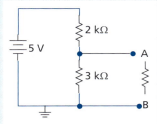

(a) What would the voltage drop between points A and B become if you placed a 1,000 Ω resistor across points A and B? Will the 3 kΩ resistor get 3 V? What is the loading error of this circuit if the goal was to produce a 3 V drop across the 1,000 Ω resistor?

The 1,000 Ω resistor will not have a 3 V drop. Calculating the equivalent resistance of the parallel 3 kΩ/1 kΩ , we get an R$_{eq}$ of 750 Ω. Using the voltage divider equation for the 2 kΩ and 750 Ω circuit, we get 1.36 V across the 1 kΩ resistor. The error in this case is 1.36 V – 3 V = –1.64 V or –55% error.

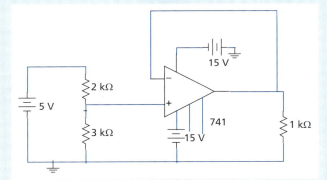

(b) Design an op amp circuit that will take the 3 V from the voltage divider and provide a 3 V output independent of loading.

This is a situation where the voltage follower is useful. It takes some input voltage and puts out that same output voltage, but because the op amp has very large resistance on the input side, whatever load one puts on the output does not load the voltage divider and hence, there really is a 3 V output. The voltage follower is shown here connected to the voltage divider circuit.

Cascading Operational Amplifiers

We have seen that a variety of relationships between input and output voltages are possible with op amps. We can combine these capabilities by cascading op amp circuits together, taking the output of one op amp and making it the input for the next. For instance, we saw in our summing amplifier Example 4.11 that we get an output voltage of –0.2 V when pressure vessels have become unsafe. We may need a larger voltage to

Example 4.13

Determine if the LED is turned on or off.

If the output voltage is 9 V, the LED will be on, if the voltage is 0 V, the LED will be off. We need to calculate the inverting and noninverting input voltages.

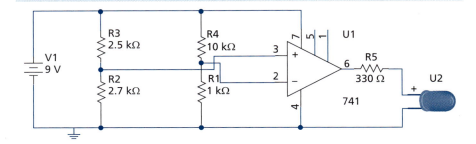

The noninverting voltage input can be found after recognizing that the resistors 2.5 kΩ and 2.7 kΩ are a voltage divider. Using the voltage divider equation, the inverting voltage is

$$V_{inverting} = \left(\frac{R_2}{R_1 + R_2} \right) V = \left(\frac{2{,}700 \ \Omega}{2{,}500 \ \Omega + 2{,}700 \ \Omega} \right) (9 \ V) = 4.67 \ V$$

Similarly, the noninverting input voltage can also be found via the voltage divider equations.

$$V_{noninverting} = \left(\frac{R_2}{R_1 + R_2} \right) V = \left(\frac{1{,}000 \ \Omega}{10{,}000 \ \Omega + 1{,}000 \ \Omega} \right) (9 \ V) = 0.82 \ V$$

Because $V_{inverting} < V_{noninverting}$, the voltage output will be 9 V and the LED will be on.

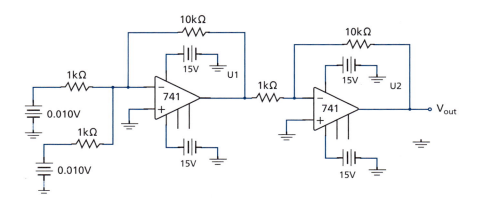

Figure 4.21 Example of cascaded op amps—summing and inverting.

turn on a warning system circuit, siren, or bell. A cascading op amp system is shown in Figure 4.21. We can use the inverting op amp and take the output of the summing amplifier as the input for the inverting op amp. The profile on the automatic titrator shows an example of cascaded op amps.

Example of a Cascaded Operational Amplifier: Instrumentation Operational Amplifier   The instrumentation op amp does something quite useful when dealing with noisy signals. It subtracts the two input voltages V_2 and V_1, and then it amplifies the result, removing any common noise to each input. The instrumentation op amp, shown in Figure 4.22, is actually three op amps with several resistors with the same resistance, R, and one resistor, R_G, which allows for easy control of gain. The output voltage is:

$$V_{out} = \left(1 + \frac{2R}{R_G}\right)(V_2 - V_1)$$ **Eq. 4.32**

where the ratio R/R_G sets the gain for the circuit.

Problem 4.20: You have five light sensors, each acting as a voltage source. The voltage range for each sensor is between 0 and 7 mV. When each sensor gets 5 mV, the device you are designing needs an output of +7 V. Design a cascaded op amp circuit to fulfill this requirement. Sketch your circuit and label the values for all relevant components.

Problem 4.21: A voltammetry experiment generates a current of 10 µA. Instead of measuring this low current, design a current to voltage op amp circuit to accomplish this goal so that a voltage of −3 V is generated as a result of the 10 µA. Sketch your circuit and label the values for all relevant components.

Problem 4.22: Sketch a circuit using a voltage comparator, an LED, and a photoresistor that will turn the LED off when light shines on the photoresistor.

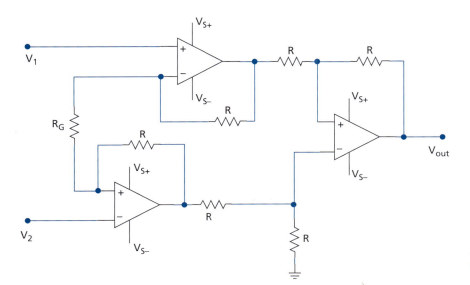

Figure 4.22 Instrumentation op amp circuit.

Electronics for an Automatic Titrator: Cascaded Operational Amplifiers and the Differentiating Operational Amplifier

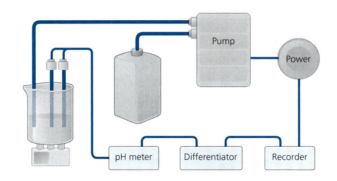

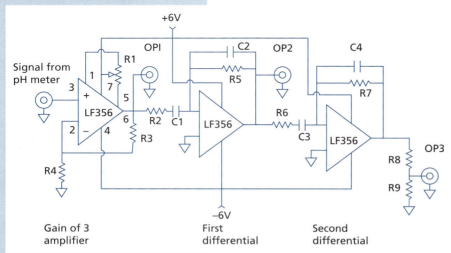

Gain of 3 amplifier

First differential

Second differential

This profile describes a powerful system[*] to help automate what can be a tedious process, titration. A sketch of the system is shown here. The pH of the sample is monitored as a pump places the titrant in the container. The pH meter voltage as well as the first and second derivatives of the pH voltage are also available as output. The circuit schematic for this device is shown here. This is an example of a *cascaded op amp* system. The first stage amplifies the voltage from a pH meter that is placed in the sample. We can now introduce the idea of the op amp differentiating configuration. The differentiating op amp has a feedback resistor in parallel with a capacitor and both are connected to the inverting input. The second stage op amp takes the output of the first stage and differentiates it. This is the derivative of the amplified pH voltage. The third stage takes a derivative of the output from the previous stage, providing the second derivative of the pH meter voltage.

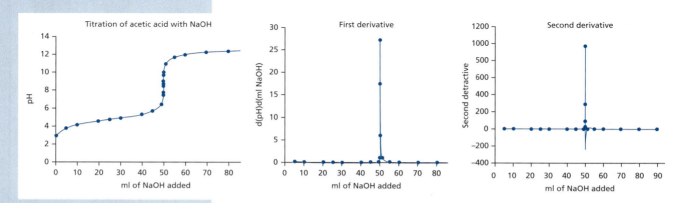

Sample output titration data for the voltage across the electrodes as well as the derivative demonstrates the power of the device. ∎

* Headrick, K. L.; Davies, T. K.; Haegele, A. N. *A Simple Laboratory-Constructed Automatic Titrator. J. Chem. Ed.* **2000**, *77*, 389–390.

4.6 Quick Survey of Electronic Components in Chemical Instrumentation

Potentiometers

The *potentiometer* or *variable resistor* is a three-terminal circuit element. The circuit symbol for the variable resistor (also known simply as a pot) is shown in Figure 4.2. In the circuit symbol, the top and bottom connections are the two terminals across which one obtains the full, maximum resistance of the pot. A connection across either the top and middle or bottom and middle terminals yields a variable resistance between near 0 and the maximum pot resistance, which depends on the rotation of the knob. By turning the knob on the pot, you vary the resistance of the pot from its minimum to its maximum rated resistance. In many cases, the volume dial on a car radio is a potentiometer.

Diodes

A diode is a device composed of an n-type semiconductor and p-type[5] semiconductor structure (known as a p/n junction). At their most basic, *diodes allow current to flow in only one direction.* This is shown in Figure 4.23. For silicon diodes, around 0.7 V is needed for current flow, as shown on the multimeter in the figure. When the battery is reversed, no current flows through the bulb.

When an AC source is in series with a diode, and a resistor, the voltage across the resistor is zero when current flow is blocked because of the diode. This is known as the *half-wave rectifier.* The AC voltage source is generally symbolized with a circuit with a sine wave inside. The negative part of the AC sine wave is blocked. We see the blocked negative part of the sine wave in the top part of Figure 4.24. When converting AC power into DC power, we prefer not to "throw away" half of our AC power. This is accomplished with a *full-wave bridge rectifier* (known as a *Wheatstone bridge rectifier*). This combination of four diodes produces an output as shown in the bottom part of Figure 4.24. Combining a capacitor in parallel with the load slows down the decrease from peak positive voltage so that the signal decreases very little between voltage maxima. The decrease in voltage from the peak describes the deviation from a true DC signal and is known as the ripple.

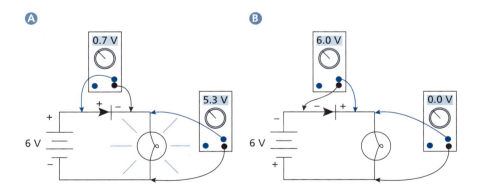

Figure 4.23 Diode and light bulb circuit with different battery orientations.

---

[5] The n-type semiconductors are so-named because the conduction is due to negatively charged particles (electrons). The p-type semiconductors are so named because conduction is due to positively charged particles (holes).

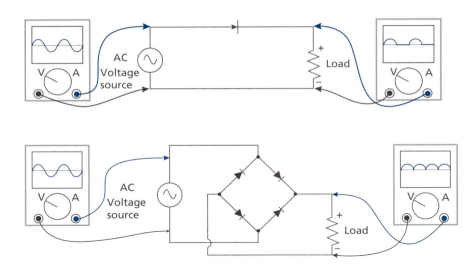

Figure 4.24 Half-wave rectifier (top) and full-wave rectifier (bottom).

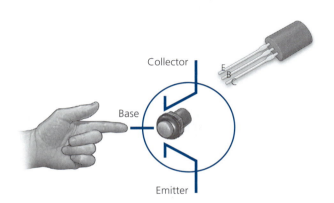

Figure 4.25 NPN transistor as switch—concept.

Transistors

The *transistor*, a three-terminal element, is used in switching and amplification applications. Transistors come in two common types: *bipolar junction* (BJT) and *field effect transistor* (FET). Our discussion will be kept to only BJTs. BJTs are either NPN or PNP, depending on whether they are a structure of n-type, p-type, n-type semiconductor (NPN) or p-type, n-type, p-type semiconductor (PNP). The terminals and labels for a BJT NPN transistor are shown in Figure 4.25. The transistor can be thought of as an open circuit across the collector emitter legs (no current flows) until a large enough current enters the base. This is also shown conceptually in Figure 4.25 for an NPN transistor where the base current "pushes the button" that allows current flow from collector to emitter. The base current that activates the transistor switch depends on a variety of factors but is generally small (<< 1 mA) and switches on a much larger collector current. The transistor obeys Kirchhoff's current law so that the emitter current is the sum of the base and collector current.

Integrated Circuits

Earlier in this chapter, we saw our first *IC*, the op amp, and learned that there are a great many ICs available with a wide range of functionalities. Because there are so many ICs,

There are a variety of inexpensive ways to construct useful instruments. Melissa A. Thal and Michael J. Samide published a paper in the *Journal of Chemical Education* that describes a simple system* for measuring absorption using inexpensive and readily available components (could be purchased at RadioShack or similar sources). The schematic for the circuit is shown here. The most expensive items are the multimeter (labeled as "ammeter" here) and the dual +15V/−15V power supply.

The left side of the schematic shows an LED, multimeter, potentiometer, and voltage source connected in series. The potentiometer allows for control of the current to the LED source. This allows for control of the LED intensity. The right side of the circuit is supplied with +15 V and is an inverting op amp, which acts as a current to voltage converter. OED is an acronym for optoelectronic detector. The authors of the paper describe the functionality of the circuit with different OED options (e.g., the photoresistor and the photodiode).

The analyte would be placed between the LED and the optical detector. Different color LEDs could be used to study different areas of the spectrum. The output voltage, V_{out}, can be related to sample absorbance. The authors demonstrate the power of this simple instrument, demonstrating Beer-Lambert linearity for absorbance between 0 and 1.5. ■

---

* *Thal, M. A.; Samide, M. J. Applied Electronics: Construction of a Simple Spectrophotometer. J. Chem. Ed.* **2001**, *78, 1510–1512.*

PROFILE

Electronics for a Simple Absorption Spectrophotometer: Operational Amplifier as Current-to-Voltage Amplifier

---

one will need to find the component data sheet online to find a description for each pin, or "leg." The data sheet will show a map of sorts, describing each pin. For example, from the data sheet, you will find the pin used for voltage to power the device, an input voltage, an output voltage, ground, or any number of other possibilities. Many data sheets will also provide a typical application for the IC.

The voltage regulator is a very common IC. In a typical application of the voltage regulator, the goal is to take a fluctuating input voltage and put out a very stable voltage. This circuit would also include several capacitors. Different voltage regulators can be chosen to provide different output voltages, depending on the needs of the circuit. The output (controlled) voltage must always be less than the input voltage. In applications with large current, a heat sink (metal structure with large surface area) is needed to dissipate heat in the voltage regulator. A voltage regulator (the LM7805) is shown in Figure 4.26 along with the pin assignments. Perhaps the most common place you will find a voltage regulator is as part of the AC to DC transformers used to provide a nearly constant voltage to charge mobile phones.

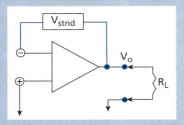

The potentiostat is the electronic hardware that is used in analytical potentiometry and analytical voltammetry.* A basic schematic is provided here. The value V_{stnd} is the voltage set by the user. The value R_L is the resistance of the load. In most cases, this would be the resistance between the working and counter electrodes in an electrochemical cell. Because the operational amplifier is wired in a feedback loop, the $V_{stnd} = V_o$. In order for V_o to $= V_{stnd}$, the current through R_L must satisfy Ohm's law ($V_{stnd} = I_L R_L$). As the resistance in RL varies, the power supply of the operational amplifier adjusts the current to maintain the relationship $V_{stnd} = V_o = I_L R_L$. ■

---

* See Chapters 18 and 19.

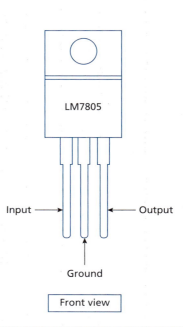

Figure 4.26 Another IC, the voltage regulator.

4.7 Analog and Digital Signals

Digital signals are generally in binary form, which is a collection of 1s and 0s, with each digit (reading right to left) representing a successively higher power of 2 (2^0, 2^1, 2^2, 2^4, and so on). Decimal number 0 is binary number 0. Decimal number 1 is binary number 1. Decimal number 2 is binary number 10. The 0 indicates that we should calculate 0×2^0, where the 2^0 is associated with the place for first digit starting on the right. The value associated with the 1 is $1 \times 2^1 = 2$. Again, this is the 2^1 place, one digit to the right from the start. These values are added ($0 + 2 = 2$) to obtain the decimal equivalent for 10, which is 2. Binary number 11 is $2^0 \times 1 + 2^1 \times 1 = 3$. Figure 4.27 shows two examples of binary numbers that are converted to their decimal equivalent. The highlighted areas show the successive power of 2 for each place in the binary number.

Signal data from sensors, whether in the form of voltage or current, are generally analog signals. *Analog* signals are continuous; they can take any value in their operating range. For example, the voltage measured via a thermistor might take on any value between 0 V and 12 V. In order to take advantage of the power of the computer for calculation and storage, analog signals are converted to digital signals. A digital signal is either on (also called high), represented by 1 in binary code, or off (also called low), represented by 0 in binary code. Although it varies, the digital 1 is often represented by +3 V (or +5 V) and the 0 signal by 0 V. Digital signals are less sensitive to noise because, unlike an analog signal, the exact signal level is not critical (because it is either on or off), so small variations in signal from noise sources are less problematic. Another example of analog versus digital signals is provided in Chapter 5, on signal processing.

The analog-to-digital converter (ADC) is an electronic device that takes in an analog signal and converts it to binary form. The ADC is specified by the number of *bits* possible. We will not focus on the particular circuit elements involved, but we will focus on *resolution*, a critical element to understanding an ADC and to understanding an instrument. An ADC will convert certain ranges of voltages (into discrete voltages (or bins). The resolution of the ADC is defined as the number of discrete values of the digital signal output. For example, an ADC might convert all analog voltages from 0 to 0.5 V to the same digital voltage output and all voltages from 0.5 to 1 V to a different digital output, and so on up to the maximum range of the voltage, perhaps 7 V. If an application requires that the digital output varies for input voltages of say 0 to 0.1 V, this ADC does not have the necessary resolution.

For example, suppose you have a 3-bit ADC. In general, the number of digital outputs is 2^n, where n is the number of bits. This device therefore allows for 8 (2^3) different digital outputs (0, 1, 10, 11, 100, 101, 110, and 111). For an input analog signal with a range from 0 to 12 V, the 12 V range would be split by the 3-bit ADC into eight different outputs. Thus, the 12 V would be divided into eight 1.5-volt units (12 V/8 = 1.5 V). This means that any voltage between 0 and 1.5 V will get the same binary output (placed in the same "bin"). Similarly, any analog signal between 1.5 V and 3 V gets the same binary output, and so on. Because any analog signal between 1.5 V and 3 V has the same binary output, there is error associated with the analog-to-digital conversion. This error is called *quantization error* or *digital resolution error*. The number of bits for the ADC fundamentally limits the precision of the conversion. The size of

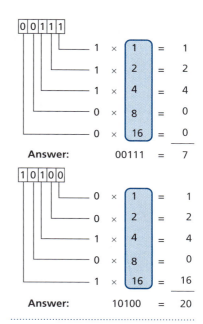

Figure 4.27 Converting from binary to decimal.

Example 4.14

Convert the binary number 1010 to decimal form.

We will start from the far right digit, which is 0. This is the 2^0 place and needs to be zero so the decimal component is $0 \times 2^0 = 0$. The next digit to the left is a 1. This is the 2^1 place and needs to be a 1 so the decimal component is $1 \times 2^1 = 2$. The next digit to the left is a 0. This is the 2^2 place and needs to be a 0, so the decimal component is $0 \times 4 = 0$. The final digit is a 1 and is in the 2^3 place. This becomes $1 \times 8 = 8$. Adding up all the components gives the decimal equivalent for 1010, which is $0 + 2 + 0 + 8 = 10$.

Example 4.15

An ADC has a bit number of 4. You plan to use it to measure a signal with a range from −3 V to +3 V. What is the size of each voltage bin?

$$\text{voltage bin size} = \frac{V_{range}}{2^n} = \frac{6}{2^4} = 0.375 \text{V}$$

the voltage "bin" for the analog input or the smallest voltage division associated with a given voltage range and bit number is:

$$\text{voltage bin size} = \frac{V_{range}}{2^n}$$

Eq. 4.33

Problem 4.23: What is the voltage resolution of a 12-bit ADC for a signal with a range of 12 V?

Problem 4.24: Is the bin size for a 16-bit ADC larger or smaller than the bin size for a 12 bit ADC? Calculate the bin size for both cases, assuming a signal with a range of 5 V.

Modern instruments typically use 8-bit to 16-bit resolution, although higher bit values are widely available. The ADC involves the comparator we described previously, which can provide a digital output, either the voltage of the supply or 0 V. There are also applications where a digital signal needs to be converted to an analog signal. The circuit for the *digital-to-analog converter* is also based on the comparator.

4.8 Further Reading

BOOKS

Alexander, C.; Sadiku, M. *Fundamentals of Electric Circuits*, 5th ed.; McGraw Hill: New York, **2012**.

Carter, B. *Op Amps for Everyone*, 4th ed.; Elsevier: Amsterdam, **2013**.

Horowitz, P.; Hill, W. *The Art of Electronics*, 3rd ed.; Cambridge University Press: New York, **2015**.

Karvinen, T.; Karvinen, K.; Valtaokari, V. *Make: Sensors: A Hands-On Primer for Monitoring the Real World with Arduino and Raspberry Pi*, Maker Media: Sebastopol, CA, **2014**.

Kybett, H.; Boysen, B. *All New Electronics Self-Teaching Guide*, 3rd ed.; Wiley: Indianapolis, **2008**.

Mims, F. *Getting Started in Electronics*, Master Publishing: Niles, IL, **2009**.

Platt, P. *Make: Electronics: Learning by Discovery*, 2nd ed.; Maker Media: Sebastopol, CA, **2009**.

Scherz, P.; Simon, M. *Practical Electronics for Inventors*, 3rd ed.; McGraw Hill: New York, **2013**.

JOURNALS

Amend, J. R.; Morgan, M. E.; Whitla, A. Inexpensive Digital Monitoring of Signals from a Spectronic-20 Spectrophotometer. *J. Chem. Ed.* **2000**, 77, 252–253.

Braun, R. D. Operational Amplifier Experiments for the Chemistry Laboratory. *J. Chem. Ed.* **1996**, 73, 858–861.

Kumar Sur, U.; Dhason, A.; Lakshminarayanan, V. A Simple and Low-Cost Ultramicroelectrode Fabrication and Characterization Method for Undergraduates. *J. Chem. Ed.* **2012**, 89, 168–172.

ONLINE RESOURCES

All About Circuits, an extensive resource for electronics, including a six-volume free textbook. http://www.all-aboutcircuits.com/ (accessed Aug 13, 2015).

4.9 Exercises

EXERCISE 4.1:

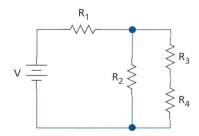

For the circuit shown here, answer the following questions as true or false:

(a) R_1 is in series with R_2

(b) R_1 is in series with R_3

(c) R_3 is in series with R_4

(d) R_2 is in parallel with R_3

(e) R_2 is in parallel with the R_3 and R_4 combination

(f) R_1 is in the combination of R_2, R_3, and R_4

EXERCISE 4.2:

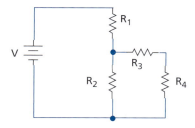

For the circuit shown here, R_1 = 330 Ω; R_2 = 500 Ω; R_3 = 800 Ω; and R_4 = 1,000 Ω. The voltage source supplies 9 V. Determine:

(a) The equivalent resistance of the circuit

(b) The current through each resistor

(c) The voltage across each resistor

EXERCISE 4.3:

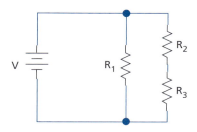

For the circuit shown here, R_1 = 1,000 Ω; R_2 = 330 Ω; and R_3 = 780 Ω. The voltage source supplies 12 V. Determine:

(a) The equivalent resistance of the circuit

(b) The current through each resistor

(c) The voltage across each resistor

EXERCISE 4.4:

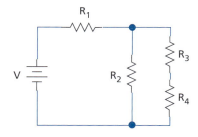

For the circuit shown here, R_1 = 780 Ω; R_2 = 500 Ω; R_3 = 330 Ω; and R_4 = 330 Ω. The voltage source supplies 15 V. Determine:

(a) The equivalent resistance of the circuit

(b) The current through each resistor

(c) The voltage across each resistor

EXERCISE 4.5: You have a circuit with three resistors in series with a 10 V DC power supply. Each resistor is 33,000 Ω. You have used a multimeter with an internal resistance of 5 MΩ to measure the voltage across one of the resistors. What will you measure? Is this an underestimate or overestimate compared to the actual voltage across the resistor?

EXERCISE 4.6: You are troubleshooting a circuit and have access to a "low impedance" multimeter that has an internal resistance of 3 kΩ. You are working on part of a circuit that is three resistors in series, each with the same resistance, 1,000 Ω. What will the low impedance voltmeter measure for the voltage across one of the 1,000 Ω resistors? Would a "high impedance" (1 GΩ) multimeter give a reading that is closer to the actual voltage across the element?

EXERCISE 4.7: You have a capacitor that has been charged so that the voltage across it is 12 V. You connect the charged capacitor to a resistor (15,000 Ω) and open the switch (all in series). You close the switch. What is the voltage across the capacitor 0.1 seconds later? What is the current 0.1 seconds later? What is the voltage across the capacitor 10 seconds later?

EXERCISE 4.8: In the section on parallel circuit elements, we said that unlike the resistors in series, each of the resistors in parallel will not necessarily get the same current. Is there a condition where three resistors in parallel with each other and a voltage supply would have the same current? Explain.

EXERCISE 4.9:

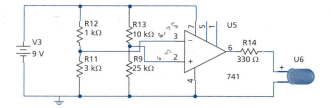

Determine whether the output voltage for the op amp comparator shown here is 0 V or 9 V.

EXERCISE 4.10:

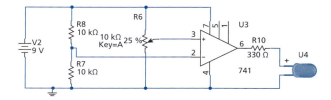

For the comparator circuit shown here, describe what happens to the LED as you vary the potentiometer from 1 to 10,000 Ω? At any point should the LED turn on? If so, at around what resistance for the potentiometer does the LED turn on? If it does not turn on, explain.

EXERCISE 4.11: You are working in a chemical manufacturing setting and have a laser that is directed into a large cuvette where a reaction is taking place where the absorption will change. You need an LED that turns on to signal when the light intensity on a detector has dropped to a certain percentage of its maximum. You choose a comparator circuit with a resistor and photoresistor in series. The photoresistor resistance varies between 1 MΩ when dark to 10,000 Ω when 10 lux is incident. You have determined that the photocell resistance is 40,000 Ω when the light intensity has reached the point where the signal is needed. What resistor would you choose to be in series with the photoresistor? Sketch your connections to the op amp.

EXERCISE 4.12: Design a high-pass filter with a cutoff frequency at 30 kHz. Sketch your design, choose the values for R and C, and label where you would measure V_{out}.

EXERCISE 4.13: A sensor voltage ranges from 0 to 3 V. A 4-bit ADC is used. Make a table that shows the binary output for the possible values of sensor voltage.

EXERCISE 4.14: A sensor voltage ranges from 0 to 6 V. A 3-bit ADC is used. Make a table that shows the binary output for the possible values of sensor voltage.

EXERCISE 4.15: Convert the following binary numbers to their decimal equivalents:

(a) 100

(b) 1111

(c) 10010

EXERCISE 4.16: For this question, refer back to the profile on the voltage divider photoresistor circuit. In the circuit, we measured the voltage across the resistor instead of the LDR and then used data on the resistance of the LDR as a function of light on the LDR (or lux) to determine the voltage across the resistor. Calculate the voltage across the LDR (not the resistor) if the lux on the LDR is 1.0.

EXERCISE 4.17: For this question, refer back to the profile on the voltage divider photoresistor circuit. Suppose you have been asked to build this circuit. Your laboratory partner has decided not to use the voltage divider. In other words, he wants to simply place the LDR in series with the voltage supply and to not use the resistor. He plans to measure the voltage across the LDR. What is wrong with this approach?

EXERCISE 4.18: This exercise is a guided derivation of the % loading error equation (Equation 4.10) that was presented in this chapter. The circuit for this derivation is from Figure 4.6. The figure shows a voltage divider circuit with a voltmeter with internal resistance, R_{int}, being used to measure the voltage across R_2.

(a) Use the voltage and current divider equations, along with the rule for determining equivalent resistance, to show that the voltage measured across R_2, taking into account the finite voltmeter resistance, is

$$V_{2, measured} = \left(\frac{R_2 R_{int}}{R_2 + R_{int}} \right) \left(\frac{V}{R_1 + \dfrac{R_2 R_{int}}{R_2 + R_{int}}} \right)$$

(b) The percent loading error is defined as

$$\% \text{ loading error} = 100\% \times \left(\frac{V_{w/meter} - V_{w/o\,meter}}{V_{w/o\,meter}} \right)$$

where $V_{w/meter}$ takes the finite resistance of the multimeter into account and $V_{w/o\,meter}$ assumes infinite resistance meter. Use this equation and your result from (a) to derive the % loading error relationship shown in Equation 4.10).

Chapter 5

Signals and Noise

An Introduction to Signal Processing

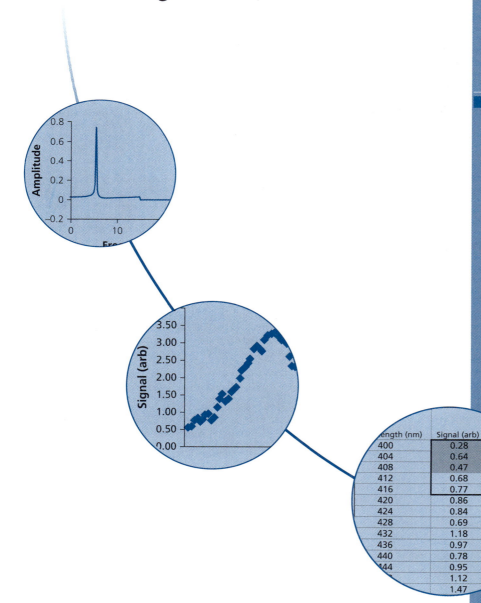

The 2014 Nobel Prize in Chemistry was awarded to William Moerner,[*] Eric Berzig, and Stefan Hell for their groundbreaking work in developing methods allowing researchers to detect and measure single molecules. Essential factors in this work included the dramatic reduction of noise associated with the signal measurement and the design of the experiment such that the signal could be exclusively detected amongst the noise that could not be eliminated.

Figure 5.1(A) shows some of Moerner's results. We are not concerned with the specifics of the experiment but the circled "peak" in Figure 5.1(A, i). This is due to a single molecule of pentacene. How could the researchers be certain the indicated feature was truly a peak rather than just more noise? They used a combination of noise reduction methods as well as a specialized laser frequency modulation technique that allowed them to observe changes in the peak that could only be explained by it being associated with a single molecule.

Since Moerner's seminal work, many researchers have used similar, as well as more advanced, techniques of signal-to-noise improvement to detect single molecules, and in some cases, even produce visual images of them. Figure 5.1(B) shows the result of fluorescence imaging of a single DNA molecule.[†]

In this chapter, we will learn to characterize signals and noise in experimental data, to find signals by altering our measurement approach, and to increase signals by working with data mathematically. A consideration of the signal strength and the noise is important as we introduce each new instrument type throughout this textbook. ∎

[*] Moerner, W. E.; Kador, L. Phys. Rev. Lett., **1989**, 62, 2535–2538.

[†] Kartalov, E. P. Single Molecule Detection and DNA Sequencing-by-Synthesis. PhD dissertation, California Institute of Technology, Pasadena, CA, **2004**.

5.1   Introduction to Signals

The vast majority of signals in modern chemical instrumentation are measurements of either voltage or current. For this reason, to understand challenges with signals, we must also think about the *electronics*. Chapter 4 provided an introduction to instrumental electronics. In order to make correct interpretations from experimental data, we need to understand the relationship between the detector response associated with the particular property of the analyte of interest (the signal) and the detector response associated with other factors not related to the particular property of the analyte (the noise). In general, we can think of the measured signal, $S_{measured}$, as the sum of the signal from the analyte, $S_{analyte}$, plus the signal associated with noise.

A signal might have components that are both direct current (DC) and alternating current (AC), although most signals associated with instruments are DC. Noise can be treated as an AC signal or signals, so understanding AC signals is also useful when working with DC signals from instruments. We can describe a single AC signal with:

$$V(t) = V_m \sin(\omega t + \phi) = V_m \sin(2\pi f t + \phi) \qquad \text{Eq. 5.1}$$

where V_m is the voltage amplitude, ω is the signal angular frequency (measured in rad/s), ϕ is the phase constant, and f is the frequency (measured in Hz or cycles/s). The inset in Figure 5.2(A) shows a combination (in this case, a mathematical sum) of three AC signals and one DC signal. The three AC signals in the inset have different amplitudes,

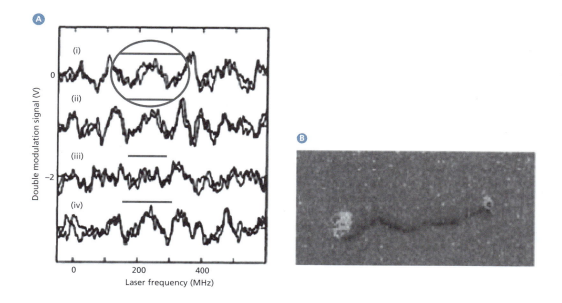

Figure 5.1 (A) Detection of a single molecule. Plots (i) through (iv) show a peak (circled), barely recognizable above the noise surrounding it. (B) Fluorescence detection of a single molecule of DNA.

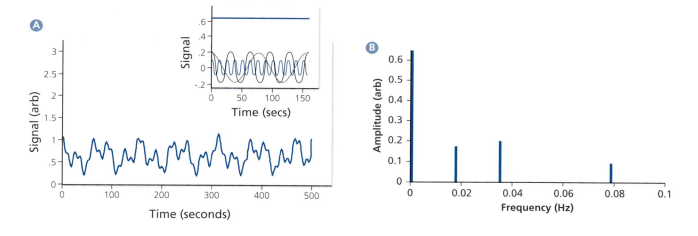

Figure 5.2 (A) Time domain spectrum of the sum of the three signals shown in the inset figure: DC signal (horizontal line) and three AC signals with different frequencies (three different sine waves). (B) The frequency spectrum.

frequencies, and phase constants. The signal with the lowest frequency (gray), has about the same amplitude as the signal of the next higher frequency (black). The frequency can be determined from the inset by determining the period of each signal (the time to complete one full cycle) and then using $f = 1/T$, where T is the period. In addition to the time domain signal shown in Figure 5.2(A), the signal can be described in the frequency domain by plotting the amplitude of each signal versus frequency. The frequency domain spectrum is shown in Figure 5.2(B). Note the DC signal at 0 Hz. We will frequently be moving between the time domain and the frequency domain as we discuss signal processing.

The individual signals in the inset can be added together to form a total signal, composed of AC and DC components. This combined signal is an example of what an instrument's detector would experience if all four signals were allowed to impinge on the detector without any discrimination. Again, our goal is to understand what aspects of this measurement are important and which are not. At times, our goal might be to only measure a DC signal. In other situations, we might only want to measure relatively high frequency signals, or conversely we might be interested in only the low frequencies components of the detector response. What constitutes the signal and what constitutes the noise is a function of the instrument and the goals for the analysis.

Problem 5.1: A measurement of voltage versus time is shown in the figure. The voltage is known to have a DC component, random voltage fluctuations of less than 0.1 V, and two AC components of different frequencies. From the figure, estimate the period for the two different AC signals and determine the frequencies. Calculate the angular frequencies. Sketch the frequency spectrum for the two AC signals.

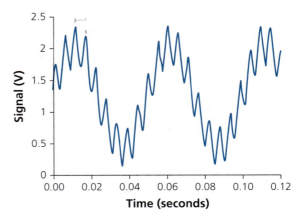

Problem 5.2: A measurement of voltage versus time is shown in the figure. The signal is known to be composed of several different frequencies. Estimate the lowest frequency contribution for this measurement. Estimate the amplitude of the lowest frequency AC component.

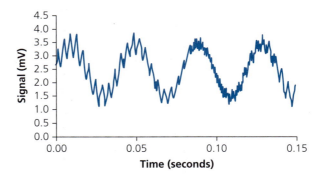

5.2 Sources and Characteristics of Noise

Noise is categorized as being either the result of the measurement system (*instrumental noise*) or the result of some chemically relevant contribution from the sample (*chemical noise*). Instrumental noise arises from, for example, temperature variations of the electronics that can alter the measurement voltage, or a poorly grounded wire that acts as an antenna for stray radiofrequency electromagnetic radiation in the vicinity. Chemical noise arises from aspects of the sample, for example, a material property that is temperature, pressure, or moisture dependent, which affects the analyte, and therefore alters the signal measured in a way that is not under the analyst's complete control or has not been accounted for during the initial calibration of the measurement.

It is important to have a sense for the different types of instrumental noise and the relative contributions of each type (e.g., detector/electronics temperature, proximity to power lines) that affect the noise. With knowledge of the type of noise relevant to a particular situation, strategies to increase signal strength and reduce noise can be used.

Different noise types have different frequency dependences. We will see shortly that two important noise types depend on the *bandwidth* of the measurement. The bandwidth, Δf, is the difference between the largest and smallest frequencies that a system measures. In particular, the bandwidth is defined as the difference in high and low frequencies at the half-power frequency. The half-power frequency is the frequency at which the power has dropped to half its maximum value. The measurement bandwidth may be defined by the measurement instrument (multimeter, oscilloscope) or by a filter that may be applied to the system that would further narrow the measurement bandwidth. For example, a multimeter that measures signals from 50 to 200 kHz has a bandwidth of $\Delta f = 200\,\text{kHz} - 50\,\text{kHz} = 150\,\text{kHz}$.

If we wish to use our instrument to take measurements as a function of time (e.g., kinetic information), we need to know how fast our instrument responds to a change in the analyte system. The relationship between the bandwidth of a measurement and the time it takes an instrument to respond to changes in the analyte system can tell us if our instrument can respond appropriately. If we are working with a rapidly changing signal, we would need an instrument that can respond rapidly as well. The instrument response time or rise time, t_r, is defined as the time it takes the instrument to increase from 10% of the input step voltage to 90% of the input step voltage. An instrument that has a very short rise time can respond very quickly to a step response input. If we model the measurement system as an effective resistor–capacitor $(\text{RC})^1$ circuit, we find that the rise time is related to the bandwidth by:

$$t_r = \frac{0.35}{\Delta f} \qquad\qquad \textbf{Eq. 5.2}$$

From Equation 5.2, we see that the larger the bandwidth of the measurement, the shorter the time response of the system. The rise time is related to the *time constant*, τ, for an *RC* circuit by:

$$t_r = 2.2\,\tau = 2.2\,RC \qquad\qquad \textbf{Eq. 5.3}$$

Thermal Noise

Thermal noise, also called Johnson or Nyquist-Johnson noise, is a voltage caused by random motion from charges due to thermal processes in electronic components. For a resistive circuit, this voltage can be written as:

$$V_{rms,Johnson} = \sqrt{4k_B TR\Delta f} \qquad\qquad \textbf{Eq. 5.4}$$

where k_B is the Boltzmann constant $(1.38 \times 10^{-23}\,\text{m}^2\,\text{kg/s}^2\,\text{K})$, T is the absolute temperature (in K) of the resistor, R is the resistance (in ohms), and Δf is the bandwidth (in Hz) for the measurement. This is a root mean square (rms) voltage. From the equation, the Johnson noise is larger for larger bandwidth measurements. Because the thermal noise is constant across all frequencies, we can think of the bandwidth dependence as simply telling us that the larger the frequency window over which we sample, the larger the thermal noise. Another way to think about this is that the thermal noise depends on frequency bandwidth but not on the actual frequency. Because of the temperature dependence, thermal noise can be lowered by lowering the temperature of the detector electronics. For example, charge coupled detector devices can be cooled to the temperature of liquid nitrogen (77K) to lower thermal noise. Another way to reduce thermal noise is to reduce the bandwidth of the measurement. Reducing the bandwidth does come with a cost. According to Equation 5.2, reducing Δf leads to a larger system rise time, t_r. Decreasing the bandwidth to reduce thermal noise therefore results in a slower instrument, potentially limiting the utility of that instrument for kinetic measurements.

---

[1] RC circuits are those that have resistors (R) and capacitors (C) in series. See Chapter 4 for a more complete discussion of RC circuits.

Example 5.1

You are looking to keep Johnson noise below 1 µV rms for a measurement. You are deciding whether to lower the detector temperature or to reduce the bandwidth measurement. You can lower the detector temperature to as low as 77K. For this measurement, it is important for the rise time to be no larger than 75 µs. The detector's effective resistance is 10,000 Ω. The bandwidth is 50 kHz.

SOLUTION −

We can solve Equation 5.4 in order to calculate the temperature that achieves this level of Johnson noise to obtain:

$$T = \frac{V^2_{rms, Johnson}}{4k_B R \Delta f} = \frac{(1 \times 10^{-6}\ \text{V})^2}{4(1.38 \times 10^{-23}\ \frac{\text{m}^2 \text{kg}}{\text{s}^2 \text{K}})(10,000\ \Omega)(50,000\ \text{Hz})} = 36.2\ \text{K}$$

Because liquid nitrogen boils at 77K, this temperature is too low for a convenient measurement. Alternatively, we can solve Equation 5.4 for the bandwidth necessary for a room temperature detector.

$$\Delta f = \frac{V^2_{rms, Johnson}}{4k_B RT} = \frac{(1 \times 10^{-6}\ \text{V})^2}{4(1.38 \times 10^{-23}\ \frac{\text{m}^2 \text{kg}}{\text{s}^2 \text{K}})(10,000\ \Omega)(300\ \text{K})} = 6.04\ \text{kHz}$$

We could combine approaches, lowering the temperature to 77 K and reducing the bandwidth. The bandwidth necessary for a 77K detector is:

$$\Delta f = \frac{V^2_{rms, Johnson}}{4k_B RT} = \frac{(1 \times 10^{-6}\ \text{V})^2}{4(1.38 \times 10^{-23}\ \frac{\text{m}^2 \text{kg}}{\text{s}^2 \text{K}})(10,000\ \Omega)(77\ \text{K})} = 23.5\ \text{kHz}$$

If we choose to reduce the bandwidth and keep the detector at room temperature, we need to check on the rise time to make sure we are still within our requirement:

$$t_r = \frac{0.35}{\Delta f} = \frac{0.35}{6,040} = 58\ \mu s$$

This is below our requirement, so we should lower the bandwidth in order to keep the Johnson noise below 1 µV.

Problem 5.3: A measurement system has an effective resistance of 1 MΩ and bandwidth of 10 MHz. Calculate the room temperature Johnson noise. What is the rise time for the measurement? How would reducing the bandwidth of the measurement to 10 kHz reduce the Johnson noise? What would the rise time be with a 10 kHz bandwidth?

Shot Noise

Shot noise results from the quantized nature of electrical charge in diodes and transistors. That is, charge is a discrete quantity, which means that current is not a continuous smooth flow but has "bursts." The associated shot noise current, $i_{rms, shot}$, is:

$$i_{rms, shot} = \sqrt{2 I e \Delta f}$$

Eq. 5.5

Example 5.2

You are going to use a multimeter with a bandwidth of 200 kHz to measure current in an instrument. Calculate the shot noise current associated with a current of (a) 1 pA and (b) 1 mA. Calculate the ratio of the shot noise current to DC current for each. For which current is the percentage of shot noise the largest?

SOLUTION –

(a) $$i_{rms,shot} = \sqrt{2Ie\Delta f} = \sqrt{(2)(1\times10^{-12}\text{ A})(1.6\times10^{-19}\text{ C})(200{,}000\text{ Hz})}$$
$$= 2.52\times10^{-13}\text{ A}$$

The ratio is $$\frac{i_{rms,shot}}{I_{DC}} = \frac{2.52\times10^{-13}\text{ A}}{1\times10^{-12}\text{ A}} = 0.25$$

(b) $$i_{rms,shot} = \sqrt{2Ie\Delta f} = \sqrt{(2)(1\times10^{-3}\text{ A})(1.6\times10^{-19}\text{ C})(200{,}000\text{ Hz})} = 8\times10^{-9}\text{ A}$$

The ratio is $$\frac{i_{rms,shot}}{I_{DC}} = \frac{8\times10^{-9}\text{ A}}{1\times10^{-3}\text{ A}} = 8\times10^{-6}$$

The shot noise for the 1 pA current is 25% of the DC current, and the much larger 1 mA current has shot noise that is 0.0008% of the DC current.

where I is the average DC current, e is the charge of an electron, and Δf is the frequency bandwidth. Unfortunately, I often is, or is related to, the desired analytical signal. Like Johnson noise, shot noise is independent of frequency but does depend on the bandwidth of the measurement. The combination of Johnson noise and shot noise are types of *white* noise. Noise that is frequency independent is referred to as white noise.

Problem 5.4: A spectrometer is used in an industrial manufacturing chamber that undergoes extreme temperature variations from −50° to 250°F. The effective resistance of the instrument is 30 kΩ and the bandwidth for the measurement is 100 kHz. Calculate the thermal noise for these temperatures. Ignore the temperature-dependent resistance that would occur at these temperatures. How would decreasing the bandwidth of the measurement to 50 kHz help reduce the thermal noise?

Problem 5.5: A system has an effective resistance of 1 MΩ. Calculate the shot noise when using three different bandwidth oscilloscopes (80 GHz, 200 MHz, and 50 MHz), assuming that the different scopes do not alter the effective resistance.

Flicker Noise

Flicker noise (also called *1/f noise*) can be a significant contribution to the total noise in an electrical measurement. Despite its importance, *1/f* noise is not completely understood. It is seen in a large range of physical systems, including traffic patterns and in biological systems. For instrumental analysis, *1/f* noise appears to be particularly important with electrical signals in semiconductors. Flicker noise has a frequency dependence of $1/f^n$, where n is in the range of 0.5 to 1.5. Given the dependence on frequency, *1/f* noise becomes larger at lower frequencies and is present at particularly detrimental levels when f is less than 100 to 200 Hz. This type of noise can also be a challenge with DC measurements. When possible, we should stay away from DC or low-frequency measurements when signal levels are low. In contrast to white noise, which has equal strength independent of frequency, *1/f* noise is sometimes called pink noise because of its particular frequency dependence.

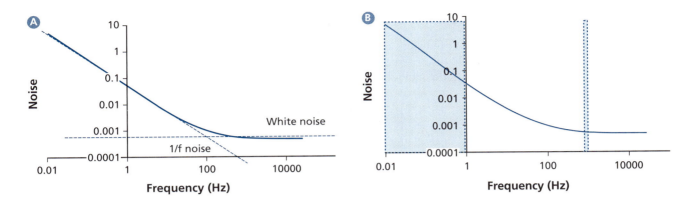

Figure 5.3 Noise voltage in electronics as a function of frequency.

Electrical noise generally has a white noise (thermal and shot) component and a pink noise ($1/f$) component. The white noise occurs across essentially all frequencies equally. There is a crossover point between the white and pink noise, very roughly around 100 to 1,000 Hz. An example plot of noise voltage versus frequency in electronics is shown in Figure 5.3(A). Note the nonlinear axes. Figure 5.3(B) shows two frequencies windows, each with a bandwidth of ~1 Hz. The total noise for each bandwidth would be the area under the curve in each window. There is clearly an advantage to moving a measurement to higher frequencies, where white noise dominates over $1/f$ noise. This point will be used later in the text where we discuss applications involving an increase in the frequency of a signal to move to a frequency region with lower $1/f$ noise.

Noise can also come from environmental sources such as the 60 Hz (in the United States; 50 Hz elsewhere) signal from electrical power lines. Cables that are used to route electronics in an instrument can actually act as an antenna receiving this and other types of radiation. A Faraday cage is an electrically grounded wire mesh box that is used to shield an instrument's electrical components from penetrating electromagnetic radiation. A simple version is shown in Figure 5.4. Many instruments with very sensitive detection electronics incorporate a Faraday cage into the instrument's design. Similarly, wiring leading to and from an instrument is often shielded; that is, it is wrapped in a grounded wire mesh.

Figure 5.4 The shielding ability of a Faraday cage can be demonstrated by simply wrapping a mobile phone in aluminum foil or putting the phone in a metal mesh cooking container. Try to call the phone. You will see that it that it does not ring. The cell signal radiation cannot pass through the metal cage.

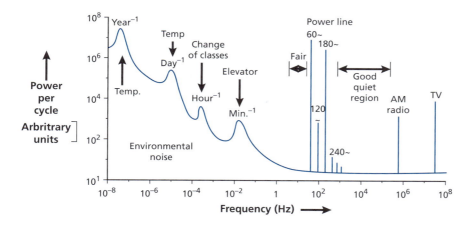

Figure 5.5 Common environmental noise spectrum. (From *J. Chem. Educ.* **1968**, *45 (7)*, A533.)

Harmonics or multiples of this 60 Hz signal also can be an issue at 120 Hz, 180 Hz, 240 Hz, and so on. A power versus frequency plot of common environmental noise sources is shown in Figure 5.5. This type of plot is also called a power spectrum.[2] Temperature variations that occur during the day can affect measurements. Even vibrations from walking in buildings or even an elevator are potential sources of noise if they can interact with a detector.

5.3 The Signal-to-Noise Ratio and Ensemble Averaging

The key figure of merit that describes the quality of a set of DC measurements is the *signal-to-noise ratio* (S/N), where S is the signal and N is the noise. The signal is taken as the mean of a number of measurements, $\bar{x}$, and the noise is the standard deviation of the mean, s, for the measurements.

$$\frac{S}{N} = \frac{mean}{standard\ deviation} = \frac{\bar{x}}{s}$$

Eq. 5.6

When comparing data sets, larger S/N ratio indicates higher quality data. A S/N ratio of 2 or 3 generally corresponds to data where the signal can be detected but cannot be quantified. For accurate quantitation, an analytical chemist usually seeks a S/N of at least 10.

In the case of white noise, the most powerful tool used to increase S/N is averaging. Because white noise involves both positive and negative values that are randomly distributed about the zero baseline, as the number of added scans increases, the average of the noise decreases and the S/N increases. The quantity $(S/N)_0$ is the signal-to-noise ratio for a single measurement or a scan (spectrum) or a single set of measurements (see Example 5.3 below). If we complete *n* sets of measurements, the S/N for all *n* is:

$$\left(\frac{S}{N}\right)_n = \sqrt{n}\left(\frac{S}{N}\right)_0$$

Eq. 5.7

In order to double the S/N, you have to increase the number of averaged scans by a factor of four.

---

[2] The power spectrum differs somewhat from the frequency spectrum. The power spectrum gives the power of different frequency components of a signal, whereas the frequency gives the amplitude of different components of a signal.

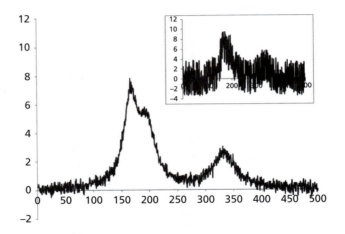

Figure 5.6 Averaging increases the S/N ratio when noise is random. The inset plot is one measurement, and the main plot is 50 measurements.

Example 5.3

For the experiment shown in Figure 5.6, we have measured the signal intensity at $x = 330$ repeatedly, under the same conditions. The data are displayed in the table here. We will assume that the variation in the data is random.

(a) What is the S/N ratio for these data?

(b) If we completed 10 sets of these measurements, how would the S/N change?

REPEATED MEASUREMENTS AT X=330
4.4
2.6
3.7
4.6
3.8
2.5
4.1
3.5

SOLUTION–

(a) The average of these measurements is 3.65. The standard deviation for the measurements is 0.77. The S/N is then 4.75.

(b) $$\left(\frac{S}{N}\right)_{10} = \sqrt{10}\left(\frac{S}{N}\right)_{0} = \sqrt{10}\,(4.75) = 15.0$$

We have improved the S/N by a factor of more than 3.

The advantage of increasing the S/N ratio when only random noise is present is shown in Figure 5.6, where we have averaged 50 measurements from an initially noisy spectrum. The inset figure shows the unaveraged signal. In this averaged plot, it is now clear that there is a third peak in the data at around $x = 200$. The type of averaging shown here, where we average several arrays of data, is called *ensemble averaging*. Averaging does increase measurement time. When averaging is done, we also are assuming that no noise level is changing consistently over each measurement. If a noise level is consistently and slowly increasing or decreasing (drift) over a set of measurements, our averaging approach would not be as beneficial. In such cases, it is incumbent on the analyst to determine the source of the biased noise and eliminate it.

Example 5.4 – A Quick Way to Estimate the Signal-to-Noise Ratio

Statistics shows us that for truly random noise, 99.7% of all data points will fall within ±3 standard deviations from the mean (see Chapter 22, Figure 22.9). So a reasonable estimate for the noise (recall we have defined this as the standard deviation for the data) can be obtained by taking the peak-to-peak baseline width as a rough estimate of six standard deviation units. Use this approximation to determine the S/N ratio for the data in the accompanying graph.

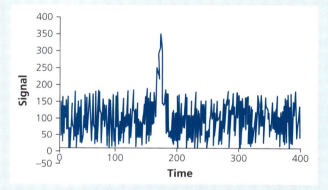

SOLUTION –

From the graph, we see that the peak-to-peak noise is 150 units. So a rough estimate of the standard deviation (s) would be to assume that 6s = 150 or $s \approx \dfrac{150}{6} \approx 25$.

The signal in this case is approximately 350 units so the $\dfrac{S}{N} \approx \dfrac{350}{25} \approx 14$.

Problem 5.6: In the table here, measurements A, B, and C correspond to repeated measurements of the same analyte under three somewhat different conditions. Assuming that the variation within each set is random, calculate the S/N ratio for each data set. Which data set has the largest standard deviation? Which data set has the largest S/N ratio?

A (mV)	B (mV)	C (mV)
12.1	3.3	023
12.4	2.6	0.25
12.3	3.1	0.22
12.9	2.8	0.24
12.5	2.9	0.25
12.2	2.6	0.24

Problem 5.7: Estimate the S/N ratio (using the 99.7% criterion) for the sharp peak in the plot shown here. Assume that the noise is random.

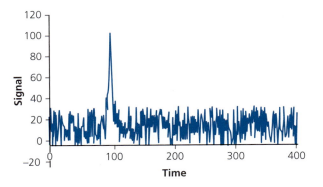

5.4　Processing Signals with Hardware and Software

Signal processing is generally categorized as either analog processing or digital processing. A common form of analog processing is through the use of *analog filters*. These filters are composed of electronic components such as resistors, capacitors, and operational amplifiers (op amps). Altering the filtering properties of an analog filter requires redesigning and rebuilding the filter circuit: for example, choosing a different capacitor or resistor. Analog filters generally alter the phase of the electronic signal. Digital signal processing involves a computer or microprocessor that alters signals using software and mathematics. Digital processing may occur after the experiment or in real time. Care should be taken with any data processing because it is possible to "create" new meaningless signals in experimental data and to obscure or distort real signals. It is important to understand how signal processing changes the data so that the operator does not draw incorrect conclusions from the processed data.

Analog Processing

Low pass, high-pass, band-pass, and band-stop passive filters are examples of *analog filters*. The low-pass filter passes low-frequency voltages and blocks high-frequency voltages. The high-pass filter blocks low-frequency signals and passes high-frequency signals. The band-pass filter allows a narrow band of frequencies to pass, and the band-stop filter blocks a narrow band of frequencies. A sketch of the ideal ratio of filter output voltage (V_{out}) to signal voltage input (V_{in}) versus frequency is shown in Figure 5.7(A) for each type of filter. Because these are passive filters, the maximum ratio of output to input is 1. An example of a signal with high-frequency noise and the result after low-pass filtering is shown in Figure 5.7(B). Note that the signal is much smoother after low-pass filtering. The specific analog circuit components that determine the filtering and bandwidth frequencies are discussed in Chapter 4.

　　The *instrumentation amplifier,* another example of analog processing, is used to subtract two voltages signals, each with *common noise* but only one with the signal of interest. The output voltage of the instrumentation amplifier is the *difference* between the two inputs. This means that the output has close to none of the common noise signal. The instrumentation amplifier is a specific configuration of op amps and other components.

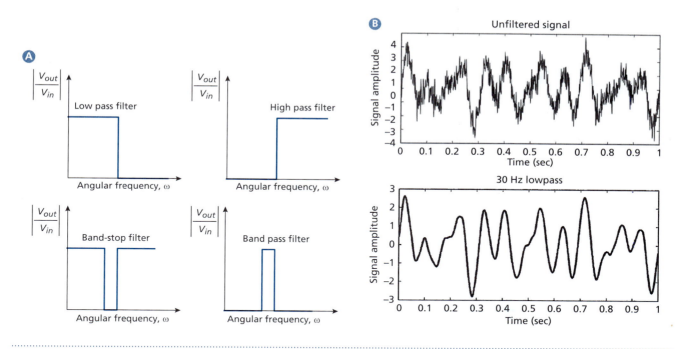

Figure 5.7 Panel (A) is the ratio of V_{out}/V_{in} for the four types of passive filters. Panel (B) is the unfiltered signal and signal after 30 Hz low-pass filter.

Op amps can be used to apply mathematical operations (such as adding, taking derivatives, and so on) to voltage input signals. More details on op amps and the instrumentation amplifier are provided in Chapter 4. The two input signals and the output signal to the instrumentation amplifier are shown conceptually in Figure 5.8, where we have a set of noisy data with a central peak. As usual, we are looking to increase the S/N ratio. The output of the instrumentation amplifier, the difference between the two inputs, is shown. The output clearly has less noise. The instrumentation amplifier is an example of an active filter because it has its own power supply such that it can increase the signal (amplify).

Problem 5.8: Compare the S/N ratio for the input signal and the signal with subtracted noise in Figure 5.8.

Boxcar Averaging with Hardware

The *boxcar averager* or *integrator* is another analog processing instrument. Boxcar averaging can be particularly useful given a time-varying signal that is periodic, but not necessarily sinusoidal, where we would like to choose a particular part of the signal (in time) to average over many periods. We can achieve *boxcar averaging* with an instrument called the gated integrator (or, often more simply, the boxcar average) to average over a user-controlled time window. The concept of boxcar averaging is shown in Figure 5.9 along with an image of the associated hardware. In this figure, the signal increases rapidly and then decays. On the instrument, the trigger delay controls where in time the gate is placed. The gate width defines the times over which the signal is averaged. These settings on the boxcar averager are chosen with the signal displayed on an oscilloscope along with a trace for the gate. The number of samples to average is also set on the instrument. The benefit of this approach is that the signal is only recorded during the time window needed, rejecting noise at other times. In this example, the gate width is set to overlap with the largest part of the peak.

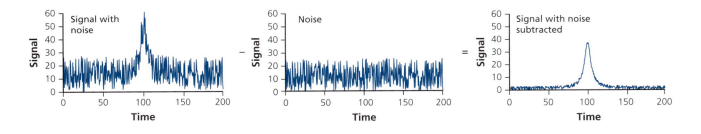

Figure 5.8 Noise subtraction instrumentation amplifier concept.

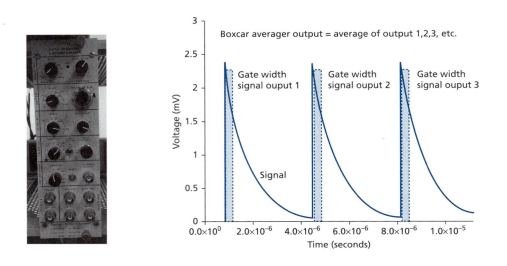

Figure 5.9 Time-varying periodic signal and boxcar gates and the gated integrator/boxcar average. (Hardware, Stanford Research Systems.)

Problem 5.9: You find that environmental noise in the form of the 60 Hz power line signal is causing problems for your measurement at 70 Hz. Which type or types of filters, if any (low-pass, high-pass, band-pass, band-stop), would you recommend for this problem? Defend your answer.

Problem 5.10: You are measuring a roughly 100 kHz signal that has a significant amount of thermal noise. You also find that $1/f$ noise is a problem at frequencies up to 200 Hz. Which type or types of filters, if any (low-pass, high-pass, band-pass, band-stop), would help with these two kinds of noise? Comment of on the effectiveness of your choices for each kind of noise ($1/f$ and thermal).

Problem 5.11: The amplitude of a signal with a frequency of 50 kHz corresponds to the amount of a contaminant in an industrial process. The amplitude is very small compared to the noise. Your system has both significant $1/f$ noise up to frequencies of around 100 Hz and significant thermal noise. Suggest ways to improve the signal to noise for these data.

Modulation and the Lock-in Amplifier

One approach to improve the S/N ratio for very low frequency signals (including DC measurements) is to artificially change or *modulate* the signal from a low frequency to a higher frequency. The modulation technique is used to increase the S/N ratio in several instruments, such as the atomic absorption spectrometer discussed in Chapter 7. It can be used effectively in other optical techniques, including fluorescence, laser absorption, and Raman spectroscopy. For spectroscopic techniques, this improvement in S/N can be accomplished by modulating the source radiation from a frequency where the *1/f* noise is high to a frequency where the *1/f* noise is significantly lower (see Figure 5.3). The white noise is also less significant if we move to a narrower bandwidth measurement, independent of the actual frequency. Band-pass filters could be also used to further reduce noise. When the DC signal is associated with or caused by an optical input (e.g., laser, lamp), the signal can be converted to AC by a mechanical chopper. The chopper is a black wheel with slits that spins in the optical path of the source radiation. The chopper's speed is controlled by a motor allowing the user to control the frequency of the source radiation that reaches the detector.[3] Different-sized chopper blades can lead to different pulse widths. Electronics can also be used to modulate a DC voltage.

The modulator can be combined with the *lock-in amplifier* to further increase the S/N ratio. The lock-in amplifier can help the analyst measure extremely small signals, as low as in the nanovolt range, that might be obscured with significantly larger noise levels. An example of the lock-in amplifier in the context of an instrument is shown in Figure 5.10(A). In this case, the laser causes emission in a sample (fluorescence). The laser is sent through a chopper, and the modulated emission from the sample is incident on a detector. The detector output is a voltage related to the emission intensity. The frequency output of the optical modulator is used as the *reference frequency* for the lock-in amplifier. The lock-in amplifier gets its name because it *locks in* to the signal at the reference frequency, ignoring essentially all signals at other frequencies. The signal output from the lock-in amplifier is an amplified DC voltage with an extremely low level of noise as compared to the signal on the detector.

A key idea is that the lock-in amplifier effectively *multiplies* the reference signal and the signal input from the detector. This is shown in Figure 5.10(B). It is not obvious that this is useful, but if we write out this product, the benefits will become clearer. In the signal voltage line, there is often some amplification before "multiplication" with the reference line.

The reference input, $V_r(t)$, from the modulator, and the signal input, $V_s(t)$, from the detector, can be written as:

$$V_r(t) = V_r \sin(\omega_r t + \phi_r)$$

$$V_s(t) = V_s \sin(\omega_s t + \phi_s) \qquad \textbf{Eq. 5.8}$$

Because the lock-in amplifier effectively multiplies these inputs, the product of these is:

$$V_r(t)V_s(t) = V_r \sin(\omega_r t + \phi_r)V_s \sin(\omega_s t + \phi_s)$$

$$= \frac{1}{2}V_r V_s \cos((\omega_r - \omega_s)t + \phi_r - \phi_s) - \frac{1}{2}V_r V_s \cos((\omega_r + \omega_s)t + \phi_r + \phi_s) \quad \textbf{Eq. 5.9}$$

---

[3] Choppers are common components of flame atomic absorption spectrometers (see Figure 7.13) and many ultraviolet-visible spectrometers.

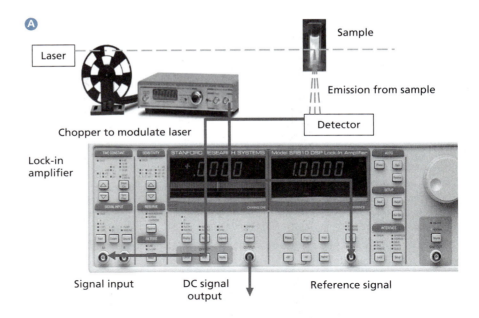

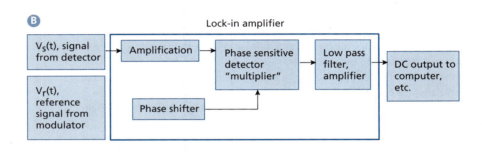

Figure 5.10 (A) Basic optical setup for using the modulator and lock-in amplifier. (B) Block diagram of lock in amplifier. (Both instruments from Stanford Research Systems.)

where we used a trigonometric identity[4] to write the product as the sum of two terms. When the reference and signal frequencies are the same $(\omega_r = \omega_s)$, the relationship simplifies to:

$$V_r(t)V_s(t) = \frac{1}{2}V_rV_s\cos(\phi_r - \phi_s) - \frac{1}{2}V_rV_s\cos(2\omega_s t + \phi_r + \phi_s) \quad \textbf{Eq. 5.10}$$

This is the output of the phase sensitive detector (also called the mixer). The phase shifter from Figure 5.10 allows the user to control the phase of the reference voltage to set $\phi_r - \phi_s$ to zero to maximize the first cosine term. The first term is DC because the phase between the two inputs is constant in time. The second term is sinusoidal with a frequency twice the input. As shown in Figure 5.9, the lock-in amplifier applies a low-pass filter to remove the second term, leaving only the DC component. The low-pass filter also removes other noise components. In particular, when the two frequencies are associated with the reference and noise, and are therefore not the same, both terms are

---

[4] Trigonometric identity:

$$\sin(A)\sin(B) = \frac{1}{2}\cos(A - B) - \frac{1}{2}\cos(A + B)$$

sinusoidal and filtering would remove both terms, leaving only the DC component proportional to signals with $\omega_r = \omega_s$. This DC component that passes the filter is then amplified and is the signal output of the lock-in amplifier. The effect is that of amplifying that lone frequency of the reference, with an extremely small bandwidth of around 0.01 Hz.[5]

Problem 5.12: Compare the contribution of white noise for two measurements that use a lock-in amplifier. For one measurement, the bandwidth associated with the lock-in is 0.1 Hz. For the other, it is 0.5 Hz. Assume that all other aspects of the measurements are the same.

Digital Signal Processing

A variety of mathematical approaches in digital signal processing are used. There are several data smoothing techniques that take a range of adjacent data points (whether in time or wavelength or some other parameter) and mathematically combine them to produce a single data point. When this is done on successive adjacent points, the result is a smoothed output compared to the initial set of data. The most straightforward of these is the rolling or running average where we take a group of data points and averaged them into one value. This can be done with three data points, five data points, or more. The three-point and five-point rolling averages are calculated according to:

$$Y_{rolling} = \frac{y_{-1} + y_0 + y_1}{3}$$

$$Y_{rolling} = \frac{y_{-2} + y_{-1} + y_0 + y_1 + y_2}{5} \qquad \textbf{Eq. 5.11}$$

The average is $Y_{rolling}$ and for the five-point average, the y_{-2} refers to the first data point, y_{-1} the second, and so on. The result of the calculation, Y, has the x value associated with the y_0 point.

The average is "rolling" because of the overlap that occurs between adjacent data sets. This is shown in Table 5.1, where a three-point rolling average is performed and shaded boxes and black lines show which intensity values are averaged for each cluster of time. The three-point averaged data point at 4 seconds in Table 5.1 is the average value of the raw data taken from 0 seconds, 4 seconds, and 8 seconds. We then shift down one position and at 8 seconds returns the average value of the raw data from

TABLE 5.1: Example Data to Illustrate Difference Between Rolling Average and Boxcar Averaging

TIME (SECONDS)	INTENSITY (ARB)	THREE-POINT ROLLING AVERAGE	INTENSITY (ARB)	BOXCAR AVERAGE
0	0.193		0.193	
4	0.525	0.432	0.525	0.432
8	0.578	0.534	0.578	
12	0.500	0.450	0.500	
16	0.272	0.349	0.272	0.349
20	0.276	0.281	0.276	
24	0.295	0.299	0.295	
28	0.327	0.308	0.327	0.308
32	0.303		0.303	

[5] Stanford Research Systems Application Notes. About Lock-In Amplifiers, Application Note #3, http://www.thinksrs.com/downloads/PDFs/ApplicationNotes/AboutLIAs.pdf.

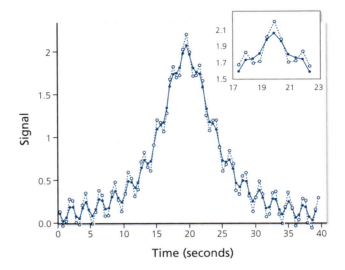

Figure 5.11 Measurement (solid) and rolling three-point average (dashed); insert shows close-up around data peak.

4 seconds, 8 seconds, and 12 seconds. The assumption made is that the noise is randomly distributed above and below the true peak in the data and as such an average of a few data points will average out the highs and lows resulting from white noise. The assumption is valid for a flat base line but near the apex of a peak, the result of a digital filter is to pull down the peak (flatten). It is also important to note that a three-point digital filter results in the loss of the first and last data point. A five-point rolling average results in the loss of the first two and last two data points.

A plot of example data and the rolling averaging smoothing using three points is seen in Figure 5.11. The inset figure shows a zoomed in look at the peak of the data. There are times when the peak value is of interest to the analyst.

The rolling average is the close relative of the boxcar average. The boxcar average takes a set of data, such as voltage from a transducer, for a narrow range of time or wavelength values and takes the average. Unlike the rolling average, the boxcar technique has no overlap of values. This is shown in Table 5.1 with the same data from the rolling average example. Generally speaking, you should only consider boxcar averaging if you have a very large set of points with small increments between them. The fewer the points you have and the greater the point-to-point increment in the x axis, the greater distortion boxcar averaging will impart.

Example 5.5

In the following experiment the signal intensity is measured as a function of wavelength. Perform a three-point and a five-point rolling average to smooth the data.

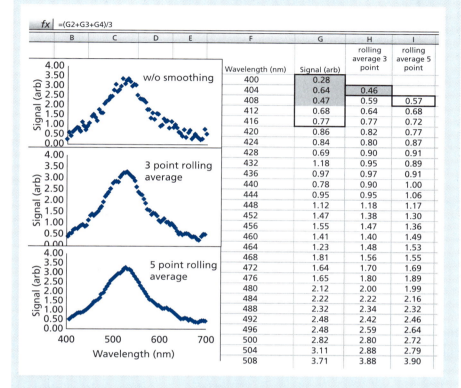

	fx =(G2+G3+G4)/3							
	B	C	D	E	F	G	H	I
					Wavelength (nm)	Signal (arb)	rolling average 3 point	rolling average 5 point
					400	0.28		
					404	0.64	0.46	
					408	0.47	0.59	0.57
					412	0.68	0.64	0.68
					416	0.77	0.77	0.72
					420	0.86	0.82	0.77
					424	0.84	0.80	0.87
					428	0.69	0.90	0.91
					432	1.18	0.95	0.89
					436	0.97	0.97	0.91
					440	0.78	0.90	1.00
					444	0.95	0.95	1.06
					448	1.12	1.18	1.17
					452	1.47	1.38	1.30
					456	1.55	1.47	1.36
					460	1.41	1.40	1.49
					464	1.23	1.48	1.53
					468	1.81	1.56	1.55
					472	1.64	1.70	1.69
					476	1.65	1.80	1.89
					480	2.12	2.00	1.99
					484	2.22	2.22	2.16
					488	2.32	2.34	2.32
					492	2.48	2.42	2.46
					496	2.48	2.59	2.64
					500	2.82	2.80	2.72
					504	3.11	2.88	2.79
					508	3.71	3.88	3.90

SOLUTION –

Equations 5.11 have been entered into fields H5 and I5 in the spreadsheet here. Comparing Equation 5.11 to the spreadsheet values, the data point 0.28 is y_{-2}, the data point 0.64 is y_{-1}, and so on. The 408 nm point gets, for the five-point rolling average value:

$$Y_{rolling} = \frac{y_{-2} + y_{-1} + y_0 + y_1 + y_2}{5} = \frac{(0.28) + (0.64) + (0.47) + (0.68) + (0.77)}{5}$$
$$= 0.57$$

The 404 nm point gets for the three-point rolling average value:

$$Y_{rolling} = \frac{y_{-1} + y_0 + y_1}{3} = \frac{(0.28) + (0.64) + (0.47)}{3}$$
$$= 0.46$$

Notice that the five-point rolling average plot provides more smoothing than the three-point rolling average, but there is a trade-off. We see that peaks tend to broaden and we lose resolution with increased data points used in a rolling average.

Problem 5.13: You are investigating a compound looking for strong evidence of a broad peak that may start at 500 nm and peak at 610 nm. In other words, you are looking for a rise in the data below, indicating the tail of a peak not shown at 610 nm. The peak at 610 nm is being used to verify that you have synthesized the compound of interest. You are not able to measure at 610 nm because of complications from the optics in the system. Use a smoothing technique of your choice and put forth a quantitative argument for whether or not the data support the existence of a 610 nm peak.

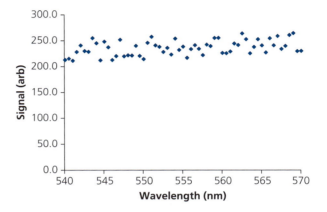

WAVELENGTH (NM)	SIGNAL (ARB)	WAVELENGTH (NM)	SIGNAL (ARB)	WAVELENGTH (NM)	SIGNAL (ARB)
540	213.0	550.5	245.9	561	228.9
540.5	215.5	551	257.5	561.5	244.3
541	211.4	551.5	240.9	562	241.4
541.5	228.3	552	238.1	562.5	263.7
542	240.9	552.5	238.2	563	252.6
542.5	230.1	553	236.0	563.5	225.4
543	228.6	553.5	223.2	564	237.7
543.5	255.0	554	253.9	564.5	252.5
544	245.2	554.5	232.0	565	240.7
544.5	212.2	555	238.4	565.5	227.0
545	248.2	555.5	216.8	566	254.4
545.5	237.3	556	233.6	566.5	240.8
546	212.8	556.5	241.1	567	258.7
546.5	220.3	557	234.1	567.5	233.8
547	251.9	557.5	221.9	568	239.7
547.5	219.9	558	242.2	568.5	260.5
548	221.8	558.5	239.3	569	264.2
548.5	221.3	559	255.0	569.5	229.5
549	240.1	559.5	255.2	570	230.0
549.5	220.7	560	226.1		
550	214.5	560.5	225.6		

TABLE 5.2: Polynomial Fitting Parameters with Quadratic Smoothing for Five Points, Seven Points, and Nine Points

	NUMBER OF POINTS IN SAVITSKY-GOLAY FILTER		
	NINE	**SEVEN**	**FIVE**
	Point Position Weighting Factor		
−4	−21		
−3	14	−2	
−2	39	3	−3
−1	54	6	12
0	59	7	17
1	54	6	12
2	39	6	−3
3	14	−2	
4	−21		

Savitzky-Golay Smoothing

There are more mathematically sophisticated methods for smoothing data that go beyond a simple rolling average of grouped data. For example, instead of smoothing with a simple three-point average, a least squares polynomial fit can be performed on a range of adjacent data points and the value for the fit at the center of the range returned as the "weighted averaged" value for that data point. The advantage of this approach over the rolling average is that there is potentially less blunting of sharp high-frequency signals with this technique. This approach can be implemented because of work by chemists Abraham Savitzky and Marcel J. E. Golay.[6] They developed a set of convolution integers that turn out to form the set of weighting factors used for polynomial least squares smoothing; this method bears their name and is known as Savitzky-Golay smoothing. Different order polynomials can be used in this approach (e.g., quadratic, quartic).

The technique puts less emphasis on large deviations in data and can do a better job of approximating peak heights than a linear rolling average. Both rolling averages and Savitzky-Golay smoothing are typically incorporated in the software that accompanies commercial instruments. As we demonstrated in Example 5.5, linear rolling average filters are most effective reducing high-frequency noise. The same is true of Savitzky-Golay filters. These types of filters can be thought of as low-pass filters. The higher the order of polynomials used (for example, quartic), the more high-frequency noise that will be part of the smoothed output.

For five-point quadratic Savitzky-Golay smoothing, we weight the five data points and calculate a single value with:

$$Y^{quadratic}_{S-G} = \frac{-3y_{-2} + 12y_{-1} + 17y_0 + 12y_1 - 3y_2}{35} \qquad \textbf{Eq. 5.12}$$

where the integers −3, 12, and so on, come from Table 5.2. The factor 35 is the sum of the weighting factors. The y_{-2} refers to the first data point, y_{-1} the second, and so on. The result of the calculation, Y, has the x value associated with the y_0 point. In other words, y_0 is the point about which the weighting occurs. For seven-point smoothing, we convert seven data points to one central point using the following equation:

$$Y^{quadratic}_{S-G} = \frac{-2y_{-3} + 3y_{-2} + 6y_{-1} + 7y_0 + 6y_1 + 3y_2 - 2y_3}{21} \qquad \textbf{Eq. 5.13}$$

where the integers −2, 3, and so on, again come from Table 5.2.

---

[6] Savitzky, A.; Golay, M. J. E. Smoothing and Differentiation of Data by Simplified Least Squares Procedures. *Anal. Chem.* **1964**, *36*, 1627–1639.

Example 5.6

In the following measurement, the peak intensity is an important part of your analysis but the data are noisy. Perform a seven-point quadratic Savitzky-Golay smoothing of these data and determine the peak signal value.

SOLUTION–

Equation 5.13 has been entered into field H5 in the spreadsheet below. Comparing Equation 5.13 to the spreadsheet values, the data point 0.28 is y_{-3}, the data point 0.64 is y_{-2}, and so on.

$$Y^{quadratic}_{S-G} = \frac{-2y_{-3}+3y_{-2}+6y_{-1}+7y_0+6y_1+3y_2-2y_3}{21}$$

$$= \frac{-2(0.28)+3(0.64)+6(0.47)+7(0.68)+6(0.77)+3(0.86)-2(0.84)}{21}$$

$$= 0.69$$

The wavelength 412 nm has the Y value of 0.69. The image here shows the equation used to calculate the Y-smoothing point based on the first seven data points. If you use the autofill feature to copy the equation from field H5 into fields H6 through the remainder of the data set, this process will repeat for each consecutive set of seven data points with the next group, starting with the wavelength 404 nm.

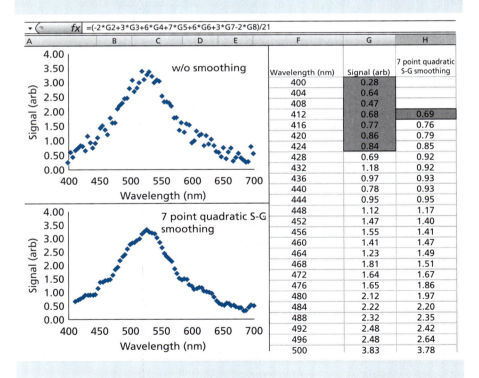

| fx | =(-2*G2+3*G3+6*G4+7*G5+6*G6+3*G7-2*G8)/21 |

Wavelength (nm)	Signal (arb)	7 point quadratic S-G smoothing
400	0.28	
404	0.64	
408	0.47	
412	0.68	0.69
416	0.77	0.76
420	0.86	0.79
424	0.84	0.85
428	0.69	0.92
432	1.18	0.92
436	0.97	0.93
440	0.78	0.93
444	0.95	0.95
448	1.12	1.17
452	1.47	1.40
456	1.55	1.41
460	1.41	1.47
464	1.23	1.49
468	1.81	1.51
472	1.64	1.67
476	1.65	1.86
480	2.12	1.97
484	2.22	2.20
488	2.32	2.35
492	2.48	2.42
496	2.48	2.64
500	3.83	3.78

We can compare the smoothing properties of the three-point rolling average, the five-point rolling average, the five-point Savitzky-Golay filter, and the seven-point Savitzky-Golay filter. These are each shown for the same set of raw data in Figure 5.12. The five-point rolling average clearly produces the smoothest data. Put another way, the Savitzky-Golay filter distorts the data to a lesser degree and provides a "softer" smoothing approach.

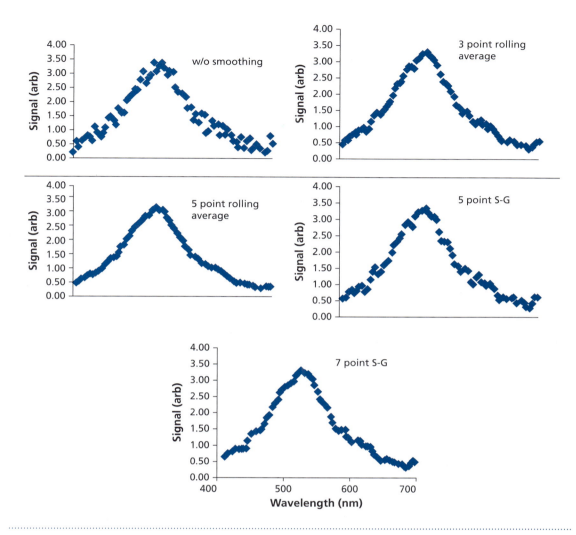

Figure 5.12 Comparison of smoothing techniques.

Fourier Filtering

The Fourier transform is a mathematical operation named after the French mathematician Joseph Fourier that takes a signal from the time domain to the frequency domain. The inverse Fourier transform takes the signal from the frequency domain to the time domain. We will see the Fourier transform used in several instrumental methods throughout this course, including infrared, Raman, nuclear magnetic resonance, and mass spectrometry. The Fourier transform can be used to perform *Fourier filtering*, which is a powerful technique that can remove unwanted AC frequency components from a signal. At this point, we will not cover the mathematical details that describe the process of calculating the Fourier transform.[7] Easily utilized functions in such software programs as Excel, Matlab, and Labview make it possible to perform Fourier transforms without these mathematical details. Many oscilloscopes can also take Fourier transforms in real time with a voltage measurement, various Fourier transform "apps" can be purchased for mobile phones, and microcontrollers such as the Arduino can be programmed to do Fourier transforms on input voltages.

In Figure 5.13(A), the signal, plotted as amplitude versus time, is composed of several AC components. The first step in Fourier filtering is to take the Fourier transform

[7] See, for example, Kreyszig, E. *Advanced Engineering Mathematics*, 10th ed.; Wiley, **2011**, or Chesick, J. P. Fourier Analysis and Structure Determination. Part I. Fourier Transforms. *J. Chem. Ed.* **1989**, *66*, 128–132.

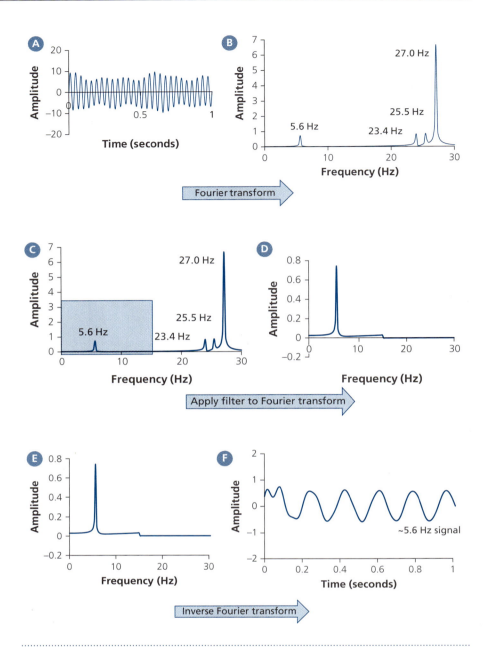

Figure 5.13 (A) Time-varying signal and its (B) Fourier transform, showing frequency components; (C, D) low-pass filter passes frequencies less than 15 Hz; (E, F) taking filtered data and taking inverse Fourier transform, with data now a function of time again.

of the time domain data. The result is the frequency domain plot of amplitude versus frequency seen in Figure 5.13(B). The Fourier transform identified four separate frequencies from the time domain data. At this stage, a filter can be applied to remove unwanted frequency components. We could apply a low-pass filter, high-pass filter, band-pass, or band-stop filter. Once we have removed the unwanted frequency components, we apply the inverse Fourier transform to recreate our time domain signal with the unwanted frequencies removed.

Suppose we know that our desired signal is less than 15 Hz for the data in Figure 5.13. We then apply a low-pass filter to the Fourier transform data. For the data here, we have applied a perfect low-pass filter, which reduces high frequencies to 0 and passes low frequencies without any loss in signal. The result shown has essentially one frequency component. Note that the 5.6 Hz signal amplitude is much less than the maximum signal value from the first data set. Once the data has been filtered, the inverse Fourier transform is applied to take the signal data back to a time domain signal. The result is a nearly sinusoidal signal at ~5.6 Hz.

Problem 5.14: The time domain signal shown here is composed of two AC components and one DC component. Roughly sketch what you think the frequency domain signal would look like.

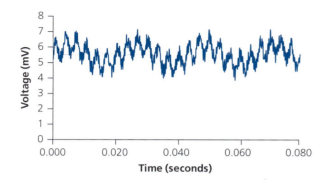

Problem 5.15: A frequency domain signal of voltage amplitude versus frequency is shown here. Roughly sketch what your think the time domain signal would look like.

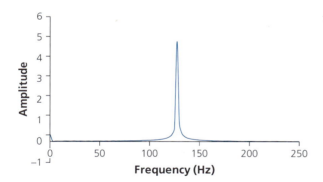

Problem 5.16: You have taken the Fourier transform for a set of data that shows narrow frequency components at 50 Hz; 100 Hz; 250 Hz; and 1,000 Hz. Your plan is to use an analog filter to remove the components that do not correspond to the aspect of the analyte in your measurements.

(a) If the component of interest is the 1,000 Hz signal, what type of filter should you use?

(b) If the component of interest is the 250 Hz signal, what type of filter should you use?

(c) If the component of interest is the 100 Hz signal, what type of filter should you use?

5.5 Sampling Rates, the Nyquist Frequency, and Aliasing

When measuring a signal, it is important to choose an appropriate sampling frequency for a given AC signal. The sampling frequency can be roughly thought of as the number of data values taken per second. In order to measure a rapidly changing signal, it is necessary to take in enough data points to adequately describe the variation in the signal. In order to correctly measure an AC signal, the sampling frequency must meet the *Nyquist condition.* The Nyquist condition (or criterion) states that *the sample rate must be at least twice the **highest** frequency contained in the signal.* The Nyquist frequency is defined as twice the signal frequency. If the Nyquist condition is not met, the frequency of the measured signal will be incorrect. This could lead to problems with analysis. To gain an intuitive sense of this effect, the blades of a helicopter serve as an interesting example. Video footage of helicopter blades sometimes makes it appear as though the blade is moving very slowly or even changing directions. The helicopter blade is moving very rapidly and the sample rate, defined by the camera and our eye, is too slow. We cannot take in enough data to adequately show the speed of the blade.

An illustration of the relationship between signal frequency, the sample rate, and the Nyquist condition is shown in Figure 5.14. The far left plot shows a single-frequency, 4 Hz sine wave (the data are plotted to look like a continuous function). This is the "true" signal, with a frequency of 4 Hz (period of 0.25 seconds). An instrument takes in or samples the signal with some controlled time between each measurement. If the measurement is taken every 0.005 seconds, the sampling rate is f = 1/0.005 seconds or 200 Hz. The middle plot (top) shows the 200 Hz sampled data. The sample rate is 50 times the frequency of the signal so the Nyquist condition is met. The data are taken at such a large sample rate that it is quite easy to determine the frequency of the signal from the plot and to see that the 200 Hz sampled data matches the 4 Hz signal.

The middle plot (center) shows the data sampled at 8 Hz, the minimum rate according to the Nyquist condition to correctly determine the frequency of the signal. Next to this plot is the sine wave that matches the data. The sine wave does have the same 4 Hz frequency as the signal. The middle plot (bottom) shows the data sampled at 5 Hz, less than what is required by the Nyquist condition. You can clearly see from the sampled data points that the sine wave produced by the instrument does not have the same frequency as the true signal of 4 Hz. To the right of the 5 Hz sample rate data, we

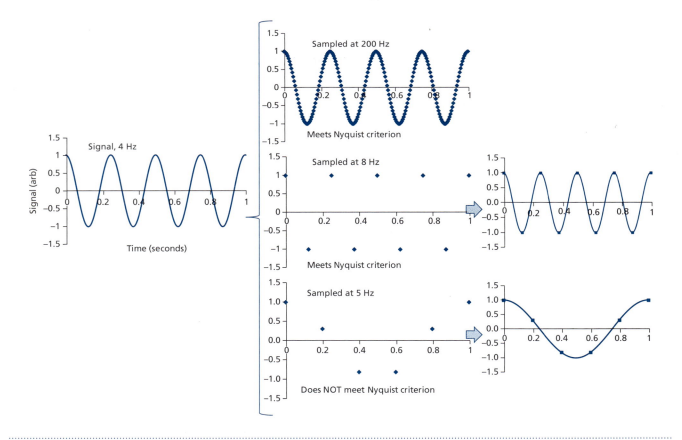

Figure 5.14 Nyquist frequency and aliasing. Left image shows 4 Hz signal. Middle images show the signal sampled at three different rates (200 Hz, 8 Hz, and 5 Hz). Right images show the sine wave constructed from the sampled data.

show the sine wave that would be produced using the 5 Hz sample rate. The frequency of the output data is 1 Hz, which is incorrect. When the Nyquist condition is not met, you see "ghost" frequencies in your output data. This effect is known as *aliasing*.

Problem 5.17: The time domain signal shown here is composed of two AC components and a DC component. Estimate the sampling rate needed to meet the Nyquist condition.

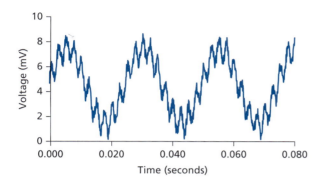

5.6 Analog-to-Digital Conversion

Signals are either *analog* or *digital*. An *analog* signal can take on any value, whereas a *digital* signal can only take on particular values or steps in a range. This concept is demonstrated in Figure 5.15, where three sinusoidal time-varying signals at 10 kHz are shown: one that is analog (looks exactly like a sine wave) and two that are digital (and look less like the sine wave and more like a series of steps). These correspond to audio signals from three different media (one analog signal—a classic vinyl record—and two digital signals—a CD and a DVD). The analog sinusoid is smooth and exactly matches a sine function. The digital signals are approximations of the sine function because the digital signals cannot take on every value of the sinusoid like the analog signal. Note that of the two digital signals, the DVD output matches the analog signal much better than the CD; the DVD output has more "allowed steps" than the CD. The number of steps for a digital output is 2^n, where n is the bit number for analog-to-digital conversion. The CD has a 16-bit processor, which results in 65,536 (2^{16}) output levels, whereas the DVD has a 24-bit processor, which results in 16,777,216 (2^{24}) output levels. Because of the higher bit number, the DVD audio can take on more values, so it can match the analog signal more closely and hence sounds better than the CD.

For an analog-to-digital converter (ADC), the voltage bin size is:

$$\text{voltage bin size} = \frac{V_{range}}{2^n}$$ **Eq. 5.14**

where V_{range} is the range of the analog input and n is the bit number for the ADC. Therefor the maximum intrinsic quantization error is half of the voltage range covered by a single bin. The key idea here is that too low a bit number could limit an instrument's ability to measure small changes appropriately.

Electronics are used to convert an analog signal to digital and to convert digital signals to analog. ADCs were briefly discussed in Chapter 4, in the introduction to instrumental electronics.

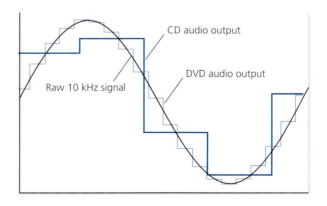

Figure 5.15 Comparison of analog signal at 10 kHz with two digital signals with different resolutions (CD and DVD).

Example 5.7

An ADC has a bit number of 8, and you plan to use it to measure a signal from −5 V to 5 V. You need to be able to measure voltage to within 0.05 V. Is this bit rate going to be adequate? If not, recommend a bit number.

$$\text{voltage bin size} = \frac{V_{range}}{2^n} = \frac{10}{2^8} = 0.039\,\text{V}$$

This should do the trick.

Problem 5.18: What is the maximum quantization error if we use a 12-bit ADC for signals with a voltage ranging between −10 to 10 V?

Problem 5.19: You are building an instrument and need to choose the ADC. The signal voltages range from −1 V to 1 V, and the instrument needs to measure voltage changes of 0.01 V in order for the instrument to measure the analyte of interest adequately. Would a 10-bit ADC be adequate? If so, support your response with a calculation. If not, suggest a bit number for an ADC that would be adequate.

5.7 Further Reading

BOOKS

Horowitz, P.; Hill, W. *The Art of Electronics*, 3rd ed.; Cambridge University Press: New York, **2015**.

JOURNALS

Blitz, J. P.; Klarup, D. G. Signal-to-Noise Ratio, Signal Processing, and Spectral Information in the Instrumental Analysis Laboratory. *J. Chem. Ed.* **2002**, *79* (*11*), 1358–1360.

Fountain III, A. Exploring Digital Signals and Noise in Instrumental Analysis: SignalsNoise.mcd. *J. Chem. Ed.* **2001**, *78*, 271.

Hoffmann, M. M. An Interactive Spreadsheet for Demonstration of Basic NMR and Fourier Transform Concepts. *J. Chem. Ed.* **2009**, *86*, 399.

O'Haver, T. C. An Introduction to Signal Processing in Chemical Measurement. *J. Chem. Ed.* **1991**, *68*, A147–A150.

Overway, K. FT Digital Filtering: Simulating Fourier Transform Apodization via Excel. *J. Chem. Ed.* **2008**, *85*, 1151.

Sengupta, S. K.; Farnham, J. M.; Whitten, J. E. A Simple Low-Cost Lock-In Amplifier for the Laboratory. *J. Chem. Ed.* **2005**, *82*, 1399–1401.

Vogt, F. Data Filtering in Instrumental Analyses with Applications to Optical Spectroscopy and Chemical Imaging. *J. Chem. Ed.* **2011**, *88*, 1672–1683.

ONLINE RESOURCES

O'Haver, T. Interactive Computer Models for Analytical Chemistry Instruction; Professor O'Haver has created an excellent set of simulation tools to understand various topics in instrumental analysis, including several tools for understand gratings. http://terpconnect.umd.edu/~toh/models/index.html (accessed Aug, 2015).

5.8 Exercises

EXERCISE 5.1: Which signal is most likely to suffer from *1/f* noise? Which signal is most likely to suffer from thermal noise?

(a)   1 Hz

(b)   10 Hz

(c)   100 Hz

(d)   1 kHz

EXERCISE 5.2: The S/N ratio for a set of 50 measurements is 10. What number of measurements would make the S/N ratio 5? What number of measurements would yield a S/N ratio of 30? If each measurement takes 3 seconds, how long would the S/N ratio = 5 and S/N = 30 measurements take?

EXERCISE 5.3: The S/N ratio for a set of measurements is 7. If you take 10 times more data, what is the S/N ratio? If you take 100 times more data, what is the S/N ratio?

EXERCISE 5.4: Statistically, 99% of all measurements should lie within ± 2.5 standard deviations of the mean. Consider Example 5.4 and repeat the S/N ratio calculation in that example using this alternate criterion. Explain any differences you observe in the calculated S/N ratio.

EXERCISE 5.5: An ultrafast oscilloscope has a bandwidth of 80 GHz. A basic oscilloscope has a bandwidth of 200 MHz.

A low-cost oscilloscope has a bandwidth of 50 MHz. Estimate the rise times for each of these oscilloscopes.

EXERCISE 5.6: You have taken the Fourier transform of a set of data that shows narrow frequency components at 400 Hz; 1,250 Hz; and 2,000 Hz. Your plan is to use an analog filter to remove the components that do not correspond to the aspect of the analyte in your measurements.

(a) If the component of interest is the 400 Hz signal, what type of filter should you use?

(b) If the component of interest is the 1,250 Hz signal, what type of filter should you use?

(c) If the component of interest is the 2,000 Hz signal, what type of filter should you use?

(d) Suppose you are interested in both the 1,250 Hz and the 2,000 Hz signals. What type of filter might you use?

EXERCISE 5.7: A plot of amplitude versus time is shown here. Does this signal have only an AC component? Roughly sketch the frequency domain version of this signal.

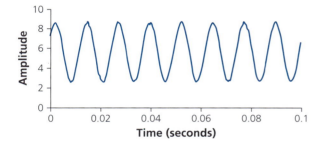

EXERCISE 5.8: You are working with an instrument with signal at 15 kHz. The usual $1/f$ noise and white noise occurs.

A band-pass filter, centered at 15 kHz, is to be used in the circuit. Discuss qualitatively, how, if at all, the bandwidth of the band-pass filter would affect the amount of $1/f$ noise and/or thermal noise?

EXERCISE 5.9: You have measured the following time-dependent voltage signal off of an optical detector. The high-frequency component of the signal is thought to be noise. Estimate the amplitude and frequency of this noise.

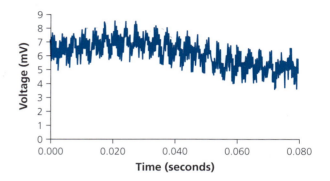

Advanced Exercises:

EXERCISE 5.10: This question makes use of content in Chapter 4 on passive filters. The crossover frequency for $1/f$ noise and white noise for a given condition and measurement system is at 150 Hz. Design a passive filter that would block these low frequencies and would have a V_{out}/V_{in} of 10% at 150 Hz.

EXERCISE 5.11: This question makes use of content in Chapter 4 on passive filters. Design a filter to pass a narrow band of frequencies between 100 and 105 kHz.

Chapter 6

Molecular Ultraviolet and Visible Spectroscopy

PROFILE
James Clerk Maxwell

The first thing a sommelier* does is swirl his or her wine in a glass to evaluate the viscosity of the wine. The second thing the sommelier does is examine the color of the wine. From an analytical perspective, the assessment of a sample based on its color is one of the oldest analytical techniques, known historically as colorimetry. Our contemporary understanding of light, energy, and color was founded by James Clerk Maxwell, called the father of modern physics. In 1864, Maxwell published the landmark paper "Dynamical Theory of the Electromagnetic Field"[†] in which he described light as mutually perpendicular oscillating electric and magnetic fields. During the presentation of his paper, Maxwell proclaimed: "We have strong reason to conclude that light itself—including radiant heat and other radiation, if any—is an electromagnetic disturbance in the form of waves propagated through the electro-magnetic field according to electro-magnetic laws." The equations outlined in this paper became what we now refer to as the Maxwell equations. Ivan Tolstoy proclaimed in his biography of Maxwell: "Maxwell's importance in the history of scientific thought is comparable to Einstein's (whom he inspired) and to Newton's (whose influence he curtailed)." ▬

James Clerk Maxwell (1831–1879)

Maxwell formulated our understanding of light as an electromagnetic wave. He is considered one of the three most influential physicists of all time.

---

* A sommelier is a trained wine steward who examines wine for quality and matches a particular wine with a specific dish.

[†] Maxwell, J. C. Dynamical Theory of the Electromagnetic Field. Phil. Trans. R. Soc. Lond., 1866, 155.

6.1 Introduction

Before Maxwell's explanation of light, the ancient Greeks had developed their own theories on the nature of light based on the four elements: fire, air, water, and earth. They believed that light emanated from the eye in much the same way as light emanates from a candle or lamp and that the colors we see are the result of the ethereal light's interaction with the fundamental elements. The Greek philosopher Empedocles expanded this idea, believing that all of the perceived colors were the result of mixing of four fundamental colors (derived from the four fundamental elements). Democritus, Empedocles' pupil, advanced this theory by postulating that the characteristics of a particular color result from the characteristics of the atoms that produce the color. The analytical result is that one can infer the elemental composition of a material by examining the material's color.[1] It is an impressive feat of logic when you consider how close these philosophers came to our current understanding of color. Regardless, the analysis of a material based on its color is one of our oldest analytical techniques. *Colorimetric techniques* rely on the analysis of visible color. Even though our understanding of the fundamental science has evolved since Democritus' time, the analysis of color still remains one of our primary analytical tools. This chapter represents a modern version of colorimetry, and so it seems fitting that ultraviolet-visible (UV-vis) light is the first of the optical instrument techniques we will study. In a modern UV-vis spectrophotometer, we have expanded our ability to "see" beyond the visible part of the electromagnetic spectrum into the ultraviolet ($\approx$ 195 nm) and infrared regions ($\approx$ 900 nm) of the electromagnetic spectrum (EMS). However, from a historical perspective, it is still correct to think of UV-vis spectroscopy as an evolved form of colorimetry.

As we learned in Chapter 2 the UV-vis spectrum of a molecule results from the quantum states that exist within a molecule. Therefore, UV-vis spectroscopy can often provide qualitative insights into the electronic structure of a molecule. However, it is safe to say that the UV-vis spectrophotometer is one of the most widely used analytical instruments for *quantitative* analysis of organic molecules (not atoms) and transition metal coordination compounds. Its usefulness comes from the direct quantitative relationship between the concentration of the species under analysis and the signal that is measured. The quantity of photons absorbed by an analyte is proportional to the number of absorbing species in the sample (see Chapter 2, sections 1 and 2). Furthering its usefulness, the UV-vis spectrophotometer is often used as the detector in other techniques such as liquid chromatography and capillary electrophoresis (see Chapters 15 and 17, respectively).

A basic schematic of a UV-vis spectrophotometer is presented in Figure 6.1. Block 1 represents the radiant source, which provides the light. The source puts out light at a range of wavelengths within the region of interest (ultraviolet and/or visible). Block 2 represents the monochromator, the function of which is to select or isolate specific

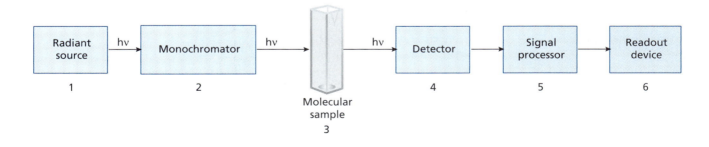

Figure 6.1 Basic schematic of a UV-vis spectrophotometer.

---

[1] Benson, J. L. *Greek Color Theory and the Four Elements: A Cosmological Interpretation,* University of Massachusetts Amherst Libraries, Amherst, MA, **2000**.

frequencies (or wavelengths) of light from the source. Block 3 represents the analytical sample. Only the wavelengths selected by the monochromator go forward to block 3 and on for measurement. Notice that in block 3, we have emphasized the molecular nature of the sample. We will discuss similar spectroscopic methods that deal with atomic samples in Chapter 7. Block 4 represents the detector, which measures the radiant power of the light signal. In the figure, the incident light (represented by the arrow hv) has emanated from the source, been filtered by the monochromator, passed through the sample, and gone on to the detector. In some instrumental designs, the detector reads the filtered signal, directly out of the monochromator, so in some spectrophotometers, the order of blocks 2 and 3 are reversed. The detector converts the incident light signal to an electrical impulse that is processed (represented by block 5) and sent to a readout device (typically a computer) represented by block 6.

When we take a UV-vis spectrum, we are measuring the transmittance of light through a material as a function of the wavelengths of light (or energies of light) that are passing through the sample. In other words, the UV-vis spectrum gives us a profile of the amount of light transmitted as a function of the light's wavelength.

It is very important for spectroscopic analysis that you have an excellent understanding of the EMS. Within the broad regions of the EMS, you should be able to quickly identify the type of energy and discuss the ways in which that form of energy interacts with matter. The visible range of the EMS is from violet (approximately 400 nm) to red (approximately 700 nm). For spectroscopic purposes, the ultraviolet spectrum begins at about 195 nm and ends at 400 nm. Although the region extends further, below 195 nm, the air (oxygen, nitrogen, and moisture) begins to absorb the radiant light and therefore background interference becomes a limiting factor for a typical UV-vis spectrometer.

Example 6.1

What is the wavelength (nm) and frequency of a photon with an energy of 3.81×10^{-19} J? In what part of the EMS does this photon belong?

STRATEGY –
The relationship between energy, wavelength, and frequency for a photon was first presented in Chapter 2 Equation 2.2 and is repeated here. Solve the equation for v and λ.

$$E = hv = \frac{hc}{\lambda}$$

where
  h = Planck's constant $(6.626 \times 10^{-34} \text{ J·s})$
  v = frequency (s^{-1})
  c = speed of light $(2.99 \times 10^{8} \text{ m/s})$
  λ = wavelength (m)

SOLUTION –

$$v = \frac{E}{h} = \frac{3.81 \times 10^{-19} \text{ J}}{6.626 \times 10^{-34} \text{ J·s}} = 6.75 \text{ (Hz)}$$

$$\lambda = \frac{hc}{E} = \frac{(6.626 \times 10^{-34} \text{ J·s})(2.88 \times 10^{8} \text{ m/s})}{3.81 \times 10^{-19} \text{ J}} = 6.009 \times 10^{-7} \text{ m} \approx 501 \text{ nm}.$$

The photon is in the blue-green portion of the visible region of the EMS.

Problem 6.1: What is the frequency and energy of a photon with a wavelength of the following:

(a) 650 nm　　　(c) 1,700 μm　　　(e) 195 nm　　　(g) 0.5 μm
(b) 50 nm　　　(d) 0.65 μm　　　(f) 700 nm　　　(h) 500 nm

Problem 6.2: Refer to Figure 6.2. To what part of the EMS do photons with the following frequencies belong:

(a) 2.99×10^7 Hz　(c) 3.21×10^{20} Hz　(e) 6.31×10^{18} Hz　(g) 4.29×10^{14} Hz
(b) 10.00 Hz　　　(d) 1.54×10^{15} Hz　(f) 7.50×10^{14} Hz　(h) 3.33×10^{14} Hz

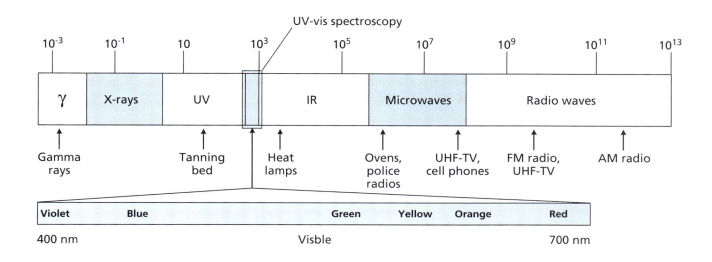

Figure 6.2 An overview of the electromagnetic spectrum.

Although it is possible to conduct spectroscopy below 195 nm, the measurement must be taken in a vacuum or in an instrument purged with helium or other nonabsorbing gases. These instruments also require specialized optics (e.g., lenses made from CaF_2). Instruments in this region do exist (they are called *far-ultraviolet spectrometers*), but they are not common and we will not emphasize them in this chapter. For the purposes of this discussion, the ultraviolet region will be limited to the range of *about 195 to 400 nm*. Similarly, on the visible end of the region, the modern UV-vis spectrometer can utilize radiation at wavelengths that are just outside the visible region, longer wavelengths up to 900 nm. Technically, light in the range of 700 to 900 nm is classified as the **near-infrared** region of the EMS and even though we cannot see this radiation with our naked eyes, when we talk about UV-vis spectrophotometry, we are talking about the full range of wavelengths that span from about 195 nm to about 900 nm.

Problem 6.3: We have stated that the typical range of a UV-vis spectrophotometer is 195 nm to 900 nm. What is this range in hertz?

6.2　Electronic Excitation and Molecular Structure

As we learned in Chapter 2, an electron within a molecule can only exist within specific energy states. When a molecule absorbs energy in the form of a photon, it does so by promoting an electron from a lower energy state (ground state) to a higher energy state (excited state). In order for the molecule to be able to absorb the photon, the photon's energy must match the difference in energy of the two electronic

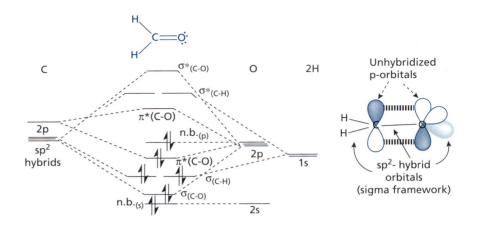

Figure 6.3 MO diagram of formaldehyde. Notice that the MO was constructed using hybrid orbitals on the carbon atom. The HOMO is a nonbonding pair of electrons originating on the oxygen atom. The LUMO is the π^* molecular orbital (left). Representation of formaldehyde using valence bond theory (right).

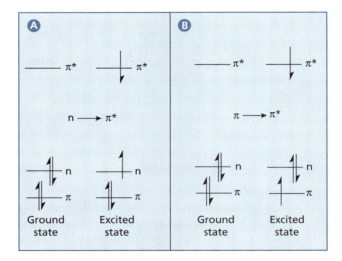

Figure 6.4 A look at two possible transitions in formaldehyde. Here we are looking at only the upper portion of the MO diagram as presented in Figure 6.3. The lower energy (sigma) portion of the MO diagram has been omitted for clarity. Panel (A) shows the $n \rightarrow \pi^*$ transition. Panel (B) shows the $\pi \rightarrow \pi^*$ transition.

states.[2] Furthermore, the energies of these electronic states are governed by molecular structure. It follows that the energy (or wavelength) of the light absorbed is governed by the *molecular structure* of the absorbing species. As a case study, let us examine the electronic structure of formaldehyde (see Figures 6.3 and 6.4).

The Lewis structure for formaldehyde is given in the top left region of the molecular orbital (MO) diagram in Figure 6.3. There are two features of this Lewis structure that are of spectroscopic interest: the existence of lone pair electrons on the oxygen atom (a spectroscopist will refer to lone pair electrons as *n-electrons*[3]) and the existence of the

---

[2] There are additional requirements that fall under a category called *selection rules*. We will discuss these later in this chapter.

[3] We will refer to lone pair electrons as *n* electrons. This term is borrowed from molecular orbital theory and stands for nonbonding electrons.

double bond between the carbon atom and the oxygen atom. We know from valence bond theory that one of the two bonds in the double bond is a σ-bond and the other is a π-bond. Because the overlap of p-orbitals in a π-bond is relatively weak, we know the gap between the $\pi \rightarrow \pi^*$-orbitals is not as large as the gap between $\sigma \rightarrow \sigma^*$ orbitals.

It happens that the energy gap for many $\pi \rightarrow \pi^*$ transitions falls within the UV-vis range of the EMS. Likewise, the energy of the n-electrons falls in between the $\pi \rightarrow \pi^*$ split and so we can expect $n \rightarrow \pi^*$ transitions in the UV-vis range as well (see Figure 6.4). Spectroscopically, we might expect to see absorption of energy associated with the transitions of electrons from the n-electrons or π-electrons to higher, unoccupied orbitals. Chemists call these features *chromophores*.[4] A *chromophore* is a structural feature (or functional group) that gives rise to an observable absorption. The word chromophore comes from the Greek "chrom," meaning color, and "phora," meaning producing. However, as we see in the case of formaldehyde, the absorptions can also occur in the ultraviolet, just outside of the visible spectrum. Chromophores are responsible for the color of a particular compound (such as the green of chlorophyll), but from the perspective of a human eye not all chromophores result in visibly colored compounds.

Figure 6.4 represents the first and second highest occupied molecular orbital (HOMO) and the lowest unoccupied molecular orbital (LUMO) of formaldehyde. The ground state is depicted in the left side of each column of Figure 6.4 and shows the π and nonbonding (n) orbitals populated. Panel (a) in Figure 6.4 shows an $n \rightarrow \pi^*$ transition resulting in the promotion of one of the n-electrons to the π^* orbital. Panel (b) in Figure 6.4 shows a $\pi \rightarrow \pi^*$ transition. The $\pi \rightarrow \pi^*$ transition's energy corresponds to a higher energy photon, a photon with a λ_{max} below 195 nm, which is outside the UV-vis region and beyond the range of most laboratory instruments.[5]

> **Problem 6.4:** Using Figure 6.3 as a guide, construct an MO diagram for ethylene (ethene). How is the MO diagram similar to the MO diagram of formaldehyde? How is the MO diagram different from the MO diagram of formaldehyde? Using your MO diagram, predict the UV-vis absorption properties of ethylene.

Structure and "Color"

It is important to develop the ability to look at a molecule's structure and make generalized predictions about the UV-vis absorption properties of that molecule and conversely to look at a UV-vis spectrum and make generalized predictions about the structural features of the molecule. We will consider organic molecules first. A cursory examination of the MO diagram in Figure 6.3 allows us to make the generalization that $\sigma \rightarrow \sigma^*$ transitions require significantly more energy than a $\pi \rightarrow \pi^*$ or $n \rightarrow \pi^*$ transition. Saturated hydrocarbons have only $\sigma \rightarrow \sigma^*$ transitions so as a general rule the electronic transitions in saturated hydrocarbons occur in the far-ultraviolet and out of the range of the typical UV-vis spectrometer. We have already seen in our examination of formaldehyde that the existence of a π-bond creates a $\pi \rightarrow \pi^*$ transition with significantly lower energy than a $\sigma \rightarrow \sigma^*$ transition, and you would be correct in making the generalization that *unsaturated hydrocarbons are more likely to have electronic transitions with absorbances that can be measured by a UV-vis spectrophotometer*. In other words, organic substances with one or more unsaturated bonds (double bonds or triple bonds) are more likely to be colored in a spectroscopic sense. Organic functional groups with lone pairs and double bonds are chromophores, as we saw with formaldehyde.

Pi-conjugation also plays an important role in the color of a molecule. When a molecule is conjugated, you can view the π-system as one large orbital that has been delocalized over many atom centers. An electron that has been promoted into the π^* orbital can correctly be thought of as a *particle in a box* (see Problem 2.10 and Exercise 2.10

Recall

A chromophore that absorbs red photons will make the material appear blue, and conversely a chromophore that absorbs a blue photon will make the material appear red.

---

[4] A term coined by the German chemist O. N. Witt (1876).

[5] Turro, N. J. *Modern Molecular Photochemistry.* University Science Books: Sausalito, CA, **1991**.

from Chapter 2 for a review of *particle in a box* energy states). Quantum mechanics tells us that the energy levels for a particle in a one dimensional box can be expressed as:

$$E_n = \frac{h^2 n^2}{8ma^2}$$

Eq. 6.1

where

h= Planck's constant

n = positive whole number and represents the quantum state

m = mass

a = length of the "box"

The important thing to recognize about Equation 6.1 is the fact that the length of the box "*a*" is in the denominator. And the "box" in this case is the conjugated π-system. As the degree of conjugation increases, the length of the box gets larger and the energy gap π → π* gets smaller. As the π → π* gap gets smaller, the photon absorbed moves from the ultraviolet (higher energy) and into the visible range (lower energy) of the EMS. In other words, as the degree of conjugation increases, we see λ_{max} shift to longer wavelengths (known as a *red shift*). The spectroscopic term for a shift to longer wavelengths is called a *bathochromic shift*.

As an example of the particle in a box concept, Figure 6.5 shows the UV-vis spectrums of anthracene and tetracene. As you can clearly see, λ_{max} for each absorption shifts to longer wavelengths (red shifts) as the length of the conjugated π-system increases. You can actually see the effect of π-conjugation with your eyes. Naphthalene, the smallest of the structures, is colorless, implying the wavelength of the absorbed photon occurs outside the visible range, in this case in the ultraviolet. This is the result of the relatively small π-conjugated system "the box" and therefore requires a high energy for the absorption. Slightly larger, anthracene is pale yellow in color, implying an absorption in the near-ultraviolet that reaches slightly into the visible spectrum. Tetracene, a bit longer, is orange in color; in this case, the absorption is fully into the visible "blue" end of the EMS. Finally, the largest structure, pentacene, is blue in color, because the absorption is much lower in energy and occurs at the "red" end of the visible spectrum.[6]

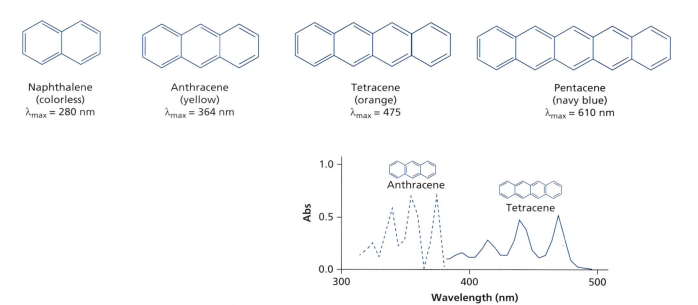

Naphthalene	Anthracene	Tetracene	Pentacene
(colorless)	(yellow)	(orange)	(navy blue)
λ_{max} = 280 nm	λ_{max} = 364 nm	λ_{max} = 475	λ_{max} = 610 nm

Figure 6.5 UV-vis spectra of anthracene and tetracene.

[6] λ_{max} values are for the lowest energy transition.

Heteroatoms

The existence of heteroatoms (atoms other than carbon and hydrogen) in organic molecules can also have a profound effect on the UV-vis spectrum, especially if they bring lone pairs into the system. Refer back to the MO diagram in Figure 6.3. The MO diagram for formaldehyde actually shows the HOMO to be one of the nonbonding "*n*" electrons. And in fact the lowest energy transition for formaldehyde is $n \rightarrow \pi^*$. The influence of lone pair electrons on the absorption properties of a material is highly dependent on the overall structure of the molecule. In the case of formaldehyde, the lone pair electrons are on the oxygen atom and the oxygen atom is also part of the π-system. *This is a key point!* The intensity of an $n \rightarrow \pi^*$ transition is related to how much spatial overlap exists between the n-orbitals and the π^*-orbitals (see Figure 6.6). If the lone pair orbitals have significant spatial overlap with the π^* orbitals, then you can expect to see a relatively intense absorption. Conversely, if the lone pair orbitals have little or no overlap with the π^* orbitals, then you would expect to see a very weak absorption or none at all. This is a direct application of the *Franck-Condon principle*, which states that electronic transitions are so fast compared to molecular vibrations that the nuclear positions remain unchanged during the transition. In order for the transition to occur, the ground state and the excited state must be connected or more precisely *the intensity of a vibronic transition is proportional to the square of the overlap integral between the vibrational wave functions of the two states that are involved in the transition.*[7] Structurally, a lone pair–bearing heteroatom must be physically bound to the π-system in order to have a significant contribution to the UV-vis spectrum. As a general rule, $n \rightarrow \pi^*$ transition will be lower in energy than $\pi \rightarrow \pi^*$, because the nonbonding state will be between the π and π^* states, thus splitting the energy gap between π and π^* (see Figure 6.3). But the $\pi \rightarrow \pi^*$ transition will be more intense because the two states share a greater spatial overlap.

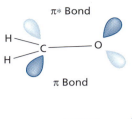

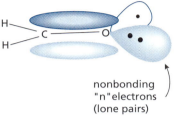

Figure 6.6 A look at the spatial orientation of the π, π^*, and n-orbitals of formaldehyde.

Problem 6.5: Benzophenone has a structure that is very similar to DPK. Using our case study of DPK as a guide, discuss in general terms how the UV-vis spectrum of benzophenone will be similar to that DPK and how will it be different. Then sketch your prediction of the UV-vis spectra of benzophenone. (See DPK: A Case Study, p. 170.)

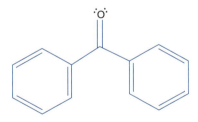

Vibronic Transitions

The peaks in the absorption spectrum of DPK in Figure 6.7 represent electronic transitions within the DPK molecule as a result of absorbed electromagnetic radiation. You might assume that electronic transitions should occur at a discrete wavelength—and that assumption is essentially correct for electronic excitation for an *atomic* species. However in a molecule, the observed transitions are occurring with bonding electrons ($\sigma \rightarrow \sigma^*$, $\pi \rightarrow \pi^*$, $d \rightarrow d^*$, and/or $n \rightarrow \pi^*$). These are all transitions between bonding states, and molecules *vibrate*, causing changes in bonding orbital energies. You also have vibrational states from the bonds superimposed on the electronic states. In addition, you have rotational states superimposed on the vibrational states (see Figures 2.11 and 2.12). For any given molecule, the electronic transition can begin from any number of

---

[7] IUPAC Compendium of Chemical Terminology, 2nd ed., **1997**.

DPK: A Case Study

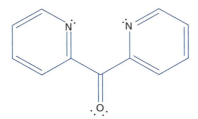

It is important for a spectroscopist to be able to correlate features of the UV-vis spectrum with structural features of the molecule under study. Figure 6.7 shows representative UV-vis spectra of di-2-pyridyl ketone (DPK) at two concentrations in dry acetonitrile. Two strong bands are observed at 238 and 270 nm, and a third very weak band is observed at 356 nm. What are the structural features of DPK that are of interest to a spectroscopist? First, we should notice the two pyridyl rings. The pyridyl rings contain a conjugated (aromatic) π-system with a lone pair of electrons on the nitrogen atom. An MO diagram of the pyridyl rings yields results similar to those seen in Figure 6.4 in which the two lowest energy transitions are n $\rightarrow$ π^* and π $\rightarrow$ π^*. We should expect to see an n $\rightarrow$ π^* and a π $\rightarrow$ π^* transition from the pyridyl chromophore. We should also expect to see an n $\rightarrow$ π^* and a π $\rightarrow$ π^* transition arising from the carbonyl π-system in a manner exactly analogous to that seen in formaldehyde.

Once we understand the number and types of chromophores, we can look at their expected relative energies and intensities. The pyridyl π-system is conjugated; thus, we should expect the energy of the π $\rightarrow$ π^* transition of the pyridyl chromophore to be lower than the energy of the π $\rightarrow$ π^* transition from the carbonyl chromophore (particle in a box argument) because the conjugated pyridyl system is much larger than the carbonyl. Likewise, because the π $\rightarrow$ π^* gap is smaller (lower in energy) for the pyridyl chromophore than it is for the carbonyl chromophore, we should also expect the n $\rightarrow$ π^* transition in the pyridyl chromophore to be lower in energy than the n $\rightarrow$ π^* transition in the carbonyl chromophore.

Now that we can place the relative energies of the possible transitions, we also want to be able to predict the relative

intensities of the possible transitions. If we consider the spatial overlap of the orbitals involved in these transitions, we can also make informed predictions about the relative intensities of the different transitions. Since the orbitals involved in the π $\rightarrow$ π^* transition share a high degree of spatial overlap, compared to the orbitals involved in the n $\rightarrow$ π^* transition, we can expect the π $\rightarrow$ π^* transition to be more intense. As discussed earlier, this is an example of the Franck-Condon Principle.

Based on what we have discussed so far, *we can assign the observed peaks in Figure 6.7.* Considering relative intensities, we can make a preliminary assignment of the two peaks at ~238 and ~270 to π $\rightarrow$ π^* transitions and the one at ~356 to an n $\rightarrow$ π^* transition. The peak at ~238 nm is higher in energy than the one at ~270, so using our particle in a box argument, we can assign this π $\rightarrow$ π^* transition to the carbonyl chromophore, and the peak at ~270 nm to the π $\rightarrow$ π^* transition from the pyridyl chromophore. Because we would expect an n $\rightarrow$ π^* transition to be lower in energy, we can once again assign the peak at ~354 nm to an n $\rightarrow$ π^* transition from the pyridyl chromophore.

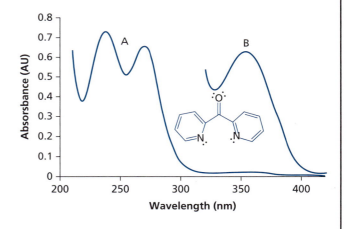

Figure 6.7 UV-vis spectrum of DPK in dry acetonitrile. The concentration of DPK is 0.0828 mM (A) and 4.17 mM. From K. D. Sienerth, R. M. Granger, et al.; *Inorg. Chem.* **2004**, *43*, 72–78.

the superimposed vibrational and rotational states, the probability of which is governed by a Boltzmann distribution of states (see Equation 2.18). Likewise, the electronic excited state also has superimposed vibrational and rotational states and the transition can terminate in any one of the superimposed states (see Figure 2.11). The result for *molecular* UV-vis spectroscopy is relatively broad peaks compared to *atomic* UV-vis spectroscopy. Because of all these overlapping transitions, when we report the wavelength of an absorption event in a *molecule*, it refers to the wavelength of maximum absorption for that peak (λ_{max}). Electronic transitions within *molecules* are called *vibronic transitions,* which is a portmanteau made from *vibr*ational and electr*onic*. Because a single atom does not have a bond and therefore no bond vibrations, we see very narrow

TABLE 6.1: Absorption Maxima (λ_{max}) for DPK in Various Solvents

SOLVENT	POLARITY INDEX*	λ_{max} (nm)		
		$\pi \to \pi^*$	$\pi \to \pi^*$	$n \to \pi^*$
Cyclohexane	0.2	236	268	358
Methyl *tert*-butyl ether (MTBE)	2.5	236	268	358
Acetonitrile	6.8	238	270	354
Dimethyl sulfoxide (DMSO)[†]	7.2	—	272	354
Water	9.0	242	276	342

* Polarity Index from *The HPLC Solvent Guide*, Wiley Interscience: New York, 2002.
[†] Note that the ultraviolet cutoff for DMSO is 260 nm.

bandwidths associated with the electronic absorption events in *atomic* UV-vis spectroscopy. We will revisit the distinction between an electronic transition and a vibronic transition in Chapter 7 when we explore atomic absorption spectroscopy.

Solvent Polarity

Analytical techniques that probe the analyte in the gas phase or as a pure substances (such as some surface techniques) benefit from a lack of background interference from the solvent. Conversely, in UV-vis spectrophotometry, the analyte is nearly always in solution, and we must consider the ways in which the solvent can perturb the energy levels of the affected orbitals in the chromophore. Recognizing that the geometry of the lone pair electron orbitals protrudes out into the solvent, it seems reasonable to imagine that the lone pair electrons would have a strong interaction with polar solvents. The dipoles of the solvent molecules coordinate about the lone pair electrons and help to lower the energy of the lone pair orbitals. If the *n*-orbital is lowered in energy, then the n $\to \pi^*$ gap becomes larger and the transition between the two states requires more energy and we observe a shift to lower λ_{max} for those absorption peaks when conducted in polar solvents (a *hypsochromic shift*; known as the *"blue shift"*). Similarly, the π^* orbital has a geometry that allows a weak interaction with the solvent molecule (see Figure 6.6), so polar solvents can slightly lower the π^* orbital's energy. If π^* is lower in energy, then the $\pi \to \pi^*$ gap becomes smaller and the transition between the two states requires less energy and we observe a shift to higher λ_{max} for those absorption peaks in polar solvents (a *bathochromic shift*). By taking the UV-vis spectrum in solvents with different polarities and observing the changes in λ_{max} for the various peaks, you can gain insight into the nature of the orbitals involved in the different transitions. As an example of the effects of solvent polarity on absorption spectra, Table 6.1 shows the λ_{max} values for DPK in various solvents.[8] The bands at ~238 nm and ~270 nm both show a bathochromic (red) shift in polar solvents, consistent with a $\pi \to \pi^*$ transition, and the peak at ~354 nm showed a *hypsochromic (blue) shift* in polar solvents, consistent with an n $\to \pi^*$ transition.

Transition Metal Coordination Compounds

Many transition metal complexes are colored. From a spectroscopic perspective, this means that transition metal complexes absorb radiation in the 400 to 700 nm or visible range. Unlike the absorptions of purely organic compounds ($\sigma \to \sigma^*$, $\pi \to \pi^*$, or $n \to \pi^*$), the color observed in the transition metal complexes are typically the result of d $\to$ d* excitations (see Figure 6.8). However, because the ligands themselves are often organic in nature, it is important to realize that ligand-based $\sigma \to \sigma^*$, $\pi \to \pi^*$, or $n \to \pi^*$

[8] Sienerth, K. D.; Granger, R. M., et al.; *Inorg. Chem.* **2004**, *43*, 72–78.

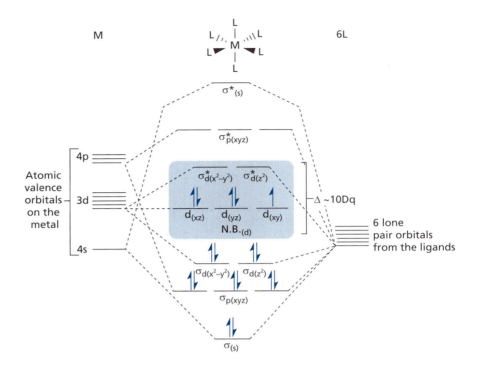

Figure 6.8 A representative MO diagram of a generic d^5 ML_6 compound. The shaded blue area shows the HOMO–LUMO gap. "M" represents a transition metal and "6L" represents the six ligands.

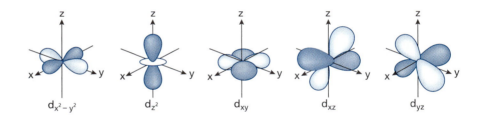

Figure 6.9 Orientation of the d-orbitals. The ligand lone pairs approach the metal center along the *x*, *y*, and *z* coordinates, thus pushing those orbital energies up. The orientation of the d_{xy}, d_{xz}, and d_{yz} orbitals prohibits any overlap with the ligand lone pair orbitals.

excitations can still occur.[9] The bonding of the ligands to the metal causes a split in the d-orbital's energy and creates a low-energy HOMO-LUMO gap that often produces a peak in the visible range of the EMS. Figure 6.8 shows an MO diagram for a generic *homoleptic* octahedrally coordinated transition metal complex[10] ML_6. Because of their orientation in space, the d_{xz}, d_{yz}, and d_{xy} orbitals are incapable of σ-overlap along the Cartesian coordinates (*x-y-z*) and therefore are incapable of forming bonds with the ligands (see Figure 6.9). This results in a HOMO derived from nonbonding d-orbitals and the LUMO derived from d* orbitals, resulting from the σ-overlap of the $d_{x^2-y^2}$ and

---

[9] A ligand is an ion or molecule that forms a *dative bond* with a metal. A dative bond is formed through the donation of one or more of the ligand's lone electron pairs. It is also correct to think of a *ligand* as the Lewis base moiety attached to the transition metal acting as a Lewis acid.

[10] The term *homoleptic* is used to indicate a coordination compound in which all of the ligands are identical.

d_{z^2} orbitals with ligand lone pair orbitals. Figure 6.8 is a very typical MO diagram for a homoleptic ML_6 complex.[11]

The HOMO-LUMO gap is represented by the shaded area in Figure 6.8. The magnitude of the HOMO-LUMO gap is the result of the interactions of the ligand's lone pair electrons with the metal's d-orbitals. An empirically derived ranking of ligand strengths can be found in the *spectroscopic series*. An abbreviated version of the spectroscopic series is presented in the margin.

Ligands on the far left side of the spectroscopic series make the HOMO-LUMO gap larger, and as such, the complex will absorb a more energetic photon (a bluer photon), thus making the transition metal complex look more red in color. Conversely, ligands on the far right side of the spectroscopic series produce smaller HOMO-LUMO gaps, and as such the complex will absorb a less energetic photon (photons nearer the red side of the visible spectrum), thus making the complex look more blue in color. A complete understanding of the absorption properties of transition metal complexes is aided by an understanding of *ligand field theory* and the use of *Tanabe Sugano diagrams*.[12]

> **The Spectroscopic Series**
>
> *CN⁻, CO > NO₂⁻ > dipy, phen >en > HN₃ > py > H₂O > OH⁻ > F⁻ > Cl⁻ > Br⁻ > I⁻*

6.3 Quantitative Measurements

Selection Rules

There are several requirements that need to be met in order for an electronic transition to occur. These requirements are termed *selection rules*. The selection rules for UV-vis spectroscopy are:

1. The *energy selection rule*, which states that the energy of the absorbed photon must match the energy gap of the transition

2. The *spin selection rule*, which states that the electronic transition cannot change the net spin multiplicity of the molecule

3. The *LaPorte selection rule*, which states that for a centrosymmetric molecule, transitions between orbitals of the same parity (symmetry) are forbidden

4. The *Franck-Condon selection rule*, which states that the symmetry of the ground state and excited state must overlap

The first of the selection rules, the requirement that the energy of the photon must match the energy gap of the transition, has already been discussed. Recall the relationship between a photon's energy and its wave properties (see Equation 6.2) that was first introduced in Chapter 2.

$$E = h\nu = \frac{hc}{\lambda} \qquad \textbf{Eq. 6.2}$$

where

E = energy of a photon (joule, J)

h = Planck's constant = 6.626×10^{-34} J·s

ν = frequency (cycles per second, s^{-1}, which is also known as hertz, Hz)

c = the speed of light = 2.99×10^{8} (m/s)

λ = wavelength (m)

---

[11] Although the general shape of this MO diagram is "generic," the number of electrons placed in the MO diagram would be dependent on the metal chosen for the complex.

[12] For more on ligand field theory and Tanabe Sugano diagrams, see Huheey, J. *Inorganic Chemistry: Principles of Structure and Reactivity*; Cotton, F. A. *Advanced Inorganic Chemistry*, 6th ed.; and Douglas, B.; McDaniel, D.; Alexander, J. *Concepts and Models of Inorganic Chemistry*.

Problem 6.6: There are three peaks in the UV-vis spectrum of DPK (Figure 6.7) with λ_{max} values of 238 nm, 270 nm, and 354 nm. Using Equation 6.2,

(a) Calculate the energy gap (in joules) associated with each of these transitions.

(b) Calculate the frequency (in hertz) of the photon absorbed for each of these transitions.

The *spin selection rule* states that the electronic transition cannot change the net spin multiplicity of the molecule. You may recall from quantum mechanics that the total spin multiplicity is:

$$M = 2S + 1 \qquad\qquad \textbf{Eq. 6.3}$$

where
$\quad$ M = total spin multiplicity
$\quad$ S = sum of the electron quantum numbers for the electrons in a molecule ($\pm\frac{1}{2}$)

Although a complete discussion of spin physics would be interesting, it is unnecessary for our purposes in understanding UV-vis spectroscopy. If you are interested in reading more about spin physics, some recommended texts are provided at the end of this chapter. Application of the spin selection rule to UV-vis spectroscopy simply states that in order for an electronic transition to be *allowed*, it cannot change the total number of spin-paired and/or unpaired electrons. Conversely, a spin *forbidden* transition is one in which the transition results in a change in the number of spin-paired and/or unpaired electrons. Spin forbidden transitions are sometimes observed but they are always very weak when compared to spin allowed transitions.

Take a look at Figure 6.10. Panel (A) shows a spin allowed transition as the total number of spin paired electrons remains the same in the ground and excited states. For every $+\frac{1}{2}$ electron, you have a $-\frac{1}{2}$ electron, and the total spin multiplicity for the ground state and for the excited state is:

$$M = 2S + 1 = 2(0) + 1 = 1$$

Systems with a total spin multiplicity of one are called singlet states.

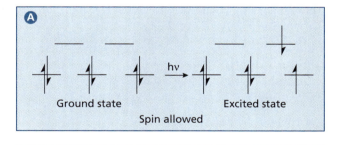

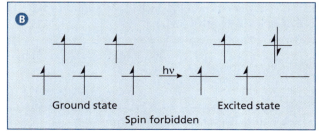

Figure 6.10 The d $\rightarrow$ d* gap for a ML_6 transition metal complex. Panel (A) shows a spin allowed transition. Panel (B) shows a spin forbidden transition.

Figure 6.10, panel (B) shows a spin forbidden transition. In order for the excited state to satisfy the Pauli Exclusion principle, the excited state electron must undergo a spin flip. In doing so, the spin multiplicity changed from:

$$M = 2(M = 2(\tfrac{5}{2})+1=6 \quad \text{to} \quad M = 2(\tfrac{3}{2})+1=4$$

Problem 6.7: Refer back to Figure 6.4. Determine the total spin multiplicity of the ground and excited states for panels (A) and (B). State if the transition is spin allowed or spin forbidden.

The third selection rule is called the *LaPorte rule*, which states that for a *centrosymmetric* molecule, transitions between orbitals of the same parity (symmetry) are forbidden. *Centrosymmetric* simply means that a molecule is symmetric about a central point, in other words it has a center of inversion (see Figure 6.11).

A complete discussion of the LaPorte rule benefits from a complete understanding of molecular symmetry and applications of group theory;[13] however, for our purposes in understanding UV-vis transitions, we can say this: for molecules that are symmetric about a central point, transitions between orbitals of the same type are forbidden. For instance, d-d* transitions are *forbidden* for a centrosymmetric molecule. However, we still observe d-d* transitions, which we will explain in the next paragraph.

In transition metal spectroscopy, the spin selection rule plays a more important role than the LaPorte selection rule in determining if a transition is observed. There are two reasons we see LaPorte forbidden transitions. The first reason is that most molecules do not have a true center of symmetry and even if they do, molecular vibrations will momentarily distort the symmetry and temporarily remove the center of inversion. This transient distortion of the symmetry makes transitions between *d subshells in transition metal complexes "momentarily" allowed*. The second reason can be understood by borrowing a concept from valence bond theory. Valence bond theory is the idea that when atomic orbitals combine to form MOs, they hybridize and form new orbitals. For instance, when a carbon atom forms four σ-bonds with the hydrogen atoms to form methane, we describe the four MOs as sp^3 hybrids. The new orbitals are neither purely *s* nor *p* in nature. The hybrid orbitals have more *p*-character than *s*-character. In a ML_6 transition metal complex, the orbital hybridization about the metal center can best be described as sp^3d^2 hybrid orbitals. In this case, a d → d* transition would be "partially" allowed since it is not purely *d* in character. In practice, applying the LaPorte rule to the interpretation of a UV-vis spectrum helps more in making predictions about how intense the transition will be than in whether or not you will see the transition. LaPorte forbidden transitions or transitions that are *mostly* forbidden are weak, whereas LaPorte allowed transitions are more intense. For instance, hexaaquocobalt(II), $[Co(H_2O)6^{2+}]$, is octahedrally coordinated and centrosymmetric (see Figure 6.11). Its d-d transitions are Laporte "forbidden," and so it exhibits a molar absorptivity of around 5 L/mol·cm. The tetrahedral $CoCl_4^{2-}$ complex, which has no center of symmetry, exhibits a LaPorte-allowed d-d transition with a molar absorptivity of more than 500 L/mol·cm.

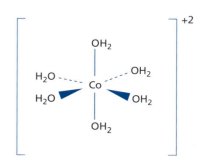

Figure 6.11 An example of a centrosymmetric molecule.

Problem 6.8: Discuss the two transitions seen in Figure 6.4 in terms of the LaPorte selection rule. Make sure you discuss how the LaPorte selection rule would affect peak intensity.

We touched on the *Franck-Condon* selection rule earlier when we were evaluating the UV-vis spectrum of formaldehyde and DPK as we explained why the n → π* transition is much less intense than the π → π* transition. The *Franck-Condon* selection rule is derived from the Franck-Condon principle, which states that an electronic transition occurs on a timescale that is essentially instantaneous compared to the time scale of

---

[13] Tsukerbla, B. S. *Group Theory in Chemistry and Spectroscopy: A Simple Guide to Advanced Usage* and Cotton, F. A. *Chemical Applications of Group Theory*, 3rd ed.

molecular vibrations. As a result, the symmetry of the ground state and excited state must overlap.[14] For the purposes of understanding UV-vis spectra, we can simplify the Franck-Condon principle and say that in order for a transition between two different MOs to occur, they must be symmetrically capable of overlap. The intensity of the transition is proportional to the degree of orbital overlap. (see Figures 6.6 and 6.7).

Problem 6.9: Draw an MO diagram of pyridine and predict the relative energies and relative intensities of the two lowest energy transitions.

Beer's Law

In 1729, the French scientist J. Bouguer published a paper in which he reported that the fraction of light absorbed by a material was directly proportional to the thickness of the material. Unfortunately, Bouguer's work was mostly forgotten until Bouguer's observation was independently rediscovered by the Swiss scientist J. H. Lambert in 1760. In 1852, J. Beer published[15] a much more complete work on the absorption of light, and in his publication he showed that there is a logarithmic relationship between the transmission of light through a material and the distance the light travels through the material. Today we summarize this relationship with Beer's law. It is also called the Beer-Lambert law, and some textbooks even call it the Beer-Lambert-Bouguer law.

At its simplest, a UV-vis spectrophotometer measures the *difference* in radiant power from the source that reaches the detector when there is no sample present (P_0), and the radiant power from the source that reaches the detector as the light passes through a sample (P). The ratio of the two power readings is called the *transmittance (T)*, which can range from 0 to 1 *(or 0 to 100%)*:

$$T = \frac{P}{P_o} \qquad\qquad \textbf{Eq. 6.4}$$

Absorbance is defined as the negative log of the transmittance.

$$A = -\log T = -\log \frac{P}{P_o} = \varepsilon bc \qquad\qquad \textbf{Eq. 6.5}$$

where
    A = absorbance (unitless)
    T = transmittance (unitless)
    P_o = initial radiant power reaching the detector
    P = radiant power reaching the detector after passing through the sample
    ε = molar absorptivity (L/mole·cm)
    b = path length of the sample (cm)
    c = concentration (M)

Beer's law is a shortened version of Equation 6.5.

$$A = \varepsilon bc \qquad\qquad \textbf{Eq. 6.6}$$

where ε has units of $cm^{-1} \cdot M^{-1}$, *b* has units of cm, and *c* has units of molarity (M).[16]

---

[14] For further reading on the Franck-Condon principle, see Turro, N. J. *Modern Molecular Photochemistry*, University Science Books: Sausalito, CA, **1991**.

[15] Beer, J. *Annalen der Physik und Chemie*, **1852**, 86, 78–88.

[16] The *molar* absorptivity, ε ($cm^{-1} \cdot M^{-1}$) has specific units, and in using it we must express concentration in molarity and path length in cm. If we use other concentration or length units, Beer's law is written in the more generic form A = abc, where a is *absorptivity* and has units related to c and b.

Derivation of Beer's Law

Equation 6.5 can be derived from Equation 6.4 as follows. A beam of radiant light passing through an infinitesimally small portion of the sample cell containing an absorbing species with a concentration of "c" will have a path length of db, the loss of radiant power will be –dP. This can be expressed as:

$$-\frac{dP}{P} = kc\,(db)$$

where k = species-specific proportionality constant

If we integrate over the entire path length, the loss in radiant power is:

$$\int_{P_o}^{P} \frac{dP}{P} = -k \int_{o}^{b} c\,db$$

and this gives:

$$ln\left(\frac{P}{P_o}\right) = -kbc$$

Converting natural log (ln) to $\log_{10}$ gives:

$$log\left(\frac{P}{P_o}\right) = -2.303kbc$$

The product of 2.303 and k gives a different constant, which we can name "ε." If we choose centimeters for our path length and molarity for our concentration. then:

$$log\left(\frac{P}{P_o}\right) = -A = -\varepsilon bc$$

and multiplying through by –1 gives us Equation 6.5.

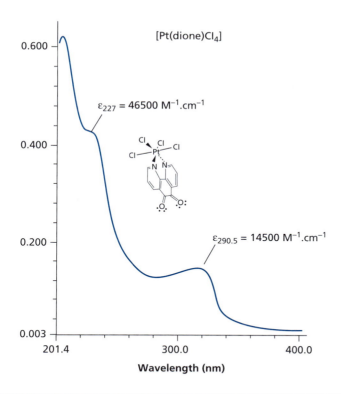

Figure 6.12 UV-vis spectrum of 1-10-phenanthroline-5,6-dione platinum(IV) chloride [Pt(dione)Cl$_4$] in dry acetonitrile.

Modern spectrophotometers can do a mathematical conversion of transmittance to absorbance. As the spectroscopist, you control the output signal (in percent transmittance or absorbance) by a flip of a switch (toggle or other instrument panel device).

The *molar absorptivity* (ε) is the proportionality constant that relates absorbance to concentration. The molar absorptivity is wavelength specific and ε must be reported for each peak in a spectrum. For example, Figure 6.12 shows the UV-vis spectrum of 1,10-phenanthroline-5,6-dione platinum(IV) chloride in dry acetonitrile.[17] The figure also shows the molar absorptivity at λ_{max} = 290.5 nm and λ_{max} = 227 nm.

Example 6.2

Use the data presented in Figure 6.12 to determine the concentration of [Pt(dione)Cl$_4$] in the test sample.

STRATEGY –
First solve Equation 6.6 for "c" and substitute in values for A, b, and $\varepsilon_{290.6}$. Check units!

SOLUTION –

$$C = \frac{A}{b\varepsilon_{290.5}} = \frac{0.145}{(1\,cm)(14500)} = 10\,\mu M$$

[17] Granger, R. M.; Davies, R.; White, P. S.; et al. *JUCR* **2005**, *2*, 47.

Problem 6.10:   Using Figure 6.12, determine ε_{205} for $[Pt(dione)Cl_4]$ (see Example 6.2 for additional data).

Problem 6.11:   What would the concentration of the solution in Figure 6.12 be if you obtained an absorbance of 0.05 at $\lambda_{max} = 290.5$ nm?

Deviations from Beer's Law

According to Equation 6.6, a plot of absorbance (A) versus concentration (c) should give a straight line with a slope of εb. In practice, the slope of A versus c is usually linear if the total absorbance is kept below 1. When the absorbance exceeds 1, the slope of the line often curves downward and flattens out. Remember that absorbance is $-\log(T)$ so at an absorbance of 1, only 10% of the radiant power is reaching the detector. For this reason, it is considered a *best practice* to keep absorbance values below 1 when conducting quantitative analysis. For quantitative analysis, we need to be sure that we are within the linear part of the Beer's law curve, the black line in Figure 6.13.

There are many reasons for a Beer's law plot to have nonlinear regions, and we will explore several of the more common reasons here. However it is still possible to conduct quantitative UV-vis analysis with nonlinear systems. All that is required is that you first construct a calibration curve using samples of known concentration. Below is a discussion of three common causes of departures from Beer's law.

Internal Screening   It is assumed that each absorbing species is "visible" to the source and detector. However, at higher concentrations, it is possible for an absorbing species to be in the shadow of another absorbing species. Figure 6.14 depicts a representation of *internal screening*. The blue wedges represent the "shadow" cast by each absorbing species.[18] In Figure 6.14, the concentration has become so high that some

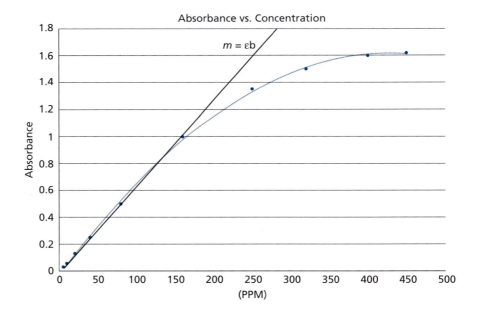

Figure 6.13 A representation of the linearity of Beer's law as a function of concentration. Beer's law is generally linear below one absorbance. Above one absorbance, the slope of the Abs. vs. Conc. line usually bends towards zero.

---

[18] A "shadow" is an imprecise analogy. An absorbing species does not "block" light in a traditional sense.

of the absorbing species lie within the "shadow" of other absorbing species and are therefore hidden from the radiant source. This scenario would lead to a negative deviation from Beer's law (i.e., the calibration plot would curve "down" as you increased concentration).

Changes in Refractive Index Another reason that you can get deviations from Beer's law stems from the assumption that the refractive index of the solution and the molar absorptivity of the solute are constant over the entire concentration range of the analysis. Usually P_o is measured with a sample cell containing only the solvent. Beer's law assumes that changes in transmittance are due to changes in the concentration of analyte species only. This is usually a good assumption; however, the assumption fails as the concentration of the solute is increased for solutes that have a strong interaction with the solvent. In these situations, the refractive index of the solution changes appreciably as the concentration of the solute is varied. For example, one might anticipate a nonlinear A versus c response if the solute is capable of hydrogen bonding and one is using water as the solvent. Likewise, if the analyte has strong intermolecular forces of attraction for itself, or if there are other chemical species in solution that can polarize the orbitals of your solute (such as a salt), then it is possible that ε will vary with concentration. It is generally accepted that below 0.01 M, the intermolecular distance between solute molecules is sufficient to preclude any solute–solute interactions. For this reason, it is considered a *best practice* to keep analyte concentrations below 0.01 M when conducting quantitative analysis. Likewise, it is important that the ionic strength of the blank be similar to that of the sample.

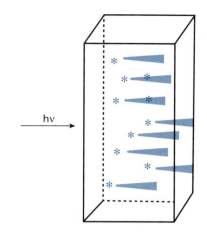

Figure 6.14 A representation of internal screening.

Bandwidth Resolution Insufficient bandwidth resolution of the monochromator can also lead to deviations in Beer's law.[19] For this discussion, let us simplify the concept of a bandwidth and just consider two closely spaced wavelengths of light. If a monochromator is passing two different wavelengths of light (λ and λ') through a sample and each wavelength of light has a different molar absorptivity (ε_λ and ε_λ') at the two wavelengths, the total absorption of light through the sample cell will be:

$$A_{total} = -\log\frac{P}{P_o} - \log\frac{P'}{P_o'} = \varepsilon_\lambda bc + \varepsilon_{\lambda'} bc \qquad \textbf{Eq. 6.7}$$

Solving for P and P′ and substitution gives us Equation 6.8.

$$P = P_o 10^{-\varepsilon_\lambda bc}$$

$$P' = P_o' 10^{-\varepsilon_{\lambda'} bc}$$

$$A_{total} = \log\frac{(P_o + P_o')}{(P_o 10^{-\varepsilon_\lambda bc} + P_o' 10^{-\varepsilon_{\lambda'} bc})} \qquad \textbf{Eq. 6.8}$$

If you program Equation 6.8 into a spreadsheet and plot A versus c, you obtain a straight line only if $\varepsilon_\lambda = \varepsilon_\lambda'$. All monochromators have a finite bandwidth, and therefore it is impossible to avoid this issue. To minimize the error in quantitative

---

[19] Christian, O'Reilly. *Instrumental Analysis,* 2nd ed. Allyn and Bacon: Boston, **1986**; Skoog, Holler, Crouch. *Principles of Instrumental Analysis,* 6th ed. Thomson, Brooks/Cole: Belmont, CA, **2007**. http://www.chem.uoa.gr/applets/appletbeerlaw/appl_beer2.html (accessed June 2015).

measurements, one should select a region of the spectrum where the $\Delta\varepsilon$ versus λ is lowest. For instance, consider Figure 6.15.[20] The λ_{max} in this spectrum is at 524 nm. Let us assume we have a very low quality monochromator with a bandwidth resolution of 6 nm.

If we take our absorption reading at λ_{max}, and we have a precision of $\pm$ 3 nm, we will introduce negligible error in our absorption value. In other words, at λ_{max}, the top of the peak is relatively flat and $\varepsilon_\lambda \approx \varepsilon_\lambda{}'$. However, if we took our absorption reading at 560 nm instead of 524 nm, the $\pm$ 3 nm precision could lead to a $\Delta A = (A_{max} - A_{min}) = (4.5 \times 10^{-3} - 2.3 \times 10^{-3}) = 2.2 \times 10^{-3}$ A, which represents an error of approximately $\pm$ 16% in your reading. In this region of the peak, $\varepsilon_\lambda \neq \varepsilon_\lambda{}'$ and Beer's law will not be linear. In addition, if the instrumental technique has a small systematic error, it will be amplified by taking the absorption reading in a region of the peak where the ε values are changing rapidly. The take-home message is this: it is considered a *best practice* to take absorption readings at the λ_{max} value when conducting quantitative measurements.

Problem 6.12: Explain why it is considered a *best practice* to keep the total absorbance below 1 when conducting quantitative work.

Problem 6.13: Explain why it is important that the ionic strength of your blank be similar to the ionic strength of your samples, even if the salt does not absorb light at the frequency of interest.

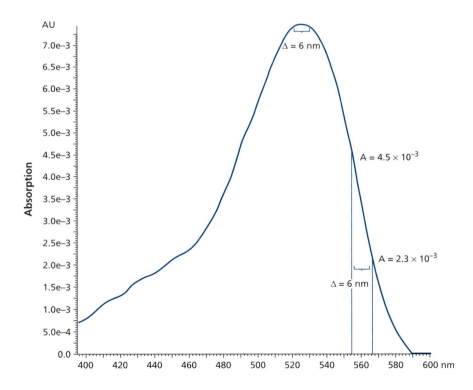

Figure 6.15 UV-vis spectrum of mavidin 3-O-glucoside. At λ_{max} a bandwidth of $\Delta\lambda$ = 6nm gives nearly the same absorption value (the top of the peak is almost flat) but at 560 nm a bandwidth of $\Delta\lambda$ = 6nm gives us a range of absorption values of $\Delta A = (4.5 \times 10^{-3} - 2.3 \times 10^{-3}) = 2.2 \times 10^{-3}$ Abs.

[20] Modified from http://commons.wikimedia.org/wiki/File:Spectre_UV_vis_malvidine3Oglc.PNG.

Absorbance vs. Concentration

Explore the effects on the relationship between total A versus c by manipulating the variables of molar absorptivity (ε_λ and ε_λ') and initial radiant power (P_o and P_o'). Construct an Excel spreadsheet with the following column headings and vary the concentration from 0.003 to 0.05 M. The values you input for P_o and P_o' are arbitrary because the unit is unspecified. Initially try using a value of 1 for each. Input initial values of ε_λ and ε_λ'. Once your spreadsheet is complete, plot A versus c and change the input variables to observe the effect on relationship of A versus c.

P_o	P_o'	ε_λ	ε_λ'	b	c
				1	0.003
					0.005
					0.007
					etc.

$$A_{total} = \log \frac{(P_o + P_o')}{(P_o 10^{-\varepsilon_\lambda bc} + P_o' 10^{-\varepsilon_\lambda' bc})}$$

6.4 Instrumentation Designs

There are three general UV-vis spectrophotometer designs. These are the fixed spectrophotometer, the scanning spectrophotometer, and the array spectrophotometer. In most cases, the fixed spectrophotometer is a low-cost, specific-use instrument and is more correctly described as a visible spectrophotometer. The source and the optics used in many low-end spectrophotometers preclude the use of the instrument for analysis in the ultraviolet range. The fixed spectrophotometer has a single light source and a single sample holder, and it monitors only one wavelength at a time. Computer-controlled scanning is not available. The scanning spectrophotometer is similar in design to that of the fixed spectrophotometer, except that the instrument can be instructed to scan through a range of wavelengths while measuring the absorption throughout the scan range. Scanning spectrophotometers can be further categorized as single beam or dual beam. We will discuss this distinction shortly. The third category is the array spectrophotometer. Even though the output data obtained from an array spectrophotometer resembles the output data obtained from a scanning spectrophotometer, the array spectrophotometer is conceptually different from the previous two designs in that it *does not scan*; instead, it measures the absorbance of multiple wavelengths of light simultaneously by dispersing the emergent light as a function of wavelength and incorporating multiple detectors, one for each bandwidth of light measured.

Fixed-Wavelength Spectrometers

The most pervasive fixed spectrophotometer on the market, which has since been replaced, was the Spectronic 20 (called "Spec 20" for short). The Spec 20 was first designed and marketed by Bausch and Lomb in 1953 (see Figure 6.16). More than 600,000 of these photometers were sold over the years, and during their long run in the marketplace, the analog display was eventually replaced by a digital display and the analog control knobs were eventually replaced by a soft touch pad. However, for more than 60 years the basic design remained the same.

With the increased sophistication of electronic devices and the decrease in the cost of computerization, even the low-end photometer market that was dominated by the Spec 20 in years past now contains spectrophotometers capable of scanning and storing entire spectra. For instance, in 2010, Thermo-Fisher Scientific announced that its Spectronic 20 would be phased out and replaced by the new Spectronic 200 (or a "Spec 200"), a rugged spectrophotometer suitable for a classroom environment yet still capable of scanning entire spectra and with USB ports for easy interface with a computer. In 2014, one could buy a Spectronic 200 for less than $1,500. As the price of low-end spectrophotometers

The HACH DR3900 is an example of a single-analysis spectrometer. It is also a single-beam, fixed-wavelength spectrophotometer. It is primarily used for water quality analysis. The spectrometer comes preprogrammed for specific field kits sold by HACH. The user simply follows the instructions on the field kit, inputs the name of the analysis being conducted, places the sample into the spectrometer, and the desired data is displayed. The system is designed to be very user friendly and requires a minimum of technical training to operate. The HACH DR3900 can be used to measure turbidity, pH, salinity, dissolved oxygen, total nitrogen, total phosphorus, and more. ■

Figure 6.16 A photograph of a basic Spec 20 spectrophotometer.

has dropped, the fixed photometer's utility in the laboratory has become more and more limited. However, there are a few areas where the fixed spectrometer still has a vital role to play, especially in conjunction with field kits and water quality kits.

Although the Spec 20 is yesterday's photometer, the basic internal design of a UV-vis spectrophotometer has not changed radically, and so it is still worth examining. Figure 6.17 shows an internal schematic of a Spec 20, and Figure 6.18 shows photographs of the internal components of a Spec 20. Figure 6.18 is a series of photographs taken of the

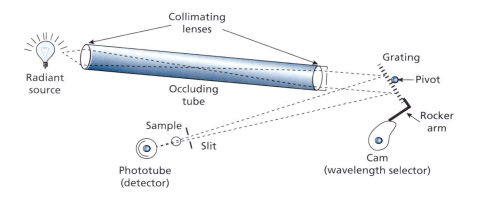

Figure 6.17 Schematic of a basic Spec 20 type spectrophotometer.

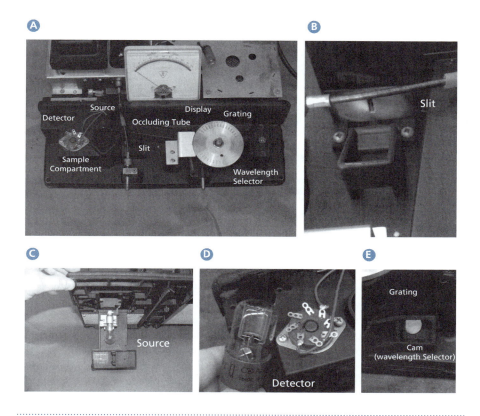

Figure 6.18 A look inside a Spec 20-type spectrophotometer.

inside of a Bausch and Lomb Spec 20 purchased in 1970. The source is a simple tungsten halogen bulb, not unlike the 5 W bulb used to power a child's nightlight. The light emitted from the source is collimated[21] using two lenses. Part of the collimating process includes an occlusion tube, which serves to absorb any stray light or noncollinear light that is present after the collimating lenses. The collimated light then passes to a grating and is dispersed as a function of wavelength.[22] The grating sits on a spring loaded pivot and its angle to the sample slit is governed by a rocker arm attached to a cam. The cam

---

[21] See Chapter 3, Section 3.2, for a review of collimating optics.

[22] See Chapter 3, Section 3.3, for a review of grating monochromators.

and rocker arm can be seen in the bottom right photograph of Figure 6.18. Turning the cam alters the relative angle the grating makes to the sample slit. You select the wavelength of light that reaches the sample slit by changing the angle at which the incident collimated light impinges onto the grating. The light then passes through the sample to a detector. Light striking the phototube creates a current and as the current passes through a resistor, it produces a voltage. Current (or voltage) is the actual *signal* that is measured by the spectrophotometer. Notice the display is a simple voltmeter. Although the simple volt meter seems antiquated, the data it provides is just as reliable as the modern digital display (and perhaps less prone to failure).

Scanning Spectrometers

The basic design of scanning spectrometers is similar to that of the Spec 20-type fixed photometer. There are some important differences—namely the source, the quality of the optical components, the sophistication of the detector, and the incorporation of computer controlled monochromators (wavelength selection). Scanning spectrophotometers fall into one of two categories: single beam and dual beam.

Single-Beam Scanning Spectrophotometers   The single-beam UV-vis spectrophotometer is conceptually identical to the Spec 20. The fundamental difference is the automation that is incorporated into the instrument. In a scanning spectrophotometer, the computer controls the wavelength selection and stores the power reading at the detector as a function of wavelength. The user places a blank into the spectrophotometer and instructs the computer to take a scan between two designated wavelengths. The computer then stores a P_o value for each wavelength in the scan range. The user then replaces the blank with the actual sample and runs the scan again, this time the computer stores P values for each wavelength. Because $A = -log\dfrac{P}{P_o}$, the computer can construct a plot of absorbance versus wavelength as the output data. Figure 6.19 is a schematic representation of a single-beam scanning UV-vis spectrophotometer. Compare Figure 6.19 to Figure 6.1. Although the Spec 20-type photometer and the single-beam spectrophotometer are conceptually similar, the single-beam scanning spectrophotometer is much more sophisticated than the fixed wavelength photometer.

For starters, the source has to be able to provide radiation that spans the range of 200 to 900 nm. We will discuss source options in the next section of this chapter. Not only does the source need to have a larger spectral range, it must be very stable (providing a constant power at each wavelength) for high-precision analysis. For this reason, the power supply that feeds the source is feedback regulated.[23]

Scanning UV-vis spectrometers also need superior optics in order to transmit ultraviolet light in addition to visible light. Recall from Chapter 3 that quartz is the material of choice for the ultraviolet spectral range. Lastly, the detector in most spectrophotometers is more sensitive than the phototube used in a Spec 20-type photometer. We will discuss detector options in Section 6.7; however the most common detector in use is very similar to the phototube. It consists of a multistage phototube and is called a photomultiplier tube (PMT).

Dual-Beam Scanning Spectrophotometers   Figure 6.20 shows a schematic of a dual-beam spectrophotometer. The dual-beam spectrophotometer takes the radiant energy emanating from the monochromator and splits it into two beams. One beam is passed through the sample cell and the other is passed through a reference cell containing solvent only or a suitable blank. In this way, the difference between P and P_o is measured directly in real time and absorbance is computed as the –log of the ratio of the power output from the two detectors.

Before you can perform an analysis with a dual-beam spectrophotometer, you must first "zero" the instrument. This entails making certain that the current output

---

[23] See Chapter 21 for a discussion of *voltage followers* and *power supplies*.

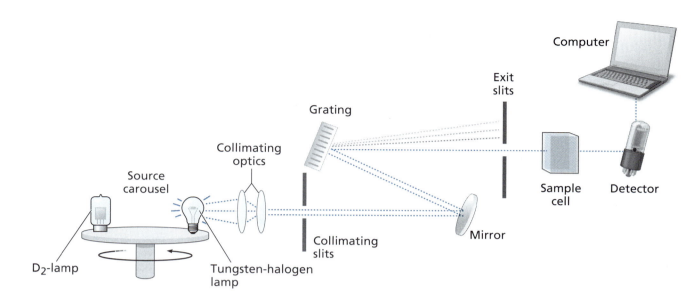

Figure 6.19 Schematic representation of a single-beam scanning UV-vis spectrometer. The source carousel is used to select between a tungsten–halogen lamp (visible) and a D_2 lamp (ultraviolet).

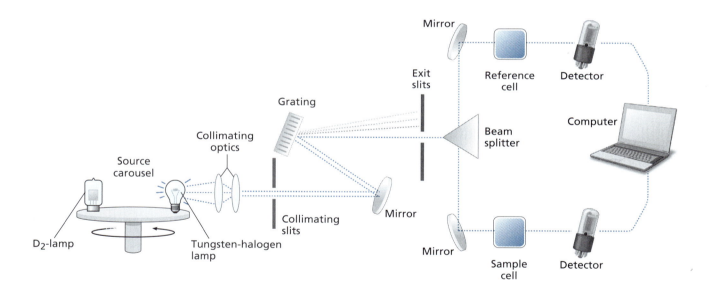

Figure 6.20 A schematic of a dual-beam spectrophotometer. From the source to the beam splitter, the dual-beam spectrometer is conceptually identical to the single-beam spectrometer.

of the two detectors is the same under the same conditions. This is accomplished by first placing a blank into *both* the sample and reference cells. Then the rheostats[24] controlling the output voltage from the detectors are adjusted until the two detectors have the same current output. The dual-beam spectrophotometer is obviously more complex than the single-beam spectrophotometer. You have additional optics, a second detector, the additional electronics associated with the second detector, the electronics needed to balance the output between the two detectors, and the additional

---

[24] A crude rheostat is the dimmer switch used to control the brightness of a light.

electronics needed to process the signal from the second detector before it can be read by the computer. So why would you want such a complicated spectrophotometer?

The dual-beam spectrophotometer is much less susceptible to $1/f$ noise (drift).[25] Because both the sample and the reference are simultaneously subjected to any fluctuations in the output from the source, corrections to $1/f$ noise are made in real time. In addition, feedback regulation of the power supply for the source is not as critical in a dual-beam spectrophotometer as it is in a single-beam spectrophotometer.

COMPARE AND CONTRAST

Single- and Dual-Beam

Either a single- or dual-beam spectrophotometer will provide satisfactory data if you are conducting a simple quick scan. The single-beam spectrophotometer has the advantage of cost and it has a smaller footprint, an important feature when laboratory space is at a premium. However, $1/f$ noise (drift) becomes a serious concern if you are conducting numerous scans against the same blank over a period of time. Depending on the stability of your instrument, you might find that you need to rescan the blank every 10 minutes or so in order to compensate for $1/f$ noise. In this scenario, the single-beam spectrophotometer becomes a much more time-consuming and labor-intensive way of obtaining your data. The dual-beam spectrophotometer is especially suited for monitoring the kinetics of a reaction, especially if the reaction takes several minutes or more to run its course. Figure 6.21 shows the UV-vis kinetic study involving the growth of gold nanoparticles using a natural leaf extract as the growth template.* The reaction was monitored by observing the absorption at 560 nm over a period of 18 hours. A dual-beam instrument is much more suited than a single-beam instrument for an experiment such as this. The operator would simply need to leave the reaction solution in the sample cell and a solvent blank in the reference cell and take a scan every 3 hours. In a single-beam instrument, the operator would need to rerun the blank for each time interval scanned.

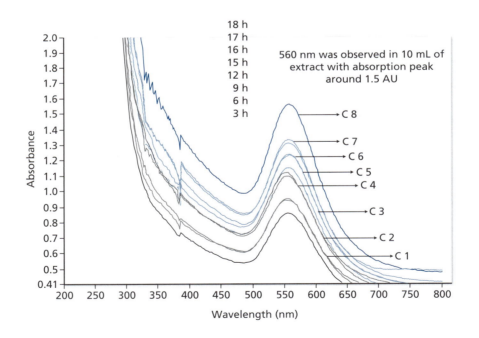

Figure 6.21 A kinetic study of the growth of gold nanoparticles.

* Elavazhagan, T.; Arunachalam, K. D. *Inter. J. Nanomed.* **2011**, 6, 1265–1278.

---

[25] See Chapter 5 for a review of noise sources.

Array Spectrophotometers

The array spectrophotometer is fundamentally different from the single- or double-beam scanning spectrophotometers. The primary difference is that the array spectrophotometer *does not scan* through a range of wavelengths. Instead, the array spectrophotometer sends the entire spectrum of radiation from the source through the sample instead of using a monochromator to select a narrow bandwidth to send through the sample. The light is then dispersed as a function of wavelength and the individual bandwidths of light are guided to a separate detector. Array spectrophotometers have existed for many years. However, the need for multiple detector components made them cost prohibitive for most applications. Fortunately, with the development of compact, robust, and inexpensive light-sensitive diodes, the cost of constructing an array detector has dropped dramatically, and the *diode array spectrophotometer* now dominates the market. The diode array consists of a series of light sensitive diodes[26] that serve as the detectors.

The technology is similar to that of a solar panel but many thousands of parallel diodes can be placed on a single "chip" that occupies a few square centimeters. Each diode monitors a narrow bandwidth of light and the output from each diode is recorded separately. Figure 6.22 represents a schematic of a diode array spectrophotometer and Figure 6.23 is a photograph of a diode array chip. In addition to diode array detectors, other common array spectrometers make use of charge coupled device (CCD) detectors and charge injection devices. Schematically, these spectrophotometers are the same with the exception of the detector type used. We will look at the different detectors in detail in Section 6.7. The primary advantage of the array spectrophotometer is speed. Quite literally, a spectrum can be obtained in the time it takes to open and close a shutter. To run an analysis, a zero light power reading is stored in the computer's memory. This is the "dark" current emitted by the diodes in the absence of any light. The analyst then places into the spectrometer a blank containing only the solvent, the shutter is opened and the power at each diode is measured, then the computer subtracts the dark current from the blank current and stores that as (P_o) at each diode. The analyst then replaces the blank with the sample cell and takes a (P) reading in the same manner. As already mentioned, the main advantages of the diode array spectrophotometers are speed of analysis and the compact size of the instrument. However, it should be noted that the diode array

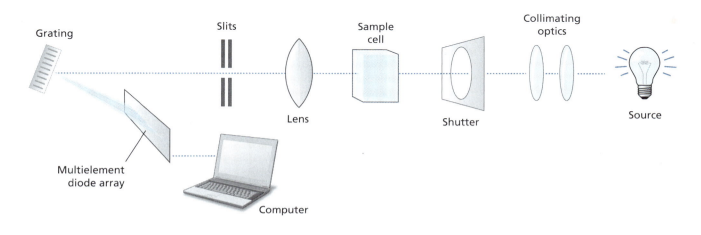

Figure 6.22 A schematic representation of a diode array spectrophotometer.

---

[26] See Photovoltaic Cells later in this chapter and Chapter 21 for more on diode detectors.

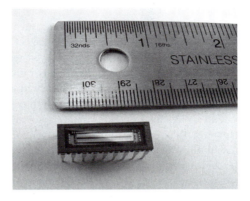

Figure 6.23 Photograph of a multielement diode array detector "chip."

instrument has significantly greater computational requirements. Each channel of the diode array detector is monitored separately, and diode array detectors routinely have 1,500 channels or more. Only in the past few decades have the price of array detectors and the price of the computing power needed to monitor the array detector allowed array spectrophotometers to become comparable in cost to that of the scanning spectrophotometers.

6.5   Monochromators: Prisms, Gratings, and the Location of Wavelength Dispersion Components

We discussed the design and construction of a prism and a grating monochromator in Chapter 3, Section 3.3. We mentioned then that a prism monochromator is less expensive but the bandwidth (resolution) is wavelength dependent. Historically, a high-quality grating of sufficient resolution was significantly more expensive than a prism monochromator. With the advance of computer-assisted laser etching, the cost of quality gratings has become comparable to a high-quality prism. Today, almost all commercial spectrophotometers use a grating monochromator, and you will most likely never encounter a prism monochromator in a commercial instrument.

Note in Figure 6.22 that the grating (wavelength dispersion) comes *after* the sample cell, whereas in Figures 6.19 and 6.20, the wavelength selection comes *before* the sample. By selecting the wavelength before passing the source radiation through the sample, you minimize the quantity of radiation passing through the sample at any given time. Most scanning spectrophotometers have the monochromator before the sample cell in order to reduce sample heating. Because scanning spectrophotometers are relatively slow, passing all of the source radiation through the sample would produce unacceptable heating during the course of the experiment. On the other hand, the advantage of an array spectrophotometer is speed, which results from the fact that it measures the absorption of all of the wavelength from the source simultaneously. So wavelength dispersion *must* occur *after* the sample. Fortunately, sample heating is usually not an issue because the sample is only exposed to the source radiation for a brief moment while the shutter opens and closes.

Building a Monochromator

One way to truly understand how something works is to build it. This activity describes how to construct a functional monochromator from commonly available parts and materials.

Materials – Your will need four razor blades, a shoebox, a can of flat black spray paint, a magnifying lens or cardboard tube, an unused CD or a grating (available through optical supply stores), a light source (a standard nightlight is enough or a white light-emitting diode (LED) bulb will also suffice), black electrical tape, a small piece of white cardboard, modeling clay or rubber cement, and a popsicle stick or pen or pencil or a thin piece of PVC pipe.

Construction – Two possible schematics of the shoebox monochromator are presented below. The first thing you need to do is to spray paint the inside of the shoebox with flat black paint. This will serve to absorb stray light. On one of the short sides of the shoebox, cut an entrance slit (approximately 1 cm × 1 cm).

Use the black electric tape to secure two razor blades on either side of your entrance slit. Try to get the edge of the blades as parallel as possible. We will cut the exit slits last.

Collimation – Next, you will need to decide if you will focus your source radiation on the grating or use an occluding tube to blank any light that is not collinear with the grating path. Ideally, you can incorporate a lens and an occluding tube. The occluding tube will help to blank any stray light from the room that enters the entrance slit. If the source radiation is sufficiently bright, an occluding tube alone will suffice. If you are using a focusing lens, you will need to find the location of the lens in the shoebox that will focus the source radiation to a penny sized spot on the opposite wall of the shoebox. Once this position is located, glue the lens in place or secure it using modeling clay. If you decide to use an occluding tube, you will want to cut a cardboard tube from a roll of paper towels so that it will fit into the shoebox without obstructing the optical path to the exit slit. A length of approximately 8 inches will suffice. If you are using the occluding tube, you will want to spray paint it black inside and out and then glue it into place or secure it with modeling clay.

Grating – Most chemistry departments have old and broken Spec 20 spectrophotometers sitting on a shelf someplace. So, ask if you can cannibalize the grating from a dysfunctional Spec 20. If a used grating is not readily available, you can purchase one for a few hundred dollars. If you want to keep your budget low, you can fabricate a reasonable grating from an unused CD. Cut a piece out of a CD about 1 cm × 6 cm near the edge. Use a razor blade to part the layers of the CD and remove the top cover plastic. This step is optional, but it will make the grating work better. Glue the grating or piece of CD to a pencil, pen, or piece of PVC pipe and poke the pencil through the bottom of the shoebox where shown. With the grating glued to the pencil and the pencil pushed through the bottom of the shoebox, place the lid over the shoebox and push the pencil through the lid. Remove the lid and with the source shining on the grating, determine the location of the projected image on the long wall, and cut the exit slit. Use black tape to secure a razor blade on either side of the exit slit so that you create a narrow (2 mm) slit with parallel and straight edges. Place the lid on the shoebox, making sure the pencil that contains the grating pokes through the top and bottom of the shoebox. Seal the edges of the shoebox with black tape. Using a piece of white cardboard, create a view screen for the exit radiation. Now that your monochromator is complete, you can rotate the pencil to direct different colors of light at the exit slit.

The only thing you would need to do to convert this shoebox monochromator into a functioning spectrophotometer is to devise a way to introduce a sample (see Figure 6.17 (schematic of a Spec 20) above), and incorporate a detector that is sensitive to light intensity (conduct an Internet search for "light meters").

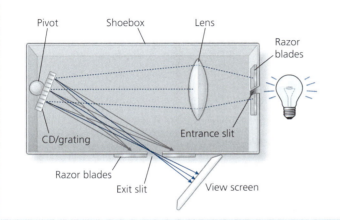

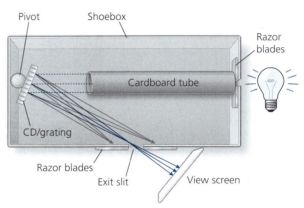

6.6 Sources

Deuterium Arc/Tungsten Halogen Bulb

There are several sources one can use for UV-vis spectroscopy. The most common sources used in commercially available instruments are actually a combination of two sources, a deuterium discharge lamp and a tungsten-halogen bulb. The ultraviolet portion of the source radiation is provided by a deuterium discharge lamp and the visible and near-infrared radiation is provided by the tungsten halogen bulb. The deuterium arc lamps operate at low pressure and produce a continuum of radiation from approximately 165 to 460 nm. Figure 6.24 shows the emission spectra of a deuterium lamp.[27] Although the bulb produces radiation below 200 nm, the optical materials (primarily fused silica or quartz) used to guide the radiation through the spectrometer limits the low-wavelength range to about 200 nm.

The deuterium discharge lamp involves a rectangular discharge chamber constructed of nickel (see Figure 6.25). An electric discharge between two tungsten electrodes serves to excite the deuterium molecule. The energy imparted to the deuterium molecule is sufficient to cause it to dissociate into its component atoms, with the concomitant emission of energy in the form of a photon (see Equation 6.9). If the D_2 molecule simply relaxed to ground state, we would expect the emitted energy to be quantized and the emission to be a line spectrum. However, kinetic energies are not quantized, and the total energy initially absorbed by the D_2 molecule is distributed among the kinetic energies of the resultant D atoms and the emitted photon. Thus, light is emitted as a continuum rather than as a line spectrum.

$$D_2 \xrightarrow{\ arc\ } D_2^* \longrightarrow D'_{KE_1} + D''_{KE_2} + h\nu \qquad\qquad \textbf{Eq. 6.9}$$

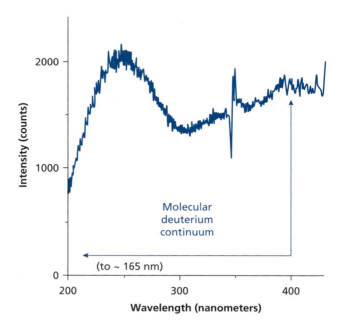

Figure 6.24 Emission spectrum of a D_2 discharge lamp.

[27] This image is being used under the GNU Free Document License agreement.

where

D$_2$ = ground state deuterium molecule
D$_2$* = excited state deuterium molecule
arc = electric discharge
D'_{KE_1} & D''_{KE_2} = deuterium atoms with different kinetic energies
hv = emitted photon

Above 450 nm, sharp atomic emission lines from deuterium are present, and slight errors in wavelength selection near the sharp lines would lead to large-intensity variations. For this reason we switch from using deuterium to the tungsten bulb above 450 nm. In some spectrometers, you can actually observe or hear the spectrometer switching lamps. If you have a scanning spectrophotometer, you may observe the scan process pause around 400 nm and you will hear a "click" inside the instrument as the source is changed.

The tungsten bulb is a black body emitter whose emission profile is directly related to the voltage applied to the tungsten filament. It is not difficult to get a tungsten bulb to emit light with wavelengths below 400 nm and reaching well into the infrared. Unfortunately, the voltage needed to get a sufficiently intense emission over the UV-vis range substantially shortens the lifetime of a tungsten filament. The addition of a small amount of iodine to the lamp extends the life span of the lamp and allows for the operation of the lamp at higher voltages. These types of lamps are collectively called halogen lamps. Most automobile headlamps use this same technology. The reason the tungsten lamp's lifetime is shortened at higher voltages is due to the sublimation of the filament. With iodine present, the sublimed tungsten atoms react to form a gaseous tungsten–iodine (W-I) compound; when the W-I compound strikes the hot filament, it decomposes and the tungsten atoms are deposited back onto the filament.

Xenon Arc Lamps

Although not as common, some commercial instruments and many research-grade instruments use a xenon arc lamp as the source. These sources consist of a quartz bulb containing xenon gas under very high pressure. An electric arc is used to excite the gas,

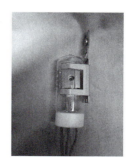

Figure 6.25 D$_2$ discharge lamp.

Figure 6.26 Xenon arc lamp.

and the high pressure results in extreme pressure broadening,[28] producing an intense continuum of radiation from approximately 250 to 700 nm (see Figure 6.26).

Light-Emitting Diodes

In the past 15 years, the cost, reliability, and availability of light-emitting diodes (LEDs) have greatly improved. An LED is a p/n junction[29] that has been forward biased. In a very crude way, you can think of an LED as a solar panel that has been forced to operate in reverse by applying an external voltage that exceeds the voltage of the p/n junction's band gap. Instead of absorbing light and producing electricity, an LED consumes electricity and produces light.

LEDs have λ_{max} outputs that tends to be very narrow (50 nm or less). A single LED would not make a useful source for a UV-vis spectrometer that was tasked with scanning a large wavelength range. However, individual LEDs do make good sources for colorimetric probes that are only tasked for a specific analysis. Furthermore, "white" LEDs are now available that cover a wide range of the visible spectrum. White LED is a bit of a misnomer, because the white light is actually generated by the secondary emission from stimulated phosphorous. The LED's housing is coated with phosphorus and the radiation from the LED causes the phosphorus to luminesce. The LEDs have a very low power draw, have a very intense output for the λ_{max} in question, and last a very long time—all good qualities for a source to be used in a handheld colorimetric probe.

Problem 6.14:   What are the two most common sources used in a UV-vis spectrophotometer?

Problem 6.15:   Why is it a common design feature to see two different sources used in a UV-vis spectrophotometer?

Problem 6.16:   Why is iodine sometimes added to tungsten lamps?

6.7   Detectors

The Photomultiplier Tube

The basic function of a detector is to convert electromagnetic energy (a photon) into electrical energy. The traditional detector for commercial instruments is the *photomultiplier tube* (PMT) and the PMT remains the detector of choice when the measurement of very low light levels is required. You can think of a PMT as a multistage phototube. The function of a phototube is based on the photoelectric effect.[30] In a simple phototube, an impinging photon strikes a negatively biased cathode and dislodges an electron. The electron is accelerated toward an anode and is collected. The current thus generated in the circuitry is a measure of the photon flux impinging on the cathode.

A PMT functions in a similar way. In a PMT the incident photon strikes a transparent film covered with a phosphorescent material. This type of device is called a scintillation counter.[31] The photon that is emitted by the scintillator strikes a photocathode dislodging an electron, but instead of collecting the electron after the first stage, the electron is accelerated toward a secondary cathode (called a dynode). As the electron

---

[28] This effect is also called *collision energy transfer* and is described in Chapter 7.

[29] See Chapter 4 for a review of diodes and p/n junctions.

[30] Einstein won the 1921 Noble Prize in physics for his explanation of the photoelectric effect.

[31] Anthracene is a common phosphor in scintillation devices.

The Jaz visible spectrometer by Ocean Optics is a handheld unit with interchangeable components. This allows for a wide range of experimental parameters, including the option to change the source. LED source options include a 365 nm, 405 nm, 470 nm, 590 nm, 640 nm, and a white LED.

Notice the narrow bandwidth of a typical LED as seen in the blue LED. Note the intense output centered at 470 nm. The emission profile of the white LED shows a continuum. The white LED has very little output in the ultraviolet or the near infrared but is more than sufficient for the visible range of the EMS. ■

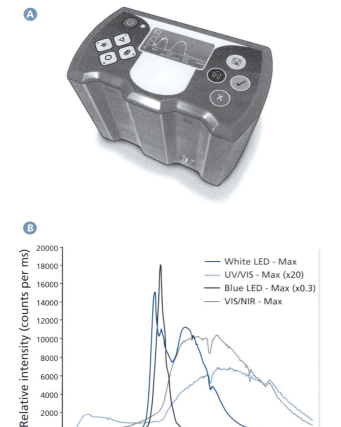

impinges on the secondary electrode, multiple electrons are dislodged. Each of these secondary electrons is accelerated toward a tertiary electrode and the process is repeated through numerous stages. Commercial PMTs are capable of a 10^8 current amplification.[32] In order to accelerate the dislodged electron toward the next stage in the cascade, it is necessary to precisely control the voltage drop between each dynode and

---

[32] *The Photomultiplier Handbook*. RCA Company, **1980**.

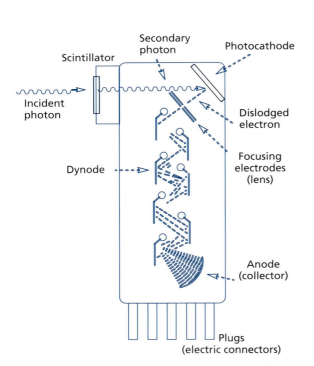

Figure 6.27 (left) A schematic of an eight-stage PMT and (right) a photograph of a PMT. The incident photon dislodges an electron from the scintillator, which initiates a cascade of ejected electrons as they proceed toward the collector.

the initial photocathode. The fluorescence of the scintillator ensures that all photons impinging on the primary photocathode fall within a uniform energy bandwidth and as a result, the initial electrons ejected from the photocathode will fall within a narrow range of kinetic energy. Figure 6.27 shows a schematic and a photograph of a PMT.

Photovoltaic Cells

Photovoltaic technology makes use of a diode (p/n junction) in which one of the semiconductors is transparent (usually the p-type semiconductor). As the photon passes through the p-type semiconductor into the n-type, an electron is promoted from the n-type into the p-type semiconductor. Because the p/n junction is a one-way current gate,[33] the excited electron can be made to pass through a wire in order to close the circuit. Figure 6.28 shows a schematic of a photovoltaic p/n junction. A single p/n junction does not provide the current gain needed to detect the low levels of light typically reaching the detector in a UV-vis spectrometer, but you can stack multiple p/n junctions onto a single chip. These types of devices are called cascade photovoltaic devices.[34] Current gains of up to 100 are easily obtained. Although the sensitivity of photovoltaic cells is inferior to that of PMTs, these devices are still used widely. The diode array spectrometers are probably the most noteworthy example. The diode detectors also have the advantage of being very small and robust. The handheld probe spectrometers and colorimeters make use of diode detectors.

---

[33] See Chapter 4 for a review of diodes, p/n junctions, and rectifiers.

[34] (left) Landis, G. A. Technology assessment of thin-film cascade photovoltaic modules. *Solar Cells,* **1988** *(25)*3, 203. (right) Yang, R. Q.; Tian, R.; Klem, J. F.; Mishima, T. D.; Santos, M. B.; Johnson, M. B. Interband cascade photovoltaic devices. *Appl. Phys. Lett.,* **2010**, *96*, 063504.

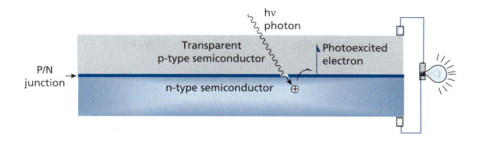

Figure 6.28 A schematic of a photovoltaic p/n junction.

Charge Coupled Devices

Charge coupled device (CCD) detectors are closely related to photovoltaic cell detectors. CCDs make use of metal oxide semiconductors (MOS) structures. The most common MOS structure is constructed of a layer of silicon dioxide sandwiched between a layer of pure silicon (the substrate) and a layer of polycrystalline silicon. The resulting structure becomes polarized in an electric field (i.e., it is a dielectric material). The CCD detector consists of three electrodes biased as seen in Figure 6.29. When light strikes the MOS at the positive electrode, electrons are "pushed" into a potential "well." The accumulated electrons are moved laterally and onto a circuit containing an operational amplifier by switching the bias on the three electrodes. In addition to their wide use in digital cameras, CCD detectors find use in imaging spectroscopy, and for techniques where low-light level detection is critical.[35]

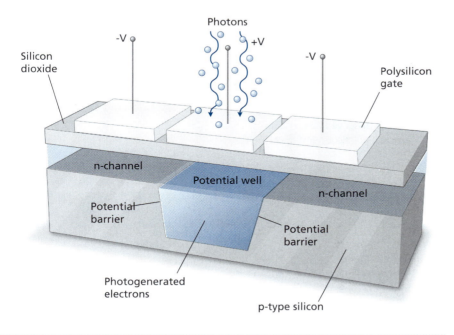

Figure 6.29 An MOS semiconductor used in a CCD detector.

[35] For additional information on CCDs, see (a) Matlina E. G.; Nakhleh M. B. How Students Use Scientific Instruments to Create Understanding: CCD Spectrophotometers. *J. Chem. Educ.,* **2003**, *80(6),* 691. (b) Giles J. H.; Ridder T. D.; Williams, R. H.; Jones D. A.; Denton M. B. *Anal. Chem.,* **1998**, *70(19),* 663A–668A. (c) Photometrics High-efficiency CCD. *Anal. Chem.,* **1995**, *67(17),* 564A–564A.

6.8 Noise

We discussed the various types of noise and signal-to-noise ratios in Chapter 5. However, it would be a good idea to review the possible types of noise with the design aspects of a UV-vis spectrophotometer in mind.

Stray Light

Recall that absorption is the negative log of the transmittance and the transmittance is the power ratio in the presence and absence of an absorbing sample. Because stray light reaches the detector regardless of the absorbing characteristics of the sample, the effect of stray light hitting the detector is an apparent loss of absorption. See Equation 6.10.

$$A_{obs} = -log\frac{(P+P_s)}{(P_o+P_s)}$$ Eq. 6.10

where

A_{obs} = observed absorbance
P_s = radiant power of the stray light

Because absorbance is a log term, the effect of stray light on the signal-to-noise ratio can be important. If the stray light is significant compared to either P or P_o, then the log term (absorbance) approaches zero. Therefore, elimination of stray light is a key design aspect of a spectrophotometer.

Detector Noise

In the absence of any light, one might expect the output current of a detector to be zero. However, this is not the case. Vacuum tube detectors have a *dark current*. In the middle of the spectral range of these types of detectors, the dark current is usually minimal when compared to the output current of the detector in normal operating conditions. However, when operating at the spectral extremes of an instrument, the source output is weak and the presence of dark current results in an apparent loss of absorption.

$$A_{obs} = -log\frac{(P+P_D)}{(P_o+P_D)}$$ Eq. 6.11

where

A_{obs} = the observed absorbance
P_D = power output from dark current

ACTIVITY

Absorbance and Stray Light

Explore the effects on the relationship between total absorbance vs. stray light by manipulating the variables of initial radiant power (P_o), sample radiant power (P), and stray radiant power (P_s). Construct an Excel spreadsheet with the following column headings and vary the amount of stray light reaching the detector. Initially try using a value of 1 for (P_o) a value of 0.5 for (P) and a value of 0.0 for (P_s). Vary the value of P_s from 0.0 to 1.0 and construct a plot of total absorbance vs. stray light power.

Excel Sheet

P_o	P	P_s	$A_{obs} = -log\frac{(P+P_s)}{(P_o+P_s)}$
1	0.5	0.0	$A_{obs} = -log\frac{(0.5+0.0)}{(1.0+0.0)} = 0.3$
1	0.5		
1	0.5		
1	0.5		
1	0.5		
1	0.5		

Shot noise is named after Walter Hermann Schottky (1886–1976) and is often seen spelled Schott noise in deference to Schottky. Schottky himself used the term *Schrot effect*, which translates from German as "a small shot effect." ■

Notice the similarity in Equations 6.10 and 6.11.

A second type of detector noise is called *shot noise*.[36]

In a PMT and diode detectors, the generated signal is the result of electrons crossing a junction,[37] which represent a discrete quantized event and are therefore subject to statistical fluctuations. The fluctuations in the output current generated by the quantized nature of the current carrier (electrons) can be described by:

$$i_{rms} = \sqrt{2Ie\Delta f}$$ **Eq. 6.12**

where

i_{rms} = root mean square of the current fluctuation
I = average output current
e = charge on an electron
Δf = bandwidth of the signal

Equation 6.12 shows us that *shot noise* can be minimized by reducing the bandwidth of the signal.[38] However, the instrument response (the time constant t_c) is related to the instrument bandwidth by:

$$t_c = \frac{1}{3\Delta f}$$

---

[36] See Chapter 2 and Equation 2.5 for further discussion of shot noise.

[37] Horowitz, P.; Hill, W. *The Art of Electronics*, 2nd ed., Cambridge University Press: New York, 431, **1989**.

[38] See Chapter 5 and Equation 5.2 for a discussion of RC time constants.

What is important to note from this equation is the fact that our ability to measure a fast event is inversely proportional to the bandwidth of radiation reaching the detector. Therefore, there are trade-offs between S/N and instrument speed one must accept in the operation of a UV-vis spectrometer if you wish to measure fast kinetic events.

Source Noise

The most prevalent type of noise generated by the source is *thermal noise* (also called *Johnson noise* or *Nyquist noise*[39]). Thermal noise is generated by the random motion of electrons found in electric conductors. This motion is present even in the absence of an applied voltage. Thermal noise is considered to be *white noise* and as such can be reduced through signal averaging. The equation describing thermal noise is:

$$i_{rms}^2 = \frac{4k_B T \Delta f}{R}$$

Eq. 6.13

where

k_B = Boltzmann constant
T = Kelvin temperature
Δf = bandwidth of the instrument
R = resistance of the device in ohms

Like shot noise, thermal noise can be reduced by decreasing the bandwidth being measured. This can be accomplished by either increasing the RC time constant or by narrowing the slit width of the monochromator. Increasing the RC time constant will limit the instruments utility for kinetic measurements, and decreasing the slit width will reduce the power reaching the detector (sensitivity). Every experimental choice is a trade-off.

> **Problem 6.17:** For each of the types of noise described so far,[40] discuss how the presence of noise affects the output data and give a physical explanation for what caused the noise, include equations (if applicable). Also, discuss methods or ways to minimize the noise.

6.9   Kinetic Ultraviolet-Visible Techniques

In addition to quantitative and qualitative analysis, the principles of UV-vis spectroscopy can also be used to study the rates of chemical reactions. In kinetic studies, the absorbance is measured as a function of time. The species being monitored can be one of the reagents, one of the products or an excited state intermediate. In the case where you are monitoring a reagent or an excited state intermediate, you record the decrease in absorbance as a function of time and in the case where you are monitoring a product, you record the increase in absorbance as a function of time. Figure 6.21 shows a kinetic study of the growth of gold nanoparticles as a function of time and Figure 6.30 shows the UV-vis data from a study where silica particles were placed in ethanol and the release of ethyl octanoate and ethyl decanoate into ethanol was monitored as a function of time.[41]

---

[39] Johnson, J. B. Thermal Agitation of Electricity in Conductors, *Phys. Rev.,* **1928**, *32*, 97; Nyquist, H, Thermal Agitation of Electric Charge in Conductors, *Phys. Rev.,* **1928**, *32*, 110. See Chapter 5 and Equation 5.4 for a discussion of thermal noise.

[40] See Chapter 5.2 for a review of the sources of noise.

[41] Restricted diffusion and release of aroma molecules from sol-gel-made porous silica particles. Veith, S. R.; Hughes, E.; Pratsinis, S. E.; *J. Controlled Release,* **2004**, *99(2)*, 315–327.

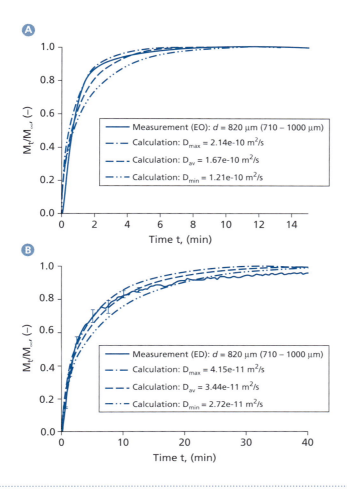

Figure 6.30 Measured release kinetics (UV-vis) of (A) ethyl octanoate (EO) and (B) ethyl decanoate (ED) in ethanol from silica particles.

In the case where you are monitoring the kinetics of a reaction between two or more reagents, it is important that the spectra of the products and reagents have features that do not overlap. One must also consider the time scale of the reaction being monitored. In Figure 6.21 the data points were separated by hours, and in Figure 6.30 the data were collected over a period of 40 minutes. Experiments that occur on these timescales can be performed in a standard laboratory grade UV-vis spectrophotometer. The user simply mixes the reagents and quickly transfers them to the sample chamber and takes a set of spectra over time. However, for reactions that occur on a shorter timescale, specialized equipment is required. We will discuss two instrumental designs for the study of short-lived reactions.

Stop Flow Ultraviolet-Visible Studies

UV-vis kinetic studies that occur on the time scale of several seconds are routinely studied using a technique called *stop flow*. In a stop flow system, a set of syringe pumps are used to rapidly deliver reagents to a mixing chamber and then onto the sample cell of a UV-vis spectrophotometer. Figure 6.31 shows a basic schematic of a stop flow UV-vis system. To study the kinetics of the reaction between reagent A and reagent B, you would place reagent A in one syringe and reagent B in the other. The syringe pumps are depressed simultaneously and the two reagents are rapidly mixed and then they displace a blank solution in the sample chamber. The mixing and flush step is accomplished in a few milliseconds. The spectrometer then measures the absorbance as a function of time. The stop syringe can be set to ensure a specific volume of reaction

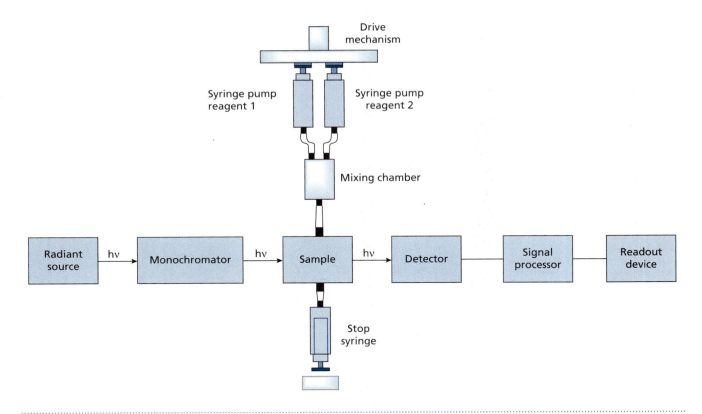

Figure 6.31 Schematic of a stop flow UV-vis system for use in studying reaction kinetics.

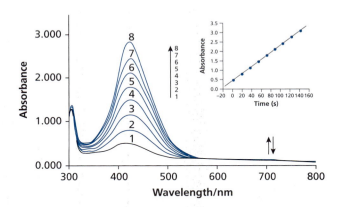

Figure 6.32 Rapid-scanning stopped-flow spectra for the reaction of OPDA and H_2O_2 catalyzed by Hb. Time interval between successive scans is 18 s. The insert picture is the kinetic curve ($A \sim t$) of the reaction monitored at 425 nm. [Hb] = 1.0×10^{-6} mol/L, [H_2O_2] = 2.0×10^{-3} mol/L, [OPDA] = 4.0×10^{-3} mol/L, 48°C.

solution is delivered. Figure 6.32 shows data collected for the rapid catalytic determination of peroxide.[42] Stop flow UV-vis accessories are commercially available and can be purchased through several different vendors.[43]

---

[42] Zhang, K; Mao, L.; Cai, R. Stopped-flow spectrophotometric determination of hydrogen peroxide with hemoglobin as catalyst. *Talanta*, **2000**, *51(1)*, 179.

[43] OLIS, Bogart, GA 30622. Applied Photophysics Ltd., Leatherhead Surrey KT22 7BA United Kingdom. Spectro Biotek Instruments, Laxmi Nagar, Delhi-110092, India.

For example, researchers used flash photolysis to study the rate that carbon monoxide (CO) binds to cytochrome cd1. A sample of cytochrome cd1-CO complex was subjected to a pulse of 460-nm light, resulting in a rapid (< 50 ns) change in the UV-vis spectrum. The absorbance change was due to the dissociation of CO from the cytochrome. By measuring the rate at which the UV-vis spectrum reverted back to its original form, a rate constant of $k = 12$ s^{-1} was determined for the combination of cytochrome cd1 with CO. ■

PROFILE

Flash Photolysis of Cytochrome cd1-CO

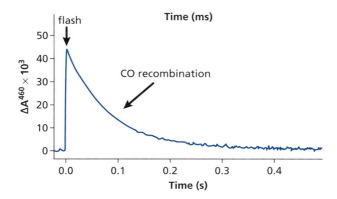

*Adapted with permission from Sjögren, T.; Svensson-Ek, M.; Hajdu, J.; Brzezinski, P. Biochemistry **2000**, 39, 10967. © (2000) American Chemical Society.*

Flash Photolysis

This technique is ideal for studying fast photophysical or photochemical processes. However, the chemical system must meet two criteria. First, the photoreaction leading to a particular intermediate species must be initiated on a timescale that is shorter than the decay process, and the photolysis must be carried out in such a manner that chemical homogeneity is obtained. Second, the decay process must have an associated change that can be observed spectroscopically. In other words, you need to be able to photoexcite your analyte with a very short burst of light, the duration of which must be shorter than the decay rate of the excited state. In addition, the excited state UV-vis spectrum must have features that are measurably different than the UV-vis spectrum of the ground state.

For instance, if the excited state molecule has a very strong absorbance at 360 nm and the ground state molecule does not, then assuming you can complete the excitation faster than the decay process, monitoring the absorbance as a function of time at 360 nm will allow you to observe the rate of decay of the excited state. Alternatively, you can also perform flash photolysis by monitoring the reoccurrence of the ground state molecule.

Flash photolysis was first introduced in 1948 by Norrish and Porter.[44] Early flash photolysis systems used a xenon arc lamp or a xenon-mercury arc lamp as the excitation source. The sample was subjected to short pulses of excitation radiation by means of a

---

[44] Norrish, R. G. W.; Porter, G. Nature **1949**, 164, 658; Porter, G. Proc. R. Soc., Sec. A., **1950**, 200, 284.

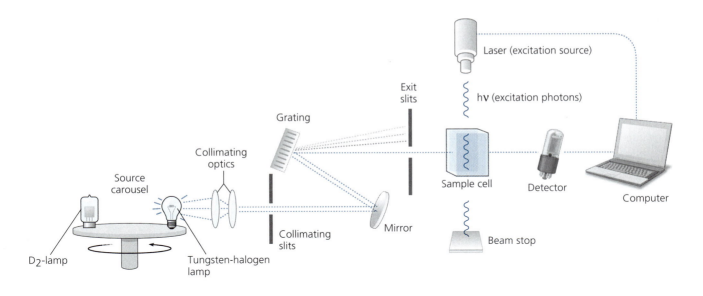

Figure 6.33 Block diagram of a laser flash photolysis spectrophotometer. The laser provides a short excitation pulse perpendicular to the optical train of the UV-vis spectrophotometer. The laser's flash is set as time = 0 and the rate of change of the absorbance is monitored versus time.

chopper or a shutter which mechanically interrupted the excitation source getting to the sample.

With the advent of pulsed lasers, which emit light in short bursts, the need of a chopper has been eliminated and the timescale of the excitation event has grown much shorter. For instance, Nd-YAG lasers are now capable of producing a pulse with a duration of less than 6 ns, allowing for the study of faster events. The narrow excitation pulse allows us to study faster events, which would have been complete and obscured under a wider pulse.

Femtosecond pulsed lasers are also now available, which should allow for studying events on the 10^{-15} second timescale. However, from an instrumental design perspective, what limits our ability to measure very fast events is the response time of the detector. Today, laser flash photolysis systems with detectors having microsecond time constants are common and research grade instruments in the subnanosecond range have been developed. Figure 6.33 shows a schematic of a laser flash photolysis spectrophotometer. The horizontal optical train is essentially the same as any UV-vis spectrophotometer. The vertical optical train serves as the excitation source for the photoreaction.

Figure 6.34 shows a representation of the type of data one would collect using flash photolysis.[45] At time equals zero the laser has fired and produced an excited state species with an absorbance different than the ground state. We then observe the decay of the absorbance as a function of time. By $\approx 25\ \mu s$, the sample has returned to the ground state. Typically, the analyst is trying to understand the kinetics (and hence the mechanism) of a reaction, so the next step would be to fit the absorbance versus time data points to a kinetic rate law equation.[46]

---

[45] Reprinted from *Laser Flash Photolysis: Photochemistry and Kinetic Studies of Palladium(I), Platinum(I), Tungsten(I) and Ruthenium(0) Binuclear Complexes.* R. M. Granger. PhD thesis: Purdue University, Lafayette IN, June **1993**.

[46] For a review of chemical kinetics see Atkins, P. W. *Physical Chemistry,* 3rd ed. Oxford University Press: New York, **1986**.

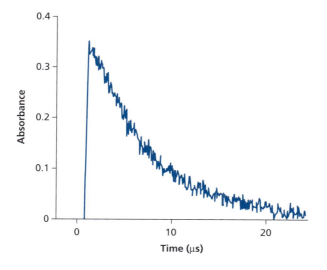

Figure 6.34 Transient absorbance decay of an excited state species.

6.10 Useful Data

Tables 6.2 and 6.3 contain data that the reader may find useful.

TABLE 6.2: Some Common UV-vis Solvents Arranged by Increasing Wavelength Cut Off

SOLVENT	λ CUTOFF* nm	POLARITY INDEX[†]	DIELECTRIC CONSTANT*,[‡]
Acetonitrile	190	6.8	37.5
Heptane	200	0.0	1.92
Water	200	9.0	78.54 (25)
Cyclohexane	210	0.2	2.023
Ethanol	210	6.2	24.30 (25)
Hexane	210	0.0	1.890
Isopropyl alcohol (rubbing alcohol)	210	3.9	19.92
Methanol	210	6.1	32.63
Methyl *tert*-butyl ether (MTBE)	210	2.5	2.6
Pentane	210	0.0	1.844
Diethyl ether	218	2.8	4.335
Tetrahydrofuran (THF)	220	4.0	7.6
Ethyl acetate	225	4.4	6.02 (25)
Dichloromethane (methylene chloride)	235	3.1	9.08[§]
Chloroform	245	4.1	4.806
Carbon tetrachloride	265	1.6	2.238
Dimethylsulfoxide (DMSO)	265	7.2	4.7
Dimethyl formamide (DMF)	270	6.4	36.7
Benzene	280	2.7	2.284
Toluene	286	2.4	2.379 (25)
Acetone	330	6.1	20.7 (25)

* Adapted from the *CRC Handbook of Basic Tables for Chemical Analysis 2nd Ed.* 2000.
[†] Polarity Index from *The HPLC Solvent Guide*, Wiley Interscience: New York, 2002.
[‡] Dielectric constants were measured at 20°C unless otherwise noted.
[§] From *CRC Handbook of Chemistry and Physics*, 47th ed. p E-54.

TABLE 6.3:   Selection Rule Effects and Relative Absorbance

TRANSITION TYPE	RELATIVE STRENGTH	TYPICAL EPSILON (ε) VALUES (M^{-1} cm^{-1})
Spin forbidden LaPorte forbidden	Very weak	< 0.1
Spin allowed (Oh symmetry) LaPorte forbidden	Weak	1–10
Spin allowed LaPorte forbidden (orbital-mixing)	Moderate	50–200
Spin allowed LaPorte allowed	Very strong	1000–1,000,000

6.11 Further Reading

BOOKS

Harris, D. C.; Bertolucci, M. D. *Symmetry and Spectroscopy: An Introduction to Vibrational and Electronic Spectroscopy.* Oxford University Press: New York, 1978 (Dover Publications: Mineola, NY, **1989**).

Hollas, J. M. *Modern Spectroscopy*, 4th ed. John Wiley & Sons: Chichester, UK, **2004**.

Milton, H. H.; Jaffe, O. *Theory and Applications of Ultraviolet Spectroscopy*; John Wiley & Sons: New York, **1966**.

Owen, A. J. *The Diode-Array Advantage*, 1st ed. Hewlett-Packard: Palo Alto, CA, **1988**.

Pavia, D. L.; Lampman, G. M.; Kriz, G. S.; VyVyan J. A. *Introduction to Spectroscopy*, 5th ed. Cengage Learning: Stamford, CT, **2015**.

Thomas M. J. K.; Ando D. J. *Ultraviolet and Visible Spectroscopy: Analytical Chemistry by Open Learning.* John Wiley & Sons, Chichester, UK, **1996.**

JOURNALS

Strelow, F. W. E.; Weinert, C. H. S. W.; Deviation from Beer's Law Caused by Change in Bathochromic Shift of Absorption Maximum. *Anal. Chem.*, **1975**, 47 (13), 2292–2293.

6.12 Exercises

EXERCISE 6.1: What is the wavelength range of a typical UV-vis spectrophotometer? Briefly discuss the physical constraints that limit a typical UV-vis spectrometer to this wavelength range.

EXERCISE 6.2: To what does the term *n-electrons* refer? What is the spectroscopic significance of *n-electrons*?

EXERCISE 6.3: Why are the peaks in molecular UV-vis spectroscopy broad relative to atomic UV-vis spectroscopy?

EXERCISE 6.4: Discuss the significance of λ_{max} when reporting molecular UV-vis absorption data. Why do spectroscopists report λ_{max} for a transition and not simply λ?

EXERCISE 6.5: What is the percent transmittance if the absorbance is

(a) 0.5   (c) 1.0   (e) 2.5

(b) 0.05   (d) 0.001   (f) 0.8

EXERCISE 6.6: What is the absorbance if the percent transmittance is

(a) 3%   (c) 30%   (e) 90%

(b) 0.5%   (d) 10%   (f) 50%

EXERCISE 6.7: If the molar absorptivity for a compound is 3500 $M^{-1}cm^{-1}$, calculate the concentration of each solution in Exercise 6.5.

EXERCISE 6.8: If the dark current from a spectrophotometer's PMT was 0.03 mA and a blank solution produced a current at the PMT of 4.45 mA, what is the absorbance of a sample if the reading at the PMT was 3.75 mA?

EXERCISE 6.9: Research the photoelectric effect and describe its importance in modern spectrophotometer design. In your discussion explain why PMT detectors are not used for infrared spectroscopy.

EXERCISE 6.10: The 3s → 3p transition in sodium is observe at about 5893 Å. What is this wavelength in nanometers? What is the frequency of the absorbed photon? What is the energy of the absorbed photon?

EXERCISE 6.11: The λ_{max} for the complex $[Fe(phen)_3]^{2+}$ is 510 nm. If the molar absorptivity is 1.89×10^4 M^{-1}cm^{-1}, what is the concentration of a $[Fe(phen)_3]^{2+}$ solution if it produced an absorbance of 0.03 in a 1 cm cell?

EXERCISE 6.12: Compare and contrast the optical components of a single-beam and a dual-beam spectrophotometer.

EXERCISE 6.13: Compare and contrast the optical components of a single-beam and a diode array spectrophotometer.

EXERCISE 6.14: What were the market forces that have allowed array spectrophotometers to dominate the UV-vis spectrometer market?

EXERCISE 6.15: Rank each of the following molecules from highest to lowest λ_{max} for the π → π* transition.

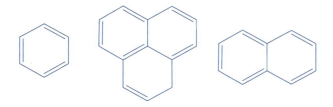

EXERCISE 6.16: During the course of this chapter, we used the term *best practice* three times. List each of these best practices and describe the objective of each of these best practices.

EXERCISE 6.17: Why is the regulation of the radiant source less critical in a dual-beam spectrophotometer than it is in a single-beam spectrophotometer?

EXERCISE 6.18: We discussed two basic classes of monochromators: prism and grating. Discuss the advantages and disadvantages of choosing one type of monochromator over the other.

EXERCISE 6.19: It has been stated that it is considered a *best practice* to take absorption readings at λ_{max}. What were the two reasons given to justify this best practice?

Advanced Exercises

EXERCISE 6.20: A plant extract was analyzed by UV-vis spectroscopy and the following data were obtained:

Concentration (M)	Absorbance
5×10^{-5}	0.051
10×10^{-5}	0.098
15×10^{-5}	0.157

What is the molar absorptivity of the chromophore?

EXERCISE 6.21: Diamond has a refractive index of 2.42. What is the frequency and wavelength of a photon with a wavelength of 622 nm as it passes through the diamond?

EXERCISE 6.22: The 340 nm peak in benzophenone is seen to shift to longer wavelengths when placed in a more polar solvent. What predictions can you make about the type of transition responsible for this peak?

EXERCISE 6.23: What are the physical characteristics of a grating monochromator responsible for determining the spectral resolution?

EXERCISE 6.24: What slit width is required to obtain a spectral resolution of 4 nm if the monochromator contains an 1,100 blazes per millimeter diffraction grating with a 50 mm focal length? Assume the angle of incidence on the grating is 0 degrees and the wavelength is 400 nm (m = 1).

EXERCISE 6.25: Given a diffraction grating with 1,500 blazes per millimeter, what is the linear reciprocal dispersion of the monochromator if the path length between the grating and the exit slit is 0.3 meters and the slit with is 0.50 mm? What is the effective bandwidth of the monochromator?

EXERCISE 6.26: What wavelength of light is diffracted at 35 degrees and at 40 degrees if light from a tungsten–halogen lamp is incident at 0 degrees on a 1,500 blaze per millimeter diffraction grating?

EXERCISE 6.27: The $\varepsilon(\lambda_{238}) = 8{,}760$ for DPK in acetonitrile. What would be the predicted absorption if a 1-ml aliquot of 2 M DPK were diluted to 1 liter?

EXERCISE 6.28: What is the physical basis of the Nyquist theory (Johnson noise)[47]?

EXERCISE 6.29: Imagine you have been hired as an outside consultant to a local paint and dye company wishing to purchase a UV-vis spectrophotometer. The spectrophotometer will be used in the quality control laboratory to test colors between different batches of dyes from the supplier. Outline the pros and cons of the various sources and detectors one might choose from, and defend your recommendation.

[47] Nyquist, H. *Phys. Rev.* **1928**, 32, 110.

EXERCISE 6.30: The UV-vis spectrum shown here is for 1,10-phenanthroline-5,6-dione.

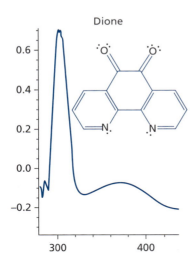

Dione

Absorption Maxima for free 1,10-phenanthroline-5,6-dione in Various Solvents[a]

SOLVENT	POLARITY INDEX[b]	λ_{max} (nm)	λ_{max} (nm)	λ_{max} (nm)
Cyclohexane	0.2	254	282(sh)	—
Ethanol	6.0	—	300	358(sh)
Acetonitrile	6.8	—	300	370
Dimethyl sulfoxide (DMSO)	7.2	262	294(sh)	—
Water	9.0	—	298	364(sh)

[a] Data from; R. M. Granger, et al. *JUCR* **2005**, *2*, 47
[b] Polarity Index from *The HPLC Solvent Guide*, Wiley Interscience: New York, 2002.

Use the data in the table to assign the peaks as $n_{ox} \rightarrow \pi^*$, $\pi \rightarrow \pi^*$ (carbonyl), $n_N \rightarrow \pi^*$, or $\pi \rightarrow \pi^*$ (aromatic).

Chapter 7

Atomic Absorption Spectroscopy

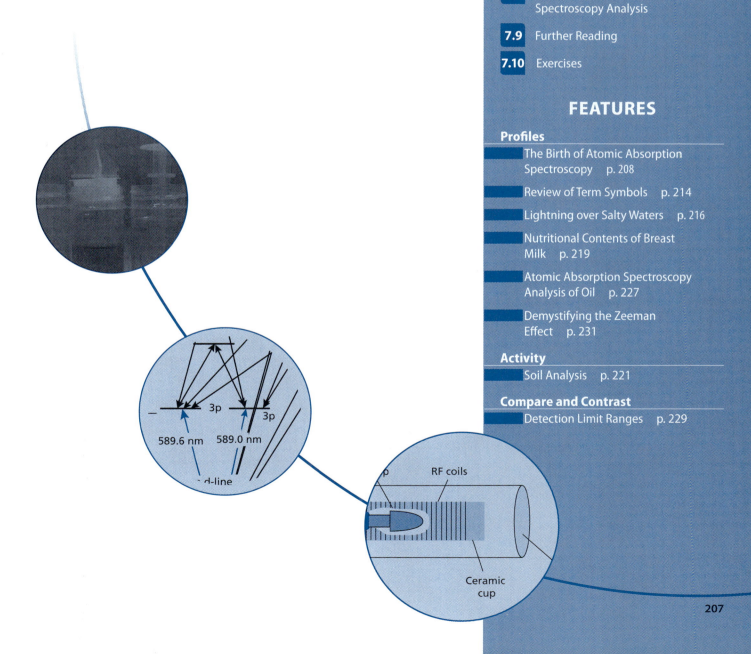

589.6 nm   589.0 nm

3p   3p

d-line

RF coils

Ceramic cup

PROFILE

The Birth of Atomic Absorption Spectroscopy

In 1859, **Robert Bunsen** and **Gustav Kirchhoff** began a collaboration studying the emission spectra of heated elements and coined the term *spectrum analysis*. We will examine the modern versions of this instrument type in Chapter 9, "Atomic Emission Spectroscopy" (AAS). In 1860 they published a paper that described the qualitative identification of alkali and alkali-earth elements, and they also described the apparatus they developed for this purpose. The citation for a translated version of that paper can be found in the footnote.* In the process, Bunsen and Kirchhoff developed many of the basic principles that underlie our understanding of modern AAS.† In that same year, they also used their "spectroscope" to discover two new elements, rubidium and cesium. The use of spectroscopes ushered in a rapid expansion of the known elements at that time.

It is often said that necessity is the mother of invention. In order to see an emission, Bunsen and Kirchhoff realized that they needed to get their samples very hot in order to generate excited state atoms. So Bunsen developed a natural gas burner that had a carburetor. Sadly, it is fair to say that Bunsen is better known for the heating element he designed (the Bunsen burner) than he is for his development of a new form of spectroscopy or his discovery of two new elements.

For more than 100 years, atomic spectroscopy would remain an obscure tool used only in the more advanced research laboratories until the company Techtron Pty. Ltd. produced the first commercially available atomic absorption spectrometer in the mid-1960s. The prototype is credited to **Sir Alan Walsh**.‡ The innovation that Walsh pioneered was to measure the absorption spectra of atomic vapor rather than the emission spectra of atomic vapor. It takes a great deal more energy to create an atomic vapor that contains elements in the excited state than it does

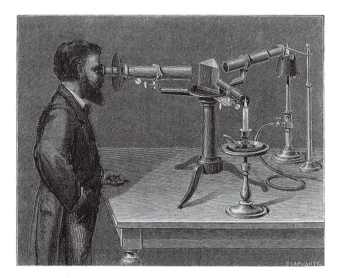

to create atomic vapor in the ground state. By measuring atomic absorption rather than atomic emission, Walsh was able to greatly expand the number of elements that could be probed by atomic spectroscopic techniques.

Note: Gustav Kirchhoff also developed what is now known as Kirchhoff's Voltage Law, which we first discussed in Chapter 4 (see Equation 4.5). Kirchhoff also developed what is now known as Kirchhoff's Law of Thermochemistry on which Josiah Gibbs based his groundbreaking work (Gibbs Free Energy). Kirchhoff also studied emissions from heated solids and coined the term *black body radiation.*[§] While studying black body radiation, Max Planck discovered the quantized nature of energy (Nobel Prize in Physics, 1918), which led to Erwin Schrödinger's development of quantum mechanics (Nobel Prize in Physics, 1933). ■

---

[*] *Kirchhoff, G.; Bunsen, R.; Chemical Analysis by Observation of Spectra. Annalen der Physik und der Chemie, 1860, 110, 161–189.*

[†] *Bunsen, R. W. Journal of the American Chemical Society, 2007, 23, 89–107.*

[‡] *Hannaford, P., Walsh, Sir Alan. (December 19, 1916.) Biographical Memoirs of Fellows of the Royal Society, 2000, 46, 534–564.*

[§] *While studying black body radiation, Max Planck discovered the quantized nature of energy.*

7.1 Introduction

Atomic absorption spectroscopy (AAS) is an *elemental analysis* technique. In other words, it is used to identify and quantify the elements present in a sample matrix. If you wanted to know the concentration of mercury in fish flesh, or of arsenic in a water source, or of lead in a paint sample, it would be quite reasonable to choose AAS as your analytical tool.[1] We will look at three other elemental analysis techniques later in this textbook: atomic emission spectroscopy, X-ray fluorescence spectroscopy, and inductively coupled plasma mass spectrometry. (We will compare and contrast these techniques at the end of Chapter 13.) Like the ultraviolet-visible (UV-vis) spectrometers we discussed in Chapter 6, AAS is fundamentally a form of UV-vis absorption spectroscopy. Unlike molecular UV-vis, AAS requires the analyte to be first converted to atomic vapor prior to analysis. As a result, all *molecular* information is *lost.* This is not to say that *atomic* AAS is inferior to *molecular* UV-vis, but rather that AAS is a tool used to answer different questions than UV-vis spectroscopy answers. AAS is used to quantify the elemental contents of a sample matrix rather than a tool used to probe the molecular nature of a pure analyte. This is an important distinction.

AAS can be used for both quantitative and qualitative determination of more than 70 elements (see Figure 7.1). The utility of the technique is in its capacity for elemental determination within complex matrices such as soil, water, petroleum, plastics, paint, and plant and animal tissue *without the need to isolate* (purify) the analyte from the matrix. As we have already mentioned, the basic principles of AAS parallel those of molecular UV-vis spectroscopy, and a basic block schematic of an AAS looks remarkably similar to a basic block schematic of a UV-vis spectrometer. Figure 7.2 is a block schematic of an atomic absorption spectrometer, and Figure 6.1 is a block schematic of

---

[1] (a) Konieczka, P; et al. Determination of Total Mercury in Fish and Cormorant Using Cold Vapor Atomic Absorption Spectrometry. *Polish J. of Environ. Stud.* **2010**, *19(5)*, 931–936; (b) Behari, J. R.; et al. Determination of Total Arsenic Content in Water by Atomic Absorption Spectroscopy (AAS) Using Vapor Generation Assembly (VGA), *Chemosphere* **2006**, *63(1)*, 17–21; (c) Marcow, P. G. Determining the Lead Content of Paint Chips: An Introduction to AAS. *J. Chem. Educ.*, **1996**, *73(2)*, 178.

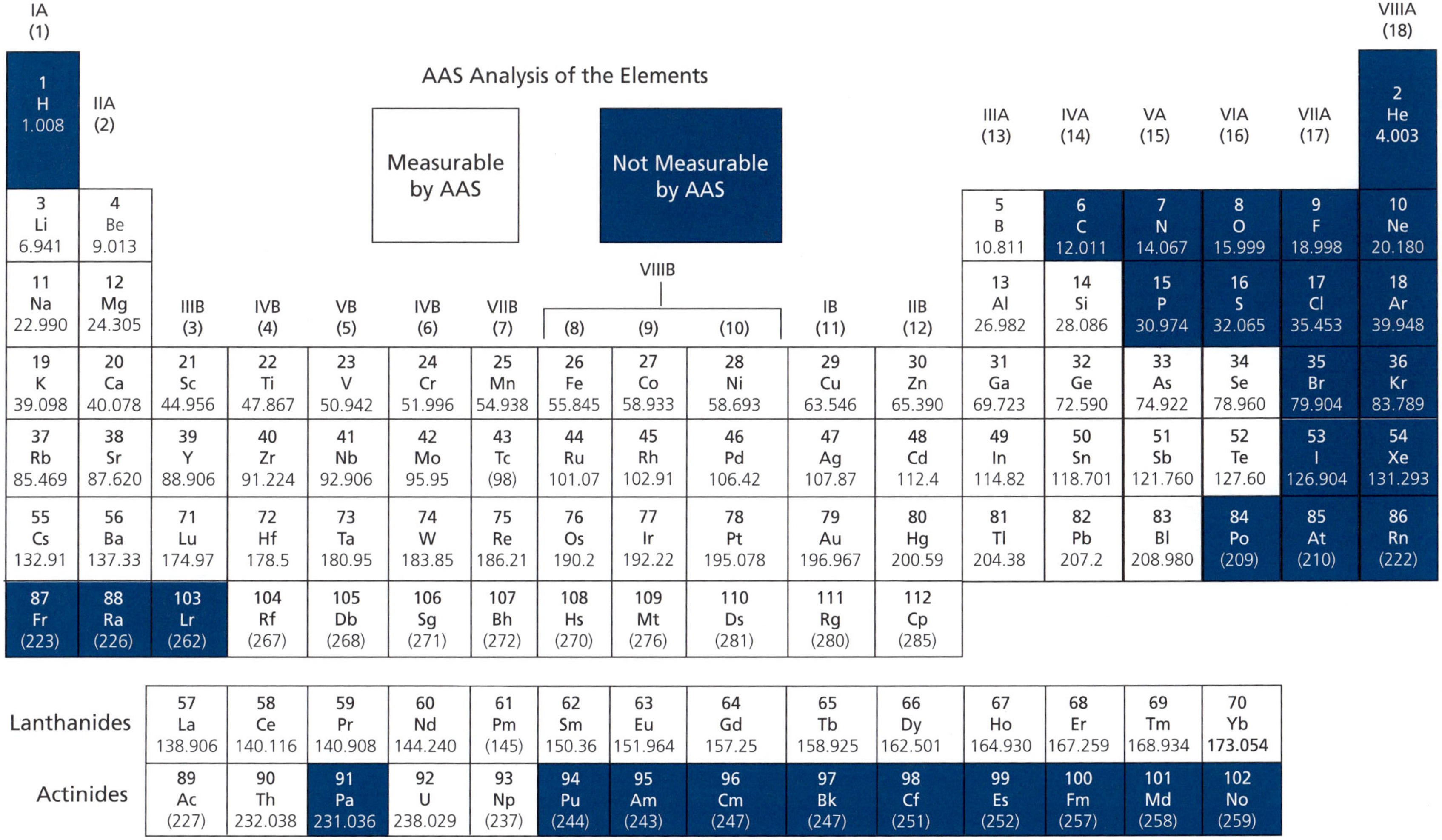

AAS Analysis of the Elements

Measurable by AAS

Not Measurable by AAS

IA (1)	IIA (2)	IIIB (3)	IVB (4)	VB (5)	VIB (6)	VIIB (7)	VIIIB (8)	VIIIB (9)	VIIIB (10)	IB (11)	IIB (12)	IIIA (13)	IVA (14)	VA (15)	VIA (16)	VIIA (17)	VIIIA (18)
1 H 1.008																	2 He 4.003
3 Li 6.941	4 Be 9.013											5 B 10.811	6 C 12.011	7 N 14.067	8 O 15.999	9 F 18.998	10 Ne 20.180
11 Na 22.990	12 Mg 24.305											13 Al 26.982	14 Si 28.086	15 P 30.974	16 S 32.065	17 Cl 35.453	18 Ar 39.948
19 K 39.098	20 Ca 40.078	21 Sc 44.956	22 Ti 47.867	23 V 50.942	24 Cr 51.996	25 Mn 54.938	26 Fe 55.845	27 Co 58.933	28 Ni 58.693	29 Cu 63.546	30 Zn 65.390	31 Ga 69.723	32 Ge 72.590	33 As 74.922	34 Se 78.960	35 Br 79.904	36 Kr 83.789
37 Rb 85.469	38 Sr 87.620	39 Y 88.906	40 Zr 91.224	41 Nb 92.906	42 Mo 95.95	43 Tc (98)	44 Ru 101.07	45 Rh 102.91	46 Pd 106.42	47 Ag 107.87	48 Cd 112.4	49 In 114.82	50 Sn 118.701	51 Sb 121.760	52 Te 127.60	53 I 126.904	54 Xe 131.293
55 Cs 132.91	56 Ba 137.33	71 Lu 174.97	72 Hf 178.5	73 Ta 180.95	74 W 183.85	75 Re 186.21	76 Os 190.2	77 Ir 192.22	78 Pt 195.078	79 Au 196.967	80 Hg 200.59	81 Tl 204.38	82 Pb 207.2	83 Bi 208.980	84 Po (209)	85 At (210)	86 Rn (222)
87 Fr (223)	88 Ra (226)	103 Lr (262)	104 Rf (267)	105 Db (268)	106 Sg (271)	107 Bh (272)	108 Hs (270)	109 Mt (276)	110 Ds (281)	111 Rg (280)	112 Cp (285)						

Lanthanides	57 La 138.906	58 Ce 140.116	59 Pr 140.908	60 Nd 144.240	61 Pm (145)	62 Sm 150.36	63 Eu 151.964	64 Gd 157.25	65 Tb 158.925	66 Dy 162.501	67 Ho 164.930	68 Er 167.259	69 Tm 168.934	70 Yb 173.054
Actinides	89 Ac (227)	90 Th 232.038	91 Pa 231.036	92 U 238.029	93 Np (237)	94 Pu (244)	95 Am (243)	96 Cm (247)	97 Bk (247)	98 Cf (251)	99 Es (252)	100 Fm (257)	101 Md (258)	102 No (259)

Figure 7.1 Elements measurable by AAS. The shaded elements are not easily measured using AAS.

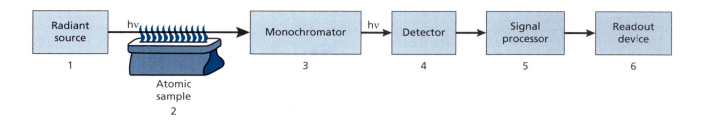

Figure 7.2 Block schematic of a *flame* atomic absorption spectrometer.

a UV-vis spectrometer. Most often the monochromator is a basic Czerny-Turner type (see Figure 3.26) and the detector is a single photomultiplier tube. Take a moment to compare Figure 7.2 to Figure 6.1. Another attractive aspect of AAS is the fact that it renders classic qualitative wet-chemical methods obsolete.

7.2 Molecular versus Atomic Absorption

Despite their similarities, there are fundamental physical differences between molecular spectroscopy and atomic spectroscopy. In Section 2 of Chapter 6 we introduced the idea of *vibronic* transitions. Molecules have bonds and bonded atoms vibrate, so when a *molecule* absorbs a photon, the electronic quantum states involved contain superimposed vibrational states. Figure 7.3(A) shows three different possible starting and ending vibrational quantum states ($v = 1, 2,$ or 3) for a transition between $n = 1$ and $n = 2$ and we are able to represent nine different energies between $n = 1$ and $n = 2$. There are obviously more than three possible vibrational states, and the result is an overall transition with a relatively broad spectral bandwidth. Broad spectral bandwidth in *molecular* UV-vis absorption spectroscopy is a contributing factor to the relatively poor resolution of UV-vis spectroscopy compared to other spectroscopic techniques. In order for molecular UV-vis spectroscopy to be useful in quantifying an analyte, one often must first purify and isolate the analyte. However, a single atom has no bonds and therefore no vibrational quantum states. So when an *atom* absorbs a photon (Figure 7.3(B)), the resulting transition occurs between discrete electronic quantum states. The result is a very narrow spectral bandwidth for that transition and a relatively high analytical specificity.

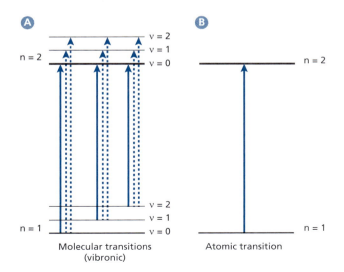

Figure 7.3 Panel (A) represents a molecular (vibronic) transition; n = principal quantum number, v = vibrational quantum number. Panel (B) represents an atomic transition; n = principal quantum number.

Analytical Specificity

Analytical specificity is defined as the ability of a technique to distinguish between two closely related analytes. In AAS, analytic specificity is obtained, in part, as a result of the narrow bandwidth of atomic transitions. The narrow bandwidths allow the analyst to select a wavelength characteristic of a specific element's absorption spectrum and be reasonably confident that there will be no overlapping absorptions from nonanalyte elements in the matrix. In AAS, overlapping absorption bands from other elements are rare. We will revisit analytical specificity when we discuss lamps (sources) later in this chapter. Relative to molecular UV-vis absorption, AAS has very high *analytical specificity*, and usually one can quantify a single element within a complex matrix without first having to isolate the element from the matrix.

On the other hand, a fundamental shortcoming of AAS compared to molecular absorption spectroscopy, is the loss of information regarding the *chemical speciation*[2] of the analyte. In other words, AAS might tell you that you have 3 ppm of cadmium in a sample but AAS cannot tell you the oxidation state of the cadmium nor if the cadmium exists as an ionic salt or as part of a larger biomolecule. Information about the speciation has to be determined by techniques other than AAS.

> **Problem 7.1:** In your own words, explain why the features in molecular absorption spectroscopy are broader than the features seen in AAS.

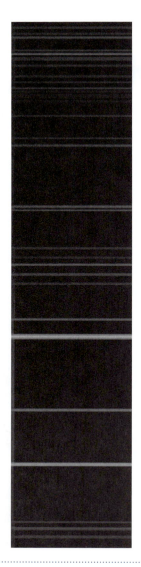

Figure 7.4 Calcium emission spectrum.

7.3 Spectral Bandwidth

From Figure 7.3 one might conclude that an atomic transition corresponds to an exact, *singular* energy transition between two discrete energy states within the atom—sometimes referred to as "infinitely narrow." In practice, we find line widths that range from 10^{-2} picometers to tens of picometers. (See Figure 7.4.). To give us some way of relating to this bandwidth, recall that the visible spectrum is only 300 nm wide (400–700 nm), so 10^{-2} pm is equal to approximately 30 ppb of the entire visible spectrum and a line width of several picometers would be equivalent to about 7000 ppb (7 ppm) of the entire visible spectrum. We can intuitively stipulate that a 10^{-2} picometer bandwidth (30 ppb) is narrow, far exceeding the limits of the spectral resolving power of some low-end monochromators; however, a spectral bandwidth of 10 picometers (30,000 ppb) seems rather large for a transition that should be "infinitely" narrow. Several factors lead to the observed bandwidth of an atomic transition:

1. Lifetime broadening

2. Magnetic field broadening

3. Pressure broadening

4. Doppler broadening

Lifetime Broadening

The first contribution to the intrinsic bandwidth is described by the *Heisenberg uncertainty principle*, which, in the form relating to electronic transitions, states that the accuracy with which you can simultaneously know both the energy and the lifetime of an excited state is limited—as follows:

$$\Delta E \Delta t \geq \frac{\hbar}{2}$$

Eq. 7.1

---

[2] *Chemical speciation* refers to the molecular or ionic form of an element in an analytical matrix.

where

$$\hbar = \frac{h}{2\pi}$$

ΔE = bandwidth for the electronic transition

Δt = lifetime of the electronic transition

Equation 7.1 shows us that as the lifetime of a transition becomes very short, the bandwidth of the transition increases. It is also important to recognize that because $E = \frac{hc}{\lambda}$ the bandwidth due to the Heisenberg uncertainty principle is also *wavelength dependent*. See Example 7.1.

Example 7.1

Using the Heisenberg uncertainty principle, calculate the minimum theoretical bandwidth associated with an electronic transition centered at 500 nm given a lifetime of 10 ns.

STRATEGY –

Solve Equation 7.1 for ΔE and recognize that the theoretical minimum bandwidth would occur when

$$\Delta E \Delta t = \frac{\hbar}{2} \, .$$

SOLUTION –

Solving for ΔE:

$$\Delta E = \frac{\hbar}{2\Delta t} = \frac{1.0546 \times 10^{-34} \text{ J} \cdot \text{s}}{2 \left(1 \times 10^{-8} \text{s}\right)} = 5.3 \times 10^{-27} \text{ J}.$$

Recall that $E = \frac{hc}{\lambda}$ and if we take the derivative we obtain

$$\frac{dE}{d\lambda} = \frac{hc}{\lambda^2} \text{ and } \therefore \Delta E = \Delta\lambda \frac{hc}{\lambda^2}$$

Solving for bandwidth $(\Delta\lambda)$ we obtain

$$\Delta\lambda = \Delta E \frac{\lambda^2}{hc} = 5.3 \times 10^{-27} \text{ J} \frac{\left(500 \times 10^{-9} \text{ m}\right)^2}{\left(6.626 \times 10^{-34} \text{ J} \cdot \text{s}\right)\left(2.99 \times 10^8 \text{ m/s}\right)} = 6.7 \times 10^{-15} \text{ m} = 6.7 \text{ fm}$$

Problem 7.2: Using Example 7.1 as a guide, demonstrate the wavelength dependence of lifetime broadening by calculating the minimum theoretical bandwidth at 200 nm, 400 nm, and 700 nm, assuming a lifetime of 10 ns for each transition. Compare your answers to Example 7.1.

Problem 7.3: Using Example 7.1 as a guide, demonstrate the lifetime dependence of lifetime broadening by calculating the theoretical minimum bandwidth for three different transitions, all centered at 500 nm transition with lifetimes of 10 µs, 10 ns, and 10 fs, respectively.

PROFILE
Review of Term Symbols

Recall that in light atoms (Z < Kr), the total spin (s) of the electrons in the valence shell combine to form a total spin (S). Similarly, the angular momentum of the orbitals (l) combine to form a total angular momentum (L), and S and L couple to form a total angular momentum J (S + L = J).

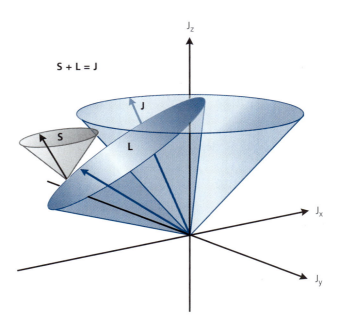

The term symbol is a shorthand method used to designate the angular momentum quantum numbers in a multielectron atom. Each energy level has its own term symbol. The convention for designating term symbols is $^{2S+1}L_J$

$$S = 2s + 1$$

L = Azimuthal quantum number (L = 0, 1, 2, 3... → S, P, D, F...)

For example*, the ground state electron configuration of sodium is Na:[Ne]3s¹. Therefore, 2s + 1 = 2, L = 0 (S) and J = 2 + 0, so the term symbol would be $^2S_2$. ■

---

* Caution! The symbol "s" in the electron configuration of sodium tells us that the single valence electron in sodium resides in an s-orbital. Unfortunately we are also using "s" in the equation 2s+1. In the latter, the symbol "s" is the sum of the spin quantum numbers of the valence electrons. In this case it is simply +1/2.

Magnetic Field Broadening

This type of broadening (also known as the *magnetic field effect* or *spin-orbit effect*) is a second factor contributing to bandwidth of an atomic transition.[3] The *magnetic field effect* occurs in multielectron elements and arises from the fact that the orbitals themselves have an angular momentum, and so the rotation of an occupied orbital generates its own magnetic field. The magnetic field created by the spin of an electron promoted through an electronic transition can be either parallel or antiparallel to the magnetic

---

[3] Griem, H. R. Spectral Line Broadening by Plasmas. *Pure and Applied Physics,* **1974**, 39, 421.

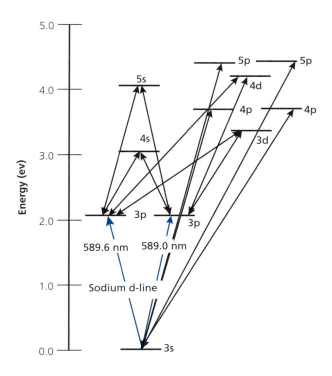

Figure 7.5 A sodium Jablonski diagram showing the degeneracy of the p-orbitals.

field generated by the angular moment of the orbital itself. This results in transitions with slightly different energies. For instance, the promotion of sodium's 3s electron into the 3p orbital can occur parallel or antiparallel to the magnetic moment of the 3p orbital and the result is a difference in energy of 3.3×10^{-16} joules, which translates into a wavelength separation of 0.6 nm. (See the sodium d-line in Figure 7.5.) Although commercial spectrometers are capable of resolving 0.6 nm for larger elements, there are many more closely spaced transitions due to the magnetic field effect. The magnetic field effect is not technically intrinsic line broadening but rather the result of the limitations of the monochromator's resolving power. Because the origin of the magnetic field effect is rooted in the physical nature of an atom, there are no adjustable parameters for the operator to exploit to correct for this bandwidth and one must accept a certain amount of bandwidth as inevitable when using an instrument incapable of resolving such closely spaced atomic lines.

Problem 7.4: Above we stated that the promotion of the sodium 3s electron into the 3p orbital can occur parallel or antiparallel to the angular momentum of the 3p orbital ($J = S + L$ or $J = S - L$). Derive the term symbols for the two possible excited states.

Problem 7.5: The 4s → 4p transition in potassium experiences a splitting of the 4p orbital in a manner similar to that seen for sodium.[4] The 4s → $4p_{1/2}$ transition occurs at a frequency of 3.89×10^{14} Hz and the 4s → $4p_{3/2}$ transition occurs at a frequency of 3.91×10^{14} Hz. Calculate the energy difference in joules and the wavelength difference in nanometers for these two transitions.

[4] Sansonetti, J. E. Wavelengths, Transition Probabilities, and Energy Levels for the Spectra of Potassium. *Journal of Physical and Chemical Reference Data*, **2008**, *37(1)*, 796.

Have you ever watched a lightning storm out over the ocean? If you have, you might have noticed that the color of the lightning strikes was a yellow-orange color instead of the white color seen over land. The yellow-orange color comes from the sodium atoms in the "salty air." The lightning strike produces excited state sodium atoms and you are seeing the fluorescence from the d-lines. See Figure 7.5. ■

Pressure Broadening

Pressure broadening is a third factor affecting the bandwidth of an atomic transition. It is also commonly referred to as *collision energy transfer* or the pressure effect.[5] Collisions between analyte atoms with other particles in the flame results in small perturbations in the ground state energy of the electrons in the analyte, and this leads to a broadening of the electronic transition. The bandwidth (Δv_p) due to pressure broadening can be calculated as:

$$\Delta v_p = \sigma N_p \frac{\sqrt{\left[2\pi RT\left(\frac{1}{M_1} + \frac{1}{M_2}\right)\right]}}{\pi}$$

Eq. 7.2

where

σ = collision cross-sectional area of the particle (cm^2)
N = number density of perturbing particles per cubic centimeter (cm^{-3})
R = gas constant
T = Kelvin temperature
M_1 and M_2 = molar mass (g/mol) of the analyte and perturbing particles, respectively

The values for Δv_p calculated from Equation 7.2 range from 1 to 10 pm and are often a significant source of line broadening in AAS.[6]

---

[5] Baranger, M. Simplified Quantum-Mechanical Theory of Pressure Broadening. *Phys. Rev.*, **1958**, *111*, 481.

[6] Lindon, J. C. (ed.) *Encyclopedia of Spectroscopy and Spectrometry*, 2nd ed., Academic Press: New York, **2010**.

Although there is not much one can do to prevent collisional energy transfer from occurring in the sample, one can control (to some degree) collisional energy transfer *in the source*. We will explore this further when we examine the construction of source lamps.

Doppler Broadening

Doppler broadening results from the thermal motion of the analyte atoms.[7] At the temperatures of a typical flame AAS, the analyte atoms are moving at $\approx 1000\,\text{m}/\text{s}$. The energy gap for a specific transition has an energy of $E = h\nu = \dfrac{hc}{\lambda}$, but the actual wavelength of light absorbed by the analyte atoms is Doppler shifted due to their thermal motion relative to the stationary source as well as the speed of the photons from the source:

$$E = \frac{h(c+u)}{\lambda_t} \quad \text{(moving toward the source)} \qquad \textbf{Eq. 7.3a}$$

$$E = \frac{h(c-u)}{\lambda_a} \quad \text{(moving away from the source)} \qquad \textbf{Eq. 7.3b}$$

where

u = *velocity of the analyte atom relative to the source*
λ_t = *photon absorbed by analyte moving away from the source*
λ_a = *photon absorbed by analyte moving towards the source*

Equation 7.3a tells us that an analyte atom moving toward the source will have to absorb a longer wavelength photon (lower energy) in order to match the energy gap of a stationary atom and likewise Equation 7.3b tells us that an analyte atom moving away from the source lamp during the absorption event will have to absorb a shorter wavelength photon (higher energy) in order to match the energy gap for the transition. The effect is proportional to the kinetic energy of the atoms in the flame and thus is more pronounced as flame temperature increases. This phenomenon is exactly analogous to the compression of sound waves one hears as a car approaches. If the car is moving faster, we hear a greater Doppler shift in the sound waves, resulting in a higher pitch to the sound.

Problem 7.6: Doppler broadening is a significant source of line broadening in atomic spectroscopy. Using a maximum velocity of 2000 m/s for an analyte atom in a flame, use Equation 7.3 to determine the line broadening ($\Delta\lambda$) associated with a transition centered at 500 nm. Report your answer in nanometers.

Problem 7.7: Show the wavelength dependence of Doppler broadening by repeating Exercise 7.6 for transitions centered at 200 nm, 420 nm, and 680 nm.

Flame temperatures in flame-AAS (FAAS) are selected to maximize atomization of the analyte.[8] However, excessive flame temperatures increase Doppler broadening unnecessarily and can lead to loss of analytical selectivity.

Note
IUPAC makes the following distinctions with respect to terms used in describing pressure broadening: if the collisions are between dissimilar uncharged particles, the preferred terms are *foreign gas broadening* or *van der Waals* broadening. If the colliding particles are the same atom type, the proper term is *resonance broadening.* IUPAC now discourages the use of the term *Lorentz broadening*; however, the term is still in common use.* In the case of AAS, what we are observing is properly termed van der Waals broadening.

---

* *IUPAC. Compendium of Chemical Terminology (the "Gold Book"), 2nd ed. Compiled by A. D. McNaught and A. Wilkinson. Blackwell Scientific Publications: Oxford, UK, 1997.*

---

[7] Rautian, S. G.; Sobel'man, I. I. The Effects of Collisions on the Doppler Broadening of Spectral Lines. *Sov. Phys. Usp.* **1967**, *9*, 701.

[8] The same considerations regarding operating temperatures should be made in graphite furnace–AAS.

Problem 7.8: Doppler broadening of atomic transitions is temperature dependent and the effect on the bandwidth ($\Delta\lambda$) compared to an atom with no velocity is $\dfrac{\Delta\lambda}{\lambda} = \dfrac{v}{c}$, where v is the velocity of an atom and c is the velocity of light. The relationship between an atom's velocity and temperature is given by the equation $v = \sqrt{\dfrac{8kT}{\pi m}}$, where k = the Boltzmann constant, T is temperature (Kelvin), and m is the mass of the atom in kilograms. Calculate the effect of Doppler broadening on the bandwidth of the calcium 422.7 nm line in an 1800°C acetylene flame.

Problem 7.9: List and explain the factors contributing to the electronic bandwidths seen in atomic transitions.

Although we must accept some Doppler broadening in the sample, we can enhance analytical selectivity by minimizing Doppler broadening at the source.

7.4 Atomic Absorption Spectroscopy Sources

The selection of a radiant source for AAS presents unique challenges that do not exist for molecular spectroscopy. In molecular spectroscopy, any continuum source of sufficient intensity over the desired absorption range would suffice. However, in AAS, the absorption bandwidth of the analyte is often only a few picometers wide, so if you used a continuum source, you would be throwing away the vast majority of your source radiation. Take, for example, a typical UV-vis spectrometer. A radiant source might span a wavelength range of 190 to 700 nm for a total bandwidth of approximately 510 nm (510,000 picometers). If a typical absorption bandwidth for an atomic transition is only 5 picometers wide, then 99.9991% of the energy is being lost or wasted (only $\approx$ 0.00091% of the radiant power is being used for probing the analyte[9]).

$$\frac{5\,pm}{550{,}000\,pm} \times 100\% = 0.00091\% \qquad \textbf{Eq. 7.4}$$

Given a 100-watt bulb for the source, you would have only 0.91 milliwatts of power probing your analyte. Assuming your monochromator could adequately prevent the remaining 99.9991% of the source radiation from reaching the detector, you would have to have an exceptionally sensitive detector capable of responding to small changes relative to 0.91 milliwatts of the radiant power reaching the detector. And if you tried to power up the source, by using for example a 1000-watt bulb, you would increase the difficulty of shielding the detector from all of the additional incident energy.

The Hollow-Cathode Lamp

The source in AAS is an element-specific line source similar in design to the schematic seen in Figure 7.6. This lamp design is referred to as a *hollow-cathode lamp* (HCL). The cathode is designed as a cylindrical cup or cavity and the cathode is made from or coated with the element to be analyzed (target element). It is the excitation of the target element that produces an element-specific line source. The HCL contains a small pressure, typically less than 30 torr, of a fill gas specifically chosen to have no spectral overlap to the target element. As electrons stream from the cathode, they collide with and excite the fill gas atoms to produce cations of the fill gas. These positive ions are accelerated toward the negative cathode, colliding with the target element on the surface of the cathode. Collision of the fill gas ions with the target element results in sputtering of excited state target atoms into the gas state. As the

---

[9] We are making a gross assumption that radiant power is constant over the entire bandwidth of the source.

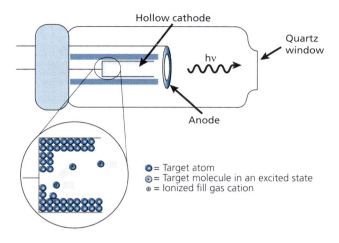

Figure 7.6 Schematic of a hollow-cathode lamp.

gaseous excited state target atoms relax, they emit the characteristic line spectrum of that element. The purpose of the hollow cathode design is to trap the sputtered target atoms in a small region of space (the cavity area), which intensifies the emission as a narrow directed beam. Trapping the sputtered atoms within the cup also increases the likelihood that the atoms will redeposit on the cathode, thus extending the lifetime of the lamp.

The design of the HCL also addresses several analytical objectives.

- First, we are using a source that is generating a line spectrum specific to our analyte, so we are no longer throwing away the vast majority of our source radiation. This greatly enhances the sensitivity of the technique and allows us to use monochromators and detectors with lower design specifications (i.e., they are less expensive).

- Second, because the HCL operates at very low pressure and at temperatures that are cooler than the flame, the HCL emissions suffer less pressure and Doppler broadening than the corresponding absorption events of the analyte in the flame. In other words, the emission bandwidths from the source are significantly narrower than the absorption bandwidths of the analyte atoms. As a result, the absorption of source radiation by the analyte atoms *can occur only* within the *narrower* bandwidth of the source emissions. So the instrument is *"blind"* to absorption events that occur due to those atoms whose absorption bandwidth has been increased due to pressure and Doppler broadening within

Monitoring the nutritional content of breast milk is important in developing nations because breast milk remains the primary source of nutrition for babies. In 2010, scientists in China reported developing multielement hollow cathode lamps that have enabled them to develop a method that allows for the AAS analysis of breast milk for Ca, Cu, Fe, Mg, Mn, and Zn with a linear analytical range of 0.01 to 10 ppm and detection limits in the low ppb range. ■■

PROFILE

Nutritional Contents of Breast Milk

Fu, W., et al. J. Agric. Food Chem. 2010, 58, 9396.

the flame. In summary, it is the narrow bandwidths generated by the HCL that preserves the analytical selectivity of AAS.

The intensity of an HCL is proportional to the voltage drop between the anode and cathode. As you increase the voltage, you create more charged fill-gas ions, which results in more target atom sputtering. However, you also increase the energy of the collisions between the fill-gas and the target element. One can equate a higher voltage in the HCL to higher temperatures in a flame, and the result is an increase in Doppler broadening as the voltage between the cathode and anode is increased. The optimum voltage drop is one that will give you a sufficiently bright source without exceeding the bandwidth of the absorbing analyte. The ideal voltage for a particular lamp used for a particular analysis is determined experimentally. The software that accompanies most commercial AAS instruments has preset voltage values for each lamp that accompanies the instrument. These preset voltages best approximate the ideal voltage for each lamp used under "normal" conditions. However, the operator should understand that one might be able to enhance the spectral purity of a particular technique by lowering the voltage in the lamp. The trade-off would be a weaker source and poorer detection limits. Under certain conditions, this is an acceptable trade-off. Conversely, the operator may be able to improve the detection limit for a particular analysis by increasing the voltage drop in the lamp; however, this decision would certainly decrease the life of the lamp and in rare cases result in a decrease in spectral purity and lower analytical selectivity. Likewise, under certain conditions, these are acceptable trade-offs.

Because, in most cases, each lamp emits characteristic lines only for a single element, it is necessary to switch lamps when measuring different species in a single experiment. Many modern instruments have arrays or turrets of lamp mounts so that several lamps can be installed in a given instrument, with computer-controlled selection of the appropriate lamp during an experiment. Furthermore, multielement HCLs exist, in which the cup is lined with an alloy of several elements instead of a single element. The elements used in a multielement lamp must be carefully selected so that there is no overlap between the elemental emissions. Also, the amount of a given element per unit area of the inner lining of the cup is diminished with respect to a single-element lamp, so to achieve a similar emission intensity, the voltage must be increased significantly, resulting in the expected decrease in life span.

Electrodeless Discharge Lamps

Some metals are relatively volatile, and the elevated temperatures of HCLs result in lamps with somewhat short life spans. Furthermore, because HCLs work on the principle of electric discharge, they are unsuitable for those metals that have low electrical conductivity. For metals that fall into either of these categories, there is another type of source lamp used in AAS: the *electrodeless discharge lamp (EDL)* (See Figure 7.7). The EDL is constructed with a quartz chamber containing an inert gas and a ceramic tube surrounded by radiofrequency coils. The ceramic tube also contains the target element or a salt containing the target element. Inert gas ions are energized by the radiofrequency radiation and impinge on the target atoms, resulting in emission of the characteristic line spectrum of the target element.

EDLs have several advantages over HCLs:

- First, because of the lower temperatures, EDLs last longer than HCLs.

- Second, by carefully selecting target elements whose emissions do not overlap, EDLs can be constructed with more than one target element with no decrease in life span because the elements are trapped within the quartz bulb and cannot diffuse away as is the case for HCLs. This allows the operator to use the same lamp for multiple element analyses.[10]

The narrow bandwidth of the source also leads to a linear Beer's Law response for absorption vs. concentration.

---

[10] Patel, B. M.; Browner, R. F.; Winefordner J. D. Design and Operation of Temperature-controlled Multiple Element Electrodeless Discharge Lamps for Atomic Fluorescence Spectrometry. *Anal. Chem., 44 (14),* 2272–2277, **1972.**

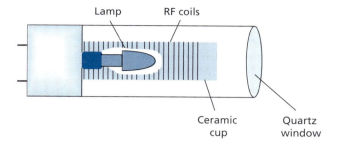

Figure 7.7 Schematic of an EDL. EDLs are used for less conducting, metals with higher volatility.

- Third, EDLs are brighter than HCLs (10 to 100 times), which results in better detection limits. This feature is perhaps the most important consideration from an analytical perspective.

EDLs also have several disadvantages compared to HCLs:
- First, they are considerably more expensive. However, this negative aspect is offset somewhat by the fact that EDLs last longer than HCLs and by the fact that EDLs can be used for the analysis of more than one element without wearing out quickly.

- Second, the intensity of the light emitted by EDLs is not as consistent. As a result, the analyst must average more data to obtain the same confidence interval. This may not be much of a concern if the analyst is looking at a single sample, but in a commercial laboratory where throughput of large batches requires analysis of multiple samples, the added analysis time can result in considerable cost.

Problem 7.10:   Why is it desirable to use an element specific source instead of a continuum source in AAS?

Problem 7.11:   Explain how the use of the HCL improves the analytical selectivity of AAS.

Problem 7.12:   Explain how the construction of the HCL extends the life of the lamp.

ACTIVITY

Soil Analysis

In a 2006 paper,* an undergraduate researcher reported on the micronutrient "health" of both the soil and petioles (leaf stems) of a local orchard and vineyard. By comparing the nutrient concentration in the soil and petiole, she was able to gauge both the gross availability of the micronutrients and their bioavailability (uptake). Copper and zinc are essential micronutrients in the development and growth of vineyards and orchards. Copper is a vital enzyme component that contributes to the color and flavor of fruits. Zinc functions to regulate enzymes, influence plant hormones, and promote starch formation. When there is a deficiency of copper or zinc in the environment, the fruit is often underdeveloped and the leaves begin to show symptoms of stress. This paper readily demonstrated why atomic absorption spectroscopy (AAS) is such a popular technique, especially for undergraduates wanting to conduct original research. A similar experiment could easily be conducted on any nearby agricultural site using AAS. In addition, in the past, arsenic-based pesticides were commonly used in apple orchards. Student researchers could easily modify the techniques in this paper to study trace toxins such as arsenic, lead, and mercury in the soil of commercial sites.

* Wilson, K. M.; Granger, R. M. Analysis of the Micronutrients Copper and Zinc in a Commercial Orchard and Vineyard by Atomic Absorption Spectroscopy. Chem. Educator, 2006, 11, 1–3.

7.5 Sample Introduction

As the name implies, samples introduced into an AAS instrument must be converted into atoms (atomized) before they can be measured. This is commonly accomplished in one of two ways: using a nebulizer that sprays a mist containing the sample into a flame or using electrothermal atomization (a graphite furnace). Many commercial instruments come equipped with interchangeable flame and furnace modules that can be moved into and out of the optical path.

Flame Atomic Absorption Spectroscopy

In *flame atomic absorption spectroscopy* (FAAS), the sample matrix is *nebulized*, converting it to an aerosol, and the mist is then mixed with a fuel (such as acetylene) and swept into a flame, where the solvent is quickly evaporated and the sample is thermally decomposed into an atomic gas. The reproducibility of the nebulization step governs the precision and quantification level of the technique. Figure 7.8 shows a pneumatic nebulizer manufactured by Perkin Elmer in line with a laminar-flow burner head. The flame produced by a laminar-flow burner appears as a line, either 5 or 10 cm in length, depending on the type of oxidant used. Typically, one uses compressed air as the oxidant with a 10 cm flame; however, some experimental methods use pure oxygen or nitrous oxide with a 5 cm flame. Within the nebulization chamber, compressed air passes over the tip of the sample capillary at a right angle, causing the sample solution to be aspirated into the nebulizing chamber. The mist containing the sample is mixed with fuel, and the

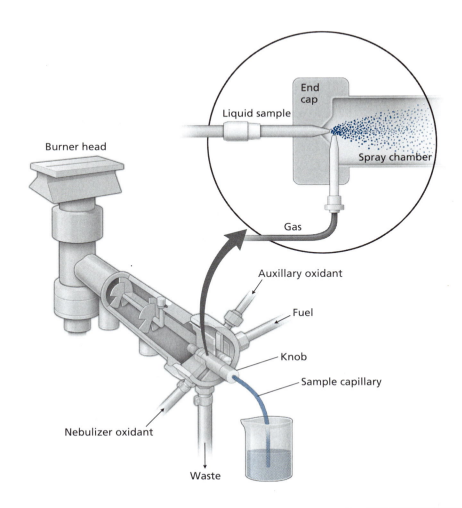

Figure 7.8 The laminar flow burner and pneumatic nebulizer from a Perkin Elmer AAS.

larger droplets are removed through impact with baffles in the expansion chamber. The remaining aerosol makes its way to the burner head where the sample is atomized in the flame. The optical beam of the spectrometer is aligned with the longest axis of the burner head in order to maximize the path length and similarly maximize absorption.

The pneumatic nebulizer is the most commonly found type for FAAS. However, this does not work as well as other techniques for samples with a high ionic strength or samples with a high viscosity. The Babington nebulizer is more suited for these types of samples. In the Babington nebulizer, the sample is allowed to run down the surface of a sphere. The oxidant gas is forced through an orifice in the sphere that jets the sample as a fine mist (see Figure 7.9). The Babington nebulizer is a better choice for viscous liquids and has been successfully used for such samples as whole blood, urine, seawater, evaporated milk, and tomato sauce.[11]

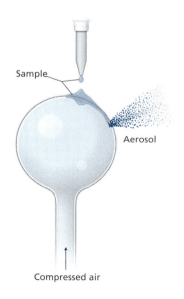

Figure 7.9 A Babington nebulizer.

Problem 7.13: Discuss the considerations made in choosing a nebulization technique.

Problem 7.14: Explain why the nebulization step is fundamentally responsible for determining the lower limit for quantitative analysis in AAS.

The Flame   When performing a flame AAS experiment, our goal is to maximize the quantity of ground state analyte atoms in the optical path of the spectrometer. Therefore, it is prudent for us to spend some time contemplating the varied events that occur in the flame and how each affects the concentration of useful analyte in the optical path of the spectrometer. When the sample matrix enters the flame, some portion of the molecular matrix is converted to an atomic vapor, but incomplete atomization of the sample matrix always occurs. If the percent atomization of the sample matrix were the only variable in determining the concentration of analyte in the optical path, we might falsely assume that our goal in flame AAS is to create as hot a flame as possible. However, as the molecules in the flame are destroyed, the resulting atoms can be produced in one of three states: (1) ground state atomic gas, (2) excited state atomic gas, and (3) atomic gaseous ions. Only ground state atomic gas is desirable in an atomic absorption experiment. Excited state atomic gas will emit at the same wavelength that the ground state atoms absorb and will lead to low absorption values; atomic gaseous *ions* will have a different line spectrum altogether.

There are two important components of the flame that the analyst controls. The first is the overall temperature of the flame. The second is the height of the optical path above the burner head. We will explore each of these variables separately.

The flame's temperature has a direct effect on the percent atomization of the sample matrix as well as the number of ground state atoms, excited state atoms, and atomic ions. By varying the type of fuel and oxidizer, the fuel's flow rate, the ratio of fuel to oxidizer, and/or the rate at which we aspirate sample into the flame, the analyst can control the temperature of the flame. The aspiration rate obviously effects the quantity of analyte in the optical path; however, if you aspirate at too high a rate, you can cool the flame somewhat. The optimal aspiration rate is determined experimentally while aspirating a standard containing the element of interest.

In practice, your instrument will most likely be set up with only one type of fuel and only one source of oxidizer, but it is important to realize that most commercial instruments allow for changes in the fuel or oxidizer. For most sample matrices, an acetylene/air flame will be sufficient to atomize your matrix. However, many metal oxides can

---

[11] (a) Bings, N. H.; Bogaerts, A.; Broekaert, J. A. C. Atomic Absorption Spectroscopy: A Review. *Anal. Chem.*, **2010**, *82*, 4653–4681. (b) Saba, C. S.; Rhine, W. E.; Eisentraut, K. J. Efficiencies of sample introduction systems for the transport of metallic particles in plasma emission and atomic absorption spectrometry. *Anal. Chem.*, **1981**, *53*(7), 1099–1103. (c) Fry, C. R.; Denton, M. B. High Solids Sample Introduction for Flame Atomic Absorption Analysis. *Anal. Chem.*, **1977**, *49*(9), 1413–1417.

TABLE 7.1: Flame Temperatures

	AIR	NITROUS OXIDE	OXYGEN
Acetylene	2,100–2,500°C	2,500–2,800°C	3,000–3,400°C
Natural gas	1,700–1,960°C		2,650–2,850°C

The John Zink Combustion Handbook, CRC Press: Boca Raton FL, 2000.

be difficult to atomize, and an acetylene/oxygen flame might be required in order to obtain the desired sensitivity. Table 7.1 shows the temperature range of various fuel/oxidizer mixtures. We also can control the flow rates of the fuel and oxidizer, allowing us to adjust the fuel/oxidizer ratio. A fuel-rich flame is a cooler, more reducing environment than an oxidant-rich flame.

As previously mentioned, it is not correct to assume that a hotter flame is always a better choice. The flame's temperature has a direct effect on the quantity of ground state and excited state analyte atoms obtained in the optical path of the spectrometer. The ratio of excited state to ground state atoms can be estimated using the Boltzmann distribution (Equation 7.5).

$$\frac{N_e}{N_g} = \frac{g_e}{g_g} e^{\left(-\frac{\Delta E}{k_B T}\right)}$$

Eq. 7.5

where

$\dfrac{N_e}{N_g}$ = ratio of excited state versus ground state atoms

ΔE = energy difference between the ground state and the excited state

T = absolute temperature

$\dfrac{g_e}{g_g}$ = ratio of degenerate states in the excited and ground states

k_B = Boltzmann constant

We can see by examining Equation 7.5 that increasing the temperature increases the ratio of excited state atoms relative to ground state atoms. However, the temperature has to be quite hot before there is significant signal loss due to the generation of excited state analyte atoms.

Example 7.2

What is the ratio of ground state to excited state atoms for sodium atoms in an oxygen/acetylene flame? Assume a flame temperature of 3,400°C and that $\dfrac{g_e}{g_g} = 1$.

STRATEGY –

Use Figure 7.5 to find ΔE for the 3s → 3p transition and Equation 7.5 to find the ratio.

ANSWER –

From Figure 7.5 we find that the transition of interest has an absorption wavelength of 589 nm, therefore:

$$\Delta E = \frac{hc}{\lambda} = \frac{\left(6.626 \times 10^{-34}\,\text{J} \cdot \text{s}\right)\left(2.99 \times 10^{8}\right)}{5.89 \times 10^{-7}\,\text{m}} = 3.36 \times 10^{-19}\,\text{J}$$

Solving Equation 7.5 we obtain the relationship:

$$\frac{N_e}{N_g} = \frac{g_e}{g_g} e^{\left(-\frac{\Delta E}{kT}\right)} = (1)e^{-\frac{3.36 \times 10^{-19} \, J}{\left(1.38 \times 10^{-23} \, J \cdot K^{-1}\right)\left(3673 \, K\right)}} = 0.0013$$

So only about 0.13% of the atoms are excited.

We want the flame to be hot enough that the majority of the sample matrix is atomized. Also, we want it hot enough to discourage the formation of oxides in the flame but not so hot that the analyte is produced as an excited state atom or ion.

Problem 7.15: Using Example 7.2 as a guide, calculate the temperature needed to create gaseous sodium atoms with 50% of the atoms in the excited state. Is 3,400°C a realistic temperature?

The Flame Height   In addition to the flame temperature, the height at which the source beam of the spectrometer passes through the flame is a key variable in determining the sensitivity of an AAS measurement. When one selects an element for AAS measurement, one of the first things the analyst will do is adjust the height of the flame relative to the optical path. This is done while aspirating a standard containing the element of interest and monitoring the signal. The goal for the analyst is to find the flame height that gives the maximum absorption value. Figure 7.10 represents a typical hydrocarbon flame showing the primary combustion zone, the interzonal region, and the secondary combustion zone. The figure is drawn as the cross-section of the narrow edge of the flame, with your eye in the vantage point of the detector, and the circle in the middle of the flame represents the HCL or EDL light source. Although most of the sample is converted to ground state atoms, some of the atoms generated in the flame are generated in an excited state and others are ionized. Also, as the atoms rise in the flame, some combine with oxygen to form stable oxides or with other species in the flame. The formation of molecular species containing our analyte atom is undesirable and effectively reduces the sensitivity of the technique. Ideally, the source will shine through the region of the flame with the highest relative abundance of atomic *ground state* atoms. The height in the flame where the highest concentration of analyte atoms occurs varies considerably by element. For instance, if your element is difficult to oxidize, you would find that the absorbance increases steadily from the base of the primary combustion zone to the edge of the secondary combustion zone. This is a direct result of your sample's spending more time in the flame and therefore generating more gas phase atoms. On the other hand, some elements display their maximum absorption at the very top of the primary combustion zone. This type of behavior is usually seen for elements that readily form stable oxides. The decrease in absorption as you move higher in the flame is a result of oxide formation. As you can surmise, most elements display a behavior in between these two extremes and the maximum absorption will typically be seen in the interzonal region.

Problem 7.16: Using your own words, write a short paragraph that describes the relationships between maximum sensitivity, flame temperature, and flame height.

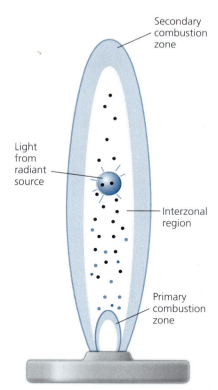

Secondary combustion zone

Light from radiant source

Interzonal region

Primary combustion zone

• Ground state analytic atoms
• Excited state analytic atoms
* Analyte ions

Figure 7.10 Profile of a typical acetylene/air flame.

Electrothermal Atomic Absorption Spectroscopy/Graphite Furnace Atomic Absorption Spectroscopy

In electrothermal AAS (also known as graphite furnace–AAS or GFAAS), the pneumatic nebulizer and burner head are replaced with a graphite furnace. The graphite furnace is a small graphite tube aligned with the optical path of the spectrometer (see Figure 7.11). Samples are introduced through a small hole in the side of the tube. Liquid samples (1–5 μL) are typically introduced using a robotic micropipetter to improve reproducibility. Solid samples can also be analyzed using a graphite furnace.

Sample atomization occurs in four steps:

- First, an electrical current is passed through the graphite tube to raise its temperature to approximately 800°C. This dries the sample so that rapid evaporation of water in the sample does not cause sample loss due to steam eruptions during the atomization step.

- Second, the electric current is increased in order to rapidly raise the temperature to between 1,200 and 1,400°C. This ashes the sample, decomposing any carbonaceous material.

- Third, the electric current is increased again, bringing the graphite furnace to a temperature above 2,000°C, resulting in a plume of gaseous atoms that are constrained in the optical path. At this time, the absorbance is read.

- Fourth, after the absorbance reading is obtained, the current is increased yet again, bringing the graphite furnace to a temperature above 3,000°C. At this point, the graphite furnace is white hot, and any residual sample matrix is "cleaned" from the furnace. Once the furnace has cooled down, the instrument is ready for another sample.

The graphite furnace is surrounded by a sleeve of argon gas that protects the graphite from oxidation, sweeps out the gases produced by the sample during the heating stages, and serves to cool the graphite tube between runs.

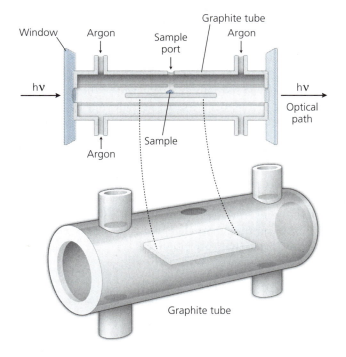

Figure 7.11 Schematic of a graphite furnace.

Scientists in the petroleum refining industry monitor the vanadium (V) and nickel (Ni) content of crude oil. These two metals can adversely affect the catalysts used in the refining process. A 2010 paper* outlines an atomic absorption spectroscopy procedure for the simultaneous determination of total Ni and V and nonvolatile Ni and V. This is a valuable development because the V/Ni ratio remains relatively constant regardless of the physical or chemical processes applied to the crude oil. Thus, an evaluation of the V/Ni ratio also allows one to determine the origin of the crude oil. ■■

Quadros, P. C. et al. Evaluation of Brazilian and Venezuelan Crude Oil Samples by Means of the Simultaneous Determination of Ni and V as Their Total and Non-volatile Fractions Using High-Resolution Continuum Source Graphite Furnace Atomic Absorption Spectrometry. Energy Fuels, 2010, 24, 5907–5911.

The confinement of the atom plume in the graphite tube during the atomization step results in higher analyte concentrations in the optical path; this, in turn, results in better detection limits. One often finds three orders of magnitude better detection limits with GFAAS compared with FAAS.

Flame versus Electrothermal Atomic Absorption Spectroscopy

GFAAS has several distinct advantages over FAAS.

- First, it uses considerably less sample than FAAS. A typical measurement using GFAAS consumes 5 to 25 µL of sample per measurement, whereas FAAS uses between 9 to 15 mL of sample per measurement.[12]

- Second, the detection limits are considerably better in GFAAS[13] than FAAS, often by as much as one thousand times.

However, FAAS has several distinct advantages over GFAAS, and one should consider all variables when choosing an analytical technique (Table 7.2).

- First, FAAS has a much higher sample throughput (it is much faster) than GFAAS. The typical FAAS measurement takes about 30 seconds per sample. Because of the thermal cycles involved in GFAAS (drying, ashing, atomization, cleaning, cooling), a typical GFAAS measurement can take up to 5 minutes per sample.

TABLE 7.2: Comparison of Flame Atomic Absorption Spectroscopy (FAAS) and Graphite Furnace Atomic Absorption Spectroscopy (GFAAS)

	SAMPLE CONSUMPTION	DETECTION LIMITS	MEASUREMENT TIME	RELATIVE ERROR
FAAS	~ 10–15 mL	~1–2 ppm	~20–30 s	~ 1–2%
GFAAS	~ 5–15 µL	~ 1–2 ppb	~ 5–8 min	~ 5–20%

[12] This assumes five replicates for each sample in electrothermal mode and three replicates for each sample in flame mode.

[13] Mickadeit, F. K.; Berniolles, S.; Kemp, H. R.; Tong, W. G. Sub-Parts-Per-Quadrillion-Level Graphite Furnace Atomic Absorption Spectrophotometry Based on Laser Wave Mixing. *Anal. Chem.*, **2004**, *76(6)*, 1788–1792.

- Second, the relative error from sample to sample is higher using electrothermal atomization (typically 5–20%) than with flame atomization (typically < 1–2%). This translates into needing a higher sample number for a given confidence limit using electrothermal atomization. Because you will need to conduct more replicate analyses for the same confidence interval using GFAAS and each replicate takes more time using GFAAS, you find that GFAAS is considerably slower than FAAS.

Many AAS instruments have autosampler trays for both modes (flame and electrothermal), and if you are content to let a batch run over lunch or through the night, the slow nature of GFAAS may not be a concern. However, if the detection limits of FAAS are sufficient for the desired measurement and you have sufficient sample, FAAS is a superior technique when high throughput is required.

Detection limits for FAAS are typically in the mg/L (ppm) range and, depending on the element and the specific technique, detection limits in the sub-µg/L (ppb) range can be obtained. Detection limits for GFAAS are typically in the µg/L (ppb) range.

Problem 7.17:   Under what analytical parameters is FAAS a more suitable technique than GFAAS? Under what analytical parameters is GFAAS a more suitable technique than FAAS?

Problem 7.18:   Assume you own a water analysis laboratory and one of the services you provide is the determination of magnesium and calcium in well water. Your overhead (e.g., utilities, rent, payroll, chemicals, supplies) is approximately $1,200/week. Using your autosampler tray, you can process 400 samples/day in flame mode and 150 samples/day using furnace mode. Operating at full capacity:

(a) What is your "breakeven" price per measurement if you ran all of your samples in flame mode at full capacity?

(b) What is your "breakeven" price per measurement if you ran all of your samples in furnace mode at full capacity?

Hydride Atomic Absorption Spectroscopy

Some elements such as antimony (Sb), arsenic (As), bismuth (Bi), lead (Pb), tin (Sn), and selenium (Se) do not readily produce atomic vapor in a flame, especially if these elements exist in a high oxidation state within the matrix. Given the fact that each of these elements can have significant health and environmental impact, a reliable method of detection and quantification is desirable. Fortunately, a technique called *hydride generation AAS* can be used to increase the detection limits for these elements by as much as a factor of a thousand. The matrix is first digested in an acidic medium and then introduced to a 3% solution of sodium borohydride and 1% sodium hydroxide. The metal analyte is reduced and converted to its hydride. The resulting hydride vapor is swept by a stream of argon into a quartz tube suspended over the burner head where the temperature is sufficient to thermalize the hydride and produce the atomic vapor of the analyte (see Figure 7.12).

A typical reaction would be:

$$4H_3SbO_4\left(aq\right)+3H^+\left(aq\right)+3BH_4^-\left(aq\right)\rightarrow 4SbH_3\left(g\right)+3H_3BO_3\left(aq\right)+3H_2O(l)$$

Recent developments in hydride techniques have allowed for the simultaneous measurement of antimony, arsenic, bismuth, and selenium.[14]

---

[14] (a) Bings, N. H.; Bogaerts, A.; Broekaert, J. A. C. Atomic Absorption Spectroscopy: A Review. *Anal. Chem.*, **2010**, *82*(12) 4653–4681. (b) Elsayed, M.; Björn, E.; Frech, W. Optimisation of operating parameters for simultaneous multi-element determination of antimony, arsenic, bismuth and selenium by hydride generation, graphite atomizer sequestration atomic absorption spectrometry. *J. Anal. At. Spectrom.*, **2000**, *15*, 697–703.

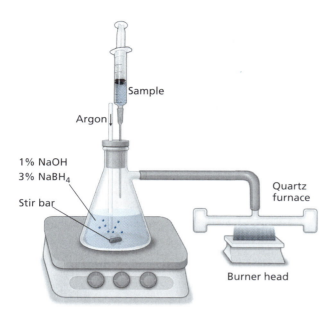

Figure 7.12 Schematic of a hydride generator for AAS. Introduction of the analyte as an acid digest to the borohydride solution produces the metal hydride gas. The gas is swept into the quartz cell where it is thermalized into atomic vapor.

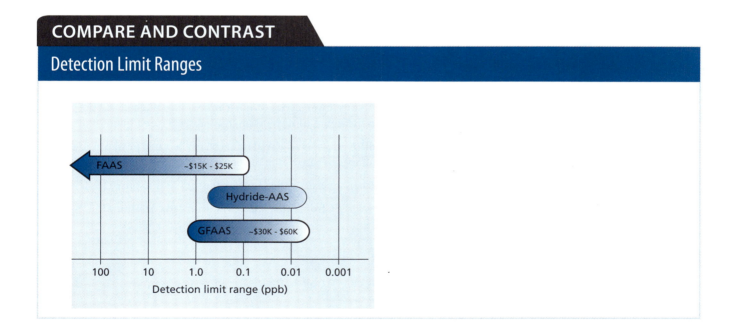

COMPARE AND CONTRAST

Detection Limit Ranges

Cold Vapor Atomic Absorption Spectroscopy

Another very toxic metal is mercury, and it is toxic at very low concentrations. However, mercury is the only metal that exhibits an appreciable vapor pressure at room temperature, and so a much more sensitive method, called cold vapor AAS, has been developed for its measurement. Sample introduction for the determination of mercury using AAS is similar to the hydride technique just mentioned, but the reducing solution is stannous chloride $(SnCl_2)$ and the mercury is produced as an elemental vapor so no flame is required.[15] If you replaced the $NaBH_4$ with $SnCl_2$ in Figure 7.12 and turned off the flame, you would have a cold vapor system for the determination of mercury. Alternate methods of mercury detection are also under investigation.[16]

7.6   Measuring Atomic Absorption

In Chapter 6, we described the absorption process for molecular spectroscopy in terms of Beer's law (see Equation 6.5 and 7.6).

$$A = -\log(T) = -\log\frac{P}{P_o} = \varepsilon bc \qquad\qquad \textbf{Eq. 7.6}$$

where

   A = absorbance
   T = transmittance
   P = radiant power reaching the detector when analyte is present
   P_o = radiant power reaching the detector when no analyte is present
   ε = extinction coefficient of the analyte
   b = sample path length
   c = sample concentration

In this chapter, we have discussed the use of HCL and EDL sources. Specifically, we mentioned how the use of these sources allows us to maintain analytical specificity. A secondary advantage of HCL and EDL sources stems from the fact that as long as the source bandwidth remains narrower than the absorption bandwidth, AAS is also subject to Beer's law. However, there are some experimental corrections that one must apply in AAS.

When considering causes of error in our absorption readings, it is easier to think about the radiant power reaching the detector (P_{net}), and we identify the following sources of error.

1. The bandwidth of the monochromator is larger than the bandwidth of the radiant source (leaking or stray light). This allows light to reach the detector that the sample is incapable of absorbing (P_L). The result is an increase in the background and thus lower signal-to-noise ratios. We talked about the effect of stray light on the signal-to-noise ratio in Chapter 6 (see Equation 6.10).

2. Some of the radiant power is scattered rather than absorbed or transmitted (P_s) due to a large number of particles in the flame.

3. The flame emits light in exactly the same wavelength as the source (P_F).

4. Background absorption from other species in the matrix or flame (P_B).

So the net power reaching the detector when analyte atoms are present can be represented as:

$$P_{net} = P_o - P + P_L - P_s + P_F - P_B \qquad\qquad \textbf{Eq. 7.7}$$

In order to obtain reliable data, we must eliminate, control, or at the very least understand erroneous power readings at the detector.

---

[15] Techniques for cold vapor detection of other elements are under investigation. See (a) Molloy, J. L.; Holcombe, J. A. Detection of Palladium by Cold Atom Solution Atomic Absorption. *Anal. Chem.*, **2006**, *78(18)*, 6634–6639. (b) Panichev, N.; Sturgeon, R. E. Atomic Absorption by Free Atoms in Solution Following Chemical Reduction from the Ionic State. *Anal. Chem.*, **1998**, *70(9)*, 1670–1676. (c) Guo, X.; Sturgeon, R. E.; Mester, Z. N.; Gardner, G. J. UV Vapor Generation for Determination of Selenium by Heated Quartz Tube Atomic Absorption Spectrometry. *Anal. Chem.*, **2003**, *75*, 2092–2099.

[16] Gil, S.; Lavilla, I.; Bendicho, C. Ultrasound-Promoted Cold Vapor Generation in the Presence of Formic Acid for Determination of Mercury by Atomic Absorption Spectrometry. *Anal. Chem.*, **2006**, *78(17)*, 6260–6264.

Background Correction

The $\mathbf{P}_L$ term is handled by the monochromator and the monochromator slit width is the variable that the analyst can change in controlling the monochromator band pass. The slit width of the monochromator is automatically set by the software that accompanies most commercial AAS instruments and is determined based on known interferences in the flame and the bandwidth of the chosen emission line from the source. Although the analyst can manually set the slit width (and there are circumstances where this is warranted), stray light through the monochromator is usually not the factor limiting detection limits or analytical selectivity. The choice of slit widths is limited and is determined at the point of instrument design. Slits are typically set to obtain a spectral resolution between 0.1 and 0.7 nm.

The $\mathbf{P}_s$ term (scatter) results from particulate matter (smoke) in the optical path. In GFAAS, the smoke is typically purged prior to the analysis. In FAAS, minimizing smoke is handled either in the sample preparation step by manipulating the matrix or by adjusting flame temperatures. In GFAAS, one also minimizes smoke by manipulating the matrix during sample preparation or by adjustments to the thermal cycles in the graphite furnace. Effects of scattering are also minimized by a technique called *Zeeman background correction* (see related Profile).

PROFILE

Demystifying the Zeeman Effect

Many chemistry departments have gas discharge tubes and gratings used to show line spectra in general chemistry laboratories. And many physics departments keep a supply of strong magnets on hand. Align two magnets positive to negative about the discharge tube and observe the line spectrum. You should notice an increase in the bandwidth of each transition in the presence of a magnetic field. If the magnet is strong enough, you can even observe splitting of the spectral line. With the magnet in place, slide a piece of polarized film in the optical path. Describe what you observe in terms of the Zeeman effect. If you do not have polarized film, a lens from a pair of polarized sunglasses will suffice. ■

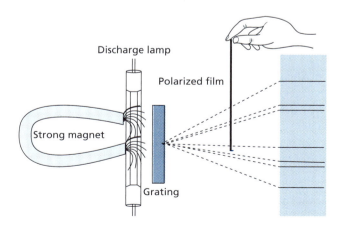

The $\mathbf{P}_F$ term results from emission from the flame and tends to be DC in nature.[17] To correct for this, the source radiation is modulated by use of a rotating wheel containing a slit (chopper). Alternatively, if the source is an EDL, modulation can be achieved by pulsing the voltage to the EDL lamp. In either case, the current generated by the detector's response to the source radiation will have an AC component. Application of *a high-pass* filter at the detector allows for the attenuation of the DC signal from the flame (see Figure 7.13). We first discussed signal modulation and high-pass filters in Chapter 5.

The $\mathbf{P}_B$ term (background) results from absorption of source radiation from species other than the analyte. These can be from the flame itself or from the formation of molecular compounds in the flame. Absorption of source radiation from the flame itself is easily corrected by first aspirating a blank and adjusting the absorption reading to zero. Correcting for absorption of source radiation from molecular compounds in the flame is more difficult. Nonanalyte molecular species formed in the flame lead to broad vibronic absorption bands. This type of interference is more common when the matrix has a high salt concentration. Background correction for molecular absorptions in the flame is done by passing a continuum light source colinear to the narrow beam generated by the HCL. The continuum source is usually a deuterium bulb. The HCL and the deuterium lamp are modulated 180 degrees out of phase. The flame is alternately exposed to the D_2 source and the HCL source. During exposure to the D_2 source, a background absorption reading is made and during exposure to the HCL, an analyte plus background absorption reading is made. A lock-in amplifier is used to subtract the two readings. We first discussed lock-in amplifiers in Chapter 5. As we saw earlier, the bandwidth of an atomic absorption is narrow compared to the bandwidth of a continuum source (see Equation 7.4). So absorption due to analyte species during exposure to the D_2 lamp is negligible (see Figure 7.13).

Problem 7.19: Using Beer's law, explain why GFAAS has lower detection limits than FAAS.

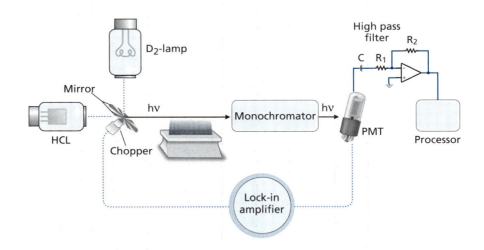

Figure 7.13 Schematic of an AAS using background correction. Modulation of the source radiation coupled with a high-pass filter attenuates emissions from the flame. Use of a continuum light source 180 degrees out of phase with the line source allows a lock-in amplifier to subtract absorptions due to molecular species in the flame.

[17] Bower, N. W.; Ingle, J. D. Jr. Precision of Flame Atomic Absorption Measurements of Copper. *Anal. Chem.*, **1976** *48(4)*, 686–692.

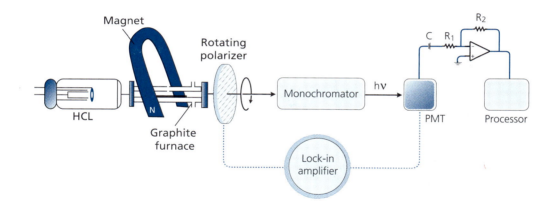

Figure 7.14 Zeeman-corrected AAS. In the presence of a magnetic field the analyte transition is split and polarized. A rotating polarizer allows for the independent measurement of analyte signal and background absorption. Additionally, the rotating polarizer serves to modulate the analyte signal and a high-pass filter is used to attenuate DC power at the detector.

Zeeman Background Correction

Electronic transitions sometimes are derived from degenerate states. In other words, there are actually several different transitions that all have the same energy. Because each transition has the same energy, they all correspond to the same spectral line. In the presence of a strong magnetic field, the degeneracy is broken due to interactions with the magnetic moments of each orbital, and the single spectral line becomes several closely spaced lines. The splitting of degenerate electronic states by a magnetic field is called the *Zeeman effect*.[18] For most transition metals, the Zeeman effect results in a single line being split into three polarized lines with the central line retaining its original position and the other two symmetrically shifted a few picometers to either side of the original line with a polarization that is 90 degrees from the central line. In Zeeman background correction, a rotating polarizer is placed in the optical path. Atomic absorption is measured when the polarizer is aligned with the transition of the analyte atom and background absorption (and scattering) is measured when the polarizer is out of phase with the analyte transition. A lock-in amplifier is used to subtract the two signals. Zeeman correction can be used with either flame or electrothermal AAS, but it is more commonly used with electrothermal AAS (see Figure 7.14). In addition to allowing the subtraction of background signal from analyte signal, the rotating polarizer also serves to modulate the analyte signal. Application of a high-pass filter allows for attenuation of direct current power (P_F) at the detector.

Smith-Hieftje Background Correction

The Smith-Hieftje background correction method is less common and is used primarily for highly volatile elements. The method is quite useful for removing absorption from interfering elements whose line spectra are close in wavelength to the analyte atom. The method utilizes a single HCL lamp and measures the absorption of the analyte during normal operation current (3–10 mA) and then measures the background absorption during a period in which the lamp is subject to a very high current (500+ mA). When the HCL is subjected to high currents, the line width broadens considerably due to collision energy transfer and Doppler broadening. In addition, a

---

[18] (a) Atkins, P. W. *Physical Chemistry*, 3rd ed. Oxford University Press: New York, **1986**. (b) Condon, E. U.; Shortley, G. H. *The Theory of Atomic Spectra* (Chapter 16). Cambridge University Press: New York, **1935**. (c) G. Schlemmer G.; Radziuk, B. *Analytical Graphite Furnace Atomic Absorption Spectrometry: A Laboratory Guide* (Chapter 4). Springer/Birkhäuser Verlag: New York, **1999**.

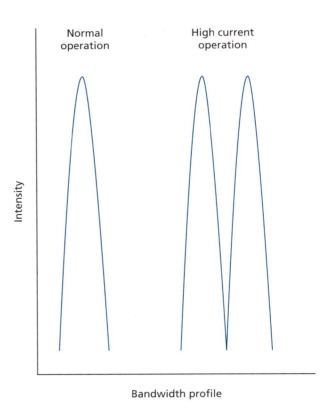

Figure 7.15 Bandwidth profile of an HCL under normal operating conditions and under high current operation.

plume of analyte atoms is generated in the center of the hollow cathode. The plume of analyte atoms serves to absorb the narrow characteristic emission seen during normal operation, and so the characteristic narrow bandwidth line is absent leaving only the expanded frequencies of the broadened line (see Figure 7.15). In high-current mode, the HCL lamp can still excite interfering atoms but cannot excite the analyte atoms. Subtracting the absorption reading during high current operation from the absorption reading during normal operation allows the analyst to correct for absorption due to closely spaced line spectra from interfering species. A notable (and costly) disadvantage of using Smith-Hieftje background correction is the fact that the high currents needed significantly shorten the lifetime of the hollow cathode lamps.

Chemical Interferences

Typically, one measures atomic absorption using the most intense line of the emission spectrum from the corresponding lamp. In the majority of samples analyzed by AAS, there will be no interfering spectral lines from other elements in the matrix. However, if you suspect there are interfering species present, it is prudent to change to a different emission line on your lamp and see if you obtain similar results for your measurement. A classic example is the measurement of aluminum by AAS. It is often better to analyze aluminum at 309.27 nm instead of using the more intense line at 308.215 nm due to interfering absorption by vanadium at 308.211 nm. Table 7.3 lists the most commonly used absorption line for several elements. The table has been ordered by absorption wavelength to emphasize those elements with similar absorption lines.

Spectral interference can also result from broadband absorption generated by molecular species formed in the flame. As we have already mentioned, these sort of broadband absorptions are usually dealt with by subtracting a blank. However, if the composition of the blank is significantly different from the sample matrix, subtraction will not yield satisfactory results. If you suspect you have an interfering absorption originating from molecular species in the matrix, there are several actions you can take to mitigate the effects the interference has on your data.

1. Chemical modification of the matrix. The user's guide that accompanies most commercial AAS instruments contains guidelines for chemical modifications that will help deal with chemical interferences in the matrix. In addition, there are several published guidebooks on the subject.[19] For instance, measuring calcium in natural waters is often hindered by the presence of phosphates. Calcium phosphate exhibits a very high lattice energy and so is difficult to decompose into its composite atoms. To alleviate this problem, it is common to add an excess of strontium or lanthanum to preferentially bind the phosphates, allowing calcium to more fully atomize in the flame. The strontium (or lanthanum) is considered a *releasing agent*. Alternatively, we could add an excess of a ligand that will preferentially bind to the analyte (Ca^{2+} in this case) to form a volatile complex that will decompose easily in the flame—that is, a *protecting agent*. The most common protecting agent is ethylene diamine tetraacetic acid (EDTA). Finally, to minimize ionization of the analyte, we might add an excess amount of *ionization buffer*—an element that exhibits a lower ionization potential than our analyte. Typical ionization buffers include alkali metals in rows 4, 5, and 6 of the periodic table.

2. Increase the temperature of the flame. If you suspect molecular species are causing serious absorptions, you can sometimes eliminate the molecular species by increasing the temperature of the flame. However, as we have mentioned above, there is a limit to how hot you want the flame.

3. Create a series of spiked matrix samples. A technique known as standard additions involves deliberately adding known concentrations (spiking) of the analyte to your matrix and measuring the absorption. The additional absorption is the result of the known amount of analyte only and not from interfering species. By creating a series of spiked samples, you can plot a line whose slope represents the instrument's response to the analyte only. By extrapolating the line back to the *x* axis, you can obtain the concentration in the original sample matrix that resulted from the analyte prior to spiking (see Example 7.3).

Another advantage of the method of standard additions is the fact that it does not require you to first make a calibration curve and thus it does not require you to first make up a series of standards. The amount of time it takes to spike a single sample is about the same amount of time it takes to make up a set of calibration standards. Therefore, the method of standard additions is particularly attractive if you are only analyzing a single (or very few) samples. The method of standard addition is also useful when you are conducting measurement at or near the limits of quantitation the technique. For instance if you were to measure the absorbance of a soil sample for barium and obtained a reading of 0.0003 A, you would have very little confidence in this reading because you are measuring absorbance at the very limits of the instrument's capabilities (you have a very low signal-to-noise ratio). However by spiking the sample with barium, you move the reading to a higher absorbance range and can therefore take your readings with a greater degree of confidence in your values (see Exercise 7.20).

TABLE 7.3: Common AAS Absorption Wavelengths (nm) for Select Elements

ELEMENT	WAVELENGTH (nm)
Ir	208.9
Zn	213.9
Te	214.3
Pb	217.0
Sb	217.6
Bi	223.1
Cu	224.8
Cd	228.8
Ni	232.0
Co	240.7
Au	242.8
Pb	247.6
Fe	248.3
Pt	265.9
Tl	276.8
Mn	279.5
Mg	285.2
Ga	287.4
In	325.6
Ag	328.1
Rh	343.5
Ru	349.9
Cr	357.9

[19] *Atomic Absorption Spectroscopy: Webster's Timeline History, 1950 – 2007*, Icon Group International. San Diego, CA, **2009**; Van Loon, J. C. *Analytical Atomic Absorption Spectroscopy: Select Methods*. Academic Press: Durham, NC, **1980**.

Example 7.3

Imagine you are performing an analysis of lead in a soil sample and you suspect that you are getting a false positive absorbance from interfering species in the soil. If you add 0.100 mL of your 1000 ppm lead standard to 100 grams of your soil sample, you will have spiked your soil by 1.00 ppm. Take a new absorption reading and repeat the process, spiking your soil sample by 1 ppm each time. Now plot the absorption versus spiked values using a spreadsheet and extrapolate the slope of the line back to the x axis. The value at the x axis is -1.00 and indicates that the original sample contained 1.00 ppm of lead.

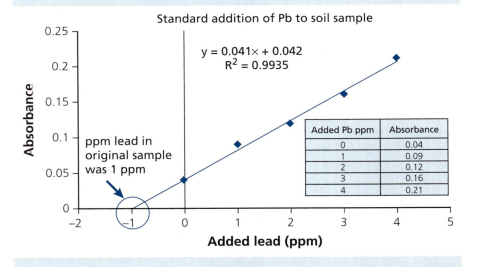

Standard addition of Pb to soil sample

$y = 0.041x + 0.042$
$R^2 = 0.9935$

Added Pb ppm	Absorbance
0	0.04
1	0.09
2	0.12
3	0.16
4	0.21

Problem 7.20:   Outline the stoichiometry showing that the addition of 0.1 mL of a 1000 ppm standard to 100 grams of soil represents a 1 ppm spike. Recall that the definition of ppm for an aqueous solution is:

$$\text{ppm} = \frac{mg}{L}$$

and the definition of ppm for a solid matrix is

$$\text{ppm} = \left(\frac{grams\ of\ analyte}{grams\ of\ matrix}\right) \times 10^6.$$

Problem 7.21:   Imagine you have performed a standard addition for the measurement of barium in soil and obtained the following data. Using a spreadsheet, plot your data and determine the original concentration of barium in the soil sample.

ADDED BA ppm	ABSORBANCE
0	0.0003
5	0.0046
10	0.0087
15	0.0123
20	0.016

7.7 Sample Preparation

Although one can introduce solids into an electrothermal atomizer, the vast majority of AAS analyte samples are introduced as a liquid. If the original sample matrix is not a liquid, then it must first be "digested" and converted into a liquid phase.

Acid Digestion

Common matrices associated with AAS samples include soils, paint, wood, clothing, animal tissue, vegetable tissue, and plastics. In addition, AAS is often used to analyze the quality of metal ores and the composition of metal alloys.[20,21] The most common way of introducing these types of solid samples into an atomic absorption spectrometer is in the form of an acid digest. One of the most widely used acid digest solutions for inorganic samples is known as aqua regia (royal water). Aqua regia, a 3:1 (v/v) hydrochloric and nitric acid, is a very efficient oxidizer of most metals and will quickly solvate metal atoms and metal ores. When such extreme measures are not needed, other acid recipes are utilized. For instance, acid digest recipes for organic samples usually contain sulfuric acid as one of the components. The Environmental Protection Agency has digestion guidelines for most common matrix materials.[22] In addition, manufacturers of AAS instruments publish handbooks that describe sample preparation methods for a variety of analytes in a variety of matrices.[23]

A typical sample preparation might involve placing about 0.5 grams of dry matrix material into a digestion vessel and adding 50 mL of concentrated acid digest solution. The ability of the acid solution to completely digest the matrix is greatly diminished by the presence of water in the matrix. For this reason, if you suspect the matrix contains significant quantities of water, it is important to dry the sample in an oven before weighing and digesting the samples. The drying step should be made as consistent as possible. The next step is to heat the digestion solution for approximately 1 hour. A common laboratory tool for performing acid digestions is a device known as a *Hot Block* (see Figure 7.16), but digestion can be accomplished with a beaker and watch glass on a hot plate.

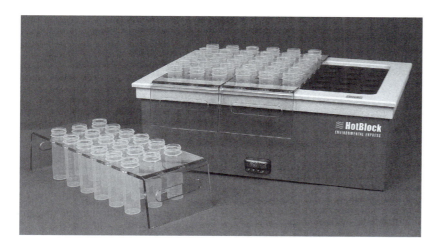

Figure 7.16 Fifty-four position *Hot Block* from Environmental Express.

---

[20] Barrett, P.; Copeland, T. R. *Monitoring Toxic Substances* (Chapter 7–Trace Metal Monitoring by Atomic Absorption Spectrometry). *ACS Symposium Series,* **1979**, *94*, 101–111.

[21] Verstuyft, A. W. *Analytical Techniques in Occupational Health Chemistry* (Chapter 14–Atomic Absorption Spectroscopy in the Occupational Health Laboratory). *ACS Symposium Series,* **1980**, *120*, 241–265.

[22] Search for "Atomic Absorption Acid Digestion" on the Environmental Protection Agency's website (www.epa.gov).

[23] Analytical Methods for Atomic Absorption Spectroscopy, Perkin Elmer, Part No. 0303–0152, **2000**.

Figure 7.17 Filter cartridge for digestion vessel. The filter is designed to fit into the digestion vessel. The filter detaches from the plunger and remains in the bottom of the vessel.

After digestion is complete, the sample solution is allowed to cool and is typically diluted 1:1 with distilled water and filtered (see Figure 7.17). The dilution step is necessary because (1) the acids can damage the nebulizer if introduced in concentrated form and (2) the addition of water lowers the viscosity and ionic strength of the solution and allows for proper aspiration and nebulization of the sample. Alternatively, after the digestion step, the sample is evaporated to dryness and redissolve using dilute nitric or hydrochloric acid.

7.8 Performing an Atomic Absorption Spectroscopy Analysis

Unless you are using the method of standard additions (spiking) then the first thing you will need to do when conducting an AAS measurement is to develop a calibration curve. This is done by analyzing a set of solutions that contain known concentrations of your analyte. Analyte standards are typically purchased as 1000-ppm stock solutions. You will need to make a series of standard dilutions starting with this stock solution.

Example 7.4

Show how you would make a series of 5 ppm, 10 ppm, 40 ppm, and 100 ppm standards, starting from a 1000 ppm stock solution. In your laboratory, only 10 mL, 25 mL, and 100 mL volumetric flasks are available.

STRATEGY −

Equation 7.7 is used for calculating a series of standard dilutions.

$$C_1 V_1 = C_2 V_2$$ Eq. 7.8

where

C_1 = initial concentration of your stock solution
V_1 = volume of stock solution you will use to make your dilution standard
C_2 = concentration of your dilution standard
V_2 = volume of your dilutions standard

In this case, C_1 is our 1000 ppm stock solution. The value C_2 is the concentration of the standard we wish to make. The value of V_2 is limited by the volumetric glassware we have available, and the value of V_1 is the volume of stock solution we must transfer to our dilution flask. Solving Equation 7.8 for V_1 we obtain:

$$V_1 = \frac{C_2 V_2}{C_1}$$

Eq. 7.9

ANSWER –

5-ppm standard:

$$V_1 = \frac{C_2 V_2}{C_1} = \frac{(5\,ppm)(100\,mL)}{1000\,ppm} = 0.50\,mL\,of\,stock\,solution$$

In order to make a 5 ppm standard from a 1000 ppm stock, you would take 0.5 mL of stock solution and dilute it to 100 mL. (**See footnote!!**[24])

10 ppm standard:

$$V_1 = \frac{C_2 V_2}{C_1} = \frac{(10\,ppm)(100\,mL)}{1000\,ppm} = 1.00\,mL\,of\,stock\,solution$$

In order to make a 5 ppm standard from a 1000 ppm stock, you would take 1.00 mL of stock solution and dilute it to 100 mL.

40-ppm standard (given the equipment available, there are two ways to make a 40 ppm standard solution.)

Scenario 1

$$V_1 = \frac{C_2 V_2}{C_1} = \frac{(40\,ppm)(100\,mL)}{1000\,ppm} = 4.00\,mL\,of\,stock\,solution$$

In order to make a 40 ppm standard from a 1000-ppm stock, you could take 4.00 mL of stock solution and dilute it to 100 mL. However you will only use about 5mL of your standard in the construction of your standard curve, leaving you with 95mL of waste to dispose (Figure 7.18).

Scenario 2: use a smaller volumetric flask to make your diluent

$$V_1 = \frac{C_2 V_2}{C_1} = \frac{(40\,ppm)(25\,mL)}{1000\,ppm} = 1.00\,mL\,of\,stock\,solution$$

In order to make a 40-ppm standard from a 1000 ppm stock, you would take 1.00 mL of stock solution and dilute it to 25 mL. Note: Scenario 2 uses less stock solution and produces less waste.

---

[24] Pneumatic pipettes such as the one shown in Figure 7.18 work on air pressure. Therefore, it is necessary to first calibrate the pipette on an analytical balance using ultrapure water prior to use. It is also a best practice to calibrate glass pipettes; however, glass pipettes do not need to be calibrated as often.

Be Flexible

Students often find the process of making a series of standards difficult, and the primary reason for the difficulty is rigid thinking. For instance, you may have sketched out the volumes you would need to generate a series of desired dilutions only to find that you do not have the appropriate sized volumetric flasks or the appropriate sized pipettes available in your stockroom. Be flexible! Looking back on Example 7.4, you should realize that it is not necessary to have exactly a 5.00 ppm, 10 ppm, 40 ppm, and 100 ppm standards. What is important is that you have four standards that span an appropriate range of concentrations *and* you need to know what those concentrations are. So use the glassware you have and make standards that span an appropriate range. So long as you know what the concentrations of your standards are, it would be perfectly acceptable to have four standards with concentrations of 4.8ppm, 9.7 ppm, 54.0 ppm, and 97.8 ppm.

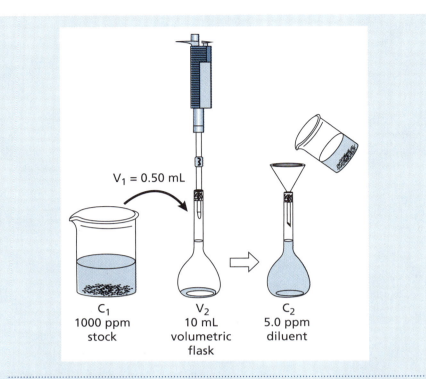

Figure 7.18 Method of standard dilutions.

100 ppm standard (given the equipment available, there are three ways to make a 100 ppm standard solution)

> Scenario 1: one could take 10 mL of stock solution and dilute it to 100 mL.
> Scenario 2: one could take 4 mL of stock solution and dilute it to 25 mL.
> Scenario 3: one could take 1 mL and dilute it to 10 mL.

$$V1 = \frac{C_2 V_2}{C_1} = \frac{(100\,\text{ppm})(10\,\text{mL})}{1000\,\text{ppm}} = 1.00\,\text{mL of stock solution}$$

Note: Scenario 3 would be the preferred method because it would generate the least amount of waste and would consume the least amount of expensive stock solution.

7.9 Further Reading

BOOKS

Hollas, J. M. *Modern Spectroscopy* 4th Ed., John Wiley & Sons: Chichester, UK, **2004**.

Meharg, A. A.; Zhao, F.-J., *Arsenic & Rice*, Springer: Dordrecht, Heidelberg, London, New York **2014**.

Stuart, B. H. *Forensic Analytical Techniques* (Chapter 5.2), John Wiley & Sons: Chichester, UK, **2013**.

JOURNALS

Bings, N. H.; Bogaerts, A.; Broekaert, J. A. C. Atomic Spectroscopy: A Review. *Anal. Chem.* **2010**, *82(12)*, 4653–4681.

Burden, S. L.; Petzold, C. J. Antacids Revisited with Modern Chemical Instruments: GCMS, AAS, and CCT. *J. Chem. Educ.,* **1999**, *76(11)*, 1544.

Kratzer, J.; Boušek, J.; Sturgeon, R. E.; Mester, Z.; Dědina, J. Determination of Bismuth by Dielectric Barrier Discharge Atomic Absorption Spectrometry Coupled with Hydride Generation: Method Optimization and Evaluation of Analytical Performance. *Anal. Chem.,* **2014**, *86(19)*, 9620–9625.

Kristian, K. E.; Friedbauer, S.; Kabashi, D.; Ferencz, K. M.; Barajas, J. C.; O'Brien, K. A. Simplified Digestion Protocol for the Analysis of Hg in Fish by Cold Vapor

Atomic Absorption Spectroscopy. *J. Chem. Educ.*, **2015**, *92(4)*, 698–702.

Legrand, M; Passos, C. J. S.; Mergler, D.; Chan, H. M. Biomonitoring of Mercury Exposure with Single Human Hair Strand. *Environ. Sci. Technol.*, **2005**, *39(12)*, 4594–4598.

Lisboa, M. T.; Clasen, C. D.; Oreste, E. Q.; Ribeiro, A. S.; Vieira, M. A. Comparison between Vapor Generation Methods Coupled to Atomic Absorption Spectrometry for Determination of Hg in Glycerin Samples. *Energy Fuels*, **2015**, *29(3)*, 1635–1640.

Markow, P. G. Determining the Lead Content of Paint Chips: An Introduction to AAS. *J. Chem. Educ.*, **1996**, *73(2)*, 178.

Pollock, E. N. Trace Elements in Fuel, Chapter 2 (Trace Impurities in Coal by Wet Chemical Methods). *Advances in Chemistry*, **1975**, *141*, 23–34.

Wilson, K. M.; Granger, R. M. Analysis of the Micronutrients Copper and Zinc in a Commercial Orchard and Vineyard by Atomic Absorption Spectroscopy: A Modern AAS Laboratory Exercise. *Chem. Educator* **2006**, *11*, 1–3.

7.10 Exercises

EXERCISE 7.1: Show how you would make a series of four standards ranging from 1 to 100 ppm given the following constraints. Your stock solution is 1000 ppm. The available pipettes have volumes of 2 mL, 3 mL, and 5 mL. The available volumetric flasks have volumes of 10 mL, 25 mL, 50 mL, and 100 mL. See the "Be Flexible" sidebar.

EXERCISE 7.2: Table 7.4 contains the data from an AAS analysis of calcium found in powdered milk. The data table contains the absorption values for four standards ranging from 5 to 90 ppm, and it also contains absorption values (in replicates of five) for three different brands of powdered milk (M1–M3). The procedure for preparing the powdered milk samples for AAS analysis involved digesting 0.05 grams of powdered milk in 50 mL of acid followed by a final dilution with water to a volume of 100 mL.

TABLE 7.4   Analysis of Calcium in Milk

SAMPLE ID	ABSORPTION
5-ppm standard	0.03
15-ppm standard	0.10
45-ppm standard	0.26
90-ppm standard	0.57
M1—Trial #1	0.37
M1—Trial #2	0.33
M1—Trial #3	0.35
M1—Trial #4	0.35
M1—Trial #5	0.36
M2—Trial #1	0.27
M2—Trial #2	0.25
M2—Trial #3	0.25
M2—Trial #4	0.26
M2—Trial #5	0.27
M3—Trial #1	0.18
M3—Trial #2	0.19
M3—Trial #3	0.18
M3—Trial #4	0.17
M3—Trial #5	0.18

(a) Use a spreadsheet to construct a calibration curve using data from your four standards. Construct a trend line (least squares regression line) and determine the equation for the slope of your line.

(b) Using the equation for the trend line found in part (a), determine the mean ppm calcium for each powdered milk sample and report the 90% confidence limit.

(c) Using your data from part (b) and your knowledge of the experimental procedure, calculate the mass percentage of calcium in each powdered (dry) milk sample.

EXERCISE 7.3: Explain why the analytical selectivity of FAAS is not affected by Doppler broadening and collision energy transfer events in the flame.

EXERCISE 7.4: What is the typical relative error from sample to sample in

(a) FAAS

(b) GFAAS

EXERCISE 7.5: In your own words, explain how the background noise from flame emissions is attenuated in AAS.

EXERCISE 7.6: Explain how the use of an HCL in AAS reduces the need for a monochromator of high spectral purity. In other words, how does the HCL allow us to get away with using a cheaper monochromator?

EXERCISE 7.7: What is the purpose of using a D_2 lamp collinear with the optical path in AAS?

EXERCISE 7.8: The 3s → 3p transition in Na is at about 589 nm. The same Mg^+ transition is at about 280 nm. Use the Boltzmann equation (Equation 7.5) to calculate the fraction of atoms or ions in the excited state for each analyte at 2,000°C.

EXERCISE 7.9: Demonstrate the lifetime dependence of Heisenberg broadening by calculating the theoretical minimum bandwidth for three different transitions, with lifetimes of 10 ms, 10 ns, and 10 fs respectively for transitions centered at

(a) 200 nm   (b) 400 nm   (c) 700 nm   (d) 900 nm

EXERCISE 7.10: Determine the ground state and first excited state term symbols for each of the following elements.

(a) Ca   (b) Mg   (c) Ti   (d) K

EXERCISE 7.11: Use Equation 7.2 to determine the pressure broadening of sodium analyte atoms in a flame AAS experiment. Assume a pressure of 1 atm, a flame temperature of 2,300°C, and an average mass of perturbing particles in the flame of 15 g/mol. *Hint*: You will need to use the ideal gas law to estimate N_p.

EXERCISE 7.12: Assuming an atomization temperature of 1,400°C in a graphite furnace AAS experiment, determine the line broadening ($\Delta\lambda$) associated with the d-lines in a sodium atom. Report your answer in nanometers. *Hint*: See Exercise 7.7.

EXERCISE 7.13: Repeat Exercise 7.12 for a flame AAS experiment assuming a flame temperature of 2,000°C. Compare the resolution of the two techniques and comment on the effect of atomization temperature on resolution as a result of Doppler broadening.

EXERCISE 7.14: Explain why the use of element specific sources such as the HCL or EDL negate the concerns exposed by Exercise 7.13.

EXERCISE 7.15: Discuss the steps involved in converting the sample matrix into analyte atoms.

EXERCISE 7.16: Discuss the benefits and disadvantages of Smith-Hieftje background correction.

Luminescence Spectroscopy

$S_0 \rightarrow S_1$

$S_1 \rightarrow S_0$

$S_0 \rightarrow S_2$

400

8.1   Introduction

Luminescence spectroscopy is an important technique used in a wide range of applications, including food science, water quality, pharmaceuticals, nanotechnology, and biochemistry. The technique is extremely sensitive and femtomolar concentrations of gas, solid, and liquid analytes have been measured.[1] Luminescence is a broad term that encompasses two types of light emission: phosphorescence and fluorescence. We will explain this difference between phosphorescence and fluorescence in the next section of this chapter, and we treat them together because the instrumentation used to measure the two types of quantum events is essentially the same.

Spectroscopic techniques that make use of luminescence benefit from a very high signal-to-noise ratio. In theory, overall noise can be limited to intrinsic detector noise (e.g., shot noise). For species that emit electromagnetic radiation (in other words, that luminesce), the technique is generally more sensitive than UV-vis absorption, essentially because there is less noise (see Chapter 5 on signal processing).

Luminescence at its most basic level involves irradiating[2] a sample with a wavelength of light that can cause electronic transitions to excited states and then measuring emission from the sample at some other wavelength as electrons relax to a lower energy state (see Chapter 2 on the quantum mechanics of spectroscopy for a review of radiative emission). Figure 8.1 shows four images of objects luminescing: quinine in tonic water excited with black light (366 nm), bananas with varying levels of ripening with black light, diamonds with a range of point defects with black light, and a jellyfish exhibiting bioluminescence. Bioluminescence (or chemiluminescence) is an emission of light due to a chemical reaction that produces a product in an excited electronic state. From an instrumental perspective, the primary distinction for bioluminescence is the

Figure 8.1 Fluorescence from quinine (excited with UV light, 366 nm), bananas with varied ripening (excited with 366 nm), diamonds with different point defects (excited with 366 nm under a black light), and bioluminescence from jellyfish.

---

[1] See the "Profile—Fluorescence Pushes the Limits of Detection: Femtomolar and Single-Molecule Detection," later in this chapter.

[2] Except in the case of chemiluminescence.

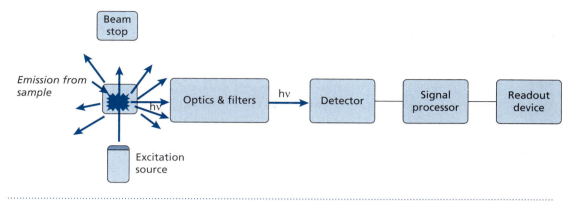

Figure 8.2 Basic setup for measuring luminescence.

fact that the excited state is not the result of irradiation but the result of a chemical reaction. For the examples in Figure 8.1, the luminescence is bright enough for our eye to be used as the detector. Figure 8.2 shows a very basic schematic of a luminescence spectrometer. We will move on to a more complete discussion of the components and design of the luminescence instrument later in this chapter.

It is important to note that not all molecules luminesce. Although this is a drawback for the technique, one may be able to find a molecule (examples include fluorescent dyes and quantum dots) that emits strongly and can be found to selectively bind to the species of interest. For example, DNA does not luminesce, but a variety of fluorescent dyes can be found that bind to various sites on DNA.

The fact that many molecules do not luminesce seems to only be a drawback, but there is actually a benefit to this. Assuming you are studying a molecule that does luminesce, you may find very little background fluorescence from other types of molecules in the sample. This lower background would lead to a better limit of detection for the fluorescent molecule. Although luminescence is a stand-alone instrumental technique, it is also commonly used as a detection scheme in conjunction with separation techniques such as liquid chromatography (Chapter 15) or capillary electrophoresis (Chapter 17).

8.2 Theory

The various radiative and nonradiative processes associated with luminescence are well described using the *Jablonski diagram*, the energy level diagram first introduced in Chapter 2, on quantum mechanics and spectroscopy, and again in Chapter 6, on molecular ultraviolet and visible spectroscopy. A Jablonski diagram is shown in Figure 8.3(A) and shows the processes of absorption, fluorescence, phosphorescence, vibrational relaxation, internal conversion, and intersystem crossing. Figure 8.3(B) shows an absorption and emission spectrum for quinine. S_0 refers to the electronic ground state, and S_1, S_2, and so on, refer to the electronic excited *singlet states*. T_1 refers to the excited *triplet state*. Each horizontal line denotes a vibrational energy level.

Fluorescence is the emission of light resulting from the relaxation of an electron from an excited singlet state to a lower energy singlet state.[3] It is quantum mechanically allowed and results in transitions that occur on the order of nanoseconds to microseconds. *Phosphorescence* is the emission of light resulting from the relaxation of an

---

[3] For a review of spectroscopic term symbols, see Douglas, B.; McDaniel, D.; Alexander, J. *Concepts and Models of Inorganic Chemistry*, pp 32–41, 442–443. John Wiley & Sons: NY, **1994**.

electron from an excited triplet state to a lower energy singlet state, and because it is quantum mechanically forbidden (because it requires a spin flip of the excited state electron), the decay takes much longer than fluorescence, typically from milliseconds to even minutes.

There are two peaks in the absorption spectrum shown in Figure 8.3(B) for quinine, one from the transition from S_0 to S_1 and one from S_0 to S_2. The absorption associated with these two transitions is represented in the Jablonski diagram in part (A) of the figure. When quinine is excited at around 350 nm, the sample emits at around 450 nm. There is one peak associated with fluorescence and no peak shown associated with phosphorescence. The transition with the associated fluorescence is shown in the Jablonski diagram. Exciting the sample at wavelengths longer than around 410 nm does not result in emission.

The absorption and fluorescence spectra, from Figure 8.3, are both plotted together. This may lead you to think that for every absorbed photon there will be a corresponding fluorescence photon. This is not the case, except for very particular molecules that convert a high percentage of absorbed photons into emitted photons. In general, many excited electrons in S_1 will relax nonradiatively. Note that in the figure that internal conversion is a nonradiative pathway from S_1 to S_0. Even for those molecules that fluoresce efficiently, instrumental factors (e.g., collection optics do not collect *all* the fluoresced photons) would make the measured number of absorbed photons greater

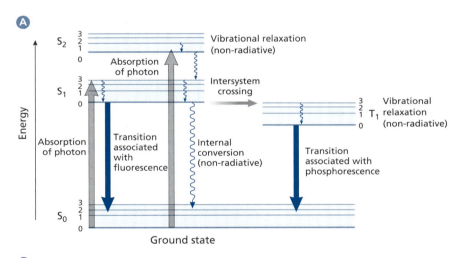

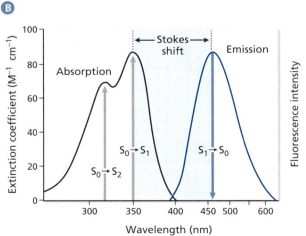

Figure 8.3 (A) Jablonski diagram for molecular spectroscopy and (B) absorption and emission spectrum from quinine.

than the measured number of emitted photons. Typically, in plots with both absorption and fluorescence spectra plotted on the same graph, the two spectra are not drawn to the same scales. Note in Figure 8.3(B) that the absorbance plot is graphed as the extinction coefficient of the analyte versus wavelength and the emission spectra is plotted as fluorescence intensity.

At first glance, it might seem like the peak emission energy for a given transition would match the peak absorption energy. But in fact, there is a difference in energy (or wavelength) from the S_0 to S_1 *absorption* peak to the S_1 to S_0 *emission* peak. This energy difference is known as the *Stokes shift*. The Stokes shift for quinine is labeled in Figure 8.3(B). Stokes shifts occur because transitions to excited vibrational states of S_1 will relax to the ground *vibrational* state of S_1 before relaxation to S_0. In other words, all relaxation events occur from the **lowest** vibrational state of S_1 to either the ground vibrational state of S_0 or excited vibrational states of S_0. Relaxation of the excited *vibrational* states of S_1 occurs prior to the relaxation of $S_1 \rightarrow S_0$ because the lifetime of the excited vibrational states of S_1 are extremely short ($\sim 10^{-15}$ seconds) compared to the lifetime excited electronic state S_1 ($\sim 10^{-9}$ seconds). We see the emission associated with the transition from the ground vibrational state of S_1 to S_0 in Figure 8.3.

A more detailed look at the transitions among the vibrational states of S_1 and the vibrational states of S_0 and the associated absorption and fluorescence spectra is worthwhile. Figure 8.4 shows the potential energy diagram between the two electronic states along with the vibrational states as well as the wave functions associated with each state. As we saw in Figure 8.3, transitions for absorption primarily start from the lowest vibrational state of S_0. This is because of the extremely low population of excited vibrational states (see Exercise 8.1) relative to the ground vibrational state of S_0 due to thermal energy. There is also a tendency for the transition to end at excited vibrational states of S_0, which means the emission wavelength is longer than that of the peak absorption, accounting for most of the Stokes shift.

The probability (or the intensity of the fluorescence) of the transition between the two electronic states S_1 and S_0 is described by the Franck-Condon factor. According to the *Franck-Condon factor*, the transition probability is proportional to the product of the vibrational wave functions between the two states. For example, the transition probability from the vibrational ground state of S_1 to various vibrational states of S_0 depends on the wave function overlap of the vibrational state of S_1 and the different vibrational wave function of the different vibrational states of S_0. The most intense fluorescence will occur when the ground state vibrational wave function of S_1 can move to a vibrational state of S_0 with the largest product[4] of the wave functions. In other words, the two states must occupy the same volume in space (overlap).

You can see that the emission transition drawn in Figure 8.4 that is associated with two vibrational wave functions with the largest product of the wave functions is also the most intense part of the spectra in part (B). The transitions from 0 to 2 (associated with absorption) and 2 to 0 (associate with fluorescence) are the most intense because of the Franck-Condon overlap. Note that the energy of the transition from a ground vibrational state of 0 (electronic ground state) to the excited vibrational state (excited electronic state) of 1 does not have the same energy as the transition from ground vibrational state (excited electronic state) to the ground vibrational state (ground electronic state).

You may have noticed that the emission and absorption spectra in Figure 8.4 are symmetric about a particular wavelength or energy. This is an example of the mirror image rule. The mirror image rule states, for molecules in which the excitation is into S_1 (not into S_2, and so on), that the absorption spectrum is the mirror image of the emission spectrum.

---

[4] The Franck-Condon factor is the volume integral of the product of the two different vibrational wave functions, $\left| \int \psi_{vib,f}^{*} \, \psi_{vib,i} \, dV \right|^2$.

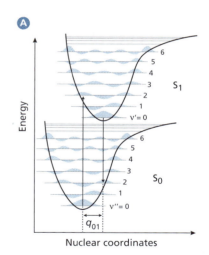

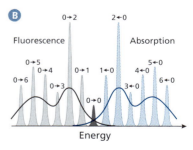

Figure 8.4 (A) Potential energy diagram showing two electronic states, ground state S_0 and excited state S_1, with vibrational levels (v) and (B) associated absorption and fluorescence spectra.

Example 8.1

In the fluorescence spectrum of a given molecule, an emission peak at 490 nm is from $S_1(v_0)$ to $S_0(v_0)$. In the absorption spectrum, the peak is near 395 nm.

(a) Calculate the difference in energy between the S_1 and S_0 states associated with emission. Determine this energy in joules.

$$\Delta E = \frac{hc}{\lambda} = \frac{1.99 \times 10^{-25}\ Jm}{490 \times 10^{-9}\ m} = 4.06 \times 10^{-19}\ J$$

We can also express the energy of this one photon in terms of the energy of one mole of these same photons. Completing this conversion:

$$\left(4.06 \times 10^{-22}\ \frac{kJ}{photon}\right)\left(6.02 \times 10^{23}\ \frac{photons}{mole}\right) = 244.4\ \frac{kJ}{mole}$$

For a sense of scale, you may recall that bond energies or bond enthalpies (for gases) are on the order of tens to many hundreds of kJ/mole.

(b) Calculate the thermal energy imparted to the molecule as a result of the absorption to emission process. The thermal energy is the difference in energy between the absorbed photon at 395 nm and the emitted photon at 490 nm.

$$E_{abs} - E_{emission} = hc\left(\frac{1}{\lambda_{abs}} - \frac{1}{\lambda_{emission}}\right) = 1.99 \times 10^{-25}\ Jm\left(\frac{1}{395 \times 10^{-9}\ m} - \frac{1}{490 \times 10^{-9}\ m}\right)$$
$$= 9.77 \times 10^{-20}\ J$$

Note the subtle difference in how we have used the equation relating wavelength and energy. In part (a), we found the energy difference between two energy states from the wavelength associated with the transition. In part (b), we found the difference between two energies, each of which has an associated wavelength.

Problem 8.1: A molecule phosphoresces with a single peak wavelength of 550 nm. The single fluorescence peak is at 500 nm. The absorption peak is at 425 nm. Sketch a rough Jablonski diagram based on this information, labeling transitions and calculating differences in each energy state in nm.

Problem 8.2: You have been given a report on luminescence measurements for an important molecule in your biochemistry laboratory. The report describes the fluorescence of the molecule with a peak at 675 nm, absorption peak at 455 nm, and phosphorescence peak at 560 nm. What is wrong with this information?

Problem 8.3: The absorption and emission spectra for quinine are shown in Figure 8.3. Does quinine obey the mirror image rule? Based on your response, what can you conclude about whether there is absorption to states above S_1?

Problem 8.4: Use the anthracene absorption and emission spectrum in Figure 8.5 to construct a combined Jablonski diagram/absorption/emission spectrum sketch, as shown in Figure 8.3. Label all states. Does the mirror image rule apply to all transitions?

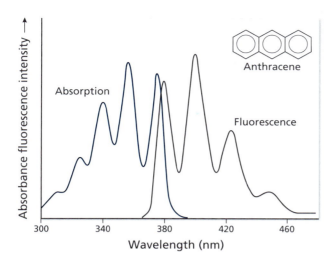

Figure 8.5 Fluorescence and absorption of anthracene.

Problem 8.5: Figure 6.5 shows the absorption spectra for anthracene and tetracene. Apply the mirror image rule and use the absorption and emission spectrum shown in Problem 8.4 as a guide to sketch an absorption and emission spectrum for tetracene. Then, construct a combined Jablonski diagram/absorption/emission spectrum sketch, as shown in Figure 8.4. Label all states and calculate transition energies in cm^{-1}.

Problem 8.6: A molecule phosphoresces with a single peak wavelength of 550 nm. The single fluorescence peak is at 500 nm. The absorption peak is at 425 nm. Based on this information, roughly sketch the Jablonski diagram, labeling transitions and calculating differences in each energy state in nm and cm^{-1}.

Fluorescence Quantum Yield and Time Decay of Fluorescence

The intensity of fluorescence is dictated by the relative pathways for radiative and nonradiative decay. The quantum yield, ϕ, is the ratio of the number of molecules fluorescing to the number of those in the excited state. If the ratio is 1, then every time an electron is in an excited state it fluoresces; if the ratio is 0, then no electrons in the excited state fluoresce. The quantum yield tells us about the competition between radiative decay and nonradiative decay. The quantum yield for several molecules is shown in Table 8.1. Each of these species is highly fluorescent. The quantum yield is notoriously difficult to measure consistently.

The quantum yield can be written as:

$$\phi = \frac{k_f}{k_f + k_{nr}}$$ Eq. 8.1

where k$_f$ is the rate constant for fluorescence and k_{nr} is the sum of various nonradiative decay channel[5] rate constants. Note that if $k_{nr} = 0$, then the quantum yield is 1. There

TABLE 8.1: Fluorophore Quantum Yields

FLUOROPHORE	QUANTUM YIELD, ϕ
Quinine sulfate	0.59
Rhodamine 6G	0.91
Fluorescein	0.89
Anthracene	0.28
Oxazine 1	0.15

From Wurth, C.; et al. Relative and Absolute Determination of Fluorescence Quantum Yields of Transparent Samples. *Nature Protocols*, 2013, 8, 1535–1550.

---

[5] Other deactivation events can be considered in the determination of quantum yield. Here we have chosen the most prevalent.

may be several nonradiative decay channels (e.g., internal conversion, external conversion, intersystem crossing; see Chapter 2 for more information on these).

Nonradiative processes can be an important part of fluorescence studies because these processes can yield information about the molecular environment, including interactions with the solvent, for the species under investigation. We will discuss the practical applications of this later in the chapter. In general, the quantum yield for fluorescence will increase with lower temperature as solvent molecules are moving more slowly and are less likely to carry energy away from the excited molecule. Similarly, the viscosity of the solvent will play a role in quantum yield for fluorescence. The more viscous the solvent, the higher the quantum yield for fluorescence.

The lifetime or decay time for fluorescence, τ, is related to the fluorescence rate and nonradiative decay rate as:

$$\tau = \frac{1}{k_f + k_{nr}}$$

Eq. 8.2

By measuring the intensity, I, of fluorescence as a function of time after excitation, one can determine the decay time, assuming a first-order rate equation, by fitting to:

$$I = I_0 e^{-t/\tau}$$

Eq. 8.3

where I_0 is the intensity of fluorescence at $t = 0$. A sketch of this single exponential decay is in Figure 8.6. In this plot, we have not shown the initial increase in fluorescence intensity as the excitation pulse is incident on the sample. Multiple exponential components are possible when the analyte contains multiple types of excited fluorophores.

We will discuss the importance of fluorescence lifetime later in this chapter when we look at time-resolved fluorescence studies. We will also see an example of time-resolved fluorescence in the Profile "Is Your $100 Bill Real? Find Out with Time-Resolved Fluorescence." First, we need to know how to determine the lifetime. If you take the natural log of both sides of Equation 8.3, you obtain:

$$\ln(I_t) = -\frac{1}{\tau}t + \ln(I_o)$$

Eq. 8.4

and a plot of $\ln(I)$ versus time will produce a slope of $-\frac{1}{\tau}$. In Figure 8.6, we estimate that $\tau \sim 10$ ns. Another way to estimate the lifetime is to find the half-life (time for intensity to decrease by half) and divide this time by $\ln(2)$. As one might expect, measuring lifetimes in the microseconds requires a less sophisticated instrument than measuring lifetimes in the picoseconds.

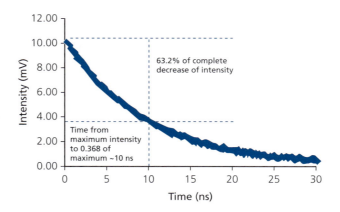

Figure 8.6 Time-resolved fluorescence measurement and explanation of fluorescence lifetime.

Example 8.2

Calculate the fluorescence lifetime for the data in Figure 8.6. With the data in a spreadsheet, we can calculate the natural log (ln) of each intensity value. This can be related to Equation 8.4. The plot $\ln(I)$ versus time and the fit of this data to a line is shown here. The fitted line's slope is the $-1/\tau$. The slope from the field yields a τ of 10.3 ns. Note that the y intercept from the fit is 2.2781, approximately $\ln(10)$, which is $\ln(I_0)$.

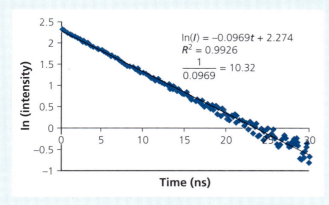

Factors that Determine Fluorescence Intensity

The amount (power) of emission, F, depends on the quantum yield ϕ, the power or intensity of the excitation source I_0, the molar absorptivity (also called molar absorption) ε (units L mole^{-1} cm^{-1}), the path length of sample b, and the molar concentration of the fluorescent species c (units mole L^{-1}).

$$F = S\phi I_0 (1 - 10^{-\varepsilon bc})$$ **Eq. 8.5**

The constant S is an instrumental factor and is the fraction of photons measured. We do not generally need to know the value of S, and we simply report fluorescence intensities, sometimes in arbitrary units, or sometimes as a voltage on a detector. Our measured amount of fluorescence is only a fraction of the total fluorescence emitted from the sample because fluorescence occurs at all radial angles but detection usually occurs at a narrow set of angles defined by the collection optics (size of lenses, slit width of monochromator, and so on). The quantity F above has units of power but is also sometimes simply referred to as the intensity of the emission because of difficulty in determining the instrumental factor S, making a true power measurement of fluorescence too challenging. Because of this, it is often common to refer to the intensity as I (especially in kinetics) instead of the actual power. For values of absorbance A, (A= εbc) less than around 0.05, we can approximate Equation 8.5 using the Taylor series (see Exercise 8.4 for the Taylor series and for details), assuming terms of A^2 and higher are negligibly small, we can approximate the amount of fluorescence as:

$$F \approx 2.303 S\phi I_0 \varepsilon bc$$ **Eq. 8.6**

PROFILE

Is Your $100 Bill Real? Find Out with Time-Resolved Fluorescence

Fluorescence spectroscopy can be used to determine the authenticity of US currency. For denominations of $5 and above, a fluorescent strip has been placed in the bill, and each denomination's fluorescent strip emits a different wavelength when exposed to ultraviolet light.*

Spectroscopy of the paper can also be useful to determine authenticity. Time-resolved fluorescence spectroscopy[†] has been shown to help characterize the paper for real currency (paper is a blend of cotton and linen fiber). Real currency paper and several types of counterfeits are well described by a two-component fluorescence decay:

$$I(t) = I_{01}e^{-t/\tau_1} + I_{02}e^{-t/\tau_2}$$

The time decay of fluorescence is shown here from authentic currency. Blue data points correspond to the time decay of emission from the authentic paper, and the black data correspond to the instrumental response from the excitation pulse. The blue line is the fit to the emission data using the two-component equation shown above. Chia and Levene found that real and counterfeit paper shared the same short lifetime ($\tau_1 \sim 160$ ps), but that the longer lifetime of the two-component fit could be used to differentiate between authentic bills ($\tau_2 \sim 2000$ ps) and counterfeits (bleached counterfeits $\tau_2 = 1552$ ps and traditional counterfeits $\tau_2 = 1725$ ps). The bar graph shows the long lifetime component with error bars for the different sample types. The counterfeit bill lifetime is statistically different than the other bill types. ■

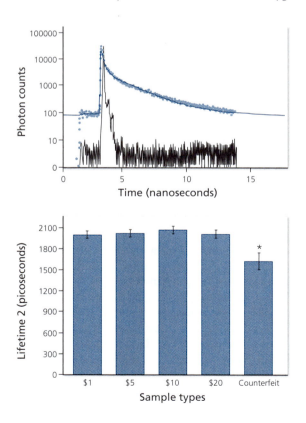

* http://www.secretservice.gov/data/KnowYourMoneyApril08.pdf.

† Chia, T.; Levene, M. Detection of Counterfeit U.S. Paper Money Using Intrinsic Fluorescence Lifetime, Opt. Express, **2009**, 17, 22054–22061.

Problem 8.7: You have taken time resolved fluorescence measurements for two samples, A and B. The table shows your data. Determine the fluorescence lifetime for each sample. Do this by plotting the ln(intensity) versus time and using the best fit line to calculate τ for each sample.

TIME (ns)	INTENSITY A (mV)	INTENSITY B (mV)
0	10.01	4.03
1	6.76	2.86
2	4.55	2.01
3	3.06	1.47
4	2.04	1.04
5	1.38	0.75
6	0.98	0.53
7	0.64	0.37
8	0.47	0.30
9	0.29	0.21
10	0.23	0.16

There is an important distinction between absorption spectroscopy and fluorescence found in Equation 8.5. In absorption spectroscopy, the intensity of the source does not affect the absorbance, absent source noise considerations. Fluorescence is fundamentally different in this regard. The intensity of fluorescence is directly related to the incident excitation power. If your fluorescence signal is weak, you may increase it dramatically by increasing the power of the excitation source. Another important distinction between absorption and emission spectroscopy is that the fraction of photons absorbed depends on the concentration (see from Chapter 5, Equation 5.5) by $10^{-\varepsilon bc}$, whereas the emission intensity from luminescence depends linearly on concentration (Equation 8.6).

Fluorescence spectroscopy can be used to do quantitative analysis. To do this, one must create a calibration curve of fluorescence intensity as a function of fluorophore concentration. In Figure 8.7, we see the fluorescence from fluorescein as a function of concentration. One drawback of this approach is that the fluorescence signal is unique to the instrument being used, so the data cannot be directly applied to another instrument.

Stern-Volmer Quenching

One way in which fluorescence can be a powerful tool to study the local environment of a molecule is through *quenching*. As the excited state molecules move, it is possible for one fluorophore to transfer energy to a different type of molecule (quenching) or even to another molecule of the same type (self-quenching). The relationship between the rate (or intensity) of a fluorescence event and the concentration of the quenching species was first described by Otto Stern and Max Volmer. Quenching due to the solvent is an example of *external conversion* and is also called *dynamic* or *collisional quenching*.

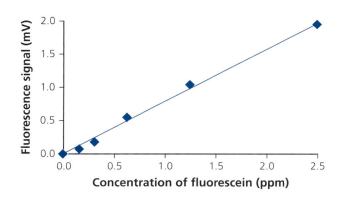

Figure 8.7 Calibration curve for concentration of fluorescein and fluorescence signal.

Example 8.3

Sample A has an absorbance of 0.02 at 532 nm and a quantum yield of 0.8. Sample B has an absorbance of 0.03 at 532 nm and a quantum yield of 0.60. Assuming samples are placed in identical instruments and have the same sample size, which would you expect to have stronger emission?

Use $F = 2.303 S\phi I_0 \varepsilon bc = 2.303 S\phi I_0 A$ and assume both situations have the same I_0 (same excitation source) and k (using identical instruments and sample sizes). We can take the ratio of fluorescence intensities:

$$\frac{F_A}{F_B} = \frac{\phi_A A_A}{\phi_B A_B} = \frac{(0.7)(0.02)}{(0.6)(0.03)} = 0.77 \text{ or } F_B = 1.3 F_A$$

We see that molecule B will have higher fluorescence intensity F.

Oxygen (O_2) is a common quencher because[6] of its triplet ground state and the electronic structure of O_2. A collision of the triplet ground state oxygen with the excited fluorophore often results in the fluorophore transferring the energy to the oxygen molecule, leading to a ground state fluorophore and an excited singlet oxygen molecule. Because of this, many fluorescence-based methods involve scrupulously removing oxygen from solution. Oxygen quenching would be an additional possible nonradiative decay mechanism to list in Equation 8.2 and would depend on the concentration of the quenching molecules colliding with the fluorophore. When energy is transferred from our excited state molecule, the molecule is no longer in an excited state and can therefore not emit light. The Stern-Volmer model is often invoked to describe oxygen quenching. In this model, a species $[Q]$ robs energy from an excited molecule preventing it from fluorescing and hence decreasing the intensity of fluorescence from the sample. Under quenching conditions, the amount of fluorescence decreases as the amount of the quenching species, Q, increases. This, then, is the basis of a quantitative method for the measurement of the concentration of the analyte Q.

In order to quantitatively describe quenching, we can return to our original equation for fluorescence quantum yield, Equation 8.1. The quantum yield without quenching can then be written as:

$$\phi_{withoutquench} = \frac{k_f}{k_f + k_{isc} + k_{ic}} \qquad\qquad \textbf{Eq. 8.7}$$

where k_{isc} is the intersystem crossing rate constant and k_{ic} is the internal conversion rate constant. We can add a rate constant associated with quenching to in Equation 8.8. With quenching added as a nonradiative option (k_{nr}), the quantum yield can be written as:

$$\phi_{withquenching} = \frac{k_f}{k_f + k_{isc} + k_{ic} + k_{quench}} \qquad\qquad \textbf{Eq. 8.8}$$

Taking the ratio of these quantum yields gives:

$$\frac{\phi_{withoutquench}}{\phi_{withquenching}} = 1 + \frac{k_{quench}}{k_f + k_{isc} + k_{ic}} = 1 + K_q[Q] \qquad\qquad \textbf{Eq. 8.9}$$

---

[6] Gouterman, M. Oxygen Quenching of Luminescence of Pressure Sensitive Paint for Wind Tunnel Research. *J. Chem. Ed.,* **1997,** *74(6),* 697.

where we replaced the quantity $\dfrac{k_{quench}}{k_f + k_{isc} + k_{ic}}$, called the Stern-Volmer constant, as K_q, and $[Q]$ is the concentration of the quenching molecules. The Stern-Volmer constant is also equal to $\tau_0 k_{quench}$, where τ_0 is the lifetime of the fluorophore in the absence of quenching. This ratio of quantum yields is effectively the same as the ratio of the amount or intensity of fluorescence with quenching (F) over without quenching (F_0). This leads us to the Stern-Volmer equation:

$$\frac{F_0}{F} = 1 + K_q [Q] \qquad \textbf{Eq. 8.10}$$

We see that for a given interaction of fluorophore and quencher, set by K_q, that as $[Q]$ increases, the amount of fluorescence F decreases.

We can also look at how the quencher concentration may alter the fluorescence lifetime. Referring back to Equation 8.2, the decay time, τ, can be written in terms of the rate constant for fluorescence and the sum of nonradiative rate constants. The introduction of quenching will increase the nonradiative rate constant contributions, increasing the denominator, and therefore reducing the fluorescence lifetime with quenching. For collisional quenching, the ratio F_0/F is also equivalent to τ_0/τ.

Example 8.4

The intensity of anthracene fluorescence was measured with zero concentration of CCl_4, a molecule that quenches anthracene emission via collisional quenching. The experimenter then added CCl_4 to the solution, measuring the fluorescence intensity with each new CCl_4 concentration. Anthracene concentration was kept constant for all solutions. The plot (often called a Stern-Volmer plot) here[7] shows F_0/F as a function of quencher concentration. Estimate K_q from these data. The plot follows Equation 8.10. The y axis is the ratio of fluorescence intensity without and with quenching and the x axis is the concentration of CCl_4 $[Q]$. A linear fit to the data gives a slope of ~14.1 liters/mole and is equal to K_q.

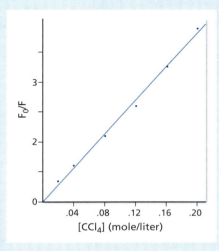

[7] Legenza, M. W.; Marzzacco, C. The Rate Constant for Fluorescence Quenching. *J. Chem. Ed.*, **1977**, *54(3)*, 183.

If we can control the amount of Q in the sample, we can measure fluorescence intensities as a function of [Q] to create a calibration curve. Using this calibration curve, we could take a sample, with the same quenching molecule but with an unknown quenching concentration [Q], to determine the amount of the quenching molecule.

We saw that in dynamic quenching, an excited molecule can lose energy and hence not fluoresce due to a transfer of energy to a quenching agent. In some cases, the fluorescent molecule does not even get to the excited state because of the quenching agent. This is referred to as *static quenching*. In static quenching, the quenching agent keeps the fluorophore from absorbing excitation light to begin with and so there is no change in lifetime due to quenching. The telltale sign of combined static and dynamic quenching is a deviation from the linear Stern-Volmer plot to a parabolic plot of F_0/F versus [Q].

Self-quenching is a term used to describe a situation where the quenching agent is the same type of fluorescent molecule as the analyte. Self-quenching will cause a deviation from the linear relationship we saw between fluorescence intensity and concentration of fluorophore from Equation 8.6. In this case, the fluorescence intensity increases with fluorophore concentration up to a point and then decreases.

Problem 8.8: A sample of a fluorescent molecule (concentration of 1.0×10^{-5} M) with no quenching agent has an emission intensity resulting in a 4.6 mV reading on the detector. An unknown amount of quenching agent is added to the solution and the fluorescence intensity is found to be 2.5 mV. Use the value of K_q = 15.0 liter/mole to determine the concentration of quenching agent.

Problem 8.9: The data table shows the ratio F_0/F for tryptophan as a function of the concentration of acrylamide quencher. What conclusion can you draw about dynamic and static quenching?

CONCENTRATION (mmol/L)	F_0/F
0	1
50	1.7
100	2
150	3
200	4.1
250	5.6

Relating Fluorescence and Molecular Structure

As a person conducting chemical analysis, it is a good idea to have an appreciation for the molecular features that lead to fluorescence and/or phosphorescence. A cursory review of the molecules known to fluoresce reveals a trend. The only organic molecules with appreciable fluorescent yields are those with rigid polyconjugated π-systems and compounds that contain an aromatic moiety. The existence of conjugation in an organic molecule creates a low energy π^* state. Population of the π^* state can occur by either a $\pi \rightarrow \pi^*$ or n $\rightarrow \pi^*$ transition and results in a π^* singlet state.[8] Relaxation of the excited state can occur through fluorescence or through nonradiative decay (collision energy transfer). Nonradiative mechanisms can be inhibited if the analyte molecule is placed in a rigid matrix (glass matrix) or if the molecule itself is structurally rigid. For example, phenolphthalein is nonfluorescent, but the more rigid fluorescein is highly fluorescent (quantum yield = 0.89). Each molecule is shown in Figure 8.8.

Another important structural feature of fluorescein is the existence of the oxygen heteroatom in the conjugated ring system. The existence of heteroatoms provides a low-energy n $\rightarrow \pi^*$ transition. However, for single-ring systems containing heteroatoms

---

[8] See Figure 6.3 for a review of n $\rightarrow \pi^*$ and $\pi \rightarrow \pi^*$ transitions.

Oxygen is a common fluorescence quencher. This allows for fluorescence to be used as a means to sense oxygen, which can be important in a variety of areas. For example, researchers* have added fluorophores (pyrene and ruthenium complexes) to paint to create a pressure-sensitive paint. The paint can be placed on structures studied in wind tunnels as a means to qualitatively and quantitatively understand air flow without obstruction due to measurement devices. When the paint is illuminated with an ultraviolet lamp, the model, an airplane or car for example, will fluoresce based on the local oxygen content, which is related to the pressure. A Stern-Volmer calibration curve for pyrene and ruthenium complex is shown. Note that in lieu of the concentration of the quencher, $[Q]$, the x axis is the pressure, p. Detection occurs with several charge coupled device cameras. Because the linear fit is much better for the pyrene, this fit would lead to a more accurate pressure measurement. ■

PROFILE

Fluorescence Quenching Helps with Aerodynamics

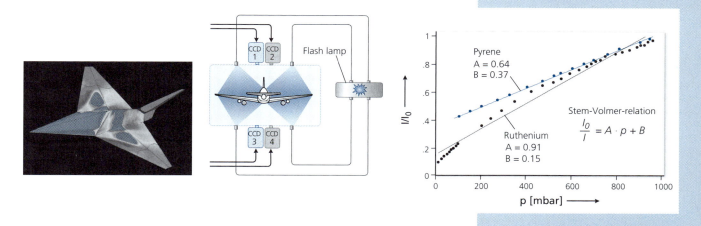

* Engler, R. H.; Klein, C.; Trinks, O. *Pressure-Sensitive Paint Systems for Pressure Distribution Measurements in Wind Tunnels and Turbomachines. Measurement Science and Technology,* **2000**, *11(7), 1077–1085.*

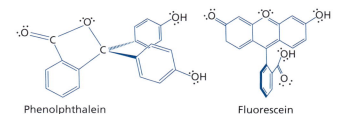

Phenolphthalein   Fluorescein

....

Figure 8.8 Molecular structure of phenolphthalein and fluorescein.

in the ring, fluorescence is generally not observed. It is believed that the $n \rightarrow \pi^*$ excited state electron rapidly converts to a triplet state, resulting in phosphorescence or deactivation through nonradiative decay rather than fluorescence. For example, pyridine does not fluoresce but quinoline and isoquinoline both do. Each molecule is shown in Figure 8.9. Again, we see that a fused ring system increases the likelihood of observing fluorescence.

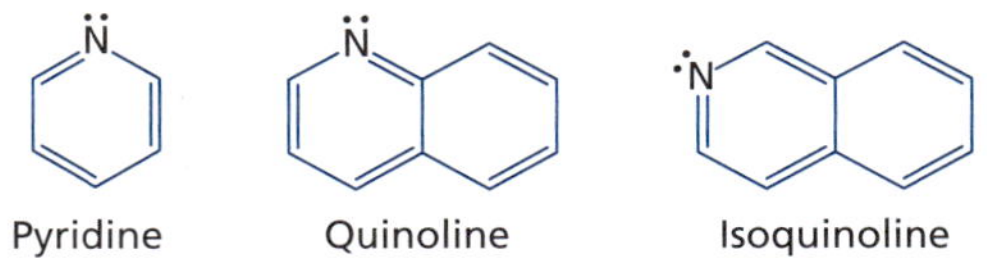

Figure 8.9 Molecular structure of pyridine, quinoline, and isoquinoline.

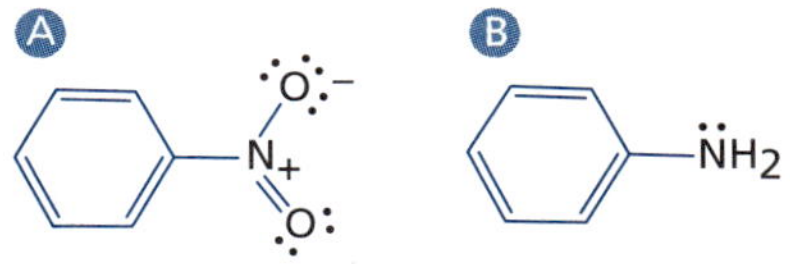

Figure 8.10 Molecular structure of nitrobenzene (A) and aniline (B).

A Quick Review of *Term Symbols*

The terms *singlet* and *triplet* come from Equation 8.11. If all of the electrons are spin-paired (± ½), then the "S" term in Equation 8.11 equals zero and 2(0) + 1 = 1 (a singlet). If an electron changes spin states (flips), the total spin of the system becomes S = (½ + ½) = 1, then 2(1) + 1 = 3 (a triplet). If you have no unpaired electrons, you have a singlet state; if you have two unpaired electrons, you have a triplet state; and if you have three unpaired electrons, you have a quadruplet state, etc.

Substituents on ring systems also affect the observance of fluorescence. For example, nitrobenzene does not have an appreciable fluorescence but aniline does. A quick look at the structures, shown in Figure 8.10, is revealing.

The nitrate group in nitrobenzene contains three very electronegative atoms and through induction, the highest occupied π-orbital in the phenyl ring is lowered, thus making the π → π* in the phenyl ring higher in energy. The highest occupied molecular orbital–lowest unoccupied molecular orbital gap is now n → π* resulting from the N=O double bond. The n → π* excited state is most likely deactivated (quenched) by collision energy transfer or other nonradiative decay mechanisms.

In Chapter 6 we introduced the spin selection rule for electronic transitions. The spin *selection rule* states that for an electronic transition the change in the total spin multiplicity of the molecule must be zero. In other words, the orientation of an electrons spin cannot change during a transition. From quantum mechanics, the total spin multiplicity, M, is:

$$M = 2S + 1$$ **Eq. 8.11**

where S is the total spin. Phosphorescence involves a spin flip (triplet to singlet) and therefore violates the spin selection rule. However, the selection rules make assumptions about the symmetry of the orbitals. These assumptions are not strictly correct, and so the probability of intersystem crossing is not zero. Looking back at our Jablonski diagram (Figure 8.3), we see that phosphorescence results from an electron flip as it passes from an excited singlet state to a triplet state. So what molecular features encourage phosphorescence? Stating the obvious, in order for an electron to pass from an excited singlet state to an excited triplet state, the triplet state must exist and it must have a similar energy to that of the singlet state. Phosphorescence is much more likely to be observed if the molecule contains a heavy atom (such as phosphorus or iodine). Atoms with high atomic numbers have many more atomic orbitals occupying a small volume of space, and this improves the probability of an existing triplet state. In addition, if we invoke orbital hybridization, we can also expect mixing of state. Mixing of states can be thought of as being analogous to the orbital hybridization we studied in general chemistry (i.e., sp³ hybridized orbitals). Mixing of states relaxes the selection rule because you no longer have a pure singlet or pure triplet state. Deactivation of the excited triplet through collision energy transfer and nonradiative decay is also possible. As a result, we also see enhanced phosphorescence if we provide a rigid environment for the excited state. Freezing the sample in a glass matrix greatly enhances the phosphorescence of the analyte.

Problem 8.10: For this question, consider benzene and naphthalene.

(a) Which molecule would you expect to have a larger quantum yield?

(b) Which molecule would you expect to have a shorter peak wavelength for fluorescence?

8.3 The Fluorescence Spectrometer

The essential components of the fluorescence spectrometer or fluorometer are similar to the components seen in a UV-vis spectrometer (Chapter 6). Take a moment to compare and contrast Figures 6.1 and Figure 8.2. Fluorescence systems may cost many tens of thousands of dollars and have a wide range of features. A basic instrument can be built for as little as $100.[9]

The fluorescence spectrometer or fluorometer has an excitation light source, a light detector or detectors, a wavelength discriminator to select the excitation wavelengths and a wavelength discriminator to select for luminescence wavelengths, filters to keep excitation light from contaminating the luminescence signal, and various optics to focus and collect light. A commercial fluorescence system is shown in Figure 8.11. We will use this as a prototypical system and will walk through each section of the instrument. This system has options for time resolved fluorescence measurements, steady-state measurements, and time-resolved phosphorescence measurements. Before we walk through the instrument, we will discuss elements common to luminescence systems: the excitation source, wavelength discriminator, and detector. Like other spectroscopic techniques, there are commercially available systems and more specialized research systems.

Steady-State Luminescence and Time-Resolved Luminescence Measurements

In steady-state mode, many times the excitation is a continuous source and the luminescence detector simply measures intensity for a set period of time for averaging. If a pulsed light source is used, steady-state measurements can still be made. In this case, the emission intensity is integrated over the lifetime of the luminescence to yield an effective steady-state measurement. Both the source and detector type may differ between time resolved and steady-state modes.

In time-resolved fluorescence mode, a short pulse excites the sample, and the emission is measured as a function of time after this excitation source is not present. An alternative technique for time-resolved measurements is to use a light source with a

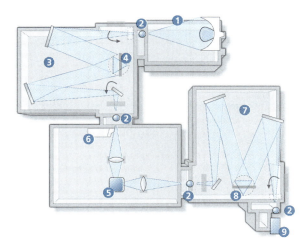

1. Xenon lamp
2. Adjustable slits
3. Excitation monochromator
4. Excitation grating
5. Sample compartment
6. Excitation correction
7. Emission monochromator
8. Emission grating
9. Detector

Figure 8.11 Commercial fluorescence and phosphorescence instrument (both time resolved and steady state), including optical layout.

[9] Wigton, B. T.; et al. A Portable, Low-Cost, LED Fluorimeter for Middle School, High School, and Undergraduate Chemistry Labs. *J. Chem. Ed.,* **2011** *88(8)*, 1182–1187.

sinusoidally varying intensity and using characteristics of the now sinusoidally varying fluorescence compared to the excitation to mathematically determine decay lifetimes. This approach does not require a short-pulse laser source. Because a tunable continuous wave lamp can be used for excitation in time-resolved studies, one has more flexibility compared to a single-wavelength, short-pulse source, in their choice of excitation wavelengths.

One very common design element of the fluorometer is that the path of the excitation light is at a right angle to the path of the luminescence to the detector. This is shown in Figures 8.2 and 8.11. This is called right-angle optics and is done to minimize the amount of excitation light that may enter the detector. Reducing this background signal from the detector will lead to better detection limit for the instrument.

Excitation Source

For steady-state fluorescence measurements, the excitation light source is often the xenon (Xe) arc lamp (see Chapter 6 for more information on how the lamp functions), a continuous wave source.

The Xe arc lamp provides broadband output from 300 to 1,000 nm; see Figure 8.12 for a typical emission spectrum. Xe lamps also show strong narrow emissions lines in the infrared. Xe arc lamps used in luminescence will have powers from 100 W and higher. The mercury arc lamp is also used in luminescence instruments. Note that the mercury lines are much more intense than the visible wavelengths available from the Xe lamp but the spectrum is not as continuous as the Xe lamp.

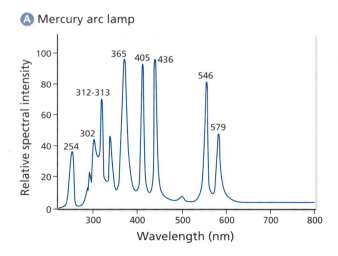

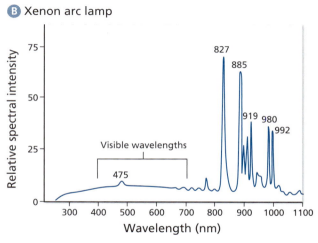

Figure 8.12 Spectrum profiles for mercury lamp and xenon lamp.

TABLE 8.2: Sample Excitation Sources

EXCITATION SOURCE	WAVELENGTH(s)	PULSE WIDTH
Q-switched Nd:YAG laser	1064 nm (532 nm, 355 nm, 266 nm, with nonlinear optics)	~ 5 ns
Mode-locked Ti:Sapphire laser	Tunable 700–1000 nm	~ 100 fs
Xe flash lamps	Broad, 200–1000 nm	~ ms
Laser diodes	Variety, ~266 nm and higher	< 100 ps
Light-emitting diodes	Variety in ultraviolet ~250–350 nm	< ns

In time-resolved fluorescence, analytes are often excited using short pulse lasers, Xe flash lamps, light-emitting diodes (LEDs), or other types of pulsed lamps. Pulses as short as 10 fs are possible from lasers. A selection of pulsed excitation sources with available wavelengths and pulse widths are in Table 8.2.

Wavelength Discrimination and Instrument Resolution

When a lamp is used as the excitation source, two wavelength discriminators are needed. One wavelength selector is needed for the excitation source (to limit excitation of the analyte to a narrow range of energies), and one is needed for wavelength selection of the luminescence. Wavelength discrimination for the excitation is usually accomplished by focusing the excitation source on a slit of a grating monochromator (often a Czerny-Turner type; see Figure 3.26). The position of the grating is controlled by computer, allowing for easy control of the output wavelength. Wavelength discrimination for the luminescence is also usually accomplished with a grating monochromator.

Compared to other techniques such as infrared absorption spectroscopy (Chapter 11) and Raman spectroscopy (Chapter 12), fluorescence instruments do not routinely require such high resolution because of the broad nature of most emission. That said, resolution can be critical for the fluorometer as well, depending on the molecular system being studied. In Chapter 3, on optics, we saw that the resolution of the instrument is limited by the number of lines/length of the grating and the distance from the grating to the detector (the focal length of the spectrometer). The more lines/length of the grating, for a given focal length, the higher the resolution will be. The longer the focal length of the spectrometer, the larger the path length will be for the dispersive element to provide spatial separation of the individual wavelengths on the detector. The wavelength resolution needed in a given measurement will depend on the analyte under investigation.

The grating monochromator (which uses diffraction gratings) does not do a perfect job of selecting only the wavelength of interest. Secondary diffraction off of the diffraction grating can cause peaks in your fluorescence spectrum that are not caused by emission from your sample. We are usually concerned with $m = 1$, first-order diffraction, for wavelength λ. However, second-order diffraction, $m = 2$, satisfies the diffraction condition for $\lambda/2$ (see Equation 3.9). This means that light at half the wavelength is reflected at the same angle as λ, albeit at a much lower efficiency. The secondary diffraction peak, often from the excitation source, could appear at twice the wavelength of the excitation source in your fluorescence spectrum. Secondary diffraction from the excitation monochromator can also cause problems. You may excite your sample with both the wavelength you have set the monochromator, as well as source radiation at half that monochromator setting. Optical filters, discussed in depth in Chapter 3, are commonly used to attenuate $m = 2$ radiation from the source from reaching the detector. In addition, optical filters are commonly used in low-cost or single-purpose systems (e.g., for the analysis of DNA, RNA, and proteins, as seen in a later Profile) in lieu of a monochromator. Filters are generally less expensive than gratings but are far inferior to monochromators with respect to wavelength bandwidth.

Example 8.5 – Revisiting Optics: Resolving Power and Diffraction Gratings

You are designing a specialized instrument to measure fluorescence. You need to buy a diffraction grating for use as a wavelength discriminator for the emission and are working with a tight budget. You are looking at three different gratings that are each 6 cm long and have 7,000 lines, 9,000 lines, and 11,000 lines. The more lines, the more expensive the grating, so do not buy a grating that has more lines than you need for this application. You are working to miniaturize your system, and as such you have only 0.09 m of path length from your grating of choice and the detector array. The light is incident at 45 degrees on the grating. You have decided that you need to have wavelengths 450 nm and 450.5 nm projected a minimum of 12×10^{-6} m (12 mm) apart on your detector.

For our three different gratings, the groove distances d (distance between grooves) are:

$$d_1 = \frac{6.0 \times 10^{-2}\, \text{m}}{7{,}000} = 8.57 \times 10^{-6}\, \text{m}$$

$$d_2 = \frac{6.0 \times 10^{-2}\, \text{m}}{9{,}000} = 6.66 \times 10^{-6}\, \text{m}$$

$$d_3 = \frac{6.0 \times 10^{-2}\, \text{m}}{11{,}000} = 5.45 \times 10^{-6}\, \text{m}$$

The distance between our projected wavelengths is (from Chapter 3):

$$\Delta x_{\text{between wavelengths}} = L\left(\tan\beta_{\lambda_1} - \tan\beta_{\lambda_2}\right)$$

$$= L\left\{\tan(\sin^{-1}(\frac{m\lambda_1}{d} - \sin\alpha)) - \tan(\sin^{-1}(\frac{m\lambda_2}{d} - \sin\alpha))\right\}$$

$$\Delta x_{\text{between wavelengths}} = 0.01m\left\{\tan\left[\sin^{-1}\left[\frac{(1)(450.5 \times 10^{-9}\, \text{m})}{d} - \sin(\frac{\pi}{4})\right]\right]\right.$$

$$\left. - \tan\left[\sin^{-1}\left[\frac{(1)(450.0 \times 10^{-9}\, \text{m})}{d} - \sin(\frac{\pi}{4}))\right]\right]\right\}$$

Completing this calculation for the three different groove spacings (d) yields:

Grating 1: distance between wavelengths 450 and 450.5 nm on the detector = 12.2 mm

Grating 2: distance between wavelengths 450 and 450.5 nm on the detector = 14.2 mm

Grating 3: distance between wavelengths 450 and 450.5 nm on the detector = 18.3 mm

Grating 1, with 7,000 lines, provides a minimum distance just larger than our requirement. We should choose the 7,000-line grating if cost is a paramount issue.

The size of the slit on the front of the grating monochromator can also be controlled to effect resolution. Increasing the slit means that more light enters the monochromator and therefore more light gets to the detector, increasing your signal. However, there is a cost to this increase in signal. The wavelength resolution of your measurement decreases as you increase the slit size because this allows light from a variety of input angles to enter the system. This idea was discussed in more detail in Chapter 3.

> **Problem 8.11:** You are setting up a fluorescence measurement using a high-power Nd:YAG laser with excitation at 355 nm. Your wavelength range of interest is 430 to 730 nm. You are using a grating monochromator to scan through the wavelength range onto a photomultiplier tube. Should you be concerned about the possibility of secondary diffraction peaks? Justify your answer.

Detectors for Luminescence

The choice of luminescence detector depends on the wavelength range of interest, the strength of the luminescence signal, and (if doing time-resolved work) the time response of the fluorescence. Common detectors are the photomultiplier tube (PMT), charged coupled device (CCD), and photodiode.

The PMT is a common detector for both steady-state and time-resolved fluorescence systems. The sensitivity (quantum efficiency) of several types of PMTs, as a function of wavelength, is shown in Figure 8.13. For time-resolved fluorescence measurements, the detector must have a sufficiently short time constant (see Equation 5.3) in order to respond to incident photons and create an electrical signal that is a true

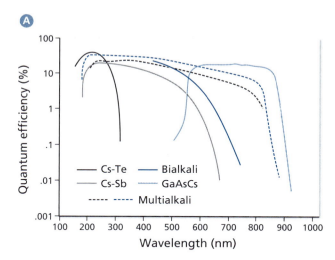

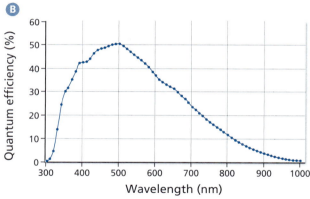

Figure 8.13 (A) Quantum efficiency as a function of wavelength for several different types of PMT photocathode materials and (B) a Si CCD.

indicator of the time-evolving fluorescence. Although it depends significantly on the type of PMT (and the voltage applied), PMTs are routinely used to study fluorescence with lifetimes less than 1 ns.

The CCD detector is essentially an array of small photodiode "pixels." It is commonly used for fluorescence imaging applications. CCD detectors are not generally used in time-resolved applications because of their slow response time compared to PMTs. As we saw in Chapter 6, on UV-vis spectroscopy, a key benefit of the array detector compared to the single-wavelength PMT detector is that the measurement of an entire spectrum can be accomplished much more quickly with an array detector than with a PMT detector. Silicon CCD arrays are sensitive to electromagnetic radiation up to a wavelength of around 1 micron, as shown in Figure 8.13. Cooling the CCD will increase the signal/noise (see Equation 5.4).

Silicon photodiodes are found in a variety of systems for instrumental response corrections (see Section 8.4) and in single-purpose systems as a detector.

These detectors can operate in two different modes: analog and *photon counting*. The photon-counting mode makes use of the particle nature of light and is used for applications with extremely low light levels. The technique is used with the PMT or the avalanche photodiode. Photon counting is more expensive to implement than analog detection but offers a higher signal-to-noise ratio. One reason for this higher ratio is the discrimination used in photon counting. This removes low-level noise pulses caused in the PMT from the photon-counting signal.

Putting It All Together: Walking Through the Luminescence System

Now that we have discussed each of the common components used in a luminescence system, we can step through the instrument. In Figure 8.11, a continuous Xe arc lamp (number 1) serves as the excitation source. Light from the Xe lamp is sent through an excitation monochromator (number 3) to select for a particular excitation wavelength. The light exiting the excitation monochromator is focused onto the sample (number 5) with a lens. The luminescence is collected with a lens to focus the emission onto the slit of the emission monochromator (number 7). The exit slit of the emission monochromator focuses the emission onto the detector (number 9).

Excitation Spectra: An Additional Measurement Mode

In order to record an excitation spectrum, the detector is set to measure a very narrow range of emission wavelengths. This wavelength is generally chosen to where the sample has the most intense fluorescence. The excitation wavelength is then scanned (via a motor connected to a diffraction grating) over the range of the absorption spectrum of the analyte, and we detect emission as a function of excitation wavelength. The data are plotted as emission intensity at one narrow range of wavelength versus excitation

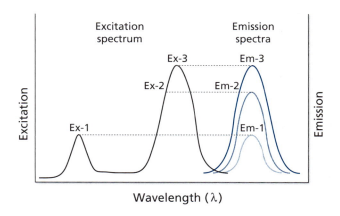

Figure 8.14 Excitation spectrum (black), and associated emission spectra caused by excitation at Ex-1, Ex-2, and Ex-3.

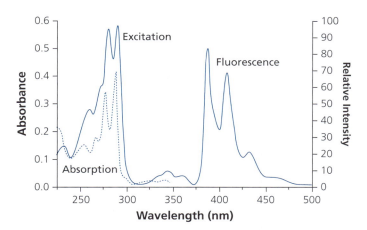

Figure 8.15 Absorption, excitation, and fluorescence spectra for benzo[a]anthracene, excitation spectrum generated at peak fluorescence wavelength.

wavelength used. A sample excitation spectrum is shown in Figure 8.14. The black line corresponds to the excitation spectrum. The different shades of blue lines correspond to the fluorescence spectra generated with three different excitation wavelengths. Note that excitation at Ex-1 causes the largest amount of fluorescence with peak at Em-1 but that excitation at all wavelengths on the excitation spectrum leads to an emission spectrum with the same shape.

Another excitation spectrum is shown in Figure 8.15. The figure shows the absorption spectrum, excitation spectrum, and fluorescence spectrum for benzo[a]anthracene, a polycyclic aromatic hydrocarbon that is of interest because of its potential negative environmental impact. The excitation spectrum is taken measuring the peak intensity of fluorescence at around 387 nm. On the plot, there is an overlaid absorption spectrum that shows the strong correlation between absorption spectrum and excitation spectrum. This relationship between absorption and excitation spectra holds when the mirror image rule holds. The excitation spectrum can be used for qualitative purposes but is often used for new types of analytes to determine the optimum excitation wavelength.

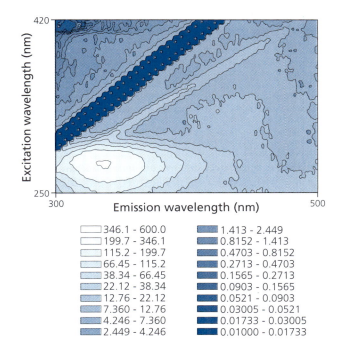

346.1 - 600.0	1.413 - 2.449
199.7 - 346.1	0.8152 - 1.413
115.2 - 199.7	0.4703 - 0.8152
66.45 - 115.2	0.2713 - 0.4703
38.34 - 66.45	0.1565 - 0.2713
22.12 - 38.34	0.0903 - 0.1565
12.76 - 22.12	0.0521 - 0.0903
7.360 - 12.76	0.03005 - 0.0521
4.246 - 7.360	0.01733 - 0.03005
2.449 - 4.246	0.01000 - 0.01733

Figure 8.16 Total luminescence spectrum for farm waste water.

Example 8.6

The absorption spectrum for the molecule fluorescein is shown here in part (A) along with the fluorescence spectrum when excited at 488 nm. How would the fluorescence spectrum change if this same sample was excited with a source at 440 nm with the same intensity as the 488 source? Roughly sketch an overlay of the two fluorescence spectra when excited at 440 nm.

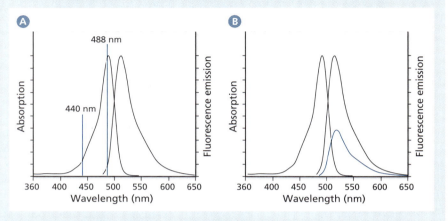

From the plot in (A), the absorption at 440 nm is much less than at 488 nm. This means that fewer electrons will be excited from the S_0 to S_1 state with 440 nm compared to 488 nm. With fewer electrons in state S_1, there will be fewer electrons that can make the radiative decay and hence less emission compared to at 488 nm. We would expect the overall shape of the fluorescence spectrum to be the same but with an overall decrease in intensity. The emission with excitation at 440 nm is shown as the blue line in (B).

We can go a step further here and create a three-dimensional plot with excitation wavelength on the y axis, emission wavelength on the x axis, and the intensity of emission on the z axis (or use shading to denote intensity). This plot is called the total luminescence spectrum. An example of a total luminescence spectrum (also called excitation emission matrix) is shown in Figure 8.16 for a water sample contaminated with animal waste.[10] The blue shading denotes intensity, with white being the most intense. The broad emission around 330 nm is associated with tyrosine and tryptophan. The black lines corresponds to where the excitation wavelength and the emission wavelength are equal.

Sample Introduction

Sample holders in fluorescence spectrometers are very similar to those in UV-vis spectrometers, except that when square cuvettes are used, all four faces of the cuvette are polished and transparent, whereas often only two polished and clear faces are required for UV-vis cuvettes. The cuvette sample holder is shown inside our commercial system in Figure 8.17.

This image also allows us to see the lens used for collecting and focusing the emission. Also notice that three different sources are available in this instrument design. Excitation light enters from the top and time resolved fluorescence is measured out of the left-hand slit and steady state and phosphorescence is measured out of the right-hand slit.

---

[10] Baker, A. Fluorescence Properties of Some Farm Wastes: Implications for Water Quality Monitoring. *Water Research,* **2002,** *36,* 189–195.

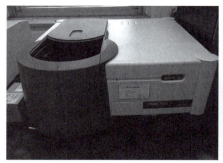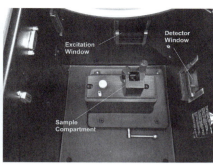

Figure 8.17 A commercial fluorometer, and sample compartment. The Cary Eclipse fluorescence spectrophotometer by Varian.

8.4 Challenges with Fluorescence Spectroscopy

Ideally, all optics, sources, and detectors would have properties that do not vary with wavelength. Unfortunately, this is not the case. For both fluorescence spectra and excitation spectra, it is important to correct the spectra for the instrumental response.

Detector Corrections

We saw in Figure 8.13 that detectors do not have constant quantum efficiency as a function of wavelength. The measured intensity at one wavelength in the emission spectrum, F_{meas}, is roughly the product of quantum efficiency of the detector and the actual intensity. The F_{meas} divided by the quantum efficiency will correct for this effect. The correction can be made experimentally by detecting a calibrated source and comparing the measured spectrum to the known calibrated spectrum. Many systems will have a correction file based on the particular detector being used.

Problem 8.12: You are using a Si CCD detector to study steady-state fluorescence in a homemade spectrometer. You measured two emission peaks at 700 nm and 800 nm that have roughly the same intensity. You have not made any correction for the detector. Roughly estimate the actual ratio of the two peaks based on the type of detector used.

Source Intensity Corrections

For excitation spectra, an additional correction is necessary. Because the instrument scans through the range of excitation wavelengths, the source does not have a constant intensity, as we saw in Figure 8.8 for the emission spectrum of the Xe arc lamp. Your excitation spectrum might make it appear as though a certain wavelength causes the most emission, but this could be because your source has a higher power at one wavelength compared to another. In order to correct for excitation wavelength dependency, a beam splitter is used to send a percentage of the excitation light to a photodiode. The photodiode records intensity data that can be used by the computer system to correct for wavelength effects (although the photodiode also has a wavelength-dependent response). In the commercial system in Figure 8.11, a photodiode is placed (number 6) just after the excitation monochromator and records source intensity as a function of wavelength for the purposes of making source corrections. In commercial systems, much of the work of correcting for source intensity variation is done by the software.

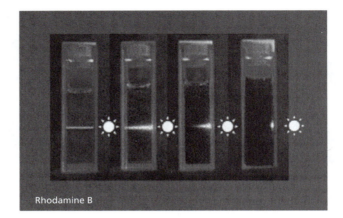

Figure 8.18 Inner filter effect for different concentrations of rhodamine B.

Challenges Associated with High Absorbance

Interpreting fluorescence data can be made more complicated for samples with higher molar absorptivities. In general, the sample is placed in the instrument so that the emission is detected from the center of the cuvette, and this works well for samples with low absorbance. However, with increasing absorbance, more emission will originate in the front of the cuvette where the excitation light enters. This results in the *inner filter* effect. This is shown in Figure 8.18. The problem is that the collection optics are configured to maximize light gathering from the center of the cuvette, not from the front of the cuvette. The measurement taken will appear as though there is less emission than there really is. To be clear, the inner filter effect is an experimental artifact and not due to quenching, where the energy of an excited molecule is transferred away. Several correction approaches have been developed.[11] One method is to correct the measured fluorescence, $F_{measured}$, using:

$$F_{corrected} = F_{measured} 10^{\frac{A_{EX} + A_{EM}}{2}}$$ **Eq. 8.12**

where A_{EX} and A_{EM} are the absorbance at the excitation and emission wavelengths.

Stray Light Contamination

Filters are used so that the excitation light does not contaminate the fluorescence spectrum. Reducing excitation light from entering the detector will also reduce the likelihood of damage to the detector from excessive photoradiation. For similar reasons, it is particularly important not to scan the emission monochromator through the excitation wavelength. See the discussion of optics in Chapter 3 for more details on optical filters. Filters can be purchased with a variety of wavelength-blocking ranges.

Photobleaching

We have so far assumed that the irradiation of our sample in a fluorescence experiment would not change the sample. There are cases when the irradiation will modify the sample. In particular, in photobleaching, the radiation causes the fluorophore to lose the ability to fluoresce. This is seen, for example, with fluorescent dyes, and the damage can be irreversible.[12] The practical result of photobleaching is that the amount of time that a dye is under irradiation (or the intensity of the excitation used) must be

---

[11] Puchalski, M. M.; Morra, M. J.; von Wandruskzka, F. R. Assessment of Inner Filter Effect Corrections in Fluorimetry. *Anal. Chem.*, **1991**, *34*, 341–344.

[12] Eggeling, C.; et al. Molecular Photobleaching Kinetics of Rhodamine 6G by One- and Two-Photon Induced Confocal Fluorescence Microscopy. *ChemPhysChem*, **2005**, *6*, 791–804.

An important aspect of chemistry is the ability to determine the presence or lack of a particular molecular species in a sample. Our ability to measure extremely low concentrations of fluorescent molecules (fluorophores*) has improved dramatically in the past 50 years, and we can now detect *single* molecules bound to a substrate. The top image shows the experimental setup used to measure fluorescence of single molecules (in this case, fluorescent dye DiI with polymer PMMA) spin coated on a microscope slide substrate. The bottom image shows fluorescence from single molecules of rhodamine 110. Although the optical setup is somewhat more complicated than the traditional luminescence instrument discussed in this chapter, the concepts are the same. The system has an excitation source, detection system, wavelength discriminators, and optics for light focusing and light collection. More conventional commercial fluorescence systems also continue to push the detection limits. Instrument manufacturers[†] have described fluorescence systems with the ability to measure fluorescence signals from 50 femtomolar (femto = 1×10^{-15}!!!) concentrations of fluorescein—a popular and strongly emitting fluorophore. ■

PROFILE

Fluorescence Pushes the Limits of Detection: Femtomolar and Single-Molecule Detection

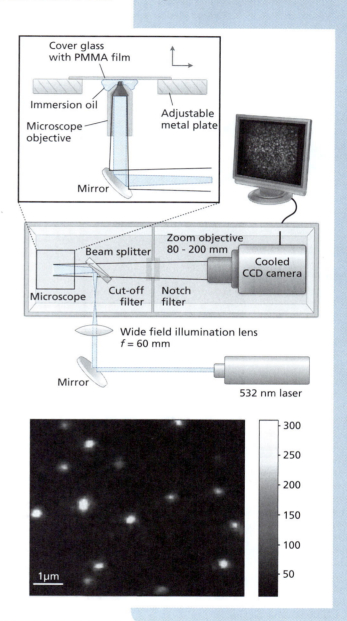

* *Fluorophores and fluorochromes (note relationship of word to chromophore) are both molecules that fluoresce.*

Top image from http://itg.beckman.illinois.edu/microscopy_suite/equipment/FCS/ Bottom image from Zimmermann, J.; van Dorp, A.; Renn, A. Fluorescence Microscopy of Single Molecules, *J. Chem. Ed.,* **2004** 81(4), 553.

[†] *For example, the Horiba FluoroLog and the PTI QuantaMaster 30.*

kept below the threshold for photobleaching. Photobleaching is more of a challenge when high intensity lasers are used for excitation.

Problem 8.13: When discussing UV-vis absorption spectroscopy (Chapter 6), we did not discuss correcting for the spectral response of the detector. Why is this correction needed when measuring an excitation spectrum in a fluorometer but not needed when measuring UV-vis absorption?

A wide range of strongly fluorescing dyes can be used to selectively bind to particular molecules. By knowing the molecule that the dye binds to, one can use the fluorescence intensity of that dye as a means to measure the quantity of that particular molecule. This approach is widely used to determine quantities of DNA, RNA, and proteins. An example of this type of small special purpose instrument is shown here. The fluorometer has a 470 nm LED and a 635 nm LED for use as excitation sources, photodiodes as the detectors, and filters for wavelength discrimination. ■■

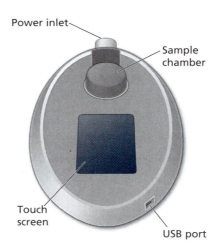

Power inlet

Sample chamber

Touch screen

USB port

8.5   Additional Fluorescence-based Techniques

There are many variants of fluorescence spectroscopy, and new techniques are always emerging. Research continues to push the limits from studying fluorescence of very small quantities (single molecule) to studying fluorescence at very short time scales (femtoseconds). In addition, there are many specialized techniques that exploit the use of a fluorescence signal.

Chemiluminescence

In *chemiluminescence*, the energy that excites the electrons in the analyte and results in luminescence is not from a light source but from a chemical reaction. Light emission from glow sticks[13] is a common example of chemiluminescence, where breaking the boundary between two containers in the stick, one containing the fluorophore and the other containing the reagent, allows the contents to mix and the resulting chemical reaction also has associated light emission. In general, the technique requires the sample of interest to be combined with another material that results in the necessary chemical reaction to cause chemiluminescence. The PMT and photodiode are common detectors for the luminometer, the name for the instrument used in chemiluminescence. Chemiluminescence has the benefit of not requiring an excitation source but the drawback that relatively few chemical reactions lead to chemiluminescence. This is overcome for certain cases where commercial reagents are available.

---

[13] Kuntzleman, T. S.; Rohrer, K.; Schultz, E. The Chemistry of Lightsticks: Demonstrations to Illustrate Chemical Processes. *J. Chem. Ed.,* **2012,** *89(7),* 910–916.

For example, quantitative analysis of ATP by chemiluminescence is used in the food industry to measure the amount of organic matter on a surface for cleanliness. Luminol (3-aminophthalhydrazide) is a common compound used to determine the presence of heme compounds, as well as many others.[14]

Fluorescence Polarization

In polarization-based fluorescence (also known as fluorescence polarization), a polarized excitation source is used or a polarizer is placed in front of an unpolarized excitation source to select for a particular source polarization. Molecules whose transition dipole moments are parallel to the polarization direction of the incident light can absorb the excitation and may fluoresce. Molecules whose transition dipole moment is perpendicular to the polarization of the excitation light will not absorb the excitation and will not fluoresce. Thus, any observed fluorescence would be polarized parallel to the polarization of the source. Because there is a finite amount of time between absorption and emission and the molecule is rotating during this time, some fluorescence will be emitted by the rotated molecule that will have a polarization different from the excitation.

The amount the molecule has rotated is a function of the size and shape of the molecule (i.e., the hydrodynamic volume), the viscosity of the solution, and temperature. The degree to which the emission polarization has rotated relative to the excitation polarization is used to characterize these effects. By using a polarizer that can be rotated, one can measure the amount of fluorescence at all polarizations and use this to determine P, the degree of polarization. The degree of polarization, P, is calculated by:

$$P = \frac{F_{\parallel} - F_{\perp}}{F_{\parallel} + F_{\perp}}$$

Eq. 8.13

where $F_{\parallel}$ is the intensity of fluorescence polarized parallel to the excitation source and $F_{\perp}$ is the intensity of fluorescence polarized perpendicular to the excitation source. This technique is extremely useful in studying protein–ligand interactions.[15]

Problem 8.14: What are the extreme ranges of possible values for degree of polarization, P? Explain what is occurring at the lowest and highest values of P.

Resonance Energy Transfer Spectroscopy

Resonance energy transfer (RET; also referred to as FRET, where the F stands for either fluorescence or Forster) is another important fluorescence technique used to study a variety of molecular systems—but particularly proteins. From an instrumental point of view, RET can be performed without any changes to a traditional fluorometer. However, RET is most often performed as a form of microscopy generating images that highlight where two specially prepared molecules are in very close proximity. For example, RET images have been used to highlight biochemical processes in living cells.[16] RET can give important information on distances in the 10 to 100 angstrom level between particular molecules. Because of this distance sensing ability, RET has been called a spectroscopic or molecular nanoruler.[17]

[14] Wu, Z. J.; Wang, C.; Ju, H.; Yan, F. Chemiluminescence Imaging Immunoassay of Multiple Tumor Markers for Cancer Screening. *Anal. Chem.*, **2012**, *84(5)*, 2410–2415.

[15] Rossi, A. M.; Taylor, C. W. Analysis of Protein-Ligand Interactions by Fluorescence Polarization. *Nature Protocols*, **2011**, *6*, 365–387.

[16] Sekar R. B.; Periasamy, A. Fluorescence Resonance Energy Transfer (FRET) Microscopy Imaging of Live Cell Protein Localizations. *Journal of Cell Biology*, **2003**, *160(5)*, 629–633.

[17] Sahoo, H. Förster resonance energy transfer—A Spectroscopic Nanoruler: Principle and Applications. *Journal of Photochemistry and Photobiology C: Photochemistry Reviews*, **2011**, *12(1)*, 20–30.

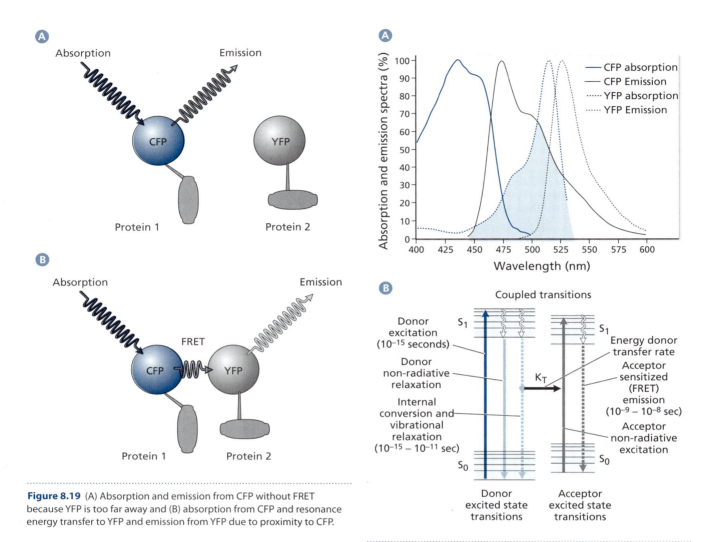

Figure 8.19 (A) Absorption and emission from CFP without FRET because YFP is too far away and (B) absorption from CFP and resonance energy transfer to YFP and emission from YFP due to proximity to CFP.

Figure 8.20 (A) Absorption and emission spectra for CFP and YFP, note the overlap of CFP emission and YFP absorption, and (B) Jablonski diagram for RET process.

A sketch of the RET process is seen in Figure 8.19. In this technique, a dye molecule[18] (the donor, cyan fluorescent protein, or CFP in the figure) is chosen to bind selectively to a relevant molecule of interest (protein 1). The donor absorbs light and is promoted to an excited state and can fluoresce. If a nearby but different type of dye molecule (the acceptor, yellow fluorescent protein, or YFP in this case) is close enough, this excited dye molecule (the donor) can transfer energy nonradiatively to this acceptor. The second dye, YFP, has been chosen to selectively bind to a different molecule of interest (protein 2). Because of the resonant energy transfer, the acceptor, now in an excited state, can then fluoresce. The acceptor dye is chosen so that the initial excitation wavelength that excites the donor cannot excite the acceptor and that the acceptor dye fluoresces at a different wavelength than the donor.

In RET, it is critical that the emission spectrum of the donor overlap spectrally with the absorption spectrum of the acceptor, as shown in Figure 8.20, because of the resonant energy transfer from the excited state of the donor to the excited state of the acceptor. A Jablonski diagram of RET process is also shown in Figure 8.20.

---

[18] Quantum dots can also be used for RET.

The distance between donor and acceptor (therefore, the distance between protein 1 and 2) plays a critical role in determining the likelihood of the acceptor emitting light. This is the key to the technique. The amount of fluorescence from the acceptor will vary as $1/r^6$, where r is the distance between the donor and acceptor molecules. The rate constant, K_T, for energy transfer from donor to acceptor in RET, is:

$$K_T(r) = \frac{1}{\tau_D}\left(\frac{R_0}{r}\right)^6 \qquad \textbf{Eq. 8.14}$$

where τ_D is the lifetime of the donor (with acceptor not present), R_0 is the Forster distance (defined as the distance where energy transfer from donor to acceptor is 50%), and r is the distance between donor and acceptor. R_0 can be thought of as the largest distance we can investigate with FRET. Distances associated with this technique tend to be those associated with biological macromolecules. The essence of this technique is that you will measure fluorescence from acceptor molecules that are at the smallest distances to the donor, generally less than 10 nm.

The nonradiative energy transfer from the donor to the acceptor in RET is a type of quenching, but it does differ from the dynamic quenching discussed earlier in this chapter. For RET, the donor and acceptor must have spectral overlap between the emission spectrum of the donor and the absorption spectrum of the acceptor. In dynamic quenching, there need not be any spectral overlap because the energy transfer is due to collisions.

Multiphoton Excitation

So far, we have assumed that one photon is associated with absorption. In other words, the energy of one photon corresponds to the difference between the ground and excited state. When a high-intensity laser source is used as the excitation source, it is possible for two photons (or more) to cause absorption from the ground state to a higher energy. The energy difference between the states is the sum of the energies of the two photons. For example, if a 532 nm high-intensity laser is incident on a sample that does not absorb at 532 nm but does at 266 nm, two-photon absorption is possible. A Jablonski diagram showing both one- and two-photon absorption is shown in Figure 8.21. Once the two photons have been absorbed, the process of decay is no different than we have discussed previously.

Two-photon–induced fluorescence is of great value when one is trying to control the spot size inducing emission. This is important when studying small-scale structures. Two-photon absorption occurs only at the highest intensities and as such the emission volume association is much smaller than one-photon absorption. This effect is seen in Figure 8.21.

Problem 8.15: In RET spectroscopy, a particular donor, CFP, and a particular acceptor, GFP, have a Forster distance of 4.8 nm. The fluorescence lifetime of CFP is 2.1 ns without the presence of GFP protein. Calculate the rate constant for a donor–acceptor distance of 6 nm.

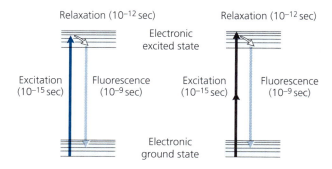

Relaxation (10^{-12} sec)   Relaxation (10^{-12} sec)

Electronic excited state

Excitation (10^{-15} sec)   Fluorescence (10^{-9} sec)   Excitation (10^{-15} sec)   Fluorescence (10^{-9} sec)

Electronic ground state

Figure 8.21 Jablonski diagrams demonstrating the distinction between a one-photon absorption (A) and a two-photon absorption (B).

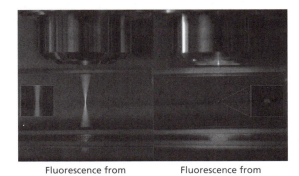

Fluorescence from out of focus planes

A

Fluorescence from focal spot only

B

Figure 8.22 Fluorescence caused by one (A) and two (B) photons. The arrow in (B) points to the extremely small and faint spot caused by two-photon absorption and fluorescence.

Problem 8.16: One advantage of multiphoton absorption is the ability to penetrate more deeply into certain types of samples—for example, living biological systems. In a particular system, a 300 nm photon is absorbed by one-photon absorption. What wavelength photon would result in this same transition as a two-photon process? How about for three photons? Sketch each process on a Jablonski diagram.

8.6   Further Reading

BOOKS

Lakowicz, J. R. *Principles of Fluorescence Spectroscopy*, 3rd ed., Springer: NY, **2006.**

Turro, N. J. *Modern Molecular Photochemistry*, University Science Books: Sausalito, CA: **1991.**

JOURNALS

Adhikary, R.; Mukherjee, P.; Krishnamoorthy, G.; Kunkle, R. A.; Casey, T. A.; Rasmussen, M. A.; Petrich, J. W. Fluorescence Spectroscopy of the Retina for Diagnosis of Transmissible Spongiform Encephalopathies. *Anal. Chem.,* **2010**, *82(10)*, 4097–4101.

Brouwer, A. M. Standards for Photoluminescence Quantum Yield Measurements in Solution (IUPAC Technical Report). *Pure Appl. Chem.,* **2011**, *83(12)*, 2213–2228**.**

Flores, R. V.; Solá, H. M.; Torres, J. C.; Torres, R. E.; Guzmán, E. E. Effect of pH on the Heat-Induced Denaturation and Renaturation of Green Fluorescent Protein: A Laboratory Experiment. *J. Chem. Ed.,* **2013**, *90(2)*, 248–251.

Jenkins, J. L.; Welch, L. E. A Mechanistic Study of Terbium Phosphorescence Quenching, *J. Chem. Ed.,* **2009**, *86(5)*, 613.

Koenig, M. H.; Yi, E. P.; Sandridge, M. J.; Mathew, A. S.; Demas, J. N. "Open-Box" Approach to Measuring Fluorescence Quenching Using an iPad Screen and Digital SLR Camera. *J. Chem. Ed.,* **2015**, *92(2)*, 310–316.

MacCormac, A.; O'Brien, E.; O'Kennedy, R. Classroom Activity Connections: Lessons from Fluorescence. *J. Chem. Ed.,* **2010**, *87(7)*, 685–686.

Rusak, D. A.; James W. H. III, Ferzola, M. J.; Stefanski, M. J. Investigation of Fluorescence Lifetime Quenching of Ru(bpy)32+ by Oxygen Using a Pulsed Light-Emitting Diode. *J. Chem. Ed.,* **2006**, *83(12)*, 1857.

Schlamadinger, D. E.; Kats, D. I.; Kim J. E. Quenching of Tryptophan Fluorescence in Unfolded Cytochrome c: A Biophysics Experiment for Physical Chemistry Students. *J. Chem. Ed.,* **2010**, *87(9)*, 961–964.

Schwarz F. P.; Wasik, S. P. Fluorescence measurements of benzene, naphthalene, anthracene, pyrene, fluoranthene, and benzo[e]pyrene in water, *Anal. Chem.* **1976** *48(3)*, 524–528.

Seetohul, L. N.; Islam, M.; O'Hare; W. T.; Ali, Z. Discrimination of teas based on total luminescence spectroscopy and pattern recognition. *Journal of Science of Food and Agriculture,* **2006**, *86(13)*, 2092–2098.

Wilczek-Vera G.; Dunbar, E.; Salin, E. D. Understanding Fluorescence Measurements through a Guided-Inquiry and Discovery Experiment in Advanced Analytical Laboratory. *J. Chem. Ed.,* **2011,** *88(2)*, 216–219.

ONLINE RESOURCES

For a list of common emission and absorption spectra, go to http://omlc.ogi.edu/spectra/PhotochemCAD/index.html

Many equipment manufacturers provide excellent resources on applications of their instruments; see, for example, the websites of Horiba (http://www.horiba.com), Bruker (https://www.bruker.com), Thermo Scientific (http://www.thermoscientific.com/), and Picoquant (https://www.picoquant.com/).

8.7 Exercises

EXERCISE 8.1: In Chapter 5, we introduced the Boltzmann distribution as we discussed the role thermal energy plays in potentially populating excited states.

(a) Use the anthracene fluorescence spectrum to estimate the energy state difference between the vibrational ground state of S_0 and the first excited vibrational state of S_0.

(b) We assumed that the first excited vibrational state was not sufficiently populated to result in absorption to S_1 (in Figure 8.4, for example, all absorption transitions originate in the ground vibrational state). Use the Boltzmann distribution to calculate the ratio of populations in these states at 273K (room temperature) and at 350K.

EXERCISE 8.2: A strong pulsed laser, $\lambda = 266$ nm, is being used to excite a sample. You find two peaks in your emission spectrum, one at 505 nm and another at 532 nm. You expected to find the 505 nm peak from your literature search. The 532 nm peak is not reported in any papers. What might be the cause? How could you test your hypothesis?

EXERCISE 8.3: You are conducting a fluorescence polarization experiment, and you have determined that the degree of polarization for your sample at room temperature to be 0.5. Would you expect this to increase, decrease, or stay the same if you did the polarization measurements at higher temperature?

EXERCISE 8.4: In our discussion of the factors that affect F, the intensity of fluorescence, we used the Taylor series for 10^{-x} to go from Equation 8.5 to 8.6, although we skipped all the details. The Taylor expansion for 10^{-x} is:

$$10^{-x} = 1 - [\log(10)]x + \frac{1}{2}[\log(10)]^2 x^2 - \frac{1}{6}[\log(10)]^3$$

$$x^3 + \frac{1}{24}[\log(10)]^4 x^4 + \ldots$$

Use the Taylor series to derive Equation 8.6.

EXERCISE 8.5: You are working on a quenching experiment and have found that the ratio of fluorescence without quenching to fluorescence with quenching increases linearly with the concentration of the quencher. You find that the ratio of lifetime without quenching to lifetime with quenching does not change as you increase the quenching concentration. What type of quenching is occurring?

EXERCISE 8.6: A particular fluorophore in a quenching solution has a K_q of 8 L/M. The fluorescence intensity is 35 mV in the absence of the quenching agents. Plot the intensity of fluorescence as a function of $[Q]$ for concentrations from 0 to 0.5 M/L.

EXERCISE 8.7: You have a set of fluorescence data for a high-absorbance sample and want to decide if the inner filter effect is resulting in more than a 20% error in your measured intensities. The optical density (OD) at the excitation wavelength is 0.1 and the OD at the emission wavelength is 0.15.

EXERCISE 8.8: Review flash photolysis in Section 9 of Chapter 6 and compare and contrast flash photolysis with time-resolved fluorescence spectroscopy. Be mindful to point out the similarities and differences in the instrumental components and the similarities and differences in the acquisition of the signal.

EXERCISE 8.9: What is the typical lifetime range of a fluorescence event? What is the typical lifetime range of a phosphorescent event?

EXERCISE 8.10: For this question, consider anthracene and tetracene.

(a) Which molecule would you expect to have a larger quantum yield?

(b) Which molecule would you expect to have a shorter peak wavelength for fluorescence?

EXERCISE 8.11: A molecule phosphoresces with a single peak wavelength of 700 nm. The single fluorescence peak is at 590 nm. The absorption peak is at 490 nm. Based on this information, roughly sketch the Jablonski diagram, labeling transitions and calculating differences in each energy state in nm and cm^{-1}.

Chapter 9

Atomic Emission Spectroscopy

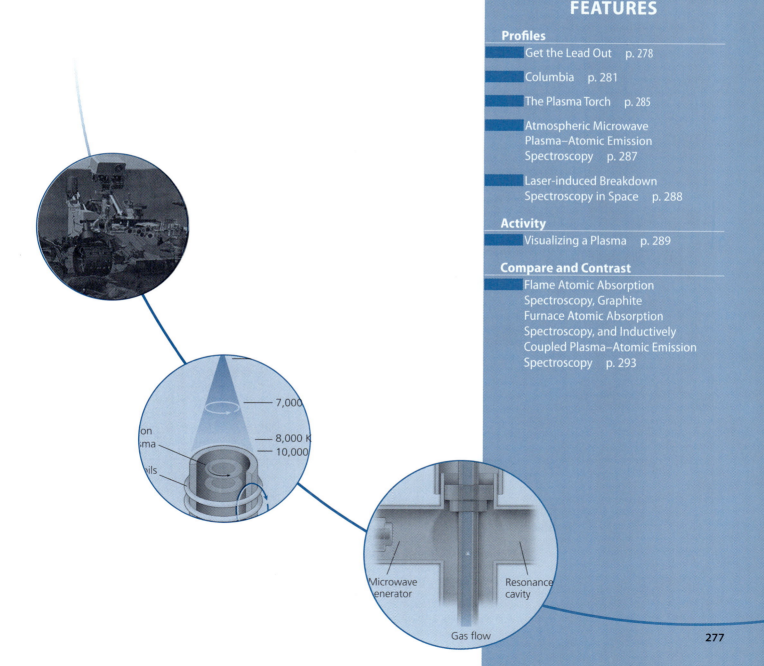

7,000

on
;ma

8,000 K
10,000

ils

Microwave
enerator

Resonance
cavity

Gas flow

PROFILE

Get the Lead Out

In 2000, an 18-month-old child in a family of Moroccan descent was brought to a Belgian hospital showing extreme lethargy and irritability. The child was diagnosed with acute lead poisoning.* Subsequent investigation found that 18 other members of the child's family were similarly intoxicated with lead but that lead was not prevalent in the normally expected sources (paint, dust, water). The Brussels Intercommunal Laboratory was charged with finding the source of the lead quickly in order to minimize the damage to residents. A fast, efficient method for the analysis of metals was needed, and the researchers at the laboratory opted for inductively coupled plasma–atomic emission spectroscopy (ICP-AES). The lead was traced to a traditional metallic teapot that the family used several times each day. Under the specific conditions of tea brewing (elevated temperature and acidity, especially when lemon juice is used), the metals are leached into the tea, and they are subsequently ingested. Multiple samples of similar teapots were collected from local vendors that sell goods from northern Africa, and it was determined that most of them leached lead, nickel, and zinc into brewed tea. The minimal sample preparation and multielement capabilities of ICP-AES led to a quick solution to what might have become a more widespread incident. ■

*Bolle, F.; et al., Food Additives and Contaminants: Part A, 2011, 28(9), 1287–1293 and Petit, D.; et al. J. de. Physique IV (proceedings), 2003, 107, 1053–1056.

9.1 Introduction

The instrumental methods discussed in the previous three chapters share a conceptual commonality with the topic of this chapter. Those chapters (Chapter 6, on molecular ultraviolet-visible spectroscopy; Chapter 7, on atomic absorption spectroscopy; and Chapter 8, on luminescence spectroscopy), along with the current chapter, present optical spectroscopic techniques that utilize the ultraviolet-visible (UV-vis) region of the electromagnetic spectrum (see Figure 9.1).

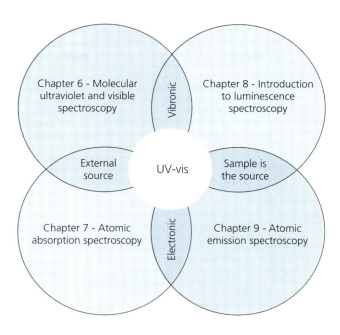

Figure 9.1 Modified Venn diagram depicting the relationships between molecular and atomic absorption spectroscopy and molecular and atomic emission spectroscopy. Note that in luminescence, just as in emission, the sample is the source of the light that serves as the *measured signal*, although many luminescence methods utilize an external source for *excitation* (see Chapter 8).

These techniques can be distinguished by the type of quantum event being studied (absorption or emission) and the nature of the sample (atomic or molecular). For example, Chapters 6 and 7 both describe absorption spectroscopy (molecular and atomic, respectively), and the only fundamental difference in the two techniques is the presence of vibronic transitions in molecular spectroscopy as opposed to purely electronic transitions in atomic spectroscopy. Both Chapters 8 and 9 describe emission spectroscopy (molecular and atomic, respectively), and the fundamental differences in the techniques involve vibronic versus purely electronic transitions. In addition, the two techniques differ in how the excited state analyte is produced. So it should not be surprising that the physical components used to construct these instruments are very similar; that is, all four spectroscopic techniques utilize similarly designed monochromators, detectors, and read-out devices. The similarity of the four techniques is represented by the central circle of the Venn diagram in Figure 9.1. The differences between these techniques can be grouped as well. For example, both atomic spectroscopic techniques have the added challenge of producing a stable atomic sample in the optical path of the spectrometer. This is represented in Figure 9.1 by the overlapping region of the bottom two circles. Other similar groupings are also in the Venn diagram.

Problem 9.1: Consider Figure 9.1 and the discussion of Boltzmann distribution in Chapters 2 and 7. Before continuing in this chapter, speculate on why a flame is an excellent atomizer for use in atomic absorption but is a poor atomizer for use in atomic emission.

Problem 9.2: Thinking in terms of quantum mechanic phenomena and instrumentation, list some similarities and differences between molecular absorption and molecular luminescence spectroscopic methods (see Figure 9.1).

9.2   The Atomizer and the Excitation Source

When we undertake experiments (absorption or emission) with molecular samples, we are typically dealing with a condensed phase and so we have the luxury of using a physical vessel to hold the sample (e.g., a cuvette). In Chapter 8, on luminescence spectroscopy, we saw that the condensed state sample could be electronically excited by an external light source and the light emitted by the sample as it relaxed to the ground state could be collected and measured by the spectrometer. Atomic emission spectroscopy (AES) is similar to molecular luminescence spectroscopy in that we are also measuring emitted light from excited state analyte. However, AES is an atomic method and as we saw in Chapter 7, atomic methods require that the sample be in the free gas state, necessitating a high-temperature atomizer that is not amenable to physical containment. Nonetheless, it is possible to produce excited state atoms using high temperatures and for this reason, AES instruments are designed such that the analyte is generated in an excited state *during* atomization. In Chapter 2, we introduced the Boltzmann distribution and learned that the number of atoms in the excited state is an exponential function of temperature, and so generally AES atomizers are much hotter than atomic absorption spectroscopy (AAS) atomizers and require special components. In summary, the atomization step in AES both produces atoms and causes their electronic excitation. A key advantage! Typically, no external excitation source is used in AES. The atomizer is also the spectroscopic source. In Chapter 8, we studied molecular fluorescence spectroscopy. Although conceptually similar to AES, a fluorometer requires an excitation source. Figure 9.2 shows a schematic of a generic atomic emission spectrometer. Notice the absence of a "source" in Figure 9.2. Figure 8.2 is a basic schematic of a fluorescence spectrometer. Take a moment to compare and contrast Figures 8.2 and 9.2.

The most common AES atomizers use plasmas maintained at several thousand Kelvins to generate atomic samples in the excited state. Samples are introduced into the plasma either (1) in the form of a nebulized aerosol mist, in the case of a liquid or powdered solid sample; or (2) by use of an electrothermal device (graphite furnace),[1] electric arc, or laser ablation, in the case of more refractory[2] solid samples. *Plasma* consists of matter in a highly energized state. It is an intimate mixture of cations and free

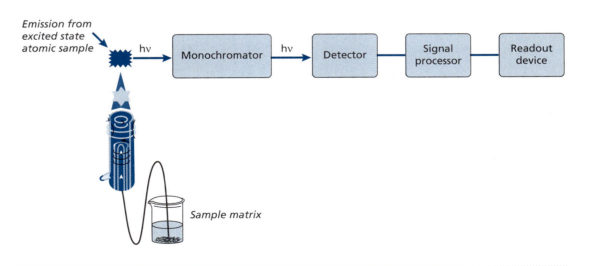

Figure 9.2 Schematic of a generic atomic emission spectrometer.

---

[1] See Chapter 7, Section 5, for a review of electrothermal atomization.

[2] A refractory substance is one that is stable at high temperatures. The term *refractory* is also sometimes used to describe a substance that is difficult to dissolve in common solvents.

electrons resulting from the ionization of a neutral species, such as argon or helium. In such a mixture, under normal conditions, the cations and electrons would most likely recombine and re-form the neutral species. To maintain the ionic mixture, the plasma must be continually energized via the introduction of energy from an external source. A neon sign is probably the most familiar artificial plasma device, and the light associated with a lightning strike is also the result of plasma. Indeed, both are essentially real-world AES experiments. The challenge in AES is to engineer an atomizer that is robust enough to withstand the hot environment needed to create excited state atoms and to do so with a stable atom flux in the optical path. The optical components needed to collimate, transfer, and detect UV-vis photons were covered in some detail in Chapters 3, 6, and 8. In the remainder of this chapter, we will explore design aspects that are unique to AES and discuss four plasma atomizer/sources that are regularly used to successfully meet the challenges of producing atoms in an excited state. The signal collected from each is essentially the same, and the differences among them lie primarily in the form of external energy applied to sustain the plasma.

On February 1, 2003, the space shuttle *Columbia* disintegrated on reentry after a successful mission in low orbit. During takeoff, debris loosened during launch damaged the heat-resistant ceramic tiles on the left wing of the shuttle. The extreme temperatures associated with reentry generated an atmospheric (probably nitrogen) plasma that flowed into the wing, melting much of the infrastructure and resulting in catastrophic failure. ■■

PROFILE
Columbia

Example 9.1

Use the Boltzmann distribution equation (introduced in Chapters 2 and 7; repeated below for easy reference) to calculate the percentage of calcium atoms that are in the first excited state in a hydrogen–air flame at 2,250K. The line associated with that transition has a wavelength of 226.51 nm.

STRATEGY –

Use the equation for the Boltzmann distribution:

$$\frac{N_e}{N_g} = \frac{g_e}{g_g} e^{\left(-\frac{\Delta E}{kT}\right)}$$

where

N_e and N_g = the number of atoms in the excited and ground states, respectively
g_e and g_g = the degeneracy (microstates) of the excited and ground states, respectively
k = Boltzmann constant
T = temperature in Kelvins

FINDING THE G-VALUE –

The number of microstates possible for an electronic configuration can be found using the formula:

$$g = \frac{2\#o!}{\#e!(2\#o-\#e)!}$$

where "$2\#o$" is twice the number of orbitals available for that state and "$\#e$" is the number of electrons in that state. So for calcium "$2\#o$" = $2 \times 1 = 2$ and "$\#e$" = 2.

$$g_g = \frac{2!}{2!(2-2)!} = 1 \ and \ g_e = \frac{6!}{2!(6-2)!} = 15$$

The percentage is just the ratio N_e/N_g multiplied by 100%. We know the values for k_B and T. We know that a photon with a wavelength of 226.51 nm results from the electronic transition of interest, so the energy of that photon is equal to ΔE, the difference in energy between the ground and excited states.

SOLUTION –

We know that:

$$\Delta E = \frac{hc}{\lambda} = \frac{\left(6.62607 \times 10^{-34} \ J \cdot s\right)\left(2.9979 \times 10^8 \ m \cdot s^{-1}\right)}{\left(226.51 \times 10^{-9} \ m\right)} = 9.03 \times 10^{-19} J$$

$$\frac{N_e}{N_g} = \frac{g_e}{g_g} e^{\left(-\frac{\Delta E}{kT}\right)} = (15)e^{-\left[\frac{6.03 \times 10^{-19} \ J}{\left(1.38065 \times 10^{-23} \ J \cdot K^{-1}\right)\left(2250 \ K\right)}\right]} = 5.57 \times 10^{-9}$$

So the percentage of calcium in the excited state is:

$$\frac{N_e}{N_g} \times 100\% = 5.6185 \times 10^{-9} \times 100\% = 5.57 \times 10^{-7} \%$$

which is approximately one out of every two hundred million atoms.

Problem 9.3: Use the Boltzmann distribution equation to calculate the percentage of magnesium atoms that are in the excited state (electronic transition from 3s to 3p, $\lambda = 285.2$ nm) under the conditions of (a) an air–acetylene flame at 2,955 K and (b) a plasma at 6,955 K. Assume that $g_e/g_g = 3$.

Problem 9.4: The Saha ionization equation allows us to estimate the ratio of gas phase atoms that will be ionized (n_i) relative to the number of neutral atoms ($n_0 = n_{total} - n_i$) under certain conditions. Assuming that the partial pressure of Mg atoms in the atomizer is 10^{-5} atm, $g_e/g_g = 3$, and the volume of the flame or plasma is 2.0 cm^3, calculate the percentage of Mg atoms that will be ionized under the two atomizer conditions given in Problem 9.3. The first ionization energy of magnesium is 737.7 kJ/mol.

$$\frac{n_i^2}{n_0} = \left(2.4 \times 10^{21} \text{ m}^{-3} K^{-\frac{3}{2}} \right) T^{\frac{3}{2}} e^{\left(\frac{-E_i}{k_b T} \right)}$$

This is a simplified Saha equation for first ionization, where E_i is the single atom ionization energy in joules.

Problem 9.5: In AES, one of the concerns of using too hot of a source is the generation of atomic ions instead of atomic atoms. The plasma source used in AES is much hotter than the atomizers used in AAS, yet we do not observe an overwhelming abundance of atomic ions in the sample spectrum. Speculate on why this is true (*Hint*: Think about the definition of and physical composition of *plasma*).

Inductively Coupled Plasma Torch

The *inductively coupled plasma (ICP) torch* takes advantage of *induction* as the source of external energy used to sustain a plasma. You are probably already familiar with the phenomenon of induction. For example, when an alternating electrical current is passed through a wire coil, an alternating magnetic field is generated. If a conductive material, such as iron, is placed within the coil, that magnetic flux will *induce* a voltage within that material, producing a current within the conductor. Because of resistance within the conductor, heat is generated during this process. This type of heating is called *inductive heating* (see Figure 9.3).

In an ICP torch, the conducting material is a collection of argon ions. Argon gas flows through concentric quartz tubes (see Figure 9.4), both longitudinally (through the center) and tangentially (through the outer tube, at an angle). The top portion of the concentric

Figure 9.3 Inductive heating of a metal bar.

Scatter

Earlier in this chapter, we talked at some length about the atomization methods used in AAS. Specifically, we spent a great deal of time talking about the importance of the temperature of flame AAS or graphite furnace AAS. If combustion is incomplete, the resulting soot (smoke) can lead to scatter of the radiant source, and this greatly reduces the absorption reading. Furthermore, if the flame (or furnace) is not hot enough, some elements readily produce refractory oxides in the flame. Refractory oxides are very efficient at scattering light and lead to very poor detection limits.

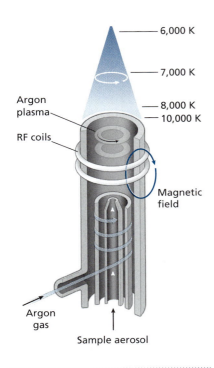

Figure 9.4 ICP torch.

quartz tubes are wrapped with a metal coil, through which a radiofrequency current is passed. This creates an alternating magnetic field, or a magnetic flux. Initially, the argon is nonconductive, but a high-voltage spark source, such as a Tesla coil, is used to ionize some fraction of the flowing argon atoms, resulting in argon ions within the volume of gas residing inside the coils. The magnetic flux triggers rapid acceleration of the argon ions within the gas. The partially ionized argon undergoes dramatic inductive heating, and collisions of the argon cations with neutral argon atoms generate additional argon ions. The collective result of these phenomena is a self-sustaining high-temperature argon plasma.

An aerosol mist of the analyte sample is carried into the torch by the argon flow through the center tube. The outermost sleeve seen in Figure 9.4 is also flushed with the carrier gas and serves two functions: first, it provides additional argon as matrix for the plasma; and second, it serves to cool the torch and prevent it from melting while in operation. The flow rate of the gas must be sufficient to keep the plasma from touching the top of the torch. A typical ICP torch operates between 6,000 K and 10,000K and consumes approximately 12 liters of high-purity argon gas per minute. The higher temperature regions are useful for thoroughly atomizing the sample and the inert environment from the argon minimizes the formation of refractory oxides and other molecular compounds often found in flame AAS experiments. The ICP torch is a very popular plasma source for AES and also finds use in elemental mass spectrometry.[3]

Direct Current Plasma Source

If you have ever seen photographs or videos of robotic welding of automobile frames, then you have observed a direct current (DC) plasma. Such robotic constructors use plasma arc welding to melt the adjoining pieces of metal, forming a continuous frame through the weld.

The first commercially available *DC plasma (DCP) atomizing sources* for chemical analysis instruments were introduced by Spectrametrics in the late 1960s. The basic design of the atomizing source has evolved over the years in the attempt to minimize stray light from the continuum plasma from entering the optical path. A schematic of a modern *DCP source* is depicted in Figure 9.5.

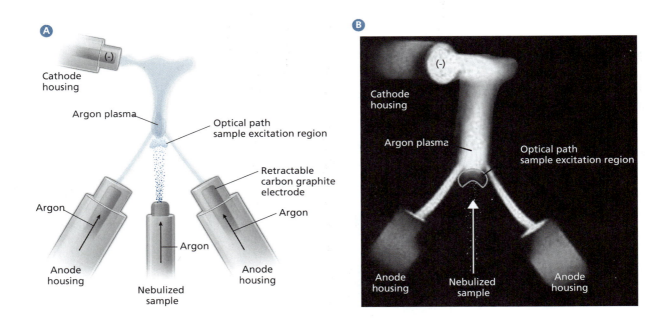

Figure 9.5 Panel (A) is a schematic of a DC plasma source, and Panel (B) is a photograph of a DC plasma source.

[3] See Chapter 13, Section 2.

Many modern welders use the plasma generated by a high-voltage arc through an inert gas to melt adjoining metal pieces and thus create the weld or cut steel. During this process, human welders are exposed to the gas-phase metal vapors that are generated, creating a health hazard. Interestingly, ICP-AES is one of the accepted Occupational Safety and Health Administration (OSHA) methods for monitoring levels of exposure to these harmful vapors (OSHA method ID-213). ■

The system consists of a tungsten cathode positioned at 90 degrees to the flow of carrier gas and sample. Two carbon anodes are positioned at 30-degree angles to the normal created by the flow path. Argon or helium serves as the carrier and plasma matrix. Initially, under a steady flow of argon, the two carbon anodes are moved into close proximity to the tungsten cathode. A high voltage (usually in the kilovolt range) is applied across the electrodes, and a spark is generated that ionizes some of the argon gas. The anodes are then retracted into ceramic sheaths to protect them from being oxidized in the high-temperature environment. A constant voltage of around 50 volts at 5 to 10 amps is maintained across the electrodes. An electron flux is generated from the tungsten cathode and accelerated toward the anodes, resulting in continuous high-energy ionization of the flowing carrier gas. Thus, the plasma is sustained by the application of external energy in the form of electricity through the electrodes. The two anodes used in conjunction with one cathode produce an inverted Y-shaped plasma plume. The inverted Y-design creates a region in the crux of the Y that is relatively free of emission from the argon plasma yet hot enough ($\approx 9,000$ K) to produce excited state atoms in the optical path of the spectrometer.

The sample is carried as a nebulized mist in argon flowing into the crux of the inverted Y. The DCP source is a bit more complicated than an ICP torch and it exhibits somewhat more difficulty in sustaining a steady plasma, but it generally consumes a lower quantity of the very expensive high-purity argon than an ICP torch (approximately 5–8 L/min).

Problem 9.6: Find the cost per volume for a tank of high-purity argon in your area. What is the cost per minute of using (a) an ICP-AES and (b) a DCP-AES?

Problem 9.7: You are the financial manager of an analytical laboratory that uses atomic emission to conduct environmental analysis. Assuming you run the instrument continuously for 6 hours on each workday, how much would you spend on argon each year (using your values from Problem 9.6)? Be sure to clearly indicate all assumptions you make to estimate the number of workdays each year.

Microwave-Induced Plasma Source

A stable plasma can also be maintained by use of a microwave generator. Although ICP remains the dominant plasma source in commercial instruments, *the microwave-induced power (MIP) source* has become more popular in the past 25 years because, in addition to argon, useful excitation plasmas can be generated using less expensive gases such as helium, nitrogen, or even nitrogen from regular air.[4] Helium plasmas can attain temperatures that are much hotter. For instance, Meubus and Parent reported a maximum temperature in a helium plasma of 24,000 K.[5] The use of helium as the plasma fuel, coupled with the use of a vacuum spectrometer, also permits the analysis of several nonmetals such as carbon and chlorine. The transition energy for these elements occurs in the vacuum UV region of the electromagnetic spectrum and require additional energy to populate. The vacuum spectrometer is required to avoid absorption of the ultraviolet emission by the optical materials in the spectrometer. An MIP source is typically constructed from a quartz tube inside a hollow metal cavity[6] that is calibrated to sustain a standing electromagnetic wave in the microwave region of the electromagnetic spectrum. A magnetron (a microwave generator) supplies microwave energy, and the standing waves build within the cavity. Support gas is supplied through the quartz tube, and an initial spark creates ions in the support gas. Microwaves produced from the magnetron cause the free electrons in the plasma support gas to oscillate. Oscillating electrons collide with atoms in the flowing support gas to create and maintain a high-temperature plasma. Analyte atoms are introduced into the MIP in much the same way as they are in ICP and DCP sources. Atomic emission spectra are measured as the excited atoms exit the microwave cavity. Figure 9.6 shows a schematic of an MIP source. It should be noted that the MIP source is often found as part of the detector system coupled in gas chromatography.[7]

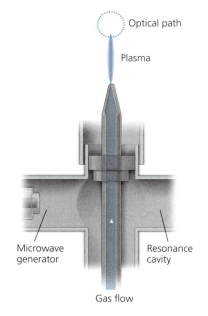

Optical path

Plasma

Microwave generator

Resonance cavity

Gas flow

Figure 9.6 Schematic of an MIP source.

[4] Long, G. L.; Motley, C. B.; Perkins, L. D. Helium High-Efficiency Microwave-Induced Plasma as an Element-Selective Detector for Packed-Column Supercritical-Fluid Chromatography, *ACS Symposium Series,* **2009,** *479,* 242.

[5] Meubus, P.; Parent, J. R. Temperature Distributions in a Helium Plasma. *Canadian Journal of Chemical Engineering,* **1969,** *47(6),* 536–542.

[6] Often referred to as a *waveguide.*

[7] See Chapter 16, Section 6.

In 2011, Agilent Technologies introduced the 4100 Microwave Plasma–Atomic Emission Spectrometer that runs on nitrogen plasma with regular air as the nitrogen source. The spectrometer was the recipient of *Research and Development Magazine*'s Top 100 award in 2012, an award that recognizes "the 100 most technologically significant products introduced into the marketplace over the past year." It boasts the advantage of eliminating the need for expensive gases and the infrastructure requirements of handling compressed gases. ■

PROFILE

Atmospheric Microwave Induced Plasma–Atomic Emission Spectroscopy

Problem 9.8: Assuming a flow rate of 2.5 L/min, determine the cost per minute and cost per year (see Problems 9.6 and 9.7) for running an MIP using (a) gas chromatography–grade nitrogen and (b) gas chromatography–grade helium.

Problem 9.9: Use the Saha equation (Problem 9.4) to compare the percentage of calcium atoms that are ionized under (a) a nitrous oxide–acetylene flame at 2,945°C, (b) an ICP argon plasma at 7,945°C, and (c) an MIP helium plasma at 20,000°C. Assume that the partial pressure of Ca atoms in the atomizer is 10^{-5} atm, $g_1/g_0 = 15$, and the volume of the flame or plasma is 1.5 cm^3. The first ionization energy of calcium is 589.8 kJ/mol.

Laser-Induced Plasma

Laser-induced plasma AES usually goes by the name of *laser-induced breakdown spectroscopy (LIBS)*. The substitution of the word *breakdown* for *plasma* was done to avoid the acronym LIPS. In a LIBS experiment, a high-power laser pulse is used to vaporize a small portion of a target sample while imparting sufficient energy to generate a plasma plume that serves as the atomizer and excitation source for the gas state atoms. Typical laser output is in the 100-MJ/pulse range, with a pulse length in the nanosecond range. The spot size is tightly focused with typical spot sizes in the 100-micron range. The cross-sectional power imparted during the pulse is immense, and spot powers in the GW/cm^2-range are typical.

Laser-induced Breakdown Spectroscopy—Selectivity

In an LIBS experiment, the relative abundance of a particular excited state atom is highly time dependent. By varying the time delay before a reading is obtained, you can influence the selectivity of the experiment toward a particular analyte.

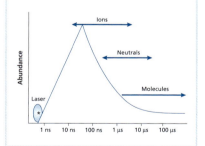

Problem 9.10: Assume that you have a laser with a power rating of 100 MJ/pulse.

(a) Determine the power (watts) of the laser pulse if the pulse width is 5 ns.

(b) Determine the cross-sectional power if the laser pulse is focused to a spot size of 100 μm.

That much energy focused on a small area is sufficient to both ablate material on the sample surface, producing a plume of analyte, and to generate the required excitation plasma (see Figure 9.7). Indeed, the temperature within a LIBS plasma plume has been estimated to exceed 100,000K in some instances, which is more than sufficient to ablate a small portion of the material and break down the material into its constituent elements in an excited electronic state. The plasma rapidly cools, and at temperatures in the 10,000K to 20,000K range, the excited state elements relax and emit their characteristic frequencies of light.

Laser-induced plasma AES enjoys two distinct advantages over the other plasma atomizers we have considered here. First, it can be used with any condensed-phase sample with no sample preparation whatsoever; there is no need to aerosolize the sample prior to introducing it into a plasma. Second, it can be used to measure samples

PROFILE

**Laser-Induced Breakdown
Spectroscopy in Space**

The Mars probe *Curiosity* uses laser-induced breakdown spectroscopy to sample and analyze soil and rocks on the planet's surface. *Curiosity* was launched in November, 2011, and it successfully landed on Mars in August, 2012. ■

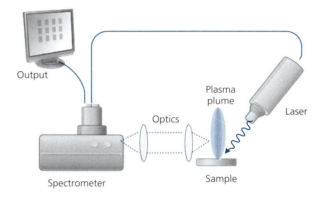

Figure 9.7 Schematic of an LIBS plasma source.

at a significant distance, up to 20 to 30 meters away from the spectrometer, which makes possible in situ measurements in virtually any environment, regardless of the hazard to humans.

The physical hardware used to conduct an LIBS experiment is very similar to the physical hardware used in laser-induced fluorescence spectroscopy and laser Raman spectroscopy. Instruments have been developed that are capable of obtaining vibrational molecular information followed by elemental analysis.[8] One of the disadvantages of the method is that it is inherently a surface technique. If the user requires a measurement of the bulk properties of the sample, it must be thoroughly homogenized prior to the acquisition of spectra.

---

[8] Sharma, S. K.; Misra, A. K.; Acosta-Maeda, T. E.; Bates, D. E. A Combined Time-Resolved LIBS-Raman System for Surface Chemical Analysis at Standoff Distances. 11th International GeoRaman Conference, **2014**; Courrèges-Lacoste, G. B.; Ahlers, B.; Pérez, F. R. Combined Raman Spectrometer/Laser-Induced Breakdown Spectrometer for the Next ESA Mission to Mars. *Spectrochimica Acta Part A: Molecular and Biomolecular Spectroscopy*, **2007**, *68(4)*, 1023–1028.

ACTIVITY

Visualizing a Plasma

This can be done either as an instructor demonstration or as an individual experiment.

Required Materials
Fresh seedless grapes
Large glass tumbler or highball glass
Microwave oven

Procedure
Carefully slice a fresh, seedless grape in half along the short axis. Take one half and slice into it, starting with the freshly cut (nonskin) side. Be careful not to cut all the way through

the skin; leave a small piece of skin connection between the two quarters. Carefully place the prepared grape on the base plate of the microwave (do not use any other dish) near the center with the cut side facing up. Place a tall glass tumbler or highball glass over the grape. Close the microwave oven, set the time for 20 seconds and then watch closely. You will usually see a spark or two, followed by one or more flashes of generated, white light–emitting plasma.

Here are links to videos demonstrating the procedure and results:
http://www.metacafe.com/watch/1011441/grape_plasma/
http://www.youtube.com/watch?v=qvSxVBalhFM

Problem 9.11: What are the properties of the ICP torch that have allowed it to become the dominant source for AES?

Problem 9.12: Summarize how each atomization method (ICP, DCP, MIP, and LIBS) achieves a plasma.

Problem 9.13: For each of the four atomization sources we have discussed, name one or two advantages it exhibits with respect to one or more of the other methods.

9.3 Sample Introduction

Sample introduction for AES is similar to that of AAS.[9] Like flame AAS, samples for AES can be introduced as an aerosol mist. In AES the plasmas exist at a very high temperatures and as a result, solvent is evaporated quite efficiently. As a result, the analytical results are less dependent on the quality of the aerosol mist in AES as in AAS.

Various methods for digesting solid sample matrices for introduction as an aerosol mist were covered in Chapter 7, Section 5, and those methods apply equally to AES. Likewise, when we discussed flame AAS, we discussed the various methods of producing a stable and reproducible aerosol mist of the sample matrix. A similar parallel can be drawn with cold vapor and hydride generation for the analysis of Hg and As. Recent developments in the area of sample introduction involve pneumatic nebulizers that operate in the submilliliter/analysis range[10] as well as the efficient use of automation.[11]

AES is also suitable for analyzing solid samples. There are several ways of introducing solid samples into a plasma source. The more common ways are electrothermal atomization[9] using a graphite furnace, an electric arc, and laser ablation. All three of these techniques produce a plume of particulate matter that is then swept into the plasma by the carrier gas. The electrothermal device is similar in design to the graphite furnace used in AAS, and although the graphite furnace is capable of producing atomic samples,

---

[9] See Chapter 7, Section 5. Note that this type of sample introduction applies to ICP, DCP, and MIP, but not usually to LIBS.

[10] Jankowski, K.; et al. Efficient Use of the NAR-1 Pneumatic Nebulizer in Plasma Spectrometry at Sub-milliliter Liquid Consumption Rates. *J. Anal. At. Spectrom.*, **2008**, *23*, 1290–1293.

[11] Bings, N. H.; et al. Atomic Spectroscopy: A Review. *Anal. Chem.*, **2010**, *82*(12), 4653–4681.

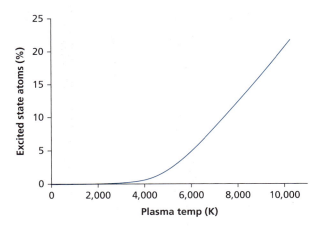

Figure 9.8 Plot of percentage of Na atoms in the excited state (3s → 3p) as a function of temperature.

that is not the goal in AES analysis. The graphite furnace simply produces an atomic plume that is then swept into the plasma.

Application

Recall in Problem 2.9, you were asked to prepare a plot of the ratio of excited state atoms to ground state atoms over a given temperature range. Your plot probably looked something like that seen in Figure 9.8. Note that at a typical flame temperature, only 0.02% of sodium atoms would be in the excited state, leaving 99.98% in the ground state. Even at the hottest flames (around 3,400K), not much difference is seen. Also, note that when an increase in percentage of excited atoms begins to be seen, the increase is nonlinear, as expected because temperature is in the exponential term of the Boltzmann equation.

At a relatively moderate plasma temperature of 6,000K, the percentage of excited state atoms increases more than 250 times that seen at 2,500K, and at a plasma temperature of 10,000K, the improvement is a factor of more than 1,200. We are now able to fully appreciate the groundwork that was laid out in Chapter 2. Flames are excellent atomizers for atomic absorption experiments because the signal we are measuring takes advantage of almost 100% of the atoms. On the other hand, flames are extremely poor atomizers/sources for AES, because the temperature is insufficient to excite many of the analyte atoms, whereas plasmas are quite good excitation sources for gas-state atoms.

Sources: Atomic Absorption Spectroscopy versus Atomic Emission Spectroscopy

As we have noted multiple times before, in order to carry out an AAS experiment, an external source of electromagnetic radiation is needed. Typically, in an AAS experiment, a unique lamp containing the element of interest (hollow cathode or electrodeless discharge) is used for each element to be measured.

In AES, the plasma atomizers essentially double as the excitation source; no external excitation light source is required. The plasma excites all elements in the sample simultaneously. So if we know the wavelength emitted by the analyte of choice, we can measure it "on the fly" without needing to change lamps as we saw in AAS.

Sample Preparation and Interferences

Recall from Chapter 7 that when conducting flame AAS experiments, it is sometimes necessary to undertake significant sample preparation to minimize chemical interferences. Protecting agents or releasing agents prevent compound formation in the flame, and the addition of an ionization buffer prevents significant ionization of elements having low ionization energies. Furthermore, there are sometimes other ubiquitous

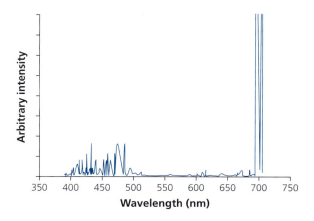

Figure 9.9 Argon plasma background emission spectrum. The emissions seen below 500 nm are primarily due to Ar⁺.

species in which we are not interested that can cause a false reading, especially when analyzing natural samples. Such interfering species must be removed or masked before a reasonable measurement can be made. On the other hand, the portion of the flame most often used for flame AAS exhibits few spectral interferences, and in most cases these can be simply removed as a part of the background measurement.

The extreme temperatures exhibited by plasmas, in most cases, allow for greatly simplified sample preparation procedures. Few chemical compounds can exist in such an extreme environment, so the need for protecting/releasing agents is largely eliminated. Furthermore, although a significant percentage of the atoms in the plasma *are* ionized, that percentage is far less than would be predicted strictly from temperature considerations due to the inherently electron-rich matrix of the plasma itself. However, plasmas do exhibit much more significant spectral interferences than do flames. Figure 9.9 shows an argon plasma emission spectrum from 390 to 710 nm; the emission lines come from excited argon atoms and Ar⁺ ions, and are always present when one is trying to measure the emission of other elements in the sample. Also, each element in the sample will exhibit a similar complex series of emissions from the excited atoms and ions produced from them. Element specificity in atomic *absorption* spectroscopy is obtained by the use of a single element lamp. Because the plasma excites all of the elements present simultaneously, plasmas exhibit a complex spectral map, and plasma instruments require very high quality monochromators. In addition, AES instruments need a very stable power supply in order to sustain a very stable plasma in the ICP torch. A standard flame AAS instrument will cost around $35,000 to $60,000, whereas a standard ICP-AES instrument will cost $60,000 to $100,000.

Zeeman Background Correction

In Chapter 7, on AAS, we introduced Zeeman background correction, and this method plays a significant role in AES as well. However, in this case, we are observing the emission event, rather than the absorption event. Recall that electronic transitions often are derived from degenerate states. However, these degenerate states reside in orbitals with different magnetic moments. So in the presence of a strong magnetic field, the degeneracy is broken and the single spectral line from the degenerate orbitals becomes several closely spaced lines.[12] Take a moment to compare and contrast Figures 7.14 and 9.10. The source and atomizer from Figure 7.14 has been replaced

---

[12] (a) Atkins, P. W. *Physical Chemistry,* 3rd ed. Oxford University Press: Oxford, UK, **1986**. (b) Condon, E. U.; Shortley, G. H. *The Theory of Atomic Spectra* (Chapter 16). Cambridge University Press: Cambridge, UK, **1935**. (c) Schlemmer, G.; Radziuk, B., *Analytical Graphite Furnace Atomic Absorption Spectrometry: A Laboratory Guide* (Chapter 4). Birkhäuser Verlag, **1999**

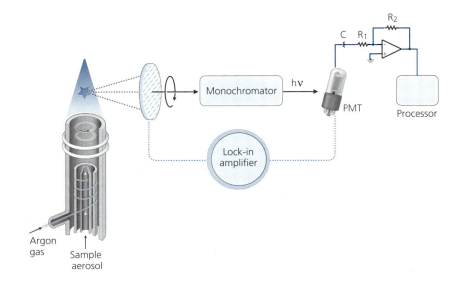

Figure 9.10 Zeeman-corrected AES. In the presence of a magnetic field, the analyte transition is split and polarized. A rotating polarizer allows for the independent measurement of analyte signal and background absorption. In addition, the rotating polarizer serves to modulate the analyte signal and a high-pass filter is used to attenuate DC power at the detector.

with an ICP torch. The remainder of the schematic is the same. Take a moment to read the explanation of Zeeman background correction in Section 7.6. The principle is exactly the same for AES.

> **Problem 9.14:** From what you understand of each atomization source for AES, rank the four sources from least expensive to operate to most expensive to operate per *sample* in an analysis. Justify your answer.

9.4 Measuring Atomic Emission

In Chapter 7 we noted that a typical flame AAS instrument uses a standard Czerny-Turner[13] monochromator with a single photomultiplier tube detector and a resolution from 0.1 to 0.7 nm. Because the flame imparts little excitation and ionization on sample components, this relatively basic system is sufficient. As noted in the above, the very complex emission spectra observed with plasmas makes it necessary to achieve much higher resolution in order to pick out the emission of the desired analyte from among the background and sample matrix emissions. Even lower-end plasma AES instruments require a monochromator with a resolution of at least 0.1 nm.

Many modern instruments now utilize an *Echelle-type dual-dispersion monochromator* (Figure 9.11) with a charge coupled device or charge injection device as a detector. The Echelle monochromator has an initial dispersing element, a prism or grating, that disperses the light in one plane. The dispersed light is then sent to another dispersing element, in this example a prism, that disperses the light in the other plane, resulting in a highly resolved two-dimensional image of all wavelengths produced by the plasma. Although this arrangement is substantially more expensive than that used in flame AAS, there are notable advantages. The dual dispersion provides for resolutions as high as ± 0.005 nm, allowing users to easily distinguish analyte spectral lines from the background.

---

[13] Refer back to Chapter 3 for a review of monochromator construction.

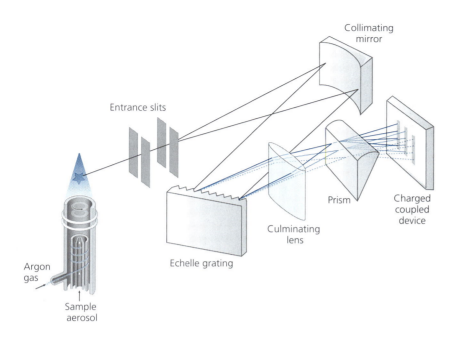

Figure 9.11 Sketch of an Echelle-type monochromator. A grating is used to disperse light by wavelength into discreet rays that comprise a bandwidth of closely spaced wavelengths. The dispersed rays from the grating proceeds to a prism where each ray is further dispersed into narrower bandwidth rays that terminate onto a charge coupled device detector where each ray is detected simultaneously.

COMPARE AND CONTRAST

Flame Atomic Absorption Spectroscopy, Graphite Furnace Atomic Absorption Spectroscopy, and Inductively Coupled Plasma–Atomic Emission Spectroscopy

TABLE 9.1:   **Comparison of Experimental Details for Atomic Spectroscopy Methods**

Feature	FAAS	GFAAS	ICP-AES
Number of elements[*]	68+	50+	73
Detection limits[†]	0.15–75000 ppb	0.002–130 ppb	0.02–10 ppb
Sample throughput	10–15 s per element	3–4 min per element	6–60 elements per min
Analytical range[‡]	10 ppb–1000 ppm	10 ppt–1 ppm	10 ppt–1000 ppm
Precision (% RSD)	0.1–1.0%	0.5–5%	0.1–2%
Spectral interferences	Very few	Very few	Some
Chemical interferences	Many	Many	Very few
Initial cost	Low	Medium	High
Running costs	Low	Medium	High
Cost per elemental analysis	Low	High	Medium

From Bart, J. C. J. *Additives in Polymers: Industrial Analysis and Applications*. John Wiley & Sons: West Sussex, UK, **2005.**

[*] This is the number of elements to which the method can be applied.
[†] From Perkin-Elmer Atomic Spectroscopy Guide, 2008–2011.
[‡] The range of concentrations over which the method is known to be applicable.

In Chapters 7 and 9, we have discussed essentially three atomization methods: flame, electrothermal, and plasma. It is worthwhile to take a moment to compare these methods in terms of analytical and practical parameters. By far, flame AAS is the least expensive option, both in terms of initial investment and maintenance costs, and it is sufficient for most routine analyses that do not involve trace-level determinations. The detection limits using flame AAS are in the ppm level on average but range from about 0.5 ppb up to the parts per thousand level for some elements. With sufficient sample size, precision better than 0.1% relative standard deviation (RSD) can be achieved without sacrificing too much throughput. Of the three methods, electrothermal provides the best detection limits, averaging about 0.25 ppb, and ranging from about two parts per trillion up to around 20 ppb, at an instrumental cost higher than that of flame AAS but much lower than ICP-AES. However, throughput is much lower than that of either flame AAS or ICP-AES, and the precision is often much worse, typically 1% to 5%, but ranging up to 20% RSD. As noted above, ICP-AES is the most expensive option of the three, and it provides detection limits between the two, but closer to electrothermal—generally around 1 ppb, ranging from 0.5 to about 20 ppb. With sufficient time given to integration, it exhibits precision on the level of flame AAS. However, the additional advantages noted in this chapter make ICP-AES one of the most viable atomic methods available today. Two additional atomic methods are yet to be discussed: *X-ray fluorescence spectroscopy* and *inductively coupled plasma-mass spectrometry*. We will cover these in Chapters 10 and 13. For the sake of comparison, it is worth noting that an ICP-MS instrument costs about 50% to 75% more than an ICP-AES instrument but provides detection limits orders of magnitude better than flame AAS or ICP-AES (even as low as 20 parts per quadrillion for some elements). Furthermore, with ICP-MS, we can quantify many elements inaccessible to the other methods. For an analytical laboratory conducting routine subtrace analyses, ICP-MS would likely be the method of choice.

Problem 9.15: Explain why it is easier to conduct multielement analysis using AES than it is using AAS.

Problem 9.16: Explain why a high-resolution monochromator is needed for AES while AAS requires a monochromator having relatively low-to-moderate resolving power.

9.5 Further Reading

BOOKS

Cremers, D. A.; Radziemski, L. J. *Handbook of Laser-Induced Breakdown Spectroscopy*. John Wiley & Sons: Chichester, UK, **2006**.

Cullen, M. *Atomic Spectroscopy in Elemental Analysis*. Blackwell Publishing: Oxford, UK, **2003**.

Hollas, Michael J. *Modern Spectroscopy*, 4th ed. (Chapter 7). John Wiley & Sons: Chichester, UK, **2004**.

Metcalfe, E. *Atomic Absorption and Emission Spectroscopy*. Prichard, E. eds. John Wiley & Sons: Chichester, UK, **1987**.

Varma, A. *CRC Handbook of Inductively Coupled Plasma Atomic Emission Spectroscopy*. CRC Press: Boca Raton, FL, **1990**.

JOURNALS

Tangen, G.; Wickstrom, T.; Lierhagen, S.; Vogt, R.; Lund, W. Fractionation and Determination of Aluminum and Iron in Soil Water Samples Using SPE Cartridges and ICP-AES. *Environ. Sci. Technol.*, **2002**, *36*, 5421–5425.

Uden, P. C., eds. *Element-Specific Chromatographic Detection by Atomic Emission Spectroscopy*; ACS Symposium Series, American Chemical Society: Washington, DC, **1992**.

Zhu, Z.; Chan, G. C. Y.; Ray, S. J.; Zhang, X.; Hieftje, G. M. Microplasma Source Based on a Dielectric Barrier Discharge for the Determination of Mercury by Atomic Emission Spectrometry. *Anal. Chem.*, **2008**, *80*(22), 8622–8627.

9.6 Exercises

EXERCISE 9.1: Why are atomic emission bandwidth so much smaller than molecular florescence bandwidths?

EXERCISE 9.2: Refer to Figure 9.1. What instrumental characteristics are shared between molecular luminescence and atomic emission methods? What would you say are the primary differences between the two methods?

EXERCISE 9.3: Thinking in terms of quantum mechanical phenomena and instrumentation, list some similarities and differences between molecular absorption and molecular luminescence spectroscopic methods (see Figure 9.1).

EXERCISE 9.4: Describe at least four ways of introducing an analyte into an ICP torch.

EXERCISE 9.5: From what you understand of each atomization source for atomic emission spectroscopy, rank the four sources from shortest to longest in terms of time required per *element* in a multielemental analysis.

EXERCISE 9.6: From what you understand of each atomization source for atomic emission spectroscopy, rank the four sources from highest to lowest in terms of cost per sample in an analysis.

EXERCISE 9.7: The primary absorbance and emission line for potassium occurs at 766.5 nm. Use the Boltzmann equation (Example 9.1 and Problem 9.3) to calculate the percentage of potassium atoms in the excited state using (a) a nitrous oxide–acetylene flame at 2,945°C and (b) an ICP plasma at 7,945°C.

EXERCISE 9.8: The method of standard addition was used to determine the amount of iron in a sample of water by DCP-AES. The data collected are presented in the table and the dark current for the spectrometer was 2.9 μA.

Added Fe (μg/L)	Detector Output (mA)
0	5.9
5	11.0
10	16.1

(a) Use a spreadsheet to plot a least squares regression line through the data. Be certain to display the x intercept on your graph. Also, display the equation for your line and the correlation coefficient on your graph.

(b) Determine the concentration of iron in the original sample and use the standard deviation of your line to determine the 95% confidence limit.

(c) If the correct answer was actually 2.81 ppb, what is the relative error in your analysis technique? Is the error most likely systematic or experimental? Explain your reasoning.

EXERCISE 9.9: In the analysis of sea water for gold content, a series of calibration standards were prepared using 4% HCl in deionized water as the diluent and then measured using ICP-AES after setting the response to zero with neat 4% HCl. A 10.00 mL aliquot of the sea water was acidified by adding 1.25 mL of concentrated (36.0%) HCl and then measured using the same method, yielding a response of 1,433 (arbitrary units).

Concentration of Calibration Standard (ng/L)	Detector Response (arbitrary units)
5.0	1,325
10.0	2,547
15.0	3,988
20.0	5,546

(a) What was the percentage of HCl in the treated sea water sample?

(b) Calculate the concentration of the gold in the original sea water sample.

(c) Calculate the standard deviation of the calculated concentration.

(d) Determine the 95% confidence interval for the measurement.

(e) Do you see any potential problems with this method as described? Explain. What would you do differently?

(f) Would your response to (e) be different if the method used were flame AAS? Explain.

Advanced Exercises

EXERCISE 9.10: Chromium is sometimes measured in abnormal concentrations near heavily traveled roads due to its presence in petroleum products (e.g., gasoline, oil, tires).

Longitude	Latitude	LIBS Response (mA)
N29.826°	W94.368°	24.25
N29.825°	W94.342°	21.57
N29.826°	W94.250°	18.11
N29.833°	W94.204°	22.93
N29.844°	W94.155°	27.89
N29.863°	W94.078°	37.65

An LIBS system was mounted on the back of a truck to sample roadside soils along Highway 73 in Texas. Prior to departure, a standard sample of 10.00 mg·kg⁻¹ Cr in clean soil was sampled using the truck-mounted system, giving a mean reading of 17.32 ± 0.13 mA for ten readings. The truck then sampled roadside soil along the highway at six locations. The locations given in the table are GPS longitude/latitude designations; you can use "loc: N29.826° W94.368°" in Google to find that location on a map.

(a) Estimate the concentration of chromium in the soil at each location.

(b) Is the Cr in the soil at the third location (N29.826°, W94.250°) significantly (statistically speaking) above the standard sample?

(c) Rationalize the trend(s) you see in the data.

(d) How accurate would you say the data are? How many significant figures do you believe you can reasonably assign to the calculated concentrations? Explain your rationale.

(e) What would you do to achieve greater accuracy in the concentration of Cr in roadside soils?

EXERCISE 9.11: What is the ratio of excited state versus ground state atoms in an 8,000K plasma, assuming the degeneracy of states is 1? (*Hint*: See Equation 4.15.)

EXERCISE 9.12: Go to the National Institute of Standards and Technology website (http://www.nist.gov/pml/data/atomspec.cfm) and look up the line spectra of magnesium and calcium. Given the fact that both calcium and magnesium are common components of soil, what recommendations can you make for suitable wavelengths for the ICP-AES determination of magnesium in soil?

EXERCISE 9.13: One of the advantages atomic spectroscopy has over molecular spectroscopy is the very narrow bandwidth of the observed transition. However the bandwidth is not infinitely narrow. In fact the Heisenberg Uncertainty principle states that $\Delta E \Delta t \geq \dfrac{h}{2\pi}$. What is the theoretical minimum bandwidth for the mercury 546.074 nm line if the lifetime of the transition (Δt) = 2 ns? *Note*: This calculation requires differentiation. Attempt this problem on your own, but refer to Chapter 7, Section 7.3 if you are struggling with it.

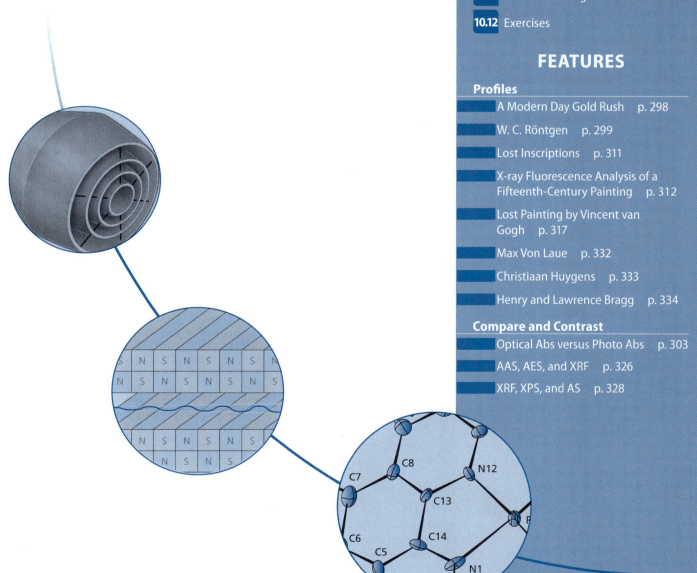

On April 24, 2013, *Bloomberg Businessweek* reported on the explosion of "We Buy Gold" businesses that seemed to have popped up on every street corner. As the price of gold surged from $300 an ounce in 2003 to $1,800 an ounce in 2012, small business entrepreneurs were looking for gold to feed the refineries any place they could find it—it seemed as if everyone was getting in on the action. If you had a drawer full of grandma's old broken jewelry, you may have, quite figuratively, been sitting on a "gold mine." But how does the shop owner know how much gold is in that bracelet you brought in? Is it 24-carat or 18-carat gold—or merely gold plated? The answer is provided by a handheld X-ray fluorescence (XRF) spectrometer (also known as a gold "gun"). For instance, Bruker Instruments markets six different models and advertises them as "handheld XRF spectrometry: a nondestructive elemental analysis technique for quantification of nearly any element from magnesium to uranium." They even have a website dedicated to "precious metal analysis." Thermo Scientific markets two XRF spectrometers specifically as precious metal analyzers. It seems that anyone can be a spectroscopist these days. ∎

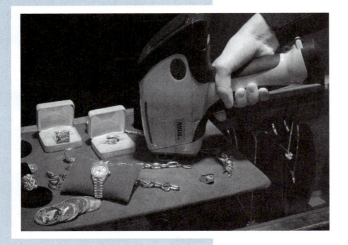

10.1 Principles of X-ray Fluorescence Spectroscopy

X-rays are high-energy photons with a wavelength ranging from 10 to 0.01 nm. The birth of X-ray spectroscopy is credited to Georg Charles von Hevesy, who in 1923 demonstrated the use of X-rays from a source to excite element specific X-ray emissions from a sample. The first practical X-ray spectrometers were developed circa 1950. There was little advancement in their capabilities until the late 1970s, when the development of

In 1895, **William Röntgen**, who lived from 1845 to 1923, observed fluorescence in $Ba[Pt(CN)_4]$ that had been placed in the path of the rays from a gas discharge tube. The tube had been covered with a blacked-out box. He correctly surmised that some previously unknown form of radiation was being produced that could penetrate the box, and he assigned the designation of X-ray (x for unknown) to his discovery. Within 2 weeks, Röntgen had produced an image of the bones in his wife's hand. Within 2 years of discovering X-rays, the use of "X-rays" in medicine had become a common practice around the world. In later experiments, Röntgen was able to deduce that X-rays were produced by the impact of cathode rays on a target anode.* We will revisit this idea when we examine X-ray sources later in this chapter. It was nearly 10 years before Max von Laue was able to show diffraction of X-rays, thus demonstrating that X-rays were a form of electromagnetic radiation and had wave-like properties. In 1901, Röntgen was awarded the Nobel Prize in Physics for his discovery of X-rays. ■

PROFILE
W. C. Röntgen

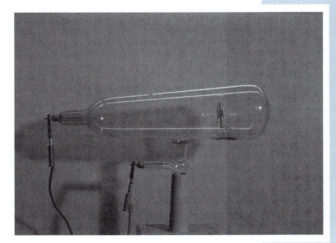
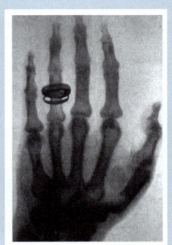

* J. J. Thompson later identified cathode rays as electrons.

http://www.deutsches-museum.de/sammlungen/ausgewaehlte-objekte/meisterwerke-ii/roentgen

advanced X-ray sources[1] allowed for the exploration of new materials and applications.[2] *X-ray fluorescence* (XRF) spectroscopy, along with all forms of X-ray spectroscopy, has seen remarkable advancements in optics and detector technology over the past decade.[3] The rate of development in X-ray spectroscopy has necessitated the publication of several good review articles and texts each year over the past decade. It is very likely that the pace will continue for the next few decades.[2]

---

[1] Beam line synchrotrons.

[2] Als-Nielsen, J.; McMorrow, D. *Elements of Modern X-ray Physics,* 2nd ed. John Wiley & Sons: NY, **2001**.

[3] Tsuji, K.; et al. X-ray Spectrometry. *Anal. Chem.* **2012**, *84*, 636–668; West, et al. *J. Anal. At. Spectrom,* **2011**, 26, 1919–1963; West, et al. Atomic Spectrometry Update-X-ray Fluorescence Spectrometry. *J. Anal. At. Spectrom,* **2010**, 25, 1503–1545; Tsuji, K.; Nakano, K.; Takahashi, Y.; Hayashi, K.; Ro, C. X-ray Spectrometry. *Anal. Chem.,* **2010**, *82*, 4950–4987; West, et al. Atomic Spectrometry Update. X-Ray Fluorescence Spectrometry *J. Anal. At. Spectrom,* **2009**, 24, 1289–1326; Tsuji, K.; Nakano, K.; Hayashi, H.; Hayashi, K.; Ro, C. X-ray Spectrometry. *Anal. Chem.,* **2008**, *80*, 4421–4454.; Erko, A.; Idir, M.; Krist, Th.; Michette, A. G. *Modern Developments in X-ray and Neutron Optics.* Springer: Berlin, Germany, **2008**; Szalóki, I; Osán, J; Van Grieken, R. E. X-ray Spectrometry. *Anal. Chem.,* **2006**, 78, 4069.

Although XRF spectroscopy is already quite a useful elemental analysis technique, it is fair to say that it is not yet a mature analytical technique. We can expect to see additional advancements for the foreseeable future.

In addition to hardware advancements, X-ray spectroscopy has been aided by advancements in software development and data analysis, which has made XRF much easier to use. As a result, XRF use has migrated out of the analytical laboratory and now finds wide use in such diverse areas as medicine (e.g., computer-assisted tomography), archaeology, anthropology (i.e., dating and sourcing antiquities), forensics (i.e., identifying art fraud[4]), and material analysis (primary metal identification in alloys[5]), as well as the development of two-dimensional and three-dimensional techniques for surface imaging. We will cover the applications of X-ray spectroscopy in the context of material and surface analysis in Chapter 20. In this chapter, we will discuss applications and instrumentation of XRF as it applies to elemental analysis; however, many of the same sources, optics, and detectors apply to both applications.

As an analytical tool, it would be correct to place atomic absorption spectroscopy (AAS), atomic emission spectroscopy (AES), and XRF spectroscopy into the same bin of your "toolbox" along with an additional technique called *inductively coupled plasma–mass spectrometry (ICP-MS)*.[6] XRF techniques are used for the routine elemental analysis of bulk materials, minerals, metal alloys, and antiquities. Although it is a form of fluorescence spectroscopy (see Chapter 8), the underlying quantum mechanics are different from all of the spectroscopic methods we have studied so far.

Without exception, all of the methods we have studied previously have involved quantum events that originate and terminate in the valence shell of the atoms or bonding orbitals of the molecules being studied. In contrast, the quantum event being exploited in XRF originates and terminates in the *core* quantum shells of the *atoms* of the sample (see Figure 10.1). When an atom is bombarded with high-energy X-rays (or gamma rays, alpha particles, protons, or high-energy electrons), it is possible to eject an inner-core electron from the analyte. The result is an ion in an *excited state* with an electron "hole" within the inner electron shell of the atom. Electrons in higher orbitals cascade into the holes as the excited state ion relaxes to the ground state. The result is the fluorescence of X-ray photons characteristic of the element in question (also known as *characteristic emissions*).

The energy of the fluoresced photon is the difference between the energies of the two corresponding orbitals. Figure 10.1 outlines the process. In panel (A), we see an incident X-ray photon from the source impinging on an analyte atom. The incident photon penetrates the outer shell of electrons and eventually collides with an inner shell electron resulting in the ejection of that inner shell electron (ejected *photoelectron*). In panel (B), we see the resulting cascade of upper shell electrons into the vacancy left by the ejected photoelectron. Panel (B) also shows two possible relaxation events. The one labeled K_α results from an electron that fell from one level above the K-shell, and the event labeled K_β results from an electron that fell from two levels above the K-shell.

X-ray Fluorescence Transitions: Terminology

When one creates an electron hole in the inner shell of an atom, the designation of the resulting X-ray photon depends on the quantum shell of the ejected electron. If the electron hole is generated in the n = 1 quantum shell (the 1s orbital), then transitions to that shell are designated as "K" transitions and the shell from which the cascading electron originated is indicated with a subscript α, β, γ, and so on. For instance, a transition from n = 2 → n = 1 is designated as K_α, where the K indicates the origin of the electron hole and the α indicates that the electron fell from one shell above the electron hole. If the cascading event is from n = 3 → n = 1, it is K_β, and so on. *All "K" transitions*

---

[4] Uffelman, E. S.; Court, E; Marciari, J; Miller, A.; Cox, L. Handheld XRF Analysis of Two Veronese Paintings. *ACS Symposium Series*, **2012**, *1103*, 51–73.

[5] dos Anjos, M. J.; Lopes, R. T.; de Jesus E. F. O.; Assis, J. T.; Cesareo, R.; Barradas, C. A. A. Quantitative Analysis of Metals in Soil using X-ray Fluorescence. *Spectrochimica Acta Part B: Atomic Spectroscopy* **2000**, *55(7)*, 1189.

[6] ICP-MS is covered in Chapter 13.

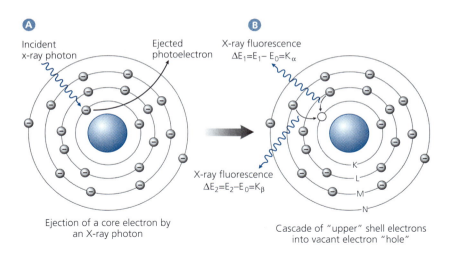

Figure 10.1 XRF resulting from the ejection of a core electron. In panel (A), an electron from the n = 1 quantum shell (the K shell) has been ejected. In panel (B), we see two possible relaxation events. The first is a cascade from the "L" quantum state resulting in a photon of ΔE_1. The second is a cascade from the "M" quantum state resulting in a photon of energy ΔE_2. Note the use of K_α and K_β to label these photons. The K-series of XRF photons results from ejection of an n = 1 electron, the L-, M-, and N-series result from the ejection of n = 2, n = 3, and n = 4 electrons, respectively.

terminate in the n = 1 quantum shell. If the original electron hole was generated in the n = 2 quantum shell, then transitions to that shell are designated as "L" transitions and n = 3 → n = 2 is designated as L_α and n = 4 → n =2 is L_β, and so on. The "L" transition is not seen for elements in the first three periods of the periodic table. In some of the very heavy elements, it is also possible to see "M" transitions in which the original electron hole was created in the n = 3 quantum shell.

Problem 10.1: The typical Jablonski diagram (i.e., Figure 8.3; also see the "Compare and Contrast—Optical Absorption versus Photoelectric Absorption") shows electronic transitions for the valence shell of an atom or molecule. Using Figure 8.3 as a style guide, construct an energy diagram in the style of a Jablonski diagram for an XRF K_α emission. As you construct your Jablonski diagram, make sure you consider the differences in atomic versus molecular transitions.

Problem 10.2: Why is it quantum mechanically impossible to observe a characteristic L_β emission for magnesium? A diagram similar to Figure 10.1 might be helpful in answering this question.

Photoelectric Absorption

Another contrast to make between XRF spectroscopy and "optical" spectroscopy (ultraviolet-visible [UV-vis] and infrared [IR], and so on) is the fact that the source radiation is not absorbed in the traditional way one would describe photon absorption in an optical spectroscopy experiment. In UV-vis and IR spectroscopy, the absorption of a photon only occurs when the energy of the incident photon matches the energy gap for the corresponding quantum events in the analyte. This is not the case for XRF spectroscopy.

For example, in UV-vis spectroscopy an electron moves from one quantum state to another in conjunction with absorption of a photon whose energy *exactly matches* the energy gap of the two electronic quantum states. In IR spectroscopy, the energy of the absorbed IR photon *exactly matches* the energy gap between two vibrational quantum states. In fact, it is this energy matching of source radiation in UV-vis and

IR spectroscopy that allows us to probe the quantum states of the analyte. Energy matching of the source radiation to analyte quantum states is not a requirement of XRF spectroscopy. In XRF, the incident photons from the source must possess a minimum threshold of energy in order to eject the core electron of the target atom. However, the incident photon may possess energy in excess of the threshold energy. Any excess energy possessed by the incident X-ray photon is carried away as kinetic energy by the *photoelectron* that is ejected from the core of the analyte nucleus[7] (see Figure 10.1).

Problem 10.3: In the orbital diagrams depicted, an ejected electron "hole" is designated with a "⊕" and the arrow represents a higher n-value electron cascading into that hole. Label each transition (i.e., K_α, L_β, and so on).

In "Compare and Contrast—Optical Absorption versus Photoelectric Absorption" (below), we have placed side-by-side a Jablonski diagram representing molecular UV-vis absorption[8] and a modified version of Figure 10.1 representing photoelectric absorption.

This is an extreme form of the *photoelectric effect*[9] and in the context of XRF, it is referred to as *photoelectric absorption*. Interestingly, the probability of ejecting an electron is maximized when the energy of the incident X-ray photon is at the minimum threshold ejection energy.

In practice, the efficiency of an X-ray source decreases as the energy of the X-ray photons exceed the threshold ejection energy of the analyte's electron. Minimum threshold ejection energies are given in Table 10.1.

Problem 10.4: Compare and contrast photoelectric absorption of source radiation with molecular UV-vis and IR absorption of source radiation. In what ways are they similar? How are they different?

Because the *wavelength* of the fluoresced photon is element specific, we can determine which elements are present in a sample (a qualitative analysis). The intensity of the fluoresced photons is proportional to the number of absorbing species. So a second goal of an XRF experiment is to measure the intensity of the fluoresced photons at each wavelength. This allows us to determine the number of atoms of each element present in the sample (a quantitative analysis).

Absorption of X-rays

In a manner similar to Beer's law for optical spectroscopy, the absorption of source X-rays in XRF has a linear absorption coefficient. We first introduced Beer's law for molecular UV-vis absorption in Chapter 6:

$$A = \varepsilon bc$$

---

[7] Measuring the kinetic energy of X-ray photoelectrons is the basis of a technique called photoelectron spectroscopy (XPS) and is related to an additional technique called auger electron spectroscopy (AES) (see Chapter 20).

[8] Jablonski diagrams were first introduced in Chapter 2 and again in Chapters 6 and 8.

[9] The photoelectric effect usually involves outer shell electrons and was first observed by Heinrich Hertz when he shone ultraviolet light on the cathode of a gas discharge tube. Albert Einstein won the 1921 Nobel Prize for his development of a theoretical explanation of the photoelectric effect. It was Einstein's work on the photoelectric effect that gave rise to the idea of a photon (a quantum of light).

TABLE 10.1: K_α Emission Energies in eV

ELEMENT		K_α	ELEMENT		K_α	ELEMENT		K_α	ELEMENT		K_α	ELEMENT		K_α
3	Li	54.3	21	Sc	4,090.6	39	Y	14,958.4	57	La	33,441.8	75	Re	61,140.3
4	Be	108.5	22	Ti	4,510.84	40	Zr	15,775.1	58	Ce	34,719.7	76	Os	63,000.5
5	B	183.3	23	V	4,952.20	41	Nb	16,615.1	59	Pr	36,026.3	77	Ir	64,895.6
6	C	277	24	Cr	5,414.72	42	Mo	17,479.34	60	Nd	37,361.0	78	Pt	66,832
7	N	392.4	25	Mn	5,898.75	43	Tc	18,367.1	61	Pm	38,724.7	79	Au	68,803.7
8	O	524.9	26	Fe	6,403.84	44	Ru	19,279.2	62	Sm	40,118.1	80	Hg	70,819
9	F	676.8	27	Co	6,930.32	45	Rh	20,216.1	63	Eu	41,542.2	81	Tl	72,871.5
10	Ne	848.6	28	Ni	7,478.15	46	Pd	21,177.1	64	Gd	42,996.2	82	Pb	74,969.4
11	Na	1,040.98	29	Cu	8,047.78	47	Ag	22,162.92	65	Tb	44,481.6	83	Bi	77,107.9
12	Mg	1,253.60	30	Zn	8,638.86	48	Cd	23,173.6	66	Dy	45,998.4	84	Po	79,290
13	Al	1,486.70	31	Ga	9,251.74	49	In	24,209.7	67	Ho	47,546.7	85	At	81,520
14	Si	1,739.98	32	Ge	9,886.42	50	Sn	25,271.3	68	Er	49,127.7	86	Rn	83,780
15	P	2,013.7	33	As	10,543.72	51	Sb	26,359.1	69	Tm	50,741.6	87	Fr	86,100
16	S	2,307.84	34	Se	11,222.4	52	Te	27,472.3	70	Yb	52,388.9	88	Ra	88,470
17	Cl	2,622.39	35	Br	11,924.2	53	I	28,612.0	71	Lu	54,069.8	89	Ac	90,884
18	Ar	2,957.70	36	Kr	12,649	54	Xe	29,779	72	Hf	55,790.2	90	Th	93,350
19	K	3,313.8	37	Rb	13,395.3	55	Cs	30,972.8	73	Ta	57,532	91	Pa	95,868
20	Ca	3,691.68	38	Sr	14,165	56	Ba	32,193.6	74	W	59,318.24	92	U	98,439

Center for X-ray Optics and Advanced Light Source, Lawrence Berkeley National Laboratory.
For additional emission lines see: http://xdb.lbl.gov/.

COMPARE AND CONTRAST

Optical Absorption versus Photoelectric Absorption

Jablonski diagram for molecular UV-vis absorption: absorption of a UV-vis photon when a bonding electron is promoted between two molecular orbitals. An exact energy match between the photon energy and molecular orbitals *is required*.

Orbital energy diagram for atomic photoelectric absorption: "absorption" of an X-ray photon results in the ejection of a core electron. The energy of the incident photon *does not* need to match the binding energy of the core electron, and any excess energy is carried by the electron as kinetic energy.

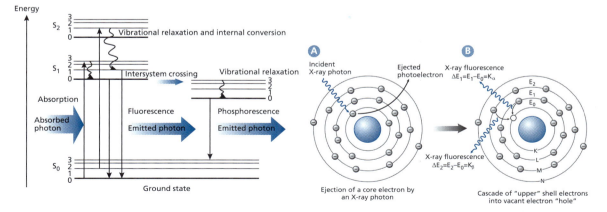

These two diagrams also point out another quantum mechanical difference in UV-vis fluorescence and XRF. In UV-vis fluorescence, an excited state electron "falls" back to its original energy state. In XRF, an electron "falls" into the vacancy left by the photoejected electron.

where
   A = absorbance (unitless)
   ε = molar absorptivity $\left(\dfrac{Liters}{(Mole)cm}\right)$
   b = path length
   c = concentration (molarity)

Imagine a beam of X-rays passing through an infinitesimally small portion of a sample with length containing an absorbing species, the loss of radiant power will be –dP. This can be expressed as:

$$\frac{dp}{d\ell} = -\eta P \Rightarrow -\frac{dP}{p} = \eta(d\ell) \qquad\qquad \textbf{Eq. 10.1}$$

where
   η = species-specific linear absorption coefficient

If we integrate over the entire path length, the loss in radiant power will be:

$$\int_{P_o}^{p} \frac{dP}{P} = -\eta \int_{o}^{\ell} d\ell$$

where
   P_o = radiant power at a distance of x = 0 (before entering the sample)

and this gives:

$$ln\left(\frac{P}{P_o}\right) = -\eta\ell$$

Converting natural log (ln) to log_{10} and multiplying through by –1 gives:

$$-log\left(\frac{P}{P_o}\right) = 2.303\,\eta\ell \qquad\qquad \textbf{Eq. 10.2}$$

The product of 2.303 and η gives a different constant, which we can symbolize as "μ." In our derivation of Beer's law in Chapter 6, we defined $-log\left(\dfrac{P}{P_o}\right)$ as absorption (A) and Equation 10.2 becomes:

$$A = \mu\ell \qquad\qquad \textbf{Eq. 10.3}$$

and in our discussion of Beer's law, we stated that the *molar absorptivity* must be determined experimentally for each molecule. Strictly speaking, one must also determine μ experimentally for X-ray absorption. However, in X-ray absorption, it is the inner-core electrons that constitute the "absorbing species," and the absorbance of X-ray photons is strongly dependent on the electron density of the elements present. Because the electron density of the core electrons in an element is dependent on the atomic number of the element (Z), in practice, the linear absorbance coefficient varies approximately as Z^4.

$$\mu \propto Z^4$$

Problem 10.5:   In what ways does the variable μ in Equation 10.3 resemble the variable ε in Equation 6.5? How is μ different from ε?

Although the parallels of Equation 10.3 to that of Beer's law are obvious, it is important to reemphasize that the analytical measurement is the absorbance of source radiation in UV-vis/IR spectroscopy and Beer's law is used to quantify the measurement. In an XRF experiment, we are not measuring absorption as a means to quantify an analyte. Instead, in an XRF experiment, the analytical measurement is composed of two variables: the wavelengths of the characteristic X-ray emissions and the intensities of the characteristic X-ray emissions. The wavelengths tell us which elements are present and the intensities are proportional to the concentration of each element. Because we are measuring emission instead of absorption, Equation 10.3 only helps us understand how efficient a source might be at inducing fluorescence in a particular sample and is not used to quantify analyte concentration.

10.2 X-ray Sources

X-ray sources for XRF spectroscopy can be categorized into three broad categories. The first is *radioisotopes* that emit X-rays as part of their natural decay process. The second is *X-ray tubes* that use high-energy electrodes to generate X-rays. The third is the use of a *particle accelerator (synchrotron radiation)* in order to achieve X-ray emission. The trade-offs one makes in choosing an X-ray source are direct. Radioisotopes are the most compact, portable, and economic of the three X-ray sources. However, they also have the lowest intensity. Synchrotron X-ray sources are massive structures that typically require government funding to operate. However, the increase in intensity over radioisotopes can be as large as a factor of 10^6. In between these two extremes, you have X-ray tubes, which are the sources found in most analytical laboratory instruments.

Radioisotopes

Because of their compact size and the absence of a power supply, many of the handheld XRF spectrometers use radioisotope sources. For example, Thermo Scientific's portable Niton XL series of handheld XRF spectrometers uses *americium-241* (see Figure 10.2). Americium-241 decays by alpha emission with gamma rays as a byproduct. Americium-241 also has a half-life of 432 years, so for all practical purposes, the source never needs to be replaced.[10] The result is a very robust source that is well suited for a handheld portable XRF spectrometer for use "in the field." These devices are capable of performing primary metal identification (PMI) of as many as 40 elements and are also widely used in geochemical research and in the antiquities marketplace. Other useful radioisotopes used for XRF sources are given in Table 10.2.

Figure 10.2 The Niton XL3t 950 GOLDD+ with integrated GPS handheld XRF analyzer utilizes americium-241 as the excitation source.

---

[10] The isotope ^{241}Am is also used in most home smoke detectors.

TABLE 10.2:   Radioisotopes Used in X-ray Fluorescence Spectroscopy

ISOTOPE	^{241}Am	^{109}Cd	^{57}Co	^{244}Cm	^{55}Fe
Energy in keV	60	23	122	18	6
K-line excitation	Ru-Er	Fe-Mo	Ba-U	Ti-Br	Al-V
L-line excitation		Yb-Pu		I-Pb	Br-I
Decay	α and γ	EC	EC	α and SF	EC
t ½ (years)	432	1.27	0.75	18.1	2.7

EC = electron capture; SF = spontaneous fission.

X-ray Tubes

In 1913, General Electric produced the first commercially successful X-ray tube, the *Coolidge tube*.[11] In the Coolidge tube, a tungsten filament served as the cathode. The filament was heated beyond the tungsten's work function (binding potential of the electrons), and the free electrons were accelerated through a potential gradient toward a target anode. The tungsten cathode was connected to a voltage source that replenished the tungsten's lost electrons, thus allowing for a steady stream of electrons toward the anode.

One of the technical problems with early X-ray tubes was the fact that the anode would get hot from the constant electric discharge, and in the low pressures of the discharge tube, the anode would sublime, greatly shortening its lifetime. In addition, the wavelength profile of the emitted X-rays would vary as the temperature of the anode changed. A design improvement incorporated into the Coolidge tube was the use of a constant water bath surrounding the anode. Cooling the anode slowed the sublimation process, giving the tube a longer lifetime and it allowed for a more stable output over time. An additional advantage of cooling the anode was the fact that you could apply greater voltages without overheating the anode, the result being a more intense X-ray beam. A schematic and a photograph of a Coolidge tube is shown in Figure 10.3.

Modern versions of the Coolidge tube incorporate a *spinning anode*. Spinning the anode avoids "hot spots" and provides for a more stable beam. However, spinning anode X-ray tubes are more expensive due to the technical challenges of spinning a tube while under a high vacuum.

Whether you are considering a traditional Coolidge tube or a Coolidge tube with a spinning anode, all X-ray tubes work on a similar principle. Electrons from the cathode are accelerated toward the anode and on striking the anode, the rapid deceleration of the

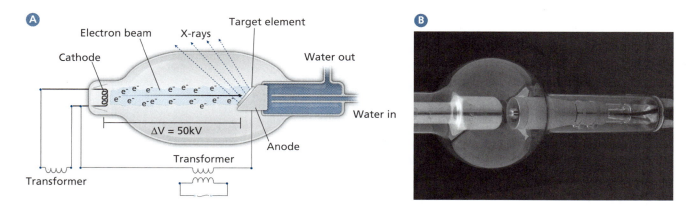

Figure 10.3 Schematic of a Coolidge tube (A) and a photograph showing the cathode and anode (B).

---

[11] The electrical engineer who developed General Electric's X-ray tube was W. D. Coolidge.

electron results in a conversion of the electron's kinetic energy into X-ray photons. This type of X-ray emission is called *bremsstrahlung* from the German word *bremsen* (to brake) and *strahlung* (radiation; braking radiation). The mechanism is fundamentally different from the element-specific characteristic emissions of XRF depicted in Figure 10.1. The bremsstrahlung portion of the emission spectrum is a continuum, which shifts to shorter wavelengths (higher energy) as the voltage gradient between the cathode and anode increases. In other words, the maximum energy of a bremsstrahlung X-ray photon is limited by the maximum kinetic energy of the electron that strikes the anode. Because the maximum kinetic energy of the electron is governed by the voltage gradient between the cathode and anode, the analyst has the ability to tune the spectral output of the source by adjusting the voltage of the X-ray tube. The highest energy photon is generated when the accelerated electron loses all of its kinetic energy in a single collision. Lower energy photons are generated when the accelerated electron loses its kinetic energy in a series of collisions within the target. The relationship between the voltage gradient and the maximum output energy of a bremsstrahlung photon is given by the *Duane-Hunt law* (Equation 10.4). The Duane-Hunt law is derived from the relationships between energy, wavelength, and frequency.

$$E_{max} = h\nu_{max} = \frac{hc}{\lambda_{min}} = eV \qquad \textbf{Eq. 10.4}$$

where

    h = Planck's constant expressed in units of eV (4.136×10^{-15} eV·s)
    c = speed of light (2.99×10^8 m/s)
    ν_{max} = maximum frequency of a bremsstrahlung photon (s^{-1})
    λ_{min} = minimum wavelength of a bremsstrahlung photon (m)
    e = charge on an electron (1.602×10^{-19} coulombs)
    V = applied voltage (volts)

If you solve Equation 10.4 for λ_{min}, combine the constants h, c, and e, and express voltage in units of kilovolts, you obtain a very useful form of the Duane-Hunt law that will allow you to quickly determine the shortest wavelength of bremsstrahlung radiation in picometers (pm).

$$\lambda_{min} = \frac{hc}{eV} = \frac{1240 \, pm}{kV} \qquad \textbf{Eq. 10.5}$$

As noted above, the bremsstrahlung radiation generated by an X-ray tube is a continuum and if the electrons striking the anode have sufficient energy, you can also induce characteristic fluorescence emissions from the anode material (see Figure 10.4). Take a moment to compare Figures 10.1 and 10.4. Some key features in Figure 10.4 are worth emphasizing. First, the maximum photon energy (shortest wavelength) seen in each trace increases as the electrode voltages increase. Second, when the accelerating voltage is high enough, we also see K_α and K_β lines from the molybdenum anode.

The bremsstrahlung radiation seen in Figure 10.4 is independent of the target metal and is only dependent on the accelerating voltage gradient of the tube. However, the K_α and K_β lines are directly related to the composition of the target material, and the target material must be considered if characteristic fluorescence from the target will interfere with an analysis. Characteristic emissions from the target anode are sometimes desirable. For increased selectivity within a specified atomic number range, some instrument designs use an X-ray tube with a high bremsstrahlung output, and the bremsstrahlung output is used to excite a secondary *target*, which gives off its own characteristic lines. The sample is excited by the emission from the secondary target only (without bremsstrahlung radiation present). The secondary target is chosen to efficiently excite elements in a certain atomic number range. For instance, the molybdenum K_α line in Figure 10.4 ranges from about 13 keV to about 14.5 keV. This could be used to eject K (1s) electrons with high intensity from bromine (binding energy = 13.5 keV) and krypton (binding energy = 14.3 keV) and with somewhat lower intensity for elements below Br. Some instruments can change secondary targets automatically during the analysis so that the low, middle, and high atomic number atoms from the sample spectrum may be analyzed during the same run.

Example 10.1 – Using the Duane-Hunt Law

(a) Determine the minimum wavelength of bremsstrahlung radiation that would be generated by a 75.00 kV X-ray tube.

(b) What would be the frequency of this radiation?

(c) What would be the energy in Joules of a single photon of this radiation?

STRATEGY –

(a) Use Equation 10.5 to determine the minimum wavelength of the bremsstrahlung radiation.

(b) Use Equation 10.4 to convert wavelength to frequency. Be careful to match your units between c and λ.

(c) You can use either the wavelength or the frequency to find the energy of the bremsstrahlung radiation using the relationships $E = \dfrac{hc}{\lambda} = h\nu$ where h = Planck's constant expressed in units of joules (6.626×10^{-34} J·s).

SOLUTION –

(a) $\lambda_{min} = \dfrac{hc}{eV} = \dfrac{1240\,\text{pm}}{kV} = \dfrac{1240\,\text{pm}}{75.00\,kV} = 16.53\,\text{pm} = 16.53 \times 10^{-12}\,\text{m}$

(b) $\nu_{max} = \dfrac{c}{\lambda} = \dfrac{2.99 \times 10^8\,\text{m/s}}{16.5 \times 10^{-12}\,\text{m}} = 1.81 \times 10^{19}\,\text{Hz}$

(c) $E = h\nu = 6.626 \times 10^{-34}\,\text{J·s} \times 1.81 \times 10^{19}\,\text{s}^{-1} = 1.20 \times 10^{-14}\,\text{J}$

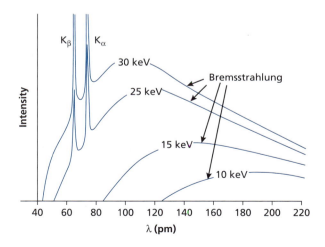

Figure 10.4 X-ray spectrum from a molybdenum target. The wavelength of bremsstrahlung radiation decreases with increasing voltage. Characteristic K_α and K_β lines for molybdenum appear at energies above 24 keV.

Problem 10.6: Use Equation 10.5 and Example 10.1 to determine the minimum wavelength, the maximum frequency and the energy in units of joules for the bremsstrahlung radiation generated by a X-ray tube with a voltage gradient of

(a) 25 kV   (b) 50 kV   (c) 100 kV

Example 10.2

What is the minimum wavelength of bremsstrahlung radiation needed to induce a K_α emission from a gold atom? What applied voltage would you need on the X-ray tube in order to obtain this radiation?

STRATEGY –

Use the Duane-Hunt law and the data from Table 10.1.

ANSWER –

From Table 10.1, the K_α line for gold has an energy of 68,803.7 eV.

$$\lambda_{min(excitation)} = \frac{hc}{eV} = \frac{(4.136 \times 10^{-15}\,eV \cdot s)(2.99 \times 10^3\,m/s)}{68,803.7\,eV} = 1.797 \times 10^{-12}\,m = 17.97\,pm$$

Substitution into Equation 10.5 and solving for V yields:

$$\lambda_{min} = \frac{hc}{eV} \quad \therefore \quad V = \frac{hc}{e\lambda_{min}} = \frac{(4.136 \times 10^{-15}\,eV \cdot s)(2.99 \times 10^8\,m/s)}{e(1.797 \times 10^{-12}\,m)} = 688.18\,kV$$

Problem 10.7: What voltage would you need to apply to the X-ray tube of an XRF spectrometer in order to induce a K_α emission from the following:

(a) Fe	(d) Ru	(g) Co
(b) Cu	(e) Zn	(h) Sn
(c) Ti	(f) Ni	(i) Ca

Synchrotron Radiation

A synchrotron is a *particle accelerator*. X-rays produced in a synchrotron can be a trillion (10^{12}) times more intense than those produced by an X-ray tube. Figure 10.5 is a schematic of a synchrotron particle accelerator. To produce X-rays in a synchrotron, electrons are accelerated to a very high energy; typically in the billions of eV (see the inner ring in Figure 10.5).[12] The high-energy electrons are then directed into a storage ring (see the outer ring in Figure 10.5). The storage ring is maintained at an ultra-high vacuum, and the electrons are bent into a circular path by very strong magnets within the storage ring. The angular acceleration of the electrons by the strong magnetic fields causes the electrons to change direction and results in the emission of X-rays. The X-rays are collimated and directed to research stations that are aligned tangentially to the storage ring.

In addition, X-ray photons can be produced by passing the electron beam through insertion devices that perturb the electron beam. The two most common devices are the *undulator* and the *wiggler* (see Figure 10.6). Both operate under similar principles. A series of magnets with alternating poles is used to perturb a perpendicular electron beam as it passes through the device. The alternating magnetic fields force the electrons to undergo

---

[12] The electrons are traveling near the speed of light.

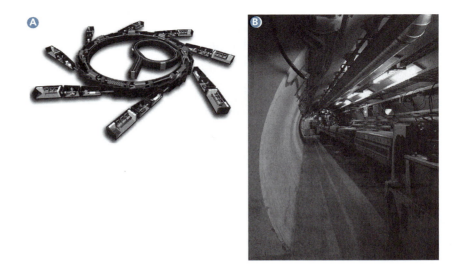

Figure 10.5 (A) Diagram of a synchrotron X-ray source. The inner ring is the particle accelerator, the outer ring is the storage ring, and the tangential buildings are associated laboratories. (B) Photo of the Large Hadron Collider.

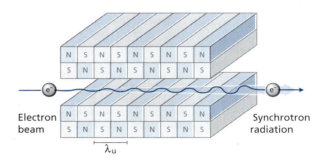

Figure 10.6 Schematic of an undulator insertion device. As the electron passes through the undulator, the electron is alternatively exposed to opposite magnetic polarities. As a result, the electron is forced to oscillate and the result is an X-ray emission. The wavelength of the emitted X-ray is related to the distance between the magnetic poles λ_u but is shortened by relativistic and Doppler effects. Typical undulator pole spacing is in the centimeter range resulting in X-rays in the angstrom (Å) range.

oscillations and thus radiate X-ray photons. Insertion devices can produce radiation that is significantly more intense than a simple bending magnet, and it is coherent. Because of the intense nature of synchrotron radiation, XRF spectroscopy using synchrotron radiation typically has detection limits in the low-ppb range. However, the use of synchrotron radiation for XRF spectroscopy is limited to a few well-funded research laboratories.[13]

10.3 X-ray Optics

As a beam of light passes from one transparent material to another, the photons change velocity and if the intersection occurs at an angle other than 90 degrees, the beam will change direction (see Figure 3.13 for a review of refraction). This is the basis of classical optics as described by *Snell's law*.[14] The ability to manipulate optical beams of light

---

[13] There are approximately 50 synchrotron XRF sources in the world.

[14] See Chapter 3, Equation 3.4 for a review of Snell's law.

was reviewed in Chapter 3. For visible light, the *refractive index (n)* of most transparent materials is in the range of 1.1 to 2 and is a function of the frequency of the light. As a result, blue light is refracted more than red light, thus allowing us to select a specific wavelength of light from a polychromatic source or focus and direct light as needed (see Figure 3.15 for a review). When dealing with X-rays, there are some key differences. When the energy of a photon correlates to an electronic transition within an atom or molecule, the refraction of light will show resonance behavior. Below the resonance energy, the refractive index (n) increases with frequency (*normal optics*) but above the resonance energy, (n) decreases. The energy of an X-ray photon lies above the resonance energy of most materials and as a result, the refractive index *is less than* 1 for optical materials in the X-ray region of the EMS $(n < 1)$. So a different approach must be used in manipulating X-rays as compared to UV-vis/IR photons. X-rays are typically manipulated using *reflection, diffraction,* and/or *interference* instead of exploiting differences in refraction.

Reflection Optics

In the case of reflective X-ray optics, most designs are based on the principle that you can only reflect X-ray photons at large *grazing angles*.[15] Otherwise, the X-ray will pass through or be absorbed by the optical material. In 1951, *Hans Wolter* designed a set of nested cylindrical mirrors made of polished metal whose surface (relative to the

---

In a paper published in the *Journal of Synchrotron Radiation,** synchrotron XRF was used to recover inscriptions from weathered Greek and Latin artifacts. Certain elements have increased concentrations in the region of the original inscription due either to deposition by the inscription tool itself, or by pigments that may have originally been used to paint the artifact. ▬

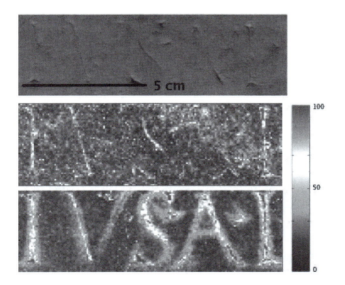

---

* *Smilgies, D. M.; Powers, J. A.; Bilderback, D. H.; Thorne, R. E. Dual-Detector X-Ray Fluorescence Imaging of Ancient Artifacts with Surface Relief. Synchrotron Rad.,* **2012**, *19, 547.*

---

[15] The grazing angle is the angle of an incident ray to the perpendicular of the surface.

PROFILE

X-ray Fluorescence Analysis of a Fifteenth-Century Painting

The photograph shown here is a painting of the Retable of the Passion of Christ. It is housed in the Museum of Fine Arts in Seville, Spain, and is believed to have been painted circa 1415. The artist is unknown. XRF spectroscopy was used to analyze the pigments used in the painting as a means of confirming its likely time of construction. The instrument used to perform the analysis was a portable XRF using a 30 kV tungsten X-ray tube and an silicon drift detector with an energy resolution of 140 eV. For example, the period-specific white pigment showed characteristic lead peaks and was confirmed as lead white, the red pigment showed characteristic mercury peaks and was confirmed as cinnabar, and the blue pigment contained copper and was confirmed as azurite. In addition to period-specific pigments, the researchers did identify areas containing more modern pigments such as zinc and barium that are consistent with a pigment known as lithopone, indicating that some restoration work had been performed on this painting. For a complete discussion of this work, see reference in the footnote.* ∎

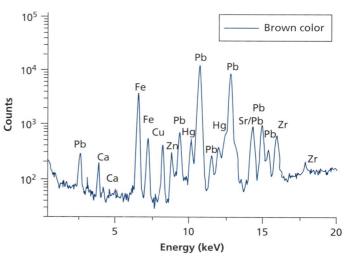

XRF spectrum of the brown color from the cross in the central table. Burned ochre (Ca, Fe) with cinnabar (Hg).

---

* Križnar, A.; et al. Non-Destructive XRF Analysis of Pigments in a 15th Century Panel Painting. The 9th International Conference on NDT of Art, Jerusalem, Israel, May **2008**.

---

incoming beam) is first parabolic and then hyperbolic (see Figure 10.7), thus keeping the X-ray beam at a grazing angle and allowing for the focusing of the beam.

The principles of *Wolter optics* can also be achieved using capillary tubes to reflect and carry the source radiation. In capillary optics, a bundle of capillary tubes is bound around a mold that gives the capillary bundle a parabolic and then hyperbolic profile (see Figure 10.8). A focused spot size in the micron range is routinely achievable with capillary optics. Many of the newer XRF spectrometers use capillary bundle optics.

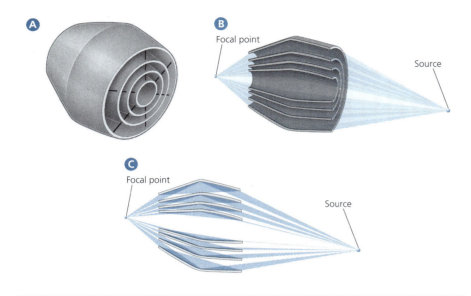

Figure 10.7 (A) A set of nested mirrors. (B) Cutaway of a Wolter optics with rays from the source to the focal point. (C) The optical path of the X-rays as they encounter the Wolter optics.

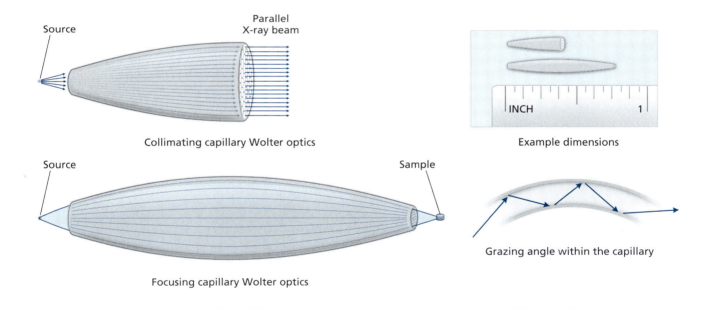

Figure 10.8 Capillary optics can be either collimating or focusing.

Diffraction Optics

In the case of diffraction X-ray optics, there are three common designs; these are *Fresnel zone plates*, *gratings*, and *single crystals/multilayer devices*.

Fresnel Zone Plates   A *zone plate* is a set of concentric rings that alternate between opaque and transparent (see Figure 10.9). X-rays striking the zone plate will diffract around the opaque rings and the zones are spaced so that the point of constructive interference is at the desired distance from the plate. The principles of Fresnel zone

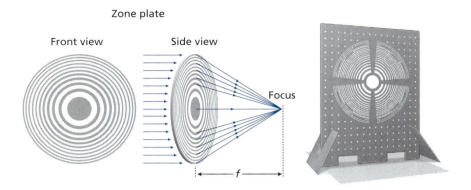

Zone plate

Figure 10.9 Schematic of a Fresnel zone plate. Image adapted from D. Attwood, *Soft X-rays and Extreme Ultraviolet Radiation: Principles and Applications* (Cambridge University Press, Cambridge, **2000**).

plates (FZPs) and gratings were discussed in Chapter 3. The critical difference in the construction of X-ray FZPs relative to UV-vis optics is in the materials used in their construction. X-rays can penetrate many materials, so the construction of X-ray FZP optics requires the opaque material to be atomically dense.[16] Gold, lead, and silver are often used for the opaque material. However, lead will form oxides over time, and silver will form sulfides over time. Therefore, gold is the material of choice if the optics will be exposed to the atmosphere. The concentric rings are produced by electrodeposition onto a semiconducting prestructured resist[17] (often Si_3N_4) that is first produced by electron or ion beam writing. A schematic of a FZP is seen in Figure 10.9.

Gratings   The gratings used in X-ray optics operate under the same principles as UV-vis gratings. However, because the wavelength of an X-ray is much smaller than a UV-vis photon, the key challenge is in the fabrication. Gratings have been manufactured for use in the soft X-ray region (10–0.095 nm); however, these gratings require as many as 5,000 lines/mm.[18] For hard X-rays (95–8 pm), diffraction is usually performed with either a single pure crystal or a fabricated nano-scale multilayer device.

Single Crystal/Multilayer Devices   Multilayer devices exploit the principles of *Bragg diffraction* to achieve wavelength specificity. Like all electromagnetic radiation, when an X-ray strikes an atom, it causes the electrons in the atom to move and the movement of the electrons reradiates the X-ray. This is known as Rayleigh scattering and was first discussed in Chapter 3. In crystalline materials and nanoscale multilayer devices, the spacing of atoms is repetitive (a lattice array), and the Rayleigh scattering of X-rays can occur at the surface or even several atomic levels

[16] Mote, R. G.; Yu, S. F.; Ng, B. K.; Zhou, W.; Lau, S. P. Near-field Focusing Properties of Zone Plates in Visible Regime—New Insights. *Optics Express*, **2008**, *16(13)*, 9554–9564.

[17] A resist is a coating used to protect the gold from adhering to undesirable areas.

[18] Heilmann, R. K.; Ahn, M.; Schattenburg, M. L. *Fabrication and Performance of Blazed Transmission Gratings for X-ray Astronomy*. Space Telescopes and Instrumentation, **2008**; Martin, E; Turner J. L.; Flanagan, K. A. *Ultraviolet to Gamma Ray*, Marseille, France: SPIE, **2008**. 701106-10. SPIE The International Society for Optical Engineering. http://dx.doi.org/10.1117/12.789176.

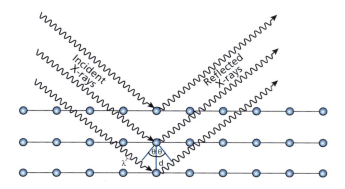

Figure 10.10 Bragg diffraction within a single crystal.

into the material. The result is a pattern of coherent and incoherent[19] scattering resulting from the differing path lengths of scattered radiation (see Figure 10.10). The effect is very similar to the wavelength dependent diffraction of visible light seen in optical gratings (Chapter 3).

Single crystal X-ray diffraction is codified by Bragg's law. *Bragg's law* is given in Equation 10.6.

$$n\lambda = 2d\sin\theta \qquad \textbf{Eq. 10.6}$$

where
  n = integer
  λ = wavelength of incident X-ray
  d = distance between the planes in the atomic lattice
  θ = incident angle between the X-rays and the plane of reflection

Bragg's Law (Equation 10.6) shows us that the angle at which an X-ray photon will experience constructive interference when reflected off of the surface of a single crystal is wavelength dependent. Therefore, by varying the incident angle, θ, single crystal X-ray diffraction allows us to efficiently disperse X-ray photons as a function of the X-ray photon's wavelength (λ).

Problem 10.8: Given a crystal lattice distance of d = 2.35Å, use Bragg's Law to calculate the (n = 1) angle required for the constructive interference of a λ = 1.54 Å X-ray beam.

Problem 10.9: For an X-ray beam of 2.01 Å experiencing constructive interference incident on a crystal surface at an angle of 45 degrees, determine the crystal lattice distance "d," assuming first-order (n = 1) Bragg diffraction.

We will revisit Bragg's Law in section 10.10 when we discuss X-ray crystallography.

Nanoscale multilayer devices work under the same principle as single crystal diffraction. The spacing in a single crystal is governed by the intrinsic atomic spacing and the orientation of the crystal to the X-ray beam. *Multilayer devices* utilize a series of

---

[19] See Chapter 3 for a review of coherence (specifically Figure 3.4).

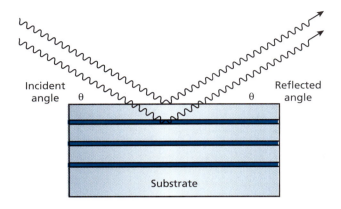

Figure 10.11   Bragg diffraction within a nanoscale multilayer device.

alternating thin films deposited onto a substrate. The spacing in a multilayer device is governed by fabrication techniques used in their manufacture. The thin films can be vapor deposited metallic films or even organic polymers (see Figure 10.11). Many of the multilayer devices outperform single crystal diffraction devices at the low atomic number range and find use for XRF analysis of lighter elements.[20]

10.4   Wavelength-Dispersive Spectrometers

Sequential and Simultaneous

There are two fundamentally different types (or configurations) of XRF spectrometers. The first type of XRF spectrometer we will examine is a wavelength-dispersive XRF (WDXRF) spectrometer. If we were to draw an analogy to spectrometers we have already studied, the WDXRF would be analogous to a UV-vis spectrometer that uses a monochromator to select a wavelength of light. The second type of XRF spectrometer we will examine is an energy-dispersive XRF (EDXRF) spectrometer. It is difficult to draw a direct analogy to other spectrometer types, but the closest analogy would be a spectrometer that generates a time domain signal such as an FT-IR spectrometer. However, EDXRF does not use the Fourier transform to create the time-dependent signal. We will revisit this topic when we discuss detector types later in this chapter. A schematic of a *sequential* and a *simultaneous* WDXRF spectrophotometer is depicted in Figure 10.12.

WDXRF spectrometers come in two basic designs: sequential and simultaneous modes. In sequential mode, a narrow bandwidth of frequencies from the source is selected by the monochromator and directed to the sample. To acquire a spectrum, the monochromator scans through a range of frequencies, sending each frequency to the sample sequentially. This is exactly analogous to a scanning UV-vis spectrometer. In this configuration, X-rays from the source are focused onto a single crystal whose angle relative to the incident beam is carefully controlled. The crystal disperses the X-ray photons according to Bragg's law, and a select bandwidth of diffracted X-rays are directed to an exit slit. Changing the angle of the crystal relative to the source radiation allows the operator to change the wavelength of X-rays reaching the exit slit. Recall from Bragg's

---

[20] Lee, P. X-ray Diffraction in Multilayers. *Optics Communications*, **1981**, *37(3)*, 159–164; Jiang, L.; Verman, B; Kim, B; Platonov, Y; Al, Z. *The Rigaku Journal*, **2001**, *18(2)*, 13–22; Lloyd, S. J.; Molina-Aldareguia, J. M.; Clegg, W. J. Structural Characterization of TiN/NbN Multilayers. *Journal of Microscopy*, **2005**, *217(3)*, 241–258; Stepanov, S. A.; Kondrashkina, E. A.; Kohler, R.; Novikov, D. V.; Materlik, G; Durbin, S. M. *Physical Review B*, **1998**, *57(8)*, 4829–4841.

In 2008, researchers at the HASYLAB[*] subjected van Gogh's painting *Patch of Grass* to XRF spectroscopic analysis in the hopes of imaging a painting believed to be hidden under *Patch of Grass*. The XRF utilized microfluorescence beamline L radiation—a second-generation synchrotron light source. The scan was conducted over 2 days with 0.5 mm² pixel resolution. The composition of the underlying pigments was analyzed and using imaging software, the researchers were able to image a women's face. The image of the woman's face is believed to be an early draft of the painting *Peasant Woman* by van Gogh, and van Gogh simply "recycled" the canvas when he painted *Patch of Grass*. ■■

PROFILE

Lost Painting by Vincent van Gogh

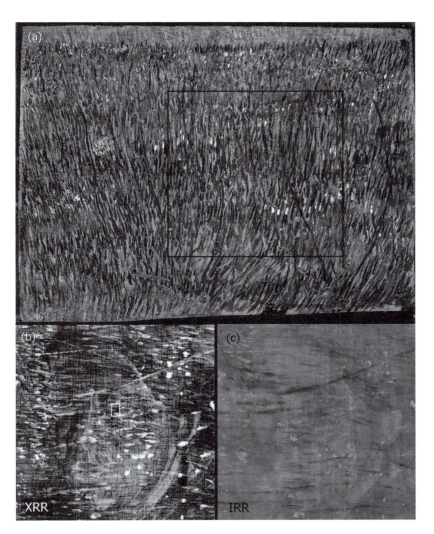

Reprinted with permission from Dik, J.; Janssens, K.; van Der Snickt, G.; van der Loeff, L.; Rickers, K.; Cotte, M. Visualization of a Lost Painting by Vincent van Gogh Using Synchrotron Radiation Based X-ray Fluorescence Elemental Mapping. *Anal. Chem.,* **2008** *80*(16), 6436–6442, American Chemical Society.

---

[*] *Hamburg Synchrotron Radiation Laboratory.*

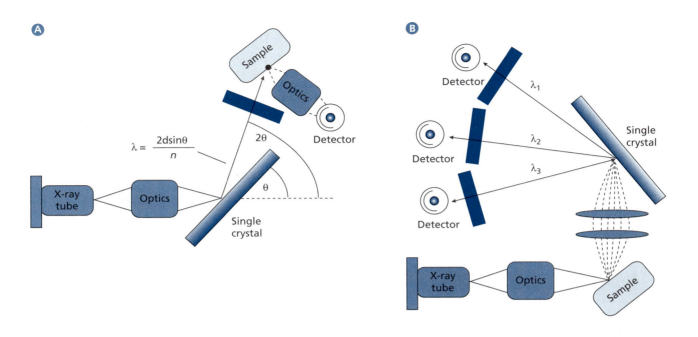

Figure 10.12 (A) Schematic of a sequential WDXRF and (B) simultaneous WDXRF.

law that the resolving power is also a function of *d* (the spacing of the crystal lattice). Therefore, spectrometers of this type often utilize multiple crystals (or multilayer devices) in order to resolve the entire X-ray region of the electromagnetic spectrum.

In a manner analogous to diode array UV-vis spectrometers, WDXRF spectrometers can also be purchased in what the industry calls *simultaneous mode*. Instead of scanning through different X-ray wavelengths by tuning the diffraction crystal, simultaneous mode spectrometers leave the crystal stationary and are configured with multiple detectors, placed so that they pick up diffracted element specific emissions from the sample over a range of angles. Simultaneous mode WDXRF spectrometers are typically used for very specific routine analysis such as metal alloy analysis or in-line quality control analysis during a manufacturing process. For example, if you were the quality control technician for a company that manufactures aluminum bronze, you might set up a simultaneous WDXRF to monitor for copper, tin, and aluminum, and you would confirm the concentrations of each element in each batch of metal produced.

Although sequential WDXRF spectrometers are slower and have more moving parts than do simultaneous WDXRF instruments, sequential mode WDXRF spectrometers allow users more analytical flexibility and are especially useful when determining the elemental composition of an unknown sample. Simultaneous WDXRF spectrometers allow for very rapid routine analysis. However, they are more expensive than sequential WDXRF spectrometers, primarily because of the additional detectors and signal processing electronics involved. Furthermore, to change elements, the user has to reconfigure the instrument, which negates many of the advantages of speed held by the simultaneous WDXRF spectrometers. WDXRF systems can routinely provide working resolutions that range from less than 5 eV to 20 eV.

Problem 10.10: It is customary to report XRF in units of eV. However, in Chapter 6, on UV-vis spectroscopy, we reported absorption wavelengths in units of nanometers. For comparative purposes, convert the range of 5 to 20 eV into units of nanometers. Compare the resolution of a typical XRF spectrometer to that of a UV-vis spectrometer in the context of the bandwidth of the X-ray and UV-vis regions of the electromagnetic spectrum.

Wavelength-Dispersive X-ray Fluorescence Detectors

There are several ways one can detect X-rays, and the choice of detectors is a trade-off between sensitivity/selectivity, cost of operation, portability, and durability. The three most common detector types are gas detectors, scintillating photomultiplier tubes, and solid-state photodiodes.

Gas Detectors   Gas detectors can be purchased in either a flow or sealed configuration. Gas detectors are best suited for the detection of longer wavelength X-rays. In both configurations, the gas is in close proximity to charged electrodes. Impinging X-rays ionize the gas and allow the gas to carry a current between the electrodes. The current is proportional to the X-ray flux. *Gas flow detectors* usually use argon as the primary carrier gas, with 10% methane added to quench any fluorescence from the recombination of argon ions and the photoejected electrons (Figure 10.13). Due to the high cost of argon, XRF spectrometers with gas flow detectors are expensive to operate. *Sealed gas detectors* differ primarily in the fact that the gas does not flow and the ionizing gas is typically krypton or xenon.

Photomultiplier Tube Detectors   The *scintillating photomultiplier tube (PMT)* was first introduced in Chapter 6 when we looked at UV-vis detectors. PMTs are designed so that the impinging X-ray photon strikes a transparent film covered with a fluorescent or phosphorescent material. The light that is emitted by the scintillating material is much lower in energy than the original X-ray that generated the emission, and it is the emitted photon that strikes the primary photocathode in the PMT, dislodging an electron. The initial dislodged electron is accelerated toward a secondary cathode. As the electron impinges on the secondary electrode, multiple electrons are dislodged. Each of these secondary electrons is accelerated toward a tertiary electrode, and the process is repeated through numerous stages. Current amplification is the primary advantage of PMTs, and their use greatly improves the detection limits of PMT XRF spectrometers. The reason we use a secondary emitted photon from the scintillator to initiate the electron cascade is due to the fact that in order to accelerate the dislodged electron toward the next stage in the cascade, it is necessary to precisely control the voltage drop between each dynode and the initial photocathode. The emission of the scintillator ensures that all photons impinging on the primary photocathode fall within a narrow bandwidth of energies and as a result, the initial electrons ejected from the photocathode will fall within a narrow range of kinetic energies. Because the construction requirements of the scintillating window, XRF spectrometers using PMT detectors are best suited for studying wavelengths below 0.3 nm. Figure 10.14 shows a schematic of a scintillating PMT.

Solid-State Detectors   A third category of WDXRF detectors includes solid-state detectors. These consist of a semiconducting material sandwiched between two charged electrodes. Figure 10.15 represents a p/n junction[21] detector. As the X-rays

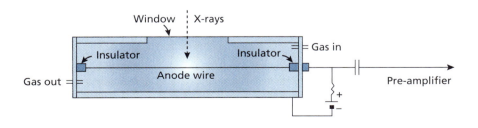

Figure 10.13 Schematic of a gas flow X-ray detector.

---

[21] See *Diodes* in Chapter 3 for a review of the principles of p/n junctions and Figure 6.28 for a schematic of a diode detector.

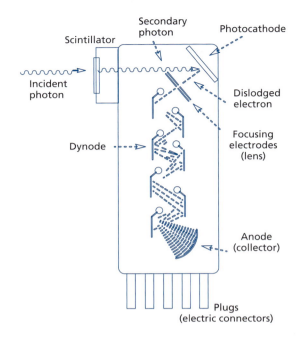

Figure 10.14 A schematic of an eight-stage scintillator PMT.

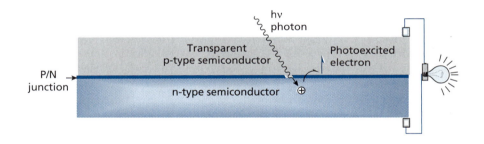

Figure 10.15 A schematic of a photovoltaic p/n junction.

impinge on the transparent p-type semiconductor, the X-rays cross the p/n junction and eject an electron across the boundary. Because the p/n junction can only conduct electricity in one direction, the electron-hole pair must migrate through the circuit in order to recombine. The resulting electric pulse is proportional to the X-ray flux. We first encountered p/n semiconductor detectors in Chapter 6 (also see Figure 6.23). The advantage of solid-state detectors is the fact that they are compact and rugged. They are also less expensive to operate, requiring less power than a PMT. However, the response time of solid-state detectors is slower than the other detector types mentioned. It is important to note that the technology of solid-state detectors is improving rapidly and their use is increasing, especially for use in portable XRF spectrometers.

10.5 Energy-Dispersive Spectrometers

EDXRF works under a different principle than does WDXRF. In general, EDXRF spectrometers are more robust, more portable, and faster than WDXRF spectrometers. The robust rugged nature of EDXRF spectrometers allows for wide use in industrial and in-the-field settings. But the primary advantage of EDXRF versus WDXRF is the analysis speed. For example, the elemental analysis of a metal alloy or a mineral can

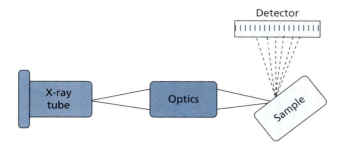

Figure 10.16 A block diagram of an EDXRF spectrometer.

be accomplished in just a few seconds. The primary disadvantage of EDXRF versus WDXRF is a decrease in resolution. A typical EDXRF has a spectral resolution of 100 eV to 300 eV or more, depending on the type of detector used.

A block schematic of an EDXRF is depicted in Figure 10.16 and shows that in EDXRF, the X-rays from the source impinge on the sample and all of the secondary fluoresced X-rays from the sample strike the detector without any wavelength discrimination. This necessitates the use of semiconducting detectors that are capable of determining both the energy of the impinging X-ray photon and the number of impinging X-ray photons. The output is a histogram of photon energy versus number of photons at each energy. The energy of the fluoresced X-ray photon allows us to determine the elements in the sample that produced it (qualitative), and the number of fluoresced X-ray photons allows us to determine the number of each atom type in the sample (quantitative).

In principle, the design of an EDXRF instrument is much simpler than a WDXRF instrument; there are no required optics between the sample and the detector and you do not need to precisely control the angle of a diffracting crystal. In fact, there are no moving parts at all. However, EDXRF spectrometers have very sophisticated detectors and require significant computing power to interpret the detector output.

Problem 10.11: In Problem 10.10, we converted the spectral resolution of WDXRF spectrometers from eV to nm. For comparative purposes, convert the spectral range of a typical EDXRF (100–300 eV) into nanometers.

Energy-Dispersive X-ray Fluorescence Detectors

As we mentioned above, EDXRF detectors must be able to measure photon counts *and* determine the energy of the photon that generated the current in the detector. There are several different designs on the market, but they all work under a similar principle. When a p-type semiconductor is in contact with an n-type semiconductor you have a *p/n junction* (also known as a rectifier).[22] Rectifiers serve as a one-way current gate. The detectors in EDXRF use PIN detectors. The "P" and the "N" in PIN stand for p-type and n-type semiconductors, respectively, and the "I" stands for intrinsic semiconductor. Intrinsic semiconductors are undoped or slightly doped. In the case of a PIN detector, population of the conduction band can be accomplished photolytically. A schematic of a PIN junction is seen in Figure 10.17. Note that the PIN in Figure 10.17 is reverse biased (the negative lead is on the p-type semiconductor and the positive on the n-type). When a p/n junction is reverse biased, the p-type and n-type semiconductors are depleted of all charge carriers, and no current flows through the device. When an X-ray photon enters the intrinsic region, it generates *electron-hole*

---

[22] See *Diodes* in Chapter 3 for a review of the principles of p/n junctions and Figure 6.28 for a schematic of a diode detector.

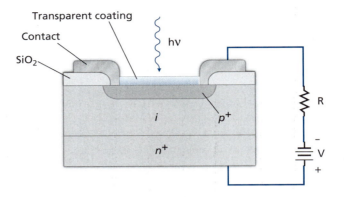

Figure 10.17 Schematic of a PIN X-ray detector.

pairs (e/h). The energy required to generate an e/h pair $(E_{e/h})$ depends on the material being used. For instance, for silicon detectors, $E_{e/h} = 3.6$ electron volts. The number (N) of e/h pairs generated is a function of the energy of the photon that impinges on the detector (Equation 10.7).

$$N \cong \frac{E_{photon}}{E_{e/h}}$$ Eq. 10.7

where

    N = number of e/h pairs
    E_{photon} = energy of incident X-ray
    $E_{e/h}$ = minimum energy required to create an e/h pair in the detector

The holes quickly migrate to the p-type semiconductor and the electrons "drift" to the n-type semiconductor. The multiple e/h pairs result in a compilation of charges, and it is the net charge (Q) that is measured in order to determine the energy of the incident photon:

$$Q = eN$$ Eq. 10.8

where

    Q = total charge generated in the detector from N e/h pairs
    e = fundamental electronic charge

$$Q \cong \frac{e\,E_{photon}}{E_{e/h}}$$ Eq. 10.9

Equation 10.9 is a combination of Equations 10.7 and 10.8, and it represents the measurement by which the instrument determines the energy of the photon. Note that Equation 10.7 (and by extension Equation 10.9) is an approximate equality, as N varies somewhat from photon to photon. This is one of the reasons that EDXRF devices exhibit poorer resolution than WDXRF ones.

Measurement of Q gives us a way to determine the energies of X-rays being emitted by the sample and so gives us our qualitative analysis. In one sense, the detector treats each photon as a unique event, and so the instrument is able to count the number of photons that generate a given range of charges (Q) over a period of integration, allowing us to quantify the amount of each element in the sample as well. However, because it is not possible to truly allow each photon "through" to the detector one at a time, EDXRF requires sophisticated software to analyze the signal generated by photons of many different energies impinging on the detector simultaneously. Transformation algorithms convert the convoluted time domain signal[23] into photon energies and photon counts. The spectrum is generated by segmenting the energy spectrum into discrete bins and counting the number of pulses registered within each energy bin.

Newer research-grade EDXRF spectrometers use the Peltier[24] cooled *silicon drift detector (SDD)*. Figure 10.18 shows a schematic of an SDD. The depletion region and the thin line seen in the depletion region in Figure 10.18 represents the charge carrier boundary as a function of photon energy within the depletion region. This line also represents the drift length that electrons travel toward the anode. When X-ray photons strike the SDD, a "plume" of e/h pairs are created within the SDD and the electrons and holes are separated under the external electric field. This event is represented by the panels (b) and (c) in Figure 10.18.

The electron holes rapidly migrate to the cathode and the electrons drift toward the anode. The bottom panel in Figure 10.18 shows the rise time profile for each discrete photon energy. The detector output is voltage as a function of time and represents the convolution of each individual rise time.[23]

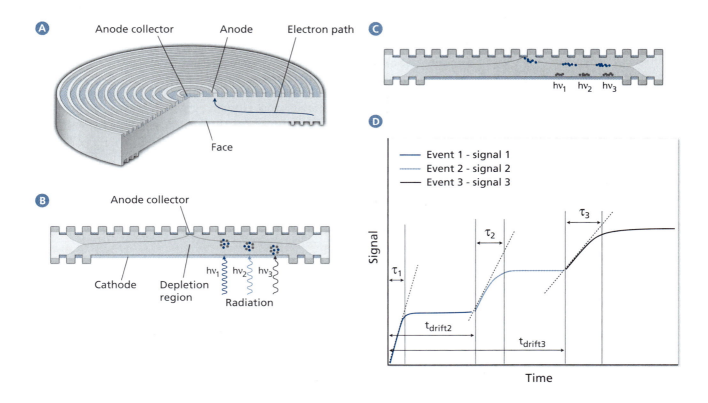

Figure 10.18 Operation of an SDD detector. Panel (A) shows a cutaway schematic of an SDD detector. Panels B and C show an expanded view of the bilayer region under bias. The darker gray regions in panels B and C represents depletion areas under bias. The blue and black dots represent e/h pairs generated by X-ray photons. In panel (B) three different energy X-ray photons strike the SDD. In panel (C), electrons "drift" to the anode. The rate of "drift" is proportional to the energy of the incident X-ray photon. Panel (D) is a graph of the signal rise time for each X-ray photon.

---

[23] We will look at time domain signals in detail when we study the Fourier transform in Chapter 11.

[24] The Peltier effect is the heating or cooling of a junction between two conducting materials (p/n). It is the same effect used in some travel coolers. These plug into a car's 12 V outlet, and a fan blows air across the cold electrode and into the travel cooler.

10.6 Direct Comparison: Wavelength-Dispersive and Energy-Dispersive X-ray Fluorescence Spectroscopy

As mentioned in the previous section, WDXRF spectrometers provide superior resolution in comparison to EDXRF spectrometers (see Figure 10.19). Depending on the quality of the instrument, the typical spectral resolution for WDXRF can vary from ~5 to ~20eV, and the typical spectral resolution for EDXRF varies from 150 to 300 eV or more. For the detailed analysis of a complex analyte, the superior resolution of a WDXRF provides fewer spectral overlaps and thus can provide superior characterization of the complex analyte. In addition, the increased spectral resolution (wavelength discrimination) of WDXRF means you also have reduced background noise (scatter) associated with each peak measured in WDXRF spectrometers. In other words, you can achieve higher signal-to-noise ratios with WDXRF than with EDXRF spectrometers. In theory, this leads to lower detection limits when compared to EDXRF. However, WDXRF requires optical components that EDXRF does not (collimators, focusing optics, and diffracting crystals).

Each of these optical components reduces the throughput of the source radiation. In practice, in order to achieve the lower detection limits promised by WDXRF, one has to incorporate a significantly more intense source. The use of a more intense source and additional optics translates into increased cost and complexity of use as compared to EDXRF spectrometers.

The virtues of EDXRF spectroscopy are mainly in the speed of analysis, ease of use, and cost of operation. In WDXRF, one scans through the X-ray region of the electromagnetic spectrum, acquiring each characteristic fluorescence sequentially as a point-by-point acquisition (time consuming) or the WDXRF can be configured with multiple detectors for simultaneous element detection (expensive). In EDXRF, a complete spectrum is acquired virtually instantaneously and can cover a large portion of the periodic table. If one is conducting routine analysis of simple materials, it is often the case that you do not need the superior resolution of WDXRF and in those cases, the rapid and relatively inexpensive EDXRF spectrometers outperform WDXRF. Two examples of EDXRF spectrometers are shown in Figures 10.20 and 10.21.

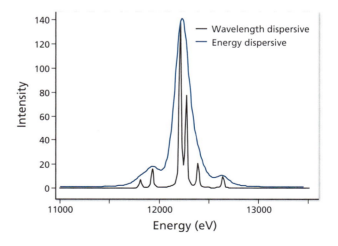

Figure 10.19 A comparison of the spectral resolution of WDXRF (black) to EDXRF (blue).

Figure 10.20 The NEX QC by Rigaku is an EDXRF analyzer utilizing a 50 kV X-ray tube as the source. It can analyze up to 81 elements with detection limits in the low percent range.

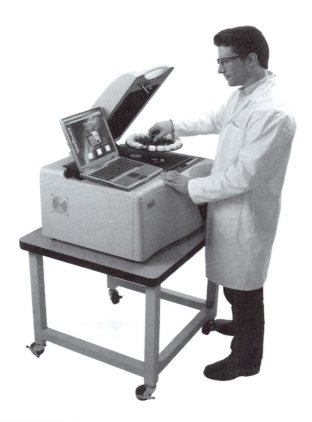

Figure 10.21 Thermo Scientific ARL QUANT'X EDXRF spectrometer showing the multisample carousel.

COMPARE AND CONTRAST

AAS, AES, & XRF

Where does XRF fits into the overall "toolbox" of the analytical chemist? In Chapter 7 we examined atomic absorption spectroscopy (AAS), and in Chapter 9 we examined atomic emission spectroscopy (AES). In a similar fashion, XRF finds wide use as an analytical elemental analysis techniques. There are some notable differences. In both AAS and AES, we had to first atomize the sample in order to remove molecular orbitals so that we could probe the quantum events of the atomic valence shells. This step also destroys the sample. Because XRF does not involve quantum events originating in the valence shell (i.e., bonding electrons) and only involves core electron shells, the energies of the fluorescent photons are unaffected by the molecular environment of the sample. Therefore, we do not need to atomize the sample before we conduct an elemental analysis. Very little (if any) sample preparation is required to use XRF, and the technique is nondestructive. In this regard, XRF has an advantage over AAS and AES techniques, especially when conducting analysis of materials that you wish to keep intact (i.e., a rare painting). On the other hand, XRF cannot achieve the low detection limits of AAS or AES.

	Cost ($1000)	Typical Sensitivity	Analysis Speed	Prep Time	Operation Maintenance Cost	Destructive
FLAME AAS*	10–20	ppm–ppb	≈ 60 s/element	Significant	Medium	Yes
GRAPHITE FURNACE AAS†	70–80	0.001–10 ppb	3 min/element	Significant	High	Yes
AES‡	40–80	<1 ppb–10 ppb	30 s/analysis	Significant	High	Yes
XRF§	75–125	1ppm–pph (%)	1–5 s/analysis	Very Little (none)	Low	No

*Agilent has a sequential mode (10 elements per analysis) that will analyze up to 10 elements at ≈ 2 min/analysis.
† Usually a graphite furnace is an added feature to flame AAS. Stand-alone graphite furnace AAS would be less expensive.
‡ AES and XRF routinely analyze multiple elements.
§ This comparison is for a research-grade instrument, not a handheld (retail-grade) instrument.

10.7 Sample Introduction

The concept of sample preparation does not really apply to XRF spectroscopy. The sample is analyzed as it is found, and with respect to the portable handheld XRF spectrometers, one simply places the nose of the "gun" above the sample and pulls the trigger (point and shoot). Likewise, the modern research grade (bench-top) XRF spectrometers do not require special sample preparation, but one does have to position the sample in the path of the excitation source within the sample chamber. Some commercial XRF spectrometers now have automated sample changers, much like the ones seen on high-performance liquid chromatography or AAS spectrometers. For example, the ARL QUANT'X Energy-Dispersive X-ray Fluorescence Spectrometer from Thermo Scientific has a 20-sample carousel (see Figure 10.21).

10.8 Total Reflection X-ray Fluorescence

Total reflection XRF (TXRF) is a surface analysis technique that we will revisit in Chapter 20, on material analysis. We briefly mention it here because of the instrumental similarities of TXRF to WDXRF and EDXRF. If a tightly collimated beam of X-rays is directed at a target at angles of less than 0.1 degrees, the reflectivity of the target increases greatly. It is therefore possible to place thin films on supports and "bounce" X-rays off of the films along with secondary fluoresced X-rays from the film. Because the reflectivity is so high, X-rays from the source do not penetrate the support, thus producing negligible secondary X-rays from the support. Detection limits exceed conventional XRF by five orders of magnitude and have been reported to be as low as 10 to 12 pg.[25] TXRF finds wide use in the semiconductor industry for the characterization of thin wafers of semiconducting materials. In addition to composition analysis, it is also possible to determine the thickness of the films by varying the angle of the incident X-ray source.[26]

Problem 10.12: Consider this scenario. You are writing a business plan in order to open a commercial laboratory that will conduct routine agricultural microanalysis of soil for the nutrients Ca, Mg, Cu, and Zn. You will need to be able to quantify these elements in the ppm range. The going rate for an analysis of this type in your area of the country is $60 per element per analysis. Your anticipated overhead will be $1500 per month,[27] which *does not* include a loan payment plan for the instrument. Use the Internet to research commercially available XRF spectrometers and complete the business plan. Justify (in your business plan) your choice of XRF spectrometer and include in your final business plan the 5-year loan payments (5% APR) for your XRF. Do not forget to include your salary. How many analyses per week will you need to conduct in order for your business to succeed? Make sure you justify your choice of XRF spectrometer against the number of analyses per week you will be conducting.

10.9 X-ray–Induced Photoelectron Spectroscopy and Auger Electron Spectroscopy

X-ray–induced photoelectron spectroscopy (XPS)[28] and Auger electron spectroscopy (AS)[29] are both forms of electron spectroscopy and are fundamentally important in the field of material surface analysis.[30] Both XPS and AS are routinely used to probe the surface *composition* of solids and viscous oils, and they can reveal the elemental composition as well as the speciation[31] of the elements present. Both techniques can "see" elements from lithium up to the heaviest of elements.

XPS and AS provide similar types of information, and the underlying principles are similar to that of XRF spectroscopy. In XRF, a core electron is ejected as the result of a collision with an X-ray photon. As higher energy electrons cascade into the vacant holes, X-ray photons are emitted (see Figures 10.1 and 10.22).

X-ray–Induced Photoelectron Spectroscopy

Ernest Rutherford deduced that the kinetic energy of the ejected electron is equal to the energy difference between the source and the electron's binding energy.[32] Therefore, it seems obvious that the energy of the ejected electron contains information about the material from which it was ejected. In XPS, you measure the energy

---

[25] Potts, Philip J.; et al. *Journal of Analytical Atomic Spectrometry,* **2001**, *16*, 1217–1237.

[26] Klockenkämper, R. *Total Reflection X-ray Fluorescence Analysis.* **1997**, John Wiley & Sons.

[27] Overhead would include office rent, insurance, utilities, mileage charged against a vehicle, and so on.

[28] XPS sometimes goes by the name electron spectroscopy for chemical analysis (ESCA).

[29] Auger was a Frenchman. The term is pronounced *O-jay* with a soft "j."

[30] Many analytical chemists use the acronym AES for Auger electron spectroscopy. However, we are already using AES for atomic emission spectroscopy, so in this textbook Auger electron spectroscopy is denoted as AS.

[31] The term *speciation* in chemistry refers to the molecular composition of a material. For instance, elemental analysis might tell that a compound contains silicon, and speciation would tell that the silicon is present as a dioxide rather than an organosilane compound.

[32] Rutherford, E.; Robinson, H.; Rawlinson, W. F. *Philos. Mag.* **1914,** *18*, 281; Rutherford, E.; *Philos. Mag.* **1914,** *20*, 305.

COMPARE AND CONTRAST

XRF, XPS, & AS

AS

The fluoresced X-ray photon ejects a secondary outer-shell electron. The energy of the secondary photo-electron is measured

XPS

The energy of the ejected core photo-electron is measured

Ionizing radiation ejects a core electron. Outer shell electrons "fall" into the vacant hole resulting in characteristic X-ray fluorescence

The energy of the fluoresced X-ray photon is measured

XRF

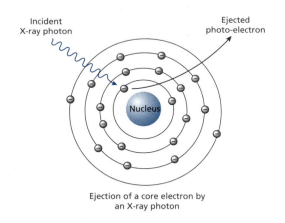

Incident X-ray photon

Ejected photo-electron

Nucleus

Ejection of a core electron by an X-ray photon

Figure 10.22 X-ray–induced photoejection of a core electron.

of the ejected *electron*. XPS makes use of the photoelectron's energy in order to study the composition of a material. As we will soon see, AS makes use of a different quantum event that also involves an ejected electron. In order to distinguish between the two photoejected electrons, we will refer to the XPS photoejected electron as the *primary photoelectron* and the AS photoejected electron as the *secondary photoelectron.*

$$KE_e = h\nu - BE_e \qquad \textbf{Eq. 10.10}$$

where

KE_e = kinetic energy of the *primary photoelectron*
$h\nu$ = energy of the incident X-ray photon
BE_e = binding energy of the *primary photoelectron*

It was not until the pioneering work of H. R. Robinson that XPS became a valuable analytical tool. In 1930, Robinson was able to resolve energy shifts (chemical shifts) in the primary photoelectron's energy caused by the chemical environment of the atoms,[33] and today XPS is routinely used to map out the surface composition of many materials with detection limits in the parts per thousand range. A typical XPS analysis involves irradiation of the surface of a sample with X-rays from a source (typically Mg-K_α or Mo-K_β), and the kinetic energy of the primary photoelectrons is measured.[34] Solving Equation 10.10 for binding energy allows you to identify the element that produced the primary photoelectron. The photoelectron spectrum is an indication of the binding energies of the atomic levels within the surface atoms. In addition, slight shifts in the binding energies allow analysts to deduce the speciation of the elements. XPS finds wide use as a tool for deducing the empirical formula, chemical state, and electronic state of the elements that exist within the first few nanometers of the surface of a material.

Auger Electron Spectroscopy

While studying X-ray induced photoelectrons, the French physicist *Pierre Victor Auger* noticed photoelectrons in the background "noise" that always had the same energy regardless of the energy of the incident X-ray photon. According to Equation 10.10, the energy of the photoelectrons should be proportional to the energy of the incident X-ray photon. Auger spent 3 years studying these anomalous photoelectrons and in 1926 published a theoretical explanation for their existence in his PhD thesis. The Auger effect (as it is now termed) is the result of inter- or intra-absorption of XRF photons resulting in a second electron ejection.

The underlying principles of AS are very similar to those of XPS. However, there is a striking difference in terms of *what* is being measured. In a manner identical to XRF and XPS, the sample is irradiated with ionizing energy in order to eject a core electron. The fluoresced X-ray that results from the cascade of upper-level electrons sometimes strikes an outer shell electron instead of escaping. The ejected outer shell electron is referred to as a *secondary photoelectron,* and it is the kinetic energy of this secondary photoelectron that is being measured in AS. This is distinctly different from XRF (where the measurand is a fluoresced characteristic X-ray photon) and from XPS (in which the measurand is the primary photoelectron). The secondary photoelectron is also known as the *Auger electron*, a term honoring Auger.

Because the characteristic XRF photon results from a cascade of core electrons and is not influenced by the chemical environment of the atom, the XRF photons have the same energy, and because it is this XRF photon that photoejects the Auger electrons, the energy of an Auger electron is a function of the energy of the XRF photon. However, the Auger electron originally resided in the valence shell so the energy of the Auger electron is also influenced by the chemical environment of the valence shell. For a particular element the Auger electrons fall within a relatively narrow band. The energy band of the Auger electron allows us to identify the element that created the Auger electron (see Figure 10.23). However, because the Auger electron is in the valence shell, its energy is shifted slightly by its chemical environment. It is this shift within the

---

[33] Robinson, H. R. *Proc. R. Soc. A.* **1923**, *104*, 455; Robinson, H. R.; Young, C. R. *Philos. Mag.* **1930**, *10*, 71.

[34] In XPS, the source radiation is monochromatic.

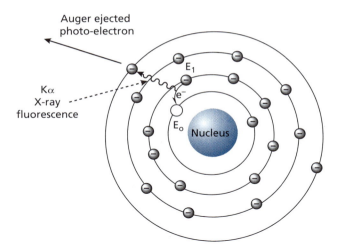

Figure 10.23 Cascade of "upper" shell electrons into vacant electron "hole" produce XRF photons. Inter- or intra-absorption of XRF photons can result in ejection of an Auger electron.

energy band that allows for speciation of the atom types observed. Furthermore, the area under the peak in an Auger experiment is proportional to the number of atoms in a given volume, so percent composition of an element type is also obtainable. Although the Auger effect was first observed using X-rays to ionize a sample, most modern Auger spectrometers use an electron gun as the ionizing source.

X-ray–Induced Photoelectron Spectroscopy and Auger Electron Spectroscopy Instrumentation

The basic components of the two instrument types are an ultra-high vacuum stainless steel sample chamber (pressure below 10^{-9} torr), an electron energy analyzer/detection system, and a source. Any form of ionizing energy can theoretically serve as the source for either experiment. However, in XPS, the source is almost always X-rays, and in AS, the source is usually an electron gun. Other devices may also be included in XPS and AS instruments such as an ion gun used to clean or etch the surface before analysis. The energy analyzer is essentially a mass spectrometer (Chapter 13) that sweeps a magnetic field range and directs electrons with a specific momentum to the exit slit (see Figure 10.24). The optics, which follow the sample, are not for light but are for electrons. What is actually differentiated is the momentum of the photoelectrons. However, because all electrons have the same mass, the momentum is proportional to kinetic energy.

Problem 10.13: Compare and contrast the physical quantum events involved in XRF, XPS, and AS techniques. How are they similar? How are they different?

10.10 Single-Crystal X-ray Diffractometry

Single-crystal *X-ray diffractometry* is often referred to as *X-ray crystallography*. Obtaining a crystal structure is probably the single most precise molecular analysis one can conduct on a solid state material. Figure 10.25 is a *crystal structure*[35] of dichloro(1,10-phenanthroline-5,6-dione)platinum(II). You will notice in the crystal structure diagram elliptical regions around each atom in the molecule. These elliptical volumes represent the 95% confidence limit for the location of each

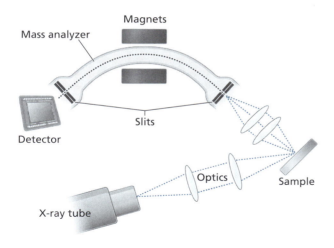

Figure 10.24 Schematic of an XPS spectrometer.

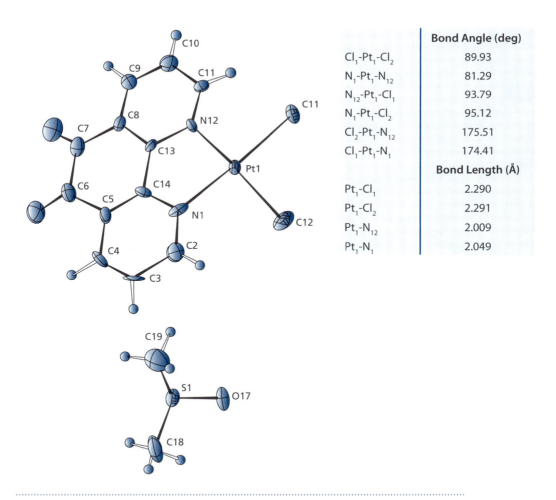

	Bond Angle (deg)
Cl_1-Pt_1-Cl_2	89.93
N_1-Pt_1-N_{12}	81.29
N_{12}-Pt_1-Cl_1	93.79
N_1-Pt_1-Cl_2	95.12
Cl_2-Pt_1-N_{12}	175.51
Cl_1-Pt_1-N_1	174.41
	Bond Length (Å)
Pt_1-Cl_1	2.290
Pt_1-Cl_2	2.291
Pt_1-N_{12}	2.009
Pt_1-N_1	2.049

Figure 10.25 ORTEP diagram showing 95% confidence ellipsoids for dichloro(1,10-phenanthroline-5,6-dione)platinum(II).

In 1906, Max von Laue (1879–1970) was an instructor at the Institute for Theoretical Physics in Berlin. He was also a research assistant to Max Planck and is credited with assisting Albert Einstein in developing his theory of relativity. However, his most notable contribution to science was his discovery that X-rays could be diffracted by a single crystal, thus demonstrating that X-rays were part of the electromagnetic spectrum. ■■

atom.[36] Unlike other techniques we have studied, a crystal structure provides detailed information on the spatial arrangement of all atoms in a chemical compound. For example, the table accompanying Figure 10.25 is a selection of key bond length and bond angles associated with this crystal structure. To emphasize the precision of the technique, note the significant figures included with each measurement. This information allows the analyst to deduce atomic connectivity, stereo conformations, bond angles, and bond distances within a prescribed confidence limit. For solid-state material analysis, the crystal structure allows us to infer the atomic density, the three-dimensional packing within the solid, stoichiometry, and the symmetry within the solid. The technique utilizes an X-ray source and Bragg's law but otherwise its underlying principles are fundamentally different from XRF, XPS, and AES spectroscopy.

In order to fully understand crystallography, a course on symmetry and group theory is a necessary prerequisite. However, by introducing a few key ideas, you can begin to understand the basics of the technique and its place in the toolbox of a practicing analytical chemist. It is the intent of this chapter to introduce the novice to the general principles of X-ray diffraction and illustrate how X-ray diffraction by a crystal can be interpreted to reconstruct the molecular structure of the molecules within the crystal.

Scatter

When the particle size (diameter) of an absorbing species approaches the wavelength of the impinging electromagnetic radiation, the probability of scattering the electromagnetic radiation increases and the probability of absorbing the electromagnetic radiation

---

[36] Oak Ridge Thermal Ellipsoid Plot (ORTEP) is a computer program that is used to plot crystal structure with ellipsoids showing desired confidence limits.

decreases. This is another example of Rayleigh scattering. As a reminder, when an X-ray (or any form of electromagnetic radiation) strikes an atom, it causes the electrons in the atom to move/oscillate, and it is the oscillation of the *electrons* that reradiates the X-ray. It is important to appreciate that it is the *electrons* in the atom that are interacting with the electromagnetic radiation. We will revisit this point later in this section. To scatter X-rays, the particle size needs to be on the order of a single atom. Because the repeating units within a crystal are on the order of 1 to 3 Å (100–300 nm), they can scatter X-rays but because the atoms in a crystal are ordered, they also produce diffraction (see Chapter 3 for a review of diffraction).

X-ray Diffraction

In Section 10.3 of this chapter we described the construction of an X-ray monochromator that made use of a single crystal in a fashion similar to an optical diffraction grating (see Figure 10.10). You may want to review that section of this chapter before proceeding. *Christiaan Huygens* (1629–1695) first demonstrated that when light waves are scattered from a point, that point can be treated as a new source of spherical waves. If light from a single source is scattered by two or more points in space, the light waves scattered from those points will be coherent (in phase) and will propagate from those points until they interfere with each other in either a constructive or destructive way. Coherent scattering of light by a periodic array is called *diffraction*. Because the repeating units within a single crystal are ordered, the scatter of X-rays occurs in the form of diffraction in a manner similar to optical gratings seen in UV-vis/IR monochromators. However, in the case of a crystal, the diffraction occurs in three dimensions and contains information about the atoms that created the scatter.

Christiaan Huygens was a pioneer in the development of light theory. He first proposed the wave properties of light to the Paris Academy of Science in 1678 and his theories were later published in 1690 as Traité de la Lumière. ■■

William Henry and William Lawrence Bragg were a father-and-son research team who studied the diffraction of X-rays by single crystals. Although von Laue is rightly credited with the discovery of X-ray diffraction, it was the Braggs who developed the theory that allowed for the use of X-ray diffraction as an analysis tool for probing the molecular structure of crystals. ▪▪

Bragg's Law

Diffraction of X-rays by crystalline material was first observed by *Max von Laue*[37] in 1912 and within the year, a father-and-son scientific team, William Henry Bragg and *William Lawrence Bragg*,[38] had worked out a simple model for understanding X-ray diffraction by crystals. We first introduced Bragg's law earlier when we described the use of a single crystal as means of constructing an X-ray monochromator (see Figure 10.26 and Equation 10.6).

Let us examine Bragg's law further. When we first described Bragg's law, we described electron diffraction using a reflection model. We imagined that the atoms within a crystal define *planes of reflection*. Although it is fundamentally incorrect, for the purposes of understanding simple two-dimensional *diffraction*, a reflection model works well. Because crystals are three dimensional, the planes of reflection stack up on top of each other.

It is important that you understand the limits of the liberties we have taken in describing Bragg's law using a reflection model. X-ray photons are not actually reflected from surfaces within a crystal but rather are scattered by the *electrons* within the molecules of the crystal. *X-ray crystallography* takes advantage of this fact to *map the electron*

---

[37] Won the Nobel Prize in Physics (1914) for the discovery of the diffraction of X-rays by single crystals.

[38] Joint winners of the Nobel Prize in Physics (1915) for their derivation of Bragg's law of diffraction and contributions to the field of X-ray crystallography.

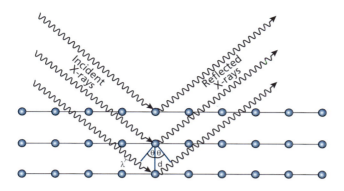

Figure 10.26 Bragg diffraction within a single crystal.

density within a crystal. However, because of the repetitive nature of a crystal lattice, thinking in terms of reflection does allow us to construct and apply Bragg's law. In Chapter 3, we learned that the angle of incidence, θ, must equal the angle of reflection (*Snell's law*) and simple geometry shows us that the extra distance traveled by the X-rays reflected by the second plane is a function of the distance between the planes (d) and the incident angle (θ) and can be calculated as d(sinθ).

Because the reflected X-ray from the second plane must travel the distance d(sinθ) twice (once entering and once exiting) before recombining with the reflected X-ray from the first plane, we can expect constructive interference anytime the total distance 2d(sinθ) is an integer value of λ (see Equation 10.6 and Figure 10.26).

We have also drawn analogies between X-ray diffraction by a crystal and diffraction of UV-vis light with a diffraction grating. Optical diffraction gratings have periodicity in only one dimension. The lattice of a crystal has periodicity in three dimensions, and as a result, the diffraction of X-rays by a crystal results in a three-dimensional diffraction pattern (a spherical pattern).

The idea of a spherical diffraction sphere was first proposed by Paul Peter Ewald and the resulting three-dimensional diffraction patterns are collectively referred to as *Ewald spheres*. The points in space where constructive interference of diffracted X-rays occurs also forms a lattice known as the *reciprocal lattice* (also known as a *reciprocal space*). For example, if your crystal had face-centered cubic symmetry (see Table 10.3), the reciprocal lattice would be body centered cubic. The crystallographer uses the concept of a Ewald sphere to determine what symmetry planes must exist in the crystal in order to obtain the observed reciprocal lattice. We will discuss symmetry planes further in the next section. Figure 10.27 is a two-dimensional projection of the reciprocal lattice for the X-ray diffraction of a tungsten crystal.

There exists an eight-volume set of crystallographic tables that contain useful applications, worked out examples, and discussions of theory known as the International Table for Crystallography. The reader is encouraged to peruse these texts.[39]

Problem 10.14: Determine the Bragg angle (θ) needed to obtain first-order (n = 1) constructive interference for an X-ray of wavelength 10 Å, given the interstitial distance (d) between lattice points in the diffraction crystal, for the following:

(a) 12 Å (c) 32 Å (e) 6.8 Å

(b) 7.5 nm (d) $6 \times 10^{-4}\,\mu$m (f) 300 pm

Figure 10.27 X-ray diffraction of tungsten.

[39] http://it.iucr.org/.

Problem 10.15: Determine the Bragg angle (θ) needed to obtain first-order ($n = 1$) constructive interference for a 200 eV X-ray, given the interstitial distance (d) between lattice points in the diffraction crystal for the following:

(a) 12 Å

(b) 7.5 nm

(c) $6 \times 10^{-4}\,\mu m$

(d) 32 Å

(e) 6.8 Å

(f) 300 pm

Problem 10.16: Determine the Bragg angle (θ) needed to obtain first-order ($n = 1$) constructive interference for an X-ray with a frequency of 3×10^{18} Hz, given the interstitial distance (d) between lattice points in the diffraction crystal for the following:

(a) 12 Å

(b) 7.5 nm

(c) 6.8 Å

(d) 300 pm

(e) $6 \times 10^{-4}\,\mu m$

(f) 32 Å

The Lattice

We need to first discuss crystal structure basics in order to understand X-ray diffraction. A crystal is composed of a specific arrangement of atoms (or molecules) whose position relative to their neighbors repeats in a periodic way. The result is a highly symmetric pattern within the crystal. The smallest repetitive grouping of atoms is called the *unit cell* and a complete crystal can be thought of as a three-dimensional stacking of unit cells. Panel (A) of Figure 10.28 shows a crystal lattice comprised from stacking unit cells. The shaded area represents a single unit cell. Panel (B) of Figure 10.28 shows some of the conventions used to describe and label a unit cell. When describing the Cartesian coordinates within a unit cell, the crystallographic convention is to use *a, b,* and *c* instead of *x, y,* and *z*. The angle formed by the intersection of the coordinates *b* and *c* is labeled α, the angle formed by the intersection of coordinates *a* and *c* is labeled β, and the angle formed by the intersection of *a* and *b* is labeled γ. The unit cell dimensions and the interaxial angles characterize the given crystalline substance.

TABLE 10.3: Lattice Systems

	ANGLES AND AXIS LENGTHS	GEOMETRY
Cubic	$\alpha = \beta = \gamma = 90°$ $a = b = c$	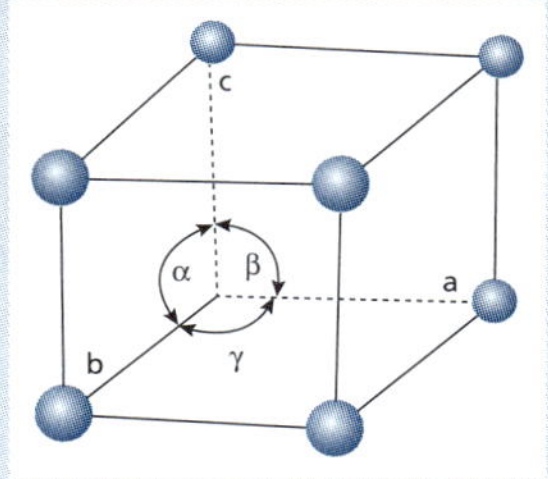
Tetragonal	$\alpha = \beta = \gamma = 90°$ $a = b \neq c$	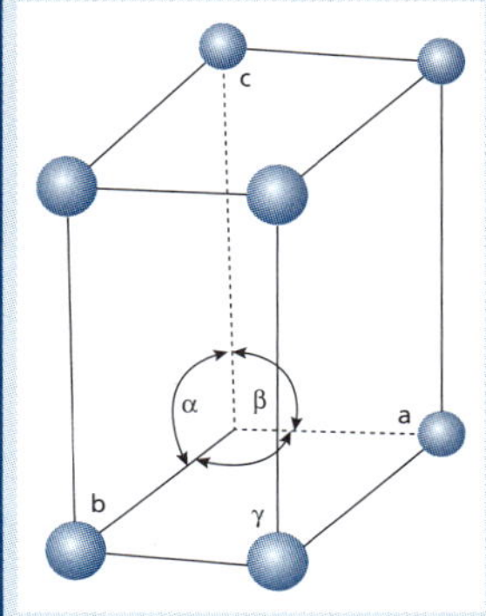

TABLE 10.3:   (*Continued*)

	ANGLES AND AXIS LENGTHS	GEOMETRY
Orthorhombic	$\alpha = \beta = \gamma = 90°$ $a \neq b \neq c$	
Rhombohedral	$\alpha = \beta = \gamma \neq 90°$ $a = b = c$	
Hexagonal	$\alpha = \beta = 90°, \gamma = 120°$ $a = b \neq c$	
Monoclinic	$\alpha = \gamma = 90° \neq \beta$ $a \neq b \neq c$	
Triclinic	$\alpha \neq \beta \neq \gamma \neq 90°$ $a \neq b \neq c$	

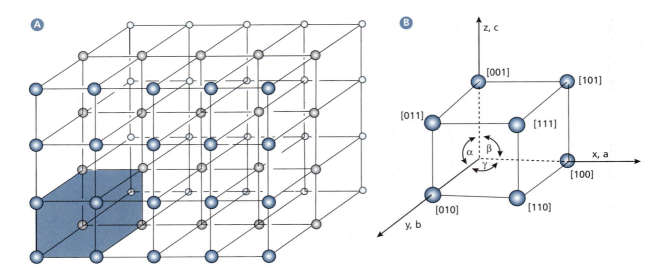

Figure 10.28 (A) A crystal lattice with a unit cell. (B) A unit cell with conventional dimensional units labeled as well as the use of Miller indices to indicate specific points within the lattice.

Depending on the relative lengths of a, b, and c and the relative angles of α, β, and γ, seven different crystallographic systems can be identified. These are summarized in Table 10.3 along with the cell geometry.[40]

Panel (B) of Figure 10.28 also demonstrates the use of *Miller indices* ([110], [010], and so on). Miller indices are used to indicate specific points within a crystal relative to a point of origin. In this case, the point of origin is the rear, left, bottom corner of the unit cell. Any point can be designated as the origin. A Miller index of [101] indicates a point from the origin that can be found by moving one unit along the a axis, zero units along the b axis, and one unit along the c axis.

Problem 10.17:   Label each of the following Miller index points within the given crystal lattice. The origin is indicated by a black dot.

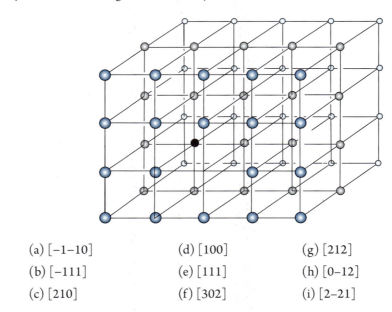

(a) [−1−10]   (d) [100]   (g) [212]

(b) [−111]   (e) [111]   (h) [0−12]

(c) [210]   (f) [302]   (i) [2−21]

---

[40] Cullity, B. D. *Elements of X-ray Diffraction*, 2nd ed.; Addison Wesley Publishing: CA, **1978**.

Example 10.3

The point of origin in the following crystal lattice is designated by a black dot. Label the point indicated by the Miller index of $[312]$.

STRATEGY –

Starting at the point of origin, count out three lattice points along the a axis, then count out one lattice point along the b axis, and then count out two lattice points along the c axis. The axis labels a, b, and c are defined in Figure 10.28.

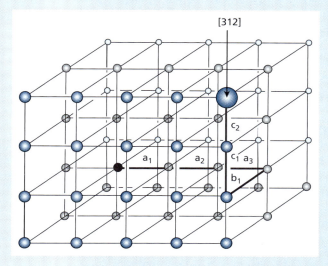

Problem 10.18: There are nine labeled points in the given crystal lattice. Provide the proper Miller index for each of these points. The origin is indicated by a black dot.

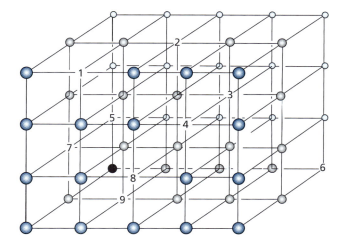

When we described Bragg diffraction, we used the analogy of planes of reflection. So it should be no surprise that the concept of lattice planes is important to crystallographers. Within a crystal, one can imagine a specific atom type within a lattice, creating lattice planes. The *Miller index system* is also used to indicate *lattice planes* within a

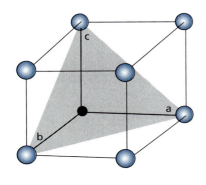

Figure 10.29 The (111) plane of a unit cell. Note the use of parentheses to designate a plane instead of the use of square brackets to indicate a point. The origin is designated by a black circle.

crystal. The convention is to enclose the Miller index in parentheses when indicating a plane and square brackets when indicating a point. It takes three points to define a plane. So the plane that is formed from a point one unit along *a*, the point that is one unit along *b*, and the point one unit along *c* would be designated as (111). Figure 10.29 shows the Miller index for the (111) plane within a unit cell. Because diffraction is the result of *"stacked planes,"* crystallographers often refer to a family of parallel planes with respect to the origin. A family of planes is indicated by enclosing the Miller index in braces, so if you wished to indicate all of the planes parallel to the (111) plane, you would write {111}.

Problem 10.19: Sketch a unit cell similar to the one depicted in Figure 10.29 and shade and label the following planes:

(a) (010)   (c) (100)   (e) (001)

(b) (110)   (d) (101)   (f) (011)

Obtaining a Crystal Structure

The diffraction pattern is a result of the atom locations within the crystal and the intensity of the diffracted X-rays (the signal) is a function of the number of electrons in an atom as well as the Bragg angle θ. The intensity values as a function of Bragg angle are tabulated as a function of the atomic number, Z. These correlation curves are referred to as *atomic scatter factors* or *atomic form factors* and given the symbol *f*.[41] Figure 10.30 shows atomic scatter factors for six atom types.

At this point, you might be thinking that a crystal structure can be determined by simply measuring the diffraction pattern to determine atom locations within the crystal and by fitting intensity data as a function of θ to an atomic scatter factor curve in order to identify the atom types that created the reflections. As a first approximation, that is essentially what is done, but in practice obtaining a crystal structure is a bit more complicated. Real data contain uncertainty due to the atoms' motion within the crystal and are affected by thermal vibrations and the rigidity of the crystal packing. The *experimental form factor* is better described as:

$$F = f_o e^{-B\left(\frac{sin^2\theta}{\lambda}\right)}$$

Eq. 10.11

where $B = 8\pi^2 \overline{\mu^2}$ and is related to the mean square amplitude of vibration $(\overline{\mu^2})$ of the atom in the crystal. In addition, we are not measuring the diffraction of a single atom but instead the diffraction from a grouping of atoms whose diffraction pattern is determined by the overall structure. So when we collect intensity as a function of θ data, what we actually measure is an experimental *structure scatter factor* (F). The observed intensities are therefore:

$$I_{obs} = K|F|^2 L_p A$$

Eq. 10.12

---

[41] Recall that it is actually the electrons in the atoms that are producing the scatter. But because the number of electrons in an atom equals the atomic number, it is correct to say that the intensities are proportional to the atomic number of the atoms that produced the scatter.

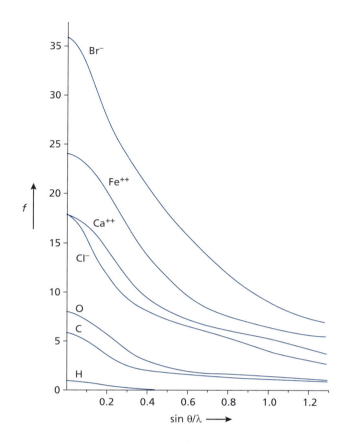

Figure 10.30 Atomic scatter factor curves as a function of $\dfrac{\sin\theta}{\lambda}$.

where

    K = scaling factor[42]
    F = structure scatter factor
    L_p = Lorentz polarization correction factor[43]
    A = absorption

But initially we do not know the structure! That is why we are performing the experiment.

The observed intensities versus θ data are converted to observed structure scatter factors (F_{obs}) that are used to identify the atoms types that created the signal. Solving the crystal structure then becomes an iterative process involving the following steps:

1. *Indexing* the diffraction pattern. Indexing requires the crystallographer to make an assumption about the crystal type (see Table 10.3), and software is used to assign Miller index numbers. Then the size and shape of the unit cell is calculated.[44]

---

[42] Empirically determined scaling factor that brings observed structure factors and calculated structure factors to the same values.

[43] If the source radiation has been made monochromatic by reflection from a crystal, it will be partially polarized and the intensity of the diffracted beam will depend on whether the electric vector is parallel or perpendicular to the reflection plane.

[44] Before the development of modern computers, solving two or three crystal structures by hand was enough to earn a PhD.

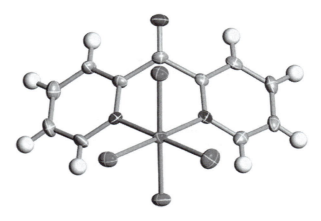

Figure 10.31 Crystal structure of [Pt(dpk)Cl$_4$]. Ellipsoids represent confidence intervals for the location of each atom and signify motion within the crystal. Selected bond angles: N(14)-Pt-N(8) 88.62°; C(3)-C(2)-C(9) 120.00°. Selected bond length: C(2)-O(1) 1.225 Å.

2. Typically, the crystallographer will have measured the density of the crystal and from the unit cell dimensions, the total number of atoms per unit cell can be calculated. This allows the crystallographer to determine the exact molecular formula.

3. The position and type of each atom in the unit cell is deduced by the F$_{obs}$, and an experimental data set is generated.

4. If the theoretical data set matches the experimental data set, you are done. If not, you go back to step 1 and start over with modified assumptions.

The last step is typically the most difficult and only yields good results if assumptions made in the first two steps are correct. The crystallographer will make use of all information available. For instance, the chemist who made the crystal will have a fairly good estimate of the product. The chemist knows what reagents were used, so the possible atom types are known and the product will have been studied by nuclear magnetic resonance, IR, and possibly elemental analysis. Figure 10.31 shows an *ORTEP diagram* of [Pt(dpk)Cl$_4$], also known as the "crystal structure diagram" of the molecule.

The Diffractometer

An X-ray diffractometer consist of three basic components: an X-ray source, a sample holder, and an X-ray detector. Early diffractometers used photographic film to capture the diffraction patterns. Today, diffractometers have movable computer-controlled X-ray detectors. X-ray sources were covered in some detail in Section 10.2 and translate directly to X-ray diffractometers. The sample holder is typically a three-axis goniometer,[45] which allows for the precise orientation of an object along all three axes. Figure 10.32 (a) is a schematic representation of a single-crystal diffractometer. Samples are mounted on thin glass fibers with epoxy or cyanoacrylate glue and the fiber is attached to a brass pin connected to the goniometer head. Adjustment of the goniometer rings allows for orientation of the *x*, *y*, and *z* orthogonal axes in order to center the crystal in the X-ray beam. The diffracted rays are then collected by the detector and the location of each diffracted "spot" relative to the incident beam is recorded by the computer along with intensity data. Modern single-crystal diffractometers use charge coupled device array detectors that can collect a great many spots simultaneously.

---

[45] The word *goniometry* is from the two Greek words, *gōnia* (angle) and *metron* (to measure). Goniometers are used to rotate an object to a desired angular position.

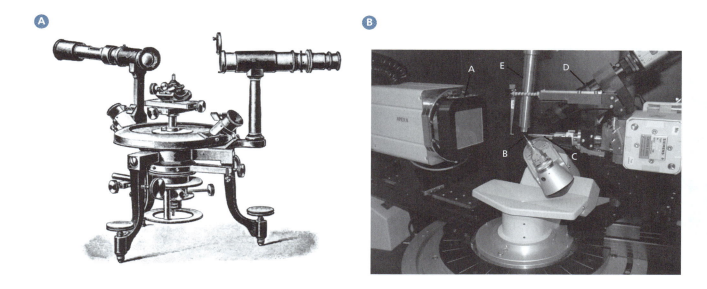

Figure 10.32 (A) A goniometer—an instrument that either measures an angle or allows an object to be rotated to a precise angular position. (B) Photograph of a Bruker Kappa Apex II diffractometer installed at the National Single Crystal X-ray Facility at Utrecht University, Netherlands: A = CCD array detector, B = Sample, C = X-ray source, D = digital camera for viewing the crystal's orientation, E = blow tube for blanketing the sample with an inert gas or for cooling the sample with cold nitrogen gas.

10.11 Further Reading

BOOKS

Buhrke, V. E.; Jenkins, R.; Smith, D. K. *A Practical Guide for the Preparation of Specimens for X-ray Fluorescence and X-ray Diffraction Analysis.* John Wiley & Sons: New York, **1998**.

Glusker, J. P.: Trueblood, K. N. *Crystal Structure Analysis: A Primer,* 3rd ed. Oxford University Press: Oxford, UK, **2010**.

Hebbar K. R. *Basics of X-Ray Diffraction and its Applications.* I. K. International Publishing House: New Delhi, India, **2007**.

Hofmann S. *Auger and X-ray Photoelectron Spectroscopy in Materials Science: A User-Oriented Guide.* (Springer Series in Surface Sciences, Book 49). Springer: Heidelberg, Germany, **2013**.

Jenkins, R. *Introduction to X-ray Powder Diffractometry.* John Wiley & Sons: New York, **1996**.

Jenkins, R.; Gould, R. W.; Gedcke, D. *Quantitative X-ray Spectrometry.* 2nd ed. Marcel Dekker: New York **1996**.

Massa W. (translated by Gould R. O.). *Crystal Structure Determination.* Springer: Berlin, Germany, **2002**.

Pecharsky, V.; Zavalij, P. Y. *Fundamentals of Powder Diffraction and Structural Characterization of Materials.* Kluwer Academic Publishers: Norwell, MA, **2003**.

Spectro X-Lab Pro Manual, circa **2000**. Spectro Analytical Instruments, Inc.

van Der Heide, P. *X-ray Photoelectron Spectroscopy.* John Wiley & Sons: Hoboken, NJ, **2012**.

Yoshio, W.; Matsubara, E.; Shindoa, K. *X-ray Diffraction Crystallography.* Springer-Verlag: Berlin, Germany, **2011**.

JOURNALS

Ferrero, J. L.; et al. Analysis of pigments from Spanish works of art using a portable EDXRF spectrometer. *X-ray Spectrometry,* **2002**, 31, 441–447.

Jenkins, R.; Gould, R. W.; Gedcke, D. Applications of X-ray Spectrometry. *Applied Spectroscopy Reviews,* **2004**, 35(1,2), 129–150.

Knoll, G. F. Radiation detectors for X-ray and gamma-ray spectroscopy. *Journal of Radioanalytical and Nuclear Chemistry,* **2000**, 243(1), 125–131.

Revenko, A. G. X-ray fluorescence analysis of rocks, soil and sediments. *X-ray Spectrometry,* **2002**, 31, 264–273.

Sieber, J. R. X-rays in research and development at the National Institute of Standards and Technology. *X-ray Spectrometry,* **2000**, 29, 327–338.

Szalóki, I.; Osán, J.; van Grieken, R. E. X-ray Spectrometry. *Anal. Chem.,* **2006**, 78, 4069.

Tsuji, K.; Nakano, K.; Takahashi, Y.; Hayashi, K.; Ro, C. X-ray Spectrometry. *Anal. Chem.,* **2008**, 80, 4421–4454.

Tsuji, K.; Nakano, K.; Takahashi, Y.; Hayashi, K.; Ro, C. X-ray Spectrometry. *Anal. Chem.,* **2010**, 82, 4950–4987.

Tsuji, K.; Nakano, K.; Takahashi, Y.; Hayashi, K.; Ro, C. X-ray Spectrometry. *Anal. Chem.,* **2012**, 84, 636–668.

10.12 Exercises

EXERCISE 10.1: Define the following terms.
a.   Characteristic emissions
b. K_β-emission
c. L_α-emission
d.   PMI analysis
e.   Photoelectric effect
f.   Photoelectric absorption
g.   Coolidge tube
h.   Grazing angle
i.   Wolter optics
j.   Fresnel zone plates
k.   Capillary optics
l.   Bremsstrahlung
m.   Duane-Hunt law
n.   Synchrotron radiation
o.   Diffraction
p.   Bragg diffraction
q.   Nano-scale multilayer devices
r.   WDXRF
s.   EDXRF
t.   TXRF
u.   Scintillator
v.   Gas detectors
w.   PIN detectors
x.   SDD
y.   Miller Index
z.   Form function

EXERCISE 10.2: Who is credited with the "birth" of XRF spectroscopy?

EXERCISE 10.3: Sketch a schematic of a Coolidge tube.

EXERCISE 10.4: The development of the Coolidge tube remedied two important technical difficulties of the early X-ray tubes. What were those technical difficulties?

EXERCISE 10.5: In this chapter, we described two different mechanisms for generating X-rays: the first was element specific *characteristic emissions* and the second was *bremsstrahlung* emissions. Describe the two mechanisms and point out how they differ.

EXERCISE 10.6: Using the graphic in Problem 10.3 as a guide, sketch an orbital diagram for each of the following transitions.

(a) K_β (d) L_β (g) K_γ
(b) N_α (e) M_α (h) L_β
(c) M_β (f) N_β (i) M_γ

EXERCISE 10.7: Use Equation 10.4 to determine the minimum wavelength of the bremsstrahlung radiation generated by a Coolidge tube with a voltage gradient of:

(a) 35 keV   (b) 65 keV   (c) 150 keV

EXERCISE 10.8: Use Equation 10.5 and Example 10.1 to determine the minimum wavelength, the maximum frequency, and the energy in units of joules for the bremsstrahlung radiation generated by a X-ray tube with a voltage gradient of:

(a) 75 kV   (b) 60 kV   (c) 125 kV

EXERCISE 10.9: What voltage would you need to apply to the X-ray tube of an XRF spectrometer in order to induce a K_α emission from:

(a) Mg   (d) Pb   (g) Ca
(b) Ti   (e) As   (h) Rb
(c) Mn   (f) Bi   (i) Cd

EXERCISE 10.10: What voltage would you need to apply to the X-ray tube of an XRF spectrometer in order to induce a K_α emission from:

(a) Os   (d) I   (g) Mo
(b) Ir   (e) Hg   (h) Ni
(c) Sb   (f) Tc   (i) Ba

EXERCISE 10.11: Why is it impractical to use refractive optics to focus X-rays?

EXERCISE 10.12: In your own words, explain the underlying principle of Wolter X-ray optics and capillary X-ray optics.

EXERCISE 10.13: Given a crystal lattice distance of d = 2.25 Å, use Bragg's law to calculate the (n = 1) angle required for the constructive interference of the following X-ray beams:

(a) $\lambda = 8.44$ Å   (b) $\lambda = 7.24$ Å   (c) $\lambda = 2.04$ Å

EXERCISE 10.14: Given a crystal lattice distance of d = 2.05Å, use Bragg's law to calculate the (n = 1) angle required for the constructive interference of a $\lambda = 1.54$ Å X-ray beam.

EXERCISE 10.15: Given an X-ray beam of 1.89 Å, incident on a crystal surface at an angle of 37 degrees, what crystal lattice distance "d" would you need to experience first-order (n = 1) Bragg diffraction?

EXERCISE 10.16: Single crystals and nanoscale multilayer devices are both used in X-ray monochromators. For which applications are nanoscale multilayers preferable to single crystals?

EXERCISE 10.17: For the spectrometer shown in Figure 10.20, what is the minimum wavelength (λ_{min}) generated by the source? Use the data in Table 10.1 to construct a list of elements that you could theoretically measure with the X-ray tube given in this instrument.

EXERCISE 10.18: Describe the atomic principles of XPS. Include the following topics in your discussion.

(a) The atomic principles underlying the technique

(b) The basic measurement apparatus

(c) The significance of the kinetic energy of the photoejected electron and the equation that describes the electron's energy

(d) The information that can be gained from an XPS measurement

EXERCISE 10.19: Describe the atomic principles of AES. Include the following topics in your discussion:

(a) The atomic principles underlying the technique

(b) The basic measurement apparatus

(c) The significance of the kinetic energy of the photoejected electron and the equation that describes the electron's energy

(d) The information that can be gained from an XPS measurement

EXERCISE 10.20: How are XRF and XPS related?

EXERCISE 10.21: What is the measurand for each of the following? Be as specific as possible.

(a) XRF   (b) XPS   (b) AES

EXERCISE 10.22: Label each of the following Miller index points within the given crystal lattice. The origin is indicated by a black dot.

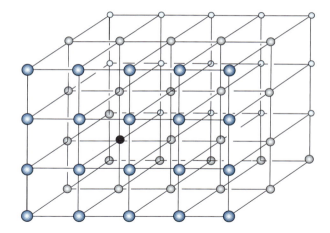

(a) [3-10]   (d) [101]   (g) [2-12]
(b) [311]   (e) [10-1]   (h) [01-1]
(c) [30-1]   (f) [1-10]   (i) [2-1-1]

EXERCISE 10.23: Explain the difference between [111], (111), and {111}.

Advanced Exercises

EXERCISE 10.24: Solve Equation 10.4 for v_{max} and combine constants in order to produce a "user friendly" version of the Duane-Hunt law that allows you to calculate the v_{max} as a function of the voltage applied across an X-ray tube. Use Equation 10.5 as a guide.

EXERCISE 10.25: Assume you are using a diamond single-crystal monochromator to select a bandwidth of X-rays from a source. The interstitial distance (d) between the planes in a diamond crystal is 3.567 Å.

(a) At what angle would a ($\lambda = 30$ pm) incident angle X-ray beam need to be in order to experience first-order diffraction (see Equation 10.6).

(b) Determine the spectral resolution (i.e., bandwidth) of a monochromator made using this crystal if the exit slit is at a distance of 8 cm from the diamond crystal and has a width of 5 μm.

Chapter 11

Infrared Spectroscopy

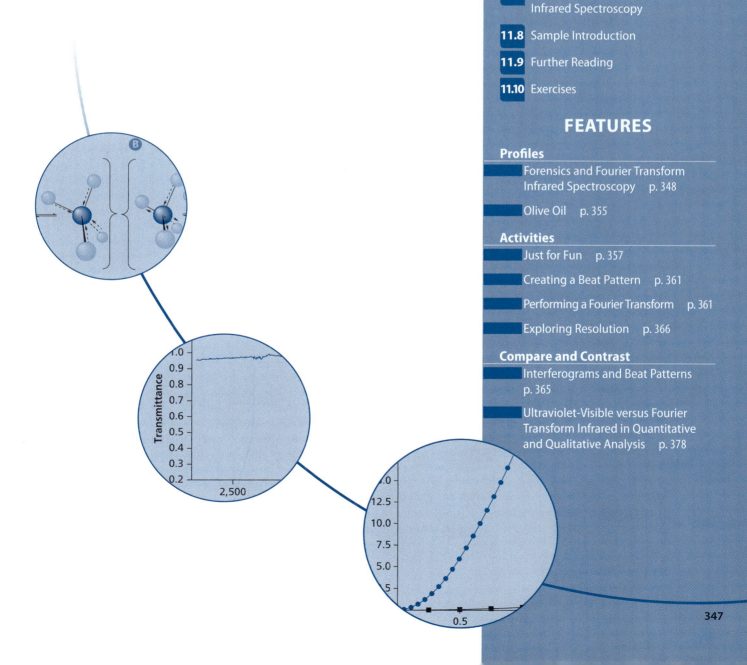

347

PROFILE

Forensics and Fourier Transform Infrared Spectroscopy

We have all watched television shows where the suspect in a crime is identified by his or her unique fingerprint patterns, patterns that are left behind in oils that are a part of our skin surface. Researchers are now looking into further analysis of those fingerprint oils. The fingerprint itself is comprised of a few micrograms of oil, and the analysis of the chemical composition of the oil can reveal the presence of caffeine, explosive materials, or a myriad of other chemicals that can give investigators valuable clues to the perpetrator's identity. In a 2007 paper published in *Analytical Chemistry*, Sergei Kazarian and coworkers* reported on the use of Fourier transform infrared (FTIR) spectroscopy to directly probe the chemical composition of latent fingerprints. Investigators can now use a gelatin tape to "lift" a fingerprint from a crime scene and, using FTIR spectroscopy, probe the chemical composition of the fingerprint oils.

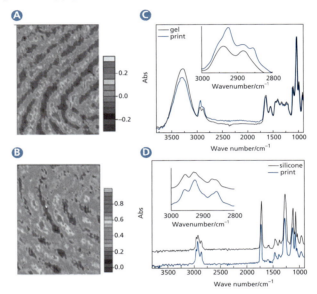

ATR FTIR spectroscopic image of latent fingerprints lifted from a door handle.

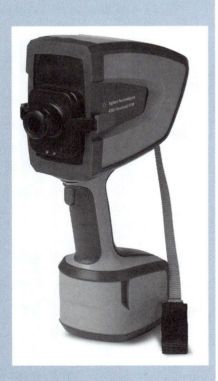

With further development, it is believed that FTIR spectroscopy will allow forensic investigators to probe a fingerprint for such things as explosive residue or the chemical composition of a particular oil, perfume, and so on. But to make this application truly field-ready, the instrument itself must become portable. In March of 2014, Agilent Technologies introduced the Agilent 4300 handheld FTIR spectrometer, "allowing the user to bring the power of infrared analysis out of the lab and into the field."

The Agilent 4300 is being marketed for use in quality-control applications, including detecting the oxidation of machined metal parts, to confirm a metal surface is free of contaminates, to verify the authenticity of polymer components, to measure the thickness and uniformity of surface coatings, and to determine the ratio of key ingredient within a chemical compound. It is not difficult to imagine the advancement of this technology leading to a first-rate forensic tool.

In later chapters, we also profile other handheld spectrometers. The next logical step in the advancement of handheld spectrometers is to combine several of these devices into a single handheld device. It is exciting to realize that something similar to the fictional "Tricorder," seen in the popular *Star Trek* series, may be a reality in the near future. ∎

---

* *Ricci, C.; Bleay, S.; Kazarian, S. G. Spectroscopic Imaging of Latent Fingermarks Collected with the Aid of a Gelatin Tape. Anal. Chem., 2007, 79, 5771–5776.*

11.1 Chemical Structure and Molecular Vibrations

In Chapter 1, we mentioned that infrared (IR) spectroscopy is used to probe molecular vibrational modes, which gives us substantive insight into the molecular structure of the analyte. A *vibrational mode* is a unique vibration within a molecule in which all atoms involved in the vibration have a sinusoidal motion and the motion of all of the atoms shares the same phase. At this point, we think it is useful to state the obvious: *in order to have a vibration between two atoms in a molecule, there must be a bond between those atoms.* One of the main uses of IR spectroscopy is to determine the types of bonds that exist within a molecule.

The *IR spectrum* can be broken down into three regions—the *near-IR* (14,000–4,000 cm⁻¹), *mid-IR* (4,000–400 cm⁻¹), and *far-IR* (400–10 cm⁻¹) regions. They are thusly named based on their anthropocentric relationship to the human eye's lower detection limit of 700 nm (~14,286 cm⁻¹). The spectroscopy described in this chapter uses the *mid-IR* region. Mid-IR radiation is used to probe the primary vibrational modes associated with the bonds within a molecule and thus allows the user to gain valuable insight into the functional groups present in a molecule. We will also discuss in this chapter how IR spectroscopy can give the analyst valuable insight into the overall symmetry of a molecule.

Recall from Chapter 2 that we modeled a molecular bond as two masses held together by a spring—the classic harmonic oscillator. The classic harmonic oscillator defines the *vibrational frequency* as:

$$\omega_{(vibration)} = \sqrt{\frac{k}{\mu}}$$

Eq. 11.1

where k is the *effective spring constant* (related to the rigidity of the chemical bond between the two atoms in the molecule) and μ is the reduced mass of the system.[1] The *reduced mass* can be found by:

$$\mu = \frac{m_1 m_2}{m_1 + m_2}$$

Eq. 11.2

where m_1 and m_2 are the masses of the atoms involved in the chemical bond.[2] Take a moment to really think about the construction of Equation 11.1. The value k in Equation 11.1 represents the type of bond involved in the molecular vibration (single, double, triple), and the value μ is derived from the masses of the atoms involved in the bond (C–C versus C–O, versus C–H, and so on).

Each exclusive bond between any two atoms will vibrate with a unique frequency. We can see this exclusivity in looking at the carbonyl functional group:

$$\underset{(R \qquad \qquad R)}{\overset{\overset{\textstyle O}{\|}}{C}}$$

All carbonyl bonds vibrate with a frequency of approximately 5×10^{13} Hz. However, the exact vibrational frequency is influenced by the local molecular environment (i.e., the remainder of the molecule) and therefore the vibration of a specific carbonyl bond in a particular molecule will be unique. It will be relatively near 5×10^{13} Hz but at a specific frequency characteristic of that particular molecule. Let us take, for instance, the molecule di-2-pyridyl ketone (DPK). We can demonstrate the effects of local molecular environment by considering the frequency of the ketone group in the DPK molecule (by itself) compared to the frequency of the ketone group in the DPK when it is bound to a metal center. The ketone vibration in DPK appears at 5.09×10^{13} Hz for the free molecule but when bound to a platinum center in a transition metal complex, it shifts to 5.14×10^{13} Hz. Although the ketone peak remains near 5×10^{13} Hz, the exact frequency is unique to the molecule containing the ketone group. Figure 11.1 shows an IR spectrum of the DPK bound to a platinum center

---

[1] For the derivation of Equation 11.1, see Equations 2.7 through 2.10 in Chapter 2.

[2] Equations 11.1 and 11.2 are the same as Equations 2.10 and 2.11.

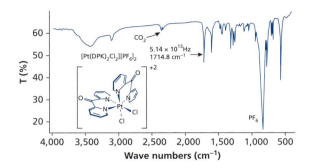

Figure 11.1 IR spectrum of $[Pt(DPK)_2Cl_2][PF_6]_2$ as a KBr pellet.

$[Pt(DPK)_2Cl_2][PF_6]_2$ taken as a suspension in solid potassium bromide (KBr pellet), where the ketone stretch is clearly visible at 5.14×10^{13} Hz.

Wave Numbers and the Infrared Spectral Region

Note that the x axis in Figure 11.1 is reported using a unit called a *wave number* instead of hertz. The unit of inverse centimeters (cm^{-1}) is called a *wave number*, and it is symbolized using $\bar{v}$ to distinguish it from frequency in units of hertz, which uses the symbol v. Because the frequency of a vibration is proportional to the inverse of the wavelength, the wave number is essentially a frequency unit. However, wave numbers are used instead of frequency to avoid the necessity of using scientific notation, keeping the abscissa from being overly cluttered when plotting IR spectra.

Recall that the relationship between the energy of a photon and its frequency or wavelength is given by:

$$E = hv = \frac{hc}{\lambda}$$ **Eq. 11.3**

where
  E = energy of a photon (joules)
  h = Planck's constant $(6.626 \times 10^{-34}$ J·s)
  v = frequency (hertz; s^{-1})
  c = speed of light $(3.00 \times 10^8$ m/s)
  λ = wavelength (meters)

Problem 11.1: Convert each of the following frequencies into wave numbers:

   (a) 4.98×10^{13} Hz   (c) 1.2×10^{13} Hz   (e) 9.0×10^{13} Hz
   (b) 6.24×10^{13} Hz   (d) 1.2×10^{14} Hz   (f) 2.4×10^{13} Hz

Problem 11.2: Determine the reduced mass for each pair of atoms:

   (a) HCl   (c) HBr   (e) CN
   (b) DCl   (d) CO   (f) CH

Problem 11.3: Convert each of the following wave numbers into (1) wavelength (in µm); (2) wavelength (in nm); (3) frequency (in hertz, s^{-1}), and (4) energy (J):

   (a) 3,000 cm^{-1} (c) 1,710 cm^{-1} (e) 800 cm^{-1}
   (b) 4,000 cm^{-1} (d) 2,100 cm^{-1} (f) 1,500 cm^{-1}

Problem 11.4: Convert each of the following wavelengths into wave numbers:

   (a) 3.3 µm   (c) 5.85 µm   (e) 12.5 µm
   (b) 2.5 µm   (d) 4.76 µm   (f) 6.7 µm

Group Frequencies

As shown above, one can expect to find a carbonyl peak[3] in the IR spectrum within a narrow range of wave numbers between 1,690 to 1,790 wave numbers. This range, or band, of wave numbers associated with a particular bond type is called a *group frequency*.

Tables of group frequencies can be found in many organic textbooks and many Web pages. One of the most comprehensive listings of group frequencies can be found in the CRC handbook.[4] Table 11.1 gives the group frequencies of several common functional groups and bond types.

TABLE 11.1:   Common Characteristic Group Frequencies (cm^{-1})* and a Sample Spectrum

BOND TYPE	$\bar{v}$	BOND TYPE	$\bar{v}$
RO-H (alcohol and phenols)	3,200–36,500	O ‖ R-C-OR (esters; C=O)	1,735–1,750
O ‖ R-C-O—H (carboxylic acids; O-H)	2,500–3,000	O ‖ R-C-OH (carboxylic acids; C=O)	1,710–1,770
R$_3$C—H (alkane)	2,840–2,960	O ‖ O ‖ R-C-H; R-C-R (aldehyde and ketone C=O)	1,690–1,745
R H \C=C/ R R (alkene; C-H)	3,050–3,100	\C=C/ (alkenes)	1,620–1,680
RC≡C-H (alkyne)	3,280–3,310	RC≡CR (alkyne)	2,100–2,260
Ar–H (aromatic stretching) (bending)	~3,300 680–860	RC≡N	2,220–2,260

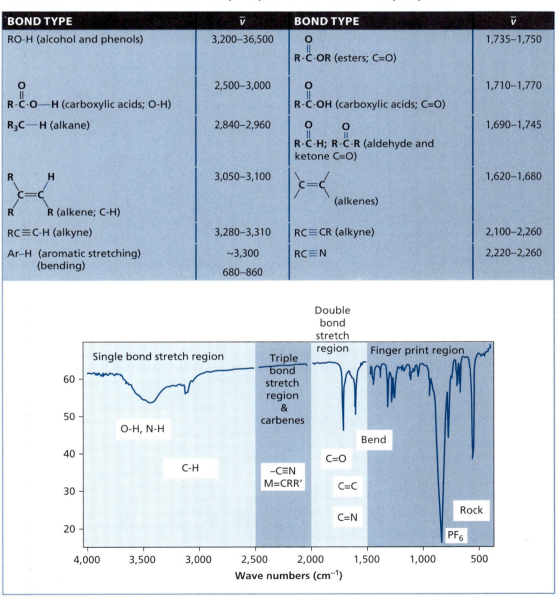

* **For an expanded list of group frequencies, see Table 11.2.**

---

[3] Aldehydes, ketones, carboxylic acids, and esters.

[4] *Handbook of Chemistry and Physics.* Vol. 91, **9**, 110–114, **2011**.

There was a time when an expert in the area of IR spectral analysis would expend significant effort in memorizing the wave number ranges of a considerable number of group frequencies. However, most modern IR spectrometers have programmed algorithms that assist the analyst in identifying the functional group regions of the observed peaks in a spectrum. Nevertheless, it is still wise to learn the most common group frequencies because you will sometimes need to interpret an IR spectrum absent from the software of the spectrometer—for instance, in a journal paper or during a lecture at a conference. If you wish to be able to use IR efficiently, you need to memorize the common group frequencies.

Example 11.1

The spectrum of acetic acid is shown. Determine the IR active functional groups and analyze the spectrum of acetic acid by assigning the peaks in the spectrum to the parts of the molecule.

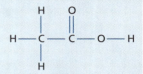

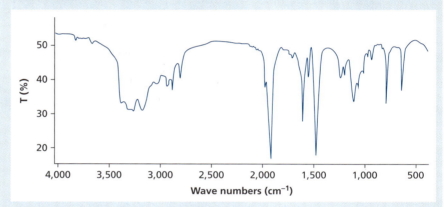

STRATEGY –

Use the data in Table 11.1 to identify the IR active groups and match those to the observed regions in the IR spectrum.

ANSWER –

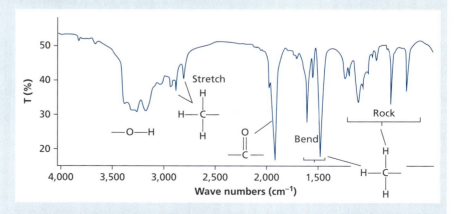

Problem 11.5: Using Table 11.1 and Example 11.1 as a guide, analyze the following spectrum:

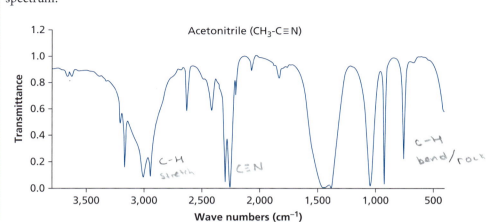

Normal Modes

The term *mode* (or *normal mode*) is used to describe the number of unique vibrations a molecule is expected to exhibit. To calculate the total number of vibrational modes in a molecule, we must consider the total degrees of freedom for the molecule. Each atom in a molecule is moving through space, therefore each atom in the molecule has three degrees of freedom (x, y, and z). Therefore, a molecule with N atoms has 3N degrees of freedom. However, because the individual atoms in a molecule are bound to each other and the *entire* molecule is moving through space (translation), three degrees of freedom are also used to describe translation of the molecule. Likewise, three degrees of freedom are needed to describe the rotation about each axis (*x, y,* and *z*) unless the molecule is linear, in which case only two degrees of freedom are needed to describe rotation. Any degrees of freedom not being used for translation and rotation are available for vibration and are referred to as *normal modes (m).* As a general rule, a nonlinear molecule has 3N–6 normal modes and a linear molecule has 3N–5 normal modes.

$$m = 3N - 6 \qquad \textbf{Eq. 11.4a (nonlinear)}$$

$$m = 3N - 5 \qquad \textbf{Eq. 11.4b (linear)}$$

Example 11.2

Let us determine the normal modes of the molecule methane (CH_4).

STRATEGY –

Apply Equations 11.4a and 11.4b to the methane molecule. Recognizing the fact that methane is nonlinear, we will use the form m = 3N – 6.

SOLUTION –

There are five atoms in the methane molecule, so there are a total of 3N = 15 degrees of freedom within the methane molecule. However, the entire molecule is undergoing translation (movement through space) within the

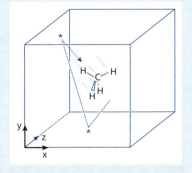

Figure 11.2 Translation of methane in three dimensions.

(Continued)

Cartesian coordinates x, y, and z, so we can assign 3 of the 15 degrees of freedom to translation. Figure 11.2 represents the translation of a methane molecule in a three-dimensional box. Likewise, the methane molecule rotates along the x, y, and z axes so we can assign 3 degrees of freedom to the rotation of methane as well. So the degrees of freedom in methane that are available for vibration are the original 15 less 3 for translation and less 3 more for rotation, thus leaving us with 9 degrees of freedom for vibration ($15 - 3 - 3 = 9$).

Problem 11.6: Determine the number of normal modes for each of the following molecules.

(a) ... (b) ... (c) H_3C—CH_2 ... CH_3 (d) $\ddot{O} = C = \ddot{O}$

Vibrational Categories

A vibration is generally described as either being a bend or a stretch. A *stretch* occurs along the axis between two atoms. A stretch is the result of an oscillation about an equilibrium bond distance. *In other words, the bond becomes longer and then shorter, longer then shorter, as it oscillates around its equilibrium bond length value.* A *bend* is the result of an oscillation about an *equilibrium bond angle*. In this case, the angle becomes smaller and then larger, smaller and then larger, with an equilibrium value in between. A vibration can be further subcategorized as a twist, scissor, rock, wag, torque, and so on, which will be explained in more detail below. It is helpful to imagine the stretching vibration by thinking about the bond as a spring. The spring compresses (becomes shorter) and lengthens, over and over again, with a periodic frequency. To visualize the bending vibration motion, imagine the two bonds (connecting three atoms) as tweezers that are opening and closing with a periodic frequency.

Let us examine the vibrational modes in the molecule methane (CH_4), and let us look at stretching modes first. There are three unique stretching modes for methane. One is the *symmetric stretch* depicted in Figure 11.3(A). All four C–H bonds are lengthening and shortening in unison. In Figure 11.3(B), we have an *asymmetric stretching* mode in which one C–H bond is shortening while the other three are lengthening and vice versa. In Figure 11.3(C), we have another asymmetric stretching mode in which two C–H bonds are lengthening while two are shortening and vice versa.[5]

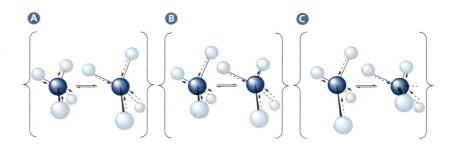

Figure 11.3 Three possible stretching modes for methane.

[5] For an animation of CH_4 stretching modes, see: http://www2.ess.ucla.edu/~schauble/MoleculeHTML/CH4_html/CH4_page.html and/or http://www.youtube.com/watch?v=zeep0q97Who.

It is estimated that in 2010, a total of 6.54×10^6 tons of olive oil were produced worldwide. The regulation of the olive oil trade has strict guidelines for the different grades of oil. As a result of price and quality differences, there is a significant economic incentive to pass off inferior olive oil for higher grades. Fourier transform infrared (FTIR) spectroscopy has been used to distinguish different grades of olive oil. In a study published in the *Journal of Food Science*, FTIR was used to correctly classify 93.5% of samples in a calibration set and 100% of samples in an independent validation set.* ▬

** Lai Y. W.; et al. Potential of Fourier Transform Infrared Spectroscopy for the Authentication of Vegetable Oils. J. Agric. Food Chem. **1994**, 42, 1154–1159.*

In Example 11.2 we calculated that there are nine normal modes (types of vibrations) for methane. The three stretches depicted in Figure 11.3 are included in the nine modes; therefore, there must be six bending motions. Let us examine those six possible bending modes for methane. Figure 11.4 shows possible bending modes for methane and identifies each one by its motion (wag, twist, torque, rock, scissoring). As you look at Figure 11.4, you want to visualize the motions in three dimensions. Imagine the central carbon as remaining fixed on the paper. Two of the hydrogens are rising above the plane of the paper (those with the wedge bonds), and two of the hydrogens are hanging below the plane of the paper (those with the dotted line bonds). The arrows indicate the direction of motion as the angle formed between two hydrogens and the carbon is increasing or decreasing, synchronously or asynchronously. As a general rule, bending modes occur at lower frequencies than do stretching modes.

Each of the vibrations described in Figures 11.3 and 11.4 are included in the *normal modes* for methane. Each of these motions has a characteristic frequency that is a

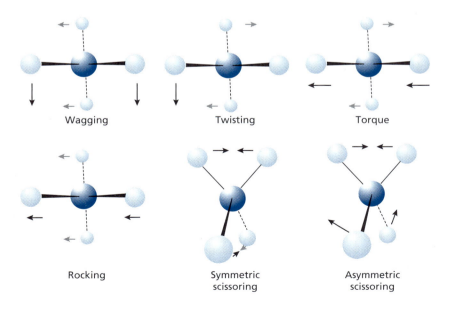

Wagging   Twisting   Torque

Rocking   Symmetric   Asymmetric
                 scissoring   scissoring

Figure 11.4 Bending modes for methane.

A Review of Bond Polarity and Dipole Moments

A polar covalent bond is a covalent bond that is partially ionic due to differences in the electronegativity of the atoms in the bond. Therefore, a covalent bond will be polar anytime the atoms involved in the bond have different electronegativities. The magnitude of the charge separation in a polar covalent bond is measured by a quantity called the dipole moment and is given by the equation p = qd, where p is the dipole moment, q is the magnitude of the charges involved, and d is the separation between the two charges. As a polar covalent bond vibrates, the distance between the separated charges changes and results in a change in the dipole moment as a function of time (an oscillating dipole moment).

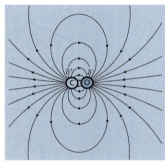

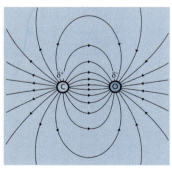

In order for a vibrational mode to be IR active, the vibration must produce a net dipole change in the molecule.

function of the bond strength (k) and the reduced mass of the atoms (μ) involved in the bond (see Equation 11.1).[6]

The Selection Rule and Molecular Symmetry

The term *selection rule* is used to describe the conditions that must be met in order for a molecule to absorb electromagnetic radiation. Selection rules are derived from the underlying quantum mechanics of a particular transition. In Chapter 2, we introduced one of the selection rules for IR spectroscopy: Δn must equal ± 1. An additional selection rule for IR spectroscopy states that *the only vibrational modes resulting in the absorption of IR radiation are those modes that produce a* net *time-dependent change in the dipole moment within the molecule.* Note the use of the word *net* in this selection rule. Only asymmetric vibrations are observed in IR spectra. This is telling us that not all of the various motions (stretches and bends) result in a peak in the IR spectrum; in other words, not all of the modes are IR *active*.

As you know from your earlier studies in spectroscopy, if the frequency of the electromagnetic energy that hits the sample equals the energy gap for the transition under investigation, the sample can respond either through absorbance or emission. IR works similarly. If the frequency of the incident electromagnetic energy exactly matches the vibrational frequency of the changing dipole moment, the electric vector of the photon can couple with the oscillating dipole moment of the vibration and absorption of radiation can occur, and the molecule may experience an increase in the amplitude of the vibration. But it is important that we understand and incorporate the implications of the IR selection rule. It is not enough that a particular bond *has* an oscillating dipole moment; the selection rule requires a vibrational mode result in a change in the *net dipole* of the *entire molecule*. This is an important point. The net dipole change for the molecule is affected by the symmetry of the entire molecule, not just a single vibrating bond. In a highly symmetric molecule, you will often find that the oscillating dipole created by one polar covalent bond is equal to but opposite to an identical oscillating dipole created by another bond on the molecule. For example, carbon dioxide has two C=O bonds that are symmetrically opposed to each other. The oscillating dipoles created by the symmetric stretch of the two C=O bonds are exactly opposite to each other and therefore create *no net* dipole change in the molecule. Take a moment to examine Example 11.3.

What we see is the fact that stretching modes that do not produce a net dipole are not visible in the IR spectrum.

Example 11.3 showed us that the symmetry of a molecule plays an important role in the IR spectrum. Because CO_2 is linear, the symmetric stretch results in no dipole change and therefore no IR absorption. In the next chapter (Chapter 12), we will look at a complementary technique called *Raman spectroscopy*. The selection rules for Raman spectroscopy are different, and symmetric stretching modes are visible in the Raman spectrum. By comparing the IR and Raman spectra, we can learn additional information about the symmetry of specific bonds in the molecule.

If we compare CO_2 to another simple three-atom molecule, H_2O, there are some important differences in structure and IR activity. Water is a bent molecule, and because of its bent geometry, it does not have a symmetric stretch with opposing dipole moment and therefore has additional IR active stretching modes when compared to CO_2. In addition, H_2O undergoes hydrogen bonding with adjacent H_2O molecules, which adds complexity to the spectrum.

As a general rule of thumb, there are fewer peaks in the IR spectrum for molecules that have a relatively high degree of symmetry (compared to those that have a low degree of symmetry), because highly symmetric molecules are more likely to have vibrational dipoles canceled out by an equal but opposite (symmetric) vibrating bond

---

[6] See Section 2.3 for a review of the quantum mechanics of molecular vibration.

Example 11.3

Determine the number of normal modes one can expect for carbon dioxide and predict how many of those modes will be IR active.

STRATEGY –

Use Equation 11.4 to determine the number of normal modes. Then analyze the possible stretching and bending modes for changes in dipole moments.

SOLUTION –

Because CO_2 is a linear molecule, the number of degrees of freedom available for vibration is $3N - 5 = (3 \times 3) - 5 = 4$. The four possible vibrations are depicted below along with the corresponding wave numbers for each vibration.

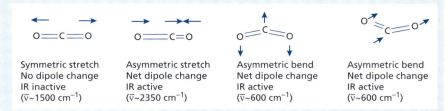

Symmetric stretch	Asymmetric stretch	Asymmetric bend	Asymmetric bend
No dipole change	Net dipole change	Net dipole change	Net dipole change
IR inactive	IR active	IR active	IR active
($\bar{v}$~1500 cm^{-1})	($\bar{v}$~2350 cm^{-1})	($\bar{v}$~600 cm^{-1})	($\bar{v}$~600 cm^{-1})

The symmetric stretching mode has the two oxygen atoms moving away from the carbon and toward the carbon. Although there is an oscillating dipole for each individual C=O bond, the bond dipoles are equal and opposite to each other; therefore, the symmetric mode does not produce a net oscillating dipole moment for the molecule overall and as such, the symmetric mode for CO_2 does not absorb IR radiation. The asymmetric stretch has the carbon atom moving toward one oxygen atom while simultaneously moving away from the other. This stretching mode does produce an oscillating dipole moment and results in absorption at approximately 2,350 cm^{-1}. The remaining two vibrations can be viewed as out of plane bending or scissoring; one in the x,z plane and the other in the y,z plane.[7] These two modes have the same energy and result in an absorption at approximately 600 cm^{-1}. Figure 11.5 shows us the gas phase IR spectrum of CO_2.

ACTIVITY

Just for Fun

Imagine a person's body as representing methane. The person's two legs represent two of the C–H bonds, and the person's two arms represent the other two C–H bonds. A stretch is when the person extends the arm or leg and then contracts it again. To demonstrate the symmetric stretch, the person would contract both legs (crouch down) and pull both arms in to the body (all compressed) as a starting position; then extend the legs and arms all together (stand up, arms out and fully extended); and then back again, over and over. How would you use your body to demonstrate the two asymmetric stretches?

---

[7] By convention, the highest symmetry axis is defined as the z axis: in this case, the linear axis of the CO_2 molecule.

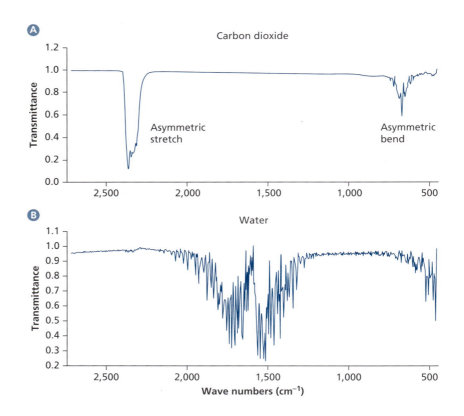

Figure 11.5 Gas phase IR spectrum of (A) carbon dioxide and (B) water.

somewhere else in the molecule. The result is the absence of a *net* dipole change for that stretching mode and a simpler spectrum overall. Figure 11.5 shows the gas phase IR spectrum of CO_2 and H_2O.

Problem 11.7: In Example 11.3, we saw how the transorientation of the two C=O bonds in CO_2 resulted in the deactivation of the symmetric vibrational mode in the IR spectrum. Based only on symmetry arguments, determine which of each pair of molecules below will have the most IR active vibrations and therefore the most complicated IR spectrum.

Vibronic Coupling

Figure 11.5 shows the IR spectrum of CO_2, and you can clearly see the asymmetric stretch at 2,350 cm^{-1} and the asymmetric bends at 600 cm^{-1}. In Table 11.1, we stated that carbonyl stretches occur between 1,690 and 1,770 cm^{-1}, and yet in the CO_2 molecule, we see the carbon–oxygen double-bond stretch at 2,350 cm^{-1}. The increase in the vibrational frequency of the carbonyl stretch in CO_2 is an example of *vibronic coupling*. Vibronic coupling occurs whenever two stretches share a common atom and it is most pronounced when the coupled vibrations are similar in energy.[8]

Here we have seen that knowledge about molecular structure and symmetry can help us to interpret IR spectra. Likewise, the IR spectra can help us determine molecular structure and symmetry. Advanced coverage of the role symmetry plays in spectroscopy is called *point group symmetry* or *group theory* and it is a topic typically covered in an advanced inorganic course or as a graduate level course.[9]

11.2 Time Domain versus Frequency Domain Spectroscopy: The Fourier Transformation

All of the spectroscopic techniques we have studied so far have used a monochromator to disperse light from the source, and the final output was a power or intensity versus frequency or wavelength plot. These types of spectrometer designs are collectively referred to as *dispersive spectrometers,* and the spectroscopy one conducts with a dispersive instrument is referred to as *frequency domain* spectroscopy. In contrast, when one conducts Fourier transform IR (FTIR) spectroscopy, all of the radiation from the source is allowed to pass through the sample and reach the detector simultaneously. The detector in an FTIR spectrometer measures the changes in total radiant power (all wavelengths simultaneously) as a function of *time.*

FTIR techniques are referred to as *time domain* spectroscopy. Let us start this discussion by imagining a spectrometer that passes only two frequencies of light through the sample. These two frequencies are shown in Figure 11.6(A) as the two oscillating sine waves in light and dark blue, having frequencies of v_1 and v_2. The two sine waves in Figure 11.6(A) can represent two different frequencies of light. Imagine these two frequencies of light passing through a spectrometer and hitting a detector. Because they are out of phase by less than 180 degrees, the degree to which they constructively and destructively interfere with each other will vary over time and the result will be a time-dependent fluctuation in the power that reaches the detector. The time-based interference pattern that would result from these two frequencies is depicted in Figure 11.6(B). The interference pattern produces its own cycle of increasing and decreasing amplitudes. This type of interference pattern is often referred to as a *beat pattern*. The beat pattern is a measure of power versus time and is referred to as the time domain plot.

The *Fourier transform* is named after the French mathematician, Joseph Fourier. It is used to isolate the individual frequencies that give rise to the observed beat pattern. Performing a Fourier transform on the beat pattern in Figure 11.6(B) yields the frequency domain plot seen in Figure 11.7. Note that Figure 11.7 represents the two sine waves from Figure 11.6(A) but as a function of frequency rather than time.

Early IR spectrometers were dispersive and similar in design to the dispersive ultraviolet-visible (UV-vis) spectrometers. However, almost all modern IR spectrometers

---

[8] See Harris, D. C.; Bertolucci, M. D. *Symmetry and Spectroscopy*; Oxford University Press: New York, **1978.** Reprinted by Dover **1989**. Also, see Cotton, F. A. *Chemical Applications of Group Theory*, 3rd ed.; John Wiley & Sons: NY, **1990.**

[9] See Cotton, F. A. *Chemical Applications of Group Theory*, 3rd ed.; John Wiley & Sons: NY, **1990.** Also, see Tsukerblat, B. S. *Group Theory in Chemistry and Spectroscopy: A Simple Guide to Advanced Usage.* Dover Publications: Mineola, NY, **2006.**

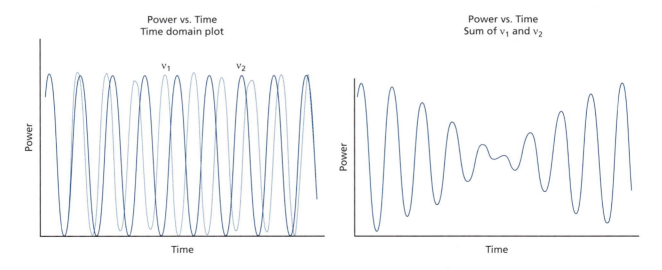

Figure 11.6 (A) A time domain plot of two (v_1 and v_2) different frequencies of light. The two frequencies have the same amplitude but are slightly out of phase. (B) A time domain plot of the interference pattern that results from adding v_1 and v_2.

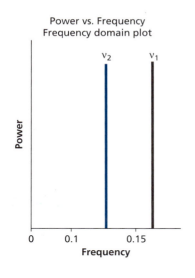

Figure 11.7 Frequency domain plot of v_1 and v_2.

are of the Fourier transform type. There are several reasons for the move away from dispersive IR spectroscopy to FTIR spectroscopy. Reasons include:

- Faster data collection times. A typical FTIR spectrum can be obtained in less than one second. Scanning IR spectrometers are slow and can take up to 15 minutes for a single scan.[10]

- Better signal-to-noise (S/N) ratios (also known as *multiplex advantage* or the *Fellgett's advantage*[11]). In Chapter 5, we learned that:

$$\left(\frac{S}{N}\right)_n = \sqrt{n}\left(\frac{S}{N}\right)_i$$

Eq. 11.5

where $\left(\frac{S}{N}\right)_i$ is the S/N ratio of a single measurement and n is the number of measurements that are being averaged. Because Fourier transform data collection is so fast, it is possible to collect many spectra in a relatively short amount of time, store them in the computer, and signal average the results. As the number of averaged spectra approaches infinity, the average value of the "white" noise approaches zero.[12] In practical terms, the S/N ratio increases as the square root of the number of spectra (n); for example, averaging 64 spectra will provide an 8× increase in S/N.

- *Better throughput* (also known as *Jaquinot's advantage*). In a dispersive instrument, only a small portion of the total energy from the source that passes through the sample reaches the detector. In a Fourier transform instrument, all of the radiant power reaches the detector. The result is better detection limits and better S/N ratios.

- *Better resolution.* The resolution of a dispersive spectrometer is related to the slit width of the monochromator. Given the fact that the energy of an IR photon is already relatively weak, narrow slit widths results in an unacceptable attenuation of the signal. To increase signal, larger slit widths are required in order to "see"

---

[10] It is semantically incorrect to refer to an FTIR spectrum as a "scan."

[11] http://en.wikipedia.org/wiki/Fellgett's_advantage.

[12] See Chapter 5 for a review of signal averaging.

ACTIVITY

Creating a Beat Pattern

Use a spreadsheet to create and plot the interference pattern generated by the sum of two sine waves of different frequencies. Repeat the exercise with three, four, and five different frequencies. Observe the changes to the beat pattern as you add in new frequencies and/or change the relative phase of the input frequencies. Spend some time playing with the spreadsheet. For example, try plotting the beat pattern of $(v_1 + v_2)$, or

$(v_1 + v_3)$, or $(v_1 + v_2 + v_3 + v_4)$, and so on. You can gain a lot of intuitive knowledge from observing how two or more sine waves interact with each other so spend some time playing around with your spreadsheet. Your column headings and cell inputs for your data might resemble the spreadsheet seen here.

A plot of column G versus A will give you a beat pattern similar to that seen in Figure 11.6(B). Save this spreadsheet as *beat pattern*.

	A	B	C	D	E	F	G
1	Time	v_1	v_2	v_3	v_4	v_5	Beat Pattern
2	1	=sin(A2)	=sin(0.9*A2)	=sin(0.8*A2)	=sin(0.7*A2)	=sin(0.2*A2)	=sum(B2:F2)
3	2	=sin(A3)	=sin(0.9*A3)	=sin(0.8*A3)	=sin(0.7*A3)	=sin(0.2*A3)	=sum(B3:F3)
4	3	=sin(A4)	=sin(0.9*A4)	=sin(0.8*A4)	=sin(0.7*A4)	=sin(0.2*A4)	=sum(B4:F4)
5	4	=sin(A5)	=sin(0.9*A5)	=sin(0.8*A5)	=sin(0.7*A5)	=sin(0.2*A5)	=sum(B5:F5)
6	↓	↓	↓	↓	↓	↓	↓
...	...	...	...	...	...	...	
257	256	=sin(A257)	=sin(0.9*A257)	=sin(0.8*A257)	=sin(0.7*A257)	=sin(0.2*A257)	=sum(B257:F257)

the sample; however, larger slit widths result in lower resolution. In an Fourier transform instrument, the resolution is dependent on the distance traveled by a moving mirror (see Section 11.3) and thus can be adjusted by the user as needed without sacrificing throughput of the source radiation.

ACTIVITY

Performing a Fourier Transform

Now let us perform a Fourier transform on the beat pattern data that you generated in the previous activity and create a frequency domain plot. Microsoft Excel can perform a Fourier transform but first you must activate the *Analysis ToolPak*. The exact sequence of steps taken to activate the Analysis ToolPak varies significantly between the various versions of Excel. If needed, use the Help function to assist you in activating the Analysis ToolPak.

Set up your Fourier transform spreadsheet. Label columns A → E with Time, Beat data, Fourier transform frequency,

Fourier transform magnitude, and Fourier transform complex number. Import your data from your beat pattern spreadsheet. If your beat pattern spreadsheet resembles the above spreadsheet from the *Activity: Creating a Beat Pattern,* then you will be importing columns A and G from your beat pattern spreadsheet into columns A and B of your Fourier transform spreadsheet. If you are using "cut and paste" functions to import your data, you will need to right click the cell A1, then select "Paste options" and then "Paste values." Otherwise, Excel will paste the equations that generated the beat pattern instead of the actual data.

Truncate your data set. The Fourier transform function requires that the number of data points must be a power of 2: 2^1 or 2^2 or 2^3, and so on (i.e., 2, 4, 8, 16, 32, 64, 128…). If you imported 100 data points, you will need to truncate your data set at 64 data points. Alternatively, you can expand your data set to 128 points* or more, but the total length of the array will need to be a power of 2. At this point, your spreadsheet should resemble Figure 11.8.

(Continued)

ACTIVITY

Performing a Fourier Transform (*Continued*)

	A	B	C	D	E
			fx		
1	Time	Data	FT Frequency	FT Magnitude	FT Complex
2	1	3.033855			
3	2	2.460373			
4	3	3.496381			
5	4	0.187216			
6	5	−0.74297			
7	6	−0.19814			
8	7	−1.01816			
9	8	1.724815			
10	9	0.499341			
11	10	0.371016			
12	11	0.637823			
13	12	−1.22593			
14	13	0.767298			
15	14	−0.61052			
16	15	−0.93716			
17	16	78267			

Figure 11.8 Spreadsheet setup for Fourier transform analysis.

Generate Fourier transform complex numbers (column E). In older versions of Excel, you will use Tools → Data analysis → Fourier analysis and in newer versions of Excel, you will select the Data tab, then Data analysis → Fourier analysis. The Fourier analysis window will pop up. Select column B as your input range and column E for your output range and select "OK." Make sure your selection has a length that is a power of 2.

Determine the Fourier transform magnitude. With the exception of the first and last data point, the Fourier analysis function returns a complex number for each of your data points. The real part is the amplitude of the cosine subcomponent, and the imaginary part is the amplitude of the sine subcomponent.

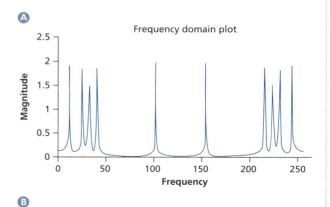

We are only interested in the absolute magnitude of the combined components. Place your cursor in cell D2 and from the formula menu, enter "=IMABS(E2)." The IMABS function returns the absolute value (modulus) of a complex number. Use the "drag" or "fill" feature to fill in the remaining cells in column D.

The Fourier transform frequency is the sampling rate of your instrument. Since the data were generated on a spreadsheet, it is the same as the time interval in column A, however, we start at zero instead of 1. Fill in column C from 0 to n.

Plot your frequency domain spectrum by constructing an x-y scatter plot of Fourier transform frequency against Fourier transform magnitude. The first thing you should notice is that the spectrum seems wrong (see Figure 11.9). In this example, there are ten peaks, and we only started with five frequencies when we generated the original beat pattern. In fact, it looks as if you could fold the plot in half and the peaks would line up. Actually, you could! From careful examination of the data in the spreadsheet, you should notice that halfway down the Fourier transform magnitude column, the data begin to repeat itself. There is no new data in the second half of the data set, and the spurious peaks seen in Figure 11.9 are known as aliasing. We first introduced the idea of aliasing in Chapter 5. To prevent aliasing, we must only use the first half of the data in the Fourier transform magnitude column. The right side of Figure 11.9 shows a frequency domain plot using only one half of the Fourier transform magnitude data.

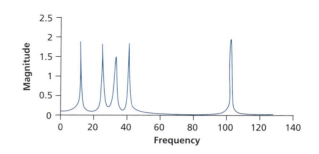

Figure 11.9 Frequency domain plots. Plot (A) shows aliasing. Plot (B) is without aliasing.

---

* *In Fourier transform nuclear magnetic resonance spectroscopy, it is a common practice to "zero fill" an array to a power of 2. Expand your data set by simply adding zeros to the end and observe how this changes your final output.*

11.3 Fourier Transform Infrared Spectroscopy and Wavelength Discrimination

The Michelson Interferometer

In 1907, Albert Abraham Michelson won the Nobel Prize for his accurate measurement of the speed of light.[13] The very precise measurements were made possible by his invention of a device we now call a Michelson interferometer. Figure 11.10 shows a block schematic of an FTIR spectrometer utilizing a Michelson interferometer. Radiation from the source (A) travels to a semitransparent beam splitter (B). A portion of the incident radiation passes through the beam splitter to a stationary mirror (C), and a portion is directed to a mirror whose position moves linearly (back and forth) about a set position (D). Both beams return to the beam splitter and pass through to the sample and detector (E).

When the stationary mirror and the pulsating mirror are the same distance from the beam splitter, the split beams recombine exactly in phase. However, when the mirrors are at different distances, there will be some degree of destructive interference on recombination of the two beams. Because the position of the moving mirror is a function of time, the degree of constructive and destructive interference of the recombined beams is time dependent. However, it is also dependent on the frequency of the beam. The difference in the path lengths of the split beams is termed the *retardation* (δ).

If the light source were monochromatic (one frequency only), then destructive interference would occur anytime the retardation were a quarter of the wavelength of the light source ($\frac{\lambda}{4}$), because the total distance traveled between the beam splitter and the pulsing mirror would be a distance $\frac{\lambda}{2}$. This would result in the combined waves being exactly 180 degrees out of phase. The mirror is pulsing at a constant velocity; thus, a time-dependent *interference* occurs on recombining at the beam splitter, and a plot of the radiant power reaching the detector as a function of δ would take the form of a cosine. This type of time-dependent power reading is called an *interferogram*. Figure 11.11 shows the relationship of power (P) to retardation (δ) for a monochromatic beam of energy (λ).

We can develop a relationship between the frequencies involved in the interferogram and the frequency of the light that generated the interferogram by recognizing that one complete cycle occurs every time the mirror moves a distance of $\frac{\lambda}{2}$ (see Figure 11.10). Because the mirror is moving at a constant velocity (u), we can designate the time it takes for the mirror to move $\frac{\lambda}{2}$ as τ. Therefore, the frequency of the detector response (f) is one divided by τ, which can be related to the velocity of the mirror and the wavelength of the incident light as:

$$f = \frac{1}{\tau} = \frac{u}{\lambda/2} = \frac{2u}{\lambda} \qquad\qquad \text{Eq. 11.6}$$

---

[13] He was the first American scientist to win the Nobel Prize.

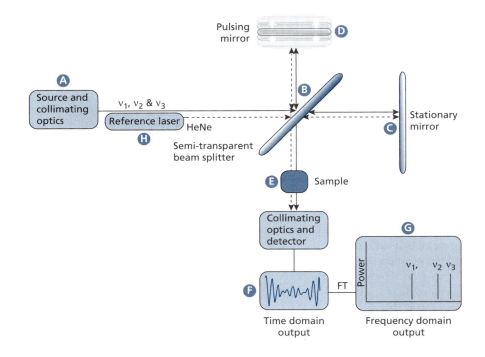

Figure 11.10 Schematic of an FTIR using a Michelson interferometer.

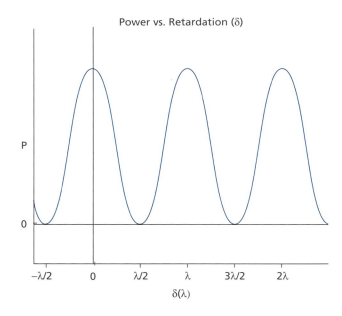

Figure 11.11 Plot of the power reaching the detector (P) as a function of the retardation (δ).

Interferograms and Beat Patterns

Although similar in concept, the interferogram generated by a Michelson's interferometer is not strictly the same thing as the "beat pattern" we described in Section 11.2 above. Let us take a moment to compare and contrast the two signals. In a beat pattern, the time-dependent power reading at the detector is the result of the constructive and destructive interference resulting from the convolution of different frequency light waves reaching the detector at the same moment. Over time, the different frequency waves come in and out of phase, thus causing a time-dependent power fluctuation at the detector. In an interferogram, you still have a convolution of different light frequencies *and* you have the constructive and destructive interference that is the result of varying the path length of a split beam of light (the pulsing mirror).

This is a key point! From an instrument design perspective, this is the reason Fourier transform infrared (FTIR) methods work. Using a Michelson interferometer, one can shift the frequency response at the detector to a much slower response time. In other words, the frequency of the beat pattern is now *within the response time of the detector.*[*] FTIR detectors typically have RC time constants in the millisecond range. If we tried to measure the beat pattern of infrared radiation directly, we would need to have a detector with an RC time constant in the femtosecond range.

---

* See Chapters 4 and 5 for a discussion of RC time constants and detector response times.

Likewise, we can relate the optical frequency of the incident light to the frequency response at the detector by using the relationship $\nu = \dfrac{c}{\lambda}$ and make appropriate substitutions into Equation 11.6:

$$f = \frac{\nu 2u}{c} \text{ or } \nu = \frac{fc}{2u} \qquad\qquad \textbf{Eq. 11.7}$$

where
   ν = frequency of the incident radiation
   f = frequency of the interferogram reaching the detector
   c = speed of light
   u = velocity of the pulsing mirror

However, what ultimately reaches the detector in a modern FTIR spectrometer is more complex than what Equation 11.7 would suggest. The source radiation is polychromatic; thus, there are a multitude of frequencies of light each generating their own interferograms, and each of these interferograms are interacting with each other to combine as a net beat pattern. A Fourier transform analysis of the beat pattern produces a set of frequencies that represent the individual interferogram frequencies. Application of Equation 11.7 to the interferogram frequencies yields the frequencies of the incident radiation from the source. In actual application, then, at least two spectra must be obtained. A blank or background spectrum is taken in the time domain to measure the maximum power of radiation at each wavelength, and a Fourier transform analysis can be performed on that spectrum to yield the background signal spectrum in the frequency domain. Then, the sample is placed in the instrument and a sample spectrum is obtained, transformed, and then transmittance is calculated as the ratio of the sample and background spectra.

Resolution

Resolution is the ability of an instrument to separate and detect two closely spaced frequencies. As mentioned earlier, in dispersive instruments, a key factor in determining resolution is the slit width. However, in a Fourier transform instrument, there is no dispersion and all wavelengths are detected simultaneously. How is resolution determined in FTIR spectroscopy? The data that ultimately reach the detector in an FTIR

ACTIVITY

Exploring Resolution

Use the same spreadsheet you created above to generate a beat pattern. This time, vary the relative frequencies of the two input sine waves.

	A	B	C	D
1	Time	v_1	v_2	Beat Pattern
2	1	=sin(A2)	=sin(0.9*A2)	=sum(B2:C2)
3	2	=sin(A3)	=sin(0.9*A3)	=sum(B3:C3)
4	3	=sin(A4)	=sin(0.9*A4)	=sum(B4:C4)
5	4	=sin(A5)	=sin(0.9*A5)	=sum(B5:C5)
6	↓	↓	↓	↓
...	...	...	...	
257	256	=sin(A257)	=sin(0.9*A257)	=sum(B257:C257)

Observe how long it takes to generate a full beat pattern as you vary the difference in the two frequencies. What you should observe is the fact that as the two sine waves become closer in frequency, it takes more time to generate a full beat pattern. Because the beat pattern in a Fourier transform infrared spectrometer is a product of the pulsing mirror and the mirror is moving at a constant velocity, the pulsing mirror must move further in order to capture a full beat pattern for two closely spaced frequencies. Resolution increases as the distance the pulsing mirror travels increases.

analysis is a beat pattern generated by the interference relationships of all impinging wavelengths of energy. In order to resolve the input frequency of a beat pattern using the Fourier transform technique, the data set must contain a full beat pattern. Take another look at the beat pattern in Figure 11.6. The data set in Figure 11.6(B) represents one complete beat. The interferogram for a specific frequency of light is generated by the pulsing mirror, and likewise, the beat pattern is a result of the mirror's motion. In other words, *in order to fully resolve two frequencies of light, the pulsing mirror must move a distance sufficient to generate the complete beat pattern associated with those two frequencies.*

Equation 11.8 gives the relationship between the spectral resolution and the distance the mirror travels in the Michelson interferometer. The value delta (δ) was defined earlier as the retardation, and the maximum δ value is twice the maximum distance (d_{max}) the mirror moves. If we measure d and/or δ in centimeters, the resolution ($\Delta \bar{v}$) can be determined in units of cm^{-1}.

$$\Delta \bar{v} = \frac{1}{\delta_{max}} = \frac{1}{2d_{max}}$$

Eq. 11.8

where
$\Delta \bar{v}$ = spectral resolution, cm^{-1}
δ_{max} = maximum retardation, cm
d_{max} = maximum mirror movement, cm

If you take another look at Figure 11.10, you will also see a reference laser (H) in the instrument schematic. In fact, if you place a piece of paper in the optical path of an FTIR spectrometer, you will notice a red point of light that looks a lot like a laser pointer. The FTIR spectrometer has a reference beam (usually a He-Ne laser) that travels collinear with the source radiation. The He-Ne laser's radiation is monochromatic; therefore, a Fourier transform of the interferogram from the He-Ne laser serves the purpose of an internal calibration and allows the computer to know the relative location of the mirror as a function of time. It is important that you do not confuse the laser with the source. The He-Ne laser simply establishes a reference frequency for determining the mirror location as a function of time.

Problem 11.8: A Michelson interferometer generates a beat pattern for a mono-chromatic radiation source. The frequency of the resulting interferogram is recorded below. Determine the frequency of incident radiation that created that interferogram given the fact that the mirror was pulsing over a distance of 3 cm at a frequency of 1 Hz.

(a) 2.40×10^4 Hz   (c) 1.38×10^4 Hz   (e) 7.20×10^3 Hz

(b) 3.60×10^3 Hz   (d) 1.20×10^4 Hz   (f) 1.08×10^4 Hz

Problem 11.9: A He-Ne laser in an FTIR instrument emits light having a wavelength of 632.8 nm. What is the expected frequency of the interferogram of the laser, assuming the mirror is pulsing over a distance of 3 cm at a frequency of 1 Hz?

Problem 11.10: What is the maximum distance the moving mirror in a Michelson interferometer would need to travel to be able to distinguish between 1,000 and 1,004 cm^{-1} in an FTIR spectrum? Between 1,000 and 999.5 cm^{-1}?

Problem 11.11: What would be the spectral resolution if the mirror in a Michelson interferometer moved a total distance of 7.5 mm?

11.4 Sources

The ideal FTIR source would be a black body emitter that emits strongly over the entire range of bond vibrational frequencies. In theory, any material can be made to emit at a usable wavelength as long as you keep the material at the appropriate temperature. But there are design characteristics that must be considered. For example, the filament in a tungsten light bulb emits a tremendous amount of IR radiation. However, in the case of the tungsten filament, the tungsten must be placed in a glass or quartz bulb to prevent oxidation of the tungsten through contact with the air and the glass bulb absorbs the majority of the IR radiation. So in practice, a primary limitation is the transparent ranges of the optical materials available.

Table 11.2 shows the transparent range of some common optical materials used in IR spectroscopy. Potassium bromide (KBr) has one of the widest transparent ranges, from 4,000 to 400 cm^{-1}. It would seem that, the ideal source for a mid-IR FTIR spectrometer would be a black body emitter with a continuum from 4,000–400 cm^{-1} (2,500–25,000 nm) that is stable in air when heated to a working temperature. For this reason, most commercial IR spectrometers use a semiconducting ceramic source. Alternatively, some IR spectrometers use a nichrome wire (NiCr) filament as the IR source.

Wien's displacement law[14] states that the wavelength of maximum emission for a black body emitter is related to the temperature by

$$\lambda_{max} T = b \qquad\qquad \textbf{Eq. 11.9}$$

where
λ_{max} = peak wavelength
T = temperature, in Kelvin
b = Wien's displacement constant $(2.898 \times 10^{-9}\, \mu\cdot K)$

---

[14] http://en.wikipedia.org/wiki/Wien%27s_displacement_law and http://hyperphysics.phy-astr.gsu.edu/hbase/wien.html.

TABLE 11.2: **Expanded Characteristic Group Frequencies (cm^{-1})**

BOND TYPE	$\bar{v}$	BOND TYPE	$\bar{v}$
RO-H (alcohol and phenols)	3,200–36,500	R-C-OR (esters), with C=O	1,735–1,750
R-C-O—H (carboxylic acids), with C=O	2,500–3,300 (stretching) 1,395–1,445 (bending)	R-C-OH (carboxylic acids), with C=O	1,710– 1,770
R-C—O-H (carboxylic acids), with C=O	1,208–1,322	R-C-H ; R-C-R (aldehyde and ketone), with C=O	1,690–1,745
R-C-O-C-R (acid anhydride), with two C=O	1,800–1,850 1,740–1,790 (two bands)	R-C-O-C-R (acid anhydride), with two C=O	890–1280
R-C-Cl (acid chloride), with C=O	1,790–1,850	R₂N—H (amines)	3,300–3,500 (one band)
C-N (amide), with C=O and H	3,300–3,500 (two bands) 1,560–1,650 (bending)	C-N (amide), with C=O and H	1,800–1,850 and 1,730–1,790
R₃C—H (alkane)	2,840–2,960 (stretching)	R—CH-R (alkane), with H	1,463–1,467 (bending)
R—CH₂ (alkane)	1,373–1,377 (bending)	R-(CH)$_n$-R (alkane), with H	718–722 (bending) n > 4
C=C (alkene), with R, H, R, R	3,010–3,100 (stretching) 1,290–1,435 (bending)	C=C (alkanes)	1,635–1,695 (isolated) 1,610–1,640 (conjugated)
R-N=O (nitro—aliphatic)	1,300–1,390 1,530–1,610	Ar-N=O (nitro—aromatic)	1,315–1,360 1,490–1,550
R₃C-X (alkyl halide) X = F	1,000–1,400	R₂P-H	2,268–2,322 (stretching)
X = Cl	538–785		810–1,090
X = Br	508–650		(bending)
X = I	480–600		
RC≡C-H (alkyne)	3,280–3,310	RC≡CR (alkyne)	2,100–2,260
RN=C=O (isocyanates)	2,268–2,272	RC≡N (nitriles)	2,220–2,260
Ar – H (Aromatic) (Stretching) (Bending)	~3,300 680 – 860	(breathing)	1,450–1,600 (two or three bands)

Figure 11.12 Nernst glower.

Regardless of the exact composition of the source, the principles governing the operation of the sources are similar. We will profile some of the more traditional IR sources in the following text.

The Nernst Glower

The early IR spectrometers used a source called the *Nernst glower* (see Figure 11.12), which consisted of a rod composed of primarily ZrO_2 and a smattering Y_2O_3 and Er_2O_3. The Nernst glower was a popular source in dispersive IR spectrometers because the source is rod shaped and could be designed to efficiently illuminate the entrance slit of a monochromator. The Nernst glower does not require any external envelope material (window) because it is not susceptible to oxidation. It is a very rugged source with a spectral output range of 25,000 cm^{-1} to 500 cm^{-1} (0.4–20 μM). Because of its early popularity and ruggedness, the Nernst glower can still be found in many IR spectrometers, so it is not yet obsolete. However, it has been largely replaced in newer spectrometers by other IR sources.

One of the oddities of the Nernst glower is the fact that it has a negative temperature coefficient, and the resistance to electric conduction decreases with increasing temperature. Also, it is an insulator at room temperature; therefore, one must preheat the Nernst glower before it will function properly. Once heated, the glower's resistance to electrical conduction maintains the high temperatures needed for black body emission. Because the glower's resistance decreases with temperature, the current running through the glower increases as the temperature rises and one must incorporate current control electronics into the design of the source. The power requirements of a Nernst glower are typically 107 V and 50 to 60 A with an operating temperature near 1,800K (2,073°C).[15]

The Globar

The *globar* is constructed of silicon carbide and is also electrically heated. The globar has a positive thermal coefficient, so it does not require a preheating stage. In some FTIR designs, the globar is housed in a water-cooled jacket that flows to a heat exchange unit, but more modern instruments often simply house the globar in a casing insulated by material similar to the thermal shield tiles used in the space shuttle. Like the Nernst glower, the globar does not require any external envelope material, and it is a very rugged source with a spectral output range of 10,000 to 250 cm^{-1} (1.0–40 μM). In comparing the spectral range of the *Nernst glower* (Δcm^{-1} = 24,500) to the *globar*, (Δcm^{-1} = 9,750), you might draw the *wrong* conclusion that the Nernst glower is the superior source because it has a larger spectral range than does the globar. In practice, FTIR spectroscopy is typically limited to a range of 4,000 to 400 cm^{-1} as a result of the optics used to guide the radiation through the spectrometer. IR optics are discussed in Section 11.7. In addition, the globar provides greater photon power output, or emissivity, than the Nernst glower below 1,700 cm^{-1} for the same energy consumption.

---

[15] LaRocca, A. *Handbook of Optics* (Chapter 11). McGraw-Hill: NY, **2004**.

Figure 11.13 Coiled wire source.

The region below 1,700 cm^{-1} is very important—it is called the fingerprint region of the IR spectrum and is extensively used in databases to positively identify unknown analytes.

Coiled Wire Sources

One of the earliest coiled wire IR sources was the nichrome wire source. *Nichrome* is an alloy of nickel that contains between 10% and 20% chromium and between 0% and 25% iron. Nichrome becomes a black body emitter at ~950°C. Portable space heaters still use this method of generating IR radiation (heat). The nichrome wire source has less emissivity than other sources and is less common in high-end research spectrometers. However, it is relatively inexpensive and finds use in teaching spectrometers and in industrial settings where sensitivity is not as large of an issue. Another popular coiled wire source for IR spectroscopy is the Kanthal wire source. *Kanthal wire sources* work under the same principle as nichrome sources; however, the alloy in this case is an iron alloy containing between 20% to 30% chromium and between 4% to 7.5% aluminum. The alloy is thermally stable to oxidation due to the formation of an aluminum oxide coating on the surface of the wire. Some compositions of Kanthal can be heated as high as 1,400°C before melting. Kanthal wire is inexpensive and widely available from many vendors, including large online shopping websites. Several companies sell proprietary alloy wires for use in FTIR spectroscopy. For example, Interspectrum carries several aftermarket IR sources for use in FTIR spectrometers. The modern alloys can be heated to 1,200°C and boast emissivities as bright as the Nernst glower and globar (see Figure 11.13).

Solid-State Sources

There are several proprietary solid-state sources used in IR spectroscopy. However, most still work on the principle of black body emission. For example, Axetris markets a source used for the IR detection of gas samples that uses a proprietary resistive heating element (ceramic) integrated onto a thin dielectric membrane,[16] which is suspended on a micromachined silicon structure for easy incorporation into electronic devices. Their sources are much smaller than the Nernst glower or the globar, but they are similarly as rugged and they consume much less space and energy. With a spectral range of 5,000 to 714 cm^{-1}, their output does not extend as deep into the IR as the Nernst glower and the globar. These types of devices find use in special purpose and portable spectrometers. Figure 11.14 is a photograph of several solid-state IR detectors manufactured by Axetris.

Figure 11.14 Solid-state IR sources by Axetris.

---

[16] A dielectric membrane is an insulator that can be polarized in an electric field.

11.5 Detectors

IR detectors fall into one of two broad categories: those that register a change in temperature on absorption of IR radiation (thermal detectors) and those that generate an electric current on absorption of IR radiation (pyroelectric and quantum well detectors).

Thermal Detectors

Thermal detectors measure the change in temperature of an absorbing body. The signal can be in the form of a change in pressure of an expanding gas (pneumatic detector), the change in resistance of a conductive metal (bolometer), the change in resistance of a semiconductor (thermistor), or the generation of a voltage (thermocouple). All of these detector designs were common in the early dispersive (scanning) IR spectrometers; however, they all have response times that are too slow for modern FTIR spectrometers. The oscillating mirror in a Michelson interferometer operates at approximately 1 Hz, so we need a detector with a response time that is in the millisecond range.[17]

Pyroelectric Detectors

One of the most common detector classes in FTIR is the pyroelectric detector. Certain crystalline materials can be fabricated with an intrinsic polarization within the crystal lattice, and absorption of an IR photon perturbs the position of the atoms in the crystal, thus changing the polarization within the crystal. The change in polarization creates a voltage fluctuation that can be amplified into a useful signal. The most common pyroelectric materials used in FTIR are deuterated triglycine sulfate, $LiTaO_3$, and piezoelectric ceramics. In all cases, the the pyroelectric material is sandwiched between two electrodes that have been coated with an IR-absorbing paint (black paint).

The electrodes measure the voltage fluctuations as a function of time. The relatively fast response time for pyroelectric detectors makes them suitable for use in FTIR instruments.

Deuterated Triglycine Sulfate Detectors
In deuterated triglycine sulfate (DTGS) detectors, the structures of the glycine moieties consists of stacked ion pairs containing $[(N^+H_3CH_2COOH)]_2[SO_4]$ spaced by protonated glycine $[(N^+H_3CH_2COOH)]^+$. The net molecular formula is $NH_2CH_2COOH)_3 \cdot D_2SO_4$. The crystals are held together by hydrogen bonding, which can be disrupted by the presence of water. As a result, DTGS detectors are very hygroscopic and must be kept enclosed in an inert gas environment. Figure 11.15 shows a schematic representation of a stacked triglycine sulfate (TGS) crystal. The spectral range of DTGS detectors is limited by the optical material used to enclose the detector. Early versions of these detectors used simply TGS, but it was found that deuterated TGS exhibits a significantly greater stability at elevated temperature, and so DTGS detectors succeeded TGS ones in the early 1980s.

$LiTaO_3$ Detectors
Pyroelectric $LiTaO_3$ detectors have sensitivities that are an order of magnitude less than DTGS detectors and operate under the same principles. $LiTaO_3$ detectors have the advantage of being insoluble in water and not prone to fogging due to hygroscopic absorption of moisture from the air. This make them more robust. In addition to being less sensitive, another disadvantage of $LiTaO_3$ detectors when compared directly to DTGS detectors is a greater degree of shot noise resulting in a lower S/N ratio.[18] It is not uncommon for the S/N of DTGS detectors to be four times greater

---

[17] See Chapters 4 and 5 for a review of resistor–capacitor time constants.

[18] Hamamatsu Technical Information SD-12. *Characteristics and Uses of Infrared Detectors*.

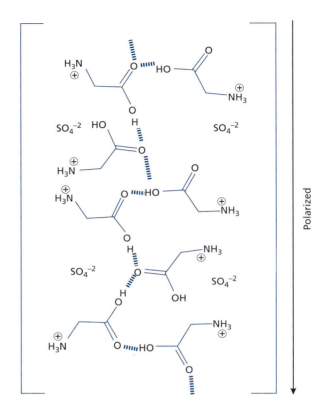

Figure 11.15 A schematic of crystalline deuterated triglycine sulfate.

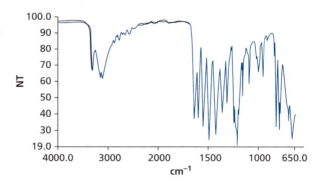

Figure 11.16 FTIR spectra of phenacetin. The black line is one scan using a DTGS detector. The blue line is an average of 16 scans using a $LiTaO_3$ detector (blue).

than that of $LiTaO_3$ detectors. As noted above, a key benefit of an FTIR compared to a scanning IR spectrometer is the ability to rapidly collect spectra and to signal average the spectra to improve S/N. However, there is a practical limit to signal averaging. As we saw in Equation 11.2, the S/N ratio improves with the square root of the sample set, so to improve the S/N ratio of an $LiTaO_3$ detector by a factor of four, we would need to signal average 16 times more scans $\left(4^2 = 16\right)$ than is required with a DTGS detector. Figure 11.16 shows two FTIR spectra taken on a Perkin Elmer Spectrum 100 FTIR. The black line in Figure 11.16 is a single scan using a DTGS detector and the blue line in Figure 11.16 is a signal-averaged set of 16 scans using a $LiTaO_3$ detector. As you can see, the S/N of the two spectra is nearly identical. As we have said several times in this textbook, most analytical decisions are a trade-off and although it takes approximately 16 times more scans to get the same S/N ratio with a $LiTaO_3$ detector compared to a DTGS detector, FTIR is such a rapid technique that many manufacturers offer their spectrometers with the $LiTaO_3$ detector in order to benefit from the robust and durable nature of the detector.

Lead/Pb Zirconium Titanate Detectors   *PZT* is an acronym that stands for lead (**P**b) **z**irconium **t**itanate, and this class of ceramics has the general formula ($Pb[Zr_{(x)}Ti_{(1-x)}]O_3$). The unit cell consists of either Zr^{+4} or Ti^{+4} in a lattice of Pb^{+2} and O^{-2} ions. The crystals assume a tetragonal or rhombohedral symmetry, resulting in a dipole moment. The voltage drop across a PZT device is temperature dependent and allows it to be used as an IR detector. PZT devices are robust and water insoluble. The response and sensitivity of PZT detectors is similar to that of $LiTaO_3$ detectors.

Photoconductive Detectors

There are quite a few materials that can function as IR-sensitive photoconductive detectors, but there is only one class of detectors that cover the entire range of the mid-IR region of the spectrum (4,000–400 cm^{-1}). For FTIR spectroscopy, the only widely used photoconduction detector is the **m**ercury **c**admium and **t**elluride[19] (MCT) detector. We have already examined a similar class of detectors when we looked at photovoltaic p/n junctions in Chapter 6 (see Figure 6.28). The MCT works on a similar principle to the p/n junction photovoltaic devices. The band gap between the valence band and the conduction band in a CdTe crystal is 12,098 cm^{-1}, and the band gap in HgTe is 0.0 cm^{-1}. The two materials cocrystallize, so by adjusting the ratio of the two materials you can tune the band gap in the MCT detector. MCT detectors are actually extensively used for both the mid- and far-IR regions of the IR spectrum. For mid-IR spectroscopy, the MCT blend is typically near 70% HgTe and 30% CdTe. For far-IR detection, the MCT detector is constructed with 80% HgTe and 20% CdTe. One of the advantages of MCT detectors is the fact that they can be cooled in order to reduce shot noise. MCT detectors have been shown to function at −50°C. However, they are susceptible to drift noise. The sensitivity of MCT detectors is related to both the composition of the detector and the operating temperature of the detector. In general, as the temperature increases, the sensitivity decreases, and the response shifts to shorter wavelengths. The sensitivity of MCT detectors at room temperature is comparable to DTGS and $LiTaO_3$ detectors, and sensitivity gains of up to two orders of magnitude can be expected by cooling the detector element.[20]

Quantum Well Detectors

Solid-state quantum well detectors are gaining popularity in many modern FTIR spectrometers because of their faster response times and greater sensitivity.[21] A quantum well detector is essentially a specialized photovoltaic cell in which large band gap semiconductors are sandwiched together to create local "wells," where the conduction band of one material is lowered with respect to the valence band of a second material.[22] The assembly of these materials is highly controlled, and the layers are on the order of

---

Lead/Pb Zirconium Titanate Ceramics

Another interesting characteristic of lead/Pb zirconium titanate (PZT) ceramics is the fact that they can produce a voltage drop across the two faces when subjected to high pressures. As a result, PZT ceramic wafers are used in the bumpers of automobiles to detect a collision. The voltage drop across the PZT device triggers the deployment of the automobile's air bag.

Mercury Cadmium and Telluride Detectors

The US military uses mercury cadmium and telluride, or MCT, detectors for night vision goggles, its aircraft (all), heat-seeking missiles, and airborne smart bombs.

---

[19] It is also common to see these types of detectors referred to as CMT detectors or MerCad Telluride detectors.

[20] Hamamatsu Technical Information SD-12. *Characteristics and Uses of Infrared Detectors.*

[21] Karunasiri, R. P. G. *Applied Physics Letters,* **1991**, *59(20),* 2588; Levine, B. F. *Applied Physics Letters,* **1988**, *53(4),* 296.

[22] You can relate the ideas of band gap theory and molecular orbital theory by considering a conduction band in a bulk material as a collection of degenerate antibonding orbitals. Similarly, the valence band is a collection of degenerate bonding orbitals.

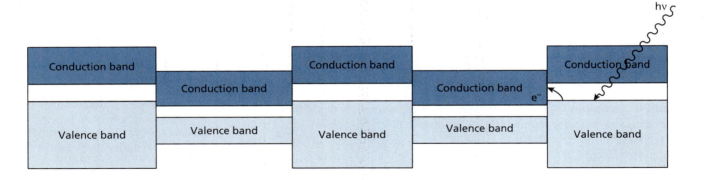

Figure 11.17 A schematic of a quantum well detector. At nanometer scales, quantum effects make the movement of electrons from the valence band of material 1 to the conduction band of material 2 forbidden. An IR photon can move the electron between the two levels.

nanometers thick. At these dimensions, quantum effects become important. A quantum barrier exists at the interface between the valence band of one material and the conduction band of the other. Even though the conduction band of material 2 is within the energy gap of the valence band of material 1, electrons do not flow unless promoted by an IR photon. These detector units can be created in a cascade mode creating high current gains (see Figure 11.17).

11.6 Spectral Output

Transmittance versus Absorbance

In contrast to UV-vis spectroscopy, it is quite common to report IR spectra using transmittance or percent transmittance instead of absorbance.

$$T = \frac{P_s}{P_o}$$ **Eq. 11.10**

where
    T = transmittance
    P_s = incident radiation reaching the detector after it has passed through a sample
    P_o = incident radiation reaching the detector with a blank in the optical path

However, if you are conducting quantitative studies and wish to apply Beer's law, modern FTIR spectrometers can also display spectra in absorbance mode (see Figure 11.18).

Problem 11.12: Speculate on the reasons why it is more common to report UV-vis spectra using absorbance, whereas it is more common to report IR spectra using transmittance (recall that absorbance = $-\log(T)$).

Quantitative Measurements and Deviations from Beer's Law

In Chapter 6 we talked about how bandwidth resolution of the spectrometer can affect adherence to Beer's law and why it was important to take measurements at λ_{max} when conducting quantitative work (see Equation 6.7). Because the peaks in IR spectroscopy are often much narrower than the peaks in UV-vis spectroscopy, it is more common to

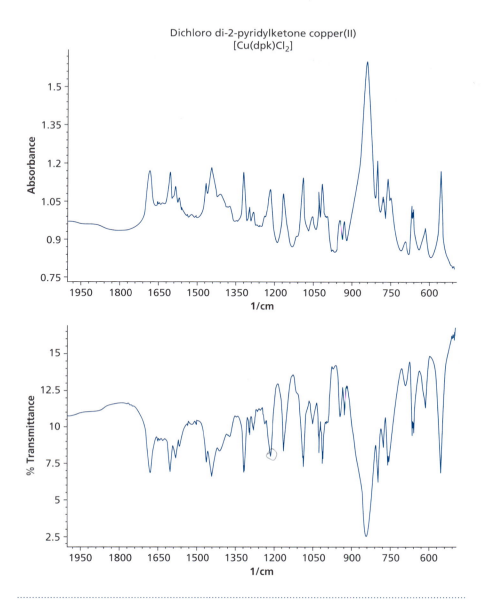

Figure 11.18 FTIR spectrum of [Cu(dpk)Cl$_2$] displayed as *absorbance* (top) and *transmittance* (bottom).

see deviations from Beer's law in IR spectroscopy, and linear calibration curves are not as commonly obtained. When conducting quantitative FTIR spectroscopy, you will often need to fit your empirical data to a nonlinear calibration curve.

Problem 11.13: Using the peak at 1,220 cm^{-1}, in the transmittance spectrum of Figure 11.18, estimate the percent error in the percent T reading if your spectrometer has a ± 2 cm^{-1} variance in its precision.

11.7 Developments: Two-Dimensional Infrared Spectroscopy

Two-dimensional IR spectroscopy is a relatively new technique that probes how the vibrations within a molecule couple with one another.[23,24] The technique involves pumping one vibrational mode and probing the effect on neighboring vibrations (pump and probe). Although the technique is primarily limited to well-funded research laboratories, commercially available instruments will likely become available in the future as the technology develops. In a two-dimensional IR experiment, you plot vibrational spectra over two axes (pump and probe), which can reveal structural and kinetic correlations that are not apparent in a one-dimensional IR spectrum. With the development of fast IR lasers capable of generating pulse width in the femtosecond range, researchers have begun developing a great many two-dimensional IR techniques. In a typical two-dimensional experiment, the first axis is the excitation axis and the second is a response axis. In other words, the plot represents a correlation showing how excitation of a specific vibration affects all other vibrations within a specified detection window.

The excitation pulse is generated by a tunable IR laser with a narrow bandwidth selected to excite a specific vibrational frequency range. The duration of the excitation pulse is usually designated as τ_1. For example, a laser with a frequency of 5×10^{13} Hz ($1,700$ cm^{-1}) could be used to excite carbonyl groups within a molecule. The probe pulse is also generated by a laser but one with a larger bandwidth encompassing a range of vibrational frequencies. The probe pulse is used to measure the absorption of a band of frequencies. For example, the probe pulse might encompass a frequency bandwidth of 4.0×10^{13} to 4.5×10^{13} Hz ($1,334$–$1,501$ cm^{-1}), which probes a section of the "fingerprint" (bending) region of the IR spectrum. The experiment continues by changing the pump frequency and probing the sample again. A compilation of all of the dynamics between the pump and probe constitutes the two-dimensional spectrum. Experiments such as this can tell the researcher a great many things. First, the coupling of vibrational modes is strongly dependent on their proximity, so an experiment such as this allows researchers to examine the through bond and through space relationships within the molecule. Second, by varying the delay between the excitation pulse and the probe pulse, researchers can examine the relaxation dynamics within the molecule. Figure 11.19 is an example of a two-dimensional IR spectrum. In Figure 11.19(A), we see a traditional one-dimensional FTIR spectrum. In the one-dimensional spectrum, we observe three peaks. If we could be assured that our sample were pure and if these three peaks had a strong correlation with peaks found in a database, we could identify our molecule from this spectrum alone. The two-dimensional spectrum of the same molecule is presented in Figure 11.19(B). The contour peaks that reside along the diagonal correspond to the peaks seen in the one-dimensional spectrum. The cross-peaks (off-diagonal) arise from the transfer of energy between coupled states within the molecule. We can see in the two-dimensional spectrum that peaks B and C are "talking to each other." In other words, energy is flowing between these two vibrational modes. They are coupled and likely reside adjacent to each other on the same molecule. Also, we observe that peak A is not coupled to either B or C. The spectra in Figure 11.19 is actually a mixture of two compounds $W(CO)_6$ and a rhodium dicarbonyl compound. In addition to readily identifying mixtures, researchers can examine the rate of energy flow between coupled states by varying the time delay between the pump and probe pulses (τ_2). This sort of information is useful for examining protein folding and contact energy quenching in large biomolecules.

[23] Baiz, C. R.; Reppert, M. E.; Tokmakoff, A. *An Introduction to Protein 2D IR Spectroscopy: Ultrafast Infrared Vibrational Spectroscopy*, edited by M. D. Fayer, CRC Press: Boca Raton, FL, **2013**.

[24] Khalil, M.; Demirdoven, N.; Tokmakoff, A. Coherent 2D IR Spectroscopy: Molecular Structure and Dynamics in Solution. *J. Phys. Chem. A,* **2003**, *107*, 5258–5279.

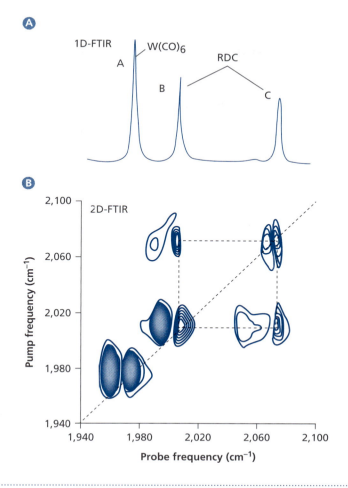

Figure 11.19 One-dimensional FTIR (A) and two-dimensional IR (B) spectra for a mixture of $W(CO)_6$ and a rhodium dicarbonyl. The cross-peaks in the two-dimensional IR spectrum show coupling between peaks B and C.

11.8 Sample Introduction

The typical way of introducing a sample into a UV-vis spectrometer is in the form of a dilute solution placed within a fused silica or quartz cuvette. Unfortunately for IR spectroscopy, solvents suitable for dissolving most analytes also absorb strongly in the IR region of the spectrum. Likewise, quartz and fused silica also absorb significant portions of the IR region of the spectrum. As a result, other methods of sample introduction have been developed for IR spectroscopy.

Optical Materials

The materials used to hold our samples and to guide IR radiation through the spectrometer must be transparent in the IR region of the spectrum—or at least transparent in the region of the IR spectrum of interest to our analysis. In addition to being IR transparent, the material must have physical and chemical properties compatible with our sample. By far the most commonly used IR-transparent optical material used in commercial instruments is potassium bromide (KBr), but there are quite a few suitable materials. Table 11.3 lists some common IR optical materials.

COMPARE AND CONTRAST

Ultraviolet-Visible Spectroscopy versus Fourier Transform Infrared Spectroscopy in Quantitative and Qualitative Analysis

The rule of thumb among chemists is that ultraviolet-visible (UV-vis) spectroscopy is very poor for qualitative analysis but excellent for quantitative analysis, and the opposite is true for infrared (IR) spectroscopy. In Chapter 1, we made the case that UV-vis spectra exhibit very few, and very broad peaks, whereas IR spectra exhibit many narrow peaks that provide a wealth of information about the chemical nature of the analyte.

But what makes UV-vis superior to IR for quantitative analysis? There are two reasons for this: (1) the broad peaks tend to lead to lower errors in measurements and (2) in general, UV-vis peaks provide significantly higher sensitivity in a quantitative measurement. In this discussion, we explore these two phenomena in more detail.

Figure 11.20(A) shows an IR absorption band centered at 1,800 cm^{-1} (5,556 nm) with a typical baseline breadth of about 4% of the nominal peak position. If our instrument's measurement of the wave number has an error as low as 0.5% in the abscissa (dotted lines), the experimental error in the signal strength is almost 6%, and a 1.0% error in the wavelength imparts a whopping 20% error in the measured signal (dark gray lines). Figure 11.20(B) shows a simulated UV-vis peak centered at 400 nm having a typical breadth of 35% of the nominal peak position (140 nm). If our instrument has an error of 0.5% in measuring the wavelength (dotted lines), the error in our signal measurement is only 0.08%. A 1% error in wavelength measurement imparts only 0.33% error in the signal (dark gray lines). Figure 11.21 demonstrates the comparison of errors for narrow IR peaks relative to broad UV-vis peaks.

The other reason UV-vis excels over IR for quantitative analysis is due to the relative peak intensities. Recall Beer's law from Chapter 6:

$$A = \varepsilon bc$$

where A is the measured absorbance (no units), ε is the molar absorptivity (cm$^{-1}\cdot$M^{-1}), b is the path length (cm) of incident light through the cell, and c is the concentration in molarity. In a quantitative analysis experiment, we would establish a calibration plot by measuring the absorbance of a series of standard solutions of known concentration and then plot absorbance versus concentration. From that point forward, we could simply measure the absorbance of any solution of unknown concentration and use the plot (or the linear regression) to determine the concentration. In this scenario, the slope of the absorbance versus concentration is referred to as the sensitivity, and a high sensitivity (steep slope) is desirable, as demonstrated in Figure 11.22. The black line represents a plot of absorbance versus concentration for the 272 nm peak in the UV-vis spectrum of di-2-pyridylketone (DPK), which exhibits a molar absorptivity of 8,400 cm$^{-1}\cdot$M^{-1}. The blue line shows the absorbance versus concentration plot of the 1,690 cm^{-1} peak in the infrared spectrum of DPK, which has ε = 380 cm$^{-1}\cdot$M^{-1}. The higher molar absorptivity seen in the UV-vis peak gives a higher sensitivity. That is, a small change in concentration will yield a relatively high change in signal, so we can better distinguish between two similar concentrations. For instance, two solutions having concentrations of 0.10 and 0.16 mM DPK will show a difference in absorbance of only 0.011 in the IR spectrum but a difference of 0.25 (2,000% larger!) in the UV-vis spectrum. For a 0.011 AU change in signal in the UV-vis spectrum, we could distinguish between a 0.10 M solution and a 0.1025 M solution, a difference of 2.5 µM. DPK is not unique; it is a general rule that molar absorptivities associated with UV-vis peaks range from 10^2 to 10^6, whereas those for IR peaks are much smaller, typically ranging from 10 to 10^3.

Gases

A typical gas FTIR cell is depicted in Figure 11.23. The cell can be either purged with the sample gas, or one can evacuate the cell and then fill the cell using Schlenk techniques.[25] Because the sample cell is often subjected to vacuum, a sturdy cell window is required. Typically, gas sample cell windows are composed of CaF$_2$; however, NaCl windows are often used if a vacuum is not required to fill the cell. The optical window is held in place by a cowl that screws over the glass chamber, and the optical window is sealed from the atmosphere using an O-ring. Cells of varying length are available, but they are usually 5 to 20 cm because the low density of gases results in low molar concentrations and thus low absorbances if typical path lengths are used.

---

[25] See http://en.wikipedia.org/wiki/Wilhelm_Schlenk.

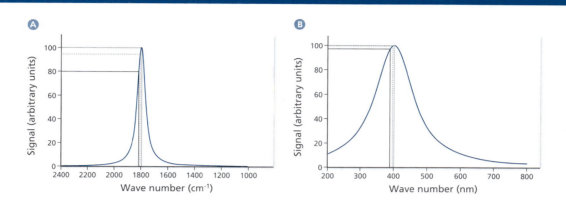

Figure 11.20 (A) Simulated IR absorbance peak at 1,800 cm^{-1} (5,556 nm) with a breadth of 4% (72 cm^{-1}). The blue lines indicate the peak position and signal. The red lines indicate the position and signal consistent with an 0.5% error in wave number (1,809 cm^{-1}). The green lines indicate the position and signal consistent with a 1.0% error in wave number (1,818 cm^{-1}).

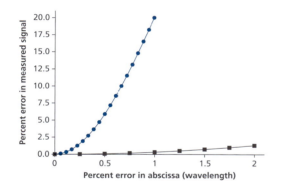

Figure 11.21 Plot of percent error in measured signal due to an error in the wavelength. The blue circles and line represent the narrow infrared peaks, and the black squares and line represent the broad UV-vis peaks.

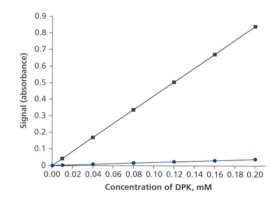

Figure 11.22 Plot of simulated signal vs. analyte concentration demonstrating the difference in UV-vis and FTIR sensitivity. The black circles and line represent UV- vis peaks; the blue squares and line represent infrared peaks.

Solution Infrared Spectroscopy

It is considerably more challenging and time consuming to conduct quantitative FTIR spectroscopy than it is to conduct quantitative UV-vis spectroscopy. However, it has value when one wants to monitor the kinetics of a reaction. To conduct quantitative spectroscopy, one needs to know the concentration of the sample, and this is most easily accomplished using a liquid FTIR cell. A typical liquid FTIR cell is more complicated in design than a gas sample cell. A schematic of a liquid FTIR cell is depicted in Figure 11.24. The optical windows are held apart by a Teflon spacer. The width of the spacer determines the path length of the cell. Typical path lengths are 1/100th to 1/10th of a mm. The optical windows are held in place by sandwiching the optical material between two stainless steel plates held together with through bolts. The sample is introduced through a stainless steel port that contains a Luer lock that allows a

TABLE 11.3: Common Infrared Optical Materials

MATERIAL	INFRARED TRANSPARENT RANGE* (cm^{-1})	PROPERTIES
KBr	40,000–400	Soft, inexpensive, water soluble, fogs
AgBr	22,000–286	Soft, inexpensive, water insoluble, darkens if exposed to ultraviolet light
AgCl	10,000–360	Soft, inexpensive, water insoluble, darkens if exposed to ultraviolet light
CaF2	77,000–1,110	Hard, relatively expensive, water insoluble, acid and base resistant, no fog, withstands high pressures
NaCl	40,000–625	Soft, inexpensive, water soluble, fogs
Diamond	25,000–33	Expensive, water insoluble, structurally strong
KRS-5 (thallium bromide-iodide)	20,000–250	Soft, good ATR material, acid insoluble, base soluble, toxic

Excerpted from Advanced Light Source (Berkley Labs), Copyright © 2012 ALS Infrared Beamlines with permission.

* **Because the optical bench of most IR spectrometers is exposed to air, the practical range of most spectrometers is limited to 4,000 to 300 cm^{-1} even if the optical material used has a larger range.**

Figure 11.23 Gas sample cell for FTIR spectroscopy. The optical window is held in place by a cowl and sealed with an O-ring.

syringe to be secured to the port. A hole in the gasket is aligned with the sample port and also with a hole in one of the optical windows. The solvated sample is injected through the sample port, floods the space between the two windows and the excess exits a second sample port. To change the optical path length, the entire cell must be disassembled and the Teflon spacer changed. Special care must be taken to align all of the components and one must not over tighten the backing plates; otherwise, the optical material will crack and be ruined.

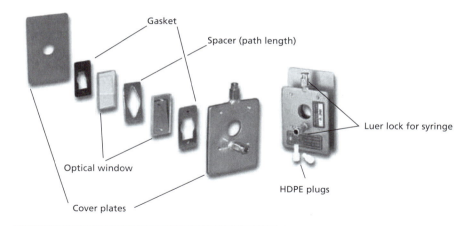

Figure 11.24 Liquid sample cell for FTIR spectroscopy.

The optical material is typically NaCl or KBr. Because both of these materials are water soluble, all solvents used must first be rigorously dried. This involves distilling the solvents over an appropriate drying agent[26] before they can be used.

The choice of solvent also requires special consideration. Because most solvents absorb in the IR region of the electromagnetic spectrum, solvents must be chosen so that they do not have strong absorption bands in the region of interest. Also, because the solvents absorb IR radiation, the path lengths are kept very small. In order to obtain an acceptable absorption by the analyte, the analyte must be very soluble in the solvent. Table 11.4 shows the useful range for various IR solvents.

If one only needs qualitative spectroscopic results, it suffices to simply place a solution of the sample between two appropriate salt plates and mount the "sandwich" in the beam path. In such uses, however, we must account for the spectral features of the solvent.

Neat Liquids

If your liquid is viscous enough and has a low vapor pressure, you can take an IR spectrum by simply smearing a thin layer of your analyte onto either a NaCl or KBr window and placing the window in the optical path of the spectrometer. This method is crude but very effective. One must still take care that the sample matrix does not contain appreciable amounts of water; otherwise, degradation of the salt window will result. Samples having relatively high vapor pressure can be measured using a sandwich cell or by placing a drop of the liquid on a sheet of polyethylene (e.g., plastic wrap) and folding the sheet over the sample to enclose it in the plastic. As with other window and solvent materials, we must be aware of the IR bands that arise from the polyethylene.

Solids

There is more than one way to take IR spectra of solids. The traditional way is to suspend a powder of the analyte in Nujol[27] and grind it in a mortar and pestle until a very thick suspension is obtained, which is referred to as a *Nujol mull*. The mull is then smeared onto either a NaCl or KBr window or a polyethylene sheet, and then placed in the optical path of the spectrometer. If you are taking routine IR spectra using Nujol mulls, you quickly learn to ignore the peaks from the Nujol.

---

[26] For a list of compatible drying agents see Gordon A.; Ford H. *The Chemist's Companion: A Handbook of Practical Data, Techniques and References*, Wiley Interscience: NY, **1972**, 445.

[27] Nujol is IR-grade mineral oil manufactured by Plough, CAS number 8012–95-1.

TABLE 11.4:   Useful Solvents for Infrared Spectroscopy*,†

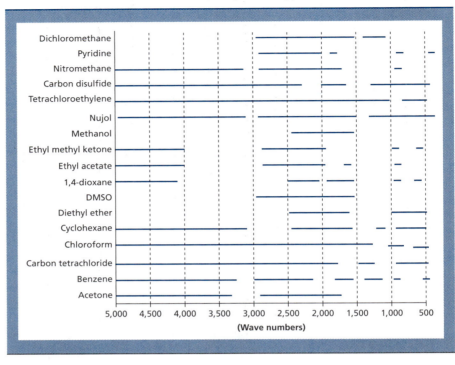

* **For a complete spectrum of each of these solvents, see Gordon A.; Ford H. The Chemist's Companion: A Handbook of Practical Data, Techniques and References, Wiley Interscience: New York, 1972, 170–176.**

† **The horizontal lines represent regions of low absorption by the solvent.**

A superior, although more time-consuming, method is to create a *KBr pellet.* Potassium bromide is transparent from 4,000 to 400 cm⁻¹ and is a very good optical material for IR spectroscopy. KBr also has another unique property. When subjected to high pressures, it will form an amorphous solid phase (glass). To form a KBr pellet, one places KBr and the solid analyte (10:1 to 100:1) in an agate mortar and pestle. The mixture is then ground into a homogeneous powder. The powder is placed in a press and subjected to high pressure, sometimes while evacuating with a mechanical pump, until a KBr window is formed (see Figure 11.25). The KBr pellet is then placed in the optical path of the FTIR spectrometer. Figure 11.1 was obtained as a KBr pellet. As you can see, the result is a very clean spectrum. Neither the Nujol mull nor the KBr pellet yield acceptable quantitative results because of the difficulty in controlling the path length ("b" in Beer's Law).

Attenuated Total Reflection

Traditionally, the biggest challenge in obtaining an FTIR spectrum was the sample preparation needed to introduce the sample into the optical path of the spectrometer. However, manufacturers of FTIR spectrometers have begun marketing attenuated total reflection (ATR) attachments for their commercial instruments that have greatly reduced the sample preparation needed to conduct IR spectroscopy.[28] In an ATR accessory, the IR radiation from the interferometer enters a crystal of high refractive index.

---

[28] Harrick, N. J. (1967). *Internal Reflection Spectroscopy.* John Wiley & Sons: NY, 1967; Perkin-Elmer. Technical Notes: FTIR Spectroscopy, http://shop.perkinelmer.com/content/technicalinfo/tch_ftiratr.pdf (accessed Aug, 2015).

Figure 11.25 Equipment for producing a KBr pellet.

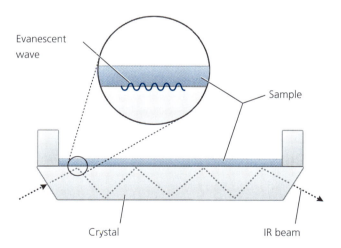

Figure 11.26 ATR. The circular cutaway view shows the incursion of the evanescent wave into the sample.

The liquid or solid sample is placed on the surface of the crystal. Because the difference in the refractive indices between the crystal and the sample are so large, the angle of reflection and the angle of refraction are equal; thus, the light undergoes total internal reflection (see Figure 11.26). At the boundary between the sample and the crystal, an evanescent wave is produced that has part of its amplitude extending into the sample.

Because the evanescent wave interacts only within a few microns of the sample, it is critical that the sample maintain good mechanical contact with the crystal. The IR radiation is allowed to bounce back and forth many times, thus increasing the path length

Crystal

Figure 11.27 Photograph of an ATR accessory. The sample (solid or liquid) is placed on the crystal face, and pressure is applied with the calipered piston.

of the IR radiation in the sample. The downside of using ATR is that the high-density ATR material attenuates the total IR beam, thus providing less signal at the detector and decreasing the net S/N ratio. Figure 11.27 is a photograph of an ATR accessory manufactured by Pike. The crystal face is the small dot visible in the center of the stainless steel ring seen in Figure 11.27. A pneumatic press holds the sample firmly against the face of the crystal (in this case a diamond) while taking a spectrum. The advent of the ATR accessory has greatly simplified sample preparation for FTIR spectroscopy. One can simply place the sample (either liquid or solid) on the crystal face, lower the pneumatic press and take a spectrum.

Problem 11.14: Define the term *evanescent wave*.

Problem 11.15: Under what conditions would an ATR accessory be ill suited? In other words, under what conditions might you get better data by using one of the traditional sample introductory modes (KBr pellet or Nujol mull)?

11.9 Further Reading

BOOKS

Cotton, F. A. *Chemical Applications of Group Theory,* 3rd ed. John Wiley & Sons: NY, **1990**.

Hamm, P.; Zanni, M. *Concepts and Methods of 2D Infrared Spectroscopy.* Cambridge University Press: New York, **2011**.

Harris, D. C.; Bertolucci, M. D. *Symmetry and Spectroscopy: An Introduction to Vibrational and Electronic Spectroscopy.* Dover: Mineola, NY, **1978**.

Larkin P. *Infrared and Raman Spectroscopy; Principles and Spectral Interpretation.* Elsevier: Waltham, MA, **2011**.

Smith B. C. *Fundamentals of Fourier Transform Infrared Spectroscopy,* 2nd ed. CRC Press/Taylor & Francis: Brooklyn, NY, **2011**.

Tsukerblat, B. S. *Group Theory in Chemistry and Spectroscopy: A Simple Guide to Advanced Usage.* Dover: Mineola, NY, **2006**.

JOURNALS

Gillie, J. K.; Hochlowski J.; Arbuckle-Keil, G. A. Infrared Spectroscopy. *Anal. Chem.*, **2000**, *72(12)*, 71–80.

Koenig, J. L.; Wang, S-Q.; Bhargava, R. FTIR Images: A New Technique Produces Images Worth a Thousand Spectra. *Anal. Chem.*, **2001**, *73(13)*, 360A–369A.

Nesbitt, D. J. High-resolution Infrared Spectroscopy of Weakly Bound Molecular Complexes. *Chem. Rev.*, 88(6), **1988**, 843–870.

11.10 Exercises

EXERCISE 11.1: Determine the number of vibrational modes for each of the following molecules.

(a) [structure]

(b) [structure]

(c) H—C≡C—F

(d) [structure]

EXERCISE 11.2: Convert each of the following frequencies into wave numbers.
(a) 5.01×10^{13} Hz
(b) 6.7×10^{13} Hz
(c) 1.5×10^{13} Hz
(d) 1.5×10^{14} Hz
(e) 8.97×10^{13} Hz
(f) 3.1×10^{13} Hz

EXERCISE 11.3: Convert each of the following wave numbers into units of hertz (s^{-1}) and then into wavelength in unit of micrometers.
(a) 2,970 cm^{-1}
(b) 43,810 cm^{-1}
(c) 1,699 cm^{-1}
(d) 2,120 cm^{-1}
(e) 756 cm^{-1}
(f) 1,597 cm^{-1}

EXERCISE 11.4: Convert each of the following wavelengths into wave numbers.
(a) 3.12 μm
(b) 2.97 μm
(c) 6.85 μm
(d) 3.76 μm
(e) 11.99 μm
(f) 5.71 μm

EXERCISE 11.5: Why are deviations from Beer's law more common in IR spectroscopy than in UV-vis spectroscopy?

EXERCISE 11.6: Determine the number of vibrational modes in each of the following molecules.
(a) SO_2
(b) NO_2^+
(c) NO_2^-
(d) SO_3^{-2}
(e) HCN
(f) CO
(g) H_2O
(h) CO_2
(i) NO_3^-

EXERCISE 11.7: In your own words, discuss the considerations one must make when selecting a suitable source for FTIR spectroscopy?

EXERCISE 11.8: Explain the relationship between the use of a Michelson interferometer and the time constant of the detector in FTIR spectroscopy.

EXERCISE 11.9: Explain why a thermal detector is not suitable for use in an FTIR instrument.

EXERCISE 11.10: Pyroelectric detectors are much less sensitive than photomultiplier tubes (Chapter 6). Why do we use pyroelectric detectors rather than PMTs in FTIR instruments?

EXERCISE 11.11: In your own words, describe how a pyroelectric detector works.

EXERCISE 11.12: Discuss some of the challenges and considerations associated with sample preparation in FTIR spectroscopy.

Advanced Exercises

EXERCISE 11.13: Construct a suitable circuit diagram for a pyroelectric detector. Be sure to include a means to drive an output reading device.

EXERCISE 11.14: The spectrum seen here was taken as a KBr pellet and belongs to a compound with the formula C_2H_6O.[29] Determine the identity of the compound.

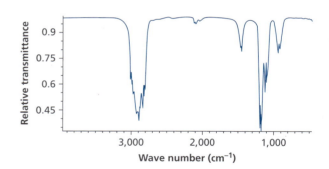

EXERCISE 11.15: Unless the solvent used is rigorously dry, the ketone group in the compound $[Pt(DPK)_2Cl_2][PF_6]_2$ undergoes hydrolysis to a stable enol (see Figure 11.1).[30] Suppose you wanted to study the kinetics of this hydrolysis

---

[29] Source: Spectral Database for Organic Compounds (SDBS).

[30] Sienerth K. D.; Granger R. M.; et al. Synthesis of Pt(dpk)Cl₄ and the Reversible Hydration to Pt(dpk-O-OH)Cl₃·H-phenCl: X-ray, Spectroscopic, and Electrochemical Characterization. *Inorg. Chem.*, **2004**, *43*, 72–78.

reaction using IR spectroscopy. Propose a suitable solvent for this study and discuss how you would introduce the sample into the spectrometer. Defend your recommendations.

EXERCISE 11.16: Review your answer to Exercise 11.5 and comment on why it is more common to use UV-vis spectroscopy for quantitative work than IR spectroscopy. You might also want to review the definition of sensitivity. See Figure 11.22.

EXERCISE 11.17: Determine the number of IR *active* vibrational modes in each of the following molecules. *Note*: You will have to use symmetry arguments to determine which vibrations result in a net change in the molecular dipole moment.

(a) SO_2

(b) NO_2^+

(c) NO_2^-

(d) SO_3^{-2}

(e) HCN

(f) CO

(g) H_2O

(h) CO_2

(i) NO_3^-

EXERCISE 11.18: Program a spreadsheet and plot the beat pattern created from the sum of each of the frequencies in Problem 11.2.

EXERCISE 11.19: Program a spreadsheet and perform a Fourier transform on the data set created by Exercise 11.18.

EXERCISE 11.20: The ketone peak in Figure 11.1 is centered at 1,714.8 cm⁻¹. What is the force constant for the C=O bond? Predict the wave number of the carbonyl peak if the carbon atom is a ¹⁴C isotope and the oxygen atom is an ¹⁷O isotope.

EXERCISE 11.21: Calculate a theoretical absorption frequency in wave numbers for a C–H bond, assuming a force constant of 4.89×10^2 N/m. How would the frequency change if you substituted deuterium for hydrogen?

EXERCISE 11.22: Compare and contrast TXRF to ATR-IR (see Chapter 10.8). What optical principle do both techniques exploit?

EXERCISE 11.23: Using Table 11.1 and Example 11.1 as a guide, analyze the following spectra.

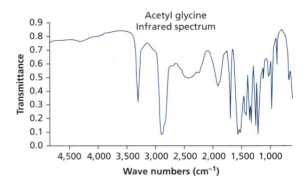

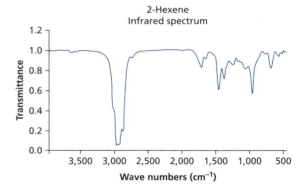

Raman Spectroscopy

Notch filter

mple   Microscope lens

α

Equilibrium position, Q = 0

zability L

PROFILE

Raman Applications in Art and Medicine

Researchers at the Massachusetts Institute of Technology have developed a blood sugar probe that can monitor blood glucose levels without having to draw blood. Because drawing blood can be painful, a method to monitor blood sugar spectroscopically without the use of a needle is of great benefit to the patient. The user

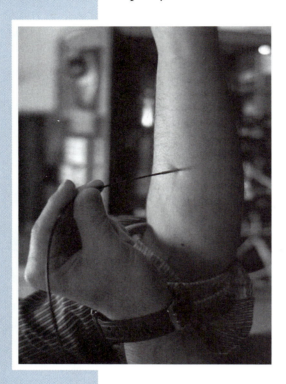
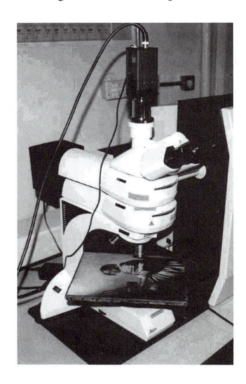

simply places the tip of a probe onto his or her skin. A fiber optic bundle directs laser light into the skin, and a second fiber optic bundle directs the Raman scattered light to a spectrometer (left).*

Raman spectroscopy can also be an important tool to help determine whether or not a piece of art is a forgery. The analyst can compare spectroscopic signatures from known pigments used by an artist or pigments known to be used during certain time periods in certain locations. The Raman fingerprint from pigments in paint allows the analyst to determine if the pigments in a piece of art are consistent with what would be expected.

The image here (right) shows a painting that was thought to be by the Dutch baroque artistst Johannes Vermeer (1632–1675). The painting is shown in a Raman microscope,[†] an instrument discussed later in this chapter. Several forgeries were thought to be in existence, warranting the analysis of the painting. In the end, it was thought to be consistent with work by Vermeer. In 2009, after the analysis, the painting was sold for $30 million. ■

---

* Barman, I.; et al. *Development of Robust Calibration Models Using Support Vector Machines for Spectroscopic Monitoring of Blood Glucose. Anal. Chem.,* **2010**, *82(14),* 6104–6114.

† Burgio, L.; Clark, R. J. H.; Sheldon, L.; Smith, G. D. *Pigment Identification by Spectroscopic Means: Evidence Consistent with the Attribution of the Painting Young Woman Seated at a Virginal to Vermeer. Anal. Chem.,* **2005**, *77(5),* 1261–1267.

12.1 Introduction

Raman spectroscopy is a form of vibrational spectroscopy and plays a role in the analyst's toolbox that is very similar to infrared (IR) spectroscopy. Raman spectroscopy has become an incredibly useful analytical technique for the identification of organic, inorganic, and biological samples. It is a nondestructive and relatively quick method for sample identification that often does not require significant, if any, sample preparation. Raman spectra can generally be measured from solids, liquids, and gases, as well as thin films and powders. Although Raman spectroscopy was once considered a technique that found only widespread use in research laboratories, new lower cost systems and new Raman-based techniques have elevated Raman spectroscopy to a level that challenges Fourier transform infrared (FTIR) analysis as one of the workhorse techniques of any well-equipped analytical laboratory. For example, low-cost, handheld Raman spectrometers are used outside the traditional laboratory in law enforcement to analyze gunpowder residue and to identify explosives and drugs. Back in the laboratory, research-grade instruments can differentiate gunpowder residue by the caliber of the ammunition used.[1] In addition to uses in law enforcement, Raman spectroscopy has been used to examine pigments, ceramics, glasses, minerals, gemstones, art, and antiquities,[2] as well as to analyze other chemical systems. Research-grade bench-top instruments are even more powerful and can be combined with an imaging system (Raman imaging) that allows for the determination of chemical composition across the surface of a sample, in some cases with submicrometer spatial resolution. An example of a Raman image[3] of human osteosarcoma (bone cancer) cells is shown in Figure 12.1. Using Raman imaging, one can investigate other small-scale features of a device or sample—for example, a silicon integrated circuit, mapping out areas of crystalline silicon, stressed silicon, and amorphous silicon.[4] Raman databases now have spectra for more than 15,000 compounds.

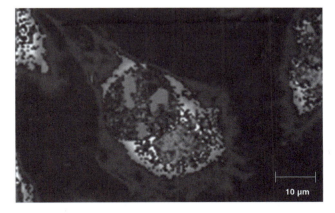

Figure 12.1 Raman image of human osteosarcoma (bone cancer) cells, showing the nuclei, nucleoli, membrane bound organelles, and the cell body.

[1] Bueno, J.; Sikirzhytski, V.; Lednev, I. K. Raman Spectroscopic Analysis of Gunshot Residue Offering Great Potential for Caliber Differentiation. *Anal. Chem.*, **2012**, *84*, 4334–4339.

[2] Vandenabeele, P.; Edwards, H. G. M.: Moens, L. A Decade of Raman Spectroscopy in Art and Archaeology. *Chem. Rev.*, **2006**, *107(3)*, 675–686.

[3] Ling, J.; Weitman, S. D.; Miller, M. A.; Moore, R. V.; Bovik, A. C. Direct Raman Imaging Techniques for Studying the Subcellular Distribution of a Drug. *Applied Optics*, **2002**, *41(28)*, 6006–6017.

[4] See, for example, the image gallery at http://www.horiba.com/.

Figure 12.2 Sir C. V. Raman.

Sir C. V. Raman, shown in Figure 12.2, discovered what is now known as the *Raman effect* in 1928.[5] By sending light from a strong mercury lamp[6] into benzene, Raman noticed that the scattered light that exited the sample contained a new wavelength, a wavelength different from the mercury source. This before-unseen scattered spectral profile is now known as a Raman spectrum. Raman used a spectroscope (a spectrometer where your eye serves as the detector) and filter to block the mercury source to determine that the wavelength of the scattered light was different than the wavelength of the mercury source. Raman went on to show that the effect occurs in many more molecules. He found that when he changed the wavelength of the excitation light that the difference in frequency between the excitation and the new light emitted remained the same for a given molecule. He also noted that the difference in frequency between the excitation and the new light emitted was unique for each molecule under investigation. This was a new type of scattering that had not been seen before, and Raman was awarded the Nobel Prize in Physics in 1930 for *"his work on the scattering of light and for the discovery of the effect named after him."*[7] Raman scattering can be thought of as a collision between a photon and the vibrational modes of the molecule, where an amount of vibrational energy of the molecule is lost from or gained by the scattered photon.

Rayleigh Scattering

A good example of light scattering can be seen at a concert during a laser light show. Without the fog or smoke in the air, you would normally not see the laser beam until it reflects off of a surface and those photons get to your eye. However, scattering through the fog allows the beam to become visible during the show. The laser beam becomes visible, as shown in Figure 12.3, despite the fact that the laser beam is not aimed at the audience. The photons you see are photons that have been scattered by particles in the air and are now directed toward and detected by your eye. The photons are scattered *in all directions*. In this example, the wavelength of the original source (laser) and the wavelength of the scattered light are *the same*. This is called *Rayleigh scattering*. In Rayleigh scattering, the difference in wavelength between the incident laser light and the scattered light is zero. This is quite different from Raman scattering, where the wavelength of the source and the wavelength (and therefore frequency and energy) of the Raman scattered light *are different*, although in practice, we will see that the two wavelengths are generally very close.

Figure 12.3 Rayleigh scattering of laser light. One can see the light as it scatters off particles and is redirected into your eye.

[5] Raman, C. V.; Krishnan, K. S. A New Type of Secondary Radiation. *Nature*, **1928**, *121*, 501–502.

[6] Raman also used filtered sunlight.

[7] From the Award Ceremony Speech presenting Sir Raman with the Nobel Prize; http://www.nobelprize.org/nobel_prizes/physics/laureates/1930/press.html.

12.2 Theory of Raman Scattering

Rayleigh[8] and Raman scattering can be discussed in terms of *conservation of energy*. In Rayleigh scattering, the incident light interacts with a material, and the scattered light leaves the material with the same amount of energy, and hence same wavelength, as it had before the interaction. Rayleigh scattering is an *elastic process* because the energy of the incident photons is the same as the scattered photons. In Raman scattering, incident light interacts with the material, and these incident photons *transfer energy to* or *gain energy from* the molecule due to changes in vibrational energy levels within the molecule. Raman scattering is an *inelastic process* because the energy of the scattered photon is not the same as the incident photons. The probability that an incident photon Rayleigh scatters is dramatically higher than the probability that an incident photon Raman scatters. When the photon loses energy to the material, it is termed *Stokes Raman scattering.*[9] When the photon gains energy from the material, it is called *anti-Stokes Raman scattering.* The terms *Stokes* and *anti-Stokes* are also used in fluorescence spectroscopy to denote emission that occurs with energies below and above the excitation source.

In order to describe Rayleigh and Raman scattering,[10] we return to our energy level view of interactions between light and matter that was developed in Chapter 4, on quantum mechanics and spectroscopy, and Chapter 8, on luminescence spectroscopy. On the left side of Figure 12.4 we see a generic energy level diagram for a molecule that has the transitions for Rayleigh, Stokes-Raman, and anti-Stokes Raman labeled. The initial state for Rayleigh and Stokes-Raman is the ground vibrational state of the molecule. The starting state for anti-Stokes Raman is an excited vibrational state of the system. As we saw in Chapter 2, these excited vibrational states have low populations at room temperature because thermal energy at room temperature is not high enough to populate the first excited state at an appreciable level.

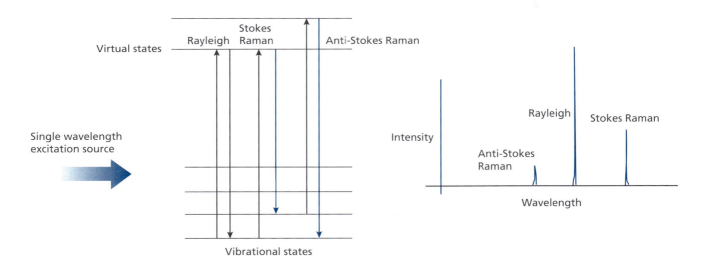

Figure 12.4 On the left, Rayleigh and Raman scattering (Stokes and anti-Stokes) as seen on an energy level diagram. An associated spectrum (to the right) is included. Note that the intensity of the Raman lines are greatly exaggerated compared to the Rayleigh line.

---

[8] Rayleigh scattering is named for Lord John Rayleigh (1849–1919), physicist and winner of 1904 Nobel Prize in Physics for isolating argon.

[9] Named for Sir George Stokes (1819–1903), mathematician and physicist.

[10] The reader interested in theory should read Long, D. A. *The Raman Effect: A Unified Treatment of the Theory of Raman Scattering by Molecules.* John Wiley & Sons, **2002**.

Because a Rayleigh scattered photon has the same energy (same frequency) as the incident photon, the length of the arrow in Figure 12.4 for Rayleigh scattering is the same pointing up (associated with absorption of the incident photon) as pointing down (associated with emission of the scattered photon emission). You will note that this is not the case for the Stokes and anti-Stokes Raman lines. The energy of the Stokes Raman scattered photon is less than the incident photon. The incident photon has transferred energy to the molecule in this case, leaving it in an elevated vibrational energy state, and a less energetic photon (shorter arrow) is the output. The energy of the anti-Stokes Raman scattered photon is more than the incident scattered photon. The incident photon has gained energy from the molecule in this case, leaving it in the ground vibrational energy state, and a more energetic photon (longer arrow) is the output.

A sketch of the intensity of the scattered light versus wavelength is shown in right side of Figure 12.4. This sketch shows that the anti-Stokes Raman scattered photons are the highest energy photons (shortest wavelength) and are shifted to a higher energy than the Rayleigh scattered light. The Stokes Raman line is shifted to a longer wavelength (lower energy) compared to the Rayleigh scattered light. In the sketch, the intensities (peak heights) of the Raman lines are exaggerated in comparison to the Rayleigh lines; Raman lines are extremely weak when compared to the Rayleigh scattered light.

In this sketch, we have plotted the Raman spectrum in terms of wavelength of the scattered photons because this may be the easiest way to think initially about Raman scattering. However, in practice, spectroscopists usually plot intensity versus Raman *shift* in units of wave numbers to be consistent with IR spectroscopy. One can convert from wavelength to wave number by taking the reciprocal of the wavelength in centimeters, as described in Chapter 11. The *Raman shift* is the difference between the wave number of the incident source radiation and the wave number of the Raman scattered light. One usually plots the Raman shift because the shift corresponds to the frequency of the vibrational mode in the molecules. For instance, in IR spectroscopy, a carbonyl stretch might absorb IR radiation at $1,700$ cm^{-1}. The Raman shift for that same carbonyl bond would also be $1,700$ cm^{-1}. This is because the incident light, with energy E_0, will be scattered to higher (anti-Stokes) or lower energies (Stokes) by coupling to that vibrational mode and will therefore occur at $E_0 + 1,700$ cm^{-1} (anti-Stokes) or $E_0 - 1,700$ cm^{-1} (Stokes). The shift, regardless of the value of the incident photon energy (E_0) will be $1,700$ cm^{-1}. That said, there are times when Raman spectra are plotted with wavelength of the scattered photons as opposed to Raman shift (wave number) of the scattered photons.

Converting between wavelength in nanometers to Raman shift in wave numbers is often necessary. This is accomplished using:

$$\Delta \tilde{v} = \left(\frac{1}{\lambda_0 [nm]} - \frac{1}{\lambda_1 [nm]} \right) \times 10^7 \, \text{nm cm}^{-1} \qquad \textbf{Eq. 12.1}$$

where $\Delta \tilde{v}$ is the Raman shift in cm^{-1}, λ_0 is the initial excitation (in nm), and λ_1 is the wavelength of the Raman scattered wavelength (in nm). 10^7 is the conversion factor from nanometers to centimeters. In a Raman spectrum, when the x axis is plotted as Raman shift in wave numbers, the Rayleigh scattered light would appear at 0 (no shift) because the shift in wave number is taken with respect to the incident laser wavelength.

Equation 12.2 is another useful conversion expression. If you know the Raman shift (in wave numbers) and the excitation wavelength (λ_0 in nanometers), you can calculate the wavelength of the Raman signal (λ_1 in nanometers) using:

$$\lambda_1 = \frac{1}{\dfrac{1}{\lambda_0} - \Delta \tilde{v} \times 10^{-7}} \qquad \textbf{Eq. 12.2}$$

Example 12.1

Using a diode laser with emission at 785 nm as a source, you measure a Stokes Raman signal at λ = 833.5 nm in a sample of gasoline. If you moved to a different laboratory where the source is an argon ion laser (488 nm), at what wavelength would you measure the Stokes Raman signal?

STRATEGY –

The Raman shift will be the same regardless of the excitation wavelength. If we measure the Raman shift using the diode laser, we can solve Equation 12.1 for λ_1 to determine the wavelength of the Raman scattered photon using the argon laser.

SOLUTION –

The Raman shift is:

$$\Delta \tilde{v} = \left(\frac{1}{\lambda_0[nm]} - \frac{1}{\lambda_1[nm]} \right) \times 10^7 \, nm \, cm^{-1} = \left(\frac{1}{785 \, nm} - \frac{1}{833.5 \, nm} \right) \times 10^7 \, nm \, cm^{-1}$$

$$= 741 \, cm^{-1}$$

We can use Equation 12.2 to solve for λ_1 because we now know the Raman shift and the new laser wavelength:

$$\lambda_1 = \frac{1}{\dfrac{1}{\lambda_0} - \Delta\tilde{v} \times 10^{-7}} = \frac{1}{\dfrac{1}{488 \, nm} - (741 \, cm^{-1} \times 10^{-7} \dfrac{1}{cm^{-1} nm})} = 506.3 \, nm$$

So λ_1 is 506.3 nm. This is the wavelength where this Stokes Raman signal would be found using the argon ion laser.

Problem 12.1: Starting with the experiment described in Example 12.1, suppose you move to a different Raman instrument and are measuring the same Stokes Raman signal using an excitation wavelength of 532 nm instead of 785 nm. Determine the wavelength of the Stokes-Raman scattered photons and the wavelength of the anti-Stokes Raman scattered photons.

We must point out one key distinction between the Raman energy level diagram in Figure 12.4 and other energy level diagrams seen in earlier chapters: the excited "state" in this energy level diagram is a *virtual state*. At this time, it is sufficient to understand that a virtual state is distinct from other quantum states we have studied in that the virtual state can exist at any energy between conventional quantum states. These virtual states exist for very short times, on the order of 10^{-13} seconds, which is 10^4 to 10^5 times shorter than the lifetimes of typical electronic excited states. It is understandable that the notion of states that can exist *at any energy* might be cause for alarm. After all, we have been careful to say that spectroscopy probes various quantum states of atoms and molecules and have stressed the resonance condition that must occur for a transition. Before becoming too concerned about the existence of states *at any energy level*, realize that (1) these virtual states exist only because of the presence of a strong electromagnetic field (the laser) and transitions between the vibrational states and virtual states occur with a very low probability (*a rule of thumb* is that 1 in 1,000,000 photons undergoes Raman scattering) and that (2) this interaction between light and matter takes place on a near-instantaneous timescale (hence, we use the term "scattering" rather than "absorption" and "emission" to describe it).

Although we have indicated in Figure 12.4 that Raman scattering is associated only with vibrational states, this is not exactly the case. The rotational states that were discussed in Chapter 2 can also play a role; however, they are not used in the standard implementations of Raman spectroscopy. Again, the rotational states are separated by energy differences much smaller than those for vibrational states.

You may be thinking that Raman scattering is similar to fluorescence because they both involve incident light and emitted light, so let us make a quick comparison. We should make a clear difference between the process of Raman scattering and the process of fluorescence described in Chapter 8. The distinction between the two has important practical implications for the technique. In fluorescence spectroscopy, the excitation source radiation had to match the difference between two electronic energy levels within the molecule. In Raman scattering, the incident light used does not need to match a particular vibrational energy level within the molecule. If a molecule is Raman active (spectroscopists use the term *Raman active* to mean that a particular vibration will result in Raman scattering), Raman scattering, although very weak, will occur for any excitation source wavelength. This is quite unlike fluorescence, where the

---

Example 12.2

In general, the metal–oxygen vibrational mode occurs between 150 and 450 cm^{-1}, depending on the particular molecule. In the molecule that you are interested in, the vibration occurs near 250 cm^{-1}. Calculate the absorption wavelength associated with this vibration for FTIR analysis. Where in the electromagnetic spectrum does this fall? By referring to Chapter 3 and the transparency range for optical materials, determine if KBr optics could be used at this wavelength for FTIR. If you use a 532 nm excitation source for a Raman experiment of this same molecule, where would the Raman line associated with this vibration fall in the electromagnetic spectrum?

SOLUTION –

We need to convert 250 cm^{-1} to wavelength in nanometers.

$$\lambda = \frac{1}{250 \, \text{cm}^{-1}} (1 \times 10^7 \, \frac{\text{nm}}{\text{cm}}) = 40,000 \, \text{nm}$$

The IR absorption at 250 cm^{-1} is equivalent to 40,000 nm. This is on the border between near and far infrared. In practical terms, the useful range of a typical FTIR is 4,000 to 400 cm^{-1}. KBr optics should not be used. The wavelength associated with the Raman line with a 532 nm excitation source is at:

$$\lambda_1 = \frac{1}{\frac{1}{\lambda_0} - \Delta \tilde{v} \times 10^{-7}} = \frac{1}{\frac{1}{532 \, \text{nm}} - \left(250 \, \text{cm}^{-1}\right) \times 10^{-7} \frac{1}{\text{cm}^{-1} \text{nm}}} = 539.2 \, \text{nm}$$

The Raman line occurs in the visible part of the electromagnetic spectrum. Note the two very different spectral regions that can be used here. Because a Raman experiment measures a "shift" in wave numbers, we could use any number of laser sources in a Raman experiment. As long as we use a sufficiently short wavelength source, so that the Raman shifted lines that fall in the ultraviolet-visible range, we can use simple quartz optics to collect and measure the scattered photons.

choice of wavelength for excitation source is critical. As we learned in Chapter 2, if the photon energy does not match the energy difference between two quantum states, a transition will not occur. Raman spectroscopy and fluorescence are different processes, but it is still very important for the experimentalist to choose the excitation source in Raman spectroscopy carefully depending on the sample under investigation. Unlike fluorescence experiments, the Raman spectroscopist actually looks to excite the sample *away from* any wavelengths where there might be an electronic absorption because the much more intense fluorescence associated with that absorption can overwhelm any Raman signal that might be present.

Raman spectroscopy is both a qualitative and quantitative method. A common qualitative approach has the analyst comparing, generally through software with the Raman instrument, the Raman spectrum (Raman shift frequency of peaks and relative peak sizes) to a wide range of Raman spectra of known molecules. By finding a match with spectra from known molecules, the analyst can determine where or not the sample contains molecules from the library. This qualitative technique is called fingerprinting. The analyst can also simply compare Raman peak frequencies from their measurement to tables[11] that list Raman band locations. Because the group frequencies discussed in Chapter 11 on IR absorption are the same vibrations (and vibrational frequencies) probed with Raman spectroscopy, we can use data such as that in Table 11.1 to make structural determinations. The Raman technique can also be quantitative. The Raman signal intensity is proportional to the concentration of the Raman active molecule. By comparing the Raman signal to a calibration curve,[12] the concentration of a Raman-active molecule can be determined.

Selection Rules

As we have seen in previous chapters, there are spectroscopic rules that govern whether a particular transition can occur. Roughly speaking, the *primary selection rule*[13] for a Raman transition is that the *molecular polarizability must change asymmetrically during the molecular vibration such that it changes from large to small or small to large*. This is a complicated statement, so let us break it down piece by piece, starting with polarizability. The *polarizability* of the molecule, α, tells us about how difficult or how easy it is for an externally applied electric field to change or distort the electron cloud (i.e., the bonding electrons) in an atom or molecule. The larger the distortion (the spatial separation of positive and negative charge in a molecule—that is, the induced electric dipole moment) caused by the application of an external electric field, the larger the polarizability of the molecule. A sketch of a diatomic molecule in the presence of a constant electric field is shown in Figure 12.5. The sketch shows how the electron cloud is shifted for different orientations of the molecule and the applied field. The gray portion shows the electron cloud without the applied field, and the blue portion shows the electron cloud with the applied field. We can relate the induced dipole moment, μ, to the external electric field, E, and the polarizability, α, with:

$$\mu_{induced} = \alpha E \qquad\qquad \textbf{Eq. 12.3}$$

The polarizability,[14] α, is a property of the molecule and has units of volume.

In general, the molecule with the larger number of electrons will have a larger polarizability compared to the molecule with fewer electrons. Larger molecules tend to have a larger polarizability compared to smaller molecules.

---

[11] Socrates, G. *Infrared and Raman Characteristic Group Frequencies: Tables and Charts*, 3rd ed. John Wiley & Sons: UK, **2004**.

[12] Sanford, C. L.; Mantooth, B. A.; Jones, B. T. Determination of Ethanol in Alcohol Samples Using a Modular Raman Spectrometer *J. Chem. Ed.* **2001**, 78(9), 1221.

[13] The selection rule we are discussing is also referred to as the gross selection rule. This is a classical (i.e., nonquantum mechanical) argument.

[14] The electric field and the dipole moment are both vector quantities. For anisotropic molecules, the polarizability is a tensor.

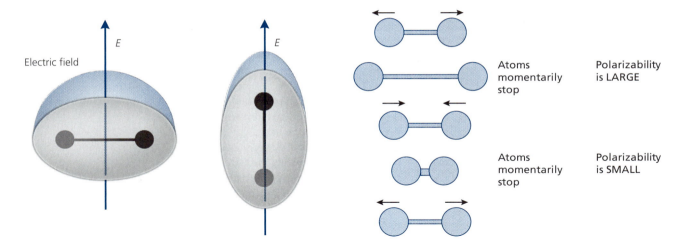

Figure 12.5 Sketch of diatomic molecule with associated electron cloud and the effect of an externally applied constant electric field for different orientations of molecule and field.

Figure 12.6 Diatomic molecule at different points in vibration with associated polarizability.

So far, the external electric field we have discussed has been constant. However, in Raman spectroscopy, the external electric field being applied is from a light source, and therefore the electric field varies sinusoidally. The sinusoidal electric field drives the molecular vibrations. In this context, the polarizability can be thought of as the sum of two components. One component describes the polarizability of the molecule in its relaxed equilibrium configuration (this is what we have discussed so far). The second component takes into account the effect on polarizability due to the *changing configuration of the molecule as it vibrates* due to external electric field.[15] We know from our initial statement of the selection rule that this second component of the polarizability is important.

We will now describe in more detail what it means for the polarizability to change *during the vibration*. In Figure 12.6, we show a generic diatomic molecule that is vibrating. The atoms stop at two configurations just before the atoms change their directions. Because we know that larger molecules have a larger polarizability, we can say that the polarizability is largest when the molecule is fully stretched (the largest) and that the polarizability is smallest when the molecule is fully compressed (the smallest). As the molecule vibrates, it moves through its equilibrium configuration (characterized by the equilibrium bond length). We can think of the vibration in terms of a coordinate that describes the molecule's departure from the equilibrium configuration. This coordinate is the distance between the atoms minus the equilibrium distance. This quantity is also called the vibrational coordinate, Q.

A plot of the polarizability versus the vibrational coordinate, shown in Figure 12.7, is a useful visual to see how the polarizability changes with vibrational coordinate. In this case, the polarizability has changed asymmetrically during the vibration. In other words, the polarizability has changed from large to small (or small to large if you prefer). The upcoming case study about CO_2 demonstrates this concept more fully. Putting this all together, using the rule that was first stated in this section, we can say that the vibration shown in Figure 12.6 is *Raman active*[16] (Raman scattering is allowed

---

[15] A more rigorous mathematical treatment of the induced dipole moment and the polarizability is given as a tutorial style homework problem at the end of this chapter.

[16] An excellent animation of how the polarizability changes with vibrational coordinate for several molecules is available at http://www.doitpoms.ac.uk/tlplib/raman/active_modes.php.

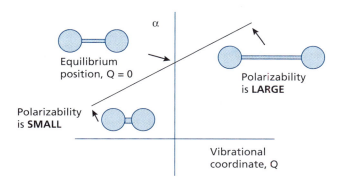

Figure 12.7 Sketch of polarizability versus vibrational coordinate for diatomic molecule vibration, showing asymmetric change in polarizability during vibration.

from the selection rule) because the polarizability changes asymmetrically during the vibration.

At this point, we should mention that FTIR and Raman analyses are often considered to be complementary vibrational techniques. In both cases, vibrational modes of a molecule are probed. We have pointed out the selection rule for Raman spectroscopy. In Chapter 11, you saw the selection rule for IR absorption. For a vibration to be IR active, the dipole moment of the molecule must change during the vibration. At this point, the distinction between changing polarizability and changing dipole moment needs some clarification. We can return to the diatomic molecule in Figure 12.5. For a homonuclear diatomic molecule (e.g., F_2, O_2 or N_2), the dipole moment is zero at all points during the vibration and therefore the dipole moment is not changing. For this reason, the vibration of the homonuclear diatomic molecule is not IR active. During the vibration of a heteronuclear diatomic molecule (e.g., HCl or CO), the dipole moment is changing. Therefore, the vibration is IR active.

Given the different selection rules for IR and Raman spectroscopic transitions, there can be vibrations that are IR active but not Raman active and vibrations that are Raman active but not IR active. The intermediate case also occurs where Raman and IR both show a particular vibration. These distinctions can help the spectroscopist determine the nature of different vibrations by comparing IR and Raman spectra.

Problem 12.2: Would you expect the vibration of HF to be Raman active?

A closer look at CO_2 will help to understand the selection rules for Raman activity and IR activity.

There are several useful *generalities* we can make concerning Raman and IR spectroscopy:

1. *Symmetric vibrations* lead to relatively strong Raman signals and no IR signals.

2. *Asymmetric vibrations* lead to much weaker Raman signals and are often quite strong in IR spectroscopy.

3. *Bending vibrational modes* lead to much weaker Raman signals and are often quite strong in IR spectroscopy.

4. A molecule can have both IR and Raman signals at the same frequency, although if the Raman signal is strong, the corresponding IR peak will be weak and vice versa.

For molecules that are centrosymmetric (symmetric about a central atom or point), generality 4 becomes a rule—centrosymmetric molecules cannot have both IR and Raman signals at the same frequency. A given vibrational mode is either Raman

Vibrations in the Linear Molecule Carbon Dioxide (CO_2): A Case Study

Like many small molecules, CO_2 has two stretching modes, symmetric and asymmetric. CO_2 also has two orthogonal asymmetric bending modes. We will discuss each so that we can decide whether each is Raman active or Raman inactive. The symmetric and asymmetric stretches, and the bending modes of CO_2 are represented in Figure 12.8. Recall that these stretching modes were first described in Chapter 11 in Example 11.3.

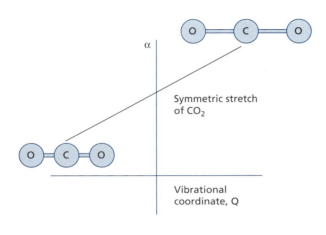

Figure 12.8 Symmetric, asymmetric, and bending modes in CO_2. The symmetric mode is Raman active, and the asymmetric mode and bending modes are not Raman active.

Considering the symmetric stretch, we need to determine if the polarizability of the molecule is different for the two different molecular configurations when the atoms pause as they change direction. For the symmetric vibration, at the point when the oxygen atoms are furthest from the carbon (and pause for an instant), the polarizbility is large. At the other point when the atoms stop, when the oxygen atoms are closest to the carbon, the polarizability is small. We can sketch this as we did previously. Because the polarizbility changes asymmetrically during the vibration, the vibration is Raman active.

Considering the asymmetric stretch of CO_2, we can think about the two points at which the molecule pauses during vibration and create the same sketch of polarizability versus vibrational coordinate. At the two points when the bonds are fully stretch and fully compressed, the polarizabilities are the same because the relative distances between the atoms are all the same. Therefore, using our Raman selection rule, the aysmmetric mode in CO_2 is not Raman active.

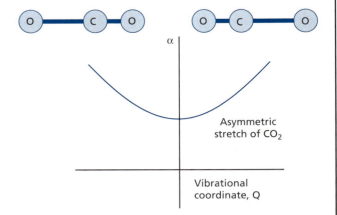

Considering the bending mode of CO_2, we can apply our argument about the relative distances between the atoms in the molecule during the bending. For this mode, we see that like the asymmetric mode, the relative distances are the same at the two points at which the molecule pauses. A sketch of the polarizability as the vibrational coordinate changes shows that the change in polarizability is symmetric. This mode would not be Raman active.

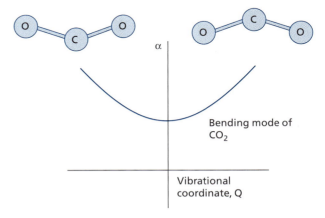

If needed, we can determine theoretical polarizibility changes using group theory and character tables.* A complete discussion of the chemical applications of group theory

---

*Cotton, F. A. Chemical Applications of Group Theory, 3rd ed.; John Wiley & Sons: NY, **1990**.

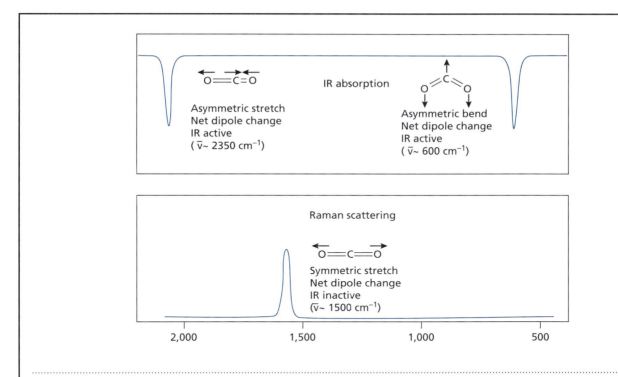

Figure 12.9 Sketch of IR and Raman spectrum for CO_2 along with associated vibrational mode.

is typically taught in an advanced inorganic course. Our empirical method of determining polarizability changes gives us a conceptual sense for why a vibrational mode will or will not be Raman active. This technique, however, is extremely difficult to implement for most cases.

As noted in Chapter 11, IR-active transitions require a change in the molecular dipole moment during the transition.

The symmetric stretch of CO_2 does not result in a change in the dipole moment, so it is not an active IR vibration. A sketch comparing IR absorption and Raman spectra is shown in Figure 12.9. The asymmetric stretch of CO_2 is IR active because the dipole moment of the molecule does change during the vibration.

active or IR active, but not both. For instance, carbon dioxide (a centrosymmetric molecule) shows a peak Raman spectra at around 1,300 cm^{-1}, and has two infrared peaks at 2,350 and 670 cm^{-1}. None of these peaks is seen in both Raman and IR spectra. This is known as the *rule of mutual exclusion*. Centrosymmetric molecules have an inversion center, a point in the center of the molecule across which each atom has a symmetrical twin.[17] Benzene and carbon dioxide, for example, are centrosymmetric molecules. We now apply our generalities to three different molecules—water, carbon tetrachloride, and molybdenum hexacarbonyl.

Problem 12.3: Argue that the vibrational modes for CCl_4 (Table 12.1) should all be Raman active by describing how the polarizability changes during the vibration.

So far, we have used our generalities to discuss Raman activity and IR activity. We can also model molecular vibrations in order to gain insight into Raman spectra. Two atoms bonded together can be modeled as two masses connected by a spring, representing the molecular bond. Although these atoms may be part of a larger molecule, we will not include effects from other parts of the molecule in our simple model. As discussed in

---

[17] An excellent animation of inversion symmetry is available at http://symmetry.otterbein.edu/tutorial/inversion. html#.

Example 12.3

Apply the four generalities we just mentioned to a qualitative description of the IR and Raman spectra of water.

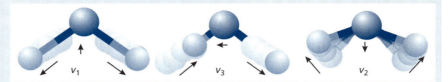

STRATEGY –

Consider the molecular structure of water as you think through the application of each generality.

SOLUTION –

H_2O has a bent structure (i.e., it is not a centrosymmetric molecule),

> Generality 1: The symmetric stretch, v_1, for water is Raman active. The symmetric stretch is also IR active.

> Generality 2: The asymmetric vibrational mode, v_3, of water will lead to strong IR absorptions. These same modes will lead to weak Raman scattering.

> Generality 3: The bending modes, v_2, of water will produce a strong IR absorption and weak Raman scattering.

> Generality 4: Water is not centrosymmetric; therefore, the rule of mutual exclusion does not apply. Strong peaks in the IR spectrum should also be seen as weak peaks in the Raman spectrum and vice versa.

TABLE 12.1: Vibrational Data for Carbon Tetrachloride (CCl_4)

MODE DESCRIPTION	FREQUENCY (CM⁻¹)	IS MODE RAMAN ACTIVE?	IS MODE IR ACTIVE?
Scissoring	218	Yes	No
Rock/umbrella	314	Yes	Yes
Symmetric stretch	459	Yes	No
Asymmetric stretch	762	Yes	Yes

Chapter 2, on quantum mechanics and spectroscopy, the angular frequency of oscillation for two masses, $\omega_{oscillator}$, with reduced mass, μ, and spring constant k can be found using:

$$\omega_{oscillator} = \sqrt{\frac{k}{\mu}} \quad \text{Or} \quad v_{oscillator} = \frac{1}{2\pi}\sqrt{\frac{k}{\mu}}$$

Eq. 12.4

or in wave numbers,

$$\tilde{v}_{oscillator} = \frac{1}{2\pi c}\sqrt{\frac{k}{\mu}}$$

The quantity $v_{oscillator}$ is the frequency of oscillation in Hz, $\tilde{v}_{oscillator}$ is the frequency of oscillation in wave numbers, and c is the speed of light. The spring constant describes

Raman Spectroscopy of the Tetrahedral Molecule Carbon Tetrachloride (CCl$_4$): A Case Study

Here we discuss the IR and Raman spectra of CCl$_4$. There are five atoms in a molecule in CCl$_4$, so using Equation 11.4 from Chapter 11 for determining the number of normal modes, m, of a molecule, gives m = 3N − 6 = 3(5) − 6 = 9 normal modes. Because of the four common Cl atoms, there are only four distinct normal mode vibrational frequencies. It turns out that all of the vibrational modes are at least partially Raman active. A representation of the Raman spectrum for CCl$_4$ is shown in Figure 12.10.

Generality 1: The symmetric stretch where the central carbon atom is fixed, and all four outer chlorine atoms move in and out is Raman active and likely to be strong. Also, following our generalities, the symmetric vibration will not be IR active because the dipole moment does not change during the vibration .

Generality 2: The asymmetric stretch, the vibration with the least symmetry, has a lower intensity Raman signal than the symmetrical stretch. The asymmetric stretch will have a strong IR signal.

Generality 3: The scissoring and rocking bending modes have a weak Raman signal in CCl$_4$ and a strong signal for IR.

Generality 4: CCl$_4$ is not centrosymmetric, so IR active bands can be Raman active.

An added complication arises with the asymmetric stretch. This complication is a quantum mechanical effect caused by the overlap between a vibrational band with a combination band. These two issues are beyond the scope of this text, although we will discuss them briefly. A combination band occurs when two excited vibrations couple, the result being a Raman line at the sum of the two vibrations. For example, in CCl$_4$, the 313.5 cm^{-1} and 458.7 cm^{-1} vibrations combine to yield a line at 772.2 cm^{-1}. It turns out that the asymmetric stretch vibration is near 776 cm^{-1}. The combined effect of the combination band being near a vibrational mode causes what is called a *Fermi resonance*. The end result of this quantum mechanical effect is that two bands appear, with an enhanced shift frequency—hence the 761.7 cm^{-1} and 790.4 cm^{-1} lines.

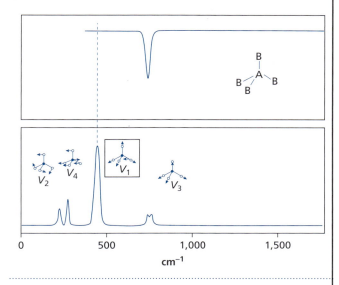

Figure 12.10 IR (top) and Raman (bottom) spectrum of CCl$_4$.

CCl$_4$ has the same tetrahedral symmetry as methane (CH$_4$). Thus, we can use Figures 12.3 and 12.4, which show the stretching modes and bending modes to determine the stretching and bending modes for methane.

the stiffness of the spring. Stiffer springs (larger spring constants) are associated with stronger bonds between the two atoms. Considering only the bond strength, we can conclude that the stronger the bond strength, the higher the frequency of vibration. Considering only the mass of the atoms, we see that higher mass atoms (leading to higher reduced mass) would lead to lower frequency oscillations. Example 12.5 demonstrates the use of this model in the analysis to Raman spectra.

Problem 12.4: Methane, CH$_4$, is a tetrahedral molecule, like CCl$_4$. Using CCl$_4$ as a model, how many Raman lines would you expect? How many IR lines? What type of vibration might have the strongest Raman line?

Problem 12.5: The symmetric stretch vibration in $^{13}C^{16}O_2$ is Raman active and occurs at 1,388.15 cm^{-1}. Estimate the Raman shift for $^{12}C^{16}O_2$.

Example 12.4

Molybdenum hexacarbonyl has the symmetric M–CO stretching mode shown in Figure 12.11. Would you expect this vibration to be IR active, Raman active, or both?

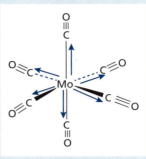

Figure 12.11 Symmetric Mo–CO stretching mode.

STRATEGY –

Apply the selection rule and our four generalities to this vibrational mode.

SOLUTION –

Symmetric vibrations do not produce a net oscillating dipole; therefore, it would not be IR active. However, symmetric stretches do produce a change in the polarizability of the molecule that is different at the two extremes of the vibration, and thus this vibrational mode is Raman active. As the CO units move toward the metal center, they move from a point of large polarizability (far from the metal) through their equilibrium point, then to a point of lower polarizability (closer to the metal center), and so on. Hence, the polarizability is changing asymmetrically during this vibrational mode. Generality 1 suggests this vibrational mode will be strong for Raman. In addition to the IR selection rule, generality 4 and the rule of mutual exclusion suggest this vibrational mode will not be IR active.

Example 12.5

In chloroform, $CHCl_3$, the vibration associated with the C–H stretch occurs at a Raman shift of 3,019 cm^{-1}. In deuterated chloroform, $CDCl_3$, how would you expect the C–D Raman stretch shift to compare to that of C–H? (Deuterated chloroform is used as a nuclear magnetic resonance solvent.)

We can model the C–H stretch as two masses connected by a spring (but not connected to the rest of the molecule). The frequency of oscillation is:

$$\tilde{v}_{oscillator} = \frac{1}{2\pi c}\sqrt{\frac{k}{\mu}}$$

where k is the spring constant and μ is the reduced mass.

We will assume that the presence of the heavy hydrogen does not alter the spring constant that describes the interaction between C–H or C–D and only affects the reduced mass. We can calculate the ratio of the frequencies for these two cases:

$$\frac{\tilde{v}_{oscillator,C-H}}{\tilde{v}_{oscillator,C-D}} = \frac{\frac{1}{2\pi c}\sqrt{\frac{k}{\mu_{C-H}}}}{\frac{1}{2\pi c}\sqrt{\frac{k}{\mu_{C-D}}}} = \frac{\sqrt{\frac{1}{\frac{m_C m_H}{m_C+m_H}}}}{\sqrt{\frac{1}{\frac{m_C m_D}{m_C+m_D}}}} = \sqrt{\frac{\frac{m_D}{m_C+m_D}}{\frac{m_H}{m_C+m_H}}} = \sqrt{\frac{\frac{2u}{14u}}{\frac{1u}{13u}}} = 1.36$$

Because the C–H stretch occurs at 3,019 cm⁻¹, we would expect the C–D stretch to occur at:

$$\tilde{v}_{oscillator,C-D} = \frac{\tilde{v}_{oscillator,C-H}}{1.36} = \frac{3019\,\text{cm}^{-1}}{1.36} = 2219\,\text{cm}^{-1}$$

Raman spectra of chloroform and deuterated chloroform from the literature[18] are shown here. The circles point out the C–H (spectrum A) and C–D frequencies (spectrum B). Our model shows good agreement with the measured Raman shift.

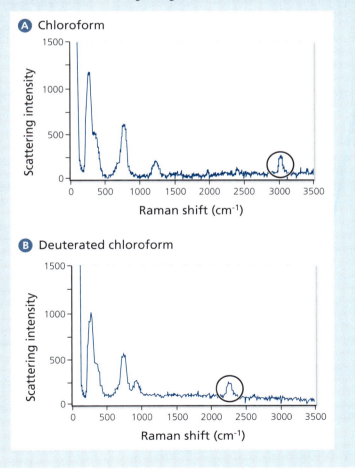

A Chloroform

B Deuterated chloroform

[18] DeGraff, B. A.; Hennip, M.; Jones, J. M.; Salter, C.; Schaertel, S. A. An Inexpensive Laser Raman Spectrometer Based on CCD Detection. *Chem. Educator,* **2002,** *7,* 15–18.

Problem 12.6: Which would you expect to have the largest Raman frequency, carbon–carbon single bonds, double bonds, or triple bonds?

In this section, we have discussed the role that molecular polarizability plays in Raman spectroscopy selection rules. This has provided us with a conceptual understanding for the requirements needed for a vibration to be Raman active. This conceptual understanding is important as we develop an intuition around molecular structure and spectroscopic activity. We should point out that in practice, commercial Raman instruments come with software libraries that can match Raman spectra with many common compounds. One can also use various resources, such as the *NIST Chemistry WebBook*,[19,20] to match Raman lines with compounds.

12.3 The Raman Spectrometer

Instrument Basics

The fundamental components of a dispersive Raman spectrometer are the radiant source, wavelength discriminator, filters, and the detector, as shown in Figure 12.12. The Raman instrument makes use of a variety of optical elements, so it is worth revisiting Chapter 3, on optics, as needed. In practice, the choices spectroscopists make in using Raman instruments are laser wavelength and power, gratings, filters, and the type of detector (and if needed, type of cooling) in the spectrometer. These choices will largely determine the type of molecular system that can be studied effectively.

We will first walk through the system in Figure 12.12 and then provide additional details on each aspect of the system. The laser excitation moves through a line filter. The line filter reduces the amount of laser light outside the narrow bandwidth of the filter. This filter "cleans up" any additional emission from the laser. The laser light is directed onto a dichroic notch filter (shown in the image as a notch filter). This optic reflects at laser wavelength and transmits other wavelengths. The laser light is the incident on a lens to focus the light onto the sample. The Raman light (and scattered laser light) will leave the sample and are collected by a lens—in this case the same lens that was used for focusing the laser light onto the sample. The collimated Raman signal is sent through the notch filter to dramatically reduce the amount of Rayleigh scattered laser light in the collected Raman signal so that it does not contaminate the measurement. The Raman light is then sent into a spectrometer that may house mirrors and one or more diffraction gratings. The grating spatially separates the various wavelengths of light, and these are directed onto an array detector.

The spectrometer/detector is generally not placed directly in line with the laser because of the amount of laser light that would enter the detector, even with laser-blocking filters. Depending on the nature of the Raman work to be done, a variety of choices must be made. We will now discuss details associated with each part of the Raman system.

Radiant Source

The radiant source for Raman spectroscopy is a laser. The laser source is used because of the high light source intensities needed to generate relatively weak Raman scattering. Common laser sources include the *argon ion laser* (wavelength 488 nm or 514 nm), *diode laser* (785 nm or 830 nm), the *Nd:YAG laser* (532 nm or 1064 nm), *frequency-doubled argon ion lasers* (244 nm), and *neon-copper lasers* (248 nm). Raman instruments generally make use of continuous wave lasers as opposed to pulsed lasers. The choice of laser source depends to some extent on the type of material you are studying. In general, the intensity of the Raman line depends linearly on the excitation power, although this can change if electronic absorption bands in the sample are close in

---

[19] *NIST Chemistry WebBook*; available at http://webbook.nist.gov/chemistry/. This resource contains a vast amount of data.

[20] Spectral Data Base (SDBS); available at http://riodb01.ibase.aist.go.jp/sdbs/cgi-bin/direct_frame_top.cgi.

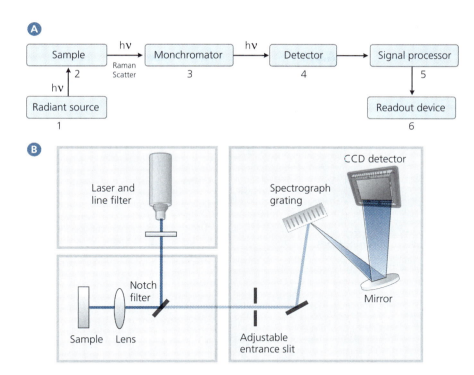

Figure 12.12 (A) Block diagram of Raman instrument. (B) Basic components of Raman spectrometer—radiant source (laser), wavelength discriminator (transmission grating), detector (CCD).

energy to the laser energy.[21] The intensity of the Raman signal, I_{Raman} (a function of the probability of a photon undergoing Raman scattering), is proportional to the wavelength of the excitation light, $\lambda_{exc,}$ according to:

$$I_{Raman} \propto \frac{1}{\lambda_{exc}^4}$$ **Eq. 12.5**

This equation tells us that shorter wavelengths of excitation light will lead to larger Raman signals.

It is important for the laser source to have a low bandwidth. That is, the laser should have a narrow range of wavelengths around the peak so that low–wave number (close to zero) Raman signals are not buried in the (more intense) Rayleigh scattered photons. Along with the choice of laser wavelength is the choice of laser power. Of course, it is critical to use a power that gives an adequate Raman signal. The Raman signal intensity will depend on the laser power used and the size of the focused laser spot. One should be careful; it is possible to damage/burn a sample during Raman analysis.

Earlier we stated that the laser should not have an emission wavelength near an electronic transition within the sample material. With shorter wavelengths, fluorescence becomes more likely, because the energy of the photon is more likely to match electronic transitions for the most common types of analytes, organic molecules. A laser wavelength that causes strong fluorescence is a problem if the fluorescence signal spectrally overlaps with the Raman signal. In this case, the fluorescence will easily overwhelm the relatively weak Raman signal. An example of the kind of spectrally broad signal largely composed of sample fluorescence is shown in Figure 12.13. Note that the signal using the 1,064 nm excitation has the most well resolved Raman spectrum.

It is important that you recognize the trade-off between different sources. A higher energy photon (shorter wavelength laser) will give you more intense Raman scattering.

---

[21] This is easily accomplished by first taking a ultraviolet-visible spectrum of the analyte in order to identify the wavelengths associated with the electronic transitions.

Example 12.6

You have access to two excitation sources with different wavelengths for a Raman investigation—514 nm (argon ion laser) and 785 nm (diode laser). Assume that (1) both sources have the same power and are focused into the sample in an identical fashion and (2) there is no sample absorption and therefore no potential fluorescence associated with either of the two potential excitation wavelengths. Which laser will result in the larger Raman signal?

The intensity of the Raman signal, I_{Raman} is:

$$I_{Raman} \propto \frac{1}{\lambda_{exc}^4}$$

If the only difference between the two lasers is the wavelength, we can take the ratio of Raman intensities:

$$\frac{I_{514nm}}{I_{785nm}} = \frac{(785)^4}{(514)^4} = 5.4$$

The intensity from the sample using the 514 nm laser is 5.4 times the intensity using the 785 nm laser. How much more powerful would you need to make the 785 nm source produce the same intensity Raman signal as the 514 nm source? If the 785 nm source was 5.4 times more powerful than the 514 nm source, then one would get the same intensity of Raman signal from these two different excitation wavelengths.

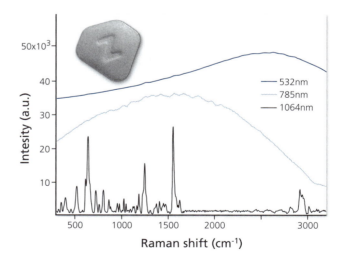

Figure 12.13 Raman spectra of the prescription drug Zantac with different excitation wavelengths demonstrating at higher excitation energy (532 nm, dark blue line) that fluorescence swamps the Raman signal. Excitation at 1,064 nm leads to the Raman spectrum with the least fluorescence.

However, if your laser source is close to an absorption band in the analyte, you risk a fluorescence event that will overwhelm your Raman signal. Ideally, you want to use the shortest possible wavelength possible without causing an electronic absorption in your analyte material. It may seem counterintuitive at first, but ultraviolet (UV)-induced Raman, using sources around 260 nm, can actually help mitigate the problem of overlapping Raman lines with fluorescence. UV excitation is likely to cause fluorescence, but the fluorescence will be at a much longer wavelength compared to the Raman signal, which is spectrally very close to the UV source wavelength. UV Raman may also benefit from increased signal levels because of resonance effects[22], a topic beyond the scope of this textbook.

Problem 12.7: A sample is known to fluoresce at excitation wavelengths below 600 nm. You have access to laser sources at 514 nm and 785 nm. Assume that both sources have the same power. What is the benefit of using the 514 nm source? What is the benefit of using the 785 nm source?

Wavelength Discrimination and Raman Spectrometer Resolution

In Raman spectroscopy, the wavelength discrimination is usually accomplished using a diffraction grating. High-resolution Raman instruments will make use of multiple gratings in order to achieve a resolution of a few cm^{-1} or less. The nature of the chemical system under investigation will determine the type of resolution needed.

In Chapter 3, on optics, we considered diffraction gratings. That discussion, in Section 3.3, is worth reviewing. We described how the groove spacing for gratings and the spectrometer focal length played an important role in wavelength resolution. The particular wavelength used will also affect the resolution a given grating and focal length will produce. For gratings in Raman instruments, groove spacing can range from roughly 1,200 lines/mm for a standard instrument to 6,000 lines/mm for a high-resolution instrument. Two types of gratings are used, conventional reflective ruled gratings and the volume phase holographic gratings. These transmission gratings are used because they allow less stray light to reach the detector. Because Raman signals are so weak, it is important to reduce stray light levels on the detector, as even a very small percentage of the excitation light would overwhelm the Raman signals.

As we saw in Chapter 3, the resolution of the instrument is limited by the number of lines/length of the grating and the distance from the grating to the detector (the focal length of the spectrometer). The more lines per length of the grating, for a given focal length, the higher the resolution will be. The longer the focal length of the spectrometer, the larger the path length will be for the dispersive element to provide spatial separation of the individual wavelengths on the array detector. For very small portable spectrometers (e.g., those that can be carried outside the laboratory), it is difficult to design the spectrometer with a long focal length. Therefore, the higher resolution gratings are needed.

Wavelength resolution is an important consideration for any spectrometer. Figure 12.14 demonstrates the role of wavelength resolution in Raman spectroscopy. The Raman peak for the symmetric stretch of CCl_4 is shown at 459 cm^{-1}. The bottom plot shows an instrument with a resolution of around 1 cm^{-1}, and the top plot shows a Raman spectrum where the instrument resolution is around 4 cm^{-1}. We see the peak at 460 cm^{-1} is actually three (possibly four) peaks with the higher resolution instrument.

Filters

Filters are often placed in front of the spectrometer to block the Rayleigh scattered laser excitation. It is important for these filters to absorb or reflect the laser light but allow the Raman signal light that may be very close in frequency to the laser light to pass into the spectrometer. Thin film *notch* or *edge filters* are commonly used for this application (Figure 12.15). Notch filters absorb strongly at a narrow wavelength range, whereas edge filters absorb strongly at wavelengths below the edge wavelength. Notch filters and edge filters can be purchased with a variety of wavelength-blocking ranges.

---

[22] When the source wavelength is near electronic transitions, the technique is called resonance Raman spectroscopy; see, for example, Strommen, D. P.; Nakamoto, K. Resonance Raman Spectroscopy. *J. Chem. Ed.,* **1977**, *54(8)*, 474.

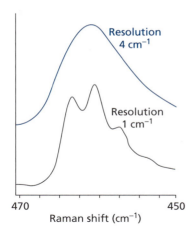

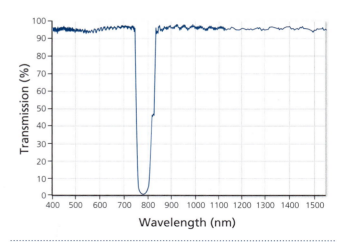

Figure 12.14 Example of two Raman spectra of CCl$_4$: (top) instrument resolution of 5 cm^{-1} and (bottom) instrument resolution of 1 cm^{-1}.

Figure 12.15 Transmission spectrum for a dichroic notch filter. Reflectivity at 785 nm is ~98%.

Dichroic notch filters are also used in Raman systems, as we saw in Figure 12.12. A transmission spectrum for a dichroic notch filter is shown in Figure 12.16. For this filter, the reflectivity at 785 nm is approximately 98%, making it a mirror for light at 785 nm. This component would be used to reflect the 785 nm laser light to the sample and would pass wavelengths between 400 and 720 nm and between 830 and 1600 nm.

Problem 12.8: The transmission spectrum shown in Figure 12.16 is for a dichroic notch filter. You found this from a data sheet in your laboratory and need to know if this optic might be appropriate for your homemade Raman system, which is configured like that in Figure 12.12. What is the approximate useful range wavelength, in nm (or wavelengths), for this optic?

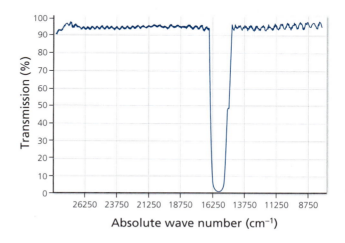

Figure 12.16 Transmission spectrum for a dichroic notch filter.

Detectors

Early Raman spectrometers operated in a scanning mode using a motorized mono-chromator to direct selected bandwidths of radiation at an exit slit. The detector was generally a *photomultiplier tube (PMT)*. Modern Raman spectrometers generally use

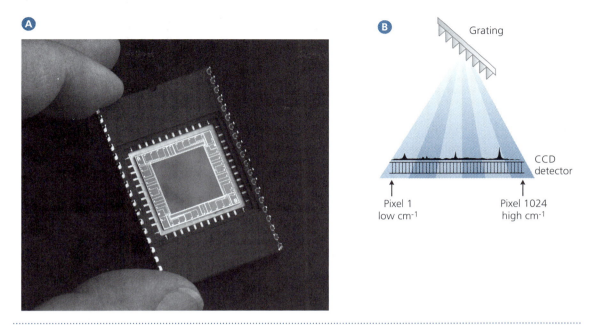

Figure 12.17 (A) Photograph of Si CCD detector array. (B) Sketch of CCD detector and reflective grating.

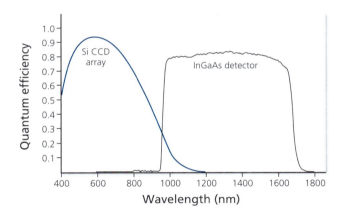

Figure 12.18 Quantum efficiency as a function of wavelength for silicon CCD arrays compared to InGaAs detectors.

array detectors. We first introduced array detectors in Chapter 6 when we discussed UV-visible spectroscopy. Light that has been spectrally separated by a grating is sent onto the array detector where each wavelength is measured simultaneously. One benefit of the array approach over a scanning single wavelength detection PMT approach is that the measurement of an entire spectrum can be accomplished in only a few seconds.

The most common detector used in modern Raman spectrometers is the silicon *charge coupled device (CCD) detector*. A CCD detector is shown in Figure 12.17. Panel (B) depicts the light diffracting off the grating and incident on the different pixels of the CCD detector. CCD detectors can be chosen with different numbers of pixels and pixel sizes. Silicon CCD arrays are sensitive to electromagnetic radiation between ~400 and 1,000 nm. The sensitivity, described here by the quantum efficiency, of the two types of detectors as a function of wavelength is shown in Figure 12.18. Cooling the CCD will reduce the thermal noise from the detector and therefore increase the signal/noise (see Chapter 5 for a review of *thermal noise*). Refer to Chapter 6 for more background on CCD detectors.

For applications between ~1,000 nm and 1,650 nm, InGaAs detector arrays are used. In Figure 12.18, we see that the InGaAs detector has large quantum efficiency from 950 to 1650 nm. The InGaAs or Ge photodiodes are also used for Fourier transform Raman spectroscopy, a topic to be discussed in Section 12.4. Because the Raman signal is measured as the shift of wavelength from the source, the choice of which detector you choose is dependent on the source you choose.

Problem 12.9: In your new job as a forensic chemist, you have been asked to study gunpowder residue. You have access to a 1,064 nm Nd:YAG laser as the excitation source for Raman spectroscopy. The Raman shift ($1,287$ cm^{-1}) you are most interested in looking for is related to the NO_2 symmetric stretching mode associated with propellant. You have access to two detectors for the experiment, a silicon CCD detector and an InGaAs detector. Based on the spectral response from these two detectors, which would be the best to use?

COMPARE AND CONTRAST

A Side-by-Side Evaluation of Fourier Transform Infrared and Raman Spectroscopy

Raman spectroscopy is often used as a complementary technique with FTIR. There are benefits and drawbacks to each technique. We provide a list of some of these in Table 12.2 below.

Table 12.2: Side-by-Side Comparison of Raman and Fourier Transform Infrared Spectroscopy

RAMAN	FTIR
Symmetric stretches are strong. Asymmetric stretches and bends are weak.	Asymmetric stretches and bends are strong. Symmetric stretches are not seen.
Standard sample holders can be used, such as glass, plastic, and so on.	IR transparent materials (salts) are generally used for sample holders. However, ATR eliminates the need for a sample holder in many cases.
Samples can be in water, and water can be the solvent.	FTIR in water is a challenge due to IR absorption, and some IR optics are water soluble.
Essentially no preparation of sample needed	Sample preparation can be challenging. However, IR variants such as ATR-FTIR and diffuse reflectance infrared FTIR spectroscopy (DRIFTS) allow for little to no sample prep.
Optics for systems can be glass or quartz.	Custom optics are required. KBr is most commonly used. KBr is hygroscopic and must be kept in a dry atmosphere.
Can be used to image small scale structures by micro-Raman	Tunable ATR-FTIR can be used to probe sample composition as a function of depth.
Can measure vibrations of homonuclear species (e.g., N_2, O_2, C_{60}, P_4)	More readily lends itself to application of group theory for the identification of molecular isomers (i.e., cis vs. trans).
Requires a laser source	Requires broadband IR source
Raman selection rules require changing polarizability.	IR selection rules require changing dipole moment.
Can use high sensitivity UV-vis detectors	Uses thermal or piezoelectric IR detectors
Wavelength discrimination is accomplished using a grating.	Wavelength discrimination is accomplished using a Michelson interferometer.

ATR = attenuated total reflection; FTIR = Fourier transform infrared; IR = infrared; UV-vis = ultraviolet-visible.

Handheld Raman Analyzers

Several companies now manufacture commercial handheld, battery-powered Raman spectrometers. Figure 12.19 shows two examples. These devices have shown promise for a variety of situations where one is performing qualitative analysis to determine whether a known compound is present. Raman handheld instruments are particularly useful when this analysis needs to be done quickly and outside of the laboratory (e.g., with detection of explosives).[23] The devices can also be used to identify unknown compounds and can be used with solids (including powders) and liquids. The measured Raman spectrum is then analyzed by onboard software, and measured spectra are compared against a sample library. The results are displayed on a liquid crystal display. As an example of the parameters of a handheld Raman instrument, the Ocean Optics product weighs only 3 pounds, uses a 785 nm diode laser as the excitation source, can measure Raman shifts of 200 cm^{-1} to 2,400 cm^{-1}, and has a spectral resolution of approximately 10 cm^{-1}.

Figure 12.19 Handheld Raman instruments (IDRaman mini and FirstGuard).

Fiber Optic Probes

For some instrumental techniques, such as UV-vis spectroscopy, the sample is generally placed in a cuvette for measurement. Instrumental approaches with increased sample flexibility can give the analyst more options. The development of fiber optic probes for Raman spectroscopy has made for more flexible sample options, beyond the cuvette, for conducting a Raman experiment. In Figure 12.21, you see an example of a probe being used to analyze a solvent within a glass bottle. In this case, the analyst does not need to pour the sample out of the container to make the Raman measurement.

[23] Moore, D. S.; Scharff, R. J. Review: Portable Raman Explosives Detection. *Anal. Chem.*, **2009**, *393(6–7)*, 1571–1578.

Identifying illegal drug components in test samples is an important application of handheld Raman instruments. Law enforcement officers who come across an array of white powders can use the handheld Raman spectrometer to determine if the powder is heroin, cocaine, or something else. Figure 12.20 shows Raman spectra of four white powder samples—heroin, methamphetamine, and two forms of cocaine. The handheld Raman spectrometers can clearly distinguish these four compounds. ■

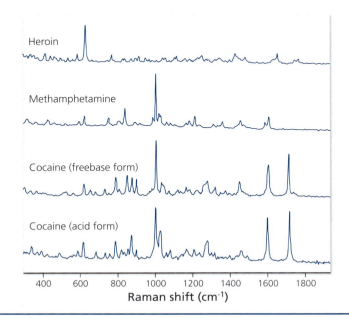

Figure 12.20 Raman spectrum showing differences between heroin, methamphetamine, and two forms of cocaine.

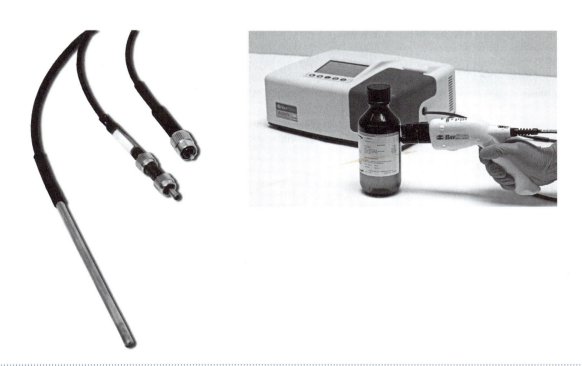

Figure 12.21 Fiber optic probe for Raman spectroscopy. Probe can be used as part of Raman system to determine identity of solution in the container without interference signal from the container.

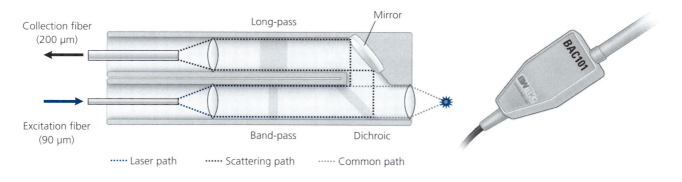

Figure 12.22 Fiber optic probe for Raman spectroscopy.

The interior configuration of a fiber optic probe for Raman spectroscopy is shown in Figure 12.22. Within the fiber optic bundle, one fiber is connected to the laser excitation source. This cable serves to direct the laser light onto the sample being analyzed. A second fiber in the probe directs the scattered radiation back to the spectrometer for analysis. The laser excitation beam moves through the optical fiber to a collimating optic. The laser light then moves through a band-pass filter. The band-pass filter in the excitation path reduces the wavelength range of the laser source to a very narrow band around the peak laser wavelength. A dichroic beam splitter on the excitation pathway passes the excitation wavelength to a lens for focusing on the sample but reflects the Raman signal wavelengths that are scattered from the sample. A second mirror directs the Raman signal to a long-pass filter.[24] The long-pass filter on the collection side blocks laser excitation from contaminating the Raman signal. Collimating optics focus the Raman signal into the collection fiber. The Raman signal is then directed to optional wavelength discriminators and then onto the detector.

12.4   Additional Raman-based Techniques

Although standard Raman spectroscopy is a very useful tool for the analyst, there are additional Raman-based approaches that can further benefit the analyst. For example, if we want to identify highly symmetric vibrations, we can use *polarized Raman* spectroscopy. If we need to identify different molecular structures across a very small sample area, we can use *Raman imaging*. If we need to investigate a Raman signal that is too weak to measure, we can use *surface-enhanced Raman spectroscopy (SERS)*. A survey of Raman-related techniques is provided below; however, this list is far from complete.

Raman Imaging

Raman imaging systems make use of focusing optics and sample stages that allow the user to measure Raman spectra as a function of sample position. These are also referred to as Raman microscopes. You may remember from the introduction to this chapter that Raman microscopes have been used to analyze works of art and their pigments for evidence of forgery. Depending on the system, spatial resolution may be in the several micron range to systems with resolution in the tens of nanometers. Raman imaging can probe on a surface and below the surface. The technique has been used, for example, to image biological cells and the distribution of chemical species in a solid matrix. An example of a Raman image comparing noncancerous breast tissue with cancerous breast tissue is presented in Figure 12.23. Although hard to see in gray scale, normally the spectra are color coded to a colored image allowing the analyst to identify the chemical composition of specific regions within the cell.

---

[24] See Figure 3.29 for a review of long- and short-pass filters.

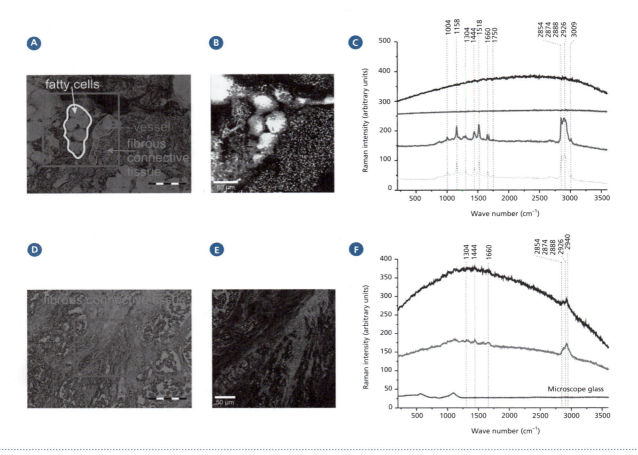

Figure 12.23 A Raman image and spectra of noncancerous and cancerous breast tissue. Typically the image and the spectra are color coded so the reviewer can identify the chemical composition of specific regions in the cell. Panels (A), (B), and (C) contain noncancerous breast tissue: (A) Microscope image, (B) Raman image (400 × 400 μm) from the region marked in (A), (C) Raman spectra. Panels (D), (E), and (F) contain cancerous breast tissue: (D) Microscope image, (E) Raman image (300 × 300 μm) from the region marked in (D), (F) Raman spectra.

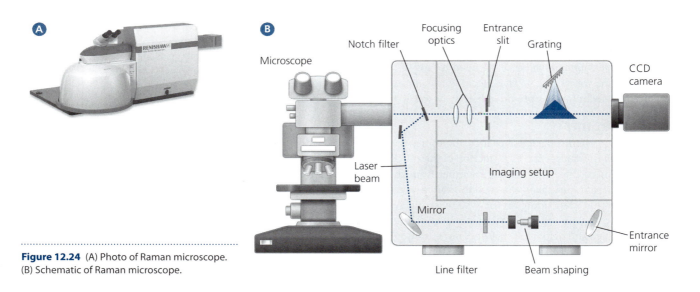

Figure 12.24 (A) Photo of Raman microscope. (B) Schematic of Raman microscope.

A commercial Raman microscope is shown in Figure 12.24. Excitation laser light enters at the bottom right and is directed by the entrance mirror to beam shaping optics and line filters. Additional mirrors direct the laser light to a notch filter and then into the microscope. The microscope focuses the laser light onto the sample. The sample

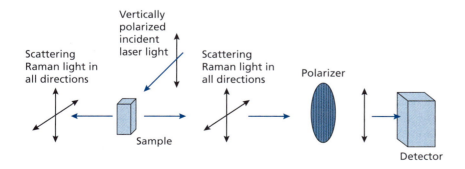

stage can be adjusted as the analyst looks into the microscope in order to choose the area of the sample to investigate. The Raman signal passes back through the microscope and the notch filter and then through focusing optics and an entrance slit. The light is directed onto a grating for wavelength selection and sent to the CCD camera, where an image of the sample is created for the analyst.

Polarized Raman Spectroscopy

Polarized Raman spectroscopy allows the spectroscopist to determine if a particular Raman signal corresponds to a highly symmetric vibration. A rough schematic for polarized Raman scattered light in its simplest form is shown in Figure 12.25. The difference between conventional Raman and polarized Raman scattered light is that the Raman signal is sent through a polarizer, which can be rotated so that the intensity of the Raman signal can be monitored as a function of polarization. The polarizer, described in Chapter 3 on optics, is a component that allows for a particular orientation of the electric field of light to be transmitted. Varying the angle of the polarizer allows the analyst to change the particular polarization state of the Raman scattered light that is being measured. This effect is described by the depolarization ratio, ρ, which is the ratio of the intensity of Raman scattered light that is polarized perpendicular to the excitation light $(I_\perp)$ and the intensity of Raman scattered light that is polarized parallel to the excitation light $(I_\parallel)$.

$$\rho = \frac{I_\perp}{I_\parallel}$$

Raman scattering from totally symmetric vibrations has a depolarization ratio of near zero. A depolarization ratio of zero means that no Raman light is polarized perpendicular to the laser source. This effect is seen in Figure 12.26 with Raman spectra of ClO_4^- ions, (A) shows the spectrum with the output polarizer set to the same direction as the input polarizer $(I_\parallel)$, (B) shows the output polarizer 90 degrees with respect to the input $(I_\perp)$. The totally symmetric vibrational mode of ClO_4^- has a depolarization ratio of close to zero because $I_\perp$ is essentially zero. We can conclude that the 900 cm^{-1} is a highly symmetric vibration. The depolarization ratio has a variety of uses and interpretations.[25]

The implementation of *polarized Raman* scattered light is slightly more complicated than our first sketch. A schematic for a commercial polarized Raman spectroscopy module is shown in Figure 12.27. This module can be added to a traditional, unpolarized Raman system. The incident laser excitation is sent through a depolarizer. A depolarizer takes linearly polarized input light (most laser sources are polarized) and

Figure 12.25 Schematic for simplest version of polarized Raman spectroscopy.

[25] Strommen, D. P. Specific Values of the Depolarization Ratio in Raman Spectroscopy: Their Origins and Significance. *J. Chem. Ed.,* **1992,** *69(10),* 803.

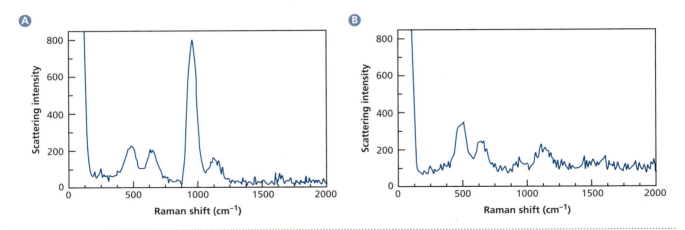

Figure 12.26 Raman spectra of ClO_4^- ions (saturated solution of $NaClO_4$) with different polarizations of excitation: (A) vertically polarized excitation with vertically oriented Raman detection and $I_{\parallel}$ (B) horizontally polarized excitation with vertically oriented Raman detection, $I_{\perp}$.

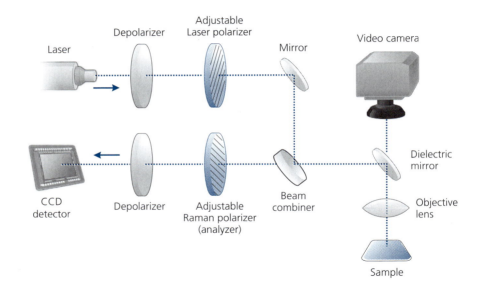

Figure 12.27 Schematic for a polarized Raman instrument.

transmits light that is roughly unpolarized. Remember that unpolarized means that there are components of the electric field associated with the light wave in all directions. This unpolarized light is then sent into a linear polarizer, so that any linear polarization of the excitation can be chosen by the experimenter. Without the depolarizer, changing the linear polarizer would dramatically reduce intensity for polarizations distinct from the initial laser polarization. The excitation light is reflected off a beam splitter and focused onto the sample. Raman scattered light from the sample is collected by the focusing lens and is transmitted through the beam splitter and then onto an adjustable linear polarizer to select for a particular orientation of Raman polarized signal light. After the polarizer, the Raman signal light moves through a depolarizer to reduce polarization-dependent effects in the detection system. If a reflective grating is used for wavelength discrimination, the grating reflectivity is dependent on the polarization of the light. This means that the depolarization ratio will be altered by the grating, not simply by the sample, causing significant problems. The depolarizer mitigates this issue. This module allows the experimenter to adjust the input excitation polarization and the polarization of the Raman scattered light.

If a depolarizer is not used for the Raman signal, another approach is to use an optical element, the half-wave plate, which rotates the initial polarization of the excitation source. This way the sample can be excited with two different polarizations (using the half-wave plate and not using it), but the same polarization state is being measured so polarization-dependent reflectivity is not an issue.

Problem 12.10: Would you expect any of the vibrational modes of CCl_4 to have a Raman depolarization ratio of close to zero? If so, which mode or modes? Recall that Table 12.1 describes the vibrational modes of CCl_4.

Fourier Transform–Raman Spectroscopy

Fourier transform–*Raman* systems measure the traditional Raman scattered light, but the wavelength resolution is accomplished with the same type of Michelson interferometer system used in the FTIR instrument discussed in Chapter 10. Commercial systems are available that combine FTIR and Fourier transform–Raman techniques.

A schematic for a Fourier transform–Raman system is shown in Figure 12.28. The excitation light is incident on the sample and the Raman scattered light leaves the sample and enters the Michelson interferometer. The moving mirror, such as in FTIR, provides the wavelength resolution of the instrument. Note that no gratings are used in this approach. Commercial Fourier transform–Raman systems generally use 1064 nm excitation, which dramatically reduces any fluorescence, as we discussed previously. The long wavelength is used because the distance the mirrors must be moved to achieve a given resolution is proportional to the wavelength being measured. This means that the distance control of the mirrors needed for UV light would be much less than that for near-IR light. It is very difficult to achieve the level of mirror control needed for the UV light. The longer excitation wavelength (1,064 nm) can be a drawback compared to shorter wavelengths because of the relationship between Raman intensity and excitation wavelength. Fourier transform–Raman excitation sources tend to operate at higher powers than dispersive Raman, so sample damage is possible. Another advantage of Fourier transform–Raman compared to dispersive instruments is that the interferometer approach can provide higher spectral resolution.

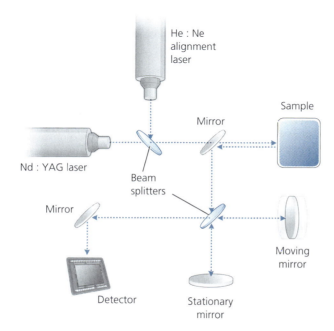

Figure 12.28 Schematic for Fourier transform–Raman spectroscopy.

The techniques we have discussed so far have required the user to have the sample close to a detector and light source in order to measure a Raman spectrum. Researchers have been developing techniques that would allow for the compound to be many meters away from the detector and excitation source. This would be of great value, especially if effects from the sample container could be eliminated. This might prove useful to screen fluids at airport security or probes on motor vehicles designed to detect ice on the road. To this end, researchers are developing stand-off spatial offset Raman spectroscopy. The experimental approach is described in Figure 12.30. A short pulsed (4.4 ns) Nd:YAG laser source 12 meters from the sample is spatially offset from a telescope that collects light scattered off the sample under investigation.* The system also used a gated detection technique,† which allows for an adjustable time delay between the laser pulse and data acquisition. This means that the detector is only measuring during a time interval when there should be a Raman signal, as opposed to measuring light intensity both during the laser pulse and after the laser pulse, when there may be fluorescence or other noise signals.

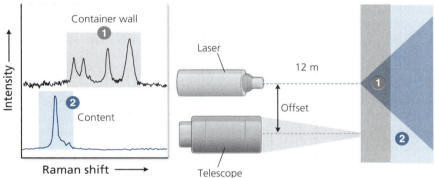

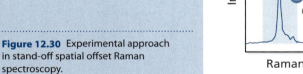

Figure 12.30 Experimental approach in stand-off spatial offset Raman spectroscopy.

The researchers demonstrated that the technique is effective at sample identification even when the sample is placed in a 1.5-mm thick opaque plastic bottle. A key benefit of the technique is that the laser interactions from the container, including fluorescence that can swamp the Raman signal, are not measured because of the spatial offset between the laser and telescope (and detector). The telescope collects light that originates from points away from where scattered or emitted light from the container signal would originate (Raman, Rayleigh, or fluorescence). The incident laser light and Rayleigh scatter from the container spreads out, and this Rayleigh scattered light can then be Raman scattered inside the container. The Raman signal from within the container is measured, as shown in the figure. As the offset moves from very close to the laser to larger distances, the telescope begins to sample deeper inside the container. ∎

---

* Zachhuber, B.; Gasser, C.; Chrysostom, T. H. E.; Lendl, B. Stand-Off Spatial Offset Raman Spectroscopy for the Detection of Concealed Content in Distant Objects. *Anal. Chem.,* **2011**, *83*, 9438–9442.

† See Chapter 3 for a review of gated detection techniques.

Surface-Enhanced Raman Spectroscopy

SERS allows the analyst to increase the Raman signal level. SERS is accomplished by placing a small volume of a liquid sample onto a metal surface, generally a silver or gold substrate. Microscope slides with a metal surface can be purchased for SERS. Interactions between the metal surface, which is rough on the nanoscopic scale, and the molecule under investigation cause an enhancement in the Raman signal. The SERS substrate and sample is sketched in Figure 12.29. The role of electric field enhancement and enhanced polarizability is not fully understood. Using the SERS technique, it is possible to increase the Raman signal dramatically, as much as 10^{14} times the normal Raman signal.[26] SERS shows promise for a variety of applications, such as environmental sensing[27] and explosives detection.[28]

Raman spectroscopy has moved from the research laboratory to becoming a nearly indispensable and highly versatile instrumental technique. New variations of Raman spectroscopy will continue to be developed. For example, recent work has demonstrated a combined approach of SERS and another variant of Raman analysis, coherent anti-Stokes Raman spectroscopy (CARS) showing dramatic enhancements of Raman intensities, even beyond those of SERS and CARS individually.[29] Raman scattering is a weak effect; thus, these improvements will help increase the number of useful ways Raman spectroscopy can be applied. Another exciting variant that is now available commercially is an instrument that will do both Raman imaging and atomic force microscopy (see Chapter 20) on the same section of a sample to generate information about surface structure and shape and chemical composition.

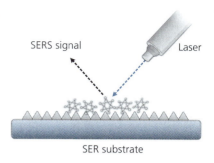

Figure 12.29 Sketch of metal substrate on nanoscale with molecule under investigation and associated laser excitation and surface-enhanced Raman signal.

12.5 Further Reading

BOOKS

Garland, C. W.; Nibler, J. W.; Shoemaker, D. P. *Experiments in Physical Chemistry*, 7th ed, McGraw-Hill: New York, **2003.**

Hollas, J. M. *Modern Spectroscopy*, 4th ed. John Wiley & Sons: Chichester, UK, **2004.**

McCreery, R. L. *Photometric Standards for Raman Spectroscopy: Handbook of Vibrational Spectroscopy*. Chalmers, J. M., Griffiths, P. R., eds. John Wiley & Sons: Chichester, UK, **2002.**

McCreery, R. L., *Raman Spectroscopy for Chemical Analysis*. Winefordner, J., ed., John Wiley & Sons: New York, **2000.**

Smith, E.; Dent, G. *Modern Raman Spectroscopy—A Practical Approach*, 1st ed. John Wiley & Sons: Chichester, UK, **2005.**

Vandenabeele, P. *Practical Raman Spectroscopy—An Introduction*, John Wiley & Sons: Chichester, UK, **2013.**

JOURNALS

Hudspeth, E. D.; et al. Teaching Raman Spectroscopy in Both the Undergraduate Classroom and the Laboratory with a Portable Raman Instrument. *Spectroscopy Letters*, **2006**, *39*, 99–115.

Mogilevsky, G.; Borland, L.; Brickhouse, M.; Fountain, A. W. III. Raman Spectroscopy for Homeland Security Applications. *International Journal of Spectroscopy*, **2012**, Article ID 808079, doi:10.1155/2012/808079.

Pavel, I. E.; et al. Estimating the Analytical and Surface Enhancement Factors in Surface-Enhanced Raman Scattering (SERS): A Novel Physical Chemistry and Nanotechnology Laboratory Experiment, *J. Chem. Ed.*, **2012**, *89(2)*, 286–290.

Wang, L.; Mizaikoff, B.; Kranz, C. Quantification of Sugar Mixtures with Near-Infrared Raman Spectroscopy and Multivariate Data Analysis. A Quantitative Analysis Laboratory Experiment, *J. Chem. Ed.*, **2009**, *86(11)*, 1322.

---

[26] Kneipp, K.; Kneipp, H.; Manoharan, R.; Hanlon, E. B.; Itzkan, I.; Dasari, R. R.; Feld, M. S. Extremely Large Enhancement Factors in Surface-Enhanced Raman Scattering for Molecules on Colloidal Gold Clusters. *Appl. Spectrosc.*, **1998**, *52*, 1493–1497.

[27] Halvorson, R. A.; Vikesland, P. J. Surface-Enhanced Raman Spectroscopy (SERS) for Environmental Analyses. *Environmental Science & Technology*, **2010**, *44(20)*, 7749–7755.

[28] Xu, Z.; Hao, J.; Braida, W.; Strickland, D.; Li, F.; Meng, X. Surface-Enhanced Raman Scattering Spectroscopy of Explosive 2,4-Dinitroanisole Using Modified Silver Nanoparticles. *Langmuir*, **2011**, *27(22)*, 13773–13779.

[29] Steuwe, C.; Kaminski, C. F.; Baumberg, J. J.; Mahajan, S. Surface Enhanced Coherent Anti-Stokes Raman Scattering on Nanostructured Gold Surfaces. *Nano Letters*, **2011**, *11(12)*, 5339–5343.

12.6 Exercises

EXERCISE 12.1: You are asked to measure a Raman Stokes signal 5 cm^{-1} from the Rayleigh scattered line. You are using the second harmonic from a Nd:YAG laser operating at 532 nm. What is the wavelength (in nanometers) of the Stokes shifted radiation?

EXERCISE 12.2: In general, the C–C vibration occurs around 800 cm^{-1}. If you use a 1,064 nm excitation source for a Raman experiment, calculate the wavelength of the Raman line associated with the C–C vibration in nanometers.

EXERCISE 12.3: C_2H_2, acetylene, is shown here.

$$H—C \equiv C—H$$

(a) Is acetylene (C_2H_2) centrosymmetric?

(b) Acetylene has several vibrational modes. One vibrational mode is shown here:

Is this vibrational mode Raman active, IR active, or both?

EXERCISE 12.4: Two normal modes for benzene (C_6H_6) are shown. The first occurs at 3,061 cm^{-1} (called the C–H stretch, where the hydrogen and carbon move in opposite directions) and the second at 992 cm^{-1} (called the ring stretch, where the hydrogen and carbons move in the same direction). Would you expect these vibrations to be Raman active? Would you expect them to be IR active?

EXERCISE 12.5: Theory shows that the ratio of Stokes Raman to anti-Stokes Raman intensity is:

$$\frac{I_{Stokes}}{I_{anti-Stokes}} = \left(\frac{\upsilon_0 + \upsilon_i}{\upsilon_0 - \upsilon_i}\right)^4 e^{-\frac{h\upsilon_i}{k_BT}}$$

where υ_0 is the Rayleigh scattered frequency, υ_i is the frequency of the Raman light, and T is the Kelvin temperature of the molecules. Use this relationship to determine the temperature of a system where the ratio of Stokes to anti-Stokes is 1.8. The excitation source is 532 nm and the Stokes emission occurs at 506.3 nm.

EXERCISE 12.6: For a molecule with Raman-active vibrations, will every Stokes line have a corresponding anti-Stokes line? Are there certain types of molecules that will exhibit Stokes but not anti-Stokes Raman scattering?

EXERCISE 12.7: In your new job as patent examiner, you have come across a patent application where a company has proposed a sensor for gaseous fuel and exhaust gases. It claims that the technique is quick and highly sensitive to H_2, N_2, and O_2. The technique makes use of the infrared absorption of these molecules. What is wrong with the technique?

EXERCISE 12.8: You have access to a 532 nm excitation source and a 785 nm excitation source for a Raman spectrum measurement. The power of the 532 nm source is 50 mW, and the power of the 785 nm source is 350 mW. Assuming that neither source will be absorbed by the sample, which source will give the largest Raman intensity signal from a sample?

Advanced Exercises

EXERCISE 12.9: You are investigating the dependence of the ratio of Stokes Raman to anti-Stokes Raman intensity on temperature using cyclohexane (C_6H_{12}). The Raman shift you are studying is 2,852.9 cm^{-1}, and you are using a Nd:YAG laser at 532 nm as the excitation source. Create a spreadsheet of the ratio of the Stokes and anti-Stokes lines as a function of temperature.

$$\frac{I_{Stokes}}{I_{anti-Stokes}} = \left(\frac{\upsilon_0 + \upsilon_i}{\upsilon_0 - \upsilon_i}\right)^4 e^{-\frac{h\upsilon_i}{k_BT}}$$

where υ_0 is the Rayleigh scattered frequency, υ_i is the frequency of the Raman light, and T is the Kelvin temperature of the molecules.

EXERCISE 12.10: The vibrational modes of CO_2 are shown.

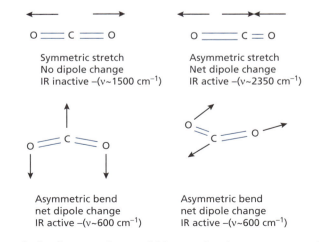

Which, if any, modes would have a depolarization ratio of near zero?

EXERCISE 12.11: A gaseous sample has strong Raman lines at Raman shifts of 3,373.7 cm^{-1} and 1,973.8 cm^{-1}. Use the online *NIST Chemistry WebBook* to determine the likely gas. Search by "vibrational energy value."

EXERCISE 12.12: Two additional Raman-based spectroscopic techniques not discussed in this chapter are CARS and resonance Raman spectroscopy. Both are valuable variants of Raman spectroscopy. Research appropriate resources and describe each of these variants. Make sure you explain how they differ and how they are similar to conventional Raman analysis.

EXERCISE 12.13: What advantage is there to using a 785 nm diode laser as compared to a 1,064 nm laser as a Raman excitation source?

EXERCISE 12.14: The excitation (black) and emission (blue) spectra for rhodamine 110 are shown here. If needed, review the concept of the excitation and emission spectrum from Chapter 8. You have access to an argon ion laser operating at 488 nm and a Nd:YAG laser operating at 532 nm with your Raman instrument. Which laser would you recommend using for rhodamine 110?

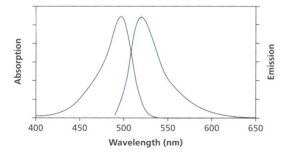

EXERCISE 12.15: You are looking to purchase a dichroic notch filter for your homemade Raman system that uses a 532 nm laser for excitation. The system is designed much like that in Figure 12.12. The transmission spectrum shown here is for a dichroic notch filter. Would this be an appropriate choice for your system? Data and image are from Semrock.com.

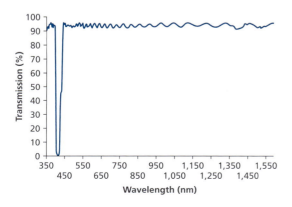

EXERCISE 12.16: Raman spectroscopy can be used to identify different allotropes of a pure element. For instance, pure carbon can form a variety of structures—for example, C$_{60}$ ("buckyballs"), carbon nanotubes, crystalline and amorphous graphite, and diamond. Imagine that you are investigating an unknown material, thought to be a form of carbon, and have found a Raman peak at a shift of 186 cm^{-1}, 1,567 cm^{-1}, and 1,591 cm^{-1}. Use the scientific literature or a Raman library/database to determine which form of carbon you have. State the source or sources you used in forming your conclusion.

EXERCISE 12.17: We see in the energy level diagram in Figure 12.4 that the initial states for Stokes Raman and anti-Stokes Raman are not the same. Based on an argument using the Boltzmann distribution described in Chapter 2, would you expect to see as much anti-Stokes Raman emission as Stokes Raman emission? Assume that the degeneracy of both of the initial states is the same.

EXERCISE 12.18: The figure shows the transmission spectrum of a notch filter used in Raman spectroscopy. If a 532 nm laser is used, use the plot to approximate the lowest Raman shift that could be measured. Would this filter be appropriate to measure both Stokes and anti-Stokes Raman spectra?

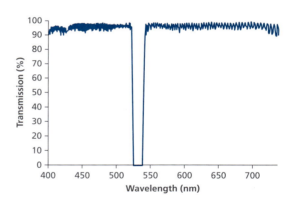

EXERCISE 12.19: The figure shows the transmission spectrum of an edge filter used in Raman spectroscopy. If a 532 nm laser is used, use the plot to determine, approximately, the lowest Raman shift that could be measured. Would this filter be appropriate to measure both Stokes and anti-Stokes Raman spectra?

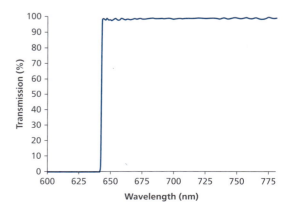

EXERCISE 12.20: In this chapter, we stated that the rule for Raman activity is that the polarizability must change asymmetrically during the molecular vibration. In this problem, you will derive this statement. We will continue to use the diatomic molecule in this exercise so that the discussion does not need to be overly general. We saw in this chapter that the induced dipole moment can be written in terms of the polarizability and external applied electric field as:

$$\mu_{induced} = \alpha E$$

After several steps in this problem, the induced dipole moment will tell us about the different scattering processes and their associated frequencies. We will start by writing out the polarizability in terms of the Taylor expansion. In general, a function f(x) can be written as:

$$f(x) = f(x_0) + \frac{df}{dx}\bigg|(x - x_0) + \frac{1}{2!}\frac{d^2f}{2!dx^2}\bigg|(x - x_0)^2 + \ldots$$

By knowing a little bit about a function (such as its derivatives), the Taylor series lets you write out an approximation for that function. When using the Taylor series, we get to pick the value for x_0, the point about which we form the series. As an example, we can write out the Taylor series for e^x about $x_0 = 0$:

$$e^x \approx e^0 + e^0\bigg|_0(x - 0) + \frac{1}{2!}e^0\bigg|_0(x - 0)^2 \approx 1 + x + \frac{x^2}{2}$$

This equation will be a good approximation for e^x for values close to 0 (check this equation against your calculator for $e^{0.1}$; it works!). Going back to our problem, we will use the Taylor series to write out the polarizability as a function of the vibrational coordinate Q, as follows:

$$\alpha(Q) = \alpha(Q = 0) + \frac{d\alpha}{dQ}\bigg|_{Q=0}Q + \frac{1}{2!}\frac{d^2\alpha}{dQ^2}\bigg|_{Q=0}Q^2 + \ldots$$

where $Q = r - r_0$, where r_0 is the equilibrium bond length. Q is the amount the diatomic molecule has stretched or compressed beyond its relaxed state. We can model the vibrations of the molecule as a simple harmonic oscillator, which has a natural frequency (associated with the molecule) of ω_0 and amplitude A:

$$Q = A\cos(\omega_0 t)$$

We can now work toward our statement of the Raman selection rule.

(a) Keeping only the first two terms of the Taylor series, write out the polarizability, $\alpha(Q)$.

(b) The external electric field associated with the laser light excitation can be written as:

$$E = E_0\cos(\omega_{exc}t)$$

where ω_{exc} is the angular frequency of the laser source and E_0 is the amplitude of the electric field. Write out the induced dipole moment using this relationship for the electric field and the result you got from part a using the first two terms of the Taylor series.

(c) After writing out the answer to part (b), you should have found the following:

$$\mu_{induced} = \alpha E \approx \alpha(Q_0)E_0\cos(\omega_{exc}t) + \frac{d\alpha}{dQ}\bigg|_{Q_0}$$
$$AE_0\cos(\omega_0 t)\cos(\omega_{exc}t)$$

The first term of the induced dipole has an associated angular frequency of ω_{exc}. This term is associated with Rayleigh scattering. The second term is more difficult to interpret because it has the product of two cosine functions with different frequencies. Use the following trigonometric identity to rewrite that second term:

$$\cos u \cos v = \frac{1}{2}\left[\cos(u - v) + \cos(u + v)\right]$$

(d) You should now have an equation for the induced dipole moment that looks like:

$$\mu_{induced} = \alpha E \approx \alpha(Q_0)E_0\cos(\omega_{exc}t) +$$
$$\frac{d\alpha}{dQ}\bigg|_{Q_0}\frac{AE_0}{2}\cos(\{\omega_{exc} - \omega_0\}t) + \frac{d\alpha}{dQ}\bigg|_{Q_0}\frac{AE_0}{2}\cos(\{\omega_{exc} + \omega_0\}t)$$

Again, the time-varying dipole moment is what produces the scattering. Look at the frequency dependence of the last two terms. These correspond to Raman scattering. Which term corresponds to Stokes and which term corresponds to anti-Stokes Raman scattering?

(e) In front of the two Raman terms in the induced electric dipole moment, we see the derivative of the polarizability with respect to Q evaluated at the equilibrium position. Going back to our sketches of polarizability versus vibrational coordinate, this is the slope of that sketch evaluated at the equilibrium point. Redraw the sketches and point out which shows an allowed Raman vibration and which does not. When that slope is zero, the derivative is zero, and we see that the two Raman terms are zero. You have now put together the basis for this selection rule!

Mass Spectrometry

Ion source → Mass (m/z) analyzer

Computer control

N$_2$

Positive ion

First dynode

- HV

e$^-$

Neutral

13.1 Basic Principles and Comparison to an Optical Spectrophotometer

A mass spectrum (MS),[1] the output of a mass spectrometer, or MS, yields information that is directly related to the structure of analyte molecules and allows us to make inferences about bond strengths and the existence of specific functional groups within the molecule. Under identical experimental conditions, each unique molecule gives a unique MS. Thus, mass spectrometry, or MS, has become one of the most important tools available for molecular identification and is often used as a molecular "finger-printing" technique to identify the unknown components of samples.

With each of the spectroscopic methods we have discussed previously, when we referred to a spectrum, the signal or output energy was on the y axis and the wavelength or wave number was on the x axis. The intent was to understand some physiochemical property of an analyte based on the energy of photons it absorbed or emitted. In MS, the analyte is ionized and the mass-to-charge ratio of the analyte (and often fragment

PROFILE

Puffer Mass Spectrometer

The next generation of airport security screening devices is based on mass spectrometry technology. The system, named the Ionscan Sentinel II™, collects samples from clothing, skin, and hair by blowing puffs of air at travelers as they stand briefly in a closed chamber. The air is collected and injected into a small quadrupole mass spectrometer, which can identify 16 different explosive compounds, detecting them in the parts-per-billion range. Earlier "puffer" devices, initially deployed in the early 2000s, used other spectrometric methods and proved to be unreliable in day-to-day use. The National Academy of Sciences stated that MS technology is the "gold standard for resolving high-consequence analysis."[1] ■

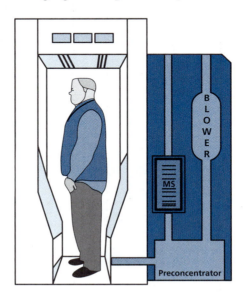

---

[1] In the chemists' vernacular, the abbreviation "MS" refers to the mass spectrum (output), as well as the instrument, the mass spectrometer, and the field of study, mass spectrometry.

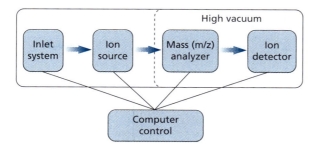

Figure 13.1 Block diagram of a generic MS. The mass analyzer and ion detector are held under high vacuum. The ion source and/or the inlet system are sometimes also under high vacuum but might be at atmospheric pressure.

ions from the analyte) is measured. Rather than an energy spectrum, such as those we have discussed previously, we obtain a mass spectrum. The term *spectrum* refers to the range of *mass-to-charge (m/z) ratios* of ions generated by the instrument from the analyte(s) in the sample, and electromagnetic radiation usually does not play a role in generating the signal.

An MS is one of the oldest of the modern instruments, having a notable history in helping chemists determine isotopic ratios of naturally occurring elements in the early twentieth century. A block diagram of an MS is seen in Figure 13.1.

Mass spectral analysis requires that the sample be volatilized and ionized. Samples are typically introduced in the gas state (at the inlet) and ionized (at the ion source), or they are volatilized energetically (using heat, light, or electricity) in the inlet and ionized in the ion source. However, MS is not restricted to only highly volatile samples. Solids can be introduced directly into the system through sputtering or thermal desorption, and several ingenious methods for introducing large nonvolatile molecules (such as proteins and other biomolecules) have been developed in the past few decades. Very often the inlet source is associated with a different instrument (such as a chromatographic system), or the inlet system and ion source are indistinguishable. This chapter will focus on describing the ion source, mass analyzer, and detector of the mass spectrometer as an independent instrument.

The ion source, as the name implies, generates ions from analyte molecules or atoms, and if the energy of the ionization step is sufficiently high, the process can also cause the molecular ion to fragment. Ions of the molecule itself, prior to any fragmentation, are called *parent ions, precursor ions,* or *molecular ions* and are useful in measuring the molecular mass of the substance. The ion fragments are called *daughter ions* or *product ions,* which are useful in discerning structure and relative bond strengths within the original molecule. The ionization step typically produces a pattern of product ions, along with the precursor ion, that is highly reproducible when obtained under identical instrumental parameters. Because the pattern is reproducible and characteristic of the analyte under study, we can think of the MS as an ion *fingerprint*. Not only does the MS determine the pattern of ions formed, but it can clearly determine the *relative number* of *each* ion formed. This pattern, having both mass information and abundance information, is the mass spectrum.

The *mass analyzer* is the MS analog of the monochromator in a spectrophotometer. The term *mass analyzer,* strictly speaking, is a misnomer. Without centrifugation or some method of size discrimination, gaseous molecules cannot easily be differentiated by size. However, gas state *ions* can be differentiated by the m/z ratio, and this is the basis of the mass analyzer. We will discuss several types of mass analyzers in Section 13.3, but the basic principle is to impart a certain kinetic energy to the ionized molecules and fragments and then separate them based on their mobility in an electrical and/or magnetic field. On separation, the detector essentially counts the number of ions of

Precursor ion = the molecular ion of the original analyte
Product ion = the ions that result from the fragmentation of the precursor ion
Base peak = the largest signal peak in a mass spectrum

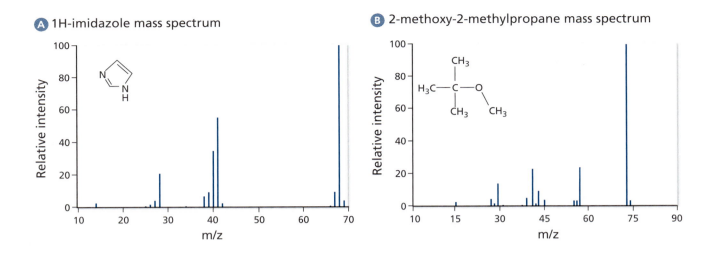

Figure 13.2 Mass spectra of selected compounds obtained using electron impact ionization: (A) imidazole ($C_3H_4N_2$, 68.08 g/mol), showing a clear precursor ion signal peak; and (B) methyl *tert*-butyl ether (MTBE) ($C_5H_{12}O$, 88.15 g/mol), in which a signal from the precursor ion is not observed.

each m/z to yield a spectrum of relative intensity versus m/z (see Figure 13.2). Although most ions will be univalent ($z = +1$), it is important to recognize that the discrimination is based on m/z, and so a mass analyzer would not be able to distinguish between, for instance, a CH_3^+ ion and a $C_2H_6^{2+}$ ion.

It is important for certain segments of the MS to be held under high vacuum. The mass analyzer and detector are always maintained at a pressure of less than 10^{-5} torr, and in some systems the ion source is also evacuated. Recall that the analyte ions are in a highly energetic state, and collisions between those particles and each other, as well as with any atmospheric gases (oxygen, nitrogen, water) could produce fragment ions that are not related to the precursor molecules.

Furthermore, the physics associated with differentiating particles of different m/z ratios within the mass analyzer are based on mobility in a vacuum, where collisions with other particles have low probability. In some cases the entire system, from inlet to detector, is under high vacuum, but in most cases the instrument uses an interface between the inlet and the ion source (or between the ion source and the mass analyzer) that allows for partial isolation of analyte species by removing solvent, carrier gas, and atmospheric gas by evacuation.

Problem 13.1:   Consider the two molecules below (molecule X and molecule Y) as analyzed by MS. Both molecules can be described as a system with three possible fragments: fragment A has a molecular mass of 10 amu, fragment B has a mass of 15 amu, and fragment C has a mass of 20 amu. Assume that (1) the charges generated on the fragments are either +1 or +2 and (2) only one bond is cleaved for each fragmentation. List all possible ions that could be generated, and determine which could be distinguished explicitly and which have m/z similar to other ions.

Molecule X: A–B–B–C

Molecule Y: C–A–C–B

13.2   Ion Sources

There are two basic types of ion sources: *hard ion sources* and *soft ion sources*. Hard ion sources typically cause significant fragmentation of molecular species. This is helpful in determining the structure of an unknown substance because it (1) provides a great deal of information about the functional groups that make up the molecule and (2) yields a "fingerprint" spectrum that can be compared to a library for unique identification. However, hard sources produce complex spectra that can be difficult to interpret and sometimes do not yield a measurable precursor ion peak.

In Figure 13.2(A), the precursor ion peak for 1H-imidazole at m/z = 68 is clearly visible, but in Figure 13.2(B), we do not see the precursor ion for methyl *tert*-butyl ether (MTBE; which would appear at m/z = 88) at all due to the relatively weak CH_3–O bond in MTBE, which breaks under hard ion source conditions. In spectra obtained with hard ion sources, it is not uncommon for the base peak to be due to another ion rather than the precursor ion. In Figure 13.2(A), the base peak ion is indeed from the precursor ion at m/z = 68, but in Figure 13.2(B) the base peak is found at m/z = 73. We can deduce that the molecular mass (M) of 88 has been decreased by 15 to form this base peak ion (M – 15 = 73). Therefore, the m/z signal at 73 represents a product ion derived from the precursor ion minus a methyl group. Again, the relatively weak CH_3–O bond in MTBE breaks under the hard ion source conditions, leaving the *tert*-butyl oxy fragment as the base.

Soft ion sources produce far fewer product ions, and spectra provide less chemical information but usually yield a strong precursor ion peak, thus allowing you to more clearly determine the molecular weight of the original compound. Figure 13.3 shows the MS of the methyl ester of decanoic acid under two ionization sources. The precursor ion peak in Figure 13.3(B), obtained with a soft ion source (chemical ionization), is clearly evident, and far fewer fragments are observed. Figure 13.3(A) shows the same analysis obtained with a hard ion source (electron ionization), in which the precursor ion peak at m/z = 186 is barely visible and a significant amount of fragmentation is seen. In spectra obtained using soft ionization, the base peak is most often due to the precursor ion. In Figure 13.3(A), the base peak is m/z = 73, representing the ion $(CH_2–C(=O)–O–CH_3)^+$, whereas in Figure 13.3(B), the base peak is m/z = 187 for an ion having a mass one unit greater than that of the precursor ion,[2] $(M + 1)^+$.

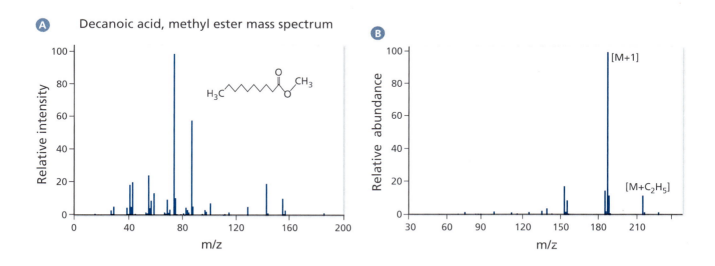

Figure 13.3 Mass spectra of the methyl ester of decanoic acid ($C_{11}H_{22}O_2$, 186.3 g/mol) obtained using (A) electron ionization (a hard ion source) and (B) chemical ionization (a soft ion source).

_____

[2] Note that Figure 13.3(B) was obtained using *chemical ionization*, which is described below. The parent ion has a mass of (M + 1) because of the fact that it was ionized by the transfer of a proton (H^+) from a reagent gas to the parent molecule.

Problem 13.2: The analytical chemist is often faced with having to make either a qualitative or quantitative determination. Given only what we have discussed so far in this chapter, which type of ion source (hard or soft) would you select for qualitative and which would you select for quantitative analysis? Explain your reasoning.

Electron Ionization

The most common ion source for laboratory-based, bench-top MS instruments (such as those used as detectors for gas chromatographs) is the *electron ionization* (*EI*; also known as the electron impact) source. As the name implies, the ionization occurs as a result of the impact of high-energy electrons with analyte molecules. Figure 13.4 depicts a typical electron ionization chamber. The filament is a high-melting metal such as tungsten or rhenium that is electrically heated to a glowing red, and a potential difference of 70 V is applied between the filament and the target electrode (also called the trap). Electrons produced by thermionic emission[3] from the filament are accelerated toward the target, creating a stream of electrons with an average kinetic energy of 70 electron volts (70 eV is equivalent to about 6.7 MJ/mole or 1.6 Mcal/mol). The motion of the high-speed electrons is shown as the downward arrow between the filament and target in Figure 13.4. The gaseous analyte sample, M(g), is introduced to the chamber and flows into the ionization region between the filament and the target.

The ionization potentials of most organic compounds are lower than 15 eV and that of inorganic substances is typically below 25 eV; therefore, the impact of a 70 eV electron is sufficient to knock out an electron from a higher molecular orbital (creating a positive ion of the precursor molecule) and also to impart enough internal energy to cause subsequent fragmentation. Because the impacts by high-energy electrons result in significant fragmentation, EI is considered a hard ionization source.

The positive ions created through the electron collisions are accelerated before being sent to the mass analyzer. The repeller is an electrode maintained at a relative

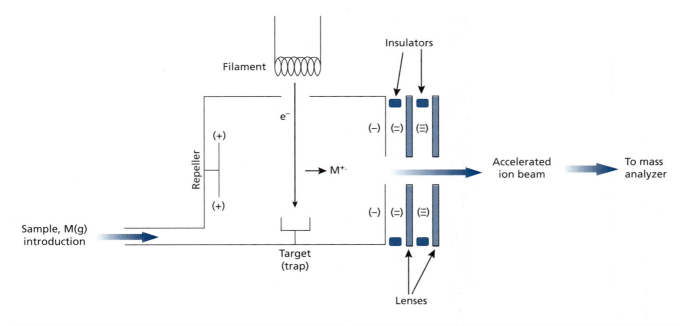

Figure 13.4 Schematic of an EI chamber.

---

[3] Thermionic emission is described in more detail in Chapter 20. For now, it is sufficient to understand that electrons are emitted from a heated filament maintained at a potential highly negative of the target electrode.

J. J. Thomson (1856–1940) was a British physicist who is mostly known for winning the Nobel Prize in physics for his discovery of the electron. However, in the field of chemical analysis, his greatest contribution was the invention of the mass spectrometer (MS). Using his MS, Thomson discovered the existence of isotopes. He was also a fantastic teacher. Some of his notable students included Niels Bohr, Ernest Rutherford, and J. Robert Oppenheimer. ■

PROFILE

J. J. Thomson

First World War.com. Licensed under Public Domain via Wikimedia Commons - https://commons.wikimedia.org/wiki/File:J.J_Thomson.jpg#/media/File:J.J_Thomson.jpg

positive potential, causing the generated positive precursor and fragment ions to move toward the opposite side of the chamber in which an exit slit is found (see Figure 13.4). Ions passing through the slit are accelerated by successively more negative potentials applied to the *lenses*, which are metal plates with slits (circular or rectangular) in their centers. The potential applied to the lenses can be varied in order to focus and collimate the ion beam. The accelerated ion beam then passes into the mass analyzer.

A few negative ions are produced during this process, but the energy of the electrons in the stream is too high in most applications to allow for efficient capturing of electrons by molecules. Any negative ions produced are drawn by the positively charged target and the repeller electrodes, and so do not make it into the mass analyzer.

The velocity of each accelerated ion is directly proportional to the accelerating voltage applied (V), which is the potential difference between the final lens (negatively polarized) and the repeller (positively polarized). The ion velocity is also proportional to the charge of the ion (usually +1) but inversely proportional to its mass. When passing through the accelerating region of the lenses, an ion assumes a specific kinetic energy:

$$KE = z \cdot e \cdot V \qquad \textbf{Eq. 13.1}$$

where z is the integer charge on the ion, e is the fundamental charge of an electron (1.602×10^{-19} C), and V is the accelerating voltage. You will recall Equation 13.2 from previous studies:

$$KE = \frac{1}{2} m \cdot v^2 \qquad \textbf{Eq. 13.2}$$

where m is the mass of the ion (in kg) and v is the velocity (in m/s). Combining Equations 13.1 and 13.2, then rearranging, we can see that the velocity of any given ion is given by Equation 13.3.

$$v = \sqrt{\frac{2 \cdot z \cdot e \cdot V}{m}} \qquad \textbf{Eq. 13.3}$$

We will see that these equations are fundamental to understanding how the most common mass analyzers function.

One can envision that only a small fraction of the analyte molecules that are introduced to the ionization chamber make it into the mass analyzer—that is, only those positive ions with a correct trajectory make it through the chamber exit slit and the lens slits. Once they pass through the lens slits, the ions move on to the mass analyzer so that the ions of different m/z can be separated and counted. What happens to the rest of the material, which is in reality the vast majority of that which was introduced? Some of it is removed by the pumping system, but much of it ends up collecting on the repeller plate, the target electrode, and the lenses themselves. Therefore, it is essential that the source components be periodically removed and thoroughly cleaned in order to maintain a well-functioning instrument.

Although the scenario we have just described is most common for EI ion sources, sometimes the nature of the analyte molecules makes the generation of negative ions more amenable to efficient analysis. Many modern EI instruments are configured to allow the user to select the *negative ion mode*. In such cases, the potentials applied to the repeller and lenses are reversed, and the potential difference between the filament and target, although not reversed, is decreased to produce electrons with much lower energies that can be captured by analyte molecules. For this to work efficiently, of course, it is desirable to have at least one large, electronegative atom in the analyte molecule. The sensitivity of negative ion EI is usually much lower than that of positive ion EI, but some advantage is gained in being able to exclude from the analysis molecules that do not easily capture electrons to form negative ions. One of the detractions from using negative ion EI is that an even greater percentage of the molecules introduced to the system are left to collect on the source components, requiring more frequent cleaning of the source.

Example 13.1 – Velocity and Accelerating Voltage

Estimate the velocity of the carbon cation after an accelerating voltage of 1.0 kV has been applied.

STRATEGY –

First, we must remember a few key definitions—namely the unit conversion between volts (V), joules (J), and coulombs (C), as well as the definition of Joules in SI units:

$$V = \frac{J}{C} \qquad \text{and} \qquad J = \frac{kg \cdot m^2}{s^2}$$

Second, note that the mass in Equations 13.2 and 13.3 is that of an individual ion, in kg, not the molar mass of the ion.

So, the C^+ ion has a molar mass of 12.011 g/mol and we must convert that to kg/ion:

$$\text{Mass of one } C^+ \text{ ion in } kg = 12.011 \frac{g}{mol} \times \frac{1 kg}{1000 g} \times \frac{1 \, mol \, ions}{6.022 \times 10^{23} \, ions}$$

$$= 1.9945 \times 10^{-26} \, kg$$

SOLUTION –

Now we can apply Equation 13.3 to solve for velocity:

$$v = \sqrt{\frac{2 \cdot z \cdot e \cdot V}{m}} = \sqrt{\frac{(2)(1.602 \times 10^{-19} \, C)(1.0 \times 10^3 \, V)\left(\frac{J \cdot C^{-1}}{V}\right)\left(\frac{kg \cdot m^2 \cdot s^{-2}}{J}\right)}{1.9945 \times 10^{-26} \, kg}}$$

$$= 1.27 \times 10^5 \, m/s$$

Problem 13.3: What would be the velocity of a methyl ion (CH_3^+) if it were accelerated across a voltage of 500 5.0×10^2 volts?

Problem 13.4: What voltage would be needed to accelerate a *tert*-butyl dication $(C_4H_9^{2+})$ to a velocity of 1.0×10^6 m/s?

Now that we have a clearer visual image of the ionization chamber, we can consider further the reason that a high vacuum is required for successful MS experiments. The main precept we must keep in mind is that in EI, we want fragmentation to take place *only* as a result of the energy imparted by collision with electrons in the ionization chamber, not as a result of collision with other molecules or ions. The formation of fragment ions due to excess internal energy imparted by collision with an electron yields information about the relative strength of the bonds in the precursor ion. Those bond energies are quantized, so we can expect the MS pattern of relative intensities between different fragments to be highly reproducible for a given analyte. It is this reproducibility of fragment ion intensities that makes MS such a valuable qualitative analysis tool. However, if fragment ions are allowed to collide with other species, collisions will be random both in number and in orientation,[4] giving rise to two variations in the fragmentation intensity patterns:

1. The size and number of fragment ions will not exhibit the high reproducibility that we rely on from MS.

2. Collisions between different fragments might produce new fragments that do not represent the actual bond structure of the original precursor molecule.

An example of the problem posed by statement 2 is demonstrated in Figure 13.5, the EI spectrum of diethyl ether, $CH_3-CH_2-O-CH_2-CH_3$. Note that the base ion is at m/z = 31, which cannot be assigned to any natural fragment of the precursor molecule. Rather, the only reasonable assignment would be CH_3O^+, but there are no CH_3 groups bonded to the oxygen in diethyl ether. The most logical explanation is that a CH_2O^+ fragment collided with either a free hydrogen atom or with another molecule or fragment which transferred a hydrogen atom to it. Undesired collisions cannot be

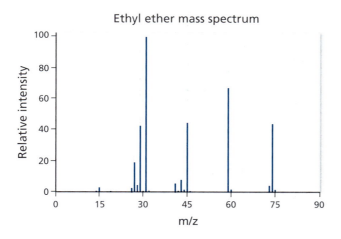

Figure 13.5 The EI MS of ethyl ether gas.

---

[4] You might recall, from when you studied kinetics in general chemistry, the concept of molecular orientation as it relates to chemical reactions. For instance, a carbon monoxide molecule might collide with another reactant via the carbon atom, the oxygen atom, or any alternate off-axis orientation. Each orientation has the potential to result in different products.

entirely avoided, but it is important to try to minimize them as much as possible in EI mass spectrometry. The parameter over which we have direct control that allows us to minimize undesired collisions is pressure; by decreasing the pressure of the system, we decrease the number of molecules and ions available for collision.

How low must the pressure be in order to achieve the desired result? If the only form of ionization were EI, then the answer might be "the lowest pressure physically possible that would still give us a measurable signal." However, we will see in the next section that some techniques, such as chemical ionization, *depend* on collisions taking place. In order to estimate the level of vacuum required for a certain level of collisional activity, we need to consider a concept called *mean free path*, which is the linear distance a particle can be expected to transit before colliding with another particle. As you should expect, the mean free path, λ, is inversely proportional to the system pressure, as shown in Equation 13.4.

$$\lambda = \frac{kT}{Pd^2 \pi \sqrt{2}}$$

Eq. 13.4

Here, k is the Boltzmann constant; T is the temperature in Kelvin; P is the pressure in pascals; and d is the diameter of the particle in meters, often taken as the mean expected diameter of particles in a system.

Example 13.2 – Pressure from Mean Free Path

Estimate the pressure, in torr and atmospheres, required for a mean free path of 1.0 m for a moderately sized organic molecule having a diameter of 5.0 Å at 25°C. What would be the maximum pressure if the molecule were being ionized in an EI source? Recall that 1 nm = 10 Å.

STRATEGY –

We will use Equation 13.4 with the units assigned in the text to find the required pressure. You will have the opportunity to ascertain that the units work out in an exercise below.

$$\lambda = \frac{kT}{Pd^2 \pi \sqrt{2}}$$

$$1.0\,\text{m} = \frac{\left(1.38 \times 10^{-23}\,\text{J/K}\right)\left(298.15\,\text{K}\right)}{P\left(5.0 \times 10^{-10}\,\text{m}\right)^2 \pi \sqrt{2}}$$

$$P_{\text{Pa}} = 0.0037\,\text{Pa}$$

$$P_{\text{atm}} = 0.0037\,\text{Pa} \times \frac{1\,\text{atm}}{101325\,\text{Pa}} = 0.0037 \times 10^{-8}\,\text{atm}$$

$$P_{\text{Torr}} = 3.7 \times 10^{-8}\,\text{atm} \times \frac{760\,\text{torr}}{\text{atm}} = 2.8 \times 10^{-5}\,\text{torr}$$

If the molecule were being ionized in an EI source, where the electrodes are closely spaced with a 70 V potential applied, we would need to reduce the pressure even further to prevent arcing across the electrodes (by a factor of two):

$$P_{\text{Torr}} = \frac{2.8 \times 10^{-5}\,\text{torr}}{2} = 1.4 \times 10^{-5}\,\text{torr}$$

We can use the mean free path equation to estimate the maximum pressure in our instrument. As a rule of thumb, a mean free path of 1 meter is used to estimate the required pressure for an MS system. If the system includes a high potential across electrodes that are relatively close together (as is the case for the filament and repeller in the EI source), the pressure must be further diminished by at least a factor of two in order to prevent arcing across the electrode gap.

Recall that we noted earlier that the pressure in the MS ionization system must be less than 10^{-5} torr. Example 13.2 shows us that for a mean free path of 1 m (for 5 Å particles), we must maintain a pressure no greater than 30 μtorr, and we would want to decrease the pressure even more, to around 15 μtorr, to prevent arcing between the filament and repeller.

Problem 13.5: Demonstrate that the units in Equation 13.4 cancel out properly.

Problem 13.6: Rewrite Equation 13.4 to relate λ to concentration of particles in molarity.

Problem 13.7: Estimate the mean free path of a water molecule (2.8 Å diameter) at standard temperature and pressure.

Problem 13.8: Assuming the water molecule in Problem 13.7 is traveling at about 610 m/s, how many collisions per second is it undergoing?

Chemical Ionization

Chemical ionization (CI) is one of the oldest and most common of the soft ionization sources. The ionization in CI results from the collision of analyte molecules with ions of a reagent gas that is introduced to the ionization chamber. Chemical ionization requires that the reagent gas be ionized by electron ionization, and many modern EI instruments are designed to be used for CI with only slight modification to prevent leakage of the reagent gas into the ion source housing (Figure 13.6).

During the experiment, a reagent gas such as methane, ammonia, or isobutane is introduced to an EI chamber at a pressure of around 750 mtorr, and the reagent gas is ionized by the EI system. Although this pressure is still relatively low (< 1 torr), it is three to four orders of magnitude higher than the partial pressure of the analyte molecules. Because electrons with normal EI energies (70 eV) have difficulty penetrating the reagent gas, the

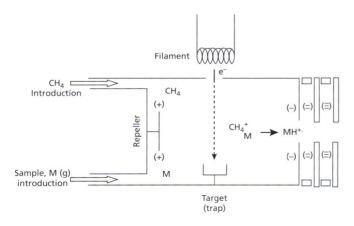

Figure 13.6 Schematic of a CI chamber.

potential difference between the filament and target electrode is increased to around 200 V. The immediate result of these modifications is twofold: (1) the probability of any given analyte molecule being struck by an electron is diminished by the relative ratio of analyte to reagent gas and (2) the probability of collision between analyte molecules and reagent gas ions is quite high (i.e., the mean free path for analyte molecules across the ionization chamber is diminished significantly). A mean free path length of about 10^{-4} m is typical for a CI experiment. Because the entire system is at relatively low pressure and moderate temperatures, collisions, although frequent, are of relatively low energy, and so reagent gas ions can cause subsequent ionization of analyte molecules but impart less internal energy than do electrons in EI. The overall result is less fragmentation in comparison to EI.

Problem 13.9: If we want the mean free path of a water molecule (2.8 Å diameter) in a CI system to be 0.10 mm, at what pressure would we need to introduce reagent gas?

Problem 13.10: Using the pressure you found in Exercise 13.9, estimate the mean free path length of the theoretical organic molecule described in Example 13.2.

Consider a system in which methane is used as the reagent gas (the most likely scenario for a typical laboratory CI MS). In this system, methane would be introduced at a pressure of 50 to 100 Pa and would undergo EI. Figure 13.7 shows the EI mass spectrum of methane gas at a pressure of about 10^{-3} torr. Note that we observe signals for the expected CH_4^+ ion, but also ones for CH_3^+, CH_2^+, CH^+, and even C^+. Furthermore, we observe a small signal at $m/z = 17$, which is assigned to the short-lived CH_5^+ ion. As the pressure of methane is increased, the intensity of the CH_5^+ increases. During the time of the experiment, then, there exists in the ion chamber, simultaneously, a mixture of a variety of cations and electrons. You may recall from Chapter 9 that this is the formal definition of a plasma. Thus, the reagent gas cloud is often referred to as the *reagent gas plasma*, even though it exists at a temperature far below that of what we might consider an ordinary plasma. As the methane pressure increases, collisions between reagent gas molecules and ions begin to produce additional species such as the $C_2H_5^+$ dimer and $C_3H_5^+$ trimer species

$$CH_5^+ + M \rightleftharpoons CH_4 + MH^+ \qquad \text{Eq. 13.5}$$

$$C_2H_5^+ + M \rightleftharpoons C_2H_4 + MH^+ \qquad \text{Eq. 13.6}$$

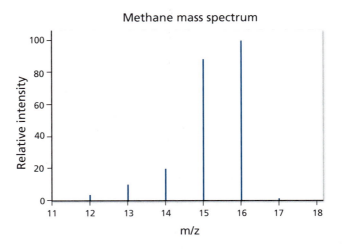

Figure 13.7 The EI MS of methane gas.

Now let us introduce the analyte molecule, M, into the reagent gas plasma. A wide variety of reactions might take place, but we will consider here only those that are prevalent. The most common type of ionization reaction is the *proton transfer* reaction, such as that seen in Equations 13.5 and 13.6, both of which result in a positive ion of the precursor molecule but with a mass of M + 1. If the analyte molecule exhibits a relatively high nucleophilic core, the relative Lewis acidity of the reagent gas ions allows them to abstract a hydride ion from the analyte, in a *hydride elimination* reaction (Equation 13.7) or a simple *charge transfer* reaction (Equation 13.8). At higher methane pressures, where dimer and trimer species become prevalent, adduct species can be produced, resulting in products with masses increased by 29 or 41 mass units, respectively (Equations 13.9 and 13.10), as seen in Figure 13.3(B).

Note that in Equations 13.7 and 13.8, the "−" is not a bond, but rather a subtraction sign, representing loss of H⁻; thus, the term $[M-H]^+$ represents the ion formed after the precursor molecule has lost a hydride to the CH_3^+. In Equations 13.9 and 13.10, the precursor molecule forms a bond with the reactive ions with which it collides, producing a new ion.

$$CH_3^+ + M \leftrightarrows CH_4 + \left[M-H\right]^+ \qquad \text{Eq. 13.7}$$

$$C_2H_5^+ + M \leftrightarrows C_2H_6 + \left[M-H\right]^+ \qquad \text{Eq. 13.8}$$

$$C_2H_5^+ + M \leftrightarrows \left[M+C_2H_5\right]^+ \qquad \text{Eq. 13.9}$$

$$C_3H_7^+ + M \leftrightarrows \left[M:C_3H_7\right]^+ \qquad \text{Eq. 13.10}$$

Unlike in EI, negative ion CI (NICI or NCI) is relatively common, although it is not strictly speaking chemical ionization in practice. That is, in common practice, the ionization of the analyte molecule does not occur as a result of a chemical reaction with the reagent gas. Rather, the reagent gas serves as a buffering medium, slowing the electrons in the flux to thermal energies such that they can easily be captured by analyte molecules. As before, the electron capture process is enhanced for molecules containing electronegative elements, and so NCI is a common method for the analysis of halogenated compounds such as pesticides and effluent from certain plastics industrial processes. The two most common ionization reactions are resonance electron capture (Equation 13.11), which results in a molecular ion, M⁻, and dissociative electron capture (Equation 13.12), which results in the formation of a fragment ion, $[M-x]^-$.

$$M + e^- \rightarrow M^- \qquad \text{Eq. 13.11}$$

$$M + e^- \rightarrow \text{product ions} \qquad \text{Eq. 13.12}$$

Problem 13.11: The mass spectrometric analysis of the common pesticide DDT ($C_{14}H_9Cl_5$) resulted in a base peak of m/z = 352, with the only other major peaks being 317, 235, and 165. Postulate about the type of ionization (EI or CI), the most logical mode of ionization (negative or positive mode); the ionization reaction responsible for the base peak; and the m/z= 317 peak.

Problem 13.12: What is the final velocity of the m/z = 235 ion in Problem 13.11 if an accelerating voltage of 1.3 kV is used in the mass analyzer?

John Fenn (1917–2010) shared the 2002 Nobel Prize in chemistry for his work on the electrospray ionization source for mass spectroscopy. Fenn received his undergraduate degree from Berea College and also took some chemistry courses at the University of Iowa and Purdue University. He received his PhD from Yale University. Dr. Fenn worked for Monsanto before taking a position on the faculty at Virginia Commonwealth University. ■

Electrospray Ionization

While multivalent ions are possible with CI and EI, the number produced is vanishingly small relative to the number of monovalent ions. *Electrospray ionization (ESI)* is a soft ionization source that makes it possible to generate a measurable number of multivalent ions. In some instances, the generation of multivalent ions can be quite important, because the mass analyzer in an MS instrument responds to m/z rather than to mass. A monovalent ion with a molar mass of 24,000 daltons (hence m/z = 24,000) could not be measured in most mass analyzers. If that same ion had a charge of +8, its m/z would be 3,000, bringing it into the range of a typical mass analyzer.

Before the development of electrospray ionization, the field of MS faced two limitations: (1) it was necessary to convert the analyte molecule to the gas phase prior to ionization, but many substances cannot be volatilized easily, and (2) mass analyzers could not be used with very large ions. We will discuss the limitations of mass analyzers in Section 13.3. Both of these limitations excluded the analysis of large organic or biochemical substances, because such species will typically undergo pyrolysis[5] before volatilizing and often have very high molar masses. As a chromatographic detector, the MS was limited to use with gas chromatography.[6] Even if a large molecule could be volatilized (by sputtering or desorption processes), the second limitation still excluded the analysis of larger biochemical species such as proteins or nucleic acids. The advent of ESI overcame these limitations, making possible the analysis of very large, nonvolatile molecules.

Furthermore, the nature of ESI made it amenable as an ion source interface between liquid-based methods (e.g., high-performance liquid chromatography [HPLC;

---

[5] Pyrolysis is the technical term for thermochemical decomposition. The process is irreversible and results in a new compound.

[6] Gas chromatography is described in detail in Chapter 16.

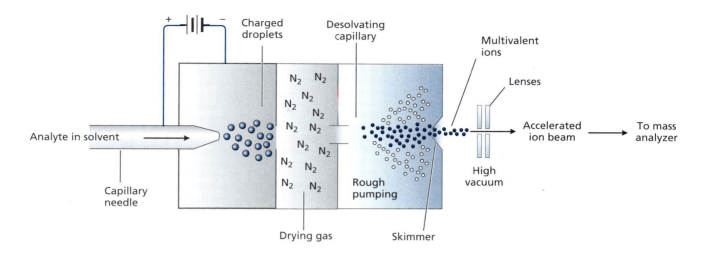

Figure 13.8 Schematic of an ESI system.

Chapter 15] and capillary electrophoresis [CE; Chapter 17]) and an MS detector. Because ESI had such a significant impact on biochemical analysis, its developer, John Fenn, was awarded a Nobel Prize in chemistry in 2002.

A schematic of an ESI source is depicted in Figure 13.8. The analyte is dissolved at low concentration in a volatile solvent, as is typically the case for HPLC or CE, and the inlet to the ESI system involves directing the solution through a metal capillary to which is applied a high voltage, typically 1 to 5 kV but sometimes up to 10 kV. The result is highly charged droplets of solvent that contain a small number of analyte molecules. To aid in this process of creating charged droplets, the mobile phase typically contains a volatile organic acid, such as acetic acid, which more easily accepts the charges and then evaporates with the solvent in the subsequent step. The droplets are desolvated by the application of a drying gas (such as dehydrated N_2), warming, and rough pumping to reduce the system pressure. As the solvent evaporates, the droplets decrease in size, resulting in an increase in the charge density as well as the surface tension. At some point (called the Rayleigh limit), the charge density becomes too high and the droplet too small to support the increased surface tension, and the droplets undergo a *coulombic explosion*. This results in the generation of individual (and thus gas state) analyte ions (see Figure 13.9). It is important to understand that the vast majority of analyte ions generated at this point are monovalent, but a measurable fraction of them do end up being multivalent. Note that normal volatilization was not needed; that is, the sample was not volatilized through heating and therefore there was no risk of pyrolysis. The gas state ions come about through other, gentler, processes. So there is little to no fragmentation of the analyte molecules, and hence, ESI is a *soft ionization method* suitable for large biomolecules. Furthermore, the existence of the multiple charges reduces the m/z ratio and allows for the use of existing mass analyzers. If, for instance, a mass analyzer had an upper limit of m/z = 3,000 Da, a protein having a mass of 60,000 Da could be measured if it were to acquire a charge of +20, which is a feasible using ESI.

Because it is impossible to strictly control the number of charges individual analyte molecules will acquire, the spectra obtained using ESI can be fairly complex. Figure 13.10 shows the ESI-MS of the enzyme carbonic anhydrase II (CA2) having a molecular mass of 29,000 Da analyzed in a quadrupole mass analyzer capable of analyzing ions up to 2,000 Da. The CA2 molecule is not fragmented but does exhibit a wide range of positive charges, from 41+ down to 19+.

This complexity can be partially resolved by using tandem MS (also called MS-MS or MS-MS-MS), in which the gas state multivalent species undergo controlled

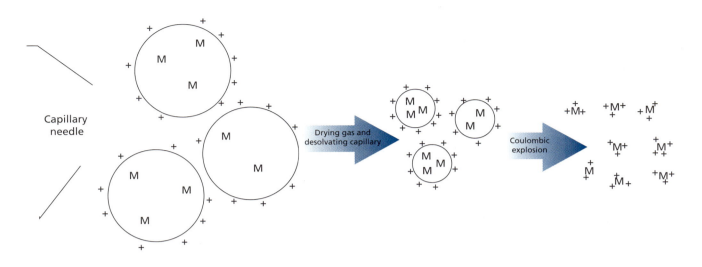

Figure 13.9 Depiction of the generation of multivalent molecular ions by ESI.

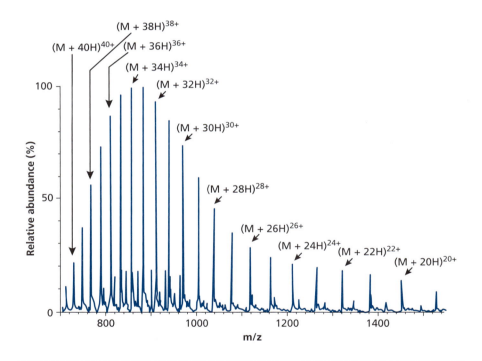

Figure 13.10 ESI-MS of the enzyme carbonic anhydrase II, 29 kDa.

fragmentation reactions between each MS measurement. The use of tandem MS methods has broadened the utility of ESI-MS into a method that can be used as a qualitative as well as quantitative technique. We will discuss some common tandem methods in Section 13.5.

Matrix-Assisted Laser Desorption Ionization

Matrix-assisted laser desorption ionization (MALDI) is another *soft ionization method* that can be used to volatilize high-boiling materials. It does not usually result in the generation of multivalent ions, and therefore its utility is limited to molecules that have a base mass within the range governed by the mass analyzer, which is typically 1,000 to 2,000 Da for standard analyzers. Because the spectra are far less complex than those seen with other ionization sources, MALDI allows for the observation of the precursor ion peak. For large biological molecules, MALDI-time-of-flight, or MALDI-TOF[7], analyzers have taken over a significant segment of the ESI market.

The basic process, depicted in Figure 13.11, involves dissolving the target analyte in a solution containing a crystal-forming compound. Droplets of the solution are deposited on a metal plate and the solvent is allowed to evaporate, leaving the analyte trapped in a crystalline matrix. A laser is directed at the crystalline mixture and the matrix absorbs the energy from the laser irradiation, resulting in desorption of the matrix and the analyte. During this process the matrix also ionizes (usually by protonation), and then the proton is transferred to the analyte by gas state collision reactions similar to those seen in chemical ionization.

Common crystal matrix-forming substances include 2,5-dihydroxybenzoic acid, sinapinic acid, and α-cyano-4-hydroxycinnamic acid. Favorable matrix compounds are typically acidic (so that they can easily transfer a proton to the analyte), contain conjugated double bonds (so that they efficiently absorb the energy from the laser), and are soluble in aqueous or acetonitrile-aqueous solutions.

As noted above, the spectra obtained are significantly simpler than those seen in ESI-MS. Figure 13.12 shows a MALDI spectrum obtained for a mixture of several proteins. Each protein is represented by only two peaks, the primary of which is the univalent precursor ion.

Secondary Ion Ionization

The standard ionization methods so far discussed are amenable only to analytes that can be isolated and, in most cases, dissolved in an appropriate solvent. However, there are instances when it is more beneficial to study a sample in bulk, such as when it is necessary to understand the chemistry of a sample surface. In 1960, the French company

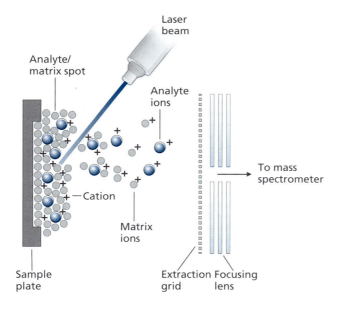

Figure 13.11 MALDI ionization.

---

[7] TOF stands for "time-of-flight"—it is a type of mass analyzer that can be used with ions of very high mass. TOF will be discussed in Section 13.3.

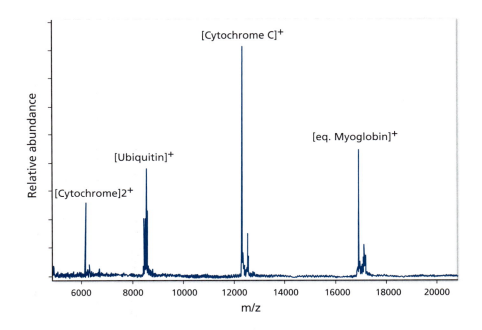

Figure 13.12 MALDI-TOF mass spectrum of a mixture of three proteins.

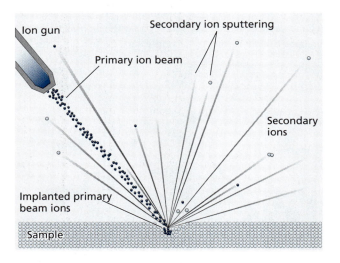

Figure 13.13 SIMS sputtering of secondary ions for analysis.

CAMECA developed a method that allowed researchers to conduct mass spectrometry on materials sputtered from surfaces. This method, *secondary ion mass spectrometry (SIMS)* has become the most sensitive technique for the analysis of surfaces, exhibiting detection limits down to the ppb range.

In SIMS, ions (e.g., Cs^+, Ar^+, O_2^+) are accelerated in a beam toward the target surface (see Figure 13.13). The ions in the beam, called primary ions, impact the surface and become embedded, while, in the process, secondary ions composed of surface molecules are sputtered off and accelerated toward the mass analyzer. The ions are not only single molecules, but also include molecular clusters; therefore, additional information about the composition of surface material can be gleaned. Because the ionization profile of sputtered ions is difficult to control, SIMS is usually considered to be a qualitative technique rather than a quantitative one.

Thermal Ionization

Thermal ionization mass spectrometry (TIMS) is the simplest and least expensive of the ion sources, but it has relatively limited use. However, it does have a powerful and important niche in the areas of geochronology, geochemistry, and cosmochemistry. A sample is deposited onto a filament that is then placed into the instrument. The instrument is sealed and brought to ultrahigh vacuum (typically $\sim 10^{-8}$–10^{-9} torr). A current is then passed through the filament, causing it to achieve a temperature of up to 2,500°C, sputtering some of the sample off of the filament and ionizing it. The ions are accelerated toward the mass analyzer.

The most common use of TIMS is to determine isotopic ratios for dating of geological samples. For instance, the mineral zircon will crystallize with uranium or thorium atoms within the structure but will reject lead atoms during crystallization. There are two known uranium-to-lead decay schemes with half-lives on the geologic time scale. The assumption made is that, during the formation of a given geologic body, only uranium would have been cocrystallized within the zircon, and any lead present is due entirely to decay of the original uranium. Therefore, the time that has passed since crystallization can be determined by standard kinetics, assuming that the initial uranium concentration is the current uranium concentration plus the current lead concentration.

> **Problem 13.13:**   Given that the half-life of the ^{238}U-to-^{206}Pb decay series is 4.47 billion years, how long ago was a zircon crystal formed if the current ^{238}U concentration is 3.97 ppb and the current ^{206}Pb concentration is 224 ppm? (*Hint:* Recall that radioactive decays follow first-order kinetics.)[8]

Inductively Coupled Plasma Ionization

We discussed the use of an inductively coupled plasma (ICP) torch for atomic emission spectroscopy (AES) in Chapter 9, where we learned that a notable proportion of the analyte atoms are also ionized in the high-energy plasma, despite the presence of a relatively large flux of free electrons. We examined ICP-AES as a way of measuring the electromagnetic radiation emitted by excited state atoms (and sometimes ions) as they decayed back to the ground state, but scientists have found a further use for the ions generated in the plasma: they can pass them into an MS and analyze them directly for quantitative and qualitative purposes. The MS provides a nearly instantaneous view of all elements present in a sample while simultaneously providing a highly sensitive quantitative elemental analysis. Indeed, *ICP-MS* is the most sensitive atomic method available, providing detection limits down to the ng/L and pg/L ranges for most elements.

To maximize sampling efficiency, the ICP torch is rotated to a horizontal position and placed directly in contact with the sampling cone of the MS[9] (see Figure 13.14). The sampling cone accepts ions streaming through the core of the plasma, because this will contain the highest concentration of analyte ions. The torch is produced under atmospheric conditions. Therefore, a series of steps must be taken in order to transfer the sample to the mass analyzer, which is maintained under high vacuum. From the sampling cone, the ions move into a short chamber that is evacuated using a rough mechanical pump, resulting in a pressure in the range of 10^{-2} torr.

From the sample cone orifice (about 1-mm diameter), materials from the torch are dispersed and directed toward another orifice (about 0.4-mm diameter) in the skimmer cone. Only a small fraction of the original plasma material passes through the

---

[8] $\ln(A)_t = -kt + \ln(A)_o$ and $kt_{1/2} = 0.693$.

[9] Students are urged to revisit Chapter 9 if they need a refresher on how the ICP torch is produced.

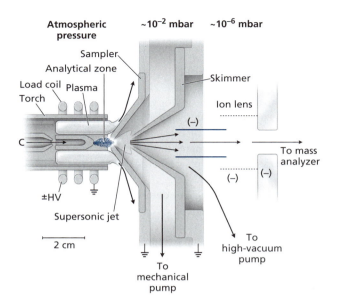

Figure 13.14 ICP-MS interface.

skimmer orifice into the high-vacuum region beyond. The ion beam is accelerated and collimated as in other ion sources, and it is then directed toward the mass analyzer.

Using an ICP as an ion source involves some unique challenges because the ion detector, which we will discuss in Section 13.4, is sensitive to ultraviolet-visible photons. With an ICP, the intense light emitted by the plasma torch produces significant background noise if the mass analyzer is a standard quadrupole (the most common type of mass analyzer) oriented in line with the detector. Light sensitivity is not an issue for most instruments (i.e., other source types), because only ions and very few photons are produced in the source. Furthermore, it is important to eliminate energetic neutrals and solid particles from the ion beam path as well, because any of those that make it through to the detector will contribute to background noise and produce a spurious signal.

Two common methods are used to eliminate such background from ICP-MS. The simplest approach is to have a beam blocker (also known as a beam stop) positioned a few millimeters behind the skimmer cone orifice. The beam blocker is a small metal disk with a positive voltage applied to it. Photons as well as neutral particles are blocked by the disk, but cations are repelled around it and refocused into a collimated beam with electrostatic lenses following the beam blocker. The other method is to use an ion guide—that is, to use electrostatics to direct ions in an indirect path, allowing photons and neutrals to travel harmlessly in a straight path. One example of this is the Perkin Elmer quadrupole ion deflector (Figure 13.15). The ICP torch is located at the bottom right of the figure, and the sampling and skimmer cones are visible in the figure. In the deflector (Figure 13.15), positive voltages are applied to the right-hand poles and negative voltages applied to the left-hand poles. Only positive ions are repelled from the right and attracted toward the left to be focused in a beam toward the mass analyzer (the quadrupole mass filter in this case).

Figure 13.15 also depicts a way to minimize other problems encountered in ICP-MS. In the energetic plasma, a variety of multielement ions are formed, such as ArO^+, CNH^+, and Ar_2^+. These ions can interfere with analytes of interest. The mass of ArO^+ is identical to that of a common analyte, $^{56}Fe^+$. Although it is possible to analyze iron using a different isotope, the difference in abundance (^{54}Fe is 5.8% abundant, whereas

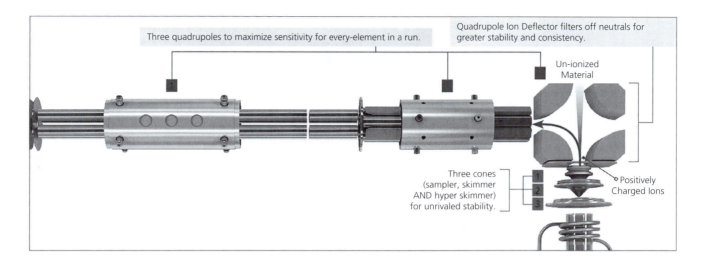

Three quadrupoles to maximize sensitivity for every-element in a run.

Quadrupole Ion Deflector filters off neutrals for greater stability and consistency.

Un-ionized Material

Three cones (sampler, skimmer AND hyper skimmer) for unrivaled stability.

Positively Charged Ions

Figure 13.15 The Perkin Elmer ICP-MS system with the quadrupole ion deflector.

^{56}Fe is 92% abundant) means we lose an order of magnitude in sensitivity. The problem is even worse for aluminum, which occurs as ^{27}Al with 100% abundance. The ion CHN$^+$ has the same m/z as ^{27}Al$^+$. Two methods have been used to minimize these interferences, and the *universal cell* depicted in Figure 13.15 can be used as either a *collision cell* or as a *reaction cell*, as described below.

A *collision cell* can be placed between the ion lenses and the mass analyzer. In this approach, an inert gas, such as He, is injected into the cell. Because molecular ion species (ArO$^+$, CHN$^+$, Ar$_2^+$) have higher cross-sectional profiles than single atom ions, they tend to collide more often with the He atoms. Each collision decreases the kinetic energy of the cation. By the end of the collision cell, the molecular ion species will have had their kinetic energy decreased significantly compared with the analyte atomic ion species. The ions then pass into another lensing region, but in this case the second lens is slightly more positive than the first lens. Ions with high enough kinetic energy (atomic ions) are able to bypass that barrier, but those with low kinetic energy (molecular ions) are rejected. The collision cell does decrease the sensitivity of the overall method but usually to a lesser degree than moving to an alternate isotope would.

Alternatively, a *reaction cell* can also be placed between the initial ion lenses and the mass analyzer. Here, a reactive gas, such as ammonia, is injected into the cell. The molecular ions tend to react more readily with the reactant gas, whereas the atomic ions tend to react less readily. Reaction of molecular ions with the gas produces either neutral species or ions of a mass that does not interfere with the analyte of interest. The reaction cell yields a smaller decrease in sensitivity than the collision cell but is less widely applicable. Although a collision cell will impact all molecular ion species, the reaction cell only impacts those that have a measurably increased reaction profile with the reactive gas.

Problem 13.14: A common ion formed in the ICP source is Ar$_2^+$. This argon ion dimer might interfere with the analysis of what atomic ion?

Problem 13.15: In wastewater analysis, the molecular ions ArCl$^+$ and ClOH$^+$ are not uncommon. These molecular ions might interfere with the analysis of what atomic ions?

Elemental Methods

Where does inductively coupled plasma-mass spectroscopy (ICP-MS) fit into the overall "toolbox" of elemental analysis techniques? Recall that in Chapter 7, we examined atomic absorption spectroscopy (AAS), and in Chapter 9, we examined atomic emission spectroscopy (AES). Both of these techniques find wide use in the elemental analysis of a wide variety of analytes. In Chapter 10, we considered X-ray fluorescence (XRF), which is very useful in examining the elemental composition of surfaces. The information one obtains from ICP-MS is similar to that of AAS or AES. As in AES, one can simultaneously analyze multiple elements at once using ICP-MS, and it would be fair to say that in this regard, ICP-MS is superior to AES. Furthermore, the detection limits of ICP-MS are typically three to six orders of magnitude smaller; however, ICP-MS instruments are considerably more costly than AAS or AES instruments.

	Cost ($1000)	Typical Sensitivity	Analysis Speed	Prep Time	Operation Maintenance Cost	Destructive
FLAME AAS*	10–20	ppm–ppb	≈ 60 sec/element*	Significant	Medium	Yes
GRAPHITE FURNACE AAS†	70–80	0.001–10 ppb	3 min/element	Significant	High	Yes
AES‡	40–80	<1–10 ppb	30 sec/analysis†	Significant	High	Yes
XRF‡	75–125§	~ppm	1–5 sec/analysis	Very Little (none)	Low	No
ICP-MS	120–180	pptr–ppb	1–5 min/analysis‖	Significant	High	Yes

* Agilent has a sequential mode (10 elements per analysis) that will analyze up to 10 elements at ≈ 2 min/analysis.

† Usually a graphite furnace is an added feature to a flame AAS. A stand-alone graphite furnace would be less expensive.

‡ AES and XRF routinely analyze multiple elements per analysis.

§ This comparison is for a research-grade instrument, not a handheld (retail) grade instrument.

‖ An ICP-MS can effectively detect all of the elements present in a material simultaneously.

<h2>13.3 Mass Analyzers</h2>

As noted in Section 13.1, the function of the mass analyzer is to separate or distribute analyte ions based on their relative m/z ratios. The mass analyzer is the mass spectrometric analog of the monochromator used in optical instruments. The separation of ions is fundamentally related to the mobility of the ions of varying m/z ratio in a controlled magnetic or electrical field.

There are two basic types of mass analyzers: single ion focusing analyzers and simultaneous analyzers. *Single ion focusing* (or *scanning*) analyzers measure only a single m/z at a given time, whereas *simultaneous* analyzers measure multiple m/z at the same time.

A spectrum is produced in single ion focusing analyzers by scanning across voltages and/or fields used in the analyzer to focus a range of m/z ions on the detector over a short period of time. This would be similar to a scanning monochromator, in which only one exit slit and one detector are used, and the optics are used to scan across the dispersed optical spectrum. Simultaneous mass analyzers allow for the detection of a wide range of m/z ions at once, analogous to the use of a diode array or charge coupled device detector in an optical system. Such analyzers can disperse ions in space (sector analyzer), disperse ions in time (time-of-flight analyzer), or allow for excited state ions to decay in time with a resulting free induction decay signal (Fourier transform–ion cyclotron resonance (FT-ICR).

R. Graham Cooks is a distinguished professor of chemistry at Purdue University and can trace his academic lineage back to J. J. Thomson. Professor Cooks received his BS (1961) and PhD (1965) from the University of Natal, South Africa, and he received a second PhD (1967) from Cambridge University, United Kingdom. He has authored more than 950 research publications and is considered one of the most highly cited living researchers in the field of mass spectrometry. Some of his accomplishments include significant contributions to the development of quadrupole ion traps, cylindrical ion traps, mini/portable mass spectrometers, secondary ion mass spectrometry, desorption electrospray ionization, and laser desorption ionization. To read more about Professor Cooks, see http://aston.chem.purdue.edu/. ▬

PROFILE
R. Graham Cooks

Resolution

One can calculate the resolving power, or resolution (R), of a mass analyzers using Equation 13.13.

$$R = \frac{\bar{m}}{\Delta m}$$ **Eq. 13.13**

where $\bar{m}$ is the mean of the two distinguishable m/z values and Δm is the difference between the two values.

If we look closely at Equation 13.13 and Example 13.3, we can see that for a given mass analyzer, the resolution will usually be different for different m/z values. We will reconsider this further in Example 13.4 below, but we first need to understand a bit about how a mass analyzer works.

In most cases, the resolution of a mass analyzer cannot be altered, but the FT-ICR, similar to the FTIR spectrometer (Chapter 11), can achieve greater resolution by measuring the free induction decay signal over a longer period of time. For scanning instruments, control of the scan parameters can provide some flexibility in resolution as a function of m/z, allowing us to avoid the limitation demonstrated in Example 13.3.

Problem 13.16: What resolution would be necessary in order to resolve peaks from univalent ions having molar masses of 1,058.0 g/mol and 1,058.5 g/mol?

Example 13.3 – Resolution of a Mass Analyzer, Part 1

A mass analyzer produces a spectrum showing two resolved peaks at m/z = 58.0 and m/z = 58.5. What is the resolution of the instrument at this m/z range? If the resolution remains constant over the mass spectrum, what would be the difference in resolvable m/z values for this analyzer when the average mass is m/z = 1,058?

STRATEGY –

We need to use Equation 13.13 in both cases.

SOLUTION –

For the first part of the question:

$$R = \frac{\bar{m}}{\Delta m} = \frac{(58.0 + 58.5)/2}{(58.5 - 58.0)} = 116.5 = 117$$

And for the second part of the question, we assume the resolution is a fixed value for the analyzer, so:

$$R = \frac{\bar{m}}{\Delta m} = \frac{(1058)}{\Delta m} = 116.5 \qquad \Delta m = 9.08$$

Going back to the question, what would be the resolved m/z values at an m/z = 1,058? By applying the calculated $\Delta m = 9.08$, this means that, at best, the instrument could resolve m/z = 1,062 (1,058 + 9.08/2) from m/z = 1,053 (1,058 – 9.08/2). This example demonstrates the difficulty of resolving m/z across a wide range when the resolution is fixed.

Problem 13.17:   Assuming the resolution of the mass analyzer considered in Problem 13.16 is fixed across the mass spectrum, what two m/z values could be resolved with respect to a mean m/z of 111?

Sector and Double-Focusing Mass Analyzer

The *sector mass analyzer* operates on the same principle that Eugen Goldstein used in 1886 to deflect the paths of canal rays produced in gas discharge tubes (see "Profile—Eugen Goldstein"). Although a variety of designs have been proposed and used, we will focus on the most common configuration, the magnetic sector single-ion focusing analyzer. In practice, sector mass analyzers are not widely used because they tend to be more expensive and have a larger bench-top footprint than most other analyzers. However, they are still commercially available, especially for methods requiring high resolution (e.g., IRMS and AMS in Section 13.5), and are still considered to be the gold standard against which other mass analyzers are measured.

Figure 13.16 represents an electron ionization source with an accelerating voltage, V. Ions leave the source with velocities related inversely to their masses (Equation 13.3). A single-ion focusing magnetic sector analyzer is designed to allow only ions of a single m/z to pass through the exit slits to the detector.

A magnetic field (B, in tesla) is applied orthogonal to the curved path, causing positive ions to bend toward the detector slit. In order for a given ion to make it through the sector, its path must be bent by the magnetic field in exactly the correct arc (r, in meters). Therefore, the force applied by the magnetic field (F_B, in newtons, Equation 13.14) must exactly equal the centripetal force (F_C, in N, Equation 13.15) required to

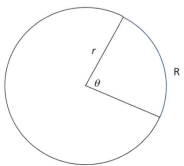

The length of the arc, r, can be found as r = θr for θ in radians.

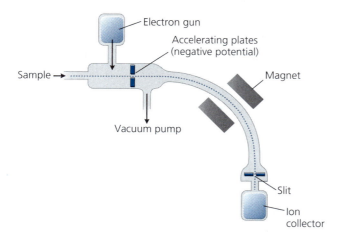

Figure 13.16 Magnetic sector mass analyzer.

curve the path of the ion. Setting these two forces equal to each other and including Equation 13.3 to substitute for velocity, we can see the experimental parameters that allow a given m/z ion to traverse the arc.

$$F_B = B \cdot z \cdot e \cdot v \qquad \text{Eq. 13.14}$$

$$F_C = \frac{m \cdot v^2}{r} \qquad \text{Eq. 13.15}$$

$$\frac{m}{z} = \frac{B^2 \cdot e \cdot r^2}{2\,V} \qquad \text{Eq. 13.16}$$

Equation 13.16 shows us that (1) for a fixed magnetic field, we can select a specific m/z ion by varying the accelerating potential, and (2) for a fixed accelerating potential, we can select a specific m/z ion by varying the magnetic field (assuming an electromagnet is used rather than a solid-state magnet). For a given V and B, only specific m/z ions will traverse through the middle of the sector. Those ions with higher mass will not have sufficient curvature imparted to their paths and thus will collide with the outer wall of the sector, whereas those of lower mass will be curved too greatly and will be lost on the inner wall.

Problem 13.18: A certain MS uses a magnetic sector mass analyzer having a fixed accelerating voltage of 2.95 kV and a radius of 0.233 m. What magnetic field (in T) must be used to focus a carbon dioxide ion $[CO_2{}^+]$ on the detector?

Problem 13.19: If the same mass analyzer as described in Problem 13.18 were used and the accelerating voltage was kept constant, what magnetic field would be required to measure a bis-imidazolium dication, $(C_7H_8N_4)^{2+}$?

In order to obtain a mass spectrum, V or B (or both) is scanned over a specified range in a very short period of time, and ions are counted at each corresponding m/z value. A full spectrum (say, about 40–4,000 Da) can be obtained in just a few seconds at fairly high resolution. We assume that the ions are being generated by the source at essentially the same rate during the time of the experiment, and for most scenarios that

Eugen Goldstein (1850–1930) was a German physicist who conducted extensive research on gas discharge tubes. His work led to the discovery of the electron by J. J. Thompson, who went on to build the first mass spectrometer. Goldstein also discovered anode rays, which were later confirmed to be protons by Ernest Rutherford. ■

is a reasonable assumption. For greater sensitivity in quantifying a given ion, the *selected ion mode* is used; that is, V and B are fixed at specific values to focus a single m/z on the detector to allow for integration of the signal over time.

Problem 13.20: Assuming the same parameters described in Problem 13.18, what magnetic field range would you need to scan in order to sweep a m/z range of 40 to 4,000 Da?

Example 13.4 – Resolution of a Mass Analyzer, Part 2

Assuming the same parameters described in Problem 13.18, what is the percentage change in magnetic field needed to isolate m/z 58.0 Da and 58.5 Da? Repeat this for m/z 1,058.0 and 1,058.5 Da.

STRATEGY –

We will calculate the field (B) for each m/z using Equation 13.16 and then calculate the percentage difference based on the lower of the two m/z values in each pair. Remember that the mass must be in kilograms *per ion*, not per mole, and a $V = J/C$ and $J = kg \cdot m^2/s^2$. Also, note that a tesla is $1\ kg/C \cdot s$.

SOLUTION –

For the m/z =58.0 Da, the mass of the ion is:

$$m = 58.0 \frac{g}{mol} \times \frac{1\,mol}{6.022 \times 10^{23}\,ions} \times \frac{1\,kg}{1000\,g} = 9.63 \times 10^{-26}\,kg$$

$$\frac{m}{z} = \frac{B^2 \cdot e \cdot r^2}{2\,V}$$

So:

$$B = \sqrt{\frac{m \cdot 2 \cdot V}{z \cdot e \cdot r^2}} = \sqrt{\frac{\left(9.63 \times 10^{-26}\,kg\right)(2)\left(2950 \frac{kg \cdot m^2}{C \cdot s^2}\right)}{(1)\left(1.602 \times 10^{-19}\,C\right)\left(0.233\,m\right)^2}} = 0.2556\,T$$

Similarly, for the m/z 58.5 Da ion, B = 0.2567, so the percentage difference is:

$$\% \, diff = \frac{\left| B_{58.0} - B_{58.5} \right|}{B_{58.0}} \times 100 = 0.43\% \text{ change in magnetic field}$$

To distinguish between m/z = 1,058.0 and m/z = 1,058.5 would require a 0.025% change in the magnetic field.

From Example 13.4, we can see that resolving m/z = 1,058.0 and 1,058.5 required a much finer control of the magnetic field than resolving m/z = 58.0 and 58.5. In other words, the change in the magnetic field to achieve a specific change in m/z decreases with the square root of the m/z value (Equation 13.16). It is far more difficult to resolve ions of high mass than those of low mass, which is why mass analyzers that rely on changing the path of ions (such as sector and quadrupole analyzers) are limited in terms of high-mass resolution. (See Table 13.1 near the end of this section.)

Because ions can also be deflected by electrical fields, a sector analyzer can be of the electrostatic (E) type as well, although for commercial instruments single-sector mass analyzers are generally of the magnetic (B) type as described in this section.

A more common type of sector mass analyzer is the double-focusing (sometimes referred to as the E-B type) instrument (see Figure 13.17). In theory, according to Equation 13.1 (KE = z·e·V), all ions of the same charge (z) should exit the accelerator with the same kinetic energy. However, the reality is that ions of each charge exit the source with a small range of kinetic energies, imparting a broadening on each m/z peak, diminishing resolution. This can be somewhat corrected, using an E-B type instrument.

In an E-B instrument, ions from the magnetic sector are passed into an electrostatic analyzer. We can treat the ion mobility in the electrostatic field in a fashion similar to that done for the magnetic sector (Equations 13.14 to 13.16). Because the ions are treated the same in the source, Equations 13.1 to 13.3 are still valid, and the force required to curve an ion's path (Equation 13.14) is also the same.

$$F_E = z \cdot e \cdot E \qquad\qquad \textbf{Eq. 13.17}$$

$$r = \frac{2\,V}{E} \qquad\qquad \textbf{Eq. 13.18}$$

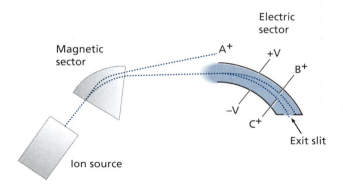

Figure 13.17 Double-focusing (E-B) type mass analyzer. V^+ and V^- represent the poles of the applied electric field in the electrostatic sector. A^+ represents an ion having sufficiently high m/z that the magnetic sector could not focus it onto the entrance slit of the electrostatic sector. B^+ represents an ion having a m/z too high for the electrostatic sector to focus it onto the exit slit. C^+ represents an ion having the appropriate m/z to be focused onto the exit slit of the instrument.

$$r = \frac{2(KE)}{zeE} \qquad\qquad \textbf{Eq. 13.19}$$

However, the electrostatic force (F_E, in N) required to impart that curvature is given by Equation 13.17, where E is the applied electrical field in V/m. If we follow the same treatment as before, we find that the radius of an ion's curvature (r) is independent of its mass and charge (Equation 13.18). In other words, in an applied electric field (E), ions of all m/z will follow the same path for a given accelerating voltage (V). Furthermore, a substitution of Equation 13.1 into Equation 13.18 shows us (Equation 13.19) that the E-type sector analyzer is one that focuses ions by kinetic energies rather than by mass. Therefore, by passing the ions from the magnetic sector into the E-type sector, the distribution of kinetic energies initially imparted by the source is diminished prior to passing the ions into the detector. The result is a decrease in peak width and thus an increase in resolution. It turns out that this double-focusing effect will work regardless of whether the electrostatic sector comes before or after the magnetic sector.

Problem 13.21: What is the magnitude of electrostatic field (E, in V/m) that would be needed in order to impart a curvature (r) of 0.233 m with a fixed accelerating voltage of 2.95 kV?

Problem 13.22: Calculate the kinetic energy of a univalent ion in the spectrometer described in Problem 13.21.

Quadrupole Mass Analyzer

The *quadrupole mass analyzer* is probably the most common type of mass analyzer found in modern chemical laboratories, in part because of its compact size, which readily allows for coupling to a gas chromatography instrument without taking up the majority of space in the laboratory. It provides the advantages of fast analysis with a very small bench-top footprint but exhibits only moderate-to-low resolution and at the sacrifice of maximum mass range ($m/z_{max} = 3,000$).

Figure 13.18 shows a schematic of a typical bench-top quadrupole mass analyzer. The poles are rods or hemispheres mounted in a square pattern. A direct current (DC) voltage is applied across each pair of diagonally opposed rods, and an alternating current (AC) voltage of radiofrequency is then superimposed on each opposing pair such that the cycles of the two opposed pairs are directly out of phase. As a "bundle" of ions enters the quadrupole, the ions will be initially drawn toward the negatively charged pair and repelled from the positively charged pair.

The nature of the *radio frequency* voltage dictates that, after half an AC cycle, the polarities of the quadrupole pairs will be reversed and positively charged ions will be drawn to the alternate pair of poles. This continues as the ions move through the quadrupole chamber. Only an ion of a specific m/z will maintain perfect resonance with a given radio frequency and will pass safely through the chamber to exit and move on to the detector. Ions of smaller m/z will be drawn too strongly initially to the negative poles and will collide with the poles quickly. Ions of larger m/z will not be able to adjust their path of curvature rapidly enough to keep up with the polarity shifts due to the radio frequency and will eventually spiral out of control and collide with the poles. For single-ion monitoring, a single radio frequency is applied so that the signal for the selected ion can be integrated over time. In order to acquire a spectrum, the DC and AC voltages are varied linearly over time, while keeping the V_{DC}/V_{AC} ratio constant, to select a range of m/z ions to send to the detector. Similar to the sector analyzer, one can acquire a spectrum in a brief period of time, typically on the order of a second or less. However, it should be noted that the mass accuracy of a quadrupole, although quite reliable for low and moderate m/z values, diminishes dramatically as the m/z increases above 1,000 amu.

Although a single quadrupole exhibits several disadvantages with respect to other mass analyzers (primarily in terms of resolution and mass range), its low cost and small footprint make it a workhorse instrument. Furthermore, the use of multiple quadrupoles in tandem can provide some unique benefits, as we will see in Section 13.5.

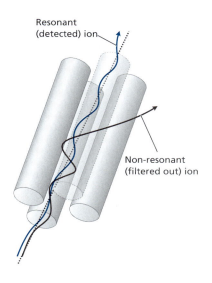

Figure 13.18 Quadrupole mass analyzer.

Time-of-Flight Mass Analyzer

The *time-of-flight (TOF)* mass analyzer works by measuring how long ions take to cross a region of space over which no magnetic or electric field is applied (the *field-free region*). Ions are accelerated in the source and then allowed to drift through the field-free region with a common start time (t_0). Ions of low m/z move more rapidly than those of higher m/z, and so the m/z spectrum is determined by measuring the time when ions reach the detector after t_0.

The primary advantage that TOF analyzers have over other types is that there is no theoretical upper limit on mass range, coupled with the ability to achieve very high resolutions (sometimes greater than 20,000). However, there is a practical limit in the mass range imparted by detectors rather than the analyzer. In general, MS detectors tend to lose sensitivity as the m/z increases, with an upper limit of around 10^6 Da. TOF analyzers are more expensive than quadrupole types but significantly less costly than sector types. TOF types also require somewhat greater expertise for proper maintenance.

As with other analyzers, an accelerating voltage is applied to the ionized analyte; therefore, Equation 13.1 is still applicable $(KE = z \cdot e \cdot V)$. In addition, Equation 13.20 describes the time (TOF) required for an ion to traverse the length of the field-free drift tube, where v is the ion velocity and L is the path length from the ion source to the detector. We can combine Equations 13.3 and 13.20 to have the mass-related TOF equation (Equation 13.21), showing that the TOF is directly related to m/z and inversely related to the accelerating voltage (V) used in the ion source. Furthermore, we can redefine the resolution in terms of flight times (Equation 13.22), where t' is the mean TOF for a given ion (center of its signal peak) and Δt is the full width of that peak at half-maximum for that peak.

$$\text{tof} = \frac{L}{v} \qquad\qquad \text{Eq. 13.20}$$

$$\text{tof} = L \cdot \sqrt{\frac{m}{z} \cdot \frac{1}{2V}} \qquad\qquad \text{Eq. 13.21}$$

$$R = \frac{t'}{2\Delta t} \qquad\qquad \text{Eq. 13.22}$$

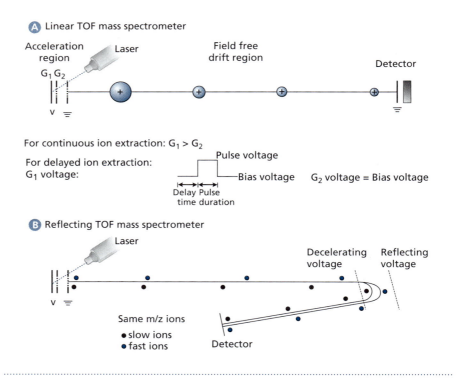

Figure 13.19 Two types of TOF mass analyzers: (A) linear and (B) reflection.

Equations 13.21 and 13.22 demonstrate that we should expect improved resolution as the drift path length increases and as the total time spent in the drift tube increases.

Figure 13.19 shows the original linear type and the more modern (and currently more common) reflection-type TOF mass analyzers. It is simpler to understand the concept of TOF analysis using the linear configuration initially, so we will start there. In Figure 13.19(A), G_1 would be the repeller described for an EI source (Figure 13.4), and the electrode designated as ground would be the final lens in the accelerating region of the source. As noted above, the ions are differentiated based on their velocities after the accelerating voltage in the ion source is applied. From Equation 13.3, we know that the velocity of a given ion is inversely proportional to its mass, and so a spectrum is developed based on when ions reach the detector after they are introduced into the field-free drift region.

Because a specific m/z determination must be made based on the measurement of the time of flight, it is imperative that the start time for a given "packet" of ions be known with a high degree of accuracy. This is typically accomplished in one of two ways.

Method 1   If ions are generated in a standard way (e.g., using a EI, MALDI, or a CI source), the ions will form throughout a certain volume of space (the volume of the source chamber), with different kinetic energies depending on how much energy remains after electron impact and fragmentation. The TOF peak for a set of ions with the same m/z will broaden based on those initial positions of ions; those ions that are near the repeller will spend more time in the accelerating voltage region and will achieve higher velocity than those ions near the exit lens. Therefore, a bundle of ions of identical m/z will have different starting kinetic energies and velocities, which means they will show up at the detector over a range of times (hence broadening the peak). This range of starting energies can be decreased by using *delayed pulsed extraction* or *delayed ion extraction*, as shown in Figure 13.19(A) and Figure 13.20.

Figure 13.20(A) represents the use of a normal ion source (continuous extraction). The black and blue dots represent two different ions of the same m/z. Because they were generated and accelerated in slightly different conditions, they reach the detector over a range of times, which the system reads as a range of m/z values

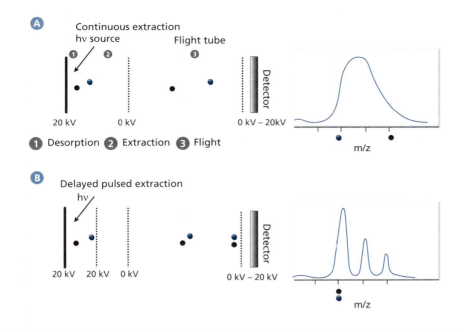

Figure 13.20 Depiction of delayed ion extraction: TOF-MS.

(a broad peak). In delayed pulsed extraction, shown in Figure 13.20(B), an additional lens is included relatively close to the repeller plate and that lens is initially held at the same positive potential, 20 kV this case, for a brief period. This is also demonstrated in Figure 13.19(A), with the pulsed program applied to lens G_2. Positive ions are trapped between the two positively polarized electrodes. During that brief period, collisions among ions allow for equilibration of initial energies and ions are more focused spatially because they are not allowed to drift throughout the entire source chamber. Then the potential of the additional lens is removed, and ions are accelerated across the 20 kV potential difference between the repeller and the accelerating lens at ground (0 V) for a timed pulse period. The result is ions accelerated as a well-defined packet across the voltage and directed into the field-free drift region. The sketched spectra shown in Figure 13.20 demonstrate the improvement in resolution using this method.

Method 2   Another method for improving TOF resolution involves the use of a reflecting field at one end of the spectrometer, visible in Figure 13.19(B). The improvement is realized due to the penetration of ions into the reflection field. Consider two ions having the same m/z. One—the blue dot in Figure 13.19(B)—initially has a slightly higher energy (velocity) than the other (black). As they move down the drift tube initially, the higher energy ion starts to out-distance the lower energy one. Left unchecked, this would result in a broad peak on detection of the ions. However, both ions encounter the reflection field. The dot, because it has higher energy, penetrates the field a bit more deeply and thus spends more time in the reflection field than does the black dot, and the black dot (lower energy) gains some ground on the white dot. After both are reflected, the higher energy ion gains back the lost ground and both encounter the detector at approximately the same time.

As noted above, resolution is improved by minimizing the dispersion in energies, and we saw that this can be accomplished by initiating the TOF in a pulsed fashion. We have discussed one way that is done above (see Figure 13.20). Another way is to ensure that all analyte ions are generated simultaneously in a pulsed fashion. For instance, if ions can be generated using a pulsed laser, the start time of the TOF can be synchronized with the laser pulses. For this reason, MALDI ionization and TOF mass analysis are naturally paired together in most MALDI instruments.

Problem 13.23: Determine the time of flight for an 800 m/z ion accelerated across 6.5 kV and allowed to drift through a length of 2 m.

Problem 13.24: If the instrument used to measure the 800 m/z ion, as described in Problem 13.23, has a resolution of 15,000, what is the expected peak width for that ion?

Fourier Transform Ion Cyclotron Resonance

Although *FT-ICR MS* is one of the less common methods used in mass analyzers, it is beginning to gain popularity on account of its ability to measure ions of essentially any mass range with exceptional resolution. It does not utilize a standard detection method; therefore, it is not limited in the upper mass ranges because of the insensitivity of normal ion detectors to high mass ions. However, this increased utility comes at a significant cost, both monetarily and in terms of the laboratory footprint. Because of the need for a high-field superconducting magnet, an FT-ICR will cost more than $200,000, and it will require, minimally, a space of 150 square feet. You will see that the measurement of the mass spectrum has many similarities to what you will learn for pulsed FT–nuclear magnetic resonance (FT-NMR) in Chapter 14.

Ions are generated using a standard ion source and then are passed through regions of successively lower pressure and temperature until a pressure of less than 10^{-10} torr and a temperature of 4K is achieved. The few ions remaining are pulse injected into the ion cyclotron cell (see Figure 13.21), which is held in a permanent magnetic field, B, generated by a 3 to 14 T superconducting magnet. The positive ions experience the Lorentz force, causing them to begin to orbit (precess) around the axis of B. The radius of the orbit and frequency of precession are dependent on the m/z ratio of the ion. These values are described by Equations 13.23 and 13.24, where r_C is the ion cyclotron resonance orbital radius (in meters), v_{xy} is the velocity (m/s) of the ion in the x-y plane (perpendicular to the z axis along which B is applied), and f_C is the precession frequency in Hz:

$$r_C = \frac{m}{z} \cdot \frac{v_{xy}}{B}$$

Eq. 13.23

$$f_C = \frac{z}{m} \cdot \frac{B}{2\pi}$$

Eq. 13.24

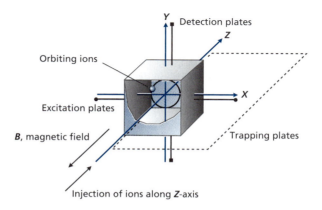

Figure 13.21 Trapping of ions in an ion cyclotron cell.

Problem 13.25: Demonstrate that the units in Equations 13.23 and 13.24 cancel each other out.

Problem 13.26: What is the resonance orbital radius of the CH_3^+ with v_{xy} of 3.0×10^4 m/s in a 7.035 T magnetic field? What is the precession frequency?

Once the ions are trapped in the ICR cell (which is sometimes called an *ion trap*) and have established equilibrium in terms of their precession, an excitation signal is imparted by applying a broadband radio frequency signal to opposing plates of the cell. If a single radio frequency is applied, only the ion whose resonance frequency matches that signal will be excited. When a broadband signal is applied, a broad range of ions become excited. The excited ions increase their resonance radius (Equation 13.25) by a factor dependent on the amplitude of the radio frequency signal (V_{p-p}) and the time of the radio frequency pulse (t_{excite}) (see Equation 13.25, where d is the distance between the excitation plates [basically the width of the ICR cell]).

$$r_{excited} = \frac{V_{p-p} \cdot t_{excite}}{2 \cdot d \cdot B}$$

Eq. 13.25

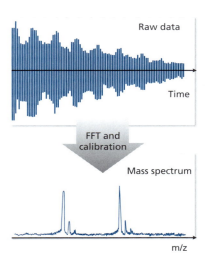

Figure 13.22 Ion cyclotron FID and the subsequent Fourier transform to the MS.

Note from Equation 13.25 that the excited state radius is independent of the m/z. Furthermore, the ions of a given m/z will be forced to cluster together in phase with the radio frequency pulse. Once the excitation pulse is stopped, the circuit to the detection plates is closed so that the signal generated by the ions as the excited state returns to the ground state in the chamber can be observed.

Charged particles moving in a circular path will generate an oscillating current within the detection circuit and produce a beat pattern consisting of a free induction decay in a manner exactly analogous to the free induction decay (FID) signal discussed in FT-NMR (see Chapter 14). Just as with the FTIR interferogram (Chapter 11) and the FT-NMR FID (Chapter 14), the ICR FID can be transformed into the frequency domain using the Fourier transform to yield the precession frequencies of ions in the cell, which can be further decoded into a mass spectrum (see Figure 13.22). Resolution can be increased experimentally by measuring the FID over a greater period of time, and signal-to-noise can be increased by repeating the excitation–decay process many times and signal averaging. Not only is FT-ICR useful for the identification of individual compounds, but its superior resolution lends to it the ability to discern components of complex mixtures.

Example 13.5 – Excitation Radius in a Fourier Transform–Ion Cyclotron Resonance Analysis

What is the excitation orbit, in cm, if a 2 V broadband radio frequency signal is applied for 1 ms to 2 cm ICR cell while ions are precessing about a 7.035 T field?

STRATEGY –
We can apply Equation 13.25 here, recalling that $1\ T = 1\ V \cdot s/m^2$.

SOLUTION –

$$r_{excited} = \frac{V_{p-p} \cdot t_{excite}}{2 \cdot d \cdot B} = \frac{(2\,V)(1 \times 10^{-3}\,s)}{(2)(0.02\,m)\left(7.035\frac{V \cdot s}{m^2}\right)} = 0.0071\,m = 0.71\ cm$$

Therefore, all ions, regardless of m/z, would achieve an excited state radius of 0.71 cm.

Problem 13.27: To avoid collision with the cell walls, what is the maximum pulse width for a 2 V excitation signal that could be used for a 2 cm cell in a 7.035 T magnetic field?

Quadrupole Ion Trap Mass Analyzer

A *quadrupole ion trap (QIT)* mass analyzer (Figure 13.23) functions similarly to a quadruple analyzer. However, two of the pairs of opposing rod electrodes are replaced with hyperbolic electrodes, called end caps, which often include apertures through which ions are introduced and ejected. The QIT replaces the other pair of opposing rod electrodes with a ring electrode that encircles the device. Thus, rather than focusing ions of a particular m/z ratio along the longitudinal z axis, as does the quadrupole analyzer, the ring electrode closes off the z axis. A combination of radio frequency and DC voltages are applied to the electrodes, resulting in the generation of three-dimensional electric field capable of trapping a group of ions within the device in a manner somewhat similar to the ion cyclotron. The ions remain trapped in a stable orbit within the QIT as long as the applied radio frequency and DC potentials remain constant. However, modifying the radio frequency will impart unstable paths to a selected set of ions having a given range of m/z ratios, causing them to be ejected through one of the end-cap apertures. This illuminates an important distinction between a quadrupole analyzer and a QIT: the two pairs of opposing electrodes in the quadrupole impart a stable path to the selected ion, allowing it to transit along the z axis to the detector. The QIT traps ions in stable paths until the radio frequency is modified, causing one m/z ion to acquire an unstable path in order to be detected.

The QIT shares many analytical characteristics with the quadrupole analyzer. Its upper limit on m/z is similar at about 4,000 Da, although in some applications that has been extended above 70 kDa. In typical use, the resolution is similar, but the resolution of the QIT can be significantly increased by slowing the change in radio frequency used to eject ions, with the sacrifice of experimental time.

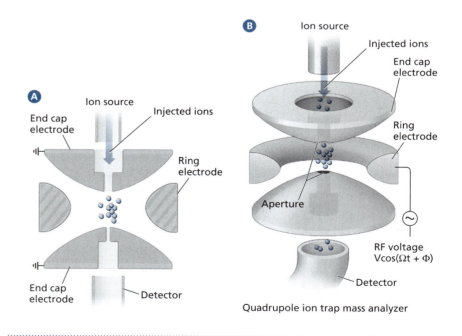

Figure 13.23 A QIT mass analyzer, in (A) a cutaway side view and (B) a three-dimensional representation.

Summary: Characteristics of Mass Analyzers

Table 13.1 provides a brief comparison of the mass analyzers discussed above in terms of a variety of important practical characteristics.

TABLE 13.1: Mass Analyzers

CHARACTERISTIC	MAGNETIC SECTOR	QUADRUPOLE	TOF	REFLECTION TOF	FT-ICR	QIT
Maximum m/z (Da)	10^4	10^3	10^6	10^6	$> 10^6$	10^3
Resolution	10^5	10^3	10^2–10^3	10^3–10^4	10^4–over 10^6	10^3
Mass accuracy	± 5 ppm	± 0.1%	± 0.1%	± 0.01%	± 10 ppm	± 0.1%*
Linear dynamic range	10^7	10^5	10^4	10^4	10^4	10^5
Operation pressure (torr)	10^{-6}	10^{-5}	10^{-6}	10^{-6}	10^{-9}	10^{-5}
Footprint	Large	Small	Small to large	Small	Very large	Very small
Cost	High	Low	Moderate	Moderate	Very high	Low to moderate

*Can achieve mass accuracy in the ppm range under certain conditions.

13.4 Detectors

For instruments other than those of the ion cyclotron type, it is necessary to have a transducer that can accept the ion flux generated in the source and separated in the mass analyzer and convert it into a measurable signal (voltage or current). If the instrument focuses the ions into a single beam, as is the case most often, only one detector is required. If the instrument produces a dispersion of ion beams, then it is necessary to utilize an array of detectors in a manner analogous to a diode array or charge coupled device (see Chapter 6, Figure 6.17). We will discuss the two most common detectors types used in modern mass spectrometers.

The most common type of detector is the *ion multiplier*, which is similar in principle to the photomultiplier tube (see Chapter 6, Figure 6.27). In a discrete dynode ion multiplier, ions exiting the mass analyzer are directed (typically off-axis to avoid interference from nonanalyte particles) to the first dynode in the electron multiplier. The dynode is coated with a material such as Cu–Be that will emit electrons when struck by an energetic particle. The electrons are focused electrostatically toward a second similarly coated dynode, and more electrons are produced during the second strike. This continues down the chain of 12 to 24 dynodes until the electrons are collected at a terminal anode, resulting in an amplification of 10^6 to 10^7 (see Figure 13.24). Another

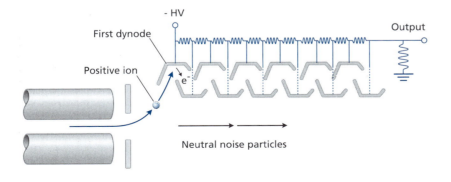

Figure 13.24 A discrete dynode ion multiplier.

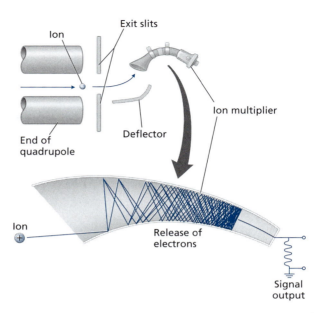

Figure 13.25 A continuous dynode ion multiplier.

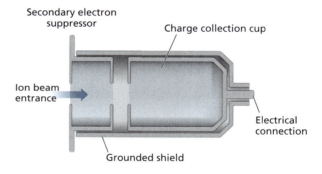

Figure 13.26 A Faraday cup detector.

configuration of the ion multiplier is the continuous dynode multiplier, which has the general shape of a curved horn, the inner wall of which is completely coated in a material that emits electrons when struck by particles (see Figure 13.25). The ion multiplier is widely used because it (1) is rugged and reliable and (2) provides internal amplification of an inherently weak signal. Its main disadvantage is that it generates a response that is dependent on the energy and mass of the incident ion: larger ions and those with greater energy will dislodge more initial electrons, which will then be amplified. In other words, the amplification is not linear with mass.

The other relatively common type of MS detector is the *Faraday cup* (Figure 13.26). The detected ion enters the cup and strikes the inner wall. The cup is shaped so as to prevent escape of the ion or any secondary ions or electrons produced on impact. Positively charged ions collect on the inner wall of the cup, and electrons flow from the grounded shield to neutralize them, resulting in a measurable current relative to the number of positive ions that enter the cup. A lens is positioned at the mouth of the cup with a slight negative polarity to suppress escape of any secondary electrons produced on impact. The Faraday cup is simple and inexpensive, but it is significantly less sensitive than an ion multiplier because it entails no internal amplification.

13.5 | Additional Techniques

Tandem Techniques

Among the more powerful mass spectrometric methods available to the modern chemist is the ability to link multiple mass analyzers in sequence (tandem) to achieve finely tuned results. It has become quite common in the literature to find references to MS-MS or MS-MS-MS (triple-quad) techniques. The application of two mass analyzers in tandem can improve resolution, provide more information about a specific ion, allow for further fragmentation in order to garner more chemical information, or improve the isolation and analysis of a specific ion of interest. Of course the trade-off is significantly increased cost, increased complexity, and increased laboratory footprint of a dual- or triple-MS instrument.

The simplest approach to tandem MS is to link two quadrupoles with a reaction cell between them (Figure 13.27). Ions generated in the source are directed to the first quadrupole (Quad 1). Ions that leave the first quad move into the reaction or collision cell. Here additional fragmentation occurs, usually by injecting a low pressure cloud of inert gas such as argon. The resulting product ions are accelerated to the second quad for final analysis. If the MS-MS is operated in the *product ion mode*, the DC/radio frequency voltages of Quad 1 are held constant to let only one m/z ion through, allowing it to refragment with analysis of the fragments conducted by scanning with Quad 2. In the *precursor ion mode*, Quad 2 is held constant while Quad 1 is scanned. This allows for the isolation of only one product ion in Quad 2 that might be generated by the further fragmentation of multiple ions separated in Quad 1. In the multiple *reaction monitoring mode*, both quadrupoles are held constant such that a single ion exits from Quad 1 and is fragmented, and then a single ion of the fragment tree is isolated in Quad 2, allowing for greater sensitivity in detection of the final ion. This is a very powerful technique for determining the identity of a specific moiety in an otherwise complex molecule.

The triple-quad (MS-MS-MS) approach is becoming more popular, especially for analysis of analytes separated using liquid chromatography. The approach is similar to that seen in Figure 13.27, except that the collision cell is replaced with a quadrupole that is typically used as an ion guide rather than as a mass analyzer. As with the plain collision cell, inert gas can be introduced into the middle quadrupole. Far greater sensitivity is achieved because fragment ions produced by collisions with the inert gas can be focused quickly into an appropriate beam path, rather than having a random distribution of trajectories as is typically the case with a simple collision cell.

The methods described above all represent *tandem-in-space* techniques. It is also possible to conduct *tandem-in-time* mass spectrometric experiments. This latter approach usually involves the use of an ion trap (similar to what we saw above for FT-ICR). An initial FT-MS measurement is made, followed by introduction of a reaction or collision gas, with a subsequent FT-MS measurement made to analyze the results of the intermediate collisional step. This can be done multiple times, giving rise to the concept of MS^n tandem experiments.

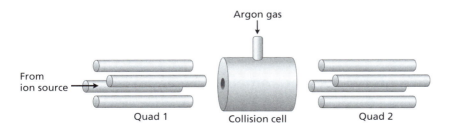

Figure 13.27 Dual quad (MS-MS) tandem MS.

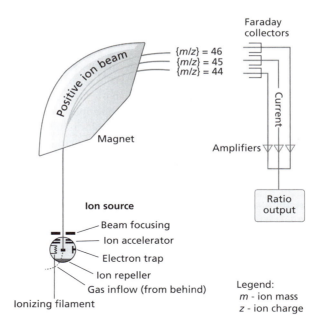

Figure 13.28 IRMS instrument.

Isotope Ratio Mass Spectrometry

One of the most important techniques used in environmental, geological, and anthropological chemistry is *isotope ratio mass spectrometry*. In your general chemistry course, you might have been required to apply a simple form of isotope ratio calculations to determine the age of rocks using the amount of various lead (Pb) isotope radioactive decay products present (^{208}Pb, ^{207}Pb, ^{206}Pb) compared to the amount of primordial ^{204}Pb. In such an assignment, you would have been given the amounts of each isotope found—but did you ever wonder how those amounts were determined? Similarly, the carbon isotope ratio in preserved human flesh remains can be used by anthropologists to deduce important food sources in prehistoric communities. The amount of ^{13}C in dissolved marine carbonate, the main source of carbon for marine life, is essentially zero, whereas the concentration of ^{13}C is significantly higher in atmospheric CO_2, the main source of carbon for terrestrial organisms. Determining the carbon isotope ratio in flesh samples from preserved human remains can allow anthropologists to deduce important food sources in prehistoric communities.

Isotope ratio mass spectrometry (IRMS) typically utilizes a sector mass analyzer, but rather than varying the field or accelerating voltage, both of those values remain fixed for a given experiment (Figure 13.28). Instead, multiple detectors are positioned at the exit port of the sector so that selected masses can be measured, with the signal integrated over time for greater accuracy. Although IRMS can be used as a detector for a gas or liquid chromatography instrument, in most cases samples are volatilized chemically and then introduced into the ionization chamber. For instance, to determine carbon isotope ratios in a sample of collagen, the sample would be burned in oxygen, converting the carbon to CO_2, and then the sector would be tuned to separate primarily in the range from 44 to 46 amu, relating to $^{12}CO_2$, $^{13}CO_2$, and $^{14}CO_2$.

Accelerator Mass Spectrometry

Accelerator mass spectrometry (AMS) is one of the least common mass spectrometric methods. Only nine facilities in the United States and just over a dozen in Europe have AMS capabilities. The technique is used only for a small number of sample types, and the cost of initiating and maintaining an AMS facility is enormous. Although it is a niche technique, it occupies a very important niche. AMS is used to conduct very

Beryllium-10 plays a unique role in geological dating because of its moderate half-life of 1.38 million years, which is fairly short on the geological time scale. There are ^{10}Be deposits in the hydrosphere through precipitation—it is soluble in slightly acidic rain water but precipitates out as the water becomes more alkaline on contact with soils. Once a layer of sediment has been covered, the ^{10}Be is no longer deposited in that soil layer, and it begins to decay to the stable isotope ^{10}B.

The analysis of ^{10}Be using accelerator mass spectrometry has led scientists to the surprising discovery that the glacial advances on the south side of Mount Everest were not synchronous with the advance of the Northern Hemisphere ice sheets. This is in contrast with the glacial advances in most of the rest of the Himalayas, which were synchronous with the ice sheet advances. This evidence indicates that the south face of Mount Everest exists in a uniquely isolated thermal zone with respect to the rest of the Himalayan range. ◼

*Finkel, R. C.; et al. Beryllium-10 Dating of Mount Everest Moraines Indicates a Strong Monsoon Influence and Glacial Synchroneity Throughout the Himalaya. Geology, **2003**, 31, 561–564.*

accurate radiocarbon dating on archaeological samples, and it has begun to find utility in geological dating in the hydrosphere using ^{10}Be. Some researchers have begun to seek uses for AMS in biomedical studies, such as determination of ^{41}Ca uptake enhanced by medication in postmenopausal women.[10]

Strictly speaking, AMS is not a single mass analysis method, but it entails a unique step in an overall mass analysis scheme. We will consider the use of AMS in terms of radiocarbon dating in order to understand its function. Note that in radiocarbon dating, the goal is to determine the concentration of residual ^{14}C in a sample. In living substances, the ^{14}C remains relatively constant due to continuous replenishment from $^{14}CO_2$ produced by cosmic radiation in the upper atmosphere. However, ^{14}C is radioactive; therefore, once a living organism dies, it no longer exchanges ^{14}C, and the amount that is in the organism begins to decay over time with a half-life of 5,730 years. By determining the ratio of ^{12}C and ^{14}C in an archeological sample and comparing the ratios to modern samples, one can determine the age of the archaeological sample with the

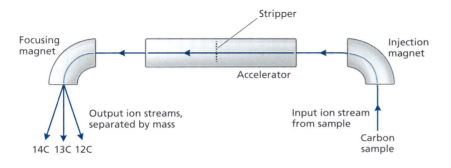

Figure 13.29 Simplified diagram of an AMS.

[10] Cheong, J. M. K.; et al. *J. Clin. Endocrinol. Metab.,* **2007**, *92(2),* 577–582.

PROFILE

Human Scent Fingerprinting

Locard's Exchange principle, proposed in the first decade of the twentieth century, states that every contact leaves a trace. This principle still serves as the foundation for all police work and forensic criminalistics. The car of a hit-and-run driver leaves paint on the damaged car. A burglar leaves fingerprints, hair, and footprints behind. Now researchers are investigating what we might leave behind without even knowing: our scent. The gas chromatography–MS sniffer method described in a paper in the *Journal of Forensic Sciences* (2010)* outperformed police canines and was able to identify individuals at a 99.74% success rate, even accounting for dietary changes over a period of days. The chart shown here pictorially demonstrates the "scent fingerprint" for five subjects sampled over a period of three days. The days of one being able to avoid specific identification by simply wearing gloves might soon be past. ∎

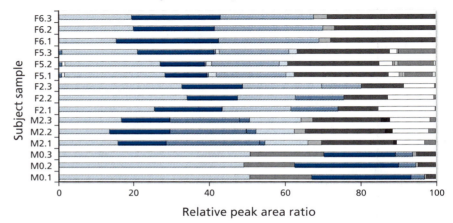

Intraday hand odor profiles of five individuals

* *Curran, A. M.; et al. The Differentiation of the Volatile Organic Signatures of Individuals Through SPME-GC/MS of Characteristic Human Scent Compounds. J. Forensic Sci., 55, 2010, 50–57.*

simple application of first-order decay kinetics. Figure 13.29 shows a simplified diagram of the AMS. In practice, the carbon sample is mounted and sputtered with Cs^+ ions, resulting in the ejection of various carbon anions (−1, −2, −3 charge) as well as other anions related to the sample. One advantage of the use of negative ions in the initial stage of AMS is that elemental isobars, such as ^{14}N, are eliminated due to the instability of their negative ions.[11] The ions are directed through a sector mass ion guide to isolate only species with m/z related to the carbon samples of interest. Unfortunately, polyatomic ions such as $^{13}CH^-$ and $^{12}CH_2^-$ can masquerade as $^{14}C^-$. Those false ions are removed in the next stage. The ions enter the accelerator, which uses an accelerating voltage of up to 1 MV. In the center of the accelerator is a stripper chamber, in which inert gas atoms collide with the ions, dissociating any polyatomic species and stripping electrons away from the anions. Thus, the negative ions are converted to cations in the stripper chamber, and for the most part only monoatomic carbon cations survive the accelerator segment. The cations are directed through another sector mass analyzer, and the three isotopes of carbon are measured quantitatively.

[11] An *isobar* is a species having the same elemental mass. For instance ^{14}N is an isobar for ^{14}C.

ACTIVITY

Selected Ion Game: A Physical Demonstration of Mass Spectrometric Principles

Preparation – Prior to the class or laboratory period in which the game will take place, the instructor and/or the students will need to accumulate the following items:

1 flat wooden plank, about 6 feet in length and no more than 2 feet in width

1 heat gun, hair dryer, or other constant air flow device

Many balls of various shapes and weights, representing "ions"

Goal – Students work in teams to try to modify the curvature of the path of their selected ion such that it reaches the detector. The plank is held at a fixed slope, and the group sends its ion (ball) down the plank. At a fixed point along the plank, the air gun is mounted and blowing a constant stream across the plank. A small bucket is placed about a foot lower on the plank than the air gun, on the opposite side.

Scenario – The slope of the plank represents the accelerating voltage in the ion source; this is the only experimental parameter over which the students have control. The air gun represents the force applied to impart curvature in the path of the ion (ball)—that is, it represents the fixed magnetic or electrostatic field. The cup represents the detector, and of course the ball represents the ion. The larger the diameter (actually, the cross-sectional area) of the ball, the larger the charge (z) of the ion, and of course the ball has a fixed "ion" mass. At the beginning of the game, each group randomly draws an "ion" from the ball bag. Group members are allowed no more than three attempts with their own ion, but they may have unlimited (within in time constraints) attempts with any other ion available to them. They can experiment with different m/z ions in order to fine tune the accelerating potential for their own ion.

13.6 Further Reading

BOOKS

Cole, R. B. *Electrospray and MALDI Mass Spectrometry: Fundamentals, Instrumentation, Practicalities, and Biological Applications*, 2nd ed. John Wiley & Sons: Hoboken, NJ, **2010**.

Dawson, P. H., editor. *Quadrupole Mass Spectrometry and Its Applications*. Elsevier/North-Holland: Amsterdam, **2013**.

de Hoffman, E.; Stroobant, A. *Mass Spectrometry: Principles and Applications*, 3rd ed. Wiley Interscience: Chichester, UK, **2007**.

Gross, J. H. *Mass Spectrometry: A Textbook*. Springer Science & Business Media: Heidelberg, Germany, **2011**.

March, R. E.; Todd, J. F. J., editors. *Practical Aspects of Trapped Ion Mass Spectrometry, Volume IV: Theory and Instrumentation*. CRC Press: Boca Raton, FL, **2010**.

Watson, J. T.; Sparkman, O. D. *Introduction to Mass Spectrometry: Instrumentation, Applications, and Strategies*

for Data Interpretation, 4th ed. Wiley: Chichester: UK, **2007**.

JOURNALS

Glish, G. L.; Vachet, R. W. The Basics of Mass Spectrometry in the Twenty-First Century. *Nature Reviews: Drug Discovery*, **2003**, *2*, 140–150.

Marshall, A. G.; Hendrickson, C. L. Fourier Transform Ion Cyclotron Resonance Detection: Principles and Experimental Configurations. *Int. J. Mass Spec.*, **2002**, *215*, 59–75.

ONLINE RESOURCES

NIST Chemistry WebBook. NIST Standard Reference Database Number 69, Eds. Linstrom, P. J.; Mallard, W. G. http://webbook.nist.gov/chemistry/.

13.7 Exercises

EXERCISE 13.1: Consider the electron ionization source depicted in Figure 13.4. What do you think would be the effect of decreasing the voltage between the filament and the target? What effect would using an unheated filament have?

EXERCISE 13.2: What would be the velocity of an imidazolium ion $(C_3H_4N_2^+)$ if it were accelerated across a voltage of 22.3 kV?

EXERCISE 13.3: What voltage would be needed to accelerate an imidazolium ion $(C_3H_4N_2^+)$ to a velocity of 4.2×10^5 m/s?

EXERCISE 13.4: Given that the half-life of the ^{235}U-to-^{207}Pb decay series is 704 million years, how long ago was a zircon crystal formed if 1.0 g of the crystal contains 4.32 ng of ^{235}U and 1.92 ng of ^{207}Pb?

EXERCISE 13.5: What would the age of the zircon crystal in Exercise 13.4 be if the amounts of ^{235}U and ^{207}Pb were reversed?

EXERCISE 13.6: Explain the function of the quadrupole ion deflector seen in Figure 13.15.

EXERCISE 13.7: What is the mass of a univalent ion if it acquired a velocity of 5.5×10^5 m/s when accelerated across a potential of 1,250 V?

EXERCISE 13.8: Which method of ionization would you suggest for each of the following samples? Explain your reasoning.

(a) Dioxin, a byproduct of the bleaching of paper by chlorine

(b) Horseradish peroxidase

(c) Steel shavings

EXERCISE 13.9: Naphthalene, a polyaromatic hydrocarbon ($C_{10}H_8$), was analyzed by MS using chemical ionization and a quadrupole mass analyzer. The base peak occurred at m/z = 129, and the largest ion mass in the spectrum was seen at m/z = 156. Explain these observances.

EXERCISE 13.10: To what excited state radius will ions in an ICR chamber be moved if a 2.5 V radio frequency signal is applied for 500 μs across a 2.3 cm cell while a 4.71 T field is applied?

EXERCISE 13.11: What is the maximum time the pulse described in Exercise 13.10 could be applied to avoid having ions collide with the cell walls?

EXERCISE 13.12: What is the maximum voltage that could be applied for 500 μs in the cell described in Exercise 13.10 to avoid having ions collide with the cell walls?

EXERCISE 13.13: Your ICM-MS spectrometer has a resolution of 175 when the average mass of univalent ions is around 120 Da. Would the molecular ion Ar–Br$^+$ interfere with the analysis of any elemental ions? Explain your answer.

EXERCISE 13.14: What resolution would be necessary in order to resolve peaks from divalent ions having molar masses of 1,000 g/mol and 1,001 g/mol?

EXERCISE 13.15: Assuming the resolution of the mass analyzer considered in Exercise 13.14 is constant across the mass spectrum, what two m/z values could be resolved around a mean m/z of 2.5×10^3?

EXERCISE 13.16: Estimate the resolution of the mass analyzer used to obtain the spectrum seen in Figure 13.2(a).

EXERCISE 13.17: Estimate the mean free path of an ethanol molecule (0.44 nm diameter) at 25°C and 1.0 torr.

EXERCISE 13.18: Assuming the ethanol molecule in Exercise 13.17 is traveling at about 555 m/s, how many collisions per second is it undergoing?

EXERCISE 13.19: If we want the mean free path for an ethanol molecule in a CI chamber to be 0.098 mm, what pressure would be required in the chamber?

Advanced Exercises

EXERCISE 13.20: Given the NMR and mass spectra shown, determine the molecular formula and structure of the compound.

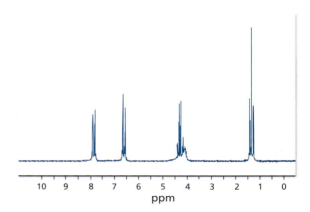

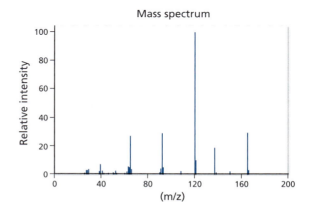

Mass spectrum

EXERCISE 13.21: Given the NMR and mass spectra shown, determine the molecular formula and structure of the compound.

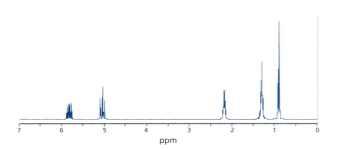

Mass spectrum

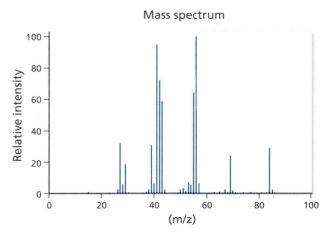

Chapter 14

An Introduction to Nuclear Magnetic Resonance Spectroscopy

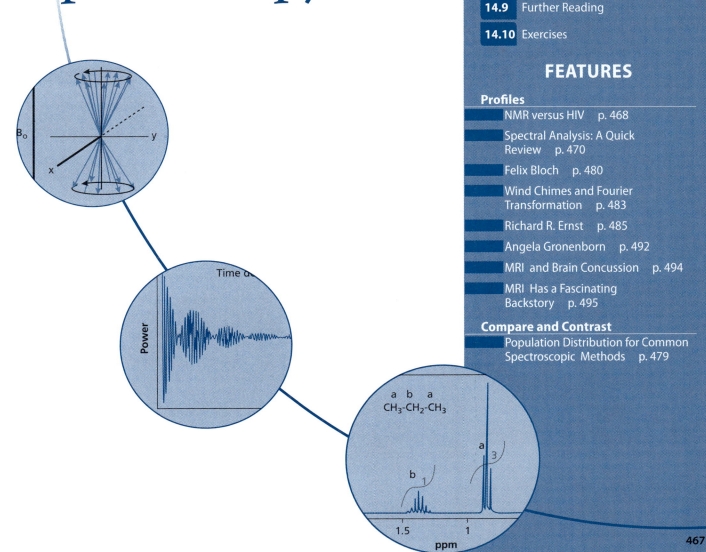

PROFILE

Nuclear Magnetic Resonance versus HIV

One of the challenges in the battle to find a vaccine against HIV (the virus that causes AIDS) is the fact that the virus mutates rapidly. However, if researchers could find a protein sequence that is common to most of the quasi-species of the HIV virus family, they could begin the process of designing a vaccine based on that protein sequence. A research team led by Leonard Spicer at Duke University used nuclear magnetic resonance spectroscopy to determine the structure of an external protein shell called the envelope membrane proximal external region of HIV-1. This HIV-1 glycoprotein (GP 41) is involved in viral–host cell membrane fusion. ■■

Reardon, P. N.; Sage, H; Dennison, S. M.; Martin, J. W.; Donald, B. R.; Alam, S. M.; Haynes, B. F.; Spicer, L. D. Structure of an HIV-1-Neutralizing Antibody Target, the Lipid-Bound Gp41 Envelope Membrane Proximal Region Trimer. Proc. Natl. Acad. Sci., **2014**, 111(4), 1391.

14.1   Introduction

Nuclear magnetic resonance (NMR) is *the* primary instrumental method for determining molecular structure. As a matter of fact, you may have some experience in using NMR and in interpreting NMR spectra, perhaps from an organic chemistry course. You may recall using an NMR spectrometer: loading your sample, in its slender NMR tube, into a large magnet and following a set of instructions to set the parameters of the experiment at the computer console. If you are like most students, you were impressed by the large magnet, housed inside its imposing shiny silver canister, and the frozen water vapor forming jets or "clouds" coming out of the top as some of the cryogenic liquid nitrogen escaped into the atmosphere. You may recall warning signs posted outside the NMR room, advising people with pacemakers or metal implants to "stay out" or telling you to leave your wallet outside the room in a safe place so as not to damage the magnetic strips on credit and ID cards. Depending on your particular NMR instrument, you may have had to climb a ladder to put your sample into the instrument! Once you had your data, you may recall looking at the NMR spectra for patterns that would provide clues to structures, and assigning peaks within the spectra to possible structures.

The underlying quantum mechanics of NMR spectroscopy were first introduced in Chapter 2, Section 7. You may want to review that material before proceeding. This chapter introduces the underlying principles of how the NMR spectrometer functions and how a signal is obtained and processed, and it reintroduces spectral interpretation.[1]

---

[1] For more detail on how NMR pulse sequences are developed for specific two-dimensional experiments, see Chapter 21.

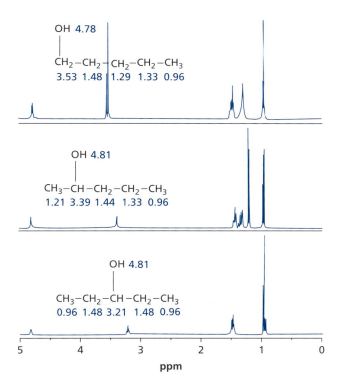

OH 4.78

$CH_2-CH_2-CH_2-CH_2-CH_3$
3.53 1.48 1.29 1.33 0.96

OH 4.81

$CH_3-CH-CH_2-CH_2-CH_3$
1.21 3.39 1.44 1.33 0.96

OH 4.81

$CH_3-CH_2-CH-CH_2-CH_3$
0.96 1.48 3.21 1.48 0.96

ppm

Figure 14.1 Structures and corresponding ^{1}H-NMR spectra of (top to bottom): 1-pentanol, 2-pentanol, 3-pentanol.

A key advantage of NMR analysis is the fact that the results provide structural information about the molecule, specifically how the individual pieces of the molecule are connected. Together with other instrumental techniques, NMR can provide the final interpretation to identify an unknown substance or to provide evidence that the compound under study is (or is not) your analyte of interest. If we think about structure determination as a puzzle to be solved, NMR data can provide the triangulation that connects mass spectroscopy (MS) data and infrared (IR) spectroscopy data. Although MS tells us about total mass and mass fragmentation, and IR spectroscopy tells us about the presence of functional groups, NMR spectra tell us how the various components are connected together to form the molecular backbone. For example, each of the isomers of pentanol (Figure 14.1) would have identical total masses, similar fragmentation patterns, and similar IR fingerprints. The MS and IR data would lead you to the general characteristics of a five-carbon structure with an alcohol group (R–OH). However, without the use of a database comparison, IR and MS spectra often cannot provide unambiguous identification. Neither IR nor MS data provide definitive information about the connectivity within the molecules. NMR, on the other hand, would provide you with specific information that details the local environment of each nucleus. Because the local chemical environment slightly shifts the net magnetic field felt by each nucleus in a molecule, NMR transitions allow us to directly probe the chemical environment of each NMR active nucleus in a molecule. In turn, we can deduce (for example) the number of –CH$_3$ groups, –CH$_2$ groups, and –CH groups in a molecule. This in turn allows us to deduce the location of functional groups such as the –OH group in our example molecule. NMR would be absolutely clear as to which of these isomers of pentanol was under investigation.

You may also be familiar with NMR in another context: *magnetic resonance imaging (MRI)*. In medicine, MRI is one of the leading diagnostic tools. It is used to image tissue. Although MRI is conducted on a much larger, human-size sample, the instrumental principles are the same as NMR. To obtain an MRI, a technician places the patient within the MRI scanner. As in the NMR, a strong magnetic field is applied around the area to be imaged. The signals are processed via computer to create a tissue density map. We will come back to MRI at the end of this chapter.

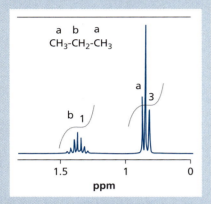

Consider the hydrogen nuclear magnetic resonance (NMR) spectrum of propane, CH_3–CH_2–CH_3, shown here. The spectrum shows a septet (7) of peaks centered around 1.4 ppm and a triplet (3) of peaks around 0.8 ppm. Recall from your introduction to NMR in organic chemistry that a hydrogen's signal is split based on how many H atoms are present on adjacent (neighbor) carbons. The splitting follows the n + 1 rule, which states that a hydrogen's signal will be split by the number of neighboring H atoms (n) on adjacent carbon atoms into n + 1 lines. Therefore, we can surmise that the triplet represents the hydrogens on the "a" carbon (the six methyl hydrogens). These a-hydrogens have a neighbor carbon with two H atoms (2 + 1 = 3; the a-hydrogens are split into a triplet).

Likewise, the septet represents the hydrogens on the "b" carbon (the methylene hydrogens). These hydrogen atoms have two equivalent neighboring carbon atoms, each with three hydrogen atoms for a total of six neighboring hydrogen atoms (6 + 1 = 7; the b-hydrogens are split into a septet). We can also integrate the area under each set of peaks, which will be proportional to the number of equivalent hydrogens at that position. The integration values—of 3 for the triplet and 1 for the septet in propane, shown in the figure in gray—indicate the empirical ratio of the number of each type of hydrogen atom (in this case a 3:1 ratio). The molecular structure has six a-hydrogens (–CH_3 groups), and two b-hydrogens (–CH_2 group), for ratio of 6:2 (empirically 3:1). ■

Problem 14.1: From the mass spectral and elemental analysis data, you have determined the formula of an unknown compound from a petroleum distillate to be C_4H_8O. You have limited the possibilities to the four different structures shown below, which all have the formula C_4H_8O.

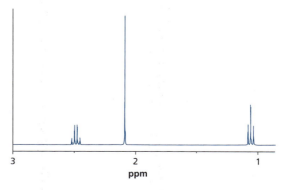

(a) Describe how an IR spectrum would aid in narrowing down your choices. Be specific.

(b) Use the hydrogen NMR spectrum of your unknown (shown) to determine which of these isomers you have.

14.2 Nuclear Magnetic Spectroscopy: All about the Nucleus

All spectroscopic techniques use electromagnetic energy to probe various aspects of structure. You select a method based on the way the energy affects the sample. The spectroscopic techniques we have studied so far have exploited energy changes between the quantum states of electrons. Relaxed electrons are excited. Excited electrons relax and quantum energy levels are assessed through absorption and/or emission. You have heard all of this before, and NMR spectroscopy *is* another spectroscopic technique. However—and this is an important *however*—unlike the other techniques you have studied, NMR is *not* about the electrons. In NMR we *are not* observing the transition of electrons between a highest occupied molecular orbital or lowest unoccupied molecular orbital (molecular ultraviolet-visible and molecular luminescence) or the electronic transition between atomic quantum states (atomic absorption spectroscopy and atomic emission spectroscopy) or the transition between the quantum states of molecular vibrations (Fourier transform [FT] IR, known as FTIR, and Raman spectroscopy). **NMR spectroscopy is all about the *nucleus.*** See Chapter 2, Section 7 for a review of nuclear magnetic quantum states.

Nuclear Quantum Numbers

Let us take a minute to think about the nucleus, the center of atomic structure. You may remember, likely from your general chemistry course, that electrons can be described (in terms of position and energy) by a set of electronic quantum numbers. Those quantum numbers have a shorthand notation in the electron configurations (e.g., $1s^2 2s^2 2p^2$). And electrons in each of those states have a spin quantum number (s) of either $+\frac{1}{2}$ or $-\frac{1}{2}$, also called *spin up* and *spin down*. Like the electrons, nuclear energy can also be described using quantum numbers. For the purposes of this chapter, we will limit our discussion to two aspects of the nuclear quantum contributions: the overall *nuclear spin (I)* and the *magnetic or directional quantum number (m).*

The rules for determining I are based on the number of protons and neutrons:

1. If the number of protons is *even and* the number of neutrons is *even*, then I = 0 and there is no overall spin.

2. If the number of protons *plus* neutrons is *odd*, then I = 1/2 or an integer value of 1/2 such as 3/2, 5/2, and so on.

3. If the number of protons is *odd and* the number of neutrons is *odd*, then I = 1 or an integer value of 1 such as 2, 3, and so on.

The nuclear magnetic quantum number (m) is derived from I. The rule states:

$$m = I, I{-}1, \ldots {-}I \qquad \textbf{Eq. 14.1}$$

Thus, m can have $(2I + 1)$ different values. In the case that I = ½, the nucleus has two possible values for m: m = $+\frac{1}{2}$ and m = $-\frac{1}{2}$. *It is the energy difference between these two states that is exploited in NMR spectroscopy.* We first introduced this idea in Chapter 2 and we will expand on it more in Section 4 of this chapter.

Nuclei that have I = ½, the "spin ½ nuclei," are very important for NMR, because the NMR signals are derived from the transitions (absorptions and relaxations) between these two states (m = $+\frac{1}{2}$ and m = $-\frac{1}{2}$). The usefulness of NMR relies heavily on the fact that hydrogen, ^{1}H, is a spin ½ nucleus, which works out well because ^{1}H is one of the most abundant nuclides on Earth.

Note

It is common in NMR vernacular to interchange the term *hydrogen* with the term *proton*. Because the most abundant isotope of hydrogen has one proton and no neutrons, the hydrogen nucleus is basically just a proton. Therefore, when NMR spectroscopists talk about hydrogen results or hydrogen experiments, it is abbreviated ^{1}H-NMR and called "proton NMR."

Note

Nuclei with I = ½ are called "spin ½ nuclei."

Problem 14.2: Determine if the nuclear spin (I) for each of the following nuclides is zero, an integer, or a multiple of ½. (See Example 14.1)

(a) ^{2}H   (c) ^{12}C   (e) ^{15}N

(b) ^{3}H   (c) ^{13}C   (f) ^{31}P

Example 14.1

What are the spin quantum states for ^1H?

STRATEGY –

See rules for determining I and Equation 14.1.

ANSWER –

^1H is the most common isotope of hydrogen and has 1 proton and 0 neutrons. The sum $(1 + 0 = 1)$ is odd. Therefore, applying the rules determining I, the spin is $I = \frac{1}{2}$. Because $I = \frac{1}{2}$, there are two values for m: $m = +\frac{1}{2}$ and $m = -\frac{1}{2}$.

Problem 14.3: Determine the nuclear magnetic quantum number (m) for each of the following nuclides.

(a) ^2H (c) ^{12}C (e) ^{15}N

(b) ^3H (d) ^{13}C (f) ^{31}P

Problem 14.4: Identify the nuclides from Problem 14.3 that are NMR active. Explain your choices.

A Nucleus in a Magnetic Field

Under the influence of an *external magnetic field*, which we call the B_o field, the two states $(m = +\frac{1}{2})$ and $(m = -\frac{1}{2})$ split in energy. In the $m = -\frac{1}{2}$ state the proton's magnetic field is aligned parallel to B_o, and in the $m = +\frac{1}{2}$ state the proton's magnetic field is aligned antiparallel to B_o. Importantly, *the magnitude of the split is proportional to the strength of the externally applied magnetic field.* Figure 14.2 shows that (1) when the B_o field is 0, the difference in energy between the two magnetic states approaches null, but (2) as the B_o field increases, moving to the right in Figure 14.2, the difference in energy between the two states (the energy gap) becomes greater.

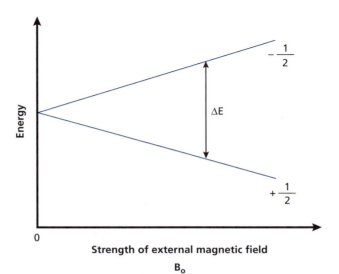

Figure 14.2 Diagram showing the energy splitting of the $+\frac{1}{2}$ and $-\frac{1}{2}$ magnetic spin states of a nucleus as a function of the applied external magnetic field.

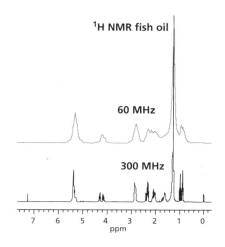

Figure 14.3 Effect of field strength on resolution. Comparison of ^{1}H-NMR spectra of fish oil taken at two magnetic field strengths, 60 MHz and 300 MHz. The convention in NMR of referring to magnetic fields in terms of frequency is introduced in the sidebar "Magnetic Field Conventions: Tesla and Megahertz."

Figure 14.4 Bruker 600 (left) and 800 (right) MHz NMR equipment.

The magnetic field dependent splitting of the energy levels is known as the *Zeeman effect*.[2]

NMR and MRI experiments are conducted by putting the sample into a strong magnetic field in order to create this energy gap between the two spin states. The stronger the B_o field, the greater the energy gap (ΔE). Larger energy gaps correspond to higher signal strength[3] and an increased level of signal energy resolution. We will talk about resolution more in Section 14.5. Thus experiments are conducted under the highest possible external field strengths (B_o).

Figure 14.3 shows the ^{1}H-NMR spectrum for a sample of fish oil taken under two different externally applied magnetic fields.[4] The lower B_o field, labeled 60 MHz, looks broader overall and shows less fine structure. The higher B_o field, labeled 300 MHz, shows many more individual proton signals resolved. You will also notice that the signal at 7.25 ppm,[5] seen in the 300 MHz spectrum, is buried in the baseline of the 60 MHz spectrum. This signal is too weak to be seen under the low field conditions of the 60 MHz B_o field.

Strong magnetic fields are produced by electromagnets (sometimes called *continuous-wave* magnets) at room temperature and are also produced by superconducting electromagnets at cryogenic temperatures (also known as *high-field* magnets). Instruments range from 60 to 900 MHz, with the most common high-field instruments in the 200 to 400 MHz range (4.7T–9.4T). Figure 14.4 shows two high-field NMR instruments located at the Ontario Cancer Institute.[6] To the left is a Bruker 600 MHz instrument, and to the right is a Bruker 800 MHz instrument.

---

[2] This was also discussed in Chapter 2, Section 7. See Figure 2.16 and accompanying text for a review of the Zeeman effect.

[3] The external magnetic field strength (B_o) is not the only factor affecting signal strength, but it is certainly a contributing factor.

[4] Process NMR Associates, 87A Sand Pit Road, Danbury, CT 06810; http://www.process-nmr.com.

[5] The description of the *x* axis in the NMR spectrum and definition of ppm is described in more detail in Section 14.5, and Equation 14.6 describes how ppm is determined.

[6] http://nmr.uhnres.utoronto.ca/ikura/nmrsuite/nmrsuite.html.

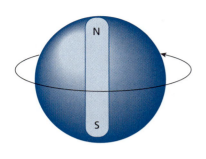

Figure 14.5 It is convenient to think of the nucleus as a spinning charged sphere having magnetic poles.

Figure 14.6 An oversimplified view of the possible orientations of the nuclear magnetic moments (small arrows) within a sample, relative to an applied external magnetic field, B_o (large arrow). In fact, the nuclear magnetic moments can adopt a variety of positions relative to the field. However, we simplify the diagram to show that we are interested in the fact that the *majority* of nuclei are aligned with the B_o field.

A spinning charged particle has an associated magnetic field; therefore, the nucleus may be conveniently modeled as a spinning bar magnet, having a north pole and a south pole. See Figure 14.5. Initially the orientation of each nucleus in a sample would vary from nucleus to nucleus throughout the sample. Once in the NMR, under the influence of an applied magnetic field (B_o), the nuclei will begin to orient themselves relative to the B_o field. This is not to imply that all of the nuclei "snap to attention" and become immediately aligned. Rather, over some period of time, the nuclei will become arranged so that the majority of nuclei are aligned in the direction of B_o. More on this, the Boltzmann distribution, and net magnetization, later.

When playing with bar magnets, you may have also noticed that there are two possible stable ways that the two magnets can be aligned: the north-to-south alignment, which is very stable (or we might say *low energy*), and a north-to-north alignment, which has a quasi-stability (or we might say a *high energy*). These quasi-stable alignments represent stable states that are physically tentative. A light "thump" with your finger, and the quasi-stable state quickly reverts to the low energy state.

For an atom inside the magnetic field of the NMR, the situation is similar in regard to stability. The two magnetic states of nucleus ($m = +\frac{1}{2}$) and ($m = -\frac{1}{2}$) represent stable and quasi-stable orientations with respect to the external field. Look again at Figure 14.2 and note that the $-\frac{1}{2}$ state is labeled as the higher energy and the $+\frac{1}{2}$ state labeled as the lower energy. These energies are derived from quantum mechanics, and this naming convention is used in NMR spectroscopy.

The magnetic field of the spinning nucleus has both direction and magnitude (i.e., a *magnetic moment*). The symbol for magnetic moment is μ, and you will often see the moment represented as an arrow. Using this symbolism, Figure 14.6 shows the direction of the external B_o field using a large arrow and shows the nuclei (the small arrows) as being aligned in one of two orientations under the influence of the B_o field. Using this symbolism, arrow up represents the more stable alignment (the $+\frac{1}{2}$ state) and arrow down represents the less stable alignment (the $-\frac{1}{2}$ state). In the NMR literature, it is typical to show the B_o field (arrow up) in a more stable alignment with the $+\frac{1}{2}$ state (lower energy state) of the nucleus (also arrow up). We say that they are "aligned." Thus the $-\frac{1}{2}$ state (higher energy state, arrow down) of the nucleus looks opposed to the B_o field. When a nucleus changes orientation with respect to B_o, we say that it has *flipped.*

You might also note that as long as nothing else is going on with the system, there are *more* nuclei that are in the lower energy state orientation (arrow up and aligned with B_o) and fewer in the higher energy orientation (arrow down and flipped with respect to B_o). The relative population of the two states[7] is dependent on B_o (see "Compare and Contrast—Population Distribution for Common Spectroscopic Methods"). Although we cannot show it here on the static page, there will be a very small number of transitions going on as nuclei "flip" from one orientation to the other under the influence of random thermal effects and so the relative number of nuclei in the two states is better thought of as an equilibrium and not a static condition.

Nuclear "flips" from $+\frac{1}{2}$ to $-\frac{1}{2}$ orientation can happen in either direction—up to down or down to up with equal probability. In one direction, we might call that absorption, whereas in the other direction, we might call that emission. In the case of the nucleus, and under the conditions in which this happens, *both processes occur* and we call this "flipping"-of-spins *resonance.*[8] It is, most precisely, *nuclear magnetic resonance.*

The external magnetic field, a key component to the NMR instrument because it provides the environment for the Zeeman splitting, is homogeneous throughout the NMR sample. Later, in Section 14.6, we will talk about how we obtain field homogeneity instrumentally. However, it is important for you to take note that, within a molecular structure, each nucleus is going to "feel" a slightly different magnetic field environment. This occurs because, from the perspective of the nucleus, its local environment

---

[7] See Equation 2.15 for a review of the Boltzmann distribution.

[8] NMR can be measured as an absorbance (typically the mode for continuous wave NMR) or as an emission (typically the mode for Fourier transform NMR).

may be shielding some of the externally applied field. This is a consequence of the presence of electrons and also other nearby nuclei. We will come back to this in Section 14.5 when we talk about *chemical shift*.

14.3 The Nuclear Magnetic Resonance Signal

In order to stimulate NMR, the system has to be excited. A common misconception is that the externally applied magnetic field (B_o) excites the nuclei in NMR measurements. This is not the case. The B_o field provides the split in energy level environments ($+\frac{1}{2}$ and $-\frac{1}{2}$). The resonance between the two states is stimulated by adding energy to the system in the form of a radiofrequency (RF) pulse. The RF pulse can be thought of as analogous to the flash lamps used to stimulate fluorescence in a fluorometer.[9]

Like other types of spectroscopy that you have studied, the excitation of the system is matched to the quantum energy difference between the two states. When the energy applied equals the energy of the gap, transitions between the states occur. The energy is absorbed into the system, the population of the higher energy state increases, and when the excited state nuclei relax, they reemit the RF energy. The frequency of the RF energy on relaxation is measured in NMR spectroscopy.[10] There are some interesting considerations along these lines in NMR, so let us examine this more fully in the context of the nucleus.

We may conveniently depict the magnetic moment with a vector, representing the direction and magnitude of the magnetic field associated with the spinning charged nucleus. If we were to look at a single nucleus, we would see that the magnetic moment of the charged particle exhibits a behavior called *precession*. You may have seen this type of physical behavior in a gyroscope or a spinning top. While the gyroscope is spinning on its axis, it is also rotating (or precessing) around the axis. As the gyroscope spins, it precesses about the vector representing the force of gravity. (See Figure 14.7 (A).)

We can envision the nucleus as having a similar behavior, which we can again represent using our arrow notation. In Figure 14.7 (B), we also see the spinning charged nucleus, like the gyroscope, aligned with the applied external field (B_o) and rotating around that axis. The magnetic moment (μ) is equal to the gyromagnetic ratio (γ) times the angular momentum (J) of the precession.

$$\mu = \gamma J \qquad \text{Eq. 14.2}$$

You might think of the *gyromagnetic ratio*, γ, as telling you what kind of gyroscope—or rather, in this case, what kind of nucleus—you are studying. The gyromagnetic ratio is an intrinsic property of the element under study and it functions to scale the *angular momentum (J)* to the measured magnetic moment. You might find it *helpful to think of the gyromagnetic ratio as being analogous to molar absorptivity* in optical spectroscopy.

Because the angular momentum (J) is a function of the applied field B_o, then that field influences not only the direction of the magnetic moment but also the frequency of precession (ω):

$$\omega = -\gamma B_0 \qquad \text{Eq. 14.3}$$

The rate of precession, ω, is known as the *Larmor frequency* of the nucleus. And, *the Larmor frequency is related to the energy difference, ΔE, between the $+\frac{1}{2}$ and $-\frac{1}{2}$ spin states.* Thus, as we saw in Section 14.2, the energy gap of the transition between the two resonance states ($+1/2$ and $-1/2$) is a function of the strength of the applied B_o field.

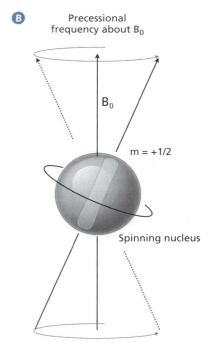

A   Precessional frequency about g

g force of gravity

Spinning top

B   Precessional frequency about B_0

B_0

m = +1/2

Spinning nucleus

Figure 14.7 Spinning top precessing about a gravitational axis (A) and spinning nucleus precessing about an applied magnetic field (B).

Note
You might find it helpful to think of the gyromagnetic ratio as being analogous to molar absorptivity in optical spectroscopy.

[9] See Chapter 8 for a review of fluorescence spectrometer design.

[10] As mentioned, NMR resonance can be measured as an absorbance (typically in continuous wave NMR) or as an emission or relaxation (typically in Fourier transform NMR). In this chapter, we focus on the more modern application of Fourier transform NMR and talk about NMR relaxation signals.

Table 14.1:   Magnetic Data for Some Common Nuclides

NUCLIDE	(%) NATURAL ABUNDANCE	SPIN (I)	MAGNETIC MOMENT (μ)*	MAGNETOGYRIC RATIO (γ)†
^{1}H	99.984	½	2.793	26.75
^{13}C	1.108	½	0.7022	6.728
^{31}P	100	½	1.131	10.84
^{19}F	100	½	2.627	25.18

* μ has units of nuclear magnetons. 1 nuclear magneton = $5.0508 \times 10^{27} JT^{-1}$

† γ has units of 10^7 rad T^{-1} sec^{-1}.

Each nuclide has its own specific magnetic moment, and thus, as shown in Equation 14.3, each nucleus also has its own Larmor frequency. The angular frequency (ω in rad sec^{-1}) in Equation 14.3 can be converted to a radiofrequency ν (in Hertz) using:

$$\nu = \omega/2\pi \qquad \text{Eq. 14.4}$$

Magnetic data for the four most commonly studied spin ½ nuclei are given in Table 14.1.

The energy for the two spins (+½ and –½) can be determined:

$$E_I = h\nu$$
$$E_I = \hbar\omega, \text{ (where } \hbar = h/2\pi) \qquad \text{Eq. 14.5}$$
$$E_I = \hbar(-m\gamma B_o), \text{ (where m is either +½ or –½)}$$

Thus the energy level difference, ΔE (taken as the –½, high state, minus the +½ low state) is equal to:

$$\Delta E = \hbar(--½\,\gamma B_o) - \hbar(-+½\,\gamma B_o)$$
$$\Delta E = ½\,(\hbar\gamma B_o) + ½\,(\hbar\gamma B_o) \qquad \text{Eq. 14.6}$$
$$\Delta E = \hbar(\gamma B_o)$$

We can see then that the energy gap, ΔE, is proportional to both γ, the identity of the nucleus under study, and the strength of the applied magnetic field, B_o (see Figure 14.8). The orientation of a particular nucleus ($\pm$ ½) will "flip" whenever the applied energy (the RF pulse) equals the energy gap as described in Equation 14.6. As discussed in Section 14.2, NMR happens in both directions, up to down and down to up. A flip from the low energy state to the high energy state is an absorption event, whereas while a flip from the high energy state to the low energy state is an emission event.

When we talk about a particular NMR instrument as being a "200 MHz" or a "600 MHz" instrument, we are actually quoting the Larmor frequency for the ^{1}H nucleus. Magnetic fields are typically measured in units of tesla. Let us see how these units correspond to each other through some sample calculations.

Figure 14.8 shows that the energy gap (ΔE) between the two energy states ($-1/2$ and $+1/2$), becomes larger with increasing B_o field, and it shows the relationship between frequency (ν) (from Equation 14.4) and B_o for the ^{1}H nucleus. The Larmor frequency for ^{1}H-NMR is used as the naming convention for the magnetic field because ^{1}H-NMR is the most commonly used NMR technique.

It is important to note that a given magnetic field will produce a different ΔE and precession frequency for different types of nuclei. Protons will have a Larmor frequency $\approx$ 100 MHz when conducting ^{1}H-NMR on a 100 MHz instrument, but the ^{13}C nuclei will have a different Larmor frequency that is dependent on the ^{13}C gyromagnetic ratio (see Figure 14.9 and Example 14.3). Holding the field strength constant, B_o = 2.35 T, the approximate resonance frequency for the four nuclides listed in Table 14.1, are shown in Figure 14.9.

Magnetic Field Conventions: Tesla and Megahertz

The tesla is the SI unit of magnetic field strength and is defined as follows: a particle with a charge of 1 coulomb and a velocity of $1\,\frac{m}{s}$ passing perpendicularly through a magnetic field of 1 tesla will experience a force of 1 newton.

The MHz frequency often quoted with an NMR spectrometer (or used as a label for the NMR) is the frequency at which a proton will precess (the Larmor frequency) about the magnetic field (B_o) of that particular instrument. Because the Larmor frequency of a spinning nucleus is also a function of the magnetogyric ratio (γ), conversions for each nuclide are needed. For example, the conversion from MHz to tesla for the ^{1}H nuclide can be derived from 100 MHz = 2.35 T and yields a conversion of 1 T = 42.55 MHz.

Example 14.2

What is the B_o field (in tesla) for a ^1H nucleus with a Larmor frequency of 100 MHz?

STRATEGY –

Use Equations 14.5 and 14.6 to relate the Larmor frequency in rad sec^{-1} to the frequency in Hz, and use the data in Table 14.1 for the physical constants. Solve for B_o.

SOLUTION –

Start with $\Delta E = \hbar(\gamma B_o)$ and $\Delta E = h\nu$. Substitution gives:

$$\nu = \gamma B_o / 2\pi$$

Add in the values for the known variables: 100 MHz $= (26.753 \times 10^7 \text{ rad T}^{-1} \text{ sec}^{-1}) B_o/2\pi$

Solve for B_o:

$$B_o = 2\pi(100 \times 10^6 \text{ rad sec}^{-1})/(26.753 \times 10^7 \text{ rad T}^{-1} \text{ sec}^{-1})$$

$$B_o = 2.35 \text{ T (see Figure 14.8)}$$

(So you have a sense of scale, a small bar magnet may have a magnetic field of around 0.01 T.)

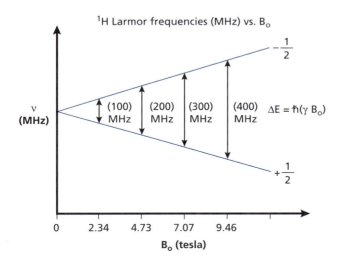

Figure 14.8 Field-splitting diagram for ^1H as a function of applied magnetic field.

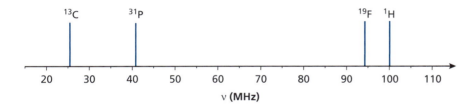

Figure 14.9 The resonance frequency of four common spin ½ nuclei (^{13}C, ^{31}P, ^{19}F, and ^1H) in a 2.35 T magnetic field.

Example 14.3

Calculate the Larmor frequency of ^{13}C in a 2.35 T magnetic field.

STRATEGY –

Use Equations 14.4 and 14.5 to convert the frequency in rad sec^{-1} to MHz.

SOLUTION –

$$\nu = \gamma B_o / 2\pi$$
$$\nu = (6.728 \times 10^{10} \text{ rad T}^{-1} \text{ sec}^{-1})2.35 \text{ T}/2\pi$$
$$\nu = 1.58 \times 10^{11} \text{ rad sec}^{-1}/2\pi$$
$$\nu = 25.3 \times 10^6 \text{ sec}^{-1} = 25.3 \text{ MHz}$$

(See Figure 14.9)

Problem 14.5:   For each nuclide below, use Equations 14.4 and 14.5 and the data in Table 14.1 to determine the Larmor frequency in a 4.7 T magnetic field.

  (a) ^1H (b) ^{13}C (c) ^{31}P (d) ^{19}F

Problem 14.6:   For each nuclide below, use Equations 14.4 and 14.5 and the data in Table 14.1 to determine the Larmor frequency in a 9.4 T magnetic field.

  (a) ^1H (b) ^{13}C (c) ^{31}P (d) ^{19}F

Problem 14.7:   Use Equation 14.6 and the equation for the Boltzmann distribution to determine the relative population distribution between m = –½ and m = +½ for a proton at room temperature for the following NMR spectrometers:

  (a) 60 MHz   (b) 200 MHz   (c) 300 MHz   (d) 900 MHz

Problem 14.8:   Determine the relative sensitivity of the spectrometers from Problem 14.7.

Problem 14.9:   Using the data from Table 14.1, derive a megahertz-to-tesla conversion for:

  (a) ^{13}C (b) $^{31}P1$ (c) ^{19}F

Problem 14.10:   The Boltzmann equation (see "Compare and Contrast—Population Distribution for Common Spectroscopic Methods") indicates that one can also affect the relative population of states by adjusting the temperature. Relative to room temperature, at what temperature would you need to be able to collect your spectrum in order to obtain the same signal enhancement achieved by switching from a 60 MHz NMR to a 200 MHz NMR? Discuss the implications.

Problem 14.11:   As of the publication of this textbook, the highest commercial field NMR used in research laboratories is a 900 MHz NMR (for ^1H?). What is the magnetic field strength in units of tesla?

COMPARE AND CONTRAST

Population Distribution for Common Spectroscopic Methods

You first saw the equation for the *Boltzmann distribution* in Chapter 2, which gives us a relative measure of species (molecules, atoms, or nuclei) in excited states and ground states as a function of temperature.

$$\frac{N_2}{N_1} = \frac{g_2}{g_1} e^{-\left(\frac{\Delta E}{k_B T}\right)}$$

Here, N_2 is the number of species in the excited state, N_1 the number in the ground state, ΔE is the difference in energy between those two states, k_B is the Boltzmann constant, and T is the temperature in Kelvin. The factors g_2 and g_1 refer to the degeneracies of the two states. Let us compare the N_2/N_1 ratio for ultraviolet-visible (UV-vis), Fourier transform IR (FTIR), and nuclear magnetic resonance (NMR). For a more direct comparison, we will use the carbonyl in benzaldehyde for the UV-vis and infrared spectra and the ^{1}H resonance from the proton on the carbon in that carbonyl in the NMR spectrum. Furthermore, because for an unsplit proton in NMR, $g_2 = g_1$, we will assume (1) that is the case for us in the UV-vis and FTIR spectra as well and (2) that all three experiments take place at room temperature (25°C). For NMR, we will assume a 300 MHz (7.05 T) instrument.

UV-Visible
The carbonyl bond exhibits a $\pi \rightarrow \pi^*$ transition at about 250 nm in benzaldehyde. We can estimate the ΔE:

$$\Delta E = \frac{hc}{\lambda} = \frac{\left(6.63 \times 10^{-34}\,\text{J} \cdot \text{s}\right)\left(3.00 \times 10^8\,\text{m/s}\right)}{250 \times 10^{-9}\,\text{m}} = 7.956 \times 10^{-19}\,\text{J}$$

We can now estimate N_2/N_1:

$$\frac{N_2}{N_1} = \frac{g_2}{g_1} e^{-\left(\frac{\Delta E}{k_B T}\right)} = (1)_e - \left(\frac{\left(7.956 \times 10^{-19}\,\text{J}\right)}{\left(1.38 \times 10^{-23}\,\text{J/K}\right)\left(298.15\,\text{K}\right)}\right)$$

$$= 1.05 \times 10^{-84}$$

This tells us that the ground state is approximately a trillion trillion trillion trillion trillion trillion trillion[11] times more populated than the excited state at room temperature! The only reasonable approach to spectroscopy is to measure absorbance (transition from ground to excited state).

FTIR
The carbonyl stretch occurs at 1,725 cm^{-1} (5.8 μm) in the FTIR spectrum of benzaldehyde. From the above approach, we find that $\Delta E = 3.429 \times 10^{-20}$ J (21 kJ/mol), and $N_2/N_1 = 2.4 \times 10^{-4}$, so the ground state is about 4,000 times more populated than the excited state. At room temperature, about 0.025% of the molecules will reside in the excited state without external excitation, but it is clear that absorbance is still the only practical approach to a spectroscopic measurement.

NMR
Equation 14.6 gives us a way to calculate ΔE in an NMR instrument. If we assume a 300 MHz NMR and use the same approach, we find that:

$$\Delta E = 1.99 \times 10^{-25}\,\text{J}\ (0.00012\,\text{kJ/mol})$$
$$\text{and}\ N_2/N_1 = 0.9999516$$

This means that we have an almost even distribution of ground and excited state populations. If we invert that and multiply by 10^6, we have 1,000,048—meaning that, out of every million nuclei, there are only 48 in excess in the ground state. When we provide an excitation source equal to the difference in energy, those 48 ppm excess nuclei are the only ones available for producing a signal. Many of the other nuclei will transition as well, but there is an equal probability of absorbance and emission for the other 999,952 nuclei, so their signals would cancel each other out. You will see later in this chapter (and in Chapter 21) that modern instruments tend to excite the excess ground state population of nuclei and then measure the signal produced as they decay back to the ground state equilibrium.

Problem 14.12: Determine the nuclear spin quantum number (I) and the nuclear magnetic quantum number (m) for each of the following nuclides. Identify the nuclides that are NMR active and explain your choices.

  (a) ^{19}F   (b) ^{119}Sn   (c) ^{16}O   (d) ^{27}Al   (e) ^{23}Na

---

Problem 14.13: What is the B_o field (in tesla) for a ^{1}H nucleus with a Larmor frequency of 800 MHz?

---

[11] A trillion is 1×10^{12}

Felix Bloch was a Swiss physicist who lived from 1905 until 1983. In 1933, he took a lecturer position at the University of Leipzig in Germany. In 1934, as Adolph Hitler rose to power, Bloch immigrated to the United States and took a position at Stanford University. During World War II, he worked on nuclear power and the US radar project. In 1946, he developed what are now known as the Bloch equations. His equations describe the magnetic resonance changes over time that characterize the spectroscopic aspect of these transitions. The Bloch equations describe the underlying theory that governs both nuclear magnetic resonance and magnetic resonance imaging technology. Bloch shared the 1952 Nobel Prize in Physics with Edward Purcell, for "their development of new methods for nuclear magnetic precision measurements and discoveries in connection therewith." Purcell was an American scientist working at Harvard University at that time. The work of Bloch and Purcell grew out of earlier discoveries by Isidor Isaac Rabi, an Austrian-born American physicist and winner of the 1944 Nobel Prize in Physics, who was the first person to demonstrate the magnetic resonance of atomic nuclei when exposed to a radiofrequency field. ■

www.nobelprize.org

Problem 14.14: What is the B_o field (in tesla) for a ^{1}H nucleus with a Larmor frequency of 500 MHz?

Problem 14.15: Using Figure 14.8 as a model, create a field-splitting diagram for ^{13}C showing the field splitting (ΔE) in units of MHz for a 2.34 T, 4.73 T, 7.07 T, and 9.46 T magnetic field.

Problem 14.16: There are two fundamental reasons why the signal strength for ^{13}C-NMR is much weaker than the signal strength for ^{1}H-NMR on any given instrument. Explain what those two reasons would be. (*Hint*: See Table 14.1 and the results for Problem 14.15.)

14.4 The Radiofrequency Pulse: Inducing Nuclear Magnetic Resonance

Recall from Section 14.3, that in order to generate a signal, we have to excite the system with radio frequency energy, called the *RF pulse*. The nuclei will undergo resonance when the frequency of the RF energy is equal to the Larmor frequency of the nuclei being studied.

In order to understand the RF pulse, let us go back briefly to the idea of the precessing magnetic moments under the influence of the external field, B_o. At equilibrium, a small majority of our precessing nuclei will be aligned in a stable sense ($+\frac{1}{2}$ or low energy state) with the B_o field. At room temperature, the number of nuclei in the more stable state will slightly outnumber those in other orientations relative to the B_o field.

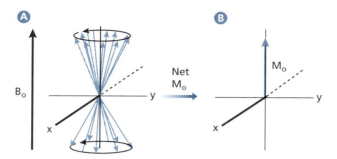

Figure 14.10 (A) Diagram showing the magnetic population imbalance of precessing nuclei in a magnetic field (B_o). (B) Diagram showing the *net* magnetization of a collection of precessing nuclei in a magnetic field (B_o). Note that M_o is aligned along the +z axis in the coordinate frame.

The signal measured is the result of the flipping of the net magnetic imbalance between the $+\frac{1}{2}$ and the $-\frac{1}{2}$ states. For the purposes of understanding how the signal is generated, it is useful to consider the population excess as a *net* magnetization of the sample.

Figure 14.10 combines the ideas presented in Section 14.2. The magnetic moments of nuclei in the sample are represented by the light blue arrows in panel A of Figure 14.10, which shows the precession of the nuclei about B_o and represents a population imbalance in which the nuclei aligned *with* B_o (up arrows) slightly outnumber those that in other orientations. If we sum all of the magnetic vectors, we are left with a single *net magnetic moment (M_o)* that represents the net magnetic imbalance of the sample. Panel B of Figure 14.10 removes the details, again showing the net magnetic moment (M_o) as a single arrow, aligned with B_o. For simplicity, as we talk about the effect of the RF pulse, we will use the net vector M_o. Note that the net magnetic moment is tremendously important. Under higher external field environments (higher B_o), the population imbalance becomes more pronounced, leading to an increase in signal strength.

Inside the magnet of the NMR spectrometer, near the sample, is a set of RF coils that transmit the pulse of RF energy and are oriented perpendicular to B_o. This RF pulse is referred to as B_1. When the RF pulse is applied, and is equal to the transition energy of the nuclear magnetic resonance, the net magnetization of the nuclei is "flipped." Figure 14.11 shows the reorientation of the net magnetization, M_o, along the $-z$ axis when an RF pulse (B_1) is applied along the x axis. This is called a *180° pulse* (or

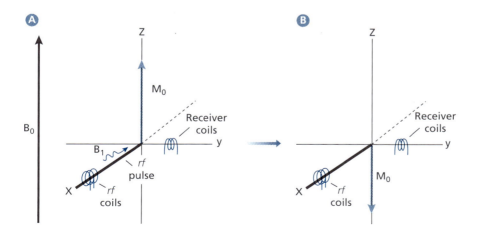

Figure 14.11 (A) Diagram showing the net magnetization aligned with B_o. This represents the equilibrium state of the net magnetization. Note that at equilibrium, the net magnetization lies along +z axis in the coordinate frame. (B) As an RF pulse is applied (B_1) along +x axis, we see reorientation of the net magnetization along –z. This is also known as a *180° or π-pulse*.

π *pulse*) because the energy applied was sufficient to change the net magnetization 180° from its original orientation.

The RF pulse is strong but short in time (measured in microseconds), and the 180° pulse is only one way in which the signals can be perturbed out of the equilibrium position. In Chapter 21, we will spend more time talking about the NMR experiment, relaxation events, pulse sequences, and coupling. Through a series of pulses and time delays, known as *pulse sequences*, the spectroscopist takes advantage of two phenomenon, *relaxation* and *coupling*, which provide information about molecular structure and environment. You are already somewhat familiar with the idea of coupling through the n + 1 neighbor rule, which describes the splitting of NMR peaks based on the number of nearest neighbors and was reviewed in Section 14.1. Likewise, the term *relaxation* should also be conceptually familiar to you from other forms of spectroscopy. Relaxation is, in short, the ways in which the excited state energy is dissipated as the system returns to equilibrium.

Before we move on, look again at Figure 14.11 and note that the receiver is positioned along the +y axis. Once the nuclear magnetic vector has been knocked out of its equilibrium position by the RF pulse, the precession of the nuclear magnetic moment is measured in the *x-y* plane and detected by a set of receiver coils (the detector). After the RF pulse has disturbed the system, the signal strength of the NMR is measured as the net magnetization vector returns to its equilibrium position, along the +z axis. The detector coils are measuring the RF energy as a function of time: a time domain signal. Each unique proton in the molecule has its own chemically shifted precession frequency because of slight differences in the magnetic field imparted by its immediate chemical environment; thus, the signal that reaches the detector is a beat pattern of constructive and destructing frequencies versus time. We first introduced the Fourier transform in Chapter 5 and again in Chapter 11. This beat pattern is very much the same sort of signal we discussed in Chapter 11 when we examined FTIR spectroscopy (Section 11.2).

Fourier Transform–Nuclear Magnetic Resonance: Time Domain versus Frequency Domain Spectroscopy and the Fourier Transformation

First-generation continuous-wave[12] NMR spectrometers operated in a scanning mode. However, today all modern high-field NMR spectrometers are time domain spectrometers that utilize the Fourier transform to generate a frequency domain spectrum. There are some excellent exercises in Chapter 11 that allow you to transform a beat pattern into a frequency domain spectrum.[13]

FT-NMR instruments use an excitation pulse that contains a *range of frequencies* spanning the Larmor frequencies of the nuclei of interest. The RF pulse contains a range of frequencies that will put *all* of the nuclei of interest into resonance at the same time. As the excited state nuclei relax, the emitted RF energy is measured in the *time domain*. For example, if you were conducting a ^{1}H-NMR experiment, you would program an RF pulse that would encompass the range of possible frequencies for ^{1}H. After the RF pulse ends, the resonance continues and ^{1}H nuclei, flipping from the m = $-\frac{1}{2}$ → m = $+\frac{1}{2}$ state, emit RF energy ($\Delta E_{\pm\frac{1}{2}}$), which is detected. The detector measures a power versus time signal (see Figure 14.12). The resulting time domain signal is a beat pattern comprised of the Larmor frequencies of all of the ^{1}H nuclei in the sample. In a manner similar to what we saw for FTIR, the Fourier transform algorithm converts the time domain signal (beat pattern) obtained with the FT-NMR into a frequency domain spectrum (see Figures 14.10 and 14.11). If you find it useful to draw analogies between optical spectroscopy and FT-NMR spectroscopy, you

---

[12] The term *continuous wave* is used to describe early NMR spectrometers that varied the excitation field (B_1) while holding the B_o field constant. In these types of experiments, the resonance signal was measured as an excitation (absorbance) rather than as an emission or relaxation. Because continuous-wave instruments have been largely replaced with FT-NMR, this text will not go into details on the continuous-wave instrument.

[13] See "Activity—Creating a Beat Pattern" and "Performing a Fourier Transform" in Chapter 11.

A gust of wind hitting a set of wind chimes is roughly analogous to the *rf* pulse that excites the molecules in nuclear magnetic resonance. The wind starts all of the chimes ringing at the same time, with each individual pipe ringing at its own frequency (or pitch). We hear the chimes, all together in time as they ring in synchrony. If you were to perform a Fourier transformation of the ringing, over time, you would mathematically deconstruct each pitch back to its original frequency.* To compare this to a scanning mode, each pipe in the set of wind chimes could be tapped in turn and measured as a function of frequency—for example, from low to high frequency. Without Fourier transformation, each one of the pipe pitches has to be measured sequentially. Some people with perfect pitch can hear a musical chord (multiple notes played together) and can tell you all of the notes in that chord. They essentially have the ability to perform Fourier transformation *in their minds.* ■■

---

* *Some digital oscilloscopes have this capability.*

could think of FT-NMR as an FT-emission technique where the electromagnetic radiation of emission (as in fluorescence spectroscopy) is in the RF range.

Free Induction Decay: The Fourier Transform–Nuclear Magnetic Resonance "Beat Pattern"

As we stated earlier, the beat pattern in an NMR experiment is analogous to the beat pattern in an FTIR experiment. However for NMR, the signal has a slightly different look. Because the nuclei are returning from the $m = -\frac{1}{2}$ to $m = +\frac{1}{2}$ ground state with different precession frequencies and decay rates, the strength of the RF energy decays exponentially with time. The result is a beat pattern with an exponential decay. This type of beat pattern is called a *free induction decay (FID)* (see Figure 14.12). Also, unlike FTIR, FT-NMR is an emission technique. In an FT-NMR experiment, the signal changes over time as the precessing nuclei *relax*.

Earlier in this chapter (see "Compare and Contrast—Population Distribution for Common Spectroscopic Methods"), we saw that the Boltzmann distribution of states in an NMR experiment is relatively small compared to other spectroscopic techniques. The implication is that NMR signals are intrinsically weak. However there are other features that help to enhance the signal. Recall that in Chapter 7, we discussed how the Heisenberg Uncertainty principle contributes to line broadening in atomic absorption spectra. We have repeated the equation for the Heisenberg Uncertainty principle here.

$$\Delta E \Delta t \geq \frac{\hbar}{2} \qquad \qquad \textbf{Eq. 14.7}$$

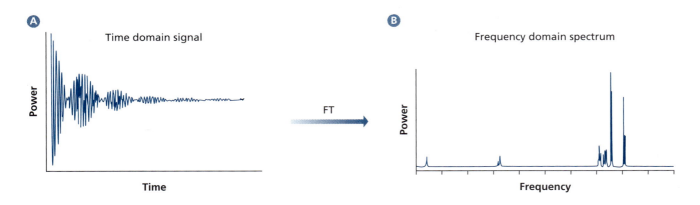

Figure 14.12 FID signal (A) comprised of many discrete frequencies. The resulting "beat pattern" diminishes over time as the system returns to equilibrium. The FID can be converted into a frequency domain signal (B) by use of the Fourier transform algorithm.

where

$$\hbar = \frac{h}{2\pi}$$

ΔE = bandwidth for the electronic transition
Δt = lifetime of the electronic transition

Equation 14.7 shows us that if Δt is large, then ΔE is small. NMR, once induced by the RF pulse, tends to relax very slowly, and long time-domain signals result in very sharp frequency domain signals. In addition, FT-NMR can take advantage of *signal averaging* and *Fourier filtering*.[14] The combination of very sharp peaks, signal averaging, and Fourier filtering leads to acceptable signal-to-noise ratios for NMR spectra.

14.5 Chemical Shift and Resolution

Now that you understand NMR, consider the simple NMR spectrum of propane again (profiled at the start of the chapter in "Spectral Analysis—A Quick Review"), shown here in Figure 14.13. Both of the signals (labeled a and b in the spectrum) represent ^{1}H nuclei (a and b as labeled on the molecular structure) are undergoing resonance. In this section we will discuss why the signals have different resonance frequencies and are spread out along the x axis.[15]

You might be thinking that, under a particular magnetic field influence, say 2.34 T, all ^{1}H nuclei resonate at the *exact same* frequency of 100 MHz. Remember, however, that the resonance frequency of a particular nucleus is shifted slightly by the chemical environment of that particular nucleus. In other words, the applied field is moderated by the local magnetic environment of the spinning nucleus ($B_{internal}$). See Equation 14.8.

$$\vec{B}_{total} = \vec{B}_0 + \vec{B}_{internal} \qquad \textbf{Eq. 14.8}$$

These *chemical shifts* are the result of the tiny magnetic fields generated by other atomic particles in the local environment. For instance, the nucleus is surrounded by electrons (bonding and nonbonding), and electrons have a spin of their own and therefore have their own associated magnetic fields. The magnetic fields generated by the electrons provide a "shielding effect" that opposes B_o and prevents the nucleus from

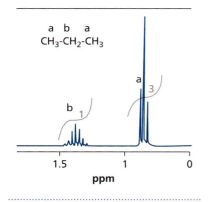

Figure 14.13 ^{1}H-NMR of propane.

---

[14] See Chapter 5 for a review of signal processing.

[15] See Chapter 21 for further discussion of coupling or why the signals are split based on their chemical environment.

Richard R. Ernst won the 1991 Nobel Prize in Chemistry for his contributions to the development of nuclear magnetic resonance (NMR) spectroscopy. Ernst's research into NMR spectroscopy led to the use of short bursts of radiofrequency energy to excite the nuclei followed by collecting a time domain signal. He then applied the Fourier transform to reestablish the frequency domain spectrum. The advent of Fourier transform-NMR allowed for signal averaging, which was necessary given the weak signals under investigation. Ernst also developed "pulse schemes" that paved the way for the introduction of two-dimensional, three-dimensional, and higher-order multidimensional NMR techniques, which will be discussed in Chapter 21. ■

feeling the full effect of B_o. The electron density about a particular nucleus is affected by considerations such as what is the atom bonded to and what kinds of bonds are involved, the electronegativity of nearby neighbors, and the three-dimensional structural positions of the atoms in the molecule.

In addition, it is usually the case that a proton will exist in close proximity to other protons, and each of those close proximity protons generate a magnetic field. The chemical shifts produced by these local variations can be exploited in the NMR experiment to elucidate chemical structures. Thus, each unique ^1H nucleus will have an actual resonance frequency that is slightly different from the other ^1H nuclei due to structural effects. This is why we find, for example, in Figure 14.13, that the ^1H nuclei in $-CH_3$ (labeled a) and the ^1H nuclei of $-CH_2$ (labeled b) can be identified separately in our propane NMR spectrum. The same can be said for the ^{13}C nuclei as well, if we were looking at a ^{13}C–NMR spectrum. Each nucleus will have its own particular chemically shifted resonance frequency, based on its local environment.

If you have studied NMR spectroscopy in organic chemistry, you most likely focused on spectral interpretation. Although spectral interpretation is obviously important, it is not the focus of this chapter. We have included several excellent texts on NMR spectral analysis in Section 14.9, Further Reading, and we continue our discussion of coupling (the splitting patterns of NMR spectra) in Chapter 21.

The Chemical Shift (ppm)

As we discuss NMR, it is important to keep in mind that there are two different magnetic influences on our analyte sample. The first is B_o, which is responsible for the split in quantum states between $m = +\frac{1}{2}$ and $m = -\frac{1}{2}$ (see Figure 14.8). The second magnetic influence is an RF pulse that initiates a transition between the $m = +\frac{1}{2}$ and $m = -\frac{1}{2}$ states. What is measured by the instrument's detector is the actual RF frequency of the transition (usually measured in MHz) within the spectral range of the RF pulse. Because the resonance frequency between the $m = +\frac{1}{2}$ and $m = -\frac{1}{2}$ states is also a function of B_o, the resonance frequency of a particular nucleus will vary between instruments with different B_o fields. If we reported NMR spectra using MHz as the unit of measure, it would be very difficult to compare spectra taken on different instruments. In order to normalize the x axis of NMR spectra, it is conventional to report the nuclear resonance frequency as a part per million (ppm) of B_o. For example, if a transition for a particular proton signal occurred at 500.000500 MHz under the influence of a B_o = 500 MHz magnet,[16] that resonance frequency would correspond to a 1 ppm *chemical shift*.

$$\left(\frac{\nu - \nu_0}{\nu_0} \right) \times 10^6 = 1\,\text{ppm}$$

Eq. 14.9

$$\left(\frac{500.000500\ \text{MHz} - 500\ \text{MHz}}{500\ \text{MHz}} \right) \times 10^6 = 1\ \text{ppm}$$

Furthermore, it will also be true that under the influence of a B_o = 400 MHz magnet, the same nuclei will resonate at 400.000400 MHz (also 1 ppm), and that under the influence of a B_o = 200 MHz, the same signal would resonate at 200.000200 MHz (still 1 ppm). Because chemical shifts in ppm are field-independent, scientists prefer to report results in ppm; however, it is easy to convert between MHz and ppm on the x axis. In fact, most high-field NMR spectrometer computer consoles have a toggle function between Hz and ppm that will do the calculations for you and will generate spectral results in the unit of your choice.

Example 14.4

Imagine that you need to determine the resonance frequency difference between two peaks in an NMR spectrum taken on a 100 MHz instrument. Let us say that you have two ^{1}H signals of interest at 1.5 ppm and 3.5 ppm. What is the frequency difference between the two peaks?

STRATEGY –

We know that at an operating field of 100 MHz, 1 ppm will equal 100 Hz. Therefore, working by proportion, 1.5 ppm will equal 150 Hz. Solve the problem by proportional reasoning.

SOLUTION –

$$3.5 - 1.5 = 2\ \text{ppm difference}$$

For the 100 MHz instrument, 100 Hz is 1 ppm. Therefore, by proportion, 2 ppm × 100 Hz/1 ppm = 200 Hz difference. There is a 200 Hz difference in the frequencies of the two signals of interest under a 100 MHz operating field.

Problem 14.17: Using Example 14.4 as a guide, determine the resonance frequency difference in Hertz between ^{1}H signals at 1.5 ppm and a 3.5 ppm as measured on an instrument with a B_o field strength of:

(a) 200 MHz   (b) 400 MHz   (c) 900 MHz

---

[16] Although it is not strictly speaking correct to report a magnetic field strength using units of MHz, this is common vernacular in "NMR-speak." As shown in the last section, 500 MHz is actually the Larmor frequency of a ^{1}H nucleus in a magnetic field of 11.75 T. The colloquial use of the term is to call the instrument a 500 MHz-NMR. Review Example 14.2 for practice in converting between these units.

Chemical Shift Reference

It is impossible to obtain the resonance frequency of a *free* hydrogen atom; therefore, chemical shifts are reported relative to a *reference compound*. For both ^{1}H-and ^{13}C-NMR experiments, the reference of choice is the compound *tetramethylsilane (TMS)*. See Figure 14.14. The 12 hydrogen atoms on TMS are equivalent, and thus its ^{1}H-NMR spectrum consists of a singlet. The chemical shift of this singlet is designated as 0.0 ppm, and the analyte chemical shifts are reported relative to TMS. Because of the electropositive nature of silicon, there is very little deshielding of the protons and the majority of analyte signals seen in a ^{1}H-NMR spectrum will be at a higher resonance frequency (also known as downfield) than the resonance frequency of TMS. TMS is also the accepted reference compound for ^{13}C-NMR and ^{29}Si-NMR. The ^{1}H signals in TMS are set to 0 ppm on the x axis; and thus all of our chemical shifts are measured relative to this reference compound.

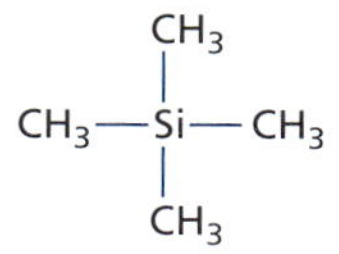

Figure 14.14 Chemical structure of TMS.

Resolution

How does NMR field strength (B_o) affect resolution? You will recall from previous chapters that the term *resolution* describes how well the signal peaks are separated across the x axis, which may, depending on the spectroscopic method, represent wavelength, time, frequency, or energy. High-resolution means that the peaks have little or no overlap, whereas poor resolution indicates that peak overlap is significant. In Example 14.5,

---

Example 14.5

What is the spectral separation in Hz of a 1.0 ppm signal and a 1.1 ppm signal obtained on a 100 MHz NMR and a 400 MHz NMR?

STRATEGY

Rearrange Equation 14.9, make appropriate substitutions, and solve for the resonance frequency of the desired signal.

SOLUTION

Rearranging Equation 14.9 and solving for the resonance frequency of each signal yields:

For the 100 MHz NMR

$$\frac{1.0\,ppm \times 100\,MHz}{1 \times 10^6} = 100\,Hz \qquad \frac{1.1\,ppm \times 100\,MHz}{1 \times 10^6} = 110\,Hz$$

For the 400 MHz NMR

$$\frac{1.0\,ppm \times 400\,MHz}{1 \times 10^6} = 400\,Hz \qquad \frac{1.1\,ppm \times 400\,MHz}{1 \times 10^6} = 440\,Hz$$

Therefore, a 1.0 ppm signal and a 1.1 ppm signal would be separated by 10 Hz (110 – 100 Hz) on a 100 MHz NMR and they would be separated by 40 Hz (440 – 400 Hz) on a 400 MHz NMR. If your detector had a baseline spectral resolution of 20 Hz, then the 1.0 ppm signal and the 1.1 ppm signal on a 100 MHz NMR would overlap by 50% (10 Hz/20 Hz). However on the 400 MHz NMR, these same two peaks, separated by 40 Hz, would be completely resolved and separated by 20 Hz.

we consider two peaks in a hypothetical experiment. One has a resonance frequency equivalent to 1.0 ppm and the other has a resonance frequency equivalent to 1.1 ppm. Will they be well resolved? The answer to this question depends on the sensitivity of the detector and the strength of the external field, B_o.

> **Problem 14.18:**   Assume you have a detector with a baseline spectral resolution of 20 Hz. How far apart (in ppm) must the two peaks be to have baseline resolution for a spectrometer with B_o equal to:
>
>   (a) 60 MHz?   (b) 200 MHz?   (c) 300 MHz?   (d) 900 MHz?

14.6 The Instrument

As a spectrometer, the high-field NMR has all of the same basic instrument components that you have studied in earlier chapters on spectrometer instrumentation. For example, an NMR spectrometer will have a source, a sample compartment, and a detector. Figure 14.15 shows a simplified schematic of an NMR spectrometer. However, unlike optical spectrometers, *the NMR source* does not emit visible light but rather emits a pulse of RF energy through a set of *RF coils*. Likewise, *the detector* in the NMR is also a set of RF coils, a receiver. In some NMR designs, the transmitter and receiver coils are physically the same set of coils; while in other designs, they are physically different coils. Figure 14.16 is a photograph of Doty Scientific's RF NMR probe. The probe has two sets of coils. The inner set of coils is the more sensitive and usually serves as the detector coils. The outer set of coils is used for advanced experiments that require the user to saturate a particular Larmor frequency (a "soft pulse"), which we will discuss in the context of decoupling experiments in Chapter 21.

In addition, the external magnet of the NMR provides a unique sample environment into which the sample is placed. The high-field cryogenic superconducting magnet is a prominent feature of the modern NMR instrument. Superconduction requires *very* low temperatures. The magnet's core is cooled with liquid helium that has a boiling point of 4.2 K. The liquid helium core is contained within an insulated container, a Dewar flask, of liquid nitrogen. The Dewar flask is a prominent feature of the NMR instrument room as seen in Figure 14.17 as the shiny silver canister to the left in the image. Nitrogen has a

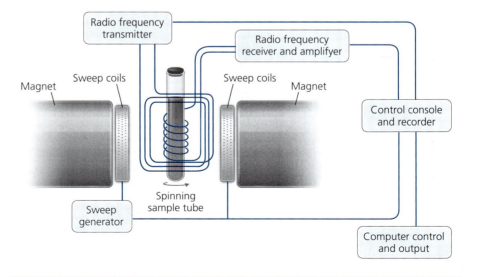

Figure 14.15 Basic schematic of an NMR spectrometer. Notice that the transmitter coils and the receiver coils are physically the same device. This is not true in all designs.

boiling point of 77 K. Because the liquid helium is expensive, the liquid nitrogen helps to minimize loss of helium due to evaporation.

The liquid nitrogen is replenished on a regular schedule to ensure the integrity of superconducting field and liquid helium is replaced as needed. Maintenance costs of an NMR spectrometer are substantial. Thus, it is common for research institutes and universities to host an NMR center where the instruments are centrally maintained and where billing/accounting for the costs are shared among multiple research groups, projects, and grants. Research laboratories typically keep at least one accessory liquid nitrogen Dewar flask on hand for intermittent fills. The storage Dewar flask is shown on the right in Figure 14.17. This storage Dewar flask is refilled on a regular schedule by a local cryogen company, and then the storage Dewar flask is used by a researcher or technician to fill the primary Dewar flask surrounding the magnet.

Figure 14.16 Photo of Doty Scientific's RF NMR probe.

Shimming

A homogeneous magnetic field maximizes signal strength and minimizes noise and signal dispersion[17] in the spectrum. NMR experiments require that the external magnetic field be homogeneous with a field variance of no more than 10 ppb. *Shimming*, a process in which the electromagnetic field is adjusted for homogeneity, is accomplished using another set of coils around the sample compartment; small adjustments to the RF voltage are applied within the shim coils in order to adjust the overall magnetic field experienced by the sample. Shimming is largely accomplished by a computer algorithm, although a skilled NMR spectroscopist can shim a magnet manually to achieve enhancements that autoshimming can miss.

Loading

The sample compartment in the NMR is inside the superconducting magnet. Therefore, samples have to be lowered into and lifted out of the magnet by a pressurized airstream. The act of putting the sample into the instrument brings the spectroscopist into close proximity to a strong magnetic field. In a high-field NMR laboratory, fields may range from 2 to 12 T. Thus, there is some concern for people with pacemakers, who should maintain a safe distance from the magnetic field at all times. Also, because the high-field magnet may desensitize magnetic card strips such as credit cards and key cards, the spectroscopist should check his or her pockets before approaching the magnet.

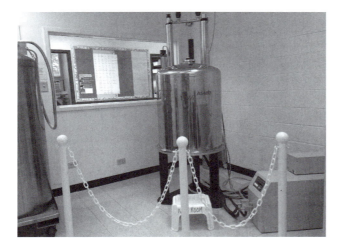

Figure 14.17 The JEOL 400 MHz NMR at Sweet Briar College. The Dewar on the right houses the superconducting magnet and the sample. The Dewar on the left is the accessory liquid nitrogen storage Dewar.

---

[17] This is essentially drift noise and is an important factor in overall resolution when performing signal averaging. If you are averaging a drifting signal, you will lose resolution.

Figure 14.18 Standard NMR tubes.

Liquid samples are contained inside a specialized sample holder, the NMR tube (see Figure 14.18). Then they are loaded into the instrument. The standard NMR tube, made of borosilicate glass, is 5 mm in diameter and comes in 17–20 cm (7–8 inch) lengths.

The NMR tube is inserted into a plastic cam called *the spinner* and lowered into the magnet using a stream of pressurized air. A jet of air is also used to rotate the spinner inside the NMR, spinning the tube at rate of about 15 to 60 Hz. Thus, the NMR tube needs to be absolutely straight and very well balanced, and the spinner cam needs to be kept clean and free of dust and other small particles. If the NMR tube wobbles inside the magnet, it will cause magnetic field inhomogeneity, resulting in peak broadening. Spinning the NMR tube has the effect of averaging out any field inhomogeneity in the *x-y* plane. This helps minimize any peak broadening and improves the spectral resolution.

14.7 Signal Processing

There are two aspects of the NMR signal that deserve further consideration and attention. First, we have already seen in this chapter that the NMR signal is measured as a *FID* in the time domain, and, as you have learned in Chapters 5 and 11, that signals in the time domain can be converted to the frequency domain using a Fourier transform. Second, we have seen that the Boltzmann distribution between the +½ and −½ spin states is very small. As well, you know that the frequency of these transitions is in the RF range, and thus, they are very low in energy. Therefore, NMR signals are inherently weak. In this section, we will discuss some of the challenges of acquiring and processing an NMR signal and how those challenges are overcome.

The signal strength is proportional to the concentration of NMR-active nuclei in the sample. In other words, the signal strength increases as the concentration of ^1H or ^{13}C in the sample increases. For an organic molecule, there are proportionally a lot of hydrogens in the molecule; 98.89% of all hydrogen atoms are the ^1H isotope (the most abundant isotope of this element). Not only are there proportionally fewer carbon atoms relative to hydrogen in an organic molecule, but only 1.108% of the carbon atoms are the ^{13}C isotope. The more abundant isotope ^{12}C has a spin of zero and is therefore not an NMR-active nuclide (see Table 14.1 for relative abundances of spin ½ isotopes). As a result, ^{13}C signals are intrinsically much weaker than ^1H signals.

The sample calculation in Example 14.6 helps demonstrate that ^1H-NMR signals are intrinsically stronger than ^{13}C-NMR signals by a factor of about 294.

Example 14.6

Using ethanol, C_2H_6O, determine the ratio of spin-active ^1H to ^{13}C nuclei for a given sample.

STRATEGY –

The ratio of hydrogen atoms to carbon atoms in ethanol is 6:2 (or 3:1). Multiply the ratio of each atom in ethanol by the relative abundance of their spin-active isotopes.

ANSWER –

$$\frac{\#\,H \times Relative\ Abundance}{\#\,C \times Relative\ Abundance} = \frac{6 \times 98.89\%}{2 \times 1.0108\%} = \frac{593.34}{2.0216} = 293.5$$

Problem 14.19: Determine the ratio of spin active ^1H to ^{13}C nuclei in each of the following molecules:

(a) Benzene (C_6H_6) (b) Hexane (C_6H_{14}) (c) Chloro(methyl)amine (CH_3NHCl)

Example 14.7

As calculated in Problem 14.19 (c), the intrinsic proton signal in chloromethylamine is nearly 400 times stronger than the carbon signal. Assuming the noise is the same for ^1H- and ^{13}C-NMR, how many additional experiments (data sets) would you need to perform in a signal averaging experiment in order to obtain the same signal-to-noise (SN) ratio in your ^{13}C-NMR as you would in your ^1H-NMR?

STRATEGY –

Signal averaging was first introduced in Chapter 5. See Equation 5.7.

ANSWER –

Substitution into Equation 5.7 and solving for N, we obtain:

$$\frac{S}{N} = \sqrt{n}\left(\frac{S}{N}\right)_o$$

$$400 = \sqrt{n}\,(1)$$

$$n = 160{,}000 \text{ experiments}$$

Using signal averaging to obtain the same SN ratio in your ^{13}C-NMR as you would in your ^1H-NMR experiment, you would need to increase the number of experiments by a factor of 160,000. ***This is obviously not practical!***

Problem 14.20: A typical ^1H-NMR experiment will incorporate n = 16 data sets into a single signal-averaged experiment. Assuming that the noise is the same and using Example 14.7 as a guide, how many signal-averaged data sets would you need to collect in order to achieve approximately the same $\frac{S}{N}$ ratio in your ^{13}C-NMR spectrum as you would in your ^1H-NMR spectrum?

Increasing the Signal-to-Noise Ratio

As Example 14.7 implies, signal averaging is a fundamental component of NMR signal processing. However, Example 14.7 also hinted at an intrinsic limitation of signal averaging. Even at the shortest practical experimental times,[18] such an approach to signal averaging would result in the experiment taking multiple days to complete. For this reason, when conducting ^{13}C-NMR, it is usually the case that the analyst will make the sample as concentrated as is reasonable for practical purposes.

Signal averaging is not the only tool used by the analyst to obtain quality data. In Chapter 5, we introduced FT-filtering, high- and low-pass filtering, and band-pass filtering. Similar techniques are applied by the software that accompanies modern NMR spectrometers. It is important to appreciate, when applying mathematical processing

---

[18] The T_1 relaxation time for a given sample can range from seconds to minutes. T_1 relaxation will be discussed in more detail in Chapter 21. If we planned for 160,000 data sets at 30 seconds per experiment, it would take approximately 56 days to complete sufficient data acquisition. If we assume the shortest possible T_1 relaxation time of only a second, the experiment would still take approximately 2 days to complete.

PROFILE

Angela Gronenborn

Angela Gronenborn is a world leader in developing methods for determining the structures of biologically important molecules. Gronenborn was educated in Germany, studying chemistry and physics at the University of Cologne. She moved to London in 1979 to pursue postdoctoral training at the National Institute for Medical Research. There she began to apply nuclear magnetic resonance (NMR), a relatively new analytical tool, to the study of enzyme structure. She discovered that proteins are able to adopt multiple stable conformations. Gronenborn's results greatly enriched understanding of protein dynamics and also further established NMR as a powerful tool with the potential to provide more than just atomic position information—NMR's capacity for elucidating interactions between protein and ligand provides solution-state structural and functional details. In 1988, Gronenborn took a position in the United States at the National Institutes of Health (NIH), where she began a line of inquiry on HIV and the development of three- and four-dimensional multinuclear NMR techniques for determining solution-state structures of large protein structures, work that provided fundamental understandings about HIV infection at the molecular level. In 2005, Gronenborn left the NIH to lead the structural biology program at the University of Pittsburgh School of Medicine. There she leads an interdisciplinary team of researchers as they collaborate on solving important problems in medicine. In pursuit of an HIV vaccine, they are applying NMR to solving one of the largest protein structures, that of the HIV-1 capsid structure. ∎

http://www.pitt.edu/research

to the FID data, that NMR signals are Lorentzian (and not Gaussian) and that certain curve-fitting applications can affect the signal shape.

Furthermore, because much of what spectral analysts do is to isolate peaks and determine the area under the peak (integration), in a process colloquially known as "peak picking," the analyst is often concerned with the peak width and peak area. An appreciation that the Lorentzian peaks have a typically broader base than a similar Gaussian peak can help the analyst obtain the best possible integration data. See Figure 14.19.

Other experimental and processing details help achieve the best possible spectrum. A few of these are described here. Background noise is reduced by using solvents in which all of the ^1H hydrogens have been replaced by deuterium (^2H). Deuterium (^2H) has a fundamental Larmor frequency outside the range of ^1H frequencies typically observed and signals from the solvent are therefore "invisible" in the spectrum. Deuterated solvents can also be used to probe the structure under analysis for exchangeable protons. Functional groups such as –OH, –NH, and –SH, which can

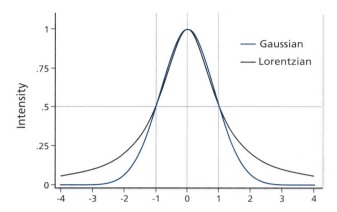

Figure 14.19 The difference between a Gaussian and Lorentzian peak is seen primarily in the width of the base of the curve.

readily exchange their ^{1}H for a ^{2}H, will "disappear" from the spectrum when a deuterated solvent is used. In older instruments, the spectral field was locked relative to TMS; however, because of the high abundance of deuterium in the deutero-solvents, modern instruments lock onto the deuterium signal of the solvent and then an off-set is applied during signal processing to create an output relative to TMS. The spectral window is set to acquire signals that fall within the expected range of frequencies around a particular "offset," which is selected at the center of the range. If the lock is not set properly or if the spectral window is too narrow, spectra can display unexpected shifts and peaks that are "folded in" from outside the window. We examined this in Chapter 11 when we discussed "*aliasing*" in the Fourier transform. Aliased peaks are peaks that fall outside the lock window and will appear as if they ran off of the end of the screen and were then "folded" back onto the opposite side of the screen. The lock ensures field-frequency stability by electronically applying a field correction whenever the signal drifts off resonance (for instance because of temperature variance). The lock helps to reduce band broadening and loss of signal.

14.8 Magnetic Resonance Imaging

This chapter began with an acknowledgment that NMR is the basis for MRI. Although NMR is the tool of choice for laboratory samples, MRI is the clinical choice where the "samples" under investigation are, more often than not, *people*. Fundamentally, however, the two techniques are derived from the same physical phenomenon and have benefitted from mutual technology development.

Whereas NMR spectroscopists go to great lengths to achieve magnetic field homogeneity in the applied (B_o) field, Lauterbur (see "Profile—Magnetic Resonance Imaging Has a Fascinating Backstory") realized that a gradient magnetic field could encode spatial distribution information, tagging each ^{1}H nucleus in three-dimensional space. The field was improved by Mansfield who developed techniques for data acquisition and processing that helped to speed the time taken to generate MRI scans from hours to seconds. The application of FT imaging principles to both NMR and MRI applications, pioneered by Richard Ernst in the mid-1970s (Ernst was awarded the Nobel Prize in Chemistry in 1991), further moved the technique ahead as Lauterbur's scanning methodologies were replaced by the highly efficient pulse techniques associated with T_1 and T_2.[19]

As the method developed throughout the 1970s, the more euphemistic term, *magnetic resonance imaging*, was adopted for clinical applications over the more established laboratory terminology, *nuclear magnetic resonance imaging* (or NMRI), due to the negative connotation of the term *nuclear* and especially, at that time, to allay fears when it

---

[19] Pulse schemes used in NMR are discussed in detail in Chapter 21.

PROFILE

Magnetic Resonance Imaging and Brain Concussion

Research on brain concussions, related to sports injuries and combat injury, has gained international attention and led to institutional changes in football regulations of neighborhood recreational leagues as well as the National Football League. Physical behaviors and symptoms may or may not be seen in all concussion events, which is why magnetic resonance imaging (MRI) can be a useful diagnostic tool. Imaging of the brain following a traumatic head injury can be used to identify the location and extent of the injury. In addition, medical scientists are learning that brain injury is characterized not

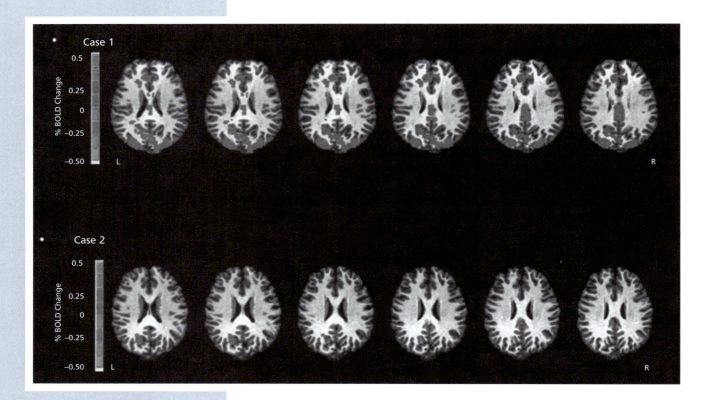

only by local and immediate physical trauma but also in terms of more subtle injury that affects the brain's neuronal functions and processes, both short term and long term. The new approach called functional MRI (*fMRI*) can provide important information related to the brain's functioning by metabolomics. Through metabolomics, scientists are studying the metabolic chemical processes that normally occur in living tissue. The researchers seek to identify and diagnose problems and diseases that result in alterations of those normal functions. The fMRI can be used to monitor tissue for changing concentrations of important metabolic intermediates. By monitoring changing ratios of key compounds, these finely tuned MRI techniques can reveal previously unidentified injury and help track recovery. MRI has been used successfully in football athletes to identify changes in the metabolomics of head injury patients and thus secure adequate and appropriate treatment for those injuries, even when more outward signs of injury are less obvious.* ■■

* Henry, L. C.; Tremblay, S.; Leclerc, S.; Khiat, A.; Boulanger, Y.; Ellemberg, D.; Lassonde, M. *Metabolic Changes in Concussed American Football Players During the Acute and Chronic Post-injury Phases. BMC Neurol.,* **2011**, *11, 105.*

Magnetic resonance imaging was first suggested in a 1971 paper published in *Science* by Raymond Damadian, a physician scientist at the Downstate Medical Center (State University of New York) in Brooklyn, New York, who discovered that tumors display different nuclear magnetic resonance (NMR) relaxation times as compared to other tissue types. Damadian wrote, "The possibility of using magnetic relaxation methods for rapid discrimination between benign and malignant surgical specimens has also been considered." Damadian's research spurred great interest in the scientific community.

Peter Lauterbur, an NMR spectroscopist and then Associate Professor of Chemistry at the State University of New York at Stony Brook, observed confirmation of Damadian's results in a comparison of normal and malignant liver tumors in an experiment carried out by Donald Hollis and Leon Saryan (Johns Hopkins University) in 1971. Lauterbur launched his own research, in search of a noninvasive way to apply NMR to a living organism, in vivo. In 1973, he published a paper in *Nature* in which he proposed a means for accomplishing in vivo imaging with NMR and demonstrated the instrumental potential by producing an image from glass capillaries of water surrounded by layers of deuterium.

The field continued to expand and by 1977, Damadian had built the first full body scanner, an instrument that his research team called "Indomitable" and which now resides at the Smithsonian Institution (see Figure 14.20).

Paul Lauterbur and Peter Mansfield were awarded the Nobel Prize in Medicine in 2003 in recognition of their contributions to the development of MRI as an in-vivo imaging technique, and worked to enhance the capabilities of the instrument with improved resolution and speed of acquisition. Key developments made in the 1970s by both Lauterbur and Mansfield were in the introduction and use of gradient magnetic fields. ■

PROFILE
Magnetic Resonance Imaging Has a Fascinating Backstory

Figure 14.20 Image of Damadian (left) with his research team.

came to analysis of human samples. Over time, as various other types of nuclear medicine have become well-established also, MRI has taken its place among the specialties that make up nuclear medicine and imaging technologies, including computed tomography (CT), positron emission tomography (PET), mammography, radiation therapy, ultrasound, and radiography.

Like the NMR, the MRI spectrometer uses RF energy to induce a nuclear spin flip in the hydrogen atoms of the patient. Hydrogen, an abundant element in the body, is found in water and lipids, two of the primary components of biological tissue. Under the influence of a high-field external magnetic field (B_o), NMR at the Larmor frequency of the nuclei is measured as the hydrogen atoms relax postexcitation. The rate of relaxation (both T_1 and T_2) is measured and, as the relaxation rates are related to the density of the tissue, a tissue density map is created via computational analysis. An MRI's image is a combination of the ^1H signals' frequency information with the phase domain information that provides location information for the ^1H signals.

Imaging data is obtained in two-dimensional slices or sections through the tissue of interest. Two-dimensional imaging is called *tomography*. A series of two-dimensional images are combined to create a three-dimensional image. In two-dimensions, the data are collected in units called *pixels*. The three-dimensional equivalent of pixel is the unit *voxel*, a portmanteau of volume pixel. In MRI, the slice is selected by applying the magnetic field gradient through the section of tissue. Under the influence of the magnetic field, the sample is also pulsed with RF energy to induce NMR in the ^1H nuclei in that section of tissue. Through the Fourier transform, the data are collected for the voxels in the section (also known as field of view) and converted into frequency and position information, which is then is used to generate an image. Figure 14.21 shows the composite nature of the various imaging techniques. Chemical contrast agents may be used to enhance particular types of images using MRI. Contrast agents are typically metal complexes that travel to particular tissues in the body and enhance the differences in the signals between tissue types. Gadolinium- and iron-based complexes are popular contrast agents and work by distorting the local magnetic field in the tissue. Use of these compounds, however, has some inherent risks due to toxicity concerns.

Another way to enhance an image from MRI is *chemical shift imaging*. This is a methodology that can highlight (or suppress) signals from either water or fat, specifically, in order to generate an image that comes from only water (fat suppression) or only fat (water suppression).

The instrumental anatomy of an MRI is, as you might expect, very similar to an NMR. Figure 14.21 shows a basic block diagram and the relative positions of sample (patient), external magnet, RF coils (which may serve as transmitters and receivers, just as in NMR) and the computer controllers that make it all work together. You will also notice the addition of gradient coils, which modify the magnetic field as needed for imaging.

The external magnetic field in the MRI is provided, just as in NMR, by a superconducting electromagnet, which is cooled by cryogens. See Figure 14.22. The Food and Drug Administration limits human exposure to 8 T for adults and children and sets limits on RF and noise exposure as well. The MRI design is such that the magnetic field strength falls off sharply as a function of distance from the magnet, thus ensuring that healthcare providers are minimally exposed and allowing for MRI centers to be placed within a healthcare facility without cause for concerns. As with NMR, magnetic swipe cards and other metal objects, including pacemakers and implants, are incompatible with MRI, and extreme care needs to be taken to prevent accidents that can kill or injure a patient or healthcare provider and/or damage the instrument.

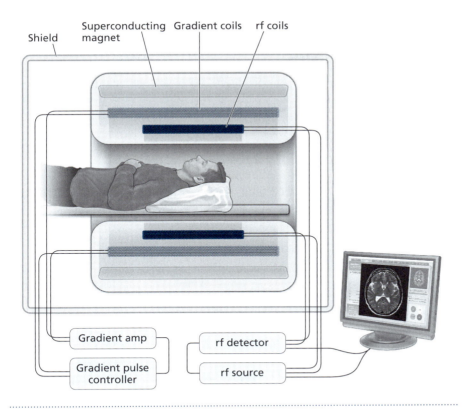

Figure 14.21 Schematic of an MRI instrument.

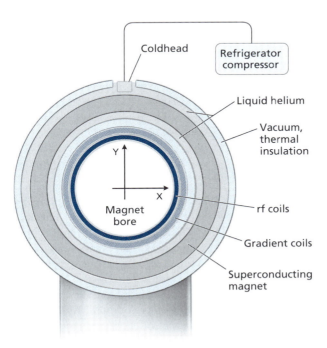

Figure 14.22 A cross-sectional view of the magnetic coils and cryogenic containment for an MRI instrument.

14.9 Further Reading

BOOKS

Berger, S.; Braun, S. *200 and More NMR Experiments.* Wiley-VCH: Weinheim, Germany, **2004**.

Dawson, M. J. *Paul Lauterbur and the Invention of MRI.* MIT Press: Cambridge, MA, **2013.**

Friebolin, H. *Basic One- and Two-Dimensional NMR Spectroscopy.* Wiley-VCH: Weinheim, Germany, **2010**.

Gunther, H., *NMR Spectroscopy: Basic Principles, Concepts and Applications in Chemistry,* 3rd ed., Wiley-VCH: Weinheim, Germany, **2013**.

Jacobsen, N. E. *NMR Data Interpretation Explained: Understanding 1D and 2D NMR Spectra of Organic Compounds and Natural Products.* Wiley: Hoboken, NJ, **2016**.

Jacobsen, N. E. *NMR Spectroscopy Explained: Simplified Theory, Applications and Examples for Organic Chemistry and Structural Biology.* Wiley-Interscience: Hoboken, NJ, **2007**.

Keeler, J. *Understanding NMR Spectroscopy,* 2nd ed., Wiley: Hoboken, NJ, **2010**.

Levitt, M. H. *Spin Dynamics: Basics of Nuclear Magnetic Resonance,* 2nd ed., Wiley: Hoboken, NJ, **2008**.

Linington, R. G.; Williams, P. G.; MacMillan, J. B. *Problems in Organic Structure Determination: A Practical Approach to NMR Spectroscopy.* CRC Press: Boca Raton, FL, **2015**.

Silverstein, R. M.; Webster, F. X.; Kiemie, D. *Spectrometric Identification of Organic Compounds,* 7th ed., Wiley: Hoboken, NJ, **2005.**

Westbrook, C.; Roth, C. K.; Talbot, J. *MRI in Practice,* 4th ed., Wiley-Blackwell: Chichester, UK, **2011**.

ONLINE RESOURCES

http://phet.colorado.edu/en/simulation/mri. A JAVA-based simulation sponsored by the University of Colorado, this is an excellent online simulation to introduce the student to concepts in NMR and MRI. It allows the user to change experimental parameters and to monitor resulting changes in the system and in various outputs.

https://www.youtube.com/watch?v=nM7jQFhrvR0. Dr. James Keeler's lectures on NMR are available on YouTube, presented by the University of Queensland and other sponsors and produced by the Australia and New Zealand Society for Magnetic Resonance (ANZMAG). The videos are divided into a series of lectures. This link is for the introduction to the series. Lecture 2 is on the vector model, and Lecture 3 is on the Fourier Transform; these lectures may be of interest to the reader.

http://chemwiki.ucdavis.edu/Physical_Chemistry/Spectroscopy/Magnetic_Resonance_Spectroscopies/Nuclear_Magnetic_Resonance. Sponsored by the National Science Foundation and the University of California, Davis, this wiki-style online text provides a textbook-like approach to NMR but in modular format with extensive illustrations, animations, and hyperlinks. Many pages are under construction, and therefore the treatment is somewhat incomplete. However, the information that is available is very accessible.

http://www2.chemistry.msu.edu/faculty/reusch/VirtTxtJml/Spectrpy/nmr/nmr1.htm /nmr/nmr2.htm. The *Virtual Textbook of Organic Chemistry* is an excellent resource that includes these two chapters on NMR by Dr. William Reusch at the University of Michigan. The site has very thorough descriptions, a mathematically rigorous treatment of background topics, animated graphics, data tables, and worked-out examples. The site is a good reference for spectral interpretation. Practice problems on spectral interpretation are interactive.

http://hyperphysics.phy-astr.gsu.edu/hbase/nuclear/nmr.html. The Hyperphysics website is sponsored by the Department of Physics and Astronomy at Georgia State University and is written and maintained by Carl (Rod) Nave. The sections on NMR and related topics provide extensive links for background work, including definitions, equations, useful constants, and diagrams.

http://www.cis.rit.edu/htbooks/nmr/inside.htm. An excellent online textbook, *The Basics of NMR,* is by Dr. Joseph P. Hornak, Professor of Chemistry and Imaging Science at the Rochester Institute of Technology (RIT) and Director of the Magnetic Resonance Laboratory at RIT. The interactive text is presented in 11 chapters, from fundamentals to more advanced topics, and it includes references, images, figures, spectra, examples, definitions, and animations.

SOME INTERESTING LABORATORY EXPERIMENTS

Gonzalez, E.; Dolino, D.; Schwartzenburg, D.; Steiger, M. A. Dipeptide Structural Analysis Using Two-Dimensional NMR for the Undergraduate Advanced Laboratory. *J. Chem. Educ.,* Oct 2, **2014** (online).

Li, Q.; Chen, Z. Yan, Z.; Wang, C.; Chen, Z. Touch NMR: An NMR Data Processing Application for the iPad, *J. Chem. Educ.,* **2014**, *91(11),* 2002–2004.

Sibbald, P. A. Nucleophilic Substitution Reactions Using Phosphine Nucleophiles: An Introduction to Phosphorus-31 NMR. *J. Chem. Educ.,* November 12, **2014** (online).

Simpson, A. J.; Mitchell, P. J.; Masoom, H.; Mobarhan, Y. L.; Adamo, A.; Dicks, A. P. An Oil Spill in a Tube: An

Accessible Approach for Teaching Environmental NMR Spectroscopy. *J. Chem. Educ.*, October 23, **2014** (online).

Williams, K. R.; King, R. W. (a series of labs):

The Fourier Transform in Chemistry. Part 1. Nuclear Magnetic Resonance: Introduction. *J. Chem. Educ.*, 1989, 66, A213.

The Fourier Transform in Chemistry. Part 2. Nuclear Magnetic Resonance: The Single Pulse Experiment. *J. Chem. Educ.* 1989, 66, A243.

The Fourier Transform in Chemistry—NMR: Part 3. Multiple-Pulse Experiments. *J. Chem. Educ.*, 1990, 67, A93.

The Fourier Transform in Chemistry—NMR: Part 4. Two-Dimensional Methods. *J. Chem. Educ.*, 1990, 67, A125;

The Fourier Transform in Chemistry—NMR: A Glossary of NMR Terms. *J. Chem. Educ.*, 1990, 67, A100.

14.10 Exercises

EXERCISE 14.1: Using the n + 1 rule reviewed in Section 14.1, explain the peak splitting pattern for the compound ethyl crotonate. The structure of trans-ethyl crotonate (ethyl trans-2-butenoate) is shown alongside the ^{1}H-NMR spectrum.

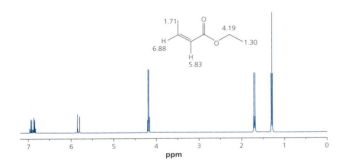

EXERCISE 14.2: From the mass spectral and elemental analysis data, you have determined the formula of an unknown compound from a petroleum distillate to be $C_3H_6O_2$. You have limited the possibilities to the four different structures shown. Each has the formula $C_3H_6O_2$.

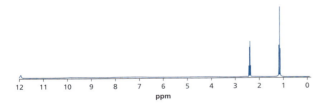

A: Glycidol   B: Methoxyacetaldehyde   C: Propionic acid   D: Methyl acetate

(a) Describe how an IR spectrum would aid in narrowing down your choices. Be specific.

(b) Use the ^{1}H-NMR spectrum of your unknown shown to determine which isomer you have.

EXERCISE 14.3: What is the Larmor frequency of ^{13}C in a 7.07 T magnetic field?

EXERCISE 14.4: What is the Larmor frequency of ^{13}C in a 19 T magnetic field?

EXERCISE 14.5: For each nuclide below, determine the Larmor frequency in a 14.14 T magnetic field.

(a) ^{1}H   (b) ^{13}C   (c) ^{31}P   (d) ^{19}F

EXERCISE 14.6: For each nuclide below, determine the Larmor frequency in a 21.21 T magnetic field.

(a) ^{1}H   (b) ^{13}C   (c) ^{31}P   (d) ^{19}F

EXERCISE 14.7: What is the magnetic field strength in units of tesla for a 600 MHz NMR instrument?

EXERCISE 14.8: What is the magnetic field strength in units of tesla for a 700 MHz NMR instrument?

EXERCISE 14.9: Using Figure 14.9 as a model, create a field-splitting diagram for ^{31}P showing the field splitting (ΔE) in units of MHz for 2.34 T, 4.73 T, 7.07 T, and 9.46 T magnetic fields.

EXERCISE 14.10: Using Figure 14.9 as a model, create a field-splitting diagram for ^{19}F showing the field splitting (ΔE) in units of MHz for 2.34 T, 4.73 T, 7.07 T, and 9.46 T magnetic fields.

EXERCISE 14.11: There are two fundamental reasons why the signal strength for ^{13}C-NMR is much weaker than the signal strength for ^{1}H-NMR on any given instrument. Explain what those two reasons would be. (*Hint*: See Table 14.1 and the results for Problem 14.15.)

EXERCISE 14.12: Determine the resonance frequency difference between two resonances in an NMR spectrum taken on a 300 MHz instrument. If the two ^{1}H signals of interest are at 2.4 ppm and 3.0 ppm, what is the frequency difference between the two peaks?

EXERCISE 14.13: Determine the resonance frequency difference between two resonances in an NMR spectrum taken on a 500 MHz instrument. If the two ^1H signals of interest at 7.75 ppm and 7.90 ppm, what is the frequency difference between the two peaks?

EXERCISE 14.14: Determine the resonance frequency difference in hertz between ^1H signals at 2.20 ppm and at 2.30 ppm as measured on an instrument with a B_o field strength of:

(a) 300 MHz   (b) 600 MHz   (c) 900 MHz

EXERCISE 14.15: Determine the resonance frequency difference in hertz between ^1H signals at 4.45 ppm and at 4.60 ppm as measured on an instrument with a B_o field strength of:

(a) 100 MHz   (b) 500 MHz   (c) 900 MHz

EXERCISE 14.16: Assume you have a detector with a baseline spectral resolution of 15 Hz, comment on the resolution of the peaks in Problem 14.17 at each field strength (a–c).

EXERCISE 14.17: Assume you have a detector with a baseline spectral resolution of 10 Hz, comment on the resolution of the peaks in Problem 14.17 at each field strength (a–c).

EXERCISE 14.18: Determine the relative population distribution between m = −½ and m = +½ for a proton at room temperature for the following NMR spectrometers:

(a) 90 MHz   (b) 400 MHz   (c) 600 MHz   (d) 800 MHz

EXERCISE 14.19: Determine the relative population distribution between m = −½ and m = +½ for a proton at a temperature of 280K, for the following NMR spectrometers:

(a) 90 Hz   (b) 400 MHz   (c) 600 MHz   (d) 800 MHz

EXERCISE 14.20: Determine the relative population distribution between m = −½ and m = +½ for a proton at a temperature of 310K, for the following NMR spectrometers:

(a) 90 Hz   (b) 400 MHz   (c) 600 MHz   (d) 800 MHz

EXERCISE 14.21: Determine the ratio of spin-active ^1H to ^{13}C nuclei in each of the following molecules:

(a) Pentane (C_5H_{12}) (b) Ethylene diamine $(C_2H_4(NH_2)_2)$
(c) Acetaldehyde (C_2H_4O)

EXERCISE 14.22: Determine the ratio of spin-active ^1H to ^{13}C nuclei in each of the following molecules:

(a) Glycerol $(C_3H_8O_3)$ (b) Heptane (C_7H_{16})
(c) Lysine $(C_6H_{14}N_2O_2)$

EXERCISE 14.23: What is the $\frac{S}{N}$ ratio enhancement when ^1H-NMR experiment incorporates n = 64 data sets into a single signal-averaged experiment rather than n = 16 data sets?

EXERCISE 14.24: What is the $\frac{S}{N}$ ratio enhancement when ^1H-NMR experiment incorporates n = 128 data sets into a single signal-averaged experiment rather than n = 64 data sets?

EXERCISE 14.25: Repeat the calculations from Exercises 14.12 and 14.13 with ^{13}C-NMR. Describe how the enhancement effect compares between ^1H and ^{13}C.

EXERCISE 14.26: If you were interested in observing nuclei other than ^1H or ^{13}C in an MRI, which of these biologically available elements would be possible candidates?

(a) Oxygen   (b) Calcium   (c) Selenium

EXERCISE 14.27: If you were interested in observing nuclei other than ^1H or ^{13}C in an MRI, which of these biologically available elements would be possible candidates?

(a) Iron   (b) Sulfur   (c) Nitrogen

EXERCISE 14.28: In an MRI, the patient is placed inside a central opening, the bore, of the magnetic field. The patient's body is aligned with the magnetic field (B_o) along the z axis. The magnetic field runs from head to toe. Is this the same relative position as an NMR tube inside an NMR? Explain using diagrams. Be sure to consider the gradient coils and their function in your answer.

EXERCISE 14.29: In an MRI instrument, there are gradient coils that modify the external magnetic field, creating gradients in the field. These gradients can be set up to image sections of the body along the sagittal, transverse, or coronal planes. Use the internet to find MRI images of tissues that represent these sections and describe the perspective of the image on the tissue. This article, from the National High Magnetic Field Laboratory at Florida State University, may help you get started; you can find it at http://www.magnet.fsu.edu/education/tutorials/magnetacademy/mri/fullarticle.html.

Liquid Chromatography

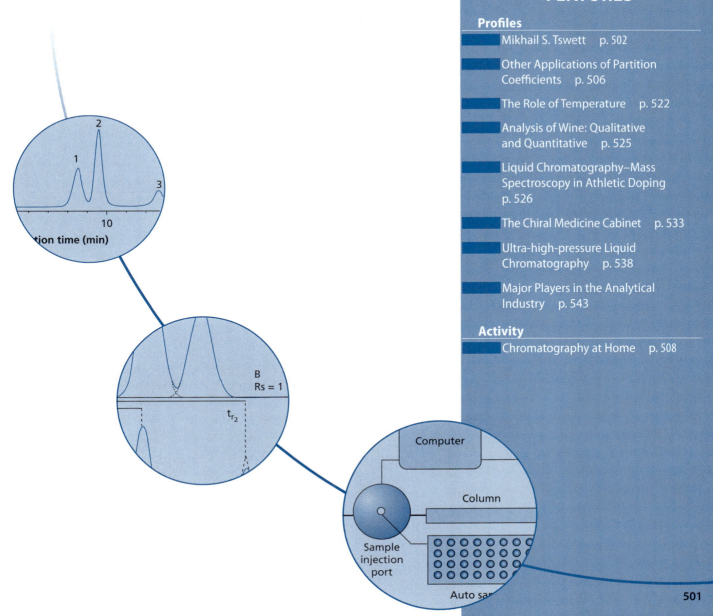

At the Warsaw Society Naturalist's convention in 1903, a 31-year-old scientist by the name of Mikhail Tswett gave a seminal lecture on a technique for the separation of solvated plant pigments by means of differential adsorption of the plant pigments on a suspended solid media.* The technique we now call column chromatography had been invented. One wonders if the attendees of this lecture fully grasped the importance of Tswett's presentation. One can arguably make the claim that his invention led to major advancements in the areas of biology, biochemistry, genetics, chemistry, pharmacology, medical research, material science, aerospace, petrochemistry, nuclear science, and much more. Furthermore, Tswett's work on differential adsorption became the foundation on which our modern understanding of distribution constants is based. Tswett was a naturalist by training and received his first PhD from the University of Geneva in 1896 at the age of 24, with his thesis "Investigations on the Physiology of Cells: Materials Leading to the Knowledge of the Movement of Protoplasma, Plasma Membranes, and Chloroplasts." He later moved to Russia, where he continued his research and, at the age of 38, defended a second PhD thesis from Warsaw University in 1910 titled "Chlorophylls in Plants and Animals."† Tswett's first application of liquid chromatography, to separate and isolate plant pigments, explains in part the etymology of the term *chromatography*, which is derived from the Greek words *graphein* (to write) and *chroma* (color)—quite literally, color writing. ∎

---

* *Tsvet, M. S. Warsaw. Obshch. Estestvoisp. Otd. Biol. (Transl. Proceedings of the Warsaw Society of Naturalists),* **1903**, *14, 20–39.*

† *For additional articles on Mikhail Tswett, see Berezkin, V. G. Chem Rev.,* **1989**, *89, 279–285; Strain, H. H.; Sherma, J. J. Chem. Ed.* **1967**, *44(4), 235–237; Williams, K. R. J. Chem. Ed.* **2002**, *79(8), 922–923.*

15.1   Introduction

Modern chromatography is used for the separation and analysis of the various components of complex mixtures. It finds wide use in analytical laboratories, crime laboratories, hospitals, and environmental laboratories. It is a powerful tool for the identification of trace materials (qualitative analysis) and also for determining the amount of a particular component in a sample (quantitative analysis). In addition, chromatography systems have been developed that can handle bulk quantities of material for the purposes of purifying a product as part of a manufacturing processes. This type of chromatography is termed *preparative* chromatography. Although the basic concepts of chromatographic separation are quite simple, the technology that has been developed around chromatography is very sophisticated. Because of its qualitative and quantitative power, chromatography is one of the workhorse techniques of analytical chemistry. In this chapter, we examine liquid chromatography (LC) and will lay out the fundamental theory. In the next chapter, on gas chromatography, we will see that many of the same concepts apply.

In LC, the sample is dissolved in a solvent (called the *mobile phase*) and the solvent carries the sample into, and/or across, and/or through the stationary phase. The *stationary phase*, as the name implies, is not mobile. As the mobile phase flows through the stationary phase, the various components of the sample are distributed between the immobile stationary phase and the flowing mobile phase. Over time, the various components of the sample become separated because of the different types of interactions each analyte has with the stationary and mobile phases. Components that spend more time associated with the stationary phase will take longer to flow through the stationary phase and will exit (*elute*) later than components that spend less time associated with the stationary phase and elute more quickly. This idea of *differential migration* is a key feature of a successful separation. Without differential migration, there is no separation.

In this chapter, we will consider the:

- Theoretical background of separations

- Instrumental considerations that provide for high performance, high resolution, and automation of modern LC

- Ways in which LC separations can be accomplished and how we take advantage of various physical and chemical characteristics to distinguish one component from another

- Need for separations and how LC supplements other types of analyses

- Applications of LC that are of current interest

15.2 Theory

Distribution Equilibrium

The principles of LC are based on equilibria. The equilibrium of significance in LC is described by a *distribution equilibrium constant*, which describes the properties of an analyte as it partitions (or distributes) between the mobile phase (solvent) and the stationary phase (solid). We will start with a review of the concept of equilibrium more generally using the familiar technique of distillation.

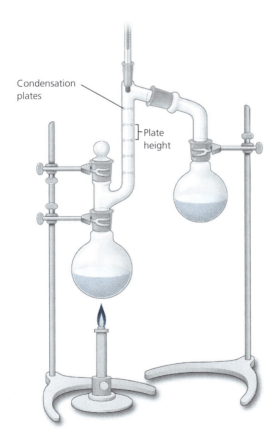

Condensation plates

Plate height

Figure 15.1 A schematic of a distillation apparatus.

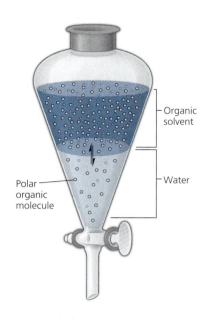

Figure 15.2 Extraction of an organic molecule from an aqueous solution.

Distillation exploits differences in the *phase equilibria* of the components in solution. During distillation, the various components alternate between different phase states (liquid and gas) as the solution is heated and cooled. A phase equilibrium between liquid and gas phases occurs at each condensation point within the distillation column. In the petroleum industry, a refinery will utilize distillation towers that have condensation plates on which different fractions are collected. Figure 15.1 is a sketch of a common distillation column one might use in an organic chemistry laboratory. As the solution is heated, vapors rise through the column, and as the temperature decreases, condensation of the less volatile components of the vapor occurs preferentially as the vapors contacts the condensation *plates*. The condensed components then fall back into solution while the vapors of the more volatile components continue to rise through the column. A phase equilibrium is established between the vapor and liquid phases at each condensation plate, and because vapors continue to rise through the column and liquids fall within the column, components of the mixture are separated based on differential equilibrium at each condensation plate. Components with low boiling temperatures rise further to the top in comparison to higher boiling temperature components. To make a distillation column more efficient at separating compounds with similar boiling points, we would need to increase the number of condensation points within the column and/or increase the length of the column. The ability to isolate two closely related compounds is quantified by the *resolving power* of the column; a concept we will discuss in detail later in this chapter. In other words, the resolution of a distillation column is a function of the number of condensation plates within the column and the *plate height* (distance between each plate).

The analogy between a distillation and chromatography is imperfect. Unlike distillation, LC does not involve an equilibrium between the liquid and vapor phases of the analyte. Instead, LC involves a dynamic equilibrium in which the analyte moves between two different preexisting media. In this regard, chromatography is similar to an extraction, where the underlying principle is a distribution or partition equilibrium between an aqueous medium and an organic medium.

You have most likely encountered partition equilibria in the laboratory when you have performed a liquid–liquid extraction using a separation funnel. In the separation funnel, the analyte distributes between two immiscible solvents (see Equation 15.1 and Figure 15.2).

$$C_{aq} \rightleftharpoons C_{org} \qquad \textbf{Eq. 15.1}$$

Typically, the two solvents self-assemble into an aqueous layer and an organic layer. A highly polar analyte will partition preferentially into the aqueous layer. Conversely, a nonpolar analyte will partition preferentially into the organic layer. Figure 15.2 is a representation of a liquid–liquid extraction in which an aqueous solution containing a solute molecule is being extracted using an organic solvent. Although the solute molecule is continuously moving between the two solvents, its higher affinity for the organic solvent results in the solute molecule becoming concentrated in the organic solvent. When the system has come to equilibrium, we can characterize the distribution of solute across the two layers using an equilibrium constant, K_D (also known as the partition coefficient K_p).

The distribution constant is the ratio of the analyte concentrations in the two phases at equilibrium. In an extraction, the convention for K_D is to report the concentration in the organic phase as a function of the concentration in the aqueous phase, as shown in Equation 15.2.

$$K_D = \frac{[C_{org}]}{[C_{aq}]} \qquad \textbf{Eq. 15.2}$$

For example, the equilibrium constant for caffeine partitioning between dichloromethane (organic layer) and water (aqueous layer) is 4.6. If Figure 15.2 represents the extraction of caffeine from an aqueous solution into dichloromethane and if the volumes of dichloromethane and water are the same, then at equilibrium, we will find the quantity

Example 15.1

If you extract caffeine from a 100 mL cup of coffee (an aqueous solution) using 250 mL of chloroform (organic solvent), how much caffeine would you find in the chloroform layer at equilibrium? Assume that the coffee initially contains 1.00 mg caffeine. The distribution coefficient for caffeine at room temperature K_D is 8.8.

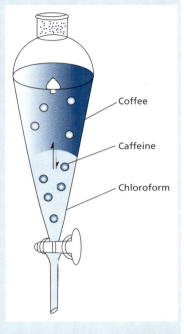

Coffee

Caffeine

Chloroform

STRATEGY –

Construct an equilibrium table and develop algebraic terms for each unknown variable in Equation 15.2. Make appropriate substitutions and solve the equation.

SOLUTION –

Step 1: Construction of the equilibrium table

	Caf (aq)	⇌	Caf (org)
Initial mass (mg)	1.00		0.00
Change (mg)	– x		+ x
Equilibrium mass (mg)	$(1.00 - x)$		x

Step 2: Express concentration of caffeine in milligrams per mL of each solvent and substitute those concentrations into Equation 15.2.

In the organic layer, $Caf_{(org)}$ will equal (x) mg in 250 mL of solvent.
In the aqueous layer, $Caf_{(aq)}$ will equal (1.00 – x) mg in 100 mL of solvent.

$$8.8 = \frac{\left(\dfrac{x}{250}\right)_{org}}{\left(\dfrac{1.00 - x}{100}\right)_{aq}}$$

Step 3: Solve the above equation for "x". The value of "x" is the milligrams of caffeine in the organic solvent.

Solving for "x" yields 0.956 mg. There are 0.96 mg of caffeine in the chloroform phase and 0.04 mg of caffeine in the aqueous phase. Caffeine partitions 96% into the organic layer and 4% into the aqueous layer at equilibrium.

of caffeine in dichloromethane would be 4.6 times larger than the quantity of caffeine in water.

Problem 15.1: Albuterol (a drug used to fight asthma) has a lipid:water K_{OW} (see "Profile—Other Applications of Partition Coefficients") value of 0.019. How many grams of albuterol would remain in the aqueous phase if 0.001 grams of albuterol initially in a 100 mL aqueous solution were allowed to come into equilibrium with 100 mL of octanol?

Of biological interest, pharmaceutical companies sometimes use octanol as a proxy for lipids when studying the fat solubility of a new drug. They use the octanol: water distribution constant $(K_D = K_{ow})$ as a measure of the relative fat solubility of the drug. Will the drug tend to find its way into the lipid layers of the body or will the drug tend to stay in the body's aqueous fluids? This is an important determination because it also provides an indication of how fast the drug will be cleared from the body. Drugs that are more lipid soluble tend to clear more slowly than do drugs that are more water soluble. Octanol:water partition coefficients* are also used in environmental science, where they can provide information about environmental sequestration (bioconcentration), likely modes of transportation through the environment, and toxicology. ■■

---

* For more information on octanol:water partition coefficients, see Sangster's text listed in the Further Reading section at the end of this chapter.

Problem 15.2:

(a) Repeat Problem 15.1, but instead of extracting the albuterol with 100 mL of octanol, extract the albuterol solution five successive times with only 20 mL of octanol per extraction. After each step, use the albuterol remaining in the aqueous phase from the previous step as the starting quantity for the next step. (*Note:* the total volume of octanol is the same in both extractions).

(b) Compare the final concentration of aqueous albuterol to that obtained in Problem 15.1. Discuss how the extraction results changed by conducting five smaller extractions instead of one large one. What is the downside to the multiple smaller volume method in this exercise compared to the single larger volume method in Problem 15.1?

Example 15.1 showed us that the efficiency of a two-phase extraction is dependent on both the K_D value and the relative volumes of the two phases. In a two-phase extraction, this separation efficiency is quantified with a term called the *capacity factor* (K_{cf}).

$$K_{cf} = K_D \frac{V_{org}}{V_{aq}}$$ **Eq. 15.3**

We will relate K_{cf} to the dynamic environment of LC.

Principles of Chromatography

The principles of chromatography are analogous to that of a two-phase extraction conducted in a separation funnel. However, in chromatography, one of the two components has been immobilized onto a solid support. As noted above, the immobilized component is referred to as the stationary phase while the other component (the mobile phase) passes over or through the stationary phase. It would be correct to view chromatography as a dynamic, multistage extraction (see Problem 15.2).

Separation of the components in the sample (a mixture) occurs because each of the components will partition differently between the mobile and the stationary phase. As we discussed in the context of the liquid–liquid extraction, the partitioning is described by the individual distribution constants for each analyte in the sample. Equation 15.4 describes K_D in terms of stationary and mobile phases. Take a moment to directly compare Equations 15.2 and 15.4.

$$K_D = \frac{C_s}{C_M}$$

Eq. 15.4

C_s is the concentration of analyte in the stationary phase and C_M is the concentration of analyte in the mobile phase. Note here that in the case of a chromatographic separation, the convention for K_D is to report the concentration of analyte in the stationary phase divided by the concentration of analyte in the mobile phase. Therefore, components with larger K_D values will spend more time on the column than those with smaller K_D values. *This is a key point!* The amount of time required for a compound to migrate through a chromatographic system is proportional to its K_D value. The term used to describe an analyte's affinity for the stationary phase is *retention*, and the time it takes for an analyte to elute from the column is its *retention time, t_r*.

It is useful to think of the compounds of interest moving past the stationary phase in a river of mobile phase, like kayakers on a river. Although some kayakers prefer to stay on the water and move along quickly, other kayakers prefer to explore the riverbank. Those who explore the riverbank the most will be retained and will move along more slowly.

Problem 15.3: Albuterol has a lipid:water K_{OW} value of 0.019. If you were to conduct a chromatographic separation using a polar solvent (i.e., methanol) and a nonpolar stationary phase, would the albuterol's distribution favor the mobile phase or the stationary phase? Explain your answer.

Problem 15.4: Caffeine has a lipid:water K_{OW} value of 0.79. If a mixture of albuterol and caffeine were to be separated using a polar solvent (i.e., methanol) and a nonpolar stationary phase, which component would elute first (having the shorter retention time)? Would the two components elute close to each other—or far apart? Explain.

Although it is possible to have compounds with K_D values that are at extremes, resulting in no retention or total retention, the majority of compounds in a well-designed protocol will distribute between the two phases in some meaningful proportion and will elute in a reasonable time. Because molecules of the same type all have the same distribution coefficient, molecules of a single type will tend to travel together in a band through the stationary phase, resulting in a characteristic retention time for each compound. Figure 15.3 shows two different analytes, A and B, traveling through an LC system. Analyte A has a higher affinity for the mobile phase and therefore is moving though the stationary phase at a more rapid pace than analyte B.

Figure 15.4 represents an idealized sample chromatogram. To a first approximation the peaks represent a normal distribution[1] about a mean retention time[2] and exhibit a peak width that is a function of the standard deviation (σ).

The time it takes for the mobile phase to pass through the column is designated t_m and is sometimes referred to as the *void time*. The amount of mobile phase that flows from the column during t_m is the *void volume*. If your analyte elutes in the void volume,

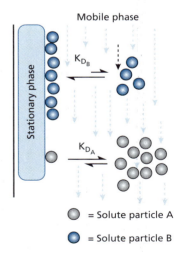

Figure 15.3 Partitioning of components A and B between a stationary and mobile phase. The distribution for B favors the stationary phase, whereas the distribution for A favors the mobile phase.

---

[1] See Chapter 22, Section 4 for additional discussion on the statistical analysis of normal distributions.

[2] Retention time can be converted to retention volume by multiplying the retention time by the flow rate.

Chromatography at Home

Black ink is usually a mixture of several different pigments. Using thin layer chromatography, we can separate the black ink into its different components. You have almost certainly conducted thin layer chromatography in organic chemistry; however, you can also perform a simple chromatographic separation with common household materials.

Materials

    Coffee filter

    Water and/or isopropanol (rubbing alcohol)

    Black felt-tipped pen or a black ballpoint pen

Method – Write an ink line 1 inch from the bottom of the coffee filter with one of your pens. Fill a glass of water with about ¾ inch of water. Place the bottom edge of the coffee filter into the water, taking care to only "wet" the coffee filter below the ink line. Capillary action will draw water up the filter paper, and the water will serve as the mobile phase. The cellulose in the filter paper serves as the stationary phase (normal phase). Over time, you will see the black ink separate into its different components.

    Try varying the experimental conditions. For example, mixing a little rubbing alcohol into the water will increase the migration rates of the less polar pigments. Or you can alter the stationary phase by using different types of paper. It is also interesting to use different manufacturers or brands of black pens or simply use a color other than black (what pigment colors do you think a green pen will reveal?). You will discover that not all ink of the same "color" is made from the same pigments.

then there was no partitioning of the analyte into the stationary phases; no retention occurred. The retention time for the void volume[3] will depend on the flow rate of the mobile phase and the volume of the stationary phase within the column. In Figure 15.4, the void volume appears on the chromatogram as the first peak.

    When using an optical detector (such as an ultraviolet-visible [UV-vis] spectrometer), the detector response is the difference in absorbance between the mobile phase and the analyte. If the solvent used to dissolve and load the analyte has a different composition than mobile phase, you will observe a peak at the void time. Depending on the nature of these differences, the void volume's peak can appear as either a positive or negative signal. Figure 15.4 shows a positive peak, and Example 15.3 shows a negative peak for the void volume.

Problem 15.5: Sometimes it is desirable to determine the retention (R) of an analyte as the difference between the observed retention time (t_r) and the void time (t_m). Using the chromatogram in Figure 15.4, determine the retention (R) of compounds A and B.

---

[3] It is designated t_m because it is the amount of time it takes for the mobile phase to pass through the column.

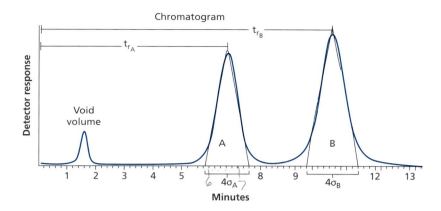

Figure 15.4 A representation of an idealized chromatogram showing the separation of two components of interest, A and B. The unique retention times for analytes A and B are labeled $t_r(A)$ and $t_r(B)$.

Problem 15.6: Sometimes you will see chromatographic peaks identified by the volume of mobile phase that was required to elute the sample instead of the time it took for the sample to elute. Using the retention times seen in Figure 15.4, calculate the retention volumes of compounds A and B assuming a flow rate of 1 mL/min. Repeat this exercise assuming a flow rate of 0.7 mL/min.

The Retention Factor

Example 15.1 and Equation 15.3 above showed us that the efficiency of a two-phase extraction is a function of both K_D and the relative volumes of the two phases. A column's ability to separate two different analyte species can be quantified in an analogous way by a term called the retention factor (k_r).

$$k_r = K_D \frac{V_s}{V_m} = K_D V_r \qquad \textbf{Eq. 15.5}$$

V_r = the ratio of the volumes of the stationary and mobile phases (the phase volume ratio).

Directly compare Equation 15.3 to Equation 15.5. Chromatography is a dynamic form of a two-phase extraction so when speaking strictly of a chromatographic separation, it is more customary to use the term *retention factor* rather than the term *capacity factor*. However, the two terms are essentially synonyms and are often used interchangeably. The retention of a particular analyte on a particular type of stationary/mobile phase is quantified using the *retention factor*. The retention factors for each analyte are often listed in published chromatography protocols. Retention factors are treated as constant for systems using the same stationary and mobile phases, and we denote them here as k_r. Therefore, one can use published retention factors to predict the suitability of a particular chromatographic system for two components in question. The retention factor for a particular analyte does not depend on the flow rate or the column length.

Because we often do not know the phase–volume ratio, the retention factor can also be found experimentally from a chromatogram using Equation 15.6:

$$k_r = \frac{t_r \cdot t_m}{t_m} \qquad \textbf{Eq. 15.6}$$

> **Note**
> A retention factor between 1 and 5 is a reasonable value for a good protocol.

In general, a retention factor between 1 and 5 is a reasonable value for a good protocol. Values of k_r that are large have long analysis times, whereas values of k_r that are small occur so quickly that no useful partitioning occurs.

Retention factor

$$k_r = \frac{t_r - t_m}{t_m}$$

Selectivity factor

$$\alpha = \frac{k_r'}{k_r''} = \frac{t_r' - t_m}{t_r'' - t_m}$$

Problem 15.7: Determine the retention factors (k_r values) for analytes A and B seen in Figure 15.4.

By comparing the retention factors for two different analytes, we can determine how difficult it will be to resolve those analytes for a particular combination of mobile phase and stationary phase. This comparison is easily accomplished by calculating the *selectivity factor*. Selectivity factors (α) are calculated by taking the ratio of the retention factors.

$$\alpha = \frac{k_r'}{k_r''} \qquad\qquad \textbf{Eq. 15.7}$$

where
k_r' = retention factor of the more strongly retained analyte
k_r'' = retention factor of the less strongly retained analyte

Substitution of Equation 15.5 into Equation 15.7 yields:

$$\alpha = \frac{k_r'}{k_r''} = \frac{K_D' V_r'}{K_D'' V_r''} = \frac{K_D'}{K_D''} \qquad\qquad \textbf{Eq. 15.8}$$

One can also derive the selectivity factor by using the relative retention times. Substitution of Equation 15.6 into Equation 15.7 yields:

Example 15.2

Determine the selectivity factor (α) for the following separation.

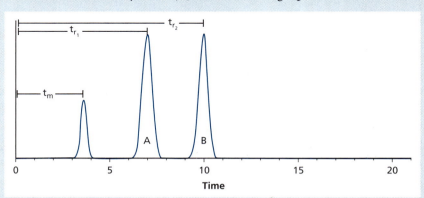

STRATEGY –
Substitute the appropriate values for t_m and t_r into Equation 15.9.

SOLUTION –

$$\alpha = \frac{k_r'}{k_r''} = \frac{t_r' - t_m}{t_r'' - t_m} = \frac{10 - 3.5}{7 - 3.5} = \frac{6.5}{3.5} = 1.86 = 1.9$$

$$\alpha = \frac{k_r'}{k_r''} = \frac{\left(\dfrac{t_r' - t_m}{t_m}\right)}{\left(\dfrac{t_r'' - t_m}{t_m}\right)} = \frac{t_r' - t_m}{t_r'' - t_m} \qquad \textbf{Eq. 15.9}$$

By definition, the selectivity factor must be greater than 1; otherwise, no separation would occur. The selectivity factor is useful in helping the analyst choose between different courses of action in order to optimize the separation of multiple components of the sample. For example, if α is close to 1, then the two analytes have about the same K_D value, and simply increasing the number of theoretical plates might not yield sufficient separation. In other words, changing the flow rate or column length would likely do very little to improve the separation. In this case, the analyst will look to change the selectivity factor by manipulating the k_r (or K_D) values of the analytes. This can be done by changing either the mobile or stationary phases, changing the temperature, or by performing chemical modifications to the sample matrix.

Problem 15.8: Calculate the retention factor for each peak in the chromatograph seen in Example 15.3 and create a results table that might be used in a report.

Resolution and Theoretical Plates

Resolution is a term used to quantify the quality of a separation. The factors that affect resolution are a combination of both thermodynamic and kinetic variables. The thermodynamic factors speak to the differences in the K_D values of the analytes. The kinetic factors speak to how rapidly the analytes are able to move between the two phases. If the kinetics of the partition process are rapid, then analytes with similar K_D values may still be resolved. However, if the rate at which the two analytes move between the two phases is very slow, even a large difference in K_D values for the two analytes may still result in poor resolution.

Example 15.3

The following chromatogram was obtained using a C18 column optimized for use in separating proteins and peptides. The separation was performed with a flow rate of 1.0 mL/min. Determine the void volume of the column and the retention volume for the Gly-Tyr dipeptide.

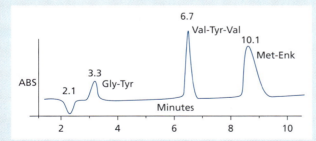

SOLUTION –

The peak at the void volume in this chromatogram is negative at 2.1 minutes. The void volume is calculated by considering the flow rate. At 1 mL/min times 2.1 minutes, the void volume is 2.1 mL. Similarly, the retention time (t_r) of the Gly-Tyr peak is found at 3.3 minutes. At a flow rate of 1 mL/min, the retention volume for Gly-Tyr peak is 3.3 mL.

Problem 15.9: Use the chromatogram from Example 15.3 to determine the retention times and retention volumes of each of the peaks in this sample and create a results table that might be included in a report.

Resolution is calculated using the retention times of two peaks and the average width of the two peaks at the base line (see Equation 15.10). Figure 15.4 demonstrates how the baseline width of the peak is measured; note that it is not the full width of the peak at the baseline but the width of the extrapolated rising and falling sides of the peak.

$$R_s = \frac{t_{r_2} - t_{r_1}}{(W_{b_2} + W_{b_1})/2}$$ **Eq. 15.10**

where

t_{r_1}, W_{b_1} retention time and base line width of the first eluting peak

t_{r_2}, W_{b_2} retention time and base line width of the second eluting peak

Note that R_s is unitless.

Assuming a normal distribution for each peak about t_r, statistics tells us that the standard deviation can be estimated as 0.424 times the width at half height (W_h) or 0.250 times the width at the baseline (W_b). See Equation 15.11 and Figure 15.5.

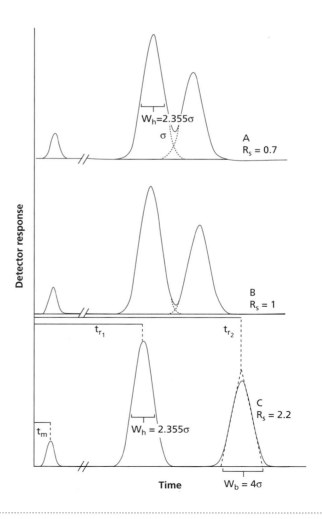

Figure 15.5 A chromatogram showing a resolution of 0.7 (A), 1.00 (B), and 2.2 (C).

$$W_b = 4\sigma \qquad\qquad \textbf{Eq. 15.11}$$

In the case when the resolution is poor and estimating the baseline width is difficult, it is often desirable to use the width at half height (W_h). For example, in Figure 15.5(A), we have two poorly resolved peaks and determining their baseline width would require complicated statistical modeling. However, the two peaks are well resolved at the half-height, so we can use Equation 15.12 to determine σ. The relationship between the standard deviation and the width at half height is:

$$W_h = 2.355\sigma \qquad\qquad \textbf{Eq. 15.12}$$

One of the goals of a chromatographic experiment is to separate a discrete analyte from a complex matrix. The *efficiency* of a separation is related to the resolution between analytes. In order for two analytes to be completely resolved, they must have different retention times and the width of the peaks must be narrow enough so that they do not overlap. Figure 15.5 shows a series of two-analyte systems having different resolutions. Panel (A) in Figure 15.5 shows a system with a resolution of 0.7. Although we can easily identify the two peaks, you can see there is significant overlap. A resolution of 0.7 is certainly sufficient for identification of a compound, but if our goal is to purify

Example 15.4

The three chromatograms below show the results for a separation of a two-component system conducted under three different experimental conditions. Determine the resolution of the system for panel (A).

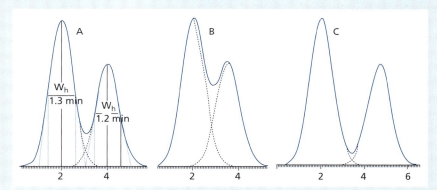

STRATEGY –

Because we would have to estimate the baseline width of the two peaks, it would be better to (1) find σ using Equation 15.12 and then (2) use σ and Equation 15.11 to estimate W_b. Determine the retention time (t_r) and the baseline width (W_b) for each peak, and substitute those values into Equation 15.10.

SOLUTION –

Using W_h from chromatogram A, we find that σ for the first peak is 0.55 and σ for the second peak is 0.51. Using these values in Equation 15.11, we obtain a W_b of 2.2 and 2.0 for the two peaks respectively. Substitution into Equation 15.10 yields:

$$R_s = \frac{(t_{r_2} - t_{r_1})}{(W_{b_1} + W_{b_2})/2} = \frac{(4.14 - 2)}{(2.2 + 2.0)/2} = 1.02$$

and isolate the two compounds, a resolution of 0.7 would be insufficient. The chromatogram in panel (B) of Figure 15.5 shows a resolution of 1.00. A resolution of 1.00 is often considered sufficient for purification if the analyst throws out the eluent at the region of overlap. This loss of sample reduces the *recovery* of the method. In panel (C) of Figure 15.5 we see a chromatogram with complete resolution (often referred to as *baseline resolution*) of the two analytes. Note, however, that extra time was spent waiting for the second component to elute. A good rule of thumb is that a resolution of 1.5 is appropriate for baseline resolution without excessive loss of time.

Problem 15.10: Use the chromatograms in panels (B) and (C) from Example 15.4 and estimate the resolution of the two-component system under each of these conditions. What challenges do you face for estimating σ for panel (B)? In terms of the value of the separation, discuss the significance of the resolution in panels (A), (B), and (C) with respect to the speed of the analysis (throughput), quantitative analysis, and qualitative analysis.

Problem 15.11: In a chromatographic experiment, two components elute with retention times of 11.52 and 12.76 minutes, respectively. The half-height widths of the peaks were determined to be 0.37 and 0.42 minutes, respectively. Assuming Gaussian peak shapes, estimate the resolution between the two peaks.

Column efficiency is the term used to describe the ability of a column to separate multiple components of a sample in a given amount of time, or mobile phase volume. Column efficiency is directly related to resolution, but it is also related to the baseline width of a peak. If band broadening is severe, it stands to reason that fewer components can be separated from a mixture in a given amount of time. On the other hand, if bands remain narrow, then many more substances can be well resolved in a given time on the system. Thus, significant band broadening is associated with poor column efficiency.

Take another look at the distillation column in Figure 15.1. The resolving power of a distillation column is a function of the number of condensation plates and the distance between the plates, or the plate height. The closer the plates are to each other, the more efficiently the column will be able to separate two different components with similar boiling points. Chromatographers borrow these concepts when comparing the resolving power of different chromatographic systems. However, a chromatography column does not actually have physically discrete plates on which partitioning occurs; thus, they talk in terms of *theoretical plates* (N_t) and the *height equivalent of a theoretical plate (H)*. The larger the number of theoretical plates, the higher the resolution. The number of theoretical plates is a quantitative way of understanding how efficiently a chromatographic system will separate the analyte from other, similar, compounds. The higher the N_t value, the more efficiently the system will perform. However, the N_t value for a particular analyte on a particular chromatography system is dependent on many variables. These include the length of the column, the flow rate of the mobile phase, the relative volumes of the mobile/stationary phases, and the temperature.

By definition, the number of theoretical plates is:

$$N_t = \frac{L}{H}$$

Eq. 15.13

where
    N_t = total number of theoretical plates
    L = length of the column
    H = height equivalent of a theoretical plate

However, we cannot know the height equivalent of a theoretical plate; therefore, in practice, N_t is determined directly from the chromatogram using Equation 15.14.

$$N_t = \frac{t_r^2}{\sigma^2}$$

Eq. 15.14

where

t_r = retention time of the analyte
σ = standard deviation of the peak
σ^2 = variance

Substitution of Equation 15.11 into Equation 15.14 gives a very useful form of the equation:

$$N_t = \left(\frac{4t_r}{W_b}\right)^2 = 16\left(\frac{t_r}{W_b}\right)^2$$

Eq. 15.15

However, when W_b is difficult to determine, it is often easier to use the width at half-height (W_h) to estimate the standard deviation of the curve. Substitution of Equation 15.12 into Equation 15.14 yields:

$$N_t = \left(\frac{2.354t_r}{W_h}\right)^2 = 5.54\left(\frac{t_r}{W_h}\right)^2$$

Eq. 15.16

Problem 15.12: Use Equation 15.15 to determine N_t for each peak in the chromatogram found in Example 15.4, panel (C).

Problem 15.13: Use Equation 15.16 to determine N_t for each peak in the chromatogram found in Example 15.4, panel (A).

In a commercial laboratory, where repetitive, routine quality control analysis of a product is monitored using LC, it is a standard operating procedure to routinely check the "health" of a column by analyzing a test injection containing well-characterized standards. If the N_t value for a particular test standard has changed significantly, the column is replaced. The concept of a theoretical plate is an attempt to relate an LC column's separation efficiency to that of a distillation column's separation efficiency. However, in an LC column, the mechanism of separation is a based on differences in K_D values. You can think of a theoretical plate as a point in the column where partitioning between the mobile and stationary phases occurs.

Problem 15.14: Relate the multistep extraction process described in Problem 15.2 to the concept of theoretical plates.

If our aim is to increase the number of plates in search of higher resolving power, then we can either lengthen the column (longer L) or we can diminish the height of the theoretical plates (smaller H). Because lengthening the column has practical and mechanical disadvantages, including longer separation times, let us look at plate height as a theoretical construct and consider how H might be made smaller.

The height equivalent of a theoretical plate is characterized by the *van Deemter equation,* an equation that provides a model for the various physical parameters that contribute to broadening of the peak. One version of this equation is shown in Equation 15.17. Interestingly, van Deemter made his original observations in 1956[4] on the contributions to band broadening in the context of gas chromatography analysis, however, it has been readily applied to LC theory as well. There are five contributions to band broadening, each of which has an effect on the height of a theoretical plate, H.

$$H = A + \frac{B}{v} + (Cv)$$ **Eq. 15.17**

where

    H = height equivalent of a theoretical plate
    A = eddy diffusion coefficient
    B = longitudinal diffusion coefficient
    v = linear flow velocity of the mobile phase
    C = rate constant for mass transport

Although theoretically descriptive, Equation 15.17 is used primarily for modeling purposes. In practice, the N_t is found experimentally using either Equations 15.15 or 15.16, and H is then calculated using an algebraic rearrangement of Equation 15.13. The remaining terms in the van Deemter equation are discussed below.

Band Broadening

There are five contributions to band broadening that we will consider in this section. They are eddy diffusion, longitudinal diffusion, and three types of mass transfer (mobile phase mass transfer, stationary phase mass transfer, and stagnant mobile phase mass transfer).

Eddy diffusion (the A term in Equation 15.17) is a process that leads to band broadening due to different paths a component may take through the stationary phase (see Figure 15.6). Eddy diffusion is minimized by having small, uniform particle size and uniform packing of the column. Smaller particle size minimizes the differences in path lengths that analytes might take in traversing the column. Columns are often packed using ultrasonic vibrations to ensure a uniform or "well-packed" composition. Note that eddy diffusion is not affected by flow velocity; fast or slow, eddy diffusion will result in band broadening. The eddy diffusion component of band broadening is directly proportional to H. Smaller stationary phase particles allow for more consistent packing, resulting in decreased eddy diffusion, and a lower A will result in a smaller H, which will improve resolution. However, the tighter packing of smaller particles in stationary phase materials requires increasing pressure to move the solvent through the columns. This helps us to understand why high pressure techniques provide high resolution.

Longitudinal diffusion (the B term in Equation 15.17) is the natural tendency for an analyte to move as a result of the motion of the molecules within the solution. This type

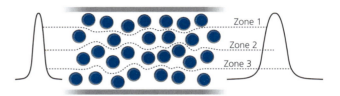

Figure 15.6 A representation of band broadening due to eddy diffusion.

[4] van Deemter, J. J.; Zuiderweg, F. J.; Klinkenberg, A. Longitudinal Diffusion and Resistance to Mass Transfer as Causes of Nonideality in Chromatography. *Chem. Eng. Sci.*, **1956**, *5*, 271–289.

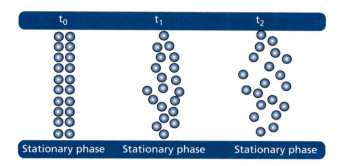

Figure 15.7 A representation of longitudinal diffusion over a period of time ranging from time t_0 to time t_2.

of motion will occur even if the flow velocity is zero. You probably already know that diffusion will occur from an area of higher concentration to an area of lower concentration; therefore, our sample will naturally tend to drift out of its band, leading to band broadening. Diffusion is a basic property of fluids and can be minimized by reducing the amount of time the analyte stays in the chromatography system. Longitudinal diffusion is also known as *time broadening* and is represented in Figure 15.7 starting at time equals zero to some time interval, t = 2. The parameters that affect longitudinal diffusion are the analysis times and the sample diffusion coefficient in the mobile phase (D_m). Analysis times can be controlled by the analyst who can adjust the length of the column and the flow velocity of the mobile phase. Diffusion coefficients are intrinsic properties of each analyte. Analytes that diffuse readily in the mobile phase solvent (higher D_m values) have enhanced longitudinal diffusion (and worse resolution because of the increase in H). In contrast, analytes that diffuse slowly in the mobile phase solvent (lower D_m values) have improved longitudinal diffusion (and better resolution due to smaller H values). Because the mobile phase composition also affects other aspects of the distribution equilibrium, consideration of its role in longitudinal diffusion may not be a primary concern. However, analysis time is readily manipulated in chromatography and can be optimized to minimize longitudinal diffusion. You should pay careful attention to the fact that there is a trade-off to be considered. Although shorter analysis times may diminish broadening due to longitudinal diffusion, a faster mobile phase flow velocity and/or a shorter column will decrease the number of times the analyte can partition during an experiment, which results in fewer theoretical plates (lower N_t) and therefore diminished resolution. Equation 15.18 is a truncated version of Equation 15.17 (the van Deemter equation) and shows this trade-off. The theoretical plate height is proportional to B in the van Deemter equation and inversely proportional to linear flow velocity (v).

$$H \propto \frac{B}{v} \qquad\qquad \textbf{Eq. 15.18}$$

Mass transfer (the C term in Equation 15.17) is a group of processes that also lead to band broadening. Think of mass transfer issues as an inhomogeneity in the way in which groups of particles move. If we go back to our kayakers on a river analogy, mass transfer would be analogous to the different ways in which the kayakers in our group explore the riverbank surface (some only superficially, others more deeply) or the way in which some kayakers may become slowed down if they happen on an area of the river that is not flowing very quickly (a stagnant area). These differences will result in the members of our group completing the trip at slightly different times, even if our overall partitioning (affinity for river or riverbank) is the same. In chromatography, we think of mass transfer in three distinct ways: *mobile phase mass transfer, stagnant mobile phase mass transfer,* and *stationary phase mass transfer.*

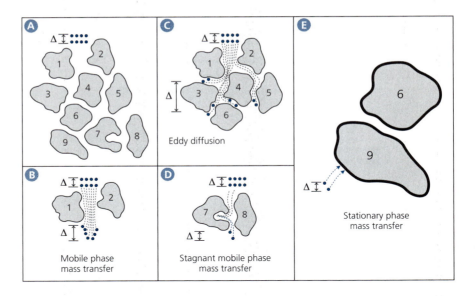

Figure 15.8 A representation of the contributions to band broadening. Panel (A) shows a plug of analyte particles at the top of a column. Panel (B) shows the contribution to band broadening due to *mobile phase mass transfer*. Panel (C) shows the contribution to band broadening due to *eddy diffusion*. Panel (D) shows the contribution to band broadening due to *stagnant mobile phase mass transfer*. Panel (E) shows the contribution to band broadening due to *stationary phase mass transfer*.

Several of the factors that contribute to band broadening are a function of how the column was constructed and the nature of the stationary phase. Considering all of the factors that relate to H and therefore improved resolution, only a few variables that affect band broadening are under the control of the analyst.* These include the flow rate of the mobile phase, the temperature, and the length of the column being used. Therefore, in practical terms, these are the parameters of interest.

*If you are packing your own columns, there are a few other issues to consider with respect to H.

Figure 15.8 is a representation of the contributions to band broadening as a function of the variables in the van Deemter equation. The relative contributions are represented by Δ. The gray shapes numbered 1 through 9 represent particles within the stationary phase. In the top of parts 15.8(A), (B), and (C), we see the injected plug represented by eight small spheres. Figure 15.8(A) shows that the analyte has been introduced to the top of the column, and the intrinsic bandwidth at this point is simply the width of the injection plug. Figure 15.8(B) shows the contribution to band broadening due to *mobile phase mass transfer*. Due to adhesion and friction, mobile phase near the edges of a channel moves slower than mobile phase in the middle of a channel. Figure 15.8(C) shows the contribution to band broadening due to eddy diffusion. Eddy diffusion arises from the fact that different channels through the stationary phase have different path lengths. Figure 15.8(D) shows the contribution to band broadening due to *stagnant mobile phase mass transfer*. As analyte migrates into stagnant mobile phase, lateral diffusion will result in analyte spending different amounts of time out of the moving mobile phase. Figure 15.8(E) shows the contribution to band broadening due to *stationary phase mass transfer*. Stationary phase mass transfer is similar to stagnant mobile phase mass transfer. If the coating of the stationary phase is too thick, differential diffusion into the stationary phase results in differential time spent in the stationary phase.

The rate constant for mass transfer (the C term in Equation 15.17) is linearly related to the linear flow velocity (v). Thus higher flow velocities tend to increase broadening due to mobile phase mass transfer because analyte molecules that are moving more quickly in the mobile phase move far beyond those retained in the stationary phase.

It is important to recognize that resolution is always at odds with time. We can see this directly in the van Deemter equation (Equation 15.17). The height of a theoretical plate is directly proportional to flow velocity in the third term (mass transfer, Cv) but inversely proportional to the flow velocity in the second term (longitudinal,

B/v). We have to find a workable compromise. There exists an optimum flow velocity at which we achieve minimum plate height (maximum efficiency), but we often operate at a somewhat higher flow velocity because the trade-off between increased plate height and time saved makes that worthwhile. As mentioned earlier, we can increase the number of theoretical plates by increasing the length of the column, but this will also increase the time the analyte spends on the column (increasing longitudinal diffusion) and will also subject the analyte to increased eddy diffusion and mobile phase mass transfer. As a rule, however, those contributions to broadening do not outweigh the advantage gained from the greater efficiency of separation produced by using a longer column.

> **Problem 15.15:** List at least three undesirable effects of band broadening.

15.3 Basic Method Development

Remember that the goal of a chromatographic separation is to achieve the best separation in a minimum amount of time. Here the idea of "best" in terms of separation really means the minimum necessary resolution (R_s). If the protocol meets your minimum resolution needs, then there is no need to pursue further method development. However, if the protocol is not sufficiently meeting your minimum resolution needs, and especially if you are developing a method that will be used for routine analysis of many samples, then the time and effort in improving the protocol will be worthwhile.

Developing the perfect chromatographic protocol is as much art as science, and finding a good, published protocol for an application that is similar to your own is the best way to get started in method development. Selecting the appropriate column and stationary phase, the solvent system, and the detector parameters are only a few of the decisions to be made. In addition, you have to consider whether you simply want to identify the existence of a component in your sample matrix or if quantification and/or isolation is a goal. If your goal is quantification then you will want to know if standards for each component are available. If not, you may have to create them yourself. In addition, the purpose of your method may be purification and recovery of the sample components. You will need to know how much sample you have to analyze, or purify, and what the expected concentration range of your analyte will be. You should also know basic information about the sample, including its volatility, thermal stability, and the number of identifiable components it contains.

Thermodynamic and Kinetic Factors

The thermodynamics of a separation speak to the differences in the K_D values of the analytes and are evaluated using the selectivity factor (α). The kinetic parameters of a chromatographic separation speak to how often the analyte changes between the two phases, which is quantified by the concept of theoretical plates (N_t).

In the initial steps of method development, the analyst will select a stationary phase that seems most likely to yield an acceptable selectivity factor (α). Then the analyst will fine-tune the protocol by manipulating the mobile phase composition and flow rate. The mobile phase can be, but need not be, a single solvent. For example, the analyst might adjust the polarity of the mobile phase by using a specific ratio of acetonitrile and water, or methanol and water, or perhaps a mixture of all three solvents. Mobile phase variability imparts great flexibility to method development but also significant complexity.

Resolution is controlled by two key parameters: changes in the selectivity factor α (which is ultimately the manipulation of the analytes' k_r values) or changes in the number of theoretical plates (N_t). Equation 15.19, which is known as the *fundamental resolution equation*, shows how these factors contribute to changes in resolution.

$$R_s = \frac{1}{4}(\alpha - 1)\sqrt{N_t}\left[\frac{k_r}{(1+k_r)}\right] \qquad \textbf{Eq. 15.19}$$

Each of the terms used in Equation 15.19 was defined previously.

The fundamental resolution equation has three, essentially independent, parts: the $(\alpha-1)$ term, the $\sqrt{N_t}$ term, and the $\left[\dfrac{k_r}{(1+k_r)}\right]$ term. We optimize resolution by optimizing each one independently, often in an iterative way.

Let us consider the $\left[\dfrac{k_r}{(1+k_r)}\right]$ term first. If the value of k_r for a particular analyte of interest is too low (i.e., it does not fall within the range of 1 to 10), then changing k_r so that it is in that range will have a good effect on resolution. Remember that if k_r is too small, the analyte is not undergoing much (if any) partitioning. Conversely, if k_r is too large, the analysis time is unnecessarily long. As the value of this term approaches 10, improvements in separation diminish. That is, an increase from a value of 1 to 4 would result in a large improvement in efficiency; increasing from 6 to 9 will result in only a small improvement. In LC, we typically manipulate k_r by changing the solvent composition; however, changing the stationary phase and/or changing the temperature are also options. It is common practice to use the term *solvent strength* in describing the way in which the solvent is changing. If your aim is to increase k_r (moving the retention time away from the void volume by making it retain more on the stationary phase), then you need to use a *weaker mobile phase*. If your aim is to decrease k_r (moving your retention time closer to the void volume by making it retain less on the stationary phase), then you need to use a *stronger mobile phase*. We need to take a moment to examine these two vocabulary words. The idea of weaker and stronger solvents speaks to the distribution coefficient of the analyte between the mobile and stationary phases. A strong solvent can be thought to "pull the analyte out of the stationary phase." Implicit in this description is the role of the stationary phase. For example, if you are trying to separate a mixture of nonpolar organic molecules, you might choose a nonpolar stationary phase. A strong mobile phase, in this case, would also be nonpolar and would compete for the analyte as it distributed between the two phases. That same solvent used in a method designed to separate polar compounds on a polar stationary phase would be a weak mobile phase.

The $\sqrt{N_t}$ term in Equation 15.19 is called the *column factor* and is simply the square root of the total theoretical plates in your column. You will recall from our earlier discussion that there are two ways to increase the number of theoretical plates: increase the column length or decrease the theoretical plate height. Because of the square root, in order to double the resolution by increasing the column length, you would have to use a column that is four times longer and your run times would be similarly four times longer. Plate height can be manipulated by adjusting the flow velocity (this is the v term in Equation 15.17). A slower flow velocity allows the analyte to partition more times in the same column length, effectively decreasing the theoretical plate height (H). A decreased flow velocity will generally result in better partitioning but at the expense of increased band broadening and longer analysis times.

Let us examine the $(\alpha - 1)$ term. By definition, the selectivity factor, $\alpha \geq 1$. Therefore, the goal in improving resolution is to make α larger; if $\alpha = 1$, the $R_s = 0$. The $(\alpha - 1)$ term is taking into consideration the relative separation of the two analytes that you are trying to resolve. Fundamentally, selectivity goes back to the K_D values for each analyte. Our goal in method development is for the K_D values to be sufficiently different in order to obtain the desired resolution between the two analytes. If the two peaks

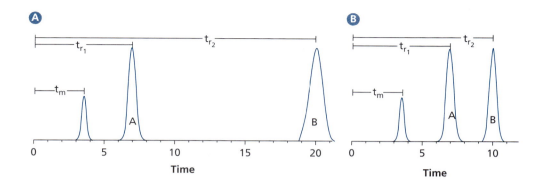

Figure 15.9 A two-component chromatogram. Panel (A) shows the results of an isocratic elution. Panel (B) shows the results of a gradient elution.

are not well resolved, their K_D values under those conditions are too similar. How we change that part of the separation is not always straightforward and typically involves some trial and error. However, changes in selectivity can allow us to improve resolution while simultaneously decreasing overall analysis time. Mobile phase composition is one parameter that can affect selectivity; but in search of a more favorable selectivity factor, you still want to maintain a range of 1 to 10 for your k_r values. Other experimental parameters that will affect selectivity include a change in stationary phase, a change in the temperature, or a mobile phase additive. For example, one might change the pH to shift an equilibrium or one might add a complexing agent that will induce a secondary equilibrium and thus change the $K_{D_{net}}$ value of one of the analytes.

Problem 15.16: Make a summary table, perhaps the size of an index card, that identifies the equations that define K_D, R_s, N_t, k_r, and α. Be sure to define the underlying terms (like t_m, W_b, H, and σ) that are used in these equations.

Isocratic versus Gradient Elution

The terms *isocratic* and *gradient* elution indicate the relative composition of the mobile phase over the time period of a chromatographic experiment. The term *isocratic* indicates that the mobile phase composition remains constant during the experiment. The term *gradient* indicates that the mobile phase composition (strength) changes as a function of time during the experiment. The use of gradient chromatography serves to shorten the analysis time of a chromatographic experiment and to improve the efficiency of separation for the late eluting peaks by lessening the time they stay on the column, thus decreasing band broadening. Figure 15.9 compares an isocratic and a gradient protocol for a two-component system. The top chromatogram shows the results of the separation under isocratic conditions. Notice how peak A eluted after 7 minutes and peak B did not elute until 20 minutes had passed. If this were a one-off analysis, the analyst might be satisfied with the results. However, if the method under development were for a routine analysis in a commercial laboratory that might involve hundreds of daily samples, waiting 20 minutes between injections would be undesirable and costly. The bottom chromatogram was obtained under gradient conditions. The analyst programmed the system to begin changing the mobile phase after 7 minutes had passed so that compound B had a higher affinity for the mobile phase. This caused compound B to elute faster while retaining satisfactory resolution. In chromatographic terminology, we would say that after 7 minutes we increased the mobile phase strength. In addition to shortening the analysis time by half, a gradient method lessens the band broadening of the later eluting peaks.

PROFILE

The Role of Temperature

Although scientists have long known that thermal effects on chromatographic separations can be highly influential, these effects have been largely ignored under traditional separation conditions. Yet the move toward miniaturization of devices and more widespread use of capillary columns in recent years has brought the temperature factor into new focus. Temperature affects both the kinetic and the thermodynamic aspects of the technique and can therefore play an important role in the overall control of the separation. Low temperature separation profiles have been shown to be effective at helping to increase resolution between enantiomers in chiral separations, an important application in the preparation of pharmaceuticals. High temperature separation profiles have been shown to have the advantage of increased efficiency because longer columns can be used and/or analysis times can be greatly shortened without a loss of resolution or peak capacity. ■■

Greibrokk, T. Heating or Cooling LC Columns. Anal. Chem., **2002***, 374A–378A.*

Problem 15.17: In the analysis of some analgesic preparations using a polar stationary phase and a mobile phase composed of 10% water and 90% methanol, four compounds were seen to elute in the following order: caffeine, acetaminophen, aspirin, ibuprofen. The last two compounds were widely separated by several minutes. How might you decrease the time required for the experiment while maintaining adequate resolution?

Problem 15.18: If a nonpolar stationary phase were used in the above experiment, what might be the elution order?

Problem 15.19: When using a nonpolar stationary phase with 10% water and 90% methanol, the first two components coelute in a single peak. How might you change the conditions of the experiment to resolve those components?

Quantitative versus Qualitative Analysis

One of the major strengths of a well-designed chromatography method is that it will be a "linear" system; that is, the response of the detector will be linearly proportional to the sample size and well suited for quantitative analysis. Because any given analyte has a retention time that does not change under the given experimental conditions, it is easily identified based on its retention time. Furthermore, the detector response can sometimes aid in identifying analytes if it provides characteristic information, such as an optical or mass spectrum. Thus, chromatography also provides excellent qualitative analysis for the identification of analytes in a sample.

In a quantitative chromatographic analysis, one uses standards to generate a calibration curve from the detector response as a function of concentration. Figure 15.10 shows a calibration curve that has been created for a chromatographic separation of acetaminophen, acetylsalicylic acid, and caffeine in commercial preparations of analgesic tablets. In this analysis, a nonpolar stationary phase consisting of an 18-carbon chain bound to silica beads (C_{18} column) was used for the separation. Detection was by mass spectrometry.

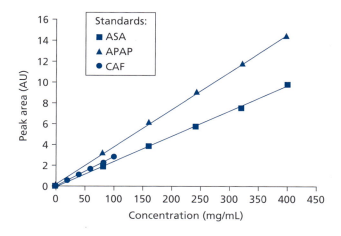

Figure 15.10 Plot of chromatogram peak area versus concentration for acetylsalicylic acid (ASA), acetaminophen (APAP), and caffeine (CAF) standards.

Problem 15.20: You have conducted an LC analysis of three different headache medicines. The column was a C_{18}, the mobile phase was 50:50 methanol:water, and the detector was a UV-vis spectrometer (output measured in AU or absorbance units). The detector response for four different standards and for the three samples is presented here. Construct a calibration curve, perform a linear regression, and determine the concentration of acetaminophen in each sample.

STANDARDS (PPM)	AU
1.00	0.098
2.00	0.211
4.00	0.402
8.00	0.699
Sample A	0.375
Sample B	0.296
Sample C	0.801

Another important feature of a linear detector response is the fact that the samples are additive. If a pure sample of an analyte compound is added to a sample matrix (a spike), the resulting chromatographic peak will have an area proportional to the new total quantity of analyte (the original plus the spike).

We take advantage of the additive feature of this linear system whenever we use a spike in the analysis of an unknown sample. We can add a spike, and if the response of the analyte peak increases due to the addition of the spike, that is evidence for the identity of the analyte. Through careful quantitative analysis, we can also use the amount of the spike that was added to determine the amount of the component in the original unknown sample matrix.[5]

For example, the chromatogram in Figure 15.11 indicates that the arsenobetaine $[(CH_3)_3As^+CH_2COO^-]$ form of arsenic was found in the muscle tissue of a rainbow trout. Peak A, the smaller peak, originates from the extracted sample, whereas peak B, the larger peak, comes from the extracted sample that has been spiked with an

---

[5] This is known as the method of standard addition. The technique is covered in detail in most sophomore/junior level analytical textbooks.

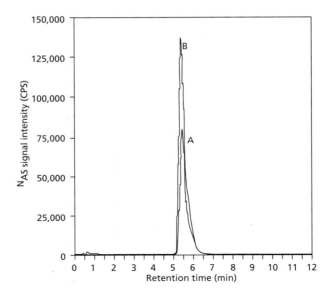

Figure 15.11 Cation exchange HPLC-ICP-MS chromatogram of muscle extract of rainbow trout: (A) Chromatogram of muscle extract of trout, and (B) chromatogram of the same sample with an arsenobetaine spike.

arsenobetaine standard. The identical retention times help identify the arsenobetaine form of the arsenic compound in the sample,[6] and the peak height (or area) can be used to determine the quantity of analyte.

The linearity of the system only holds when the sample size is small in comparison to the mobile phase volume. The maximum quantity of analyte that can be analyzed by LC while still retaining proper peak shape and symmetry is called the capacity. In determining the system capacity, we can consider the injection of the sample in terms of its volume or bandwidth. In an analytical separation, the injection volume is typically very low, a "pulse." However, in a preparative separation, the injection volume becomes more significant, a "plug." The width of the plug (in terms of volume) will be proportional to the width of the peak as long as the equilibrium process of the partitioning is not disturbed. In plain English, if the analyte concentration is too high, all of the available adsorptions spots on the stationary phase will be occupied, and any remaining analyte in the mobile phase will no longer participate in the partition equilibrium and will be swept along at the flow velocity of the method. In that situation, asymmetric peaks will be observed in the chromatogram.

Problem 15.21: Assuming a flow rate of 1 mL/min, determine the retention volume of peak #6 from the profile "Analysis of Wine: Qualitative and Quantitative".

Problem 15.22: Describe how a gradient method could be used to shorten the run time for the chromatograms seen in the profile "Analysis of Wine: Qualitative and Quantitative". Be specific! Where in the chromatogram would you change the mobile phase and how would you change it?

---

[6] Ciardullo, S.; et al. Bioaccumulation Potential of Dietary Arsenic, Cadmium, Lead, Mercury, and Selenium in Organs and Tissues of Rainbow Trout (Oncorhyncus mykiss) as a Function of Fish Growth. *J. Agric. Food Chem.* **2008**, *56*, 2442, 2451.

The three chromatograms in the figure show the results of a reverse-phase liquid chromatography separation of phenolic compounds in wine using ultraviolet (UV) detection at 280, 320, and 365 nm. The three panels demonstrate detection of the eluent compounds at three different wavelengths, using a UV-visible photodiode array detector. The characteristic retention times and spectral properties allow the researchers to identify specific components in the wine. In this analysis, the major peak with a retention time near 17 minutes (labeled 6) is identified as caftaric acid, a tartaric ester of hydroxycinnamic acid. The researchers determined that the hydroxycinnamic acids are, as a class, the major phenolic constituents of these wines and that the relative proportions of these compounds vary by grape variety (in this analysis, chardonnay was compared to pinot noir) and also by growing season. This information is useful to viticulturists who are interested in understanding how growing conditions and processing practices affect these phenolic components. ■

PROFILE

Analysis of Wine: Qualitative and Quantitative

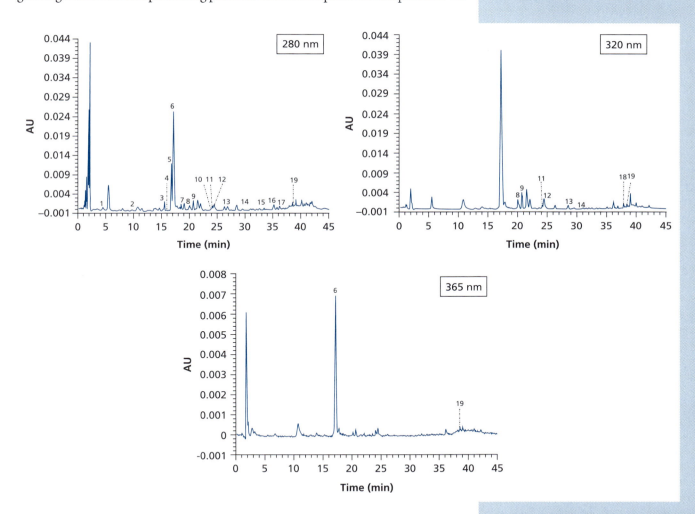

*Chamkha, M; Cathala, B.; Cheynier, V.; Roger Douillard, R. Phenolic Composition of Champagnes from Chardonnay and Pinot Noir Vintages, J. Agric. Food Chem., **2003**, 51, 3179–3184.*

15.4 Stationary Phase Materials and Modes of Separation

Chromatographic techniques are frequently described by the physical nature of the mode of separation. There are fundamentally four types of separation mechanisms used in modern LC analyses. Partition chromatography is one of these, using a liquid mobile phase and (typically) a bound–liquid stationary phase (*liquid–liquid chromatography*). Another mode of separation is *adsorption chromatography*. You might think of adsorption chromatography as an extreme. The analyte prefers the stationary phase so much, in comparison to the mobile phase, that it binds to the stationary phase and there is no "partitioning." In adsorption chromatography, depending on the analyte and the mobile phase composition, the analyte of interest is retained by a solid stationary phase through surface interactions between the stationary phase particle and the analyte of interest. When it is time to elute the sample, the mobile phase is changed in order to reverse the favor in binding and release the analyte from the stationary phase. This is sometimes referred to as *liquid–solid chromatography* or *solid-phase extraction*.

Within each of these groups, the mode of separation will be determined by essential features of the stationary phase, the mobile phase character, and the way in which the analyte interacts with both. These include separation by polarity, separation by charge, separation by size, and separation based on shape-specific interactions.

Normal Phase Chromatography

Normal phase chromatography is typically used to separate analytes that are relatively polar in nature. In *normal phase* chromatography, the stationary phase is initially more polar than the mobile phase, and the physical property used to drive the partitioning of the analyte species is a difference in polarity. See Figure 15.12. Typical normal phase

---

PROFILE

Liquid Chromatography–Mass Spectroscopy in Athletic Doping

Liquid chromatography (LC) coupled with mass spectroscopy (MS) detection is widely used in antidoping laboratories by scientists and technicians tasked with screening athletes for performance-enhancing drug use. Performance-enhancing drugs are very difficult to identify because they can be very similar or identical to naturally produced molecules in the body. Drugs of choice used to boost athletic prowess include anabolic steroids, hormones, and stimulants. The World Anti-Doping Agency (WADA) is an international organization that sets detection standards that are used for establishing drug use benchmarks. Independent analytical laboratories are certified for testing validity. Inspectors, independent of the laboratories, collect samples in the form of urine and/or blood from the athletes, initiating a chain of custody for the sample through analysis. Analytical chemists work hard to develop methods that stay one step ahead of those who seek to cheat the system. In 2007 an innovative method using LC-MS was developed at the German Sport University Cologne by Mario Thevis and colleagues, which can determine the slight variances between the composition of natural human insulin and the synthetic version of the hormone used to boost glycogen production. ■

See Mukhopadhyay, R. Out! Catching Doping Athletes. Anal. Chem., 2007, 79(15), 5522–5528.

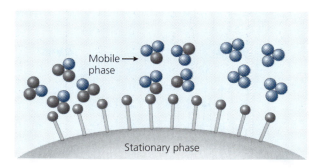

Figure 15.12 Schematic of a normal phase separation. The gray color indicates a "more" polar moiety. The blue color represents a less polar moiety. Analyte molecules with polar groups (gray spheres) will partition preferentially into the stationary phase in comparison to those analyte molecules that are nonpolar (blue spheres).

column packing material is silica or a silica-based support derivatized (or "bonded") with amino, diol, or cyano functional groups.[7]

In a normal phase experiment, the mobile phase gradient will begin with a relatively nonpolar organic solvent, such as hexane or heptane. For polar analytes, this promotes a higher affinity for the stationary phase. As the analysis progresses, the mobile phase gradually increases in polarity by the addition of miscible solvents such as isopropanol, ethyl acetate, or chloroform. By doing so, the affinity the analytes have for the mobile phase is increased during the analysis. A key consideration in selecting the mobile phase constituents is their "transparency" or ultraviolet cutoff. If you are detecting your sample using a UV-vis spectrophotometer, your solvents must be transparent within the region of the spectrum where you are detecting. For example, hexane has an ultraviolet cutoff of 201 nm, and therefore it is suitable for applications where the detection wavelength(s) is greater than 201 nm. On the other hand, ethyl acetate has an ultraviolet cutoff of 256 nm, and therefore it is not suitable unless the detection wavelength is longer than 256 nm.

Applications for normal phase chromatography involve either polar compounds that are soluble in organic solvents, such as fat-soluble vitamins, saccharides, and polyalcohols, or polar compounds that are unstable in aqueous solvents, such as organophosphates. Normal phase chromatography finds wide use in pharmaceutical analysis, as well as food purity analysis and food extract preparation.

Figure 15.13 shows a normal phase chromatogram of an extract of olive oil.[8] In this analysis, the components are eluted by polarity, with the least polar components (the ceramides, cerebrosides, and phosphatidyl cholines) eluted first under a mobile phase of acetonitrile and methanol (70:30). As the separation continues, the more polar components (digalactosyldiglycerides and phosphatidylethanolimines) are eluted as the mobile phase is switched to pure methanol. And, in the final stage of the separation, the most polar components (unidentified peak in fraction 25 between 50 and 55 minutes) are eluted, with 100% water as the mobile phase.

---

[7] With the later development of stationary phases that incorporated different partitioning schemes, the traditional polar stationary phase was considered the "normal" way of conducting chromatography; thus the term "normal phase" became widely used.

[8] Karantonis, H. C.; et al., *J. Agric. Food Chem.,* **2002**, *50*, 1150–1160.

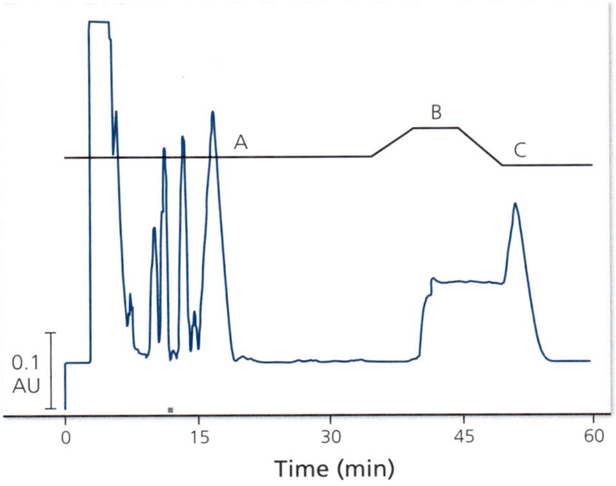

Figure 15.13 Gradient HPLC of total polar lipids from olive oil on a normal phase $-NH_2$ column. Region (A) 70:30 acetonitrile/methanol, (B) 100% methanol, and (C) 100% water.

Problem 15.23: An aqueous sample contained traces of the compounds listed in Table 15.1. The octanol–water distribution coefficients are tabulated as pK_D values. Assume you analyzed an aliquot of the aqueous sample using high-performance LC with a mobile phase that was 75% methanol and 25% water on a cellulose column. Predict the elution order and explain your answer.

TABLE 15.1: Octanol–Water Distribution Coefficients: $pK_D = -\log(K_D)$

COMPOUND NAME	pK_D
n-Pentane	3.45
2,2-Dimethylpropane	3.11
3-Methylpropane	3.60
2,2-Dimethylbutane	3.82
2,3-Dimethylbutane	3.85

Source: Sangster, J. Octanol-Water Partition Coefficients of Simple Organic Compounds. *J. Phys. Chem. Ref. Data,* 1989, *18*(3), 1111.

Problem 15.24: Based solely on molecular polarity, predict the most likely order of elution for each set of three compounds in a normal phase chromatographic experiment.

A

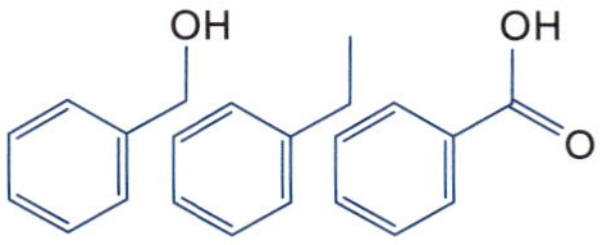

B

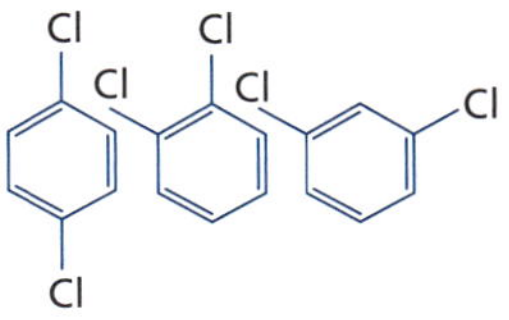

Reversed Phase Chromatography

Reversed phase chromatography (RPC) is typically used to separate analytes that are relatively nonpolar in nature or at the very least contain a substantial organic moiety as part of their structure. The term *reversed* is used with respect to normal phase chromatography. In RPC, the stationary phase is hydrophobic in nature. As in normal phase chromatography, the separation mode of the analytes is by polarity. However, in RPC, unlike in normal phase chromatography, the least polar analytes are typically retained the best. Typical RPC stationary phase packing materials are composed of an octane (C_8) or octadecyl (C_{18}) carbon chain bonded to silica beads. However, any nonpolar material that can be uniformly packed into a column can be used. For example, poly(styrene-divinylbenzene) is a reversed phase material used when the pH required is too extreme for the more common C_{18} stationary phase (see Figure 15.14). Hydrophobic analytes in the mobile phase will preferentially associate with the hydrophobic stationary phase. As the analysis continues, the mobile phase strength is increased (for example by decreasing the proportion of water). For an identical sample analyzed under conditions of identical mobile phase, the order of elution in reversed phase chromatography is opposite to that found in a normal phase chromatography.

Bonded phase RPC columns, such as C_8 and C_{18}, continue to function through surface interactions, or like very thin layers. Figure 15.15 is a schematic representation of a C_8 column. If the stationary phase becomes thick, or more three-dimensional, then additional band broadening will be the result due to increased mass transfer. Derivatization of the support with chains longer than C_{18} are likewise undesirable. The longer chains will fold back and develop surface structures. As the layer deepens, mass transfer becomes more problematic.

In a reversed phase experiment, the mobile phase gradient will begin with a highly polar solvent, typically an aqueous mixture having a small percentage of a water-miscible organic substance (e.g., methanol or acetonitrile). For nonpolar analytes, this

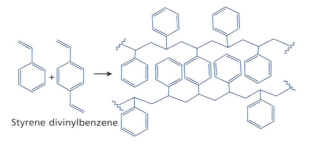

Styrene divinylbenzene

Figure 15.14 A schematic of poly(styrene-divinylbenzene).

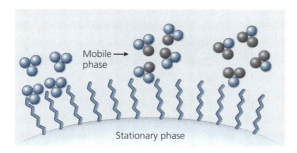

Mobile → phase

Stationary phase

Figure 15.15 Schematic of a C-8 reverse phase separation. The gray color indicates a "more" polar moiety. The blue color represents a less polar moiety. Nonpolar analyte molecules (blue spheres) will preferentially partition into the stationary phase in comparison to those analyte molecules that are more polar (gray spheres).

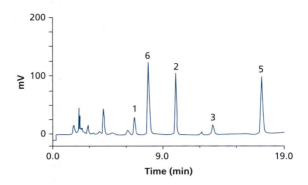

Figure 15.16 Reverse phase HPLC chromatogram of a postmortem whole blood sample. Peaks: benzoylecgonine (1), 2'-methylbenzoylecgonine (2), cocaine (3), 2'-methylcocaine (5), and 3,4-methylenedioxy-N-ethylamphetamine (6).

promotes a relatively high affinity for the stationary phase. As the analysis progresses, the mobile phase gradually decreases in polarity as the concentration of organic solvent increases. Note as with normal phase chromatography, the solvents in the mobile phase must be miscible during the changeover period, and the transparency or ultraviolet cutoff for each solvent must be considered if you are using UV-vis detection.

Figure 15.16 shows the use of reversed phase high-performance LC to detect the presence of cocaine and cocaine metabolites in the presence of human blood samples.[9] The sample was taken postmortem. The chromatographic procedure used a C_{18} reversed phase column and a mobile phase gradient. The more polar compound (benzoylecgonine; labeled 1 in the figure) was eluted first and, as the percentage of organic solvent was increased, the less-polar compounds eluted.

Problem 15.25: Assume that an aqueous sample contains equal trace amounts of the compounds listed in Table 15.1. If you extract the aqueous sample with an aliquot of octanol:

(a) List the compounds in order of increasing concentration in the octanol phase at equilibrium.

(b) Assume that you analyzed an aliquot of the aqueous sample using high-performance LC with a mobile phase that was 75% methanol and 25% water on a C_8 column. Make a sketch of the predicted chromatogram showing the expected elution order. Label each peak. Compare and contrast your answer with the one you gave in Problem 15.21.

Problem 15.26: Based solely on molecular polarity, predict the most likely order of elution for each set of three compounds for a reversed phase chromatographic experiment. Compare your answer with the one you gave in Problem 15.22.

[9] Karine, M.; Clauwaert, J. F.; Van Bocxlaer, W. E. L.; De Leenhee. A. P. Liquid Chromatographic Determination of Cocaine, Benzoylecgonine, and Cocaethylene in Whole Blood and Serum Samples with Diode-Array Detection. *Journal of Chromatographic Science,* **1997,** 35(7), 321–328.

Ion Exchange Chromatography

Ion exchange chromatography (IEC) is used to separate analytes that are ionic in nature. In IEC, the physical property used to drive the partitioning of the analyte species is ionic charge density. IEC is based on charge–charge interactions between the charged analyte and oppositely charged groups that are bonded to the stationary phase. IEC is divided into two types: *anion exchange chromatography* and *cation exchange chromatography*. These are further described by the strength of the interaction—strong cation exchange (SCX), weak cation exchange (WCX), strong anion exchange (SAX), and weak anion exchange (WAX). The strength of the ionic interactions depends on the ionization in the stationary phase. The chromatography literature typically uses these acronyms exclusively. You should take a moment to learn these acronyms.

The separation mechanism in IEC involves electrostatic interactions between the analyte and the stationary phase. The stationary phase is at first associated loosely with an appropriate counter ion. These counter ions are displaced by analyte as the sample moves through the column in the mobile phase in a stoichiometric displacement.

Note that the mode of separation, either anion exchange or cation exchange, refers to the charge on the analyte—those that are exchanging with counter ions—and not to the charge on the stationary phase (see Figure 15.17).

Just like normal and reversed phase techniques, the purpose of a gradient analysis is to change the relative affinity for the mobile and stationary phases during the analysis. In IEC, the analyst has more options. One way to create a gradient is to change the ionic strength of the mobile phase during the analysis. As the ionic strength increases, the ionic sites on the stationary phase become occupied and less likely to interact with the analyte. In the cases of weak acids or bases, the analyst can also adjust the pH of the mobile phase in order to change the percent ionization of the analyte. Figure 15.18 shows a gradient elution of several proteins using IEC.

Hydrophilic Interaction Chromatography

When a polar stationary phase is used in a chromatographic method to separate relatively nonpolar analytes, the mechanism is referred to as *hydrophilic interaction chromatography* (HIC) or sometimes hydrophilic interaction liquid chromatography (HILIC). The mobile phase is typically a nonprotic polar organic solvent with a small amount of water dissolved in it. Acetonitrile is commonly used as the polar organic solvent both because of its miscibility with water and its transparency in the UV-vis range from 190 to 900 nm. The mechanism in HIC is believed to involve the formation of a mobile phase bilayer in which the hydrophilic nature of the stationary phase produces a water-dense boundary at the surface of the stationary phase and a water-depleted

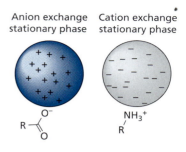

Anion exchange stationary phase   Cation exchange stationary phase

Figure 15.17 Ion exchange resins. This schematic shows the affinity of an ion for the stationary phase. The spheres represent silica beads with bonded ionic moieties.

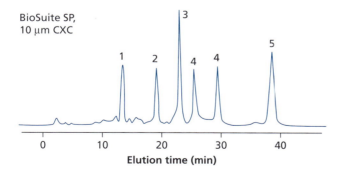

BioSuite SP, 10 μm CXC

Figure 15.18 Cation exchange separation on a Waters Biosuite SP column, a strong cation exchange column. Separation was accomplished using a linear gradient elution that increases, from 0 to 100%, in eluent B (B is 20 mM sodium phosphate pH 7.0 with 0.5 M sodium chloride) while it decreases, from 100 to 0% in eluent A (A is 20 mM sodium phosphate pH 7.0). Results show (1) trypsinogen, (2) ribonuclease, (3) alpha-chymotrypsinogen, (4) cytochrome-c, and (5) lysozyme.

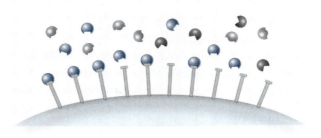

Figure 15.19 Schematic of stationary phase affinity ligand binding to a shape-specific analyte.

organic phase. Partitioning then occurs at the bilayer boundary between these two regions (liquid–liquid chromatography). In the case where the analyte is sensitive to pH, a second equilibrium involving protonation and/or deprotonation within the aqueous-rich layer occurs and ion exchange with the stationary phase can be exploited as part of the separation scheme. This allows the analyst to use pH as another variable to control resolution. Solvent gradients as well as pH gradients are often used to maximize the separation. In these circumstances, it is important that the organic solvent be aprotic. Other polar organic solvents have been used with HIC, the criterion being the ability to solvate a small amount of water and a suitable λ cutoff. See Table 6.2 for a list of solvents along with their Snyder polarity index and their λ cutoff.

Affinity Chromatography

Affinity chromatography is mostly used in biological applications where a particular analyte (molecule or molecule-type in the sample mixture) has a preference, or affinity, to bind another substance (see Figure 15.19). An affinity column will take advantage of this binding preference.

An example of this type of affinity is the preference of a glucose-binding protein for binding to glucose. The stationary phase contains an inert solid support attached to or derivatized with the agent of interest, called the ligand. The ligand could be, for instance glucose, or an antigen, enzyme, metal ion, nucleic acid, or biochemical receptor. The binding of analyte to ligand must be reversible so that the analyte can be eluted from the column. The stationary phase should have good flow characteristics, and the ligand must be accessible. Frequently the attachment of the ligand to the solid support is accomplished through a linker in order to keep the ligand from laying on the surface of the solid support. Gradients can be created by varying the pH of the mobile phase, by varying the ionic strength of the mobile phase, or by changing the polarity of the mobile phase.

Chiral Chromatography

Chiral chromatography is used to isolate an optically pure enantiomer from a racemic mixture. As you may recall, the l and d enantiomers of a racemic mixture have identical boiling points, melting points, vapor pressures, and so on, so the isolation of a pure enantiomer can often be very challenging. Fortunately, one property in which the two enantiomers in a chiral pair differ is in their affinity toward an enantiomerically pure substrate. In chiral chromatography, an optically pure (single) enantiomer of a substance is immobilized on a stationary phase such as silica gel. The resulting chiral stationary phase (CSP) allows racemic analytes in the mobile phase to undergo differential partitioning, with the CSP resulting in efficient separation of the two enantiomers.

Chiral chromatography finds its most prevalent applications in the pharmaceutical industry. It is quite often the case that a biologically active molecule will exist as an

According to an article in 2004 in *Chemical & Engineering News*, the active ingredients in the top four blockbuster drugs are comprised of a single enantiomer.* ▬

	GLOBAL REVENUE ($BILLIONS)	ACTIVE COMPOUND	THERAPY
Lipitor	$12.0	Atorvastatin	Cholesterol
Zocor	$5.9	Simvastatin	Cholesterol
Plavix	$5.0	Clopidogrel	Clotting
Nexium	$4.8	Esomeprazole	Ulcers

*Rouhi, A. M. C&EN, **2004**, 83(36), 49–53.

enantiomer. The single enantiomers of the chiral active molecules often have different pharmacological effects, with the biologically "inactive" isomer being toxic.

Size Exclusion Chromatography

There are two types of size exclusion chromatography (SEC):

1. *Gel permeation chromatography* (GPC). GPC uses a hydrophobic stationary phase material and an organic mobile phase solvent. GPC finds wide use in the separation of organic polymers—polymers that are soluble in polar organic and nonpolar solvents. One could think of GPC as the SEC analog to reversed phase chromatography.

2. *Gel filtration chromatography* (GFC). GFC uses a hydrophilic stationary phase material and an aqueous mobile phase solvent. GFC finds wide use in the separation of biopolymers—polymers that are water soluble. One could consider GFC as the size exclusion analog to normal phase chromatography.

The mode of separation in SEC is by molecular size. The stationary phase material in this technique is porous. It might be useful to imagine the packing material akin to microscopic sponge particles. As the analytes in the mobile phase pass through the stationary phase, analytes that are too large to fit into the pores are excluded and must take a path around the particle. Meanwhile, analytes that are small enough to fit into the pores will do so and will explore the solvent volume in the pore (see Figure 15.20). The pore size of the stationary material determines the extent of separation in this technique. Small molecule analytes are swept into the pores and explore a larger volume of the column and mobile phase. This behavior retains the smaller molecules on the column for a longer time. Meanwhile, larger molecule analytes can be excluded from the pores in the stationary phase and therefore move more quickly through the column (see Figure 5.21). Through exclusion, the large analytes will elute from the column first while the smaller analytes are retained longer in the pores of the stationary phase and elute later (see Figure 15.21).

Technically speaking, applications that use a size exclusion stationary phase are not chromatographic techniques. This distinction is due to the fact that the mode of separation is not the partitioning of an analyte between phases but more accurately involves the ability of the stationary phase to physically trap molecules below a given size (the exclusion limit) within pores. SEC is included in this chapter because the equipment and instrumentation used in SEC is nearly identical to the other chromatographic techniques profiled here and differs only in the properties of the stationary phase.

Size of solute substances determines whether they enter or not

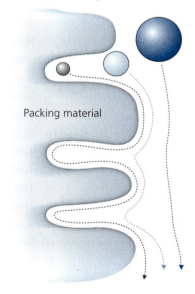

Packing material

Principle of size exclusion mode

Figure 15.20 Expanded schematic demonstrating the mechanism of separation for SEC.

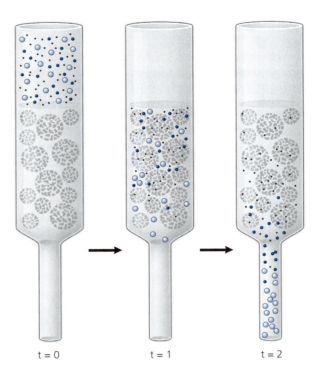

t = 0   t = 1   t = 2

Figure 15.21 Schematic of a size exclusion separation. The column labeled t = 0 represents the analyte mixture before separation. In this image, the analyte mixture contains three different analyte particle sizes. As the analyte passes down the column, the largest analyte particles are excluded from the stationary phase pores and are eluted first. The smallest of the analyte particles penetrate the deepest into the stationary phase pores and are eluted last.

The size of the largest molecule that is not excluded from the pores is the exclusion limit. Molecules larger than the exclusion limit will all be excluded and will elute together. There is no separation of analytes that are above the exclusion limit.

We have mentioned that in SEC the key property of the separation mechanism is the "size" of the analyte molecules. It is important to appreciate that when we talk about the size of a molecule in terms of SEC results, we are talking about the volume of space the molecule occupies in the solvent, or the "hydrodynamic volume," of the molecule. As molecules tumble in the mobile phase, they sweep out a volume that can sometimes be disproportionate to their molecular mass, depending on the molecular shape. Lower mass molecules can appear large if they are more rod-like, whereas higher mass molecules can appear small if they are very tightly packed and spherical. In Figure 15.22, we have two molecules with the same molecular mass, however the deca-1-ene-3,5,7-triyne on the left side of Figure 15.22 is very rigid and will sweep out a much larger volume of space compared to the naphthalene molecule on the right.

In general, for molecules of similar shape, the retention time (or elution volume) is inversely proportional to the $\log_{10}$ of the molecular weight (see Figure 15.23).

$$t_r \propto \frac{1}{\log_{10} MW} \qquad \textbf{Eq. 15.20}$$

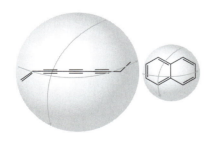

Figure 15.22 Both of these molecules have the formula $C_{10}H_8$. However, naphthalene (right) has a much smaller hydrodynamic volume and will therefore appear smaller in an SEC experiment.

Early stationary phases consisted of dextran or starch gels, agarose, or polyacrylamide. However, in order to operate under higher pressures, many of the commercial stationary phases are composed of polymers with carefully designed pore size

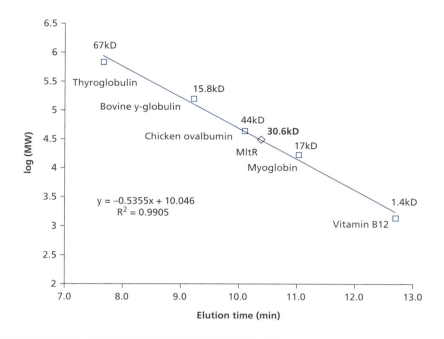

Figure 15.23 A calibration curve of molecular weight versus the elution time (or volume) for an analytical gel filtration experiment.

distributions and are available in series such that one can select a packing material, based on a predetermined calibration curve, that is best suited to the application at hand. GPC stationary phases are typically made of polymethacrylate or polystyrene. GFC stationary phases are more commonly silica based, but polymethacrylate is also used.

15.5 Instrumentation

Overview

In this section, we will look at the system components of a chromatography instrument. For the purpose of comparison, we will first discuss column chromatography, a system very similar to that first developed by Tswett. In column chromatography, a glass tube is filled with cellulose, alumina, or silica, which serves as the stationary phase. The stationary phase is wetted with solvent (mobile phase,) and a small volume of sample is loaded at the top of the column.

Additional solvent is placed on top of the column and the sample and solvent are allowed to migrate down the column under the force of gravity. Tswett's samples were highly colored and could be seen visually; thus, the human eye served as the detector.[10] Figure 15.24 shows a block diagram of the system components for a simple LC system. This type of chromatography system is often referred to as simply *column chromatography*.

From an instrumental perspective, column chromatography is not an instrument at all but rather a technique. However, you are likely to be familiar with the technique from your earlier studies and experiences in organic and inorganic chemistry, and it provides us with a good representation of the basics of the instrumental techniques (high-performance LC and ultra-high-performance LC) on which we will spend

---

[10] Sometimes a black light can be used to induce fluorescence in polyconjugated analytes, thus allowing for visual detection of otherwise colorless samples.

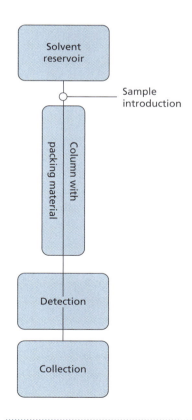

Figure 15.24 Block diagram of a simple LC system.

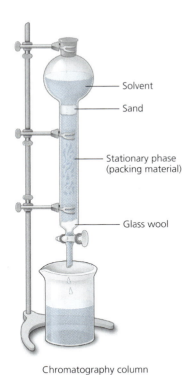

Solvent

Sand

Stationary phase
(packing material)

Glass wool

Chromatography column

Figure 15.25 A simple setup for performing column chromatography.

more time in this text. In addition, column chromatography is still commonly used to purify small quantities of material in the laboratory. As in Tswett's column LC system, the stationary phase, a finely powdered solid, is typically packed into a column (see Figure 15.25).

The bottom of the column is typically plugged with glass wool to prevent the stationary phase from escaping from the bottom of the column, the top of the column is typically capped with a few centimeters of sand, and the column is wetted with an appropriate solvent. To load the sample onto the column, the solvent is allowed to migrate down the column so that the top of the solvent is in the region of the sand cap when the flow is stopped. The sample is then loaded into the top of the sand cap, where it will form a layer in the sand. Finally, the solvent reservoir above the sand is filled with the solvent. The sand prevents the sample from being dispersed while the solvent reservoir is being filled. The flow is restarted and as the mobile phase migrates down the column, the sample components partition between the two phases. The flow in column chromatography is typically gravity fed; however, in some situations the top of the solvent reservoir may be connected to an air jet on the laboratory bench to provide a little bit of added air pressure to help push the solvent through the column. This technique, known as *flash* chromatography, helps speed up column separations that are typically slow.

Problem 15.27: Describe schematically how you might incorporate a detection system in the column chromatography system depicted in Figure 15.25. (There is more than one correct answer.)

High-Performance Liquid Chromatography Components

In the modern LC system, you can still identify the basic components of Tswett's original system. However, in order to present a large surface area to the mobile phase, modern LC systems use highly engineered packing materials, usually support beads of 3- to 5-μm diameter with a bonded stationary phase. This greatly enhances the separation efficiency of the procedure. However, it also requires that the mobile phase be under very high pressures in order to push the mobile phase through the packed stationary phase. These types of LC systems are often referred to as *high-performance LC*, or sometimes high-pressure LC (HPLC).

In order to achieve the high performance found in a modern chromatography system, it must be expanded to include a number of other important components. Figure 15.26 shows a block diagram of a typical HPLC system. Although there is significant engineering that has gone into the design of a modern HPLC system, the basic technology is the same as in Tswett's original system. In all configurations of an LC, the mobile phase is stored in a reservoir and is directed into the column. The liquid mobile phase flows through the column, and after separation, the analytes are detected in the eluent.

In an HPLC instrument, a pump draws solvent from a solvent reservoir or, in the case of gradient chromatography, multiple reservoirs. Two-, three-, and four-reservoir systems are very common; these are known as binary, ternary, and quaternary systems, respectively. Because the column packing in a high-performance system is very dense, the mobile phase must be forced through the column using a liquid pumping system. These pumps are capable of achieving very high pressures, which are typically measured in pounds per square inch, with pressures in the thousands. Recall that 14.7 pounds per square inch is equal to one atmosphere of pressure. It is common for an HPLC system to operate at pressures in the hundreds of atmospheres, and newer "ultra-high pressure" pumps can work at 1,000 atmospheres. A computer regulates the flow of each of the mobile phase components into the mixer, and it also controls the pump. The mobile phase flow is directed from the pump through the injector port and then on through

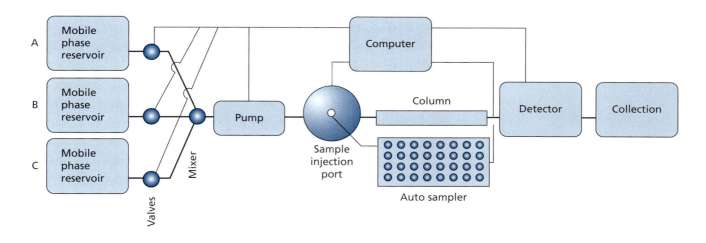

Figure 15.26 Block schematic of a tertiary HPLC system that uses three pumps and autosampling capability. Using computer-operated valves, the amount of each type of mobile phase drawn into the mixer is controlled precisely. The mixer ensures solvent uniformity of the mobile phase drawn into the pump.

the column. The eluent, still under pressure, is passed into the detector,[11] and the signal from the detector is sent back to the computer, which processes the detector response. The eluted sample can be directed to an automated fraction collector or collected by hand if the purpose of the separation was to obtain purified fractions of the original solution, or the flow can be directed to waste if sample collection is not an objective of the experiment. Although an HPLC system looks much more complicated than Tswett's original column chromatography, the "heart and soul" of HPLC is still the partitioning of an analyte between a stationary and mobile phase.

Mobile Phase

In the selection of mobile phase, purity is a crucial consideration. Solvents are purchased in quality groups that include reagent-grade, spectroscopy-grade, and HLPC-grade. The HPLC-grade solvents are the highest quality and contain minimal UV-vis absorbing contaminants. These should always be used in chromatography systems that utilize spectroscopic detectors. Also, solvents need to be filtered prior to use. Filtering protocols typically use a 0.2-micron filter in order to remove particles such as dust and bacteria. Bacterial growth can clog the solvent intake line and/or the column. In addition, bacteria can potentially create bubbles in the system, which can cause pressure problems. Furthermore, bacteria may also absorb light or produce proteins that absorb light, which will interfere with the detection of analytes. Fresh, daily filtration will keep bacterial growth under control. Some protocols may call for the addition of an antibacterial agent, such as azide, to prevent bacterial growth. Some HPLC systems have filtration accessories directly in line with the solvent uptake from the reservoir.

Solvents naturally dissolve gases from the air, and the dissolved gases can come out of solution, or "degas," as the mobile phase passes through the system under pressure. Therefore, it is important to degas the mobile phase solvents before they get into the chromatography system. "Outgassing" within the system can cause bubbles through the column or detector volumes, as well as unwanted pressure fluctuations. Because different solvents have different solubilities for air, outgassing can be more problematic when the solvents are first mixed and then pressurized. For systems such as this, sparging of the solvents prior to mixing is essential. Sparging involves connecting a tank of helium gas to the system and forcing or bubbling helium into the solvents. The helium will displace other dissolved gases; however, note that the solubility of helium in most

---

[11] The most common detection systems are UV-vis, infrared, electrochemical, and mass spectroscopy.

In 2004, Waters introduced the first commercial ultra-high-pressure (performance) liquid chromatography system, the Aqcuity UPLC™. They simultaneously incorporated advances in particle size technologies (1.7–1.8 μm) and innovations in hardware design that allowed for the use of much denser packing materials operating at much higher pressures. The result was a dramatic improvement in both the resolution and throughput. The Acquity can be purchased with ultraviolet-visible (UV-vis), fluorescence, and mass spectroscopy detectors.

Since the introduction of the Acquity, several manufacturers now offer UHPLC systems. For example, PerkinElmer markets the Flexar, which can be operated at pressures as high as 18,000 psi ($\approx$ 1,225 atm)* and comes available with UV-vis and fluorescence detectors. ▄▄

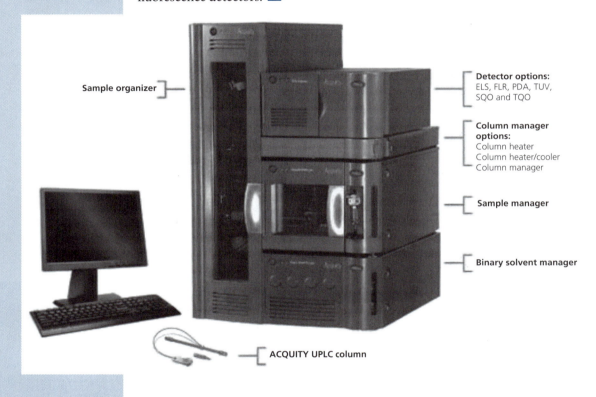

Sample organizer

Detector options:
ELS, FLR, PDA, TUV, SQO and TQO

Column manager options:
Column heater
Column heater/cooler
Column manager

Sample manager

Binary solvent manager

ACQUITY UPLC column

The deepest part of the ocean is estimated to have a pressure of 1,072 atm.

solvents is minimal. Sparging with helium at a very low flow velocity throughout the chromatographic analysis will help maintain a "gasless" solution. For systems in which the solvents are first pressurized and then mixed, a sparging system is not required. Because of the increase in the cost of helium, vacuum degassing is also popular. Vacuum degassing involves pulling a hard vacuum on the solvent for a period of time. Outgassing under vacuum can also be assisted by sonicating the solvent while pulling a vacuum. Warming the mobile phase will also help degas. Finally, for safety's sake, mobile phase solvents should be used and maintained in an appropriate environment that includes appropriate ventilation to minimize exposure to solvent fumes.

Columns

Because of the high pressures used in HPLC, the stationary phase is usually contained in a stainless steel column. Because we need the conditions experienced by the mobile phase to be uniform as it passes through the stationary phase, the "packing" of the stationary phase into the column must be done in a highly reproducible and uniform manner. The first consideration is the particle size of the stationary phase. Manufacturers of HPLC columns take great care to ensure that the particle sizes are within very narrow tolerance ranges in order to minimize eddy diffusion and mass transfer effects. The stationary phase surface area available to solutes in manufactured columns can vary—from 2 to 500 m^2/g of packing material.

Packing material can also limit performance in other ways. For example, hydrolysis of silica-based supports can occur at pH values above 8. Likewise, the rate of degradation can be increased with increasing temperature. Therefore, column operation conditions must also be consistent with the application.

One must also be aware of changes to packing material over time and through use. Oxidation and derivatization can alter the behavior of the stationary phase, and changes to the packing can cause channeling. These types of effects should be minimized over the lifetime of the column by following the manufacturer's guidelines on column tolerances and normal operating parameters.

Obviously, the chemical nature of the packing material also plays a critical role in the partitioning mechanisms used. The manufacturer's websites contain a wealth of information regarding the bonded phase, particle size, and applications for their columns. We have included the links to several prominent column suppliers in Section 15.7 and we strongly encourage you to familiarize yourself with these resources. For instance, on the Waters site, you can search for columns by the bonded stationary phase (e.g., 2-ethylpyridine, amide, amino, anion/cation exchange, C_1, C_6, C_8, C_{18}, cyano, fluorophenyl), particle size (2.5–55 μm), column length (5–300 mm), diameter (320 μm–50 mm) or mode (e.g., normal phase, reversed phase, ion exchange). Also, under the resource tab on the Waters website, you will find a support library and users guide along with webcasts and how-to videos. Similarly, the Restek website has a searchable technical library that allows you to search application notes for specific types of separation methods. The site also includes a "ChromaBLOGraphy" to which you can subscribe.

Because packed columns resist mobile phase flow, we are limited in column length: the longer the column, the greater the pressure needed to force mobile phase through it. Typical HPLC columns are 5 to 25 cm in length and exhibit efficiencies (N_t) on the order of 10,000 plates.

Injectors

As discussed earlier, the sample is introduced onto the column in a single pulse or injection. In HPLC, the sample is loaded onto an injector using a syringe. This can be done manually or in the case of routine analysis by means of an autosampler. Figure 15.27 shows the flow diagram for an injector in two positions, load and inject.

In the load position, the flow of mobile phase is directed directly into the column, bypassing the sample loop. This allows the researcher or technician ample time to load the sample properly and ensure that the system is ready. The sample is loaded into an injector "loop." Various loops of specific sizes are available. The operator will select a loop size that is appropriate for the system and method. The loop size limits the volume of sample injected. For example, if the system is fitted with a 2-microliter loop, and the technician tries to load 5 microliters, 2 microliters will fill the loop and 3 microliters will overflow into an overflow container outside of the chromatographic system. When doing analytical work, where careful quantitative analysis is required, loop overfilling is recommended. This guarantees that a consistent and exact injection volume is obtained in each successive experiment. It is recommended to overfill by two times the loop volume to ensure the injection volume is quantitative and that the loop is fully flushed of any preexisting contents. In our example, with a 2-microliter loop, one would load the loop with 4 microliters to guarantee an overfill, knowing that exactly 2 microliters will be injected. When ready, the injector is turned to the inject position, which realigns the flow so that the sample is moved along into the mobile phase and onto the column. It is also possible to underfill the loop, or end up with partial loop filling. In this case, the amount of sample injected is wholly dependent on the precision of the syringe and the

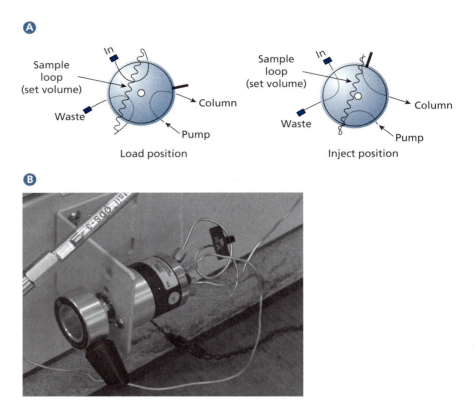

Figure 15.27 (A) Schematic of an HPLC injector. While in the load position (left), the mobile phase is bypassing the sample loop. The liquid sample matrix in loaded into the *In* port while the injector assembly is in the load position. The sample matrix enters the sample loop (represented by the wavy line) with any excess spilling out into waste. In the inject position (right), the injector has been rotated so that the mobile phase is pumped into the sample loop where it will carry the sample along into the column. (B) Photo of an LC47LG injection valve with a 20 microliter sample loop. Valve by Rheodyne.

technician who is loading it. Because of this variance, it is not recommended for quantitative uses. However, it may be more convenient. For example, if the loop size is 5 microliters, but you only want a 2-microliter injection, you can inject a 2-microliter sample using your syringe to measure the volume. This may be more convenient than changing the loop on the injector and, if the quantitative aspect of your experiment is not essential, it is a perfectly fine method of injection.

If your system is equipped with an autosampler, the injection is timed and rotation of the injector is computer controlled. The use of an autosampler is very useful in commercial laboratories where routine analysis of large batches occurs.

Pumps

Solvent pumps are an essential part of the HPLC system. Without pumps, the mobile phase cannot pass through the tightly packed stationary phase. In addition, the pumps need to operate such that they provide a continuous and constant flow, without pulsing and without changes in flow rate. Take another look at Figure 15.27, which demonstrates schematically that the solvents in the mobile phase are drawn up by the pump and pushed forward to the column. A closer look at the functioning parts of the pump is shown in Figure 15.28. Solvents are drawn in through the pump's inlet valve. The piston of the pump is made of a sapphire rod, and as it is mechanically drawn back, solvent is pulled in. To prevent the solvent from flowing backward (back to the reservoir), the inlet manifold is equipped with two one-way valves. Ruby balls float in flowing solvent, to allow for one-way flow. If the flow tries to reverse, the balls seat and stop the flow. As the piston is pushed forward, the solvent is pressurized and flows out through the outlet valve. A high-end

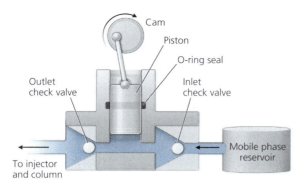

Figure 15.28 Schematic of a generic HPLC pump.

HPLC pump has two pump heads to minimize pulsing. As one pump head reaches its maximum draw, the other pump head is already applying pressure to the solvent.

Pumps have a rapid refill stroke, whereas the pressurization stroke is regulated very precisely at the set flow rate. Computer control allows the pumps to be individually regulated so that each pump is contributing the exact amount of the desired solvent as required for the chromatographic protocol. The pressurized solvents coming out of the pumps are combined into a single stream before the mobile phase passes into the injector and onto the column.

Detectors

To be a useful detector in LC analysis, the detector should be somewhat selective. In other words, it should respond to some analytes very well and not at all for others. However, because chromatographic techniques, by their nature, separate and purify the components of a complex matrix, selectivity is not always a crucial consideration. It would also be useful if the signal energy responds quantitatively to the amount of analyte detected, and also if the linear dynamic range is large. We also want the limit of detection (LOD) to be very low. In LC we need the detector to have a fast time constant such that the detector will respond very quickly as the analyte passes by and quickly diminishes once the analyte has passed. Lastly, we want an acceptable signal-to-noise ratio. Sensitivity is also a consideration.[12] Sensitivity describes the response of the detector to the analyte as a function of the concentration of the analyte. Ideally, the slope of signal-to-concentration calibration curve will be very steep; however, all analytes need not have the same sensitivity profile. Finally, the ideal LC detector needs to be stable. Both noise (short-term random fluctuations) and drift (longer timescale changes in signal) need to be minimal. Noise and drift will diminish the LOD.

Problem 15.28: Using Figure 15.29, make a plot of the instrument's response as a function of the concentration of compound A. The concentrations used to obtain the chromatograms were 1.0, 2.0, and 3.0 mM respectively. This type of plot is called a sensitivity curve.

Problem 15.29: Using Figure 15.29 and the plot you made in Problem 15.28:

(a) Estimate the minimum detection quantity for compound A.

(b) Estimate the instrument's response to a 4 mM sample of compound A.

(c) Estimate the concentration of a sample that exhibited a retention time of 4.25 minutes and a peak height of 0.366 AU.

---

[12] Recall that sensitivity is defined as the slope of the calibration curve.

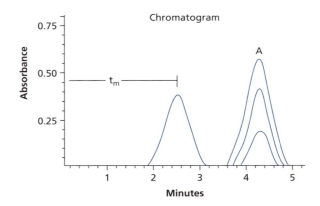

Figure 15.29 A detector response to various concentrations of analyte A.

The most common detectors are actually instrumental techniques in their own right, and the two most common detectors are the UV-vis and fluorescence spectrophotometers. These two instrumental techniques were discussed in Chapters 6 and 8, respectively. In addition, as we mentioned in Chapter 13 it is not uncommon to find an HPLC utilizing a mass spectrometer as a detector. For completeness, it should be noted that HPLC–Fourier transform infrared (HPLC-FTIR) instruments do exist. However, HPLC solvents tend to absorb strongly in the infrared region, thus limiting the utility of HPLC with detection in the infrared region.

The tandem nature of HPLC with these other spectroscopies combines to create very powerful analytical tools. Not only can you achieve separation of a complex matrix into specific analytes, but you can also obtain spectral information on each isolated fraction from the matrix. There are a great number of "detectors" used in HPLC. A casual perusal of the literature will turn up refractive index detectors, electrochemical detection, radioactive detection, and more.

Ultraviolet-Visible Detection   Figure 15.30 shows a Z-cell. The Z-cell is the most common configuration in absorbance detection. It is an example of a flow detector cell. In this particular cell, the mobile phase eluent passes into the cell from the small diameter line at the bottom left corner of the cell and exits the cell at the top right. The mobile phase flow moves from bottom to top to prevent bubbles from being trapped.

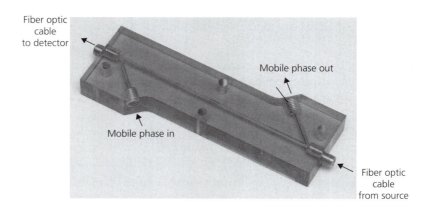

Figure 15.30 Z-cell interface for HPLC-UV-vis.

The Z-cell configuration provides a larger path length than simply using a cross-section of the tubing, without compromising peak breadth. From Beer's law we know that the signal (absorbance) is directly proportional to path length (b). The horizontal channel in the Z-cell is the optical path in the UV-vis spectrometer and two larger diameter cables in the horizontal plane contain optical fibers that guide the light into and out of the Z-cell.

Fluorescence Detection   In the case of a fluorescence detector, the Z-cell has only one optical cable, which serves to deliver the excitation pulse. The emission is collimated through traditional optics and is read perpendicular to the excitation source.

Mass Spectrometry Detection   The most common interface for HPLC–mass spectroscopy is called an atmospheric pressure ionization (API) source. The commercialization of the API source in the late 1980s brought the interface into widespread use. API sources include both electrospray ionization (ESI) as well as atmospheric pressure chemical ionization (APCI). The ESI is similar to the Babington nebulizers discussed in Chapter 7. The mobile phase eluent is drawn into a capillary under vacuum of a mass spectrometer and nebulized to create an aerosol mist that enters the mass spectrometer. In APCI, the eluent is nebulized into a heated tube where the solvent is vaporized and swept into the mass spectrometer.

Conductivity Detection   The most common type of detector used in ion chromatography is the conductivity detector. In a conductivity detector, two electrodes are positioned a fixed distance apart (typically 1 cm) within a flow cell, and they are electrically connected as one arm in a Wheatstone bridge circuit.[13] Mobile phase flows across the electrodes and an alternating current is passed through the circuit, which is adjusted to produce a null signal. As different solutes elute from the column, the conductivity within the cell will vary, causing the Wheatstone bridge to no longer be balanced, resulting in a signal.

Chromatography, in the form of high-performance liquid chromatography (HPLC), dominates the analytical instrument industry, with billions in revenue annually. The major players in the industry include Agilent Technologies, Amersham Pharmacia Biotech, Beckman Coulter, Bioanalytical Systems, Gilson, Hitachi, PerkinElmer, Shimadzu Scientific, Varian, and Waters Corporation. The use of "hyphenated" technologies, wherein the detection is accomplished through a second instrumental technique (such as LC-MS, LC-ICP-MS, and LC-FTIR, see Appendix for meaning of these terms) continues to expand at a rapid rate, and these types of systems contributed to the industry's move toward better integration of components and software across systems.* ■■

*Erickson, B. Anal. Chem., **2000**, 72, 711A–716A.*

PROFILE

Major Players in the Analytical Industry

---

[13] See Chapter 4 for a brief discussion of the Wheatstone bridge circuit.

15.6 Further Reading

BOOKS

Ackman, R. G. Errors in the Identification by Gas—Liquid Chromatography of Conjugated Linoleic Acids in Seafoods. In *Omega-3 Fatty Acids: Chemistry, Nutrition, and Health Effects*; Shahidi, F., Finley, J. W. eds.; ACS Symposium Series 788; American Chemical Society: Washington, DC, **2001**, 788, 235–242.

Bartle, K. D.; Clifford, A. A.; Myers, P.; Robson, M. M.; Seale, K.; Tong, D.; Batchelder, D. N.; Cooper, S. Packed Capillary Column Chromatography with Gas, Supercritical, and Liquid Mobile Phases. In *Unified Chromatography*; Parcher, J. F.; Chester, T. L., Eds; ACS Symposium Series, 748; American Chemical Society: Washington, DC, **1999**, 142–167.

Beck, T. L.; Klatte, S. J. Computer Simulations of Interphases and Solute Transfer in Liquid and Size Exclusion Chromatography. In *Unified Chromatography*; Parcher, J. F., Chester, T. L., eds; ACS Symposium Series, 748; American Chemical Society: Washington, DC, **1999**, 67–81.

Beesley, T. E.; Scott, R. P. W. *Chiral Chromatography (Separation Science Series)*. John Wiley & Sons: New York, **1998**.

Donnelly, C.; Pollock, A.; Heidtmann, Y.; Marley, E. Development of an Immunoaffinity Column for the Determination of T-2 and HT-2 Toxins in Cereals Using Liquid Chromatography with Fluorescence Detection. In *Food Contaminants*; Siantar, D. P., Trucksess, M. W., Scott, P. M., Herman, E. M., eds.; ACS Symposium Series, 1001, **2008**, 276–284.

Jiang, D.; Huang, Y.; Peterson, D. G. Application of Two-Dimensional Orthogonal Reversed-Phase Liquid Chromatography for Taste Compound Analysis. In *Recent Advances in the Analysis of Food and Flavors*, Toth, S., Mussinan, C., eds.; ACS Symposium Series 1098, **2012**, 137–144.

Manz, A.; Dittrich, P. S.; Pamme, N.; Lossifidis, D. *Bioanalytical Chemistry*, 2nd ed. Imperial College Press: London, **2015**. Distributed by World Scientific Publishing Co, Singapore.

Meyer, V. *Practical High-Performance Liquid Chromatography*, 5th ed. John Wiley & Sons: Hoboken, NJ, **2010**.

Sangster, J. *Octanol-Water Partition Coefficients: Fundamentals and Physical Chemistry*. John Wiley & Sons: Chichester, UK, **1997**.

Shabir, G. *HPLC Method Development and Validation in Pharmaceutical Analysis: Handbook for Analytical Scientists*. LAMBERT Academic Publishing, **2013**. Saarbrücken, Germany.

Snyder L. R.; Kirkland J. J.; Dolan, J. N. *Introduction to Modern Liquid Chromatography*, 3rd ed. John Wiley & Sons: Hoboken, NJ, **2010**.

Weber, D.; Lau, B. P.-Y.; Godefroy, S. B. Characterization of Gluten Proteins in Grain Flours by Liquid Chromatography-Tandem Mass Spectrometry Using a Quadrupole Time-of-Flight Mass Spectrometer. In *Food Contaminants*; Siantar, D. P., Trucksess, M. W., Scott, P. M., Herman, E. M., eds.; ACS Symposium Series, 1001, American Chemical Society: Washington, DC, **2008**, 482–500.

JOURNALS

Arnaud, C. H. Speeding Up Separations: Ultra-high-pressure Liquid Chromatography Improves Separations and Cuts Run Times by as Much as 90%. *Chem. Eng. News Archive*, **2010**, *88(24)*, 40–44.

Casoli, A.; Mangia, A.; Predieri, G.; Sappa, E.; Volante, M. Applications of High-performance Liquid Chromatography to Metal Carbonyl and Cluster Chemistry. *Chem. Rev.*, **1989**, *89* (2), 407–418.

Dorsey, J. G.; Dill, K. A. The Molecular Mechanism of Retention in Reversed-Phase Liquid Chromatography. *Chem. Rev.*, **1989**, *89(2)*, 331–346.

Pirkle, W. H.; Pochapsky, T. C. Considerations of Chiral Recognition Relevant to the Liquid Chromatography Separation of Enantiomers. *Chem. Rev.*, **1989**, *89(2)*, 347–362.

ONLINE RESOURCES

Hamilton (http://www.hamiltoncompany.com/HPLC/)

PerkinElmer (http://www.perkinelmer.com/catalog/category/id/hplc%20columns)

ProteCol (http://www.sge.com/products/protecol-lc-columns)

Restek USLC (www.restek.com)

Sigma-Aldrich (http://www.sigmaaldrich.com/analytical-chromatography/hplc/columns.html)

Thermo Scientific (http://www.dionex.com/en-us/products/columns/lc/lp-71678.html)

Waters (www.waters.com)

15.7 Exercises

EXERCISE 15.1: Define the following terms:

a. Exclusion limit
b. Isocratic
c. Gradient
d. Mobile phase strength
e. Distribution constant
f. Stationary phase
g. Reverse phase
h. Normal phase
i. Ion exchange
j. Size exclusion
k. Eddy diffusion
l. Selectivity factor

m. Retention time
n. Elute
o. Resolution
p. Theoretical plates
q. Plate height
r. Void volume
s. van Deemter equation
t. W_b
u. W_h
v. Column efficiency
w. Hydrodynamic volume
x. Baseline resolution

EXERCISE 15.2: Assume that an aqueous sample contained traces of the following compounds listed in the table and that you extracted the aqueous sample with an aliquot of octanol. The distribution coefficients are given as pK_D values.

Octanol-Water Distribution Coefficients

$$pK_D = -\log(K_D)$$

COMPOUND NAME	pK_D
Benzene	2.13
Toluene	2.73
o-Xylene	3.12
m-Xylene	3.20
p-Xylene	3.15

Sangster, J. J. Phys. Chem. Ref. Data, 1989, 18, 1111.

(a) List the compounds in order of increasing concentration in the octanol phase at equilibrium.

(b) Assume that you analyzed an aliquot of the aqueous sample using HPLC with a mobile phase of 75% methanol and 25% water on a C8 column. Predict the elution order and make a sketch of the predicted chromatogram. Label each peak.

EXERCISE 15.3: Assume that an aqueous sample contained traces of the following compounds listed in the table and that you extracted the aqueous sample with an aliquot of octanol. The distribution coefficients are given as pK_D values.

(a) Assume that you analyzed an aliquot of the aqueous sample using HPLC with a mobile phase of 75% methanol and 25% water on a C8 column. Make a sketch of the predicted chromatogram and label each peak.

(b) How would the chromatogram change if you repeated the experiment using a mobile phase of 50% methanol and 50% water?

Octanol-Water Distribution Coefficients

$$pK_D = -\log(K_D)$$

COMPOUND NAME	pK_D
2-Methylfuran	1.85
Tetrahydropyran	0.82
Methyl t-butyl ether	0.94
Cis-2,5-Dimethyltetrahydrofuran	1.22
Trans-2,5-Dimethyltetrahydrofuran	1.34

* Naseem, R. *Biochemical Journal*, **2006**, *395*, 529–535.

EXERCISE 15.4: Purification of a fragment of BRCA1 (amino acids 230–534) produced the chromatogram shown.

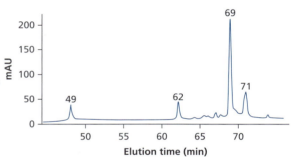

(a) Report retention times and retention volumes for each numbered peak assuming a flow rate of 1 mL/min and a flow rate of 0.3 mL/min.

(b) Assuming a t_m value of 3 minutes, use the half-height method to estimate the resolution between peak 69 and 71. What inferences can you draw from this resolution value?

EXERCISE 15.5: One form of "gradient" chromatography one can use in an ion exchange separation is to alter the pH of the mobile phase over the course of the experiment. Likewise, you can select a pH that will render a weak acid either mostly ionized or mostly in its associated form. Use pKa values for the following weak acids to predict the percent ionization of each of the weak acids in a mobile phase with pH of 5. (*Hint*: The Henderson-Hasselbalch equation will prove very useful in solving this problem.)

Some Common Biological Acids

NAME	FORMULA	STRUCTURE	pK_A
Acetic acid (vinegar)	CH_3COOH		4.76
Succinic acid (used in sour candy)	$C_4H_6O_4$		$pKa_1 = 4.21$ $pKa_2 = 5.60$
Benzoic acid (a natural preservative)	C_6H_5COOH		4.19
Formic acid (found in ant venom)	CH_2O_2		3.75
Acetylsalicylic acid (aspirin)	$C_9H_8O_4$		3.52
Citric acid	$C_6H_8O_7$		$pKa_1 = 3.15$ $pKa_2 = 4.77$ $pKa_3 = 6.40$
Oxalic acid	$C_2H_2O_4$		$pKa_1 = 1.27$ $pKa_2 = 4.2?$

Blue indicates the acidic proton.

EXERCISE 15.6: Determine N_t and H for each numbered peak in the chromatogram.

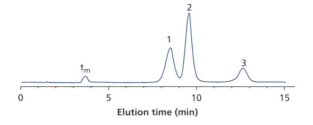

EXERCISE 15.7: Calculate the resolution and retention factor (k_r) for each peak in the chromatogram found in Exercise 15.6.

EXERCISE 15.8: For each of the parameters given on the right-hand side of Equation 15.19, indicate what effect increasing the parameter has on the expected resolution between two closely spaced peaks.

EXERCISE 15.9: Describe at least two advantages of a gradient chromatography method compared to an isocratic chromatography method.

EXERCISE 15.10: Use one of these manufacturer's websites (Agilent Technologies, Amersham Pharmacia Biotech, Beckman Coulter, Bioanalytical Systems, Gilson, Hitachi, PerkinElmer, Shimadzu Scientific, Varian, Waters Corporation) to find the chemical composition and structure of at least two normal phase stationary materials.

EXERCISE 15.11: Describe various ways one might alter the chromatographic experiment in order to improve the resolution of the two, circled peaks seen in the following chromatogram. Justify your suggestions.

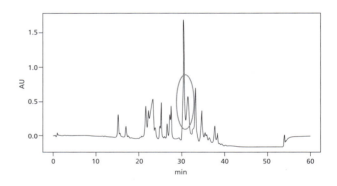

HPLC chromatogram of J'Adore perfume (aqueous), an example of a complex mixture analysis. Separation was conducted on a C18 column using a linear 5 – 100% acetonitrile-water gradient over 60 minute run time.

EXERCISE 15.12: In the HPLC analysis of caffeine in an energy drink by standard addition, the sample was prepared by pipetting 5.00 mL of the drink into a 50 mL volumetric flask, adding a fixed volume of a 1.205 mg/mL caffeine standard solution, and then diluting to volume using the HPLC mobile phase. The table summarizes the sample preparation protocol and the results obtained from the three injections:

High-Performance Liquid Chromatography Data for an Energy Drink

VOLUME OF ENERGY DRINK ADDED (mL)	VOLUME OF CAFFEINE STANDARD ADDED (mL)	FINAL DILUTION VOLUME (mL)	HEIGHT OF CAFFEINE PEAK IN CHROMATOGRAM (ARBITRARY UNITS)	AREA OF CAFFEINE PEAK IN CHROMATOGRAM (ARBITRARY UNITS)
5.00	0.00	50.00	24.79	408.3
5.00	0.50	50.00	46.77	831.7
5.00	2.00	50.00	80.40	1278.6

(a) Use the peak height data to estimate the original concentration of caffeine in the energy drink.

(b) Use the peak area data to estimate the original concentration of caffeine in the energy drink.

(c) Of the two values calculated in (a) and (b), which would you consider more reliable? Explain your answer.

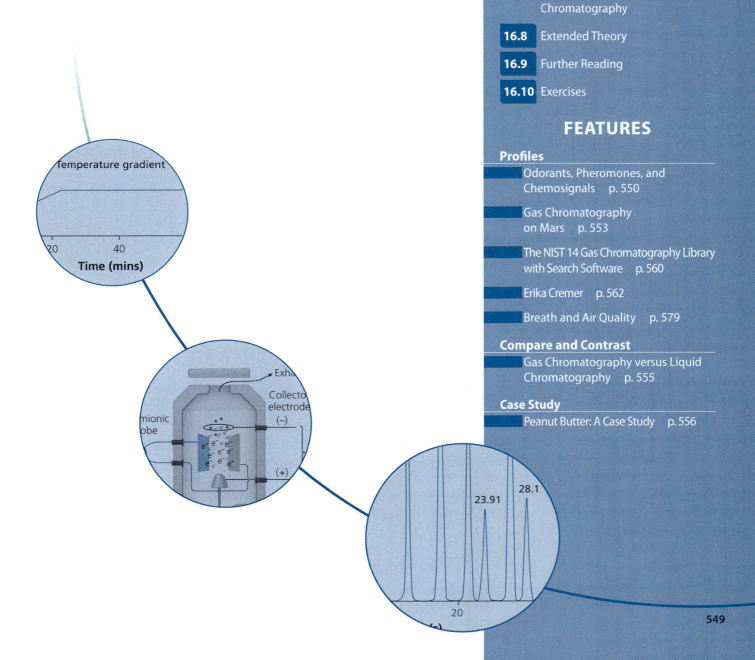

Chapter 16

Gas Chromatography

Temperature gradient

20   40

Time (mins)

Exha

Collecto electrode

mionic obe

(−)

(+)

23.91   28.1

20

Although science does not have a good grasp of these volatile compounds, nature certainly does. Many animals identify each other by scent, locate their homes and other locations using odorant molecule tracers, and interact with plant species using odor-based chemical signaling. Gas chromatography (GC) is perfectly devised as an analytical tool for analyzing these vapor-based trace compounds. In an interesting 2005 paper in *Analytical Chemistry*,[*] Dr. Alan Willse from the Pacific Northwest National Laboratory collaborated with researchers from the Monell Chemical Senses Center and the University of Pennsylvania to use gas chromatography-mass spectroscopy (GC-MS) to identify volatile compounds in the urine of mice and to use the results to identify the mice by population. Using two genetically distinct groups of mice, which differed only in genes of the major histocompatibility complex (MHC), the researchers were successful in using the chemosignals to identify the differences in the pheromones produced by the MHC genes of the two groups from among the thousands of compounds present in the samples. In another interesting example, researchers from the University of Vienna used GC to discover that the orchid *Ophrys sphegodes* releases pheromone compounds to, essentially, seduce pollinator bees to come to the flower.[†] The flower not only emits the same chemical compounds as the female bees but impressively emits them in the same proportions. The scientists used a nontraditional detector, "electroantennographic detection," which relies on the male bees' antennae as chemoreceptors. As a final example, Polish scientists used GC-MS to better understand the canine capacity for identifying cancer in human breath samples.[‡] As published in 2012, Buszewski and colleagues found positive correlations between GC-MS and dog indications for two trace volatile organic compounds implicated with the presence of lung cancer. Work such as this may lead to identifying biomarkers that have the potential for clinical use in monitoring human health. ◼

---

[*] *Willse A.; et al. Identification of Major Histocompatibility Complex-Regulated Body Odorants by Statistical Analysis of a Comparative Gas Chromatography/Mass Spectrometry Experiment. Anal. Chem.* **2005,** *77, 2348–2361.*

[†] *Schiestl, F. P.; et al, Orchid Pollination by Sexual Swindle. Nature,* **1999,** *399, 421–422.*

[‡] *Buszewski, B.; et al. Identification of Volatile Lung Cancer Markers by Gas Chromatography–Mass Spectrometry: Comparison with Discrimination by Canines. Anal. Bioanal. Chem.,* **2012,** *404, 141–146.*

16.1 Introduction

Gas chromatography (GC) is one of the most well-established analytical separations techniques. After studying liquid chromatography (LC) in Chapter 15, GC will seem very familiar to you. We will focus on important areas of comparison and important differences, both in fundamentals and in applications. As you may already suspect from the name, the primary difference between LC and GC is the physical state of the mobile phase.[1] Although LC separates components of a mixture in a liquid mobile phase, the

---

[1] *Mobile phase,* you will recall from Chapter 15, is a term used in chromatography to describe the part of the separation technique that is moving. A chromatographic separation is accomplished by differential migration of a set of analytes in a mobile phase as they move through (or across) a *stationary phase*. In GC, the mobile phase is a carrier gas, whereas in LC, the mobile phase is a liquid.

GC technique separates components in the gas phase. Thus, instead of solvents and solvent pumps as in LC, the GC techniques use pressurized gas cylinders to provide carrier gas and valves to control flow rates.

In Chapter 15, we profiled Mikhail Tswett as the innovator of chromatography. However, Tswett's discovery was underutilized until it was discovered again in the 1930s and developed more thoroughly in the 1940s.[2] Archer John Porter Martin and his colleague Richard Laurence Millington Synge received the 1952 Nobel Prize in Chemistry "for their invention of partition chromatography" and pioneering work in chromatography.[3] Martin and another colleague, Anthony T. James, introduced gas–liquid chromatography (GLC) in the early 1950s. "Gas chromatography provided a major improvement in the separation of volatile compounds, eclipsing established methods at the time. It remains the most widely used chromatographic technique for the fast and efficient separation of thermally stable and volatile compounds today."[2]

Applications of GC run the gamut of human ingenuity. Of historical significance are applications in fragrances, foods, and petroleum materials. Modern areas of interest and research include chiral separations in the pharmaceutical industry, organochlorine compounds, and pesticides, as well as problem-based applications such as the detection of potential explosive materials, extraterrestrial exploration, breath analysis, and air quality and atmospheric monitoring.

GC is a useful technique for any compound that is volatile within the thermally stable range of the column. Typical GC analyses occur in the 40°C to 400°C range. However, room-temperature GC analysis is common for extremely volatile analytes. But the limiting factor for whether or not a compound is suitable for GC analysis tends to be decomposition rather than volatility. If a compound breaks apart under the typical injection parameters of GC, that tendency may render it unsuitable for some applications. On the other hand, it is possible to take advantage of decomposition via a sampling technique called *pyrolysis*. In this case, thermal decomposition is done prior to chromatography, and the resulting chromatogram is an analysis of the decomposition products.

The general utility of GC techniques, like the LC analyses that you studied in Chapter 15, involves both *qualitative* as well as *quantitative* analysis. That is, GC techniques are used for answering questions not only about what substances are present in a sample but also how much or at what concentration those substances are found. GC provides excellent separations and characterization based on the components' differential intermolecular interactions with the column, their retention[4] behaviors in the column, and the method of detection. Methods for analyzing peak areas in GC are akin to those described for LC. The chapter is organized to walk you through the instrument, in an analogous way to the LC system—from carrier gas and injection port, through the column and column interactions, and on to the detector.

Over the past decade, papers published in the area of GC analysis have been primarily focused on applications of the technique. There have been many fewer papers published on the pure science fundamentals of the technique, with some interesting exceptions in areas related to miniaturization and multidimensional analysis. These frontier areas will be described in the last section of this chapter. Major innovations in the past decade relate to bringing the instrument to the sample site (i.e., portability) and to increasing the speed of analysis—or more precisely the throughput (i.e., how many samples can be analyzed in a given amount of time). Take, for example, problems faced

Note

Ionic compounds are generally not suitable for GC analysis because they are nonvolatile at most operating temperatures.

---

[2] Poole, C. F. *The Essence of Chromatography.* Elsevier: Amsterdam, The Netherlands, **2002**.

[3] The Nobel Prize in Chemistry 1952, http://www.nobelprize.org/nobel_prizes/chemistry/laureates/1952/ (accessed Aug 14, 2015).

[4] *Retention* is the term used in chromatography to describe the interaction of analytes with the stationary phase. Analytes that interact preferentially to the stationary phase or adsorb to the stationary phase are retained more strongly in comparison to analytes that prefer to move along in the mobile phase.

Example 16.1

In GC analysis, it is important that the analyte be volatile at the operating temperatures of the experiment. These temperatures, as we will see in the upcoming sections of this chapter, are in part limited by the type of column or stationary phase used. Consider the organic compounds listed along with their boiling points (BP = boiling point; MTBE = methyl *tert*-butyl ether; THF = tetrahydrofuran).

Suppose you had four GC columns in your laboratory. Column A has a maximum operating temperature of 200°C; column B has a maximum operating temperature of 220°C; column C has a maximum operating temperature of 280°C; and column D has a maximum operating temperature of 300°C. Which column would give you the best chance of successfully separating a mixture of all six compounds? What concerns might you have?

COMPOUND	BP (°C)
Cyclohexane	80.7
Anthracene	340
MTBE	55.2
Naphthalene	218
THF	66
Acetonitrile	81.3

EXPLANATION –

Ideally you would want to choose a column whose operating temperature exceeds the boiling point of your least volatile analyte. In this case, your least volatile analyte is anthracene; however, none of the available columns have operating temperatures that exceed the boiling point of anthracene. But the maximum operating temperature of column D does approach the boiling point of anthracene. It would be reasonable to try column D using a method that starts at a relatively cool temperature and ramps the temperature over time to a maximum of 300°C. A concern one might have in using column D is that the run times needed to elute anthracene would be exceedingly long, making the separation method untenable for routine analysis. Also, as we learned in Chapter 15, long retention times lead to peak broadening.

by the US Department of Homeland Security; the sample site might be an airport and the analysis might involve sweeping baggage contents for potentially explosive material.

Not only does the instrument need to be portable, but the throughput needs to be high in order to avoid unnecessary delays. In 1979, Terry, Jerman, and Angell published a seminal paper describing the first GC composed of microfabricated components.[5] Since then, the extent to which miniaturization has been the growing direction of the field has only continued to expand.

The widespread use of GC across nearly all areas of chemistry demonstrates not only its utility but also its value in terms of both qualitative and quantitative results. Analysts find GC to be useful, reliable, and robust. Thus, we find that GC as a method of instrumental analysis is nearly ubiquitous across laboratories, whether industrial, governmental, field-based, and/or academic.

Much of what we learned about partitioning[6] in LC (Chapter 15) is generally applicable to GC separations as well. The distribution of the analyte occurs between the mobile phase (gas) and a stationary phase that is either a solid (*gas–solid chromatography [GSC]*) or a liquid that is immobilized through adsorption or bonding to a solid support (GLC). The gas mobile phase is also called the *carrier gas*. In most applications

---

[5] Terry, S. C.; Jerman, J. H.; Angell, J. B. A Gas Chromatographic Air Analyzer Fabricated on a Silicon Wafer. *IEEE Trans. Electron Devices*, **1979,** 26, 1880–1886.

[6] *Partitioning* is a term used in chromatography that describes the physical distribution of analyte between two phases (mobile phase and stationary phase). This concept was discussed in Chapter 15.

In the summer of 1976, Viking 1 landed on Mars carrying a gas chromatograph (GC). A few months later, a second GC arrived via Viking 2. The design of these instruments, as well as the data and knowledge that were obtained, remain very relevant today. In fact, we can understand quite a bit about the nature of scientific inquiry through the discourse and controversies of the investigation. The discussion tells an interesting story about the importance of instrumental sensitivity and instrument design.

The two gas chromatography-mass spectrometer (GC-MS) systems probed the Martian surface using a technique in which the sample is heated in the absence of oxygen (pyrolysis). Organic materials decompose into simpler volatile compounds that are then swept into the GC via a stream of hydrogen gas. The sample ovens could be heated quickly, up to 500°C within 8 seconds. The instrument design, absolutely state of the art for its time, incorporated many new and important features, including a palladium-based separator to remove excess hydrogen gas, and the system was powered by a small nuclear generator.*

Scientists have expressed concern that conclusions drawn from the Viking data have been overstated—namely, that the results of soil analysis there do not show the presence of organic material. Questions about the use of pyrolysis for sampling led, in 2000–2001, to a reevaluation of the limits of detection. Also, other experimental parameters, including the temperature and the iron-rich content of soil, further called into question whether the instruments could have conceivably found evidence of organics.

Experts assert that the instruments performed in exactly the way they were designed to perform and represent a milestone of achievement in scientific instrumentation. The goals of the Viking mission did not, in fact, include a search for biological evidence, and the instruments were not designed for that purpose. Thus, extrapolation is inappropriate. Yet, the public is interested in the possibility of life

PROFILE
Gas Chromatography on Mars

NASA's space lab Curiosity, a Mars rover. Curiosity carries a GC-MS as part of its instrument panel.

on Mars, and the media are eager to make the connection. Scientific debate and peer review, although an accepted part of scientific inquiry, are often seen by those outside the scientific community as flaws in the scientific process. As public confidence waxes and wanes, important project funding is influenced.

However, critical evaluation of the 1976 data helps us make important decisions about new projects. For instance, NASA's space lab Curiosity, a Mars rover, carries a GC instrument as part of its *sample analysis at Mars* (SAM) unit.

The rover's GC includes extended capacity for detection of a broad range of compounds, and higher temperature (up to 1000°C). Pyrolytic extractors were included in the design, and upgrades include a more robust detection system in search of organic substances and isotopic measurements of carbonaceous materials.

SAM's ability to extract materials from and under the surface has also been diversified and includes the capacity for wet chemistry derivatization. Dr. Paul Mahaffy, the principal investigator of the SAM project, writes, "This extraction method has been utilized in our laboratory to analyze organics from samples of the highly oxidized, organic-poor soil from the driest part of the Atacama Desert. This wet extraction method avoids the potential issues associated with the transformation by oxidation of these organic compounds during pyrolysis."[†] The Curiosity landed on Mars on August 6, 2012 and is roving about the Gale crater, powered by the same type of nuclear generator as the 1976 Viking landers. For more information, consult the NASA website: http://ssed.gsfc.nasa.gov/sam/curiosity.html. ■

---

[*] *Mukhopadhyay, R. Analytical Chemistry, Oct 1, **2007**, 7249–7256.*

[†] *Paul Mahaffy, Geochemical News, October 2009, vol. 141.*

(although not all), the carrier gas does not interact with the analyte molecules; that is, it does not act as a solvent but merely serves to move the analytes along the column. Partitioning of the components in the sample tends to be based on physical adsorption of the analytes on the solid stationary phase (in GSC) or distribution between the gas and the liquid phases (in GLC). Both adsorption and partitioning distributions were discussed in more detail in Chapter 15, and you may want to review those concepts before proceeding. Stationary phase selection and method development in GC will be discussed in Section 16.3. Despite widespread adoption of GC analysis, the literature reveals that there remains much to discover about the fundamental mechanisms of separation, how various elution conditions affect selectivity, how various stationary phases behave and interact with eluents under chromatographic conditions, and so on. In general, GC results for various classes of compounds are often compiled empirically, with ongoing research efforts focused on generating data sets that may be useful for making predictions of retention times. However, for many eluents, there exists no way to predict retention time deductively.

Research is ongoing in thermodynamic modeling and "quantitative structure property relationship" approaches, but the practicing GC analyst usually depends on database analysis to predict retention times. There are several commercial vendors of GC databases.

COMPARE AND CONTRAST

Gas Chromatography versus Liquid Chromatography

Take a moment now to make a comparison of the block diagram of the GC instrument from Figure 16.1, comparing it component by component, to the block diagram of the high-pressure LC system from Chapter 15 (Figure 15.26). As you read through this chapter, identify the components that are the same or similar in the two systems and note the components that are not the same in the two systems. As we move on in this chapter, annotate your notes with further descriptions. Be prepared to answer the following questions:

1. How does the term *gradient elution* in LC differ from its use in GC? What instrumental components are the most relevant to this discussion?

2. How are the types of detectors used in LC and GC limited by the phase (liquid or gas) in the two systems?

3. How do column technologies differ in GC as compared to LC?

16.2 Basic Gas Chromatography Instrument Design

As mentioned in the introduction, the GC instrument is somewhat simpler than the LC instruments studied in the last chapter. Because the GC has no liquids to pump, the gas flow is simply controlled by a cylinder valve (also known as a regulator) and a flow monitor. If you will follow along in Figure 16.1, you will see that the sample is introduced into the GC via a gas tight syringe. The gas flow from the cylinder carries the sample into the column and then on to a detector. Each of these will be considered in more detail in later sections of this chapter. Pay particular attention to the part of the figure that is labeled as "oven." In GC, the separation is controlled in part by temperature. Thus, the column[7] is ensconced inside a temperature-controlled oven that is carefully and precisely regulated. As the sample components elute at the end of the column, they pass on to the detector where a signal is measured. A computer controls the entire operation and records the signals from the detector.

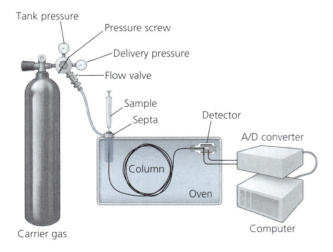

Figure 16.1 Basic schematic of a GC instrument.

---

[7] The term *column* is used synonymously with *stationary phase* in a colloquial sense. Column chromatography, distinct from other physical forms of chromatography such as thin layer chromatography, utilizes a stationary phase that is held inside a column. The mobile phase flows through the column and the analytes are separated in the column. As we will see in later sections of this chapter, the column technology of GC is very different from LC. Review the basics of column chromatography in Chapter 15.

16.3   Method Development

GC methods are as diverse as the applications to which they are applied. However, they all have a common set of parameter choices that are part of the experimental design. These include column type, detector choice, carrier gas flow rate, and temperature program.

In order to understand the GC experiment, we are first going to look at a case study. By close reading of the information provided, you will be able to understand the entire GC analysis. We will then discuss each of the instrumental components and experimental options. Figure 16.2 is a gas chromatogram that shows the separation of fatty acid methyl esters (also known as FAMEs) from peanut butter. The author of this paper was investigating the levels of *trans*-fats found in peanut butter preparations that contain hydrogenated vegetable oils. Let us take a look at the method used to acquire these data, stopping along the way to discuss each bit of information provided in the paper's experimental description (see "Peanut Butter: A Case Study").

What can we learn about gas chromatography from this description?

Peanut Butter: A Case Study

Fatty acid methyl ester analysis was performed on a Hewlett-Packard 5890 gas chromatograph (Hewlett-Packard Co., Palo Alto, CA) equipped with a flame ionization detector and a fused silica capillary (100 m × 0.25 mm) column with biscyanopropyl polysiloxane as the stationary phase (SP-2560, Supleco, Inc. Bellefonte, PA). The carrier gas was nitrogen at 12 cm/s, and the column was programmed to heat from 145° to 220° at 4°C/min with an initial hold time of 4 min. After [a] 35-min hold at 220°C, the temperature was increased to 240°C at 4°C/min and held for 10 min.

The injector and detector were set at 250°C. The injector was split with a split flow of 24.8 ml/min. Data were collected using a Hewlett-Packard model 3396 series II integrator. Fatty acid percentages were calculated using response factors for the methyl esters identified. Identifications were made on the basis of retention time compared to known reference standards purchased from Ne-Check Prep, Inc. (Elysian, MN). Qualitative accuracy was verified daily by analysis of previously characterized soybean oil and spike recoveries.

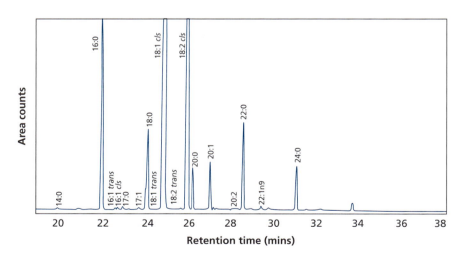

Figure 16.2 Representative GC of fatty acid esters from peanut butter.

From: Sanders, T. H. Non-detectable Levels of trans-Fatty Acids in Peanut Butter
J. Agric. Food Chem., **2001**, *49, 2349–2351.*

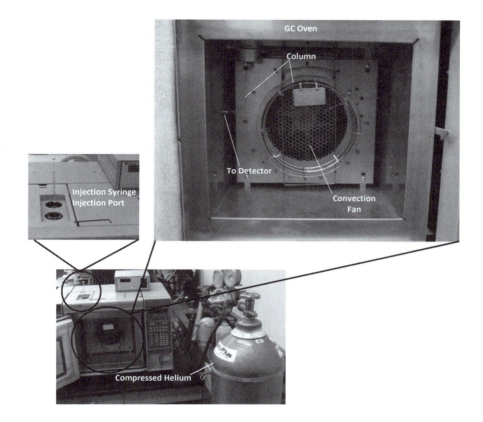

Figure 16.3 Hewlett Packard 5890 GC with a flame ionization detector and a fused silica capillary (100 m × 0.25 mm) column.

The Hewlett-Packard 5890[8] is an older model, but it is a workhorse instrument, still very commonly found in working laboratories today. The lower left photograph in Figure 16.3 shows the GC alongside the compressed helium tank that serves as the carrier gas. What is not seen in this image is the computer that is used to run the GC. As the column leaves the oven, the eluent is passed onto the detector. The detector on this particular unit is a *flame ionization detector* and sits adjacent to the oven. As we open the front oven door of the 5890 (same figure, top right), we are able to see the column inside the oven. This photograph shows the column connected to both the inlet port (connected to the injector and gas inlet) and the outlet port that leads to the detector. Notice the thick walls of the oven and insulation on the door, which are important for temperature control. You can also see, at the back of the oven, the convection fan that circulates the air throughout the oven to maintain an even distribution of heat all around.

As described in the method, the "fused silica capillary (100 m × 0.25 mm) column with biscyanopropyl polysiloxane as the stationary phase" describes the type and dimensions of the column and stationary phase used in the peanut butter analysis experiment. The analyst must choose a column that is likely to interact with the analyte in the desired manner. Usually one wants a column that will promote partitioning between the column's surface and the carrier gas. Note the length of the column: 100 meters! You will see how that length is accommodated in the bench-top GC by looking into the oven in Figure 16.3. GC columns (unlike LC columns) are coiled to minimize space requirements. Their flexibility is aided by the fact they are also very narrow; in the column described, the inside diameter of the capillary column is just 0.25 millimeters. The

---

[8] Hewlett-Packard (HP) is also known as HP-Agilent. Agilent was formed as a spin-off company from HP in 1999.

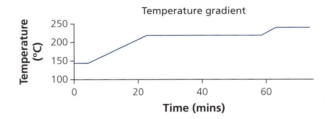

Figure 16.4 A plot of the temperature gradient used in the analysis of fatty acids in peanut butter.

stationary phase[9] material, biscyanopropyl polysiloxane, is a highly polar nonbonded polymer that was designed for the separation of FAMEs, particularly for isomer applications in terms of separating *cis/trans* isomers.

In addition, we are told that the injection was "split." The idea of a "split" injection is a result of the fact that the Hewlett-Packard 5890 can be fitted with a capillary column, as described here, or a column with a larger sample capacity (a packed column). By splitting the sample, the analyst is controlling the amount of sample that goes into the column to avoid overloading the column. With larger packed columns, the injected volume is in the microliter range. However, volume of the column decreases with r^2 (r = the radius of the column), so a typical capillary column requires injection volumes in the nanoliter range. We do not have nanoliter syringes, and so we rely on splitting off a portion of the injected volume. Columns and injectors will be discussed more thoroughly in Section 16.4.

In terms of the method profile for the peanut butter analysis, the flow rate of the nitrogen carrier gas (12 cm/sec) is given along with the temperature gradient profile. The flow rate is one of the experimental variables controlled by the analyst and represents the average linear distance a molecule of the carrier gas moves as a function of time. Another experimental variable controlled by the analyst is the temperature. Figure 16.4 shows the experimental profile for the temperature gradient for the peanut butter analysis, as described in the paper. Review the paper's description carefully, and take note in this figure how the temperature gradient profile is constructed.

The temperature gradient profile also provides us information about the fatty acids under analysis. Note that the fatty acids in peanut butter are undergoing analysis as FAMEs; that is, they have been derivatized into methyl esters for analysis. Because fatty acids have relatively high boiling points and would undergo significant degradation at these temperatures, GC analysis is not well suited for the direct analysis and detection of fatty acids. However, a simple derivatization generates the more volatile FAMEs, which are well suited for GC analysis.

Problem 16.1: In the development of a GC method, list three experimental variables under the control of the analyst.

In the case study about peanut butter, the last part of the description ("Data were collected using a Hewlett-Packard model 3396 series II integrator…") tells us about how the instrument detector (FID) and computer software processed the data. As you can tell from the chromatogram (Figure 16.2), the signal is given on the *y* axis in terms of "area counts." This is a relative signal that can be used quantitatively. In general, larger peaks indicate higher concentrations, although detector response may be influenced by molecular weight. In this experiment, standards were used to correlate area counts to actual concentrations of the respective fatty acids. Retention times (on the *x* axis, in minutes) for each of the FAMEs were matched with known reference standards, thus allowing positive identification of each fatty acid in the peanut butter.

---

[9] *Stationary phase* is the chromatographic term for the portion of the separation technique (in contrast to the mobile phase) that is not moving during the analysis.

Before we move on to a more generalized description of the modes of separation used in GC analysis, take a few minutes to review the three questions in "Compare and Contrast—Gas Chromatography versus Liquid Chromatography" at the end of Section 16.2. Then update your notes in terms of what you notice in this particular GC analysis that is the same as, and different from, the LC analysis we studied in the Chapter 15.

16.4 Modes of Separation

Isothermal versus Temperature Gradients

As you may have surmised by now, one important way in which GC differs from LC is that the mobile phase *composition* does not change during the analysis. Recall that LC methods using gradient elution have changing solvent composition. For example, the polarity of the mobile phase solvents (during reverse phase or normal phase experiments) is changing over time or the ionic strength of the mobile phase solvents (during ion exchange experiments) is changing; this is *not* the case with GC methods. In a GC analysis, the carrier gas remains the same throughout the analysis. Rather than manipulate the thermodynamics of the separation, the GC techniques manipulate the kinetics of the separation by using temperature gradients.[10]

In an *isothermal* separation, the temperature remains constant from start to finish. This is the simplest scheme and requires only that the temperature remain constant over time. As we saw in the case study above the temperature of the experiment can be changed as a function of time. This is termed a *temperature gradient* and is more common than isothermal methods.

In order to generate a temperature gradient, the column temperature needs to change homogeneously; that is, all parts of the column should be (ideally) at the same temperature. Thus, temperature changes need to be quick and uniform. The convection oven is the most common design and is very effective. The oven has a thermostat control that is interfaced to the computer. The computer controls the temperature changes, which are programmed into the method. Temperature ramps (up or down) can be linear or nonlinear over time. In the case study discussed earlier, there are two ramps, both are linear, and both ramps have the same slope (4°C/min). This does not have to be the case. Recall that decomposition is a consideration that every gas chromatographer needs to keep in mind; therefore, you need to be aware of the temperature limits of the column as well as the decomposition properties of your analytes.

> **Note**
> Isothermal separation in GC is analogous to an isocratic separation in LC analysis.

Problem 16.2:

(a) Use your knowledge of partition kinetics to speculate on how an analyte might behave if the temperature in a GC analysis was set too low.

(b) How might you exploit a "too low" temperature to your advantage in a GC analysis?

The Column

In terms of mode of separation, GC analysis is usually cataloged by the configuration of the column. Columns are described as being either *packed column* or *open tubular column*. The packed column is rarely seen in an analytical application and is mostly used for *preparative GC*. Recall from Chapter 15 that preparative chromatography is used to purify bulk quantities of a chemical species.

---

[10] You may want to review the theoretical principles behind chromatographic separations in terms of kinetics and thermodynamic contributions, which were outlined in Chapter 15.

PROFILE

The NIST 14 Gas Chromatography Library with Search Software

The National Institute of Standards and Technology, or NIST, periodically updates their gas chromatography (GC) database. The 2014 version contained a library of 82,337 chemical compounds analyzed by GC, including 385,872 GC methods and retention index citations from the literature. Users are able to search based on the name, formula, or Chemical Abstract Service number. Search results can be sorted by degree of molecular similarity, column type (packed or capillary), stationary phase type (polar or nonpolar), or the GC temperature conditions (isothermal or temperature programmed). ■

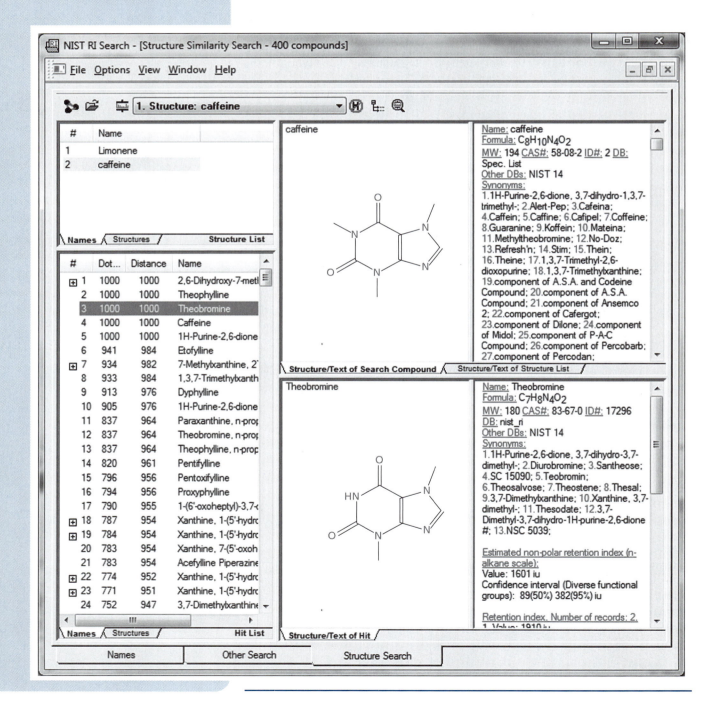

A packed column (see Figure 16.5, part D) is similar, conceptually, to an LC column. It contains tiny particles that are packed tightly into a glass, silica, Teflon, aluminum, or stainless steel column, or porous cross-linked polymers that take up the internal volume of the column. The solid particles or polymer may act as the stationary phase (*GSC*), or they may be coated with a high boiling liquid, in which case the liquid acts as the stationary phase (*GLC*). A typical internal diameter for a packed GC column is very narrow, on the order of 2 to 4 mm. Applications using packed columns are less prevalent in the literature than applications using open tubular columns, and we will therefore spend the remainder of our discussion on open tubular columns.

All of the configurations depicted in Figure 16.5 are constructed on *fused silica open tubular (FSOT)* columns. These columns have smaller internal diameters than glass or metal columns, and they are coated on the outside with a protective polymer coating that makes them both flexible and easy to handle. A typical capillary column will have an internal diameter of less than 1 mm, and this is often given in units of μm (1 mm = 1,000 μm). Because of the narrow diameter of the capillary, the capillary walls can act as the stationary phase (a type of GSC). However, many applications choose to use capillary columns in which the walls inside the capillary are coated with either a thin film of liquid (wall-coated open tubular; *WCOT*), a thin layer of a porous solid stationary phase (porous layer open tubular; *PLOT*), or a thin film of liquid-coated solid packing material (support-coated open tubular; *SCOT*). Figure 16.5 shows a cross-section of a WCOT, PLOT, and a SCOT capillary column as well as a packed capillary column. Table 16.1 provides a summary of the most common terms used to describe FSOT capillary column technology. These thin films are on the order of tenths of microns in thickness. Typically, the inside of the capillary will be coated with a thin layer of stationary phase in order to increase the partitioning selectivity for a selected class of analyte molecules. The inner walls may also be coated with a thin film of packing material (PLOT) that is not liquid-coated.

The analyte will interact with the stationary phase through adsorption/desorption partitioning, in a manner conceptually similar to what we saw in LC, as analytes partitioned between the moving and the stationary phases. Analytes that spend more time in or on the immobilized stationary phase will be retained longer than analytes that spend more time in the mobile phase and are being swept along in the carrier gas. The specific column that you choose for your analysis is going to be based on the application. In GC, we do not have the same separation mode categories as are found in LC (reversed phase, normal phase, ion exchange, affinity, size exclusion, and so on). Instead, the stationary phase material is selected by its known suitability for a particular application and columns are cataloged by analyte type. The peanut butter analysis described in the case study is a good example of this approach. The application called

TABLE 16.1: Capillary Column Terminology

FSOT	Fused silica open tubular	GSC
WCOT	Wall-coated open tubular	GLC
PLOT	Porous layer open tubular	GSC
SCOT	Support-coated open tubular	GLC

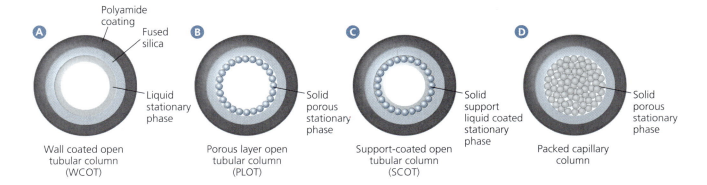

Figure 16.5 Cross section of (A) a WCOT capillary column; (B) a PLOT capillary column; (C) a SCOT capillary column; and (D) a packed capillary column. Note! Not drawn to scale.

PROFILE
Erika Cremer

In 1921, a recent high school graduate named Erika Cremer (**1900–1996**) decided to attend Lyzeum Boretius in Berlin, Germany, as a student of chemistry. At that time in history, there were very few women at the university. While at the university, she had the opportunity to attend lectures from famous scientists such as Max Planck, Albert Einstein, Fritz Haber, Walther Nernst, Max von Laue, and her thesis advisor Max Bodenstein. She eventually went on to develop kinetic models for the detonation of hydrogen-containing gases; contribute to the understanding of the energetics of nuclear fission; and most importantly, contribute to the fundamental advancement of the understanding of gas phase diffusion and adsorption properties of gases. In 1938, Cremer received a postdoctoral research position with limited lecture privileges from the University of Berlin. The university dean told her that this would be as high as she could go because women could not hold a formal professorship. The beginning of World War II drew many men away from the University of Berlin, and Cremer's professional opportunities suddenly improved. She took a professorship position at the University of Innsbruck in Austria and while there began research on the hydrogenation of acetylene. She needed a way to quantify the components of her gas phase reactions and began work on the selective adsorption/desorption of gas phase analytes. She visualized the separation process as a collection of boats that are carried by a river of gas, with the boats spending a characteristically specific amount of time ashore. She developed mathematical formulations between adsorption enthalpies and retention times, and, in 1944 she submitted a paper describing the birth of gas chromatography. Sadly, the printing house was bombed and the presses destroyed. Her paper would not actually be in print for another 30 years. ■

for a column to separate *cis* and *trans* isomers of FAMEs, which is exactly the advertised function of the biscyanopropyl polysiloxane column.

Although it is conceptually simple to think of GC as analogous to LC, there is a fundamental difference in the underlying principles of GC and LC separations that should be appreciated. In LC, we manipulated the mobile phase strength to effect the partitioning of the analyte between the mobile and stationary phases. In GC, we manipulate the temperature to the same effect. Because we can change the temperature of the GC environment during analysis, there can be phase changes that occur in GC but not in LC.

As we saw in the peanut butter case study analysis in section 16.3, the injector temperature in a GC experiment is often set at a temperature much hotter than the initial temperature of the column. This is done to ensure that all of the analytes in the sample matrix are volatilized and swept onto the column. For instance, the injector temperature in our case study was 250°C. However, the initial column temperature was only 145°C. Therefore, components of a sample that are less volatile may undergo a phase change, undergoing condensation at the top of the column. While the volatile components of the sample continue on down the column and partition between the mobile and stationary phase, the less volatile components will migrate through the column much more slowly, spending much of their time adsorbed to the column walls until such time as the temperature of the experiment is sufficiently hot to volatilize them.

When programing a temperature gradient for a GC analysis, it is useful to pay close attention to the boiling points of your analytes. From our earlier discussion on temperature gradients, you should be thinking that a careful consideration of the boiling points of the analytes is needed as a part of the overall method development in GC. In our case study, Sanders ended the method with "After [a] 35-min hold at 220°C, the temperature was increased to 240°C at 4°C/min and held for 10 min." This last step in a GC method is called a thermal clean. After each injection, the column is heated to near its maximum to volatize any "junk" left on the column from the previous injection.

Because open tubular columns are not packed, the resistance to gas flow is much lower than what is seen in packed columns. The carrier gas and the analyte can pass through the column at a much greater flow rate, thus minimizing the time the analyte spends on the column, which leads to less longitudinal diffusion and better resolution.[11] As a result, in GC, it is possible to use much longer columns with a capillary system compared to a packed system. It is common to have capillary columns as long as 100 meters compared to packed columns with a typical length of only 2 to 3 meters (see Figure 16.6). Because of their greater length, capillary columns have many more theoretical plates[12] (N is often on the order of ~10^5) as compared to packed column, which have N values typically ranging from approximately 10^2 to 10^3. In other words, with these columns we can achieve superior separation in a relatively short time, which accounts for the popularity of the column capillary systems.

The fused silica (FSOT) columns predominate the literature in comparison to other options. The FSOT columns have both practical and analytical advantages: smaller plate height, which translates into more plates per given column length; greater efficiency; speed of analysis; flexible design; and inertness. Their major disadvantage is that they have very low capacity, with appropriate sample sizes in the nanogram range. In addition, GLC-type separations are far more common than are GSC separations. The primary reason for this is that GSC separations tend to suffer from loss of analyte due to semipermanent surface adsorption and tailing, caused by adsorption effects, and also "memory," or analytes that were adsorbed in a previous experiment eluting in a subsequent experiment. GLC, as a partitioning separation, gives more acceptable results over a longer time.

In the ideal cases, the stationary phases used in GC columns have high boiling points (100°C or higher than the highest operating temperature for the column), thermal stability (they do not break down under high temperatures), and they are chemically inert. The ways in which the various stationary phases interact (cause partitioning behavior) with the analytes affect the *retention factor* (k_r) and the *selectivity factor* (a), which then determine the classes of compounds for which the column is best suited.

Most manufacturers produce a column selection guide that provides ample information about the stationary phase, type of bonding or cross-linking used, and applications. Restek, for example, has an online selection guide that starts by grouping columns into five general application-based categories: Environmental; Foods, Flavors, and Fragrances; Clinical and Forensics; Pharmaceutical; and Petroleum and Petrochemical. Columns are referenced, also, to specific analytical techniques, such as Environmental Protection Agency (EPA) protocols and American Society for Testing and Materials methods. For example, Agilent's VF-1701ms, 30 m × 0.25 mm, 0.25 μm (part number CP9151) is listed as a recommended column for EPA method 515.4 for the "determination of chlorinated acids in drinking water by liquid-liquid extraction, derivatization, and GS with electron capture detection." Likewise, Sigma-Aldrich (the parent company of Supelco) and Restek also have online selection guides for their GC columns. A list of GC column suppliers is found in Table 16.2. You would likely find browsing their websites informative and useful.

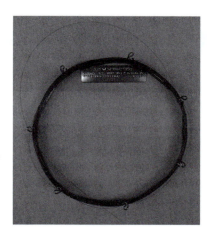

Figure 16.6 GC columns can be as much as 100 m long. They are typically wound about a metal frame and the frame is hung on a bracket in the GC oven.

---

[11] *Longitudinal diffusion* describes loss of resolution due to band spreading along the length of the column. Longitudinal diffusion was first described in Chapter 15 in the context of LC; more information in relation to GC can be found in the references at the end of this chapter.

[12] *Theoretical plates* are a measure of the column's separating ability. The *height equivalent to a theoretical plate* is an important factor in the determination of the resolving power of the column. See Chapter 15 for a complete introduction to this topic and refer to the references at the end of this chapter for more information in relation to GC.

TABLE 16.2: **Gas Chromatography Column Manufacturers**

COMPANY NAME	WEBSITE	TECHNICAL RESOURCE WEBSITE
Restek Corporation	www.restek.com	www.restek.com/Technical-Resources
PerkinElmer	www.perkinelmer.com	Searchable selection guide on website
Sigma-Aldrich	www.sigmaaldrich.com/analytical-chromatography/gas-chromatography.html	Column section guides and application notes
Crawford Scientific (J & W Scientific) (Agilent)	www.crawfordscientific.com/	Searchable selection guide on website; can also download a selection guide in PDF format
SGE Analytical Science	www.sge.com/	Online selection guide
Interchim	www.interchim.eu/pp/106/gas-chromatography.html	Website shows chemical structures of bonded phases
Phenomenex	www.phenomenex.com/gc-column	GC "learning center" tab; includes selection guide, video clips for installation, column prep guide, and derivatization guide
Cronus	www.labhut.com/	GC application library and learning center
Thermo Scientific	www.thermoscientific.com/	Several product selection guides organized by analyte polarity
Grace	grace.com	Several selection and application guides
Vici	www.vici.com	Support link leads to numerous categories

When there is no existing reference, consider the following when selecting a stationary phase. The two primary things that influence the retention of an analyte in GC are the boiling point (volatility) and the affinity the analyte may have with the stationary phase. If you wish to separate polar compounds on a nonpolar stationary phase, you will not benefit from any partitioning mechanisms with the stationary phase, and your separation will be governed only by the boiling points of the analytes. Your analytes will simply elute in order of increasing boiling point. A temperature gradient will be very beneficial in this scenario. Likewise, if you wish to separate nonpolar compounds on a column with a polar stationary phase, your separation will be governed by the boiling points of your analytes. The choice of the stationary phase becomes critically important when you wish to separate analytes that have similar boiling points. For instance, if we have two analytes of similar boiling points but one is more polar than the other, we can use a polar stationary phase to add selectivity to the separation. In general, if you wish to separate analytes that have similar boiling points, select a polar stationary phase for polar analytes or a nonpolar stationary phase for nonpolar analytes.

Some things to remember when selecting a stationary phase are:

- If the stationary phase is nonpolar, then the primary mode of interaction between the analyte molecules and the stationary phase will be van der Waal forces (i.e., dispersive forces). Recall from general chemistry that van der Waal forces increase with molecular weight. If the only mode of interaction with the stationary phase involves dispersive forces, then analytes of similar boiling points will generally elute according to molecular weight, with heavier molecules eluting later than lighter ones.

- If your analyte has aromatic rings, you may want to select a stationary phase that also has aromatic rings to take advantage of π-stacking interactions with the stationary phase.

- If your analyte molecules contain polar functional groups or heteroatoms, then you can take advantage of the many polar stationary phases available. In this case, partitioning will be governed by polarity.

Example 16.2

How would you choose a GC stationary phase for a particular separation?

The chart, which lists GC columns available from Sigma-Aldrich, ranks the columns by the polarity of the stationary phase. Included in the column's name is the maximum operating temperature for that stationary phase.

We have also included a similar list of molecules to those we examined in Example 16.1. However, this time we have also included the molecules' Snyder polarity index numbers. Notice that 2-butanone, cyclohexane, and acetonitrile all have similar boiling points. Fortunately, for a GC separation, their polarities are different. Let us think about how our choice of column will affect the elution order for a GC separation. To a first approximation, we should expect elution order to follow boiling points, especially if we are running a temperature gradient that starts below 55°C. However, at temperatures above 81.3°C, the partitioning

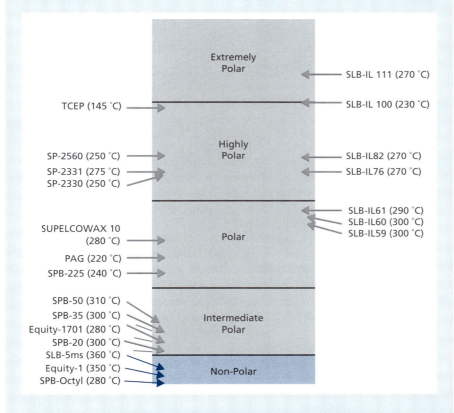

COMPOUND*	POLARITY†	BP (°C)
MTBE	2.9	55.20
THF	4.2	66
2-Butanone	4.7	79.64
Cyclohexane	0	80.74
Acetonitrile	6.2	81.3

BP = boiling point; MTBE = methyl *tert*-butyl ether; THF = tetrahydrofuran
* Most GC column vendors have similar data on their websites. See Table 16.2 for a list of vendors.
†Snyder polarity index: see Walker, N.; Stuart, D. *Acta Crystallogr.* 1983, *A39*, 158.

of the analyte between the gas phase and the stationary phase will dominate the separation. If our stationary phase is nonpolar, we should expect the most polar of those three compounds to elute first and the least polar last (acetonitrile → 2-butanone → cyclohexane). Likewise, if our stationary phase is highly polar, we should expect to see the least polar of those three compounds to elute first and the most polar last (cyclohexane → 2-butanone → acetonitrile). In practice, predicting the elution order of a complex mixture is complicated by the fact that you have two variables affecting the elution order. If the temperature of the column is cooler than the boiling point, those molecules will be well retained and will move fairly slowly through the system. If your analytes have significantly different boiling points, a temperature gradient can be used to great effect to control the elution order. A reasonable method for separating all five of these compounds would be to choose a column in the intermediate polarity range (i.e., SPB-20) and run a temperature gradient that started close to 55°C and ramped above 66°C and stayed there until the THF had eluted. During the initial period of the run, we would expect MTBE to elute first because it is least polar and has the lowest boiling point, followed by THF. Then once THF has eluted, quickly ramp the temperature above 81°C (100°C would be a reasonable choice). We would then expect the elution to follow polarity.

Example 16.3

Describe how you would design a method for the separation of aniline, benzene, naphthalene, toluene, o-xylene, and p-xylene.

STRATEGY –

With the exception of aniline (aminobenzene), all of these compounds are nonpolar aromatics. So you would want to pick a column that is relatively nonpolar, but a slight amount of polarity might help aniline partition between the phases. Ideally, you also want a column with a maximum operating temperature above the boiling point of your least volatile analyte. You would then want to program a temperature gradient that started slightly above the boiling point of your most volatile analyte (T_i) and ramp the temperature gradient above the boiling point of your least volatile analyte (T_f). Keep your temperature at T_f until all of your analyte species have eluted and then ramp the temperature just below the maximum operating temperature of the column in order to thermally clean the column before the next experiment. This last step is important so as to avoid cross-contamination from previous experiments.

SOLUTION –

The boiling points of your analytes are benzene (80.1°C), toluene (111°C), p-xylene (139°C), o-xylene (144°C), aniline (184°C), and naphthalene (218°C). Any of the intermediately polar columns from Example 16.2 would be appropriate. Let us choose SP-35. We would want to start our temperature gradient above the boiling point of benzene, so a T_i of 95°C would be appropriate. We then would want to ramp the temperature above the boiling point of naphthalene, so a T_f of 250°C would be appropriate. A ramp rate of 2 to 4°C/min would be appropriate. After the naphthalene peak has eluted, we would want to ramp the temperature again but stay below the maximum temperature of the column. A clean temperature of 280°C would be appropriate, and we would want to hold that temperature until we were confident that all volatile species had eluted from the column.

When selecting a column, in addition to the type of column, you will also have to make some decisions about the column's size (internal diameter and length), as well as perhaps film thickness of the stationary phase if there are options available. The smaller the inner diameter of the column, the greater the efficiency and better the resolution, but the capacity (how much sample you can inject) is generally lower. With a larger diameter column and a thicker film, the capacity increases. In any case, overloading the column will always lead to poor resolution. Also, take into consideration the detector. Some detectors have capacity limits as well, and it is undesirable for your column capacity to exceed your detector capacity. The manufacturer's selection guides will provide advice and recommendations per the application.

Problem 16.3: Using Examples 16.1, 16.2, and 16.3 as guides, propose a method for the separation of caffeine, catechin, and phenol. Justify your choice of columns and your temperature gradient.

Problem 16.4: Go online and use one of the GC column supplier's "How to Select a Column" guides and select a column for the analysis of gasoline. (See Table 16.2 for a list of GC column suppliers.) Explain why you selected that particular column, and justify your selection based on chromatographic principles.

Problem 16.5: Go online and use one of the GC column supplier's "How to Select a Column" guides and select a column for the analysis of metabolic steroids. Explain why you selected that particular column, and justify your selection based on chromatographic principles.

16.5 Carrier Gas and Injector

Now that you have a good idea of what GC methods are all about, let us go back to Figure 16.1 and take a more in-depth look at each of the instrumental components. In each case, we will provide a general description and some examples of the most commonly used options.

Carrier Gases

As you can imagine, using gases instead of liquids for the mobile phase requires different instrumental hardware and techniques. Unlike LC, both the pressure and temperature of the carrier gas play a large role in analyte retention in GC. As a result, in many cases data in GC are reported as *retention volumes, V_R,* (rather than retention times, t_r, as in LC). It is possible to characterize compounds based on the value of V_R because V_R depends only on the partitioning coefficient (K), the density of the liquid (in the case of GLC), and the temperature; however, the V_R values tend to vary experimentally and are not used definitively to identify compounds. A further discussion of V_R is continued in Section 16.8.

The flow rate of the carrier gas is controlled roughly at the cylinder (usually via a two-stage gas regulator) and more finely at the instrument using electronic pneumatic flow controllers. Pressures, typically in the range of 10 to 50 psi, are much lower than the LC counterparts discussed in the last chapter. However, flow rates in GC are much higher than in LC, on the order of 1 to 500 mL/min. Flow rates are typically measured and reported in units of cm/sec (linear flow rate) or mL/min (volumetric flow rate).

Ideally, the mobile phase gas selected for a GC experiment is inert within the parameters of the experiment. Thus, nitrogen is a reasonable option because of its low cost; however, it has a relatively narrow range of useful flow rates. Other commonly used gases include helium, argon, hydrogen, and carbon dioxide, depending on the application. Carrier gases are purchased from a vendor in a pressurized cylinder at a suitably high quality and are connected to the instrument via a cartridge that removes water (and/or other impurities) so that the carrier gas is dry as it enters the column.

Problem 16.6:   As noted above, nitrogen gas exhibits a fairly low range of useful flow rates as a carrier. Hydrogen exhibits the widest range of useful flow rates among common carrier gases. Postulate on the relative advantages and disadvantages of using H_2 versus N_2 to explain why N_2 is more commonly used than H_2.

Samples are typically injected into the GC with a 10 μL GC syringe. The sample (typically a few microliters in volume) is delivered by needle injection through a rubber or silicone septum into a heated injector port, generally set about 50°C above the highest boiling point of the components in your sample, or 0 to 25°C above the highest temperature in the temperature gradient of your analysis. The heated injector ensures that all of the components in the sample are vaporized and that they all move into the column at the same time. However, care should be taken not to overheat the sample at the injector by setting the port temperature too high, which can add noise due to septum bleed (contaminating peaks from septum breakdown). Ideally, just as in LC, the sample needs to be introduced to the column in as small a time frame as possible, resulting in a very narrow "plug" at the column head. Once injected, band broadening will only decrease resolution. Thus, the first step in achieving good resolution is at the injector. Poorly maintained injectors can cause other types of band broadening and tailing (asymmetry in peak shape); therefore, replacing the septa, port liners, and seals on a regular basis is important. Figure 16.7 is a schematic of a typical GC injector port.

For applications that require very high resolution, or for instruments that are designed for very high resolution purposes, band broadening at the injector is one of the major limitations. Instrument manufacturers have come up with innovations to help minimize extra-column band broadening caused by, for example, "dead volume" or incomplete flow and also due to temperature gradients within the injector port (e.g., preheating the carrier gas).

Most commercial injectors also have the capacity for what is termed *split injection* or *split flow*. In a split-flow injector (see Figure 16.7), the sample and carrier gas are mixed in the injector and the majority of sample is pulled off ("split") and does not go to the column. The split ratio can be controlled, but in general, only about 1% of the sample goes onto the column. The primary use for a split injector is when the sample is likely to overload the column, which is very easy to do with capillary GC because the capacities are low.

As presented in Section 16.4, we see that capillary columns have several key analytical advantages over packed columns. First, they have better resolution (many more theoretical plates per meter). Second, they have less resistance to flow, so they can be much longer than packed columns without creating too much resistance to flow. However, the small diameter of a capillary column (0.05–0.5 mm) means that it is quite easy to overload the column (see Figure 16.8). In order to avoid having to dilute the sample before injection, the injector can be programmed to "throw away" a portion of each injection. Most modern GC instruments have a programmable split injector. If the analyst notices overloaded peaks, he or she would increase the split ratio in the method.

Autosamplers provide the ability to automate sample injection for routine analyses and are commonplace in laboratories where throughput is a significant priority. Figure 16.9 shows a photograph of an autosampler manufactured by CTC Analytics. Samples are loaded into a sample tray, and the instrument is programmed to inject the samples in sequence on command. Autosamplers provide speed and eliminate variability. Some autosamplers also have the capability to do preparatory steps (e.g., derivatization, precolumn heating, and mixing) on command prior to injection, further automating the analysis. These features help streamline high-throughput work and minimize researchers' chemical exposure. Because it is very difficult to reproducibly inject a specific volume using the typical GC syringe, quantitative analysis is possible with standard GC

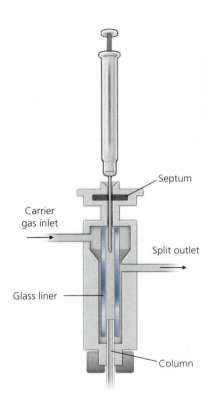

Figure 16.7 Schematic of a GC injection port.

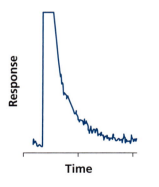

Figure 16.8 Example of an overloaded GSC column. When a GSC columns are overloaded, peaks have a tailing edge that does not return to baseline. Conversely, an overloaded GLC column will typically have peaks with a leading edge.

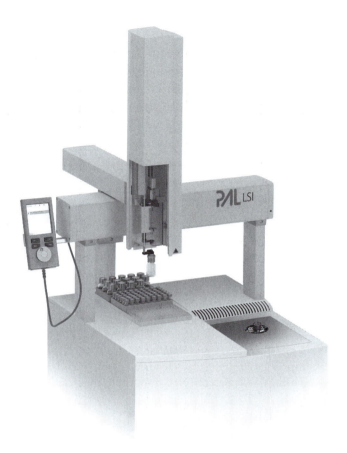

Figure 16.9 This photograph shows an autosampler, the "PAL GC-xt" manufactured by CTC Analytics. This autosampler can be programmed to inject from 0.1 to 500 mL and can hold up to 294 samples.

instruments only by using an internal standard. However, the use of an autosampler greatly diminishes the variability between injections and makes feasible quantitative analysis without internal standards (see footnote 12). On the other hand, for high-quality quantitative work, *best practices* include the addition of an internal standard of known concentration to each sample.

16.6 Detectors

A quick search of the chemical literature will show a great number of methods used to detect analytes in the carrier gas of a GC instrument. However, only a select number of methods are routinely found in commercially available instruments. In this section, we will examine the more common methods. GC detectors can be grouped into five categories:

- *Ionization detectors*: The analyte is converted into ions, and the change in conductivity between two electrodes is related to analyte concentration.

- *Optical detectors*: Optical absorption/emission is related to analyte concentration.

- *Thermal conductivity detectors*: The analyte is detected based on comparative conductivity to the carrier gas.

- *Electrochemical detectors*: The analyte reacts with hydrogen or oxygen and is then solvated to form a species that can be detected by a change in electrical conductivity.

- *Tandem instrument detection*: This involves the use of a second "stand-alone" instrument as the detector (e.g., mass spectrometry).

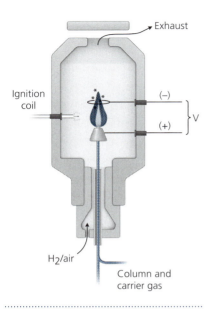

Exhaust

Ignition coil

(−)

(+)

V

H_2/air

Column and carrier gas

Figure 16.10 Schematic of an FID.

Ionizing Detectors

The carrier gases used in GC instruments are typically inert, and at the high temperatures found in GC instruments, these gases have very low electrical conductivity. The basic principle of ionizing detectors is to use the carrier gas as an insulator between two electrodes. As the analyte passes between the two electrodes, the presence of the analyte is detected as a change in the conduction between the two electrodes. The principal difference between the various ionizing detectors is the mechanism used to ionize the analyte. Some of the most prominent ionizing detectors are the *flame ionization detector (FID)*, the *thermionic ionization detector (TID)*, the *electron capture detector (ECD)*, and the *photoionization detector (PID)*. We will discuss each of these individually.

Flame Ionization Detectors   The FID is the most common detector found on GC instruments,[13] and with very few exceptions, it functions as a universal detector.[14] The FID consists of a hydrogen–oxygen flame positioned between two electrodes. Ionization of the analyte is accomplished when the carrier gas passes through the hydrogen flame and the resulting ions generate a current between the two electrodes. For hydrocarbon species, the ionization of the sample is proportional to the number of carbon atoms in the gas stream, and the FID has lower detection limits of pg/sec for hydrocarbon analytes.[15] The response to compounds containing heteroatoms is significantly less. For example, the detection limits for nitrogen oxides and small sulfur-containing molecules is inferior by several orders of magnitude and tend to be in the tens of ng/sec range. The FID is generally considered a poor choice for small nitrogen- and phosphorus-containing compounds. Similarly, fully oxidized gases (such as CO_2 and H_2O) are not easily detected using a FID. Figure 16.10 shows a schematic of an FID.

Thermionic Detectors   The term *thermionic* applies to devices that produce a flow of ions by heating a material beyond its work function[16] in the presence of an electric field.[17] The TID was first introduced in 1971 by Kolb and Bischoff for the detection of nitrogen- and phosphorus-containing compounds.[18] The basic design has not changed much since it was first introduced (see Figure 16.11). The key component of a TID is the use of an electrically heated ceramic cylinder. The ceramic cylinder is doped with an alkali metal (rubidium or cesium).[19] Just prior to elution, the carrier gas is combined with hydrogen gas as a carrier and the carrier gas is heated just below the flash point for the mixture. As the carrier gas elutes from the column, it passes over the alkali metal and ignites as a "cold" plasma. Under thermionic conditions, electrons flow between the anode and cathode based on the work function of the alkali metal. The presence of excited state nitrogen or phosphorus reduces the work function of the alkali metal, resulting in an increase in current—often by as much as a factor of 10^4. One significant drawback of the TID is the fact that the alkali metal bead is consumed over time, and must be replaced regularly.[20]

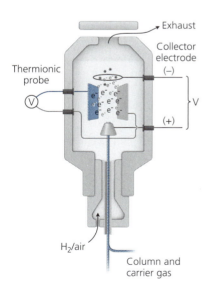

Exhaust

Collector electrode

(−)

Thermionic probe

V

V

(+)

H_2/air

Column and carrier gas

Figure 16.11 Schematic of a TID. Instead of using a flame to ionize the eluent sample, the TID uses a stream of high-energy electrons ejected from a filament. The ionized sample is represented by an asterisk (*). The analytical signal is a current generated at the collector electrode.

[13] Morgan D. J. Construction and Operation of a Simple Flame-Ionization Detector for Gas Chromatography *J. Sci. Instrum.* **1961,** *38,* 501–503.

[14] Eiceman, G. A.; Hill, Jr., H. H.; Gardea-Torresdey, J. Gas Chromatography, *Anal. Chem.* **1998,** *70,* 321–340; Poole, C. F. *The Essence of Chromatography.* Elsevier; Amsterdam, The Netherlands, **2002.**

[15] David, D. J. *Gas Chromatographic Detectors.* Wiley-Interscience: New York, **1974;** Hill, H. H.; McMinn, D. G. *Detectors for Capillary Chromatography.* Wiley: New York, **1992.**

[16] The work (energy) needed to remove an electron from an atom at the surface interface of a solid. The work function is similar in concept to the ionization energy of an atom in a vacuum but work functions apply to solid surfaces.

[17] We used thermionics to produce a stream of electrons when we examine the Coolidge X-ray tube in Chapter 10.

[18] Kolb, B.; Bischoff, J. A New Design of a Thermionic Nitrogen and Phosphorous Detector for GC. *J. Chromatogr. Sci.* **1971,** *12,* 625–629.

[19] David, D. J. *Gas Chromatographic Detectors.* Elsevier: Amsterdam, **1986.**

[20] Patterson, P. L. Recent Advances in Thermionic Ionization Detection for Gas Chromatography *J. Chromatogr. Sci.* **1986,** *24,* 41–52.

Electron Capture Detectors   The ECD was introduced in 1957[21] and remains a very popular detector for pesticides, organometallic compounds, and nitro and nitrile compounds.[22] Its popularity stems from its unsurpassed sensitivity toward halogenated compounds and compounds containing strongly electronegative moieties. As an example, Agilent Technologies markets an ECD detector with 6 fg/mL detection limits.[23] In an ECD detector, the eluent from the column enters a radioisotope chamber where it is combined with a *make-up gas* (usually nitrogen). The ECD utilizes a β-emitting radioisotope to introduce a stream of electrons into the make-up gas where the β-particles collide with the make-up gas and produce a plasma of secondary free electrons and support gas cations and radicals. The secondary electrons are accelerated toward a positively charged collector electrode and collected as a background current. If the carrier gas contains analyte molecules with strong electronegative moieties, the analyte can *capture* the electrons and the result is registered as a decrease in the current collected. The majority of ECD detectors utilize ^{63}Ni as the β-source because it can be produced as a thin foil and it is thermally stable to 400°C. Figure 16.12 shows a schematic of an ECD.

Photoionization Detectors   The PID utilizes ultraviolet (UV) light to ionize organic eluents in the carrier gas, and the resulting current is read in a manner similar to the other ionizing detectors we have examined. The PID has a clear advantage over the other ionizing detectors in that it does not utilize a support or make-up gas for its operation, and as a result, the it finds wide use in portable GC instruments (see Figure 16.13). Also, the PID ionizes only a small percentage of the eluent analyte, so it is generally considered nondestructive. This makes the PID a good choice of detectors for instruments using multiple detection modes. Some disadvantages of the PID arise from the fact that the sensitivity of the detector is concentration dependent and as such, the detector response is sensitive to flow rate.

In addition, the sensitivity of the PID increases with the number of oxidizable carbons in the analyte. One must create separate calibration curves for each analyte. Unlike the FID, the PID is not considered a universal detector. However, this feature of the PID can be exploited to create a very selective analysis. By adjusting the output of the UV lamp, the PID can be made very sensitive to specific analytes while simultaneously being "blind" to others. It has been shown that the PID can be up to 50 times more sensitive than the FID to certain analytes while simultaneously being unresponsive to non-analyte species.[24] A GC instrument with a PID in tandem with an FID can be a very versatile instrument.

Optical Detectors

Optical absorbance detectors are not as widely used in GC systems as they are in LC instruments. In general, most volatile organic compounds have UV absorption maxima less than 200 nm, which are outside the transparent range of common optical materials. As such, absorbance detectors are not considered "universal" detectors in GC. However, several optical detectors provide key advantages for specific types of analysis. We will examine the most common one here.

Flame Photometric Detectors   The flame photometric detector (FPD) is a detector that is well suited for the analysis of phosphorus- and sulfur-containing compounds. The FDP is similar in concept to an atomic emission spectrophotometer

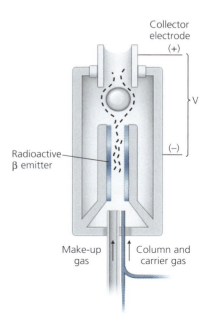

Figure 16.12 Schematic of an ECD. The ionizing source is a radioactive β emitter. Unlike the previous two ionizing detectors, the ECD generates anions that are detected at the collector electrode. ECD detectors are particularly useful for compounds containing large, electronegative elements such as chloride, bromide, and iodide.

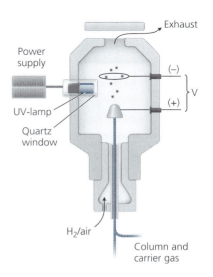

Figure 16.13 Schematic of photoionization detector. In a PID, the ionizing source is an ultraviolet lamp.

[21] Lovelock, J. E. A Sensitive Detector for Gas Chromatography. *J. Chromatogr. A.,* **1958,** *1,* 35–46.

[22] Krejči M.; Dressler M. Selective Detectors in Gas Chromatography. *Chromatographic Reviews,* **1970,** *13,* 1–59.

[23] Smith, D.; Lynam, K. GC/µECD Analysis and Confirmation of PCBs in Fish Tissue with Agilent J&W DB-35ms and DB-XLB GC Columns. Agilent Technologies Application Note, http://www.chem.agilent.com/Library/applications/5990–6236EN.pdf (accessed Aug 14, 2015).

[24] Driscoll, J. N.; Krull, I. S. Recent Advances in Gas Chromatography Instrumentation: An Historical Perspective. *CRC Crit. Rev. Anal. Chem.,* **1986** *17(2),* 193.

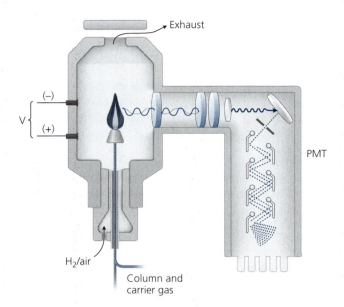

Figure 16.14 Schematic of an FPD.

(see Chapter 9). However, the flame used in FPD is not intended to atomize the eluent. The carrier gas and eluting analytes from the GC are fed into a two-stage hydrogen–oxygen flame assembly where the analyte is decomposed in the first flame and then the fragments are energized into an excited state in the second flame (see Figure 16.14).[25] As the excited state fragments relax, the emitted photon is detected using a scintillator/photomultiplier tube (PMT).[26] Although the FPD is primarily used for the detection of phosphorus- and sulfur-containing compounds, it has been shown that the FPD can be used to detect compounds containing other atoms, such as arsenic, selenium, germanium, tin, antimony, and bromine. Furthermore, many transition metal complexes are suited for emission detection.[27] The detection limit of a FPD for phosphorus-containing compounds tends to fall in the 10^{-13} to 10^{-14} g/sec range and for sulfur-containing compounds in the 10^{-12} to 10^{-13} g/sec range.

Atomic Emission Detectors

We first studied atomic emission spectroscopy (AES) in Chapter 9, where we examined the use of inductively coupled argon plasma or direct current argon plasma torches as both the atomizer and excitation source. It would seem rather straightforward to simply feed the eluent from a GC column into the argon stream of a plasma torch. However, AES is best suited for the study of transition metal compounds, and GC is primarily used to study organic compounds. The argon plasma torches do not provide sufficient excitation energy to yield atomic carbon in an excited state. In the late 1980s, Hewlett-Packard developed a helium plasma torch specifically designed for use as a GC detector.[28,29] The helium plasma is created in a microwave chamber and, as the eluent from the column passes through the chamber, it is simultaneously atomized and the atoms are created in an excited

[25] Patterson, P. L. Recent Advances in Thermionic Ionization Detection for Gas Chromatography *J. Chromatogr. Sci.*, **1986**, *24*, 41–52.

[26] See Chapter 6 for a description of scintillator photomultiplier tubes.

[27] Jing, H. W.; Amirav, A. Pulsed flame photometric detector—a step forward towards universal heteroatom selective detection *J. Chromatogr. A*, **1998**, *805*, 177–215; Aue, W. A.; Sun, X. Y.; Millier, B. Inter-elemental Selectivity, Spectra and Computer-Generated Specificity of Some Main-Group Elements in the Flame Photometric Detector. *J. Chromatogr. A*, **1992**, *606*, 73–86.

[28] Hewlett-Packard split off their analytical instrument division in 1999 as Agilent Technologies.

[29] Handley, A. J.; Adlard, E. R. (editors). *Gas Chromatographic Techniques and Applications*. **2001**, Sheffield Academic Press, Sheffield, UK.

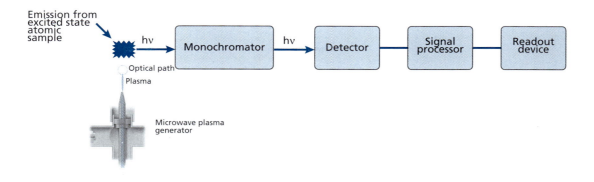

Figure 16.15 Schematic of an AED using a microwave helium plasma generator as the atomization and excitation source.

state. A key advantage of the atomic emission detector (AED) is its ability to simultaneously quantify the analytes eluting from the column while determining the elemental composition of each eluent. A schematic of an AED detector is shown in Figure 16.15. The AED is comprised of an interface for the incoming capillary GC column to the microwave-induced plasma chamber, a cooling system for that chamber, a diffraction grating and associated optics, and a position-adjustable photodiode array interfaced to a computer. The AED is considered a universal element–specific detector with lower detection limits in the sub-pg/sec range. Although the performance specifications of the AED are impressive, it is a very expensive detector to purchase and to operate. Depending on the analyte, the operator must select different "scavenger" gases, and the detector requires helium, which is expensive, as well as other make-up gases.

Thermal Conductivity Detectors

Typically, the carrier gas in GC is either hydrogen or helium. Both of these gases have relatively high thermal conductivity. The *thermal conductivity detector (TCD)* works by measuring the change in thermal conductivity of the column eluent as the analyte passes through the detector. Because most compounds have a lower thermal conductivity than the carrier gas, the TCD is considered a universal detector.

The TCD contains two separate electrically heated filaments,[30] each in a thermally insulated chamber. One chamber is the reference chamber and has a steady flow of carrier gas passing through the chamber. The other chamber is the test chamber and has a steady flow of the column eluent gas passing through that chamber. Under conditions where no analyte is passing through the chambers, there is a steady heat flow from the filament to the chamber body. As the analyte passes through the test chamber, the thermal conductivity of the chamber gas decreases and as a result, the filament becomes hotter. The filament's resistance increases, and as a consequence, the voltage drop across the resister also increases (Ohm's law). The TCD is often constructed as part of a classic Wheatstone bridge circuit (see Chapter 4), which measures the voltage change between the two filaments (see Figure 16.16).

The TCD has several advantages over other detector types. First, it is nondestructive and can be used in tandem detector designs. Second, most organic compounds have similar thermal conductivity, and as such, the TCD gives very similar responses to a wide range of analytes. Generally, this allows the TCD to be used without calibration and relative compositions can be determined by comparing the area under each peak. The main disadvantage of the TCD compared to other techniques is its higher (poorer) detection limits. The typical lower detection limit of the TCD is in the 10^{-6} to 10^{-8} g/sec.

---

[30] Less expensive models (and less sensitive) use a single filament.

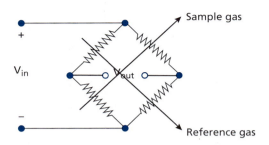

Figure 16.16 Schematic of a TCD. Sample gas passes from the lower left to the upper right while a reference gas passes from the upper left to the lower right. Temperature dependent changes in resistance are measured as a voltage differential across V_{out}.

Electrochemical Detectors

In *electrochemical detectors (ELCDs)*, the eluent is combined with hydrogen and then passed through a nickel-plated, heated tube that catalyzes the conversion of (1) halogen-containing compounds into HX (HF, HCl, HBr, HI) and (2) nitrogen-containing compounds into ammonia (NH_3). A schematic of an ELCD is depicted in Figure 16.17. If the ELCD is being used to detect sulfur-containing compounds, the make-up gas is changed to oxygen and the sulfur compounds are converted into SO_2. In either case, the eluent from the nickel catalyst chamber is passed into a conductivity flow cell containing an appropriate solvent: water (~5% v/v *tert*-butyl alcohol) for ammonia, 1-propanol for HX, and methanol (~3% v/v water) for SO_2. The ELCD is a very selective detector with lower detection limits in the sub-pg/sec range. The advantages of the ELCD lie in its ability to detect halogen- and sulfur-containing compounds at very low detection limits. Although this can also be accomplished with an AED, the ELCD is much less expensive to purchase and operate. However, the ELCD is prone to fouling. Scrubbers must be used to avoid buildup of carbon deposits in the catalyst tube, and sulfur oxide deposits can also decrease the sensitivity of the detector.

Tandem Instrument Detection

The two most common tandem instrument designs are the *GC-Fourier transform infrared (GC-FTIR)* and the *GC-mass spectroscopy (MS)* systems. The GC is a very powerful tool for separating complex mixtures, but by itself, it cannot identify the individual

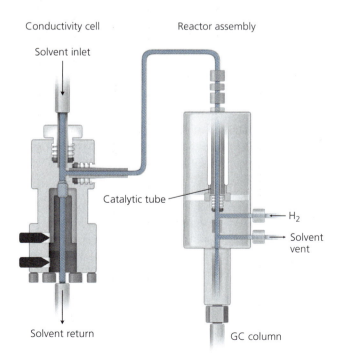

Figure 16.17 Schematic of an ELCD.

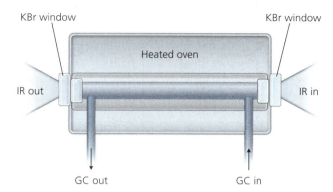

KBr window   KBr window

Heated oven

IR out   IR in

GC out   GC in

Figure 16.18 Schematic of a GC-FTIR light pipe.

components. On the other hand, infrared spectroscopy is not able to isolate components of a mixture but is an excellent tool for elucidating the bonding structure of molecules; if one uses a database of spectra, the FTIR can be used as a "fingerprint" in the positive identification of the individual components of the mixture. Likewise, the mass spectrometer (MS) is not able to isolate components of a mixture but is an excellent tool for elucidating structural components of a compound, and as with FTIR, one can use of spectral databases to positively identify each component.

The operation and design of FTIR instruments was covered in detail in Chapter 11, and the operation and design of MS instruments was covered in detail in Chapter 13. Here we will focus our attention on the design of the *interface* for each of these techniques.

Gas Chromatography-Fourier Transform Infrared   Because the eluent in a GC experiment is a gas, the interface of a GC-FTIR is simply a modified gas cell (see Figure 11.15 for a photograph of a standard gas cell). The eluent from the GC flows through a temperature-controlled "light pipe" that is placed in the optical path of the FTIR spectrometer. Figure 16.18 shows a schematic of a GC-FTIR light pipe. GC-FTIR accessories are sold by several manufacturers.

Gas Chromatography-Mass Spectroscopy   Both GC and MS techniques operate in the gas phase, so you would be correct to think that that coupling these two techniques is straightforward. With the very small diameter of some modern capillary columns, direct insertion of the column into the MS is possible. However, early GC columns were much larger in diameter and often contained packing material. The technical challenge of coupling a GC to an MS was the pressure differential between the two instruments. GC instruments are run under high pressures (> 760 mm Hg), whereas MS instruments operate under very low pressures (10^{-5} to 10^{-6} mm Hg). The instrumental design challenge was how to drop the pressure by seven or eight orders of magnitude while still transferring the bulk of the eluent from the GC into the MS. There have been several different interface designs developed over the years, but the most important one was the *jet separator* (see Figure 16.19). The jet separator isolates the organic eluent from the carrier gas by exploiting the difference in the diffusion rates of the carrier and the eluent. The carrier gas is typically a light molecule (helium or hydrogen) with a diffusion coefficient that is much larger than the analyte under study. As the analyte nears the end of the GC column, it passes through a heating element called the *transfer line*. A typical temperature of the transfer line is approximately 250°C. The heated transfer line adds kinetic energy to the eluting gasses. The eluting gases are sprayed through a small orifice into a partially evacuated chamber (about 10^{-2} torr), where the high diffusion rate of the carrier gas allows the carrier gas to make a wide spray angle. Because the organic analytes are heavier than the carrier gas, they have much lower diffusion rates and their spray angle is much narrower. By collecting the middle section of the spray cone with a skimmer, much of the carrier gas is removed from the analyte before it is passed to the mass spectrometer.

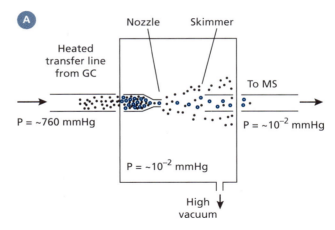

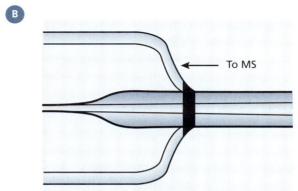

Figure 16.19 (A) Schematic of a GC-MS jet separator. (B) Rendering of a drawn-down capillary comprising the "jet" of the jet separator.

The jet separator serves two functions. First, it allows a step down in pressure from about 760 to about 10^{-6} mm Hg, and second, it effectively concentrates the analyte by removing much of the carrier gas. The jet separator is typically made by drawing down a glass capillary to a fine point and then sealing the capillary into a glass bulb that contains a line to a vacuum. The fine point of the capillary is aligned with a second capillary that functions as the skimmer.

Problem 16.7: Do you think using CO_2 as a carrier gas in an instrument with an FID would be problematic? Explain.

Problem 16.8: In your job as a chemist for a regional analytical laboratory, you are expecting a large number of samples that you will need to test for potential herbicide contamination. Prepare a plan (including explanation) for which type of detector you will need for the trace-level GC analysis of (a) 2,4-D, (b) atrazine, and (c) RoundUp. *Use an online search engine to find the structures of each of these substances.*

Problem 16.9: You will be analyzing urine samples from patients taking low levels of thiazide diuretics. What type of GC detector would you choose? Explain.

Quantitative and Qualitative Considerations

Peak heights, integrated peak areas, or area counts in GC provide quantitative results in a way analogous to LC analysis, and the details of the analysis depend primarily on the type of detector used in the analysis. Through the use of a set of external standards, creation of a calibration curve, or by using an internal standard, the amount of each analyte in the sample may be quantified. Under spectroscopic analysis (where Beer's law applies), the absorbance of the sample by the detector is converted to a concentration. On the other hand, under photonic analysis (counting phenomenon), a quantitative determination can be made and applied to the sample in comparison to a set of standards of known amounts. LC techniques rely heavily on concentration analyses and utilize primarily nondestructive detection, but GC techniques rely predominantly on mass analyses and utilize detectors that destroy the sample, although we saw in the previous sections some exceptions to this generalization.

Retention times under the chromatographic conditions of the investigation provide for a check on the identity of the analyte of interest. As discussed in the first section of this chapter, the results of the separation in terms of retention behaviors are largely only empirically understood, and most analysts rely on the literature to guide the design of the experiment and the interpretation of the results; the use of databases are also increasing in popularity for their utility in making peak identification.

Although chart recorders were the industry standard for decades, today's modern instruments send signals from the detector to a computer for analysis. At the computer, an analog-to-digital conversion is made and the digitized signal is used for computer processing, storage, and reporting. Many aspects of the signal are dependent on the type of detector used, including sensitivity, limit of detection, dynamic and linear ranges, response time, and noise. Signal processing is accomplished by the computer through peak integration, baseline adjustment, and autoscaling. As with LC, the data collecting and processing features depend on the sampling rate and range of the analog-to-digital converter. Especially in some fast GC techniques, and as new methods emerge in multidimensional GC of complex mixtures, the rate of sampling can approach the limit of the technique and will limit the resolution and performance of the analysis overall. Ways of overcoming these challenges are at the forefront of methodological advances in terms of computer interfaces and data-handling capacity.

16.7 New Developments and Directions in Gas Chromatography

Although the majority of papers published in the field of GC in the past decade are reports of applications in which the GC was used as an analytical device to obtain results on a system under study, some very interesting new developments in the modern literature are worth following. These breakthrough areas of research and new technologies are greatly advancing this area of science.

Multidimensional Gas Chromatography Techniques

For the analysis of very complex mixtures, *multidimensional GC* (MDGC) is used, in which eluting analytes from one GC analysis are directed into a second GC for further separation and characterization (see Figure 16.20). There are two ways in which the samples are directed from the first GC to the second. The older, more mature technique, known as *heart-cutting two-dimensional GC*, or *heart-cutting 2D GC*, uses an interface that follows the first GC column.

At a time just before the poorly resolved analytes of interest elutes from the first column, the interface is switched, directing the effluent to the second column. The interface remains open for only a short time, just long enough to allow for the analyte of interest to enter the second column. Once on the second column, the sample is separated once more. Typically, in heart-cutting 2D GC, the two columns are different specificities and separation is accomplished in both dimensions. Heart-cutting 2D GC has been around since 1958 and has been drawing less attention recently compared to the newer MDGC technique known as $GC \times GC$, first introduced in 1991 by Liu and Phillips.

As in heart-cutting 2D GC, in $GC \times GC$ there is an interface between the two columns that regulates and directs the flow from the outlet of the first column to the inlet of

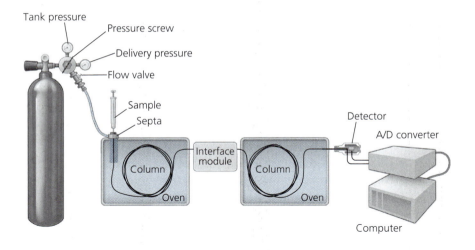

Figure 16.20 Schematic of a MDGC system, including an interface between the two columns that directs flow of the effluent from the primary column to the secondary column.

the second column. In GC × GC, however, the interface pulses the effluent across from the first column to the second in volumes that are approximately the width of the peaks. This sampling rate is called a *modulation period*. Each pulse of sample is very quickly analyzed on the second column—typically in a fast, high-temperature isothermal a-nalysis. Developments related to modulation and column performance are of great in-terest. The number of applications published using GC × GC has been increasing and is especially seen as a powerful technique for the new field of metabolomics, which seeks to identify and characterize all of the small molecule metabolites of a particular cellular system for the purpose of understanding both normal function and identifying disease states. Applications are abundant in the environmental, petroleum, and pharmaceutical fields. Currently, the most popular detectors for GC × GC techniques are MS instru-ments, because of their fast detection timescales. However, further developments are warranted, especially with regard to data-processing schemes that can handle the abun-dance of data collected from this technique.

Miniaturization, Portability, Speed, and Throughput

As mentioned in Section 16.1, major innovations in gas chromatography today are related to the conjoined ideas of portability (taking the instrument *to* the site of the analysis) and throughput (superfast analysis so that many samples can be tested in a short amount of time). These twin goals are in service to many important applications, including Department of Homeland Security procedures and extraterrestrial explora-tion. In order to achieve the goal of practical portability, innovations in miniaturiza-tion must be accomplished, and advances in this area are astounding. Not only are new advances in miniaturization relevant to engineering challenges such as power needs, size constraints, and cost, but research in this area has already led to new advances in column technology and performance, which promises advancement in fundamental understandings related to separation mechanisms and particle behaviors. Technologies stemming from the miniaturization of computers, including advances in the produc-tion of semiconductors, computer memory, and microprocessors, have been critical to the development of newer, faster, cheaper, and smaller instrumental devices. In terms of the particular advances in miniaturization of GC systems, research is being pursued in column technology and manufacturing, detector size and sensitivity, and precon-centrating devices for the purpose of enhancing detection limits under microsampling conditions. The development of detector arrays is another way in which increased throughput is being pursued.

In an interesting article on indoor air monitoring, author Sun Kyu Kim and coworkers at the University of Michigan demonstrated what they report is the "first field demonstration of [microfabricated] GC technology for automated, near-real-time, selective volatile organic compound monitoring at low- or sub-ppb levels."[*] The portable instrument was able to selectively identify and quantify a potential air contaminant, trichloroethylene (TCE), which is known to seep into buildings when the soil and or groundwater underneath the building is contaminated. One of the challenges of this analysis is the low levels at which the contaminant may be found. To further complicate the analysis, there are many other volatile organic compounds (known as VOCs) that are also commonly found in indoor air and at similarly low levels. Kim and team tested two houses: one with previously documented TCE vapor intrusion stemming from a gap between the basement wall and the concrete floor and one in which there was an intentionally hidden source of TCE. In the first house, the researchers were able to confirm the previous findings and were able to document transient changes in TCE levels over time. Furthermore, they were able to map the changes in TCE levels across the home as a function of the distance from the source of the vapor intrusion point. In the second house, the researchers were able to use the micro-GC to find the hidden source of the TCE (in a closet) by mapping the concentrations of TCE around the house. Their micro-GC prototypes are 44 cm wide by 25.5 cm deep by 14.5 cm high and weigh 4.5 kg, and they are operated by a laptop computer. Detection is accomplished by a chemiresistor array, utilizing four microsensors. In a related application, Vereb and coworkers from Virginia Tech wrote a review article on the analysis of human breath (breathalyzers) to test for a person's possible exposure to environmental toxins.[†] In the article, they review new microfabricated devices, sensors, and preconcentrators, which have been used in miniaturized, handheld GC devices for this application. Future developments may lead to the use of the micro-GC as a diagnostic tool in the physician's office. ■

---

[*]*Kim, S. K.; et al. Environmental Science and Technology, **2012**, (4), 6073.*

[†]*Vereb, H.; et al. Environmental Science and Technology, **2011**, (45), 8167.*

PROFILE
Breath and Air Quality

The miniaturization of the column has led to some remarkable developments and discoveries. At this very small scale, achieving uniformity and reproducibility are significant challenges, as is developing a satisfactory means of connecting the miniaturized column to the injection preconcentrator[31] at the inlet side and to the detector at the outlet side without introducing dead volumes and increasing band broadening. Diminishing these "extra column" effects is essential in these ultra-low volume applications. Of particular interest is the development of miniaturized "high aspect ratio columns." These are columns that are not columnar in the geometric sense; the high aspect ratio columns are produced using microchip technology taken from the computer industry, in which the channels produced have a rectangular cross-section.

---

[31] Preconcentration is any process that concentrates a sample prior to analysis, the purpose being to improve the detection limits of trace analytes.

TABLE 16.3: Symbols and Their Definitions

SYMBOL	DEFINITION	COMMENT
V_g	Retention volume	$V_g = t_r F$
V_R°	Adjusted retention volume	$V_R^\circ = j t_r F_c$
V_N	Net adjusted retention volume	$V_N = j t_R' F_c$
V_m	Hold-up volume	$V_m = t_m F_c$
V_m°	Adjusted hold-up volume	$V_m^\circ = j V_m$
V_G°	Specific retention volume	$V_G^\circ = V_N / w$ w = weight of stationary phase
t_r	Retention time	$t_r = V_g / F$
t_m	Hold time	Similar to void volume in LC
t_r'	Retention time less hold time	$t_r' = t_r - t_m$
F	Carrier gas flow rate	$F = V_g / t_r$
F_c	Average carrier gas velocity	$F_c = F_0 \left(\dfrac{T_c}{T_a} \right) \times \left(1 - \left(\dfrac{P_{H_2O}}{P_0} \right) \right)$
P_i	Inlet pressure	$P_i = P_0 + P_G$
P_0	Ambient pressure or outlet pressure	Pressure of the atmosphere on location
P_G	Column head pressure	Pressure measured at the gas regulator
P	Relative pressure	$P = \dfrac{P_i}{P_0}$
P_{H_2O}	Vapor pressure of water	Taken from a table of physical constants
j	Compressibility factor	$j = \dfrac{3}{2} \times \dfrac{(P^2 - 1)}{(P^3 - 1)}$
T_c	Column temperature	
T_a	Ambient temperature	
k_r	Retention factor	$k_r = \dfrac{t_r'}{t_m} = \dfrac{t_r - t_m}{t_m}$

These columns are "high aspect" in the sense that they are taller in one dimension and narrower in another dimension. The narrow dimension provides for the excellent chromatographic resolution that we expect from capillary GC columns, whereas the taller dimension allows for improved flow characteristics. Interestingly, Golay[32] predicted enhanced performance characteristics for rectangular cross-section tubing in his early work on GC, and he also identified some of the band broadening challenges that are introduced with the rectangular shape of the new column. Researchers have been challenged to find satisfactory ways of evenly coating the high aspect ratio columns, where the corners typically attract a thicker layer of the film. Because the corners of the tubing create more drag on the mobile phase, mass transfer effects are increased.

---

[32] Marcel J. E. Golay (1902–1989) was one of the early pioneers of GC and the inventor of capillary columns.

16.8 Extended Theory

Chromatographic separation theory was presented in Chapter 15 in the context of LC. Those general theories are applicable to GC as well. Here we will explore some specific aspects of chromatographic theory as they apply to GC. For the following discussion, Table 16.3 may prove to be a useful summary.

Evaluation of the Gas Chromatography Separation

In simplest terms, we can describe the *retention volume, V_R*, as it relates to t_r and the flow rate (F) of the carrier gas (see Equation 16.1):

$$V_R = t_r F \qquad \text{Eq. 16.1}$$

In GC, this relationship is not as straightforward as it may seem because the flow rate depends on a variety of experimental conditions as well as the compressibility of the carrier gas. In LC, because liquids are not greatly affected by pressure, the compressibility of the mobile phase is generally ignored; however, for the carrier gases in GC, compressibility must be taken into consideration. Furthermore, the pressure of the gas at the head of the column, where sample and carrier gas are introduced, (P_i) is not necessarily the same as the pressure of the gas at the end of the column (P_o). The difference $(P_o - P_i)$, or pressure drop, must also be taken into consideration. The gas *compressibility correction factor* (j) is determined by:[33]

$$j = \frac{3}{2} \times \frac{(P^2 - 1)}{(P^3 - 1)} \qquad \text{Eq. 16.2}$$

where P is the relative pressure and is determined by:

$$P = \frac{P_i}{P_o} \qquad \text{Eq. 16.3}$$

The inlet pressure, P_i, can be determined by the sum of the ambient pressure (P_o) and the pressure at the column head (P_G):

$$P_i = P_o + P_G \qquad \text{Eq. 16.4}$$

Knowing the compressibility correction factor, j, allows us to compensate for the pressure drop across the column as we calculate the average carrier gas velocity (F_c), and the average $V_R°$ for the component of interest.

Going back to Equation 16.1, we account for the compressibility to determine the average retention volume, $V_R°$ (see Equation 16.4) with the addition of the j term:

$$V_R° = j t_r F_c \qquad \text{Eq. 16.5}$$

Similarly, we can determine the *adjusted hold-up volume, $V_m°$*. This value accounts for the volume of the empty column. Like the void volume in LC, the hold-up volume accounts for the volume of carrier gas that will pass through the column from the start of the chromatogram (time = 0) until the first unretained peak would elute (see Equation 16.6), where t_m is the column's hold-up time and F_c is the average carrier gas velocity. The hold-up volume in GC is equivalent to the volume of the gas in the column at average temperature and average pressure across the column's

---

[33] Martire, D. E.; Locke, D. C. Compressibility Factor for Nonideal Carrier Gases in Gas Chromatography. *Anal. Chem.,* **1965,** *37,* 144–145.

length. Equation 16.7 applies the correction factor, j, to the calculation of the hold-up volume.

$$V_m = t_m F_c \qquad \text{Eq. 16.6}$$

$$V_m^\circ = j V_m \qquad \text{Eq. 16.7}$$

Problem 16.10: Calculate the compressibility correction factor, j, as a function of changing ambient pressures. Assume that the column head pressure remains constant at 70 mm Hg and the ambient pressure (in mm Hg) is:

(a) 760.0	(b) 767.3	(c) 752.6	(d) 749.7

To calculate the average flow rate across the column, F_c, we need to account for both temperature gradient as well as gas pressure. Equation 16.6 gives the generalized equation for determining F_c. The value for P_{H_2O} is determined from a physical constant table and represents the *vapor pressure of water* at the ambient temperature of the experiment. The ratio of P_{H_2O} to P_o corrects for the difference in pressure of the dry carrier gas in the column and the gas in a bubble meter (often used to measure volumetric flow rate), which is water based. In this equation, T_c is the column temperature and T_a is the ambient temperature.

$$F_C = F_o \left(\frac{T_C}{T_a} \right) \times \left(1 - \left(\frac{P_{H_2O}}{P_o} \right) \right) \qquad \text{Eq. 16.8}$$

Example 16.4

Your GC experimental parameters are such that your carrier gas flow at the column outlet measures 25.0 mL/min. The column temperature is 250°C. The ambient temperature is 23°C. Ambient pressure is 758.8 mm Hg. The column head pressure is 68.4 mm Hg. Determine j, the compressibility correction factor.

STRATEGY –

First start with Equation 16.4 and calculate the pressure at the column head, P_i. Use your value of P_i to calculate the overall pressure P and put that into Equation 16.2 to find j.

SOLUTION –

Solving for P_i:

P_o = 758.8 mm Hg

P_G = 68.4 mm Hg

$P_i = P_o + P_G = 758.8 + 68.4$ mm Hg = 827.2 mmHg

Solving for P:

$P = P_i / P_o = 827.2$ mmHg / 758.8 mmHg = 1.090

Solving for j:

$$j = \frac{3}{2} \times \frac{(P^2 - 1)}{(P^3 - 1)} = \frac{3}{2} \times \frac{(1.09^2 - 1)}{(1.09^3 - 1)} = 0.958$$

Example 16.5

Let us continue our calculations from the previous worked-out example and calculate the average flow rate across the column, F_c. Your carrier gas flow at the column outlet measures 25.0 mL/min. The column temperature is 250°C. The ambient temperature is 23°C. Ambient pressure is 758.8 mmHg. The column head pressure is 68.4 mmHg. Also, we have calculated j = 0.9580 mm Hg. We look up P_{H_2O} at 23°C, and the value is 21.068 mmHg.

STRATEGY –

Use Equation 16.8. Using all of the relevant terms, solve algebraically for F_c.

SOLUTION –

$$F_C = F_o\left(\frac{T_C}{T_a}\right)\times\left(1-\left(\frac{P_{H_2O}}{P_o}\right)\right)$$

$F_o = 25.0$ mL/min
P_{H_2O} at 23°C = 21.068 mmHg
$P_o = 758.8$ mmHg
$T_c = 250°C = 523K$
$T_a = 23°C = 296K$

$$F_C = 25\,mL/min\left(\frac{523K}{296K}\right)\times\left(1-\left(\frac{21.068\ mmHg}{758.8\ mmHg}\right)\right)=43\,mL/min$$

Problem 16.11: Given the following GC experimental parameters, your carrier gas flow at the column outlet measures 25.0 mL/min. The ambient temperature is 23°C. Ambient pressure is 758.8 mm Hg. The partial pressure of water at 23°C is 21.068 mm Hg. The column head pressure is 68.4 mm Hg. Calculate F_c for the following column temperatures:

    (a) 150°C   (b) 200°C   (c) 300°C   (d) 385°C

Problem 16.12: Suppose we change the flow rate from 25.0 mL/min to 30.0 mL/min. Recalculate F_c for each temperature in Problem 16.11.

Problem 16.13: Imagine you are running GC analysis of gases trapped in ice core samples in a lab in the Arctic Circle. The lab is kept at a temperature of 12°C. Your other GC parameters include carrier gas flow at the column outlet of 25.0 mL/min. Ambient pressure is 762.8 mm Hg and the column head pressure is 78.4 mm Hg. Calculate F_c for the following column temperatures: (*Hint*: You will need to look up the partial pressure of water at 12°C.)

    (a) 150°C   (b) 200°C   (c) 300°C   (d) 385°C

$V_R°$ represents the *adjusted retention volume* for the peak of interest, after correcting for the compressibility of the carrier gas.

We have one more set of adjustments to make: to determine the *net adjusted retention volume* (V_N) by accounting for the hold-up volume. In this set of calculations, we will use $t'_{R'}$ which is the difference between t_r and t_m (Equation 16.9), the retention

Example 16.6

Next, use the calculated F_c (from Example 16.5) to determine $V_m°$ and $V_R°$ for experimental data in which t_r, the retention time for the peak of interest is 3.25 min and the column hold-up time (t_m) is 0.20 min.

STRATEGY –

Use Equations 16.6 and 16.7 to determine $V_m°$. Then use 16.5 to determine $V_R°$.

SOLUTION –

$V_m = t_m F_c$

$\qquad V_m = 0.20 \text{ min} \, (43.0 \text{ mL/min})$

$\qquad V_m = 8.6 \text{ mL}$

$V_m° = jV_m$

$\qquad V_m° = 0.9580 \, (8.6 \text{ mL})$

$\qquad V_m° = 8.2 \text{ mL}$

$V_R° = jt_r F_c$

$\qquad V_R° = 0.9580 \, (3.25 \text{ min}) \, (43.0 \text{ mL/min})$

$\qquad V_R° = 133 \text{ mL}$

Example 16.7

Determine V_N for the peak $(t_r = 3.25 \text{ min})$ described Example 6.3.

STRATEGY –

Use Equation 6.9 to find t_R'. Then use t_R', j (determined in Example 16.4), and F_c (determined in Example 16.5) in Equation 16.10 to calculate V_N.

SOLUTION –

$t_R' = (t_r - t_m)$

$\qquad t_R' = 3.25 \text{ min} - 0.20 \text{ min}$

$\qquad t_R' = 3.05 \text{ min}$

$V_N = jt_R' F_c$

$V_N = (0.9580)(3.05 \text{ min})(43.0 \text{ mL/min}) = 126 \text{ mL}$

times for the peak of interest, and the hold-up time for the column. Equation 16.10 shows the calculation for the net adjusted retention volume corrected for compressibility. Note that V_N differs from $V_R°$ in the use of t_R' instead of t_r in its calculation.

$$t_R' = (t_r - t_m) \qquad \qquad \textbf{Eq. 16.9}$$

$$V_N = jt_R' F_c \qquad \qquad \textbf{Eq. 16.10}$$

V_N represents the net retention volume for the peak of interest, after correcting for the compressibility of the carrier gas.

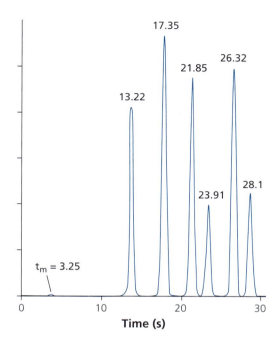

Figure 16.21 A GC chromatogram showing retention times. Carrier gas flow at the column outlet measures 28.1 mL/min. The column temperature is 285°C. The ambient temperature is 24.8°C. Ambient pressure is 760.2 mm Hg. The column head pressure is 70.4 mm Hg. The column used was 30 meters in length.

The Relationship Between V_N, k_r, and Selectivity

Now that we have learned how to determine adjusted retention values to account for the column environment, let us review (as this is similar to LC) the relationship between the retention parameters of partitioning and selectivity. Recall from our earlier introduction to partitioning (see Section 15.2) that the retention factor (k_r) is a ratio of the retention times (see Table 6.3 and Equation 16.11, and Equation 15.6) describing the proportion of time the analyte spends on the stationary phase to the time spent in the mobile phase. In addition, recall that the separation factor (α), also known as the selectivity factor, gives us a description of how far two peaks in a chromatogram are separated as the ratio of the retention factors for the two peaks. Equation 15.7 shows the general equation for calculating the separation factor for two peaks, where k'_r is the retention factor for the more strongly retained analyte and k''_r is the retention factor of the less strongly retained analyte. Therefore α is always greater than one. We can use α in GC the same way we use α in LC.

Problem 16.14: Determine $V_R{}^\circ$ for each peak in Figure 16.21.

Problem 16.15: Determine V_N for each peak in Figure 16.21.

The General Elution Problem: Review of Resolution and Band Broadening

GC is not immune to the same types of resolution problems that were discussed in the context of LC in Chapter 15. Although peaks may be separated and have better resolution through longer columns, the more time an analyte spends on the column, the more band broadening that analyte will experience. Thus, the general elution problem of chromatography is also very relevant in GC techniques. Be sure to review the concepts of resolution and band broadening from Section 15.2, especially the section on longitudinal diffusion.

Recall from Chapter 15 that the height equivalent of a theoretical plate is modeled using the van Deemter equation, an equation which accounts for the various physical parameters, both enthalpic and entropic, that contribute to broadening of the peak. The *van Deemter equation* is given in Equation 16.11 (and also in Equation 15.17):

$$H = A + B\!\!\big/\!\!_{v} + (Cv)$$ **Eq. 16.11**

where

H = height equivalent of a theoretical plate

A = eddy diffusion coefficient

B = longitudinal diffusion coefficient

v = linear velocity of the mobile phase

C = rate constant for mass transfer

The eddy diffusion term (A) in the van Deemter equation is omitted for well-packed and open tubular columns in GC. Thus in GC, plate height (H) is calculated via a modified version of the van Deemter equation, known as the *Golay equation* (see Equation 16.12):

$$H = \frac{B}{v} + C$$ **Eq. 16.12**

Of the various factors that affect peak width, longitudinal diffusion (the B term in the van Deemter equation) is *potentially* the most problematic in GC, primarily due to the high diffusion rates of analytes within the carrier gas as opposed to diffusion of analytes within liquids in LC. When longitudinal diffusion effects are not controlled, plate heights increase and resolution suffers. At slower flow rates, band broadening due to the longitudinal diffusion term is greatest, whereas at higher flow rates, longitudinal diffusion is minimized and the predominate cause of band broadening is due to the mass transfer contribution. In order to minimize plate height, a compromise must be reached that keeps longitudinal diffusion to a minimum while also minimizing mass transfer broadening. In GC, because the diffusion effects are much greater than the mass transfer concerns, it is often true that the flow rate can be increased without seeing a significant loss of resolution; thus reducing the time of the analysis and increasing the potential throughput (number of samples per time) of the process.

Although α describes relative peak position (as a ratio of retention times), it falls short as a good measure of resolution because it does not take peak *widths* into consideration (see Equation 15.10 for a review of resolution). To account for peak overlap, the peak width is an essential consideration. Again, these factors in GC are analogous to those in LC. Peak width is also a major determinant of the peak capacity of a column under the conditions of analysis. Peak capacity is the number of peaks that can be well resolved on a column under optimal experimental conditions. Because later eluting peaks are broader than early eluting peaks, the general elution problem emerges. Although longer run times and longer columns allow for greater separation (favorable effects on H and N), the longer run times lead to increased peak broadening and peak overlap diminishes the resolution and therefore the peak capacity as well. A proper calculation of peak capacity involves a differential equation to account for the changing peak width over time (see Exercise 16.23).

Temperature in GC methods is an important part of the separation design. An increase in temperature will speed up the kinetic effects of the partition equilibrium, thus modifying the distribution equilibria and also how fast the components reach equilibrium between stationary phase and mobile phase. At higher temperatures, analytes will elute more quickly and peaks will be sharper. However, increased temperatures will have an adverse effect on resolution because similar analytes do not have enough time to be fully resolved from each other. As noted earlier in this chapter, carefully

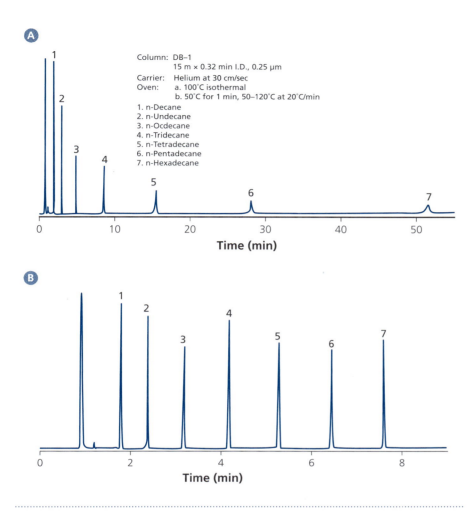

Figure 16.22 Use of a temperature gradient to sharpen peaks and shorten run times.

controlling the temperature over the time of the analysis, a process called *temperature programming*, specific temperature gradients can be designed to maximize resolution at both short and longer retention times. Furthermore, temperature gradients can be so successfully applied, that the column length in GC can be increased significantly. The longer column provides for more partitioning and better separation of analytes, and the increased temperature will speed the time of the analysis and keep peaks narrow. Figure 16.22 shows how a temperature gradient can be applied to sharpen peaks, improve resolution, and increase peak capacity. Once a suitable column has been chosen, the control of temperature in a GC method is a key design consideration. In a temperature gradient, the temperature is changed as a function of time over the analysis. Temperature changes need to be nearly instantaneous, exact to within ± 5°C, well controlled in both the heating and cooling segments of the time course, and held constant during whatever preset times are desired. Temperature gradients are analogous to mobile phase gradients in LC. Over the course of the analysis, the chromatographer will adjust the conditions to maximize resolution and minimize time, thus trying to minimize the general elution problem.

Problem 16.16: Compare and contrast the ways a chromatographer performs a gradient separation in LC and GC.

16.9 Further Reading

BOOKS

Dettmer-Wilde, K.; Engewald, W. (editors). *Practical Gas Chromatography: A Comprehensive Reference.* Springer-Verlag: Berlin, **2014**.

Hites, R. A. *Handbook of Instrumental Techniques for Analytical Chemistry.* Settle, F. (editor). Prentice Hall: Upper Saddle River, NJ, **1997**, 609.

McNair, H. M.; Miller J. M. *Basic Gas Chromatography,* 2nd ed. John Wiley & Sons: Hoboken, NJ, **2009**.

Poole, C. F. *Gas Chromatography.* Elsevier: Oxford, UK, **2012**.

Poole, C. F. *The Essence of Chromatography.* Elsevier: Amsterdam, The Netherlands, **2002**.

JOURNALS

Ahn, H.; Brandani, S. Analysis of Breakthrough Dynamics in Rectangular Channels of Arbitrary Aspect Ratio. *AIChE Journal,* **2005**, *51*, 1980–1990.

Dorman, F.; Whiting, J. J.; Cochran, J. W.; Gardea-Torresdey, J. Gas Chromatography (Review). *Anal. Chem.,* **2010**, *82(12)*, 4775–4785.

Eiceman, G . A. Gardea-Torresdey, J.; Dorman, F.; Overton, E.; Bhushan, A.; Dharmasena, H. P. Gas Chromatography. *Anal. Chem.* **2006**, *78*, 3985–3996.

Eiceman, G. A.; Gardea-Torresdey, J.; Overton, E.; Carney, K.; Dorman, F. Gas Chromatography. *Anal. Chem.,* **2002**, *74(12)*, 2771–2780.

Seeley, J. V.; Seeley, S. K. Multidimensional Gas Chromatography: Fundamental Advances and New Applications. *Anal. Chem.,* **2013**, *85*, 557–578.

ONLINE RESOURCES

Manura, J. J. Improving Sensitivity in the H.P. 5971 MSD and Other Mass Spectrometers—Part II of II, SISWEB Tech. Note D, Scientific Instrument Services, Ringoes, NJ. http://www.sisweb.com/referenc/articles/sensit-2.htm (accessed Aug 14, 2015).

Several column selection guides are available, including:

- http://www.restek.com/Chromatography-Columns/GC-Columns (accessed Aug 14, 2015)

- http://www.sge.com/support/training/columns/capillary-column-selection-guide (accessed Aug 14, 2015)

- http://www.sigmaaldrich.com/etc/medialib/docs/Supelco/General_Information/t407133.Par.0001.File.tmp/t407133.pdf (accessed Aug 14, 2015)

16.10 Exercises

EXERCISE 16.1: Define each of the following terms:

a. Eluate

b. Isothermal elution

c. Thermal gradient

d. Temperature programming

e. Selectivity factor

f. Retention factor

g. Carrier gas

h. Retention volume

i. Compressibility correction factor

j. van Deemter equation

k. Split injection

l. Hold-up volume

m. Adjusted retention volume

n. NET adjusted retention volume

EXERCISE 16.2: Calculate the theoretical plate height, H, and the number of theoretical plates, N, for peak 4 in Figure 16.21.

You might need to review Section 15.2 to answer this question. It is acceptable to estimate the values of the variables you need for the calculation from the axes of the chromatogram.

EXERCISE 16.3: In designing a separation method, we use the theoretical plate as a representation of the separation efficiency of the column. So it might seem reasonable to assume that more theoretical plates is always better than fewer theoretical plates. Having said that, why do we not automatically increase the column length anytime we need to improve separation?

EXERCISE 16.4: Describe the effect of each variable in Equation 16.11 on column efficiency.

EXERCISE 16.5: Rank the following compounds in terms of the expected elution order for a capillary GC separation run under isothermal conditions.

(a) Ethanol

(b) n-Propanol

(c) Methanol

(d) n-Pentanol

(e) n-Butanol

EXERCISE 16.6: Sketch a schematic for a split flow injector and in your own words, describe each component.

EXERCISE 16.7: Sketch a schematic of an FID and in your own words, describe how it functions.

EXERCISE 16.8: Sketch a schematic of a TID and in your own words, describe how it functions.

EXERCISE 16.9: Sketch a schematic of an ECD and in your own words, describe how it functions.

EXERCISE 16.10: Sketch a schematic of a PID and in your own words, describe how it functions.

EXERCISE 16.11: Sketch a schematic of an FPD and in your own words, describe how it functions.

EXERCISE 16.12: Sketch a schematic of an AED and in your own words, describe how it functions.

EXERCISE 16.13: Sketch a schematic of a "light pipe" interface for a tandem GC-FTIR and in your own words, describe how it functions.

EXERCISE 16.14: Sketch a schematic of a jet separator for a tandem GC-MS and in your own words, describe how it functions.

EXERCISE 16.15: Imagine the following two peaks coeluted shortly after the hold-up volume. Which of the following actions would be the *"best"* way to improve the resolution? Defend your answer and explain why you rejected the other options.

(a) Use a longer column.

(b) Use a faster carrier gas velocity.

(c) Use a slower carrier gas velocity.

(d) Use a higher column temperature.

(e) Use a cooler column temperature.

EXERCISE 16.16: Which of the various detector types discussed in this chapter is the detector of choice for the detection of halogenated organic compounds?

EXERCISE 16.17: In all chromatographic experiments, controlling the injection volume is an important consideration. Compare and contrast the methods used to control injection volume in LC and GC.

EXERCISE 16.18: Both the TCD and FID are considered universal GC detectors, but the TCD is considered "more" universal than the FID. If the TCD is so much more universal, why use a FID at all?

EXERCISE 16.19: Compare and contrast the operation of an atomic emission spectrometer (Chapter 9) with the operation of an AED.

Advanced Exercises

EXERCISE 16.20: Compare and contrast the features of the mobile phase in GC with that in LC. Describe the important properties of the mobile phase in each technique and its impact on the efficiency of the separation.

EXERCISE 16.21: Answer the following questions for the chromatogram given in Figure 16.21.

(a) Calculate the selectivity factor and resolution for the first two peaks
(see Section 15.2 for a review on calculating resolution).

(b) Calculate the number of theoretical plates for the first and last peak.

EXERCISE 16.22: Which term of the van Deemter equation is most influential in a GC separation? Explain how this factors into the general elution problem.

EXERCISE 16.23: The peak capacity is the total number of observable peaks in a chromatogram under a given set of experimental conditions. The equation for determining the peak capacity is:

$$n_C = 1 + \int_{t_m}^{t_R} \left(\frac{\sqrt{N}}{4t} \right) dt$$

where n_C is the peak capacity, t is the separation time, t_m is the hold-up volume, and t_R is the retention time of the last eluting peak. If you ignore the fact that the number of theoretical plates varies with the retention time of an individual peak, determine the peak capacity of a system given, $t_m = 6$ min., $t_R = 40$ min, and $N = 52,000$.

EXERCISE 16.24: What type of detector would you select in the routine quantitative determination of Prozac in urine samples by GC?

EXERCISE 16.25: In the analysis of a series of n-alkanes, would you prefer to use a GC-FTIR or a GC-MS instrument? Explain.

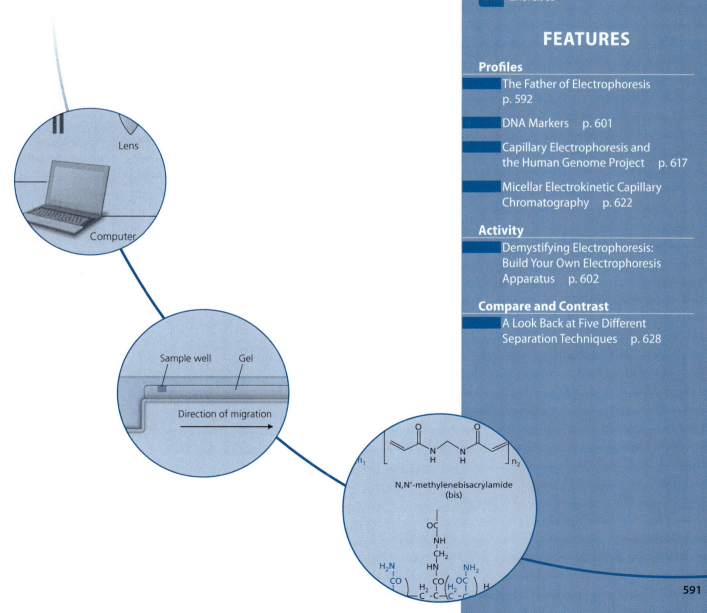

Lens

Computer

Sample well   Gel

Direction of migration

N,N'-methylenebisacrylamide
(bis)

17.1 | Introduction

The analytical technique of electrophoresis involves a very simple instrument, or what could be described in its most basic form as an apparatus. *Electrophoresis* is used to separate and identify macromolecular species based on the differential migration of charged molecules in an electric field. A particular form of electrophoresis, capillary electrophoresis, has developed over the most recent quarter century into a truly instrumental technique, offering fast run times, excellent resolution, automation, and options for obtaining quantitative results. It now occupies a significant place in the *analyst's toolbox*. Because of the prominent role that electrophoresis has in modern forensics and other analytical applications, it is important that a thorough discussion of modern electrophoretic techniques include the basics of electrophoresis. Furthermore, because the fundamentals of electrophoretic separation are distinct from other separation techniques, it is important that the analytical chemist understand the basis of electrophoresis, its applications, and results.

Like chromatography (Chapters 15 and 16), electrophoresis is a separation technique. It can be used *analytically,* to characterize the compositions of mixtures, and it can be used *preparatively*, to isolate components of mixtures that are of interest. Unlike chromatography, electrophoresis is a one-phase system. Although chromatography depends on the partitioning of analytes between two phases (e.g., liquid–gas partitioning), electrophoresis involves components that are in the same phase. There is no partitioning. Furthermore, unlike chromatography, which uses a range of techniques for separations by a variety of molecular properties (e.g., polarity, charge, affinity, size), electrophoresis is primarily limited to the separation of ionic analyte mixtures. In the last section of this chapter, we will discuss a method by which electrophoresis can be used to separate neutral analytes. But for the vast majority of applications, if the analytes have no charge, they will not participate in an electrophoretic separation. However, you should not draw the conclusion that electrophoresis is an inferior technique! Because both protein and nucleic acids (DNA and RNA) are macromolecules (or biopolymers) having extensive charge density across the polymer, electrophoresis has demonstrated widespread applicability and unprecedented power for separations of biological samples. DNA polymers are polyanionic, and protein samples can be made positive, negative, or neutral by adjusting the pH of the buffer. As a result, electrophoresis is one of

Arne Tiselius (1902–1971) won the 1948 Nobel Prize in Chemistry "for his research on electrophoresis and adsorption analysis, especially for his discoveries concerning the complex nature of the serum proteins." Tiselius was a Swedish biochemist who had begun his interest in electrophoretic separations as a PhD candidate, when he used an electric field to separate large proteins in solution. His apparatus consisted of a U-shaped tube filled with buffer solution. Electrodes were immersed in both ends and a voltage applied. This type of electrophoresis is now called *moving-boundary* or *free-boundary electrophoresis*. The proteins would migrate to either electrode depending on their charge, and detection was made by measuring the change in refractive index at the protein boundaries. Moving boundary electrophoresis was largely supplanted in the 1950s with the introduction of *zone electrophoresis*, which utilizes a solid matrix such as paper or gel. ■

the primary methods of DNA and protein analysis. Therefore, electrophoretic results are in the news continuously in the form of paternity questions, genetic testing, and archaeological and historical reckoning of artifacts.

The principles of electrophoresis can be traced as far back as the 1800s as scientists explored the movement of charged species in an electric field.[1] However, the technique did not gain prominence until the mid-twentieth century when the structures of protein and DNA were better understood.[2] The advancement of electrophoresis was accelerated during the late 1980s and into the 1990s in the quest to sequence the human genome, as electrophoretic techniques were expanded and developed.[3]

17.2 Fundamental Principles

In *zone electrophoresis*, charged components in a mixture are compelled to move through a matrix[4] by the force of an electric field. The matrix serves to minimize diffusion of the analytes such that the analyte ions of the same size-to-charge ratio will migrate together within a narrow zone of the matrix. The various components in the mixture are separated into zones (also known as bands), as a result of the difference in electric mobility of the analyte particles[5] within the matrix. Under the influence of an electric field, particles that are positively charged will move toward the more negative electrode (the cathode), and negatively charged particles will move toward the more positive electrode (the anode). See Figure 17.1.

Vocabulary Review

Electrode An electrically conducting material used to make contact with a nonmetallic part of a circuit

Electrolysis The use of direct current to drive a nonspontaneous redox reaction (i.e., recharging a battery)

Galvanic cell A spontaneous redox reaction that can produce an electric current between two electrodes (i.e., a discharging battery)

Cathode The electrode to which *cat*ions will migrate in an electric field. In a spontaneous cell, the cathode is the positive terminal; in an electrolysis cell, it is the negative terminal.

Anode The electrode to which *an*ions will migrate in an electric field. In a spontaneous cell, the anion is the negative terminal; in an electrolysis cell, it is the positive terminal.

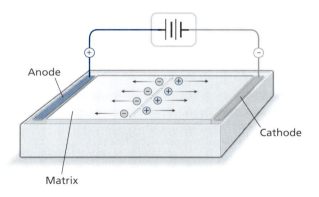

Figure 17.1 Schematic showing the electrophoretic migration of an ion in an electric field. The arrows indicate the direction each ion will migrate in the electric field. Cations travel through the matrix toward the cathode and anions travel through the matrix toward the anode.

---

[1] Michael Faraday (1791–1867) was an early pioneer in the understanding of electrical conduction. Svante August Arrhenius was awarded the Nobel Prize in Chemistry (1903) for his work on the electrolytic conduction of aqueous salt solutions.

[2] Tiselius developed the first device for electrolytic separation of charged species. See Tiselius, A. A New Apparatus for Electrophoretic Analysis of Colloidal Mixtures. *Trans. Faraday Soc.*, **1937**, *33*, 524–531.

[3] Zubritsky E. How analytical chemists saved the human genome project...or at least gave it a helping hand. *Anal Chem.*, **2002**, *74(1)*, 23A–26A.

[4] In earlier chapters, we used the term *matrix* to describe the portion of an analytical sample other than the analyte. For instance, in the analysis of lead in soil, the soil would be the matrix. However, in electrophoresis, the term *matrix* describes the medium in which the separation is conducted. In electrophoresis, the term *matrix* is most analogous to the chromatographic term *stationary phase* (see Chapter 15).

[5] We will use the term *particles* here (rather than analyte, molecules, or ions) because electrophoresis is widely applied to large biopolymers (such as protein and DNA) and even larger complexes of these macromolecules. Although electrophoresis can be accomplished on smaller molecules, and in fact, you may have had some experience with small molecule electrophoresis in one of your undergraduate laboratory sessions, the technique's primary analytical uses are in separation and characterization of biopolymers. Thus, the term *particle* is sufficiently descriptive.

Field Strength

Because the total length of the matrix in an electrophoresis experiment is typically much shorter than a meter, it is conventional to calculate the electric field strength (E) in units of $\left(\dfrac{Volts}{cm}\right)$.

The electric field is used as the motive force, pushing/pulling the charged particles through the matrix. Recall Equation 17.1a from your physics course, where F is the force exerted on the charged particle, q is the charge on the particle, and E is the field strength. If you care to draw an analogy to liquid chromatography, the electric field would be analogous to the high pressure flow of the mobile phase provided by the pumps in a liquid chromatography instrument. The electric field strength (E) is calculated as the ratio of the applied voltage over the length of the matrix (Equation 17.1b). From Equation 17a and 17b, we can derive the equation that defines the force experienced by ionic species in an electric field (Equation 17.1c).

$$F = qE \qquad\qquad \textbf{Eq. 17.1a}$$

$$E = \frac{V}{L} \qquad\qquad \textbf{Eq. 17.1b}$$

$$F = q\frac{V}{L} \qquad\qquad \textbf{Eq. 17.1c}$$

F = force exerted on a charged particle (newton)
q = charge on the particle (coulomb)
E = field strength $\left(\dfrac{Newton}{Coulomb} \text{ or } \dfrac{Volt}{meter}\right)$
V = applied voltage (volts)
L = length of matrix (meters)

Example 17.1

Calculate the electric field strength, E, in units of V/cm, for the following experimental design: 100 volts applied to a 12 cm × 12 cm matrix. Would the electric field strength increase or decrease with each of the following experimental changes?

    A. Applied voltage is increased to 200 V

    B. The matrix is changed to a 6 cm (wide) × 12 cm (length) gel

    C. The matrix is changed to a 24 cm × 24 cm gel

    D. The matrix is changed to 24 cm × 24 cm and the applied voltage is increased to 1,000 V

SOLUTION –

In the original experiment, E = 100 V divided by 12 cm (length) = 100/12 = 8.3 V/cm

    Modification A: increasing the voltage from 100 V to 200 V will increase E to 200/12 = 16.6 V/cm

    Modification B: there will be no change in E. The electric field strength is not affected by the width of the matrix. Because neither the applied voltage nor the length of the matrix were changed, there is no change in E.

    Modification C: increasing the length of the matrix from 12 cm to 24 cm will decrease E to 100/24 = 4.2 V/cm

    Modification D: doubling the applied voltage from 100 V to 200 V and doubling the length of the matrix will result in the same E as the original experiment, 200/24 = 8.3 V/cm

Problem 17.1: Calculate the force exerted on a singly charged protein in an electrophoresis experiment if the matrix used is 15 cm long and the applied voltage across the matrix is 20 V.

The field strength is one variable that determines the rate at which the particles migrate through the matrix, so E is one of the operator-controlled variables that affects the separation of analyte particles. Low-voltage experiments are typically in the range of 6 to 20 V/cm, whereas high-voltage experiments[6] can reach up to 200 V/cm. We will discuss the advantages and disadvantages of high field strength versus low field strength a little later in this chapter.

As mentioned previously, the *matrix* is an important component of the electrophoresis experiment and is typically composed of either paper or gel, generally on the order of 8 cm to 1 meter in length, saturated with buffer. Although you may be tempted to think of the matrix as analogous to the stationary phase in chromatography, remember that the electrophoresis matrix does not facilitate partitioning as in chromatography. Instead, the matrix accomplishes several other analytical objectives. *First*, it acts functionally as a resistor (an area of low conductivity) so that the voltage gets moderated across the distance of the matrix. This prevents the apparatus from functioning as an electrolysis cell across the matrix. *Second*, the matrix also serves to moderate mass transfer (mixing) of the buffer as a result of convection. Convection occurs anytime you have uneven heating within a fluid. Warm fluids rise and cool fluids sink. Heating occurs in an electrophoresis experiment due to the resistance to the flow of current in the matrix. Left unchecked, convection would severely increase the width of the zones in a zone-electrophoresis experiment. The matrix inhibits this kind of mixing, which gives zone electrophoresis much greater *resolution* than seen in free boundary electrophoresis.[7] *Third*, the large surface area of the matrix helps dissipate heat during the separation. *Fourth*, the matrix also serves as an adsorption support medium and/or a sieve through which the particles are pulled. As a sieve, the matrix will help differentiate and separate the particles based on size, shape, and mobility, given similar charge distributions.

Problem 17.2: Water will undergo electrolysis at approximately -1.5 V according to the following half reactions:

$$4OH^-(aq) \rightarrow O_2(g) + 2H_2O(l) + 4\,e^- \quad \text{Anode (oxidation)}$$
$$4H_2O(l) + 4e^- \rightarrow 2H_2(g) + 4OH^-(aq) \quad \text{Cathode (reduction)}$$
$$\overline{2H_2O \rightarrow 2H_2 + O_2 \qquad\qquad\qquad \text{Net reaction}}$$

Suppose you are performing an electrophoresis experiment using an aqueous buffer solution across a 30 cm matrix. What is the maximum voltage you should apply if you want to avoid electrolysis of the solvent buffer?

Electrophoresis is based on the physical principles that govern the motion of charged particles under the influence of an electric field. Therefore, the strength of the field (E) and the charge on the particle (q) will be important contributing factors for the rate of migration of the analyte through the matrix. However, the rate of migration is also influenced by the resistive force of friction as the particles move through the matrix. The force of viscous friction depends on the *frictional coefficient* (f) and the velocity of the particle. The viscous drag will be influenced by the mass and shape of the

Resolution

In Chapter 15 we used Equation 15.10 to define resolution for liquid chromatography. Resolution between any two bands in electrophoresis experiments is defined as:

$$R_s = \frac{d_{1,2}}{(W_1 + W_2)/2}$$

where $d_{1,2}$ is the distance between any two observed bands and W_1 and W_2 are the widths of band 1 and band 2, respectively. Compare and contrast this equation to Equation 15.10.

---

[6] Manz A. *Bioanalytical Chemistry.* Imperial College Press: London, **2004**.

[7] See opening profile titled "The Father of Electrophoresis" for a discussion of free boundary electrophoresis.

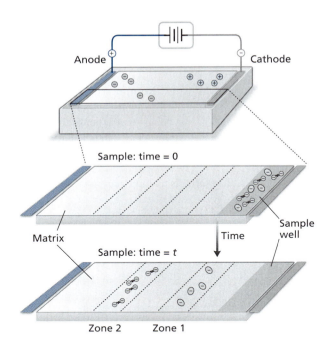

Figure 17.2 Electrophoretic migration of negatively charged particles through a matrix. The middle panel represents a single sample well at time = 0 before the electric field is turned on. The analyte mixture is loaded into the sample well. The bottom panel represents the same well after the electric field has been turned on and some time has been allowed to pass. The sample mixture has migrated under the influence of an electric field and after some time, the smaller, more mobile charged particles have traveled faster and further through the matrix compared to the larger, slower charged particles. Note that in this example, the negatively charged particles were loaded into a well at the negative (–) end of the gel and are moving toward the positive (+) end of the field.

ion. Using Newton's second law and assuming the speed of the particle is constant, the velocity or migration rate (v) of a charged particle in an electric field is therefore the result of the counterbalancing of these forces and is described by Equation 17.2. Simply put, as the electric field strength increases, the rate at which a particle migrates through the gel also increases. Similarly, as the charge on the particle increases, the rate of that particle's migration through the gel increases. Larger, bulkier particles will have higher frictional coefficients and move more slowly than smaller particles with the same charge. See Figures 17.2 and 17.3.

$$Eq = vf \qquad \textbf{Eq. 17.2}$$

where

E = electric field strength, $\left(\dfrac{N}{C} \text{ a.k.a } \dfrac{V}{m}\right)$

q = charge on a particle, (C)

v = velocity of the particle, $\left(\dfrac{m}{s}\right)$

f = frictional coefficient (a.k.a. viscous drag), $\left(\dfrac{N \cdot s}{m} \text{ a.k.a } \dfrac{Kg}{s}\right)$

Example 17.2

In a typical electrophoresis experiment, the electric field strength (E) will be a constant. Consider two particles, one with twice the charge as the other. Assuming that both particles have similar size and shape (and therefore similar frictional coefficients), demonstrate the effect of charge on the velocity of the two particles.

SOLUTION –

Solve Equation 17.2 for v.

$v = qE/f$

Let the first particle have charge q_1 and let the second particle have charge $2q_1$.

v of the first particle $= q_1E/f$

v of the second particle $= 2q_1E/f$

Because E/f is the same, we can see that v for the second particle will be two times faster than for the first particle. The velocity is directly proportional to charge, assuming all other effects are constant.

Problem 17.3: Using a similar approach as demonstrated in Example 17.2, show for two particles, both having the same charge, what the effect on velocity is if one particle is larger, having twice the frictional coefficient of the smaller particle.

You can also affect the migration rate by manipulating the matrix. For instance, you can increase the viscous drag, f, on the particles by increasing the density of the matrix. In the case of a gel matrix, the degree of cross-linking (average pore size) in the gel affects the density. Intrinsically, viscous drag on the particle will be a function of the pore size of the gel. Electrophoretic separation exploits the relationship between electromotive force and viscous drag and allows us to use electrophoresis to separate ions based on differences in mobility (size). The combination of effects is represented by the particle's *electrophoretic mobility* (μ_{ep}):

$$\mu_{ep} = \frac{v}{E} = \frac{q}{f}$$ **Eq. 17.3**

where μ_{ep} = electrophoretic mobility $\left(\dfrac{m \cdot C}{N \cdot s} = \dfrac{C \cdot s}{Kg}\right)$

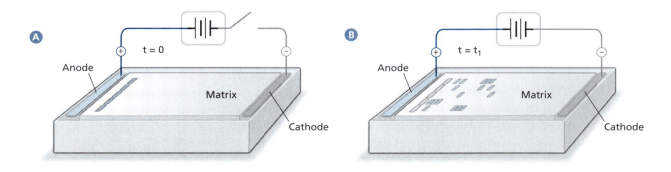

Figure 17.3 Panel (A) represents a set of sample mixtures loaded onto into the wells (dark blue) of an electrophoresis gel at time = 0 before the electric field is applied. Panel (B) represents the same system under the influence of an electric field after some time t = t₁ has elapsed. The sample mixture has separated into distinct bands or zones (light blue). Each band has a characteristic electrophoretic mobility μ which is specific to the experimental conditions.

Problem 17.4: Electrophoresis experiments conducted in solution—without the use of a matrix—are known as *moving boundary* or *free boundary electrophoresis*. Explain what problems would be expected with this experimental design and propose a few reasons why this technique is rarely used today.

Electrophoretic mobility (μ_{ep}) is determined as the velocity (v) per unit electric field strength and is proportional to the charge on the ion and inversely proportional to the friction coefficient. The friction coefficient (Equation 17.4) for a specific ion is proportional to the hydrodynamic volume of the ion (r) and the viscosity of the medium (η).

$$f = 6\pi r \eta \qquad \textbf{Eq. 17.4}$$

Substitution of Equation 17.4 into Equation 17.3 yields

$$\mu_{ep} = \frac{q}{6\pi r \eta} \qquad \textbf{Eq. 17.5}$$

The hydrodynamic volume (r) is a unique characteristic of each ion. Electrophoresis separates particles based partially on differences in hydrodynamic volume (see Figure 17.3). The analyst can manipulate the experimental environment by:

- Adjusting the electric field strength (E)

- Adjusting the pH, which in turn often affects the charge on the ion (q)

- Adjusting the viscosity of the matrix (η)

Of course, this is a simplistic evaluation. Many other physical factors influence the rate of migration; ultimately, the experiment is characterized by a variety of contributing variables that include pH, ionic strength, and also temperature. In practice, these influences are combined into an experimentally determined value for the electrophoretic mobility, μ, which is characteristic for a species under the precise conditions of the experiment. The expectation that each species in a mixture will have a unique mobility in an electric field—influenced by its mass-to-charge ratio, size, shape, and flexibility—is the basis for characterization by electrophoresis.

Problem 17.5: Finish calculating the electrophoretic mobilities, μ_{ep}, for bands B–E in Example 17.3 and create a table of results. (See Example 17.3 on p. 599.)

Of the various experimental parameters that affect electrophoretic results, temperature is the most concerning. *Joule heating* is produced any time current flows through a resistor. The matrix serves as a resistive medium; thus, one of the experimental realities of electrophoresis is Joule heating. Heating of the solution buffer can cause convection currents as well as other mechanical problems such as the denaturing of proteins and/or, in the case of a gel matrix, the dissolution of the matrix. Because temperature fluctuations can also cause changes in the viscosity (especially for gel matrix) and density, temperature changes will cause band broadening and reduced resolution. Distortion of zones, seen as curved bands, is an indication of inefficient cooling. The ionic strength of the buffer can also influence resolution (contributions to Joule heating and also counter ion effects), and therefore experiments are typically run at the lowest practical ionic strength.

Example 17.3

You have just completed an electrophoresis experiment and are creating a data table of electrophoretic mobility (μ) values from your experimental results. You have five bands (A–E), which you measure at 5 mm, 14 mm, 24 mm, 150 mm, and 400 mm from the center of the loading well. Your gel was 8 cm long and your experiment was conducted at an applied voltage of 200 V with a run time of 25 minutes. Calculate the μ_{ep} value for band A.

SOLUTION –

Calculate the electric field strength, E, for this experiment using Equation 17.1c:

$$E = V/L = 200 \text{ V}/8 \text{ cm} = 25 \text{ V/cm}$$

Next, calculate the velocity for band A (the band at 5 mm). Velocity is a measure of distance per time. Therefore, band A, traveling 5 mm in 25 min has a velocity of 5 mm/25 min, or:

$$v = 5 \text{ mm}/25 \text{ min} = 0.2 \text{ mm/min}$$

Finally, to determine the m value for band A, plug the values into Equation 17.3:

$$\mu = v/E \text{ where } v = 0.2 \text{ mm/min and } E = 25 \text{ V/cm}$$

Before solving, resolve the unit difference between mm and cm by converting the velocity to cm/min (10 mm = 1 cm) so that $v = 0.02$ cm/min:

$$\mu = (0.02 \text{ cm/min})/25 \text{ V/cm}$$

$$\mu = 8 \times 10^{-4} \text{ cm}^2/\text{V}\cdot\text{min}$$

17.3 The Basic Apparatus

The basic electrophoresis instrument is very simple and inexpensive.[8] It requires a power source, two electrodes, a conductive solution, and a matrix (typically gel or paper). The two electrodes (the positive anode and the negative cathode) are usually platinum wires, which form the electric boundary at either end of the matrix. Application of an electric potential across the two electrodes (turning the power source on) produces an electric field, and the movement of ions through the field produces a current. Figure 17.4 shows a side view schematic of simple gel (A) and paper (B) electrophoresis apparatuses, not including the power supply. Both systems are set up to separate a mixture of anionic analytes as indicated by the fact that the sample has been loaded onto the matrix at the cathodic end of the matrix. In the schematic of the gel apparatus (Figure 17.4[A]), the gel is floating in a reservoir of buffer solution. In the paper matrix setup (Figure 17.4[B]), the paper is saturated with solvent and kept moist by the wicking effect of the two ends dipped into the solvent reservoirs. Figure 17.5 shows a photograph of a typical gel electrophoresis apparatus. In some situations, the gel may be held between two glass plates—either for stability (if the gel is very thin) or to maintain moisture (paper matrix). The electrodes are connected to the power source, and the total voltage drop across the matrix is determined at the power source. Electrons

---

[8] In 2016, you could buy an electrophoresis apparatus such as the one in Figure 17.5 for less than $200 and the power supply for less than $600. For less than $1,000, you can buy a functioning electrophoresis unit or you can build your own for about $20 (see the Activity "Demystifying Electrophoresis: Build Your Own Electrophoresis Apparatus").

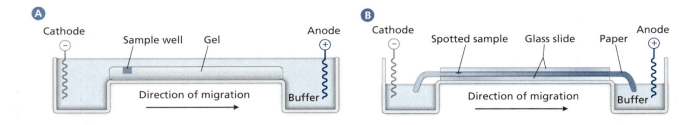

Figure 17.4 Schematic of a basic gel (A) and paper (B) electrophoresis apparatus. Analyte is either loaded into the sample well (gel) or spotted onto the paper. Application of an electric field causes anions to migrate toward the anode and cations to migrate toward the cathode. In both (A) and (B), the analyte is anionic.

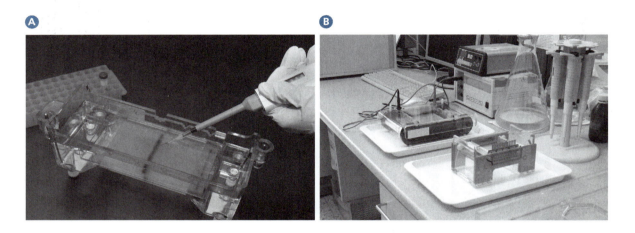

Figure 17.5 Panel (A) shows a sample and loading dye being loaded into preformed wells. Panel (B) shows the power supply and the comb used to make the sample wells.

flow from the power source to the electrodes and ions in the buffer flow through the matrix to complete the circuit. Cations flow toward the cathode and anions flow toward the anode. Likewise, ionic analyte ions are pushed or dragged through the matrix under the influence of the electric field. To separate a cationic sample mixture, you load the sample on the anodic side of the matrix. If your sample contains both cationic and anionic analytes, you would load the gel in the middle as shown in Figure 17.5(A).

Figure 17.5 shows a photograph of an electrophoresis apparatus. In panel (A) we see the analyst loading sample into indentations in the gel using a pipette. These indentations are called *wells*. The analyst also used a *loading dye* mixed with the sample to aid in seeing the sample. In addition, the loading dye is a charged molecule (often bromophenol blue), which is chosen to migrate at a rate that is just a little faster than the anticipated rate of the smallest DNA fragment. It is pushed through the gel ahead of all but the smallest DNA fragments. When the dye nears the end of the gel, the user knows it is time to turn off the power supply. In panel (B) of Figure 17.5, we also see a picture of the power supply as well as the comb used to create the sample wells when the gel was poured.

These wells typically are about 30 to 50 microliters in volume. With a paper matrix, the sample is "spotted" onto the paper using a capillary tube (in a manner similar to spotting a plate in thin-layer chromatography). The goal is to get the sample loaded in the smallest possible band. Ultimately, the resolution in a simple electrophoresis experiment is limited in part by the width of the loaded sample band. We will talk more about resolution in Section 17.8.

Zone techniques can be used in either analytical or preparative modes. The goal of an analytical experiment is to identify the various components in the sample. In an analytical experiment, a small amount of sample is loaded onto the matrix. The resulting

The term DNA marker can have several meanings. However, in the case of an electropherogram a DNA marker can be thought of as an internal standard used to calibrate the molecular mass of the observed DNA fragments in an electropherogram. A common molecular mass marker used to calibrate DNA mass is lambda DNA (sometimes called a *Lambda Ladder*). Typically when conducting an electropherogram several wells will be filled with the Lambda DNA marker and subjected to the same electrophoresis conditions as the experimental DNA. The figure seen here is lambda DNA (*cI857 ind* 1 *Sam*7). Each DNA fragment is 48.5 kilobases larger than the previous fragment. The smaller fragments migrate the furthest during an electropherogram. If one of your experimental DNA fragments traveled the same distance as the 242.5 kilobase fragment, then you know that your experimental DNA fragment is approximately 242.5 kilobases in size.

DNA markers are generally created in one of two ways. The first is to use a known piece of DNA and subject that DNA to restriction enzymes. The restriction enzyme will cut the DNA at specific sequences creating a series of fragments of differing sizes. The advantage of using restriction enzymes is the ease in which one can create more marker. The disadvantage is that the fragment sizes are determined by the enzyme. A second approach is to connect fragments of known size by use of a phosphodiester bond. This approach has the advantage of creating a linear scale (ladder) within the gel. The disadvantage is cost and complexity in creating the marker DNA. ▬

PROFILE
DNA Markers

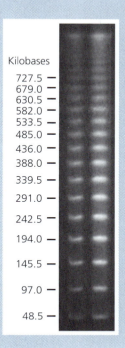

Kilobases
727.5 —
679.0 —
630.5 —
582.0 —
533.5 —
485.0 —
436.0 —
388.0 —
339.5 —
291.0 —
242.5 —
194.0 —
145.5 —
97.0 —
48.5 —

electropherogram shows discrete bands of sample components, usually with excellent resolution. The sample components are identified by comparing the banding pattern (which corresponds to different μ_{ep} values) of the analyte against known standards (also known as ladders or markers). We will look at examples of these in Sections 17.7 and 17.8.

The goal of a preparative experiment is to purify and recover a relatively large quantity of analyte. Compared to an analytical experiment, the analyst in a preparative experiment loads a larger amount of sample onto a larger volume of matrix, and the resulting electropherogram has sample bands that constitute a band across the entire matrix. The purified sample components are then recovered from the matrix.

As you increase the quantity of matrix used, the volume of the matrix will increase faster than the surface area. One of the instrumental challenges of the preparative electropherogram is the dissipation of Joule heating. A preparative experiment will get warmer than an analytical experiment run under the same conditions of matrix, buffer, and voltage. In order to keep the matrix at the desired temperature, preparative electrophoresis instruments are often water cooled. Figure 17.6 shows a preparative gel electrophoresis chamber manufactured by Bio-Rad Laboratories. The matrix is held in a gel tube jacketed by a water bath, which is circulated by a peristaltic pump.

17.4 Paper Electrophoresis

The simplest example of electrophoresis, paper electrophoresis, uses a matrix of either cellulose (paper), or cellulose acetate, a derivative paper in which the hydroxyl groups on the paper have been acetylated. Paper has the particular distinction that it does not have molecular sieving properties; that is, paper does not have pores that the particles have to travel through and therefore the separation is based mainly on the mass-to-charge ratio. Heavier particles will migrate more slowly than lighter particles of the same charge, and more highly charged particles will have larger velocities than similarly

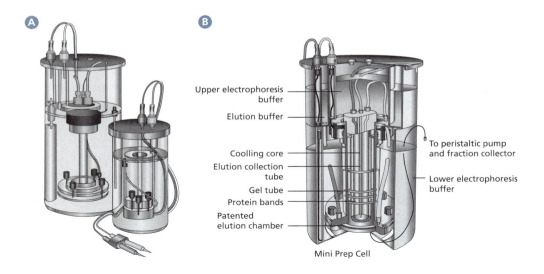

Figure 17.6 An example of a preparative electrophoresis chamber, Model 491 Prep Cell from Bio- Rad, for isolating protein molecules in complex biological matrices. Panel (A) is an overall schematic of the unit. Panel (B) is a schematic showing an expanded view of the gel column with analyte bands represented as gray disks.

massed particles with a lower charge. Without the confounding contribution of size or shape, paper electrophoresis benefits from having a simpler mode of separation than does gel electrophoresis. Therefore, the interpretation of paper electropherograms is straightforward.

ACTIVITY

Demystifying Electrophoresis: Build Your Own Electrophoresis Apparatus

The construction of a basic electrophoresis apparatus is very simple. All you need is a container to hold your gel (a plastic food container), two electrodes (aluminum foil or graphite pencil lead), and a power supply (a series of 9-volt batteries). You can cast* a gel using agar or pectin† obtained from the local grocery store. When you are casting your gel, you need a mold that will create the sample wells. A hair comb with wide teeth will work. You can use food coloring for the analyte. Many food coloring dyes are ionic and colorful, and therefore they will work nicely without the need for a detector or visualization step. We recommend loading your dyes in the middle of the gel because you do not yet know if the dye is cationic or anionic. (A step-by-step guide for this activity can be found at http://howtosmile.org/record/4255.)

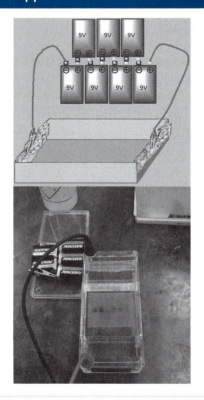

*Casting is the act of pouring a liquid into a mold and allowing it to set (harden).

† Pectin is a polysaccharide primarily found in citrus fruit. It is used by home canners when making jelly and jams to help the fruits jell. It is sold under such brand names as Ball RealFruit Pectin, Certo Fruit Pectin, Sure Jell, and more.

On the other hand, paper has the disadvantage that the hydroxyl groups on the surface of the cellulose have adsorptive properties toward polar analytes. This creates a partitioning separation mode, adding a dimension of paper chromatography to the electrophoresis experiment. The competing separation modes tend to result in poor resolution. Acetylation of the hydroxyl groups (cellulose acetate) removes the hydroxyl groups, making cellulose acetate the preferred matrix for some applications.

Samples can be retrieved out of the paper by cutting the paper at the appropriate zone and soaking in a small volume of buffer. These samples can then be further analyzed by other methods, such as nuclear magnetic resonance or Fourier transform infrared spectroscopy.

Despite its simplicity, paper electrophoresis is not commonly used today, being superseded by the more efficient gel techniques (Section 17.5). However, as we will see in Section 17.8, there are some very powerful two-dimensional electrophoresis techniques that use a paper matrix.

17.5 Gel Electrophoresis

Applications of gel electrophoresis, using either polyacrylamide (PA) or agarose as the gel matrix, are widely used today for the separation of RNA, DNA, and proteins. Both PA and agarose can best be described as porous, tangled webs of long polymers through which the particles in the sample must negotiate as they are pushed and pulled by an electric field force. You may want to envision the matrix as a three-dimensional net and the particles as travelers who are being pushed and pulled through the net. Gel offers enhanced resolution, compared to paper, because of its molecular sieving functionality.

As described in Section 17.2, separation of the particles in the sample mixture is a function of both size and charge, and the various components in the mixture will separate into bands or zones because the analyte particles have different overall electrophoretic mobilities through the matrix. *Gel electrophoresis*, a type of zone electrophoresis, will result in the separation of the mixture's components, which are resolved into zones or bands having similar characteristics in terms of charge, size, and shape. Both Figures 17.2 and 17.3 show a schematic of zone separation and the profile titled "DNA Markers" shows an agarose gel electropherogram. As with paper electrophoresis, the zones formed in a gel can be isolated at the end of the separation by cutting the gel and soaking the section of gel in buffer to retrieve the sample for further analysis.

One important difference between electrophoresis and liquid chromatography is that electrophoresis utilizes a continuous matrix (a solid polymer mass of interconnecting strands), whereas liquid chromatography utilizes a particulate solid phase (ultrasmall beads). A second important difference is the motive force. In electrophoresis the particles are being pushed or pulled by an electric field and travel through the separation matrix, whereas in chromatography the particles are pushed by a pressurized liquid or gas and travel in the mobile phase and partition between the mobile and stationary phases. As important distinction to understand is that in gel electrophoresis, large particles move very slowly (or not at all) because they cannot move through the matrix easily while smaller particles move more rapidly through the matrix as a result of their relatively lower frictional coefficient. This is exactly the opposite result of size exclusion chromatography (see Chapter 15).[9]

Polyacrylamide Gel Electrophoresis

The pore sizes typically found in *polyacrylamide gel electrophoresis (PAGE)* make it the most common form of gel electrophoresis experiment for protein analysis. These types of gels are formed from acrylamide monomers and a "cross-linker," typically N,N'-methylenebisacrylamide (also known as 'bis'). Together with an initiator and a catalyst, the polymerization reaction works to form the gel. Pore size will vary, but

---

[9] Recall that in size exclusion chromatography, the larger particles are excluded from the stationary phase and are eluted quickly.

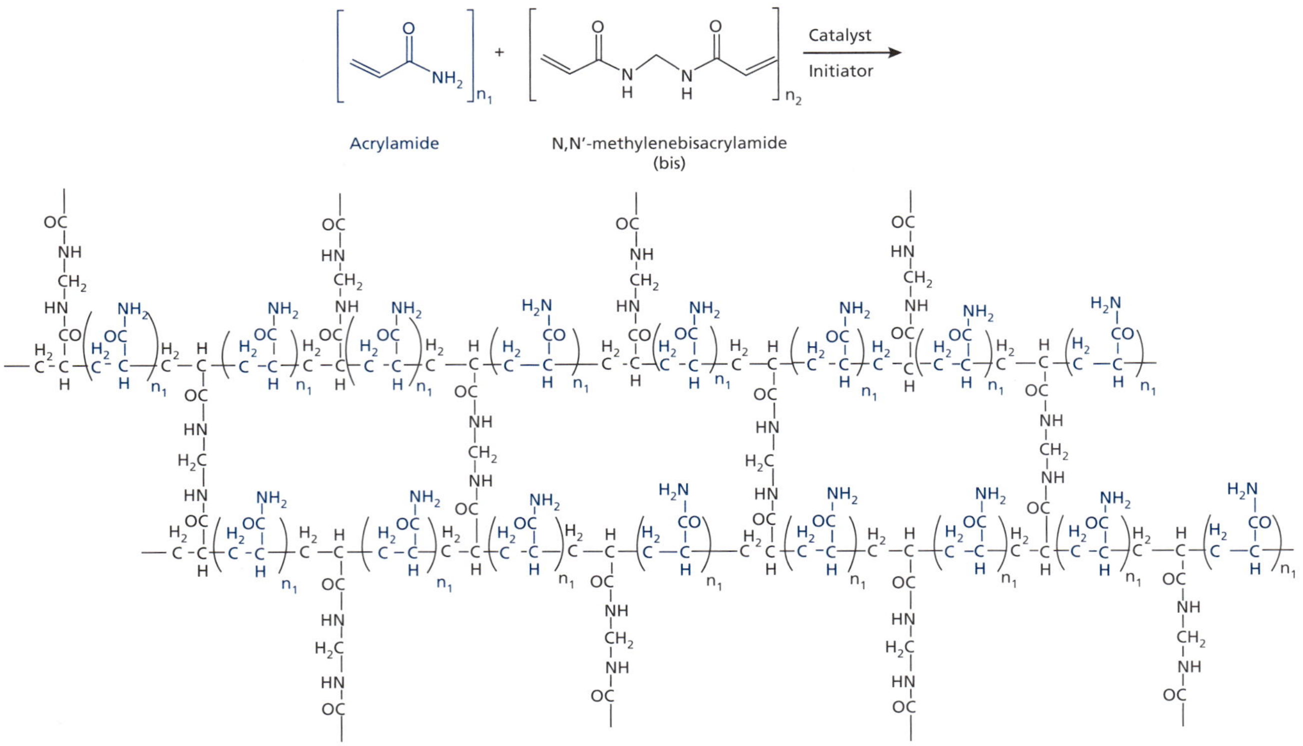

Figure 17.7 Schematic showing the polymerization of PA cross-linked with N,N'-methylenebisacrylamide.

TABLE 17.1: Polyacrylamide Gel Separation Ranges

PERCENT ACRYLAMIDE (T)	TYPICAL MOLECULAR WEIGHT RANGE
3–5	Greater than 100,000
5–12	20,000–150,000
10–15	10,000–80,000
15+	Less than 15,000

From Andrews, A. T. Electrophoresis: Theory Techniques and Biochemical and Clinical Applications, 2nd ed. Oxford University Press: New York, 1986.

gels are characterized by the average pore size, which is set at the time that the gel is cast. The average pore size is controlled by the amount of total monomer, the percent cross-linker in the gel, and the amount of catalyst present that affects the lengths of the polymer chains (see Figure 17.7).

Because of the highly toxic nature of the acrylamide monomer, these gels are typically purchased in the polymerized form. When purchasing PAGE gels, you need to specify the monomer density *and* the percent cross-linker. A standard nomenclature has been adopted in which one designates a T-value and a C-value. The value of "T" represents total concentration (m/v) of the gel (acrylamide + bis) and the "C" value represents the percentage of the total monomer that is the bis cross-linker. A PA gel with the designation of $T_{12}C_4$ would be a gel that is 12% polymer by mass and 4% of the polymer is comprised of the cross-linker. (See Table 17.1 for further information.)

Problem 17.6: Describe the composition of the following PA gels.

(a) T_9C_2 (b) T_9C_4 (c) $T_{15}C_7$ (d) T_8C_2

Problem 17.7: Use Table 17.1 to determine which of the gels in Problem 17.6 would be best suited for the separation of relatively small particles. Which would be best suited for the separation of relatively large particles?

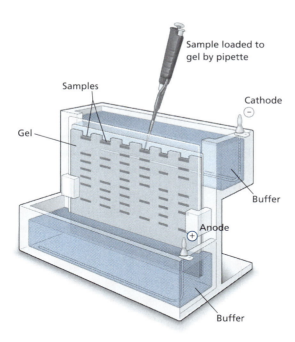

Figure 17.8 Vertical PAGE. Sample is loaded into wells at the top of the gel. Application of an electric field creates a force on the sample that pushes the sample through the gel. Separation is accomplished as a function of mobility through the gel.

Problem 17.8: What would be the correct way to label a gel that is 10% polymer by mass that has 2% cross-linker?

PAGE is typically performed using a vertical gel orientation (see Figure 17.8) in which the gel is very thin (~10 mm less) and cast between two glass plates. The samples are loaded into wells or indentations at the top of the gel and the samples move into the gel because of the applied electric field. In this case, the apparatus is set up so that the wells are cast at the negative end of the gel (cathodic end), and therefore the samples of interest travel in one direction, toward the positive anode. PA gels can also be used in horizontal setup; however, they must be thicker in order to have reasonably deep wells that can be loaded in that orientation.

Problem 17.9: If samples are loaded at the negative end of a gel, what happens to positively charged particles in a vertical PAGE experiment?

Take a moment to compare Figure 17.8 with Figure 17.4. The two apparatuses are fundamentally the same. The key distinction in vertical electrophoresis comes from the need to support the gel between the two glass plates. The thin PA gel, sandwiched between two glass plates, is suspended between the cathode and anode. The buffer solution at the top is in contact with the cathode, the buffer solution at the bottom is in contact with the anode. In this experiment, the buffer is set to a slightly alkaline pH so that the proteins will have a net negative charge. Once the electric field is turned on, the protein sample is pulled into the gel toward the positive electrode at the bottom of the apparatus. As you can see, control of the pH is an important analytical parameter. Typically, one of the wells of the gel is used for a set of protein standards of known molecular weight (markers), which can be used to benchmark the electrophoretic mobility (μ) of each analyte band by size. Having multiple wells across the same gel allows for simultaneous analysis of multiple samples under the same electrophoretic conditions, and therefore side-by-side comparisons can be made in terms of electrophoretic mobilities.

Sodium Dodecyl Sulfate Polyacrylamide Gel Electrophoresis

Another common practice in PAGE is to denature the protein samples using sodium dodecyl sulfate[10] (*SDS*). The SDS is a surfactant (i.e., a detergent) with a 12-carbon alkyl group with a sulfate at the end of the carbon chain. The interaction of SDS with the proteins disrupts the hydrogen bonding that gives proteins their specific folding and linearizes the protein. Furthermore, the SDS binds to protein at a ratio of 1.4 g of SDS per gram of protein (or one SDS per two amino acids), making the SDS-protein moiety polyanionic. The result is that the proteins all have the same charge-to-mass ratio, and all have the same (random) shape. SDS-PAGE thus allows for the electrophoretic separation to occur based on mass only (akin to gel filtration chromatography) without the confounding effects of particle shape contributing to the mobility (see Figure 17.9).

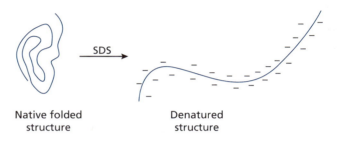

Figure 17.9 Schematic of SDS-denatured protein for analysis by SDS-PAGE.

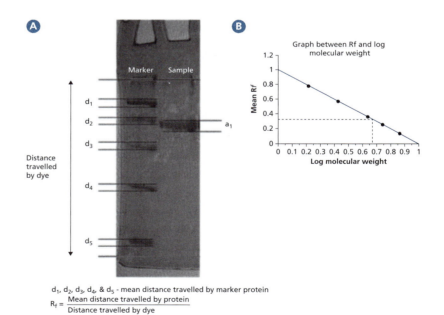

$d_1, d_2, d_3, d_4,$ & d_5 - mean distance travelled by marker protein

$$R_f = \frac{\text{Mean distance travelled by protein}}{\text{Distance travelled by dye}}$$

Figure 17.10 Schematic of SDS PAGE results (A) and the molecular weight calibration curve (B) for several proteins. The relative migration (R_f) is determined by dividing the distance traveled by each protein by the distance traveled by the marker dye. The calibration curve is constructed as the relative migration (R_f) versus log molecular weight for a set of standard proteins. The molecular weight of an unknown protein can be taken from the graph or determined from the formula for the trend line.

---

[10] SDS is the main ingredient in many household detergents.

The mobility of the particles varies by the log of the molecular weight. SDS-PAGE gels are run with molecular weight markers in one lane in order to determine electrophoretic mobility by mass. If one plots the $\log_{10}$ of the molecular mass of a protein against the mobility, you obtain a linear relationship (see Figure 17.10).

Problem 17.10: Using the graph in Figure 17.10, determine the molecular mass of a protein fragment that has a relative mobility of 0.75.

Unit Review

A dalton is defined as of the mass of one ^{12}C atom. One gram equals 6.022×10^{23} daltons. The abbreviation kD represents a kilodalton.

Agarose Gel Electrophoresis

Larger ions (> 200 kD) tend not to work well with PA gels. In order to get the large ions to move at all, the gel must have relatively large pore sizes. To get these large pores in a PA gel requires the gel to be very weakly cross-linked—so weak that it is not practical to deal with in terms of handling (too runny, too thin, too fragile) even with the support of glass plates. Therefore, for larger proteins, and especially for DNA applications, gel electrophoresis requires agarose as the matrix because of its relatively larger pore size, which makes it possible to use it for ions up to 50,000 kD in size. *Agarose* is a linear polysaccharide composed of d-galactose and 3,6-anhydrogalactose that is extracted from the cells of agarophyte, a red alga. Agarose, a natural product made from seaweed, is nontoxic and is considered a "green" material. Therefore, it is the matrix of choice for educational purposes, and you may have used agarose in one of your laboratory courses.

A typical agarose gel experiment begins with the analyst casting the gels, although they can be purchased premade. In contrast to PAGE, the pore size in an agarose gel is determined by the net percent agarose (m/v) in the gel, where 0.2% to 3% is typical. Agarose gels are usually used in a horizontal orientation and include preformed wells that may be cast at either end of the gel or in the middle. Position of the wells will depend on whether the sample mixture includes particles of interest that are all negative, all positive, or a mixture of positive and negative particles. Thus, particles can be separated in either direction (toward the positive or negative poles) or both directions if the sample contains a mixture of cations and anions of interest. Because nucleic acids are polyanions, gels for DNA analysis are always cast at one end of the gel and placed with the wells at the negative (cathodic) side of the field so that they will migrate toward the positive end of the gel.

Problem 17.11: Compare and contrast the physical parameters of the matrix used in PAGE and agarose electrophoresis. How does the analyst control those physical parameters?

17.6 Ending the Analysis: The Time Factor

In all of the experiments described, time plays an important role. The particles will continue to move in the electric field as long as the electric field exists. In order to stop the electrophoresis, the field is turned off. The time that the particles are exposed to the electric field is therefore one of the key experimental parameters. If the experiment is too short, the particles will not move far enough to obtain the desired resolution. On the other hand, diffusion increases with time and band spreading occurs throughout the entire run time of the experiment. Diffusion can make observation and/or recovery of the band difficult. Furthermore, if the experiment is too long, the particles can actually migrate off the end of the gel and be lost in the buffer solution. In determining the optimum time for the experiment, the analyst seeks to find a time that fully resolves each desired band while maintaining an acceptably narrow sample band in a reasonable amount of laboratory time. Typically, experiment times are in the 20- to 90-minute range. Higher voltages can be used to make run times shorter, but this risks overheating the gel.

Typically, one adds a tracking dye to the loading buffer[11] that serves as a visual clue as to how far the particles have been able to travel in a given time, although a reference experiment is required in order for this to be useful. If a sample contains a mixture of compounds across a wide range of mobilities, it may be necessary to use one gel for best separation of the smaller components, and another gel for best separation of larger components.

17.7 Gel Sample Detection

Visualization

In most applications of electrophoresis, the particles under analysis are not visible to the naked eye. Therefore, detection is the final step in the analysis. Staining is the most common and facile method of detection; it involves removing the matrix from the electrophoresis apparatus, soaking it in an appropriate stain, and then rinsing and visualizing by eye or with an appropriate light and/or lens. Table 17.2 lists a few examples of electrophoresis stains. Commonly used stains for protein samples include silver stain, Coomassie blue, and fluorescamine (also known as fluram). Both silver stain and Coomassie blue result in colored bands that can be seen with the unaided eye, with silver stain being the more sensitive of the two treatments. Fluorescamine reacts with primary amines in the protein, resulting in a fluorescent derivative that is visualized by placing the gel on an ultraviolet light table and observing the fluorescence. Ethidium

TABLE 17.2:   Stains for Gels

ANALYTE	STAINING REAGENT	COMMENTS
Amino acids, peptides, and proteins	Ninhydrin 2,2-dihydroxyindane-1,3-dione **CAS#: 485–47-2**	Very sensitive stain for amino acids, either free or combined in polypeptides; used after paper electrophoresis
Proteins	Amido back 10B 4-amino-5-hydroxy-3-[(4-nitrophenyl)azo]-6-(phenylazo)-2,7-naphthalene disulfonic acid, disodium salt **CAS#: 1064–48-8**	Binds to cationic end of proteins; adsorbs onto cellulose, giving high background staining with paper and dehydration and shrinkage of polyacrylamide gels Most appropriate for transferred membrane blots
	Coomassie brilliant blue R250: **CAS#: 6102–59-2** G250: **CAS#: 6104–58-1**	Originally developed for the textile industry, these dyes are now common stains for protein analysis. G250 has two additional methyl groups compared to R250. The dyes bind to basic groups on proteins and also by nonpolar interactions.
	Silver stain $(AgNO_3)_{aq}$	Silver are reduced to silver metal by the proteins. Detection in the ng range is possible.
	Ponceau S **CAS#: 6226–79-5**	A very common dye for western blotting. The stain is reversible with water washes.
	Dansyl chloride **CAS#: 605–65-2**	Fluorescent when bound to primary amino acid groups
	Fluorescamine CAS#: 38183–12–9	Fluorescent when bound to primary amino acid groups
Glycoproteins	Alcian blue **CAS#: 33864–99-2**	Used to stain acidic polysaccharides and glycoproteins

(Continued)

---

[11] For example, xylene cyanol, Chemical Abstracts Service (CAS) number 2650–17-1, is used as a tracking dye in DNA separations on either PA or agarose gels; cresol red, CAS number 1733–12-6, is used as a tracking dye in PAGE of proteins and in agarose gels of DNA or RNA; and bromphenol blue, CAS number 34725–61-6, is also used for both PA and agarose applications of protein or nucleic acid.

TABLE 17.2:   (Continued)

ANALYTE	STAINING REAGENT	COMMENTS
Polynucleotides	Acridine orange *N,N,N',N'*-Tetramethylacridine-3,6-diamine **CAS#: 494–38-2**	Intercalating fluorescent dye When bound to DNA: excitation = 502 nm, emission = 525 nm (green) When bound to RNA: excitation max = 460 nm, emission max = 650 (red)
	Pyronin Y (or G) **CAS#: 92–32-0**	A popular RNA stain; appears red to the eye and fluoresces at 302 nm
	Ethidium bromide (EtBr) **CAS#: 1239–45-8**	Intercalating fluorescent dye; a very popular nucleic acid stain. The fluorescent intensity of EtBr increases 10-fold when intercalated with DNA. Absorption max = 285; emission max = 605.

Adapted from Melvin, M. *Electrophoresis* (Analytical Chemistry by Open Learning). John Wiley & Sons: Chichester, UK, 1987.

bromide is the most commonly used stain for nucleic acids. Ethidium bromide intercalates into the DNA backbone and it is highly fluorescent when exposed to ultraviolet light (see Figure 17.11). However, ethidium bromide is classified as a mutagen, and instructional laboratories will often use a colored dye such as bromocresol green. A scanner can also be used to visualize—and even quantify—the band intensities. If the samples undergoing analysis are radiolabeled, the visualization process can involve the use of X-ray or photographic film, referred to as autoradiography. Alternatively, radiolabeled samples can be quantified by scintillation counting, in which the matrix is sliced into lanes and then rows and each area is transferred into its own vial, soaked in scintillation fluor, and counted in a scintillation counter.

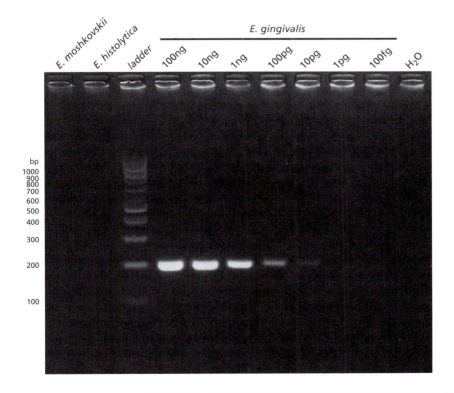

Figure 17.11 Agarose gel electropherogram containing λ-ladder markers and PBR 322 plasmid DNA. Gel was soaked in a 2% ethidium bromide solution for 20 minutes and then 20 minutes in a buffer wash solution. It was then placed on an ultraviolet light table.

Flashback

We first encountered scintillation counters when we introduced the photomultiplier tube in Section 6.7. In electrophoresis, you can use radioactive-labeled isotopes in the DNA or protein. The sample is then soaked in a solution containing a fluorescent or phosphorescent compound (the fluorophore), and the emitted light is detected using a photomultiplier device.

Blotting

In the case of gel electrophoresis, visualization can also be made more specific by transferring or *blotting* the sample bands from the gel onto a heartier substrate. The most well-known of these techniques, the *Southern blot*, is named after its developer, Edwin Southern. A Southern blot transfers DNA fragments from an agarose gel onto a nylon or nitrocellulose membrane in order to analyze the sample bands for the presence of a specific DNA sequence of interest. Transfer of the DNA to the membrane is accomplished by capillary action. A dry sponge is placed in buffer solution, and the gel is place on top of the sponge. The membrane is placed on top of the gel. A wicking medium (typically paper towels) is place on top of the membrane and a weight is placed on the very top of the stack (see Figure 17.12). Buffer is wicked vertically through the gel and transfers the DNA to the membrane. The nylon membrane is removed from the stack and soaked in a hybridization solution in which small radioactively labeled, sequence-specific DNA probes are allowed to bind to their complementary sequence. If the DNA sequence of interest exists among the bands of DNA components on the membrane, the radiolabeled complementary sequence will remain embedded in the membrane after the membrane is rinsed. In this way, the radioactive-labeled complement strand acts as a sequence specific probe. Visualization of the bound probe is then a positive indicator that the sequence of interest is among the DNA in the analysis. This technique is especially useful for the analysis of crime scene samples, paternity, and genetic disease.

Problem 17.12:   Compare and contrast the common detection methods used in liquid and gas chromatography with those used in paper and gel electrophoresis.

Quantitative Electrophoresis

For paternity testing and other qualitative experiments, it is sufficient to simply observe the patterns in an electropherogram for the presence or absence of a band. However, if

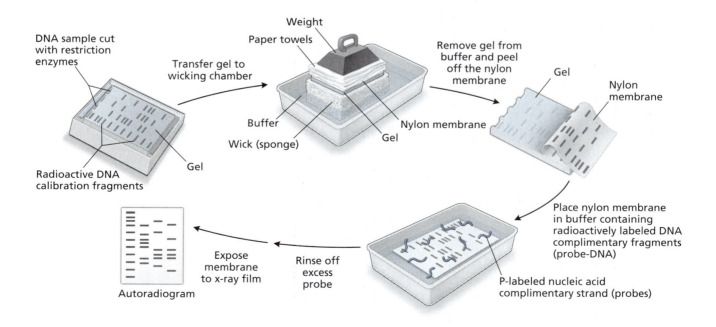

Figure 17.12 Schematic of a Southern blot electrophoresis experiment.

you are using electrophoresis to study protein metabolism kinetics, for example, you must have the ability to measure the relative quantity of protein within each band. Early efforts at quantifying bands in an electrophoresis experiment were accomplished using a device called a *densitometer*. The densitometer is essentially an ultraviolet-visible (UV-vis) spectrometer with a light source and a photodetector. The name *densitometer* comes from the fact that it is the custom to report the absorbance of an electrophoretic band in units of optical density. The *optical density* (OD) is defined as 1 absorbance unit if the path length is 1 cm. The gel would be placed between the source and the detector, and the amount of light reaching the detector would be measured. It was a tedious and time-consuming process that involved manually scanning the gel by sliding the gel over the detector and measuring OD as a function of the position on the gel. Over time, several companies developed densitometers that would automate this scanning process, and these sort of devices dominated the market for quantitative electrophoresis work. The traditional densitometers are still in use in many laboratories today. However, that technology is rapidly being taken over by digital imaging devices. For example, several companies market "densitometers" that are actually based on scanner technology. These devices are more correctly referred to as *molecular imagers*. The gel is scanned, much like you would scan a document on a photocopy machine, and the grayscale pixel density is analyzed by software and displayed as a peak. The pixel density is proportional to the OD so the output looks similar to the traditional densitometer. Many of these scanners are actually quite sophisticated and can switch sources, utilizing laser to stimulate fluorescence if needed and incorporating a variety of different optical filters to enhance contrast. The user inputs the stain or dye being used and the scanner automatically optimizes the source and optics combinations.

The height of the peak is proportional to the concentration, and the total area under the peak allows you to determine the total quantity of analyte in the band. These devices also have the advantage of allowing you to archive digital copies of your electrophoresis data. Figure 17.13 shows an electrophoresis scanner marketed by Bio-Rad. Calibration standards are commercially available from several sources. In addition to the use of scanner-like molecular imaging devices, very sensitive charge coupled device digital cameras have been used as molecular imaging devices. Cameras are primarily used for smaller gels, and the image can be directly imported to a computer using a USB cable. Analysis of the image is conducted using software in a manner similar to the scanner software.

Other Blotting Techniques
Other, derivative, techniques have been developed to accomplish similar goals with RNA, proteins, and antibodies. The Northern blot is used to transfer RNA from agarose onto a nitrocellulose sheet for similar sequence-specific analysis. The Western blot is used to transfer proteins from gel to nitrocellulose paper for identification by antibody recognition.

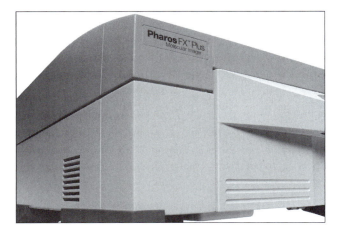

Figure 17.13 The PharosFX molecular imager by Bio-Rad.

17.8 Enhancing Resolution

Resolution is a key concern in zone electrophoresis. The resolution of a band in zone electrophoresis is defined as the ratio of the distance between bands to the average width of the two bands in question. For a sense of scale, the distance between bands and the width of the bands are usually measured in millimeters:

$$R_s = \frac{d_{1,2}}{(w_1 + w_2)/2}$$

Eq. 17.6

where

$d_{1,2}$ = distance between any two bands

W_1 and W_2 = width of bands 1 and 2, respectively

Problem 17.13: For the lane labeled "Marker" in Figure 17.14, determine the resolution between the two bands labeled:

(a) LD_2 and LD_3

(b) LD_{30} and LD_{29}

(c) LD_{22} and LD_{17}

(d) LD_{21} and LD_{18}

(e) LD_{29} and LD_{28}

Figure 17.14 shows many well-resolved bands. However, there are some poorly resolved bands in the lanes labeled "February" and "September." Ultimately, the distance between bands is influenced by the width of the bands, and the theoretical minimum width of a sample band is determined by the width of the loading well. In other words, if the analyte began its journey at the top of the gel in a region (sample well) 2 mm wide, then the narrowest sample band you could hope to achieve after electrophoresis is greater than or equal to 2 mm. Because diffusion occurs within the gel, we find that width of the bands generally increases with time. Therefore, when one is performing electrophoresis, sample solutions are made to be as highly concentrated as possible, in order to load very small amounts of sample into the wells. Typical sample volumes, including the loading buffer, are in the range of 10 to 30 microliters. Other techniques that can be used to improve resolution include *discontinuous electrophoresis, isoelectric focusing,* and *two-dimensional electrophoresis.*

Discontinuous Electrophoresis

Discontinuous electrophoresis[12] is often abbreviated as "disc electrophoresis." Instead of pipetting the sample into the wells of the running gel, samples are loaded from a *stacking gel* into the running gel by using an electric current to load the sample into the running gel (see Figure 17.15). The stacking gel is cast such that it has very large pores and a pH that is lower than the pH of the running gel, which is more basic. The stacking gel is typically made with a buffer having a weak acid component (such as glycine, pH = 9.6), and the samples are loaded into the stacking gel. The stacking gel is "stacked" on top of the running gel. When the electric field is turned on, the weakly acidic ions in the top of the running gel move into the stacking gel, where they become neutral (by losing a proton at the interface between the stacking gel and the running gel) and they stop moving. The loss of charge carriers causes the resistance of the gel to increase greatly

---

[12] Shuster, L. Preparative Acrylamide Gel Electrophoresis: Continuous and Disc Techniques. *Methods Enzymol.,* **1971**, 22, 412–433.

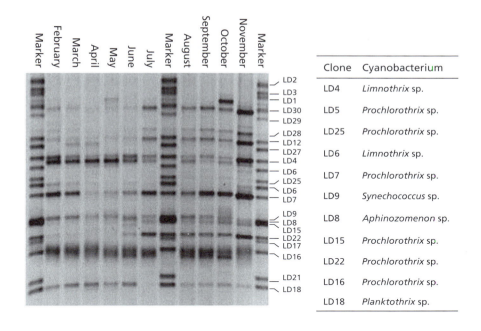

Clone	Cyanobacterium
LD4	*Limnothrix* sp.
LD5	*Prochlorothrix* sp.
LD25	*Prochlorothrix* sp.
LD6	*Limnothrix* sp.
LD7	*Prochlorothrix* sp.
LD9	*Synechococcus* sp.
LD8	*Aphinozomenon* sp.
LD15	*Prochlorothrix* sp.
LD22	*Prochlorothrix* sp.
LD16	*Prochlorothrix* sp.
LD18	*Planktothrix* sp.

Figure 17.14 Resolution is the ratio of the distance between bands and the average width of the bands in question.

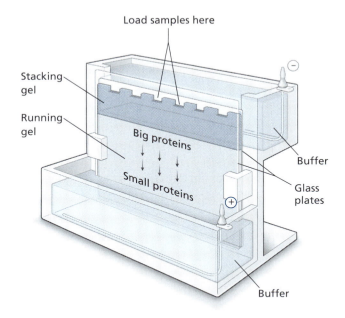

Figure 17.15 Schematic of discontinuous electrophoresis. The stacking gel is used to rapidly load samples in ultra-narrow bands onto the running gel.

at the boundary between the two gels. As a result, per Ohm's law (Equation 17.7), the voltage drop (V) at the boundary becomes very large:

$$V = IR \text{ (Ohm's law)} \qquad \textbf{Eq. 17.7}$$

A locally high electric field moves the sample ions to the running gel very quickly, forming a tight band (approximately 0.01 mm thick) and thereby ensuring much better overall resolution as compared to a manual loading of the sample.

Problem 17.14: The electrophoretic mobility μ is similar in principle to the k_r value in liquid chromatography (see Chapter 15). Compare and contrast the techniques of chromatography and electrophoresis in the context of the coefficients μ and k_r.

Isoelectric Focusing

Isoelectric focusing is a different type of electrophoresis experiment that is a differential steady-state technique that allows for high resolution of protein samples. In isoelectric focusing, the gel is cast in such a manner that the pH within the gel varies with length. As a protein moves through the gel, it experiences these different pH environments. In the case of a positively charged protein, it will deprotonate when the pH rises above the pKa; the proteins will become neutral and they will stop migrating in the gel. The pH at which the protein becomes neutral is called its *isoelectric point* and is given the symbol pI. If the particles diffuse out of their isoelectric band, they will regain charge and be forced back into the band. The result is very narrow bands yielding high resolution. Thus, the resulting electropherogram displays the protein components of the sample as a function of pI (see Figure 17.16).

The gels used to conduct isoelectric focusing experiments are seeded with special polymers called *polyampholytes*.[13] Polyampholytes are polyelectrolytes that contain both cationic and anionic monomeric units with amphipathic[14] amino and carboxy groups. The existence of these acidic and basic groups varies throughout the gel and are responsible for the pH gradient in an isoelectric focusing gel. These carrier ampholytes assemble into a pH gradient under the influence of an electric field and can be made to generate gradients that vary by as little as 0.05 pH units between zones.

Isoelectric focusing techniques are critical analytical tools. They are used for the identification of isozymes (enzymes that have different structures but the same

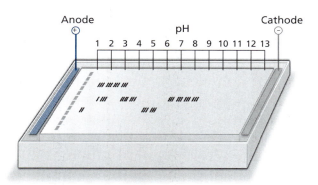

Figure 17.16 Schematic representation of an isoelectric focusing gel. At low pH all of the protein fragments are cationic. As the protein reaches a point in the gel where the pH of the gel is equal to the pKa of the protein, the protein becomes neutral and stops migrating.

---

[13] Dobrynin, A. V.; et al. Polyampholytes. *J. Poly Sci. Part B: Poly Phys.*, **2004**, *42*(*19*), 3513–3538.

[14] Amphipathic molecules have both hydrophilic and hydrophobic functionality. The most common examples are soap and detergent.

function); enzymopathies (changes in enzyme structure and function); and the diagnosis of protein-based diseases, such as hemoglobin mutations.

Problem 17.15: How is the separation mechanism in an isoelectric focusing experiment fundamentally different from other electrophoresis techniques?

Two-Dimensional Electrophoresis

A simple paper electrophoresis experiment, even at high-voltage, may not yield the desired resolution. Therefore, two-dimensional electrophoresis–electrophoresis and two-dimensional electrophoresis–chromatography techniques (also known as "fingerprinting" experiments) have been developed. In a two-dimensional technique, the first dimension refers to a paper electrophoresis experiment as has been described. The second dimension refers to a second experiment conducted perpendicular to the first by turning the paper 90 degrees. The second dimension can be either a second electrophoresis experiment or a paper chromatography experiment. In a two-dimensional electrophoresis–electrophoresis experiment, the particles are separated by charge in both dimensions; therefore, the particles that move the farthest (in both dimensions) will end up farthest from the starting point on a diagonal and complex mixtures can become better resolved. In a two-dimensional electrophoresis–chromatography experiment, the particles are separated in the first dimension by charge and in the second dimension by polarity; thus, the resulting "fingerprint" contains a multiplicity of information about the sample components.

Problem 17.16: One of the most powerful gel electrophoresis techniques is accomplished by a two-dimensional technique in which a protein mixture is separated by isoelectric focusing in the first dimension, followed by SDS-PAGE in the orthogonal dimension. Describe the resulting information that this experiment provides.

There are other variations of two-dimensional electrophoresis in use, some of which involve changing the direction of the electric field during the course of the experiment. These two-dimensional electrophoresis techniques involving time-dependent changes to the electric field are known as *pulsed field gel electrophoresis* (PFGE) experiments.

In PFGE, the sample is loaded onto the matrix, and electrophoresis is run for a designated amount of time (a pulse) in a single direction. After the pulse, the electric field is reoriented, so that when the next pulse occurs, the particles in the sample have to reorient themselves to align with the new field direction. PFGE is a valuable tool for separating and isolating very large DNA fragments (~20,000 to 2 million bases) because the technique enhances the resolution and separation. Because DNA fragments are long, thin polymers, the reorientation process occurs on a timescale relative to the length of the fragment. Smaller fragments reorient more quickly and begin again to move in the field in its new direction, whereas larger fragments take a longer time to reorient and begin moving in the new direction at a delayed rate. Pulse schemes (length of pulses and orientation angle changes) have been designed for various applications, with the result of excellent resolution of DNA fragments of masses well beyond the 50,000 kD cutoff for normal agarose electrophoresis.

Figure 17.17 shows two different pulse schemes. On the left side of Figure 17.17, the electric field is momentarily reversed at set time intervals. The ability of analyte particles to reorient in the reversed field becomes an additional separation variable. On the right side of Figure 17.17, the gel is rocked back and forth (left to right) in the electric field and the particles are forced to undergo continual reorientation in the electric field.

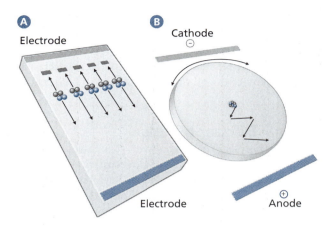

Figure 17.17 Panel (A) shows a field inversion gel electrophoresis experiment. The field is periodically reversed for a brief moment throughout the duration of the run (+/– to –/+). Panel (B) shows a rotating gel electrophoresis experiment. The field is held constant while the gel rotates first in one direction and then the other direction in a continuous manner throughout the duration of the experiment.

17.9 Capillary Electrophoresis

Introduction to Capillary Electrophoresis

Capillary electrophoresis (CE) incorporates aspects of traditional gel-matrix electrophoresis (also known as slab electrophoresis), gas chromatography, and liquid chromatography techniques. As the name implies, CE is electrophoresis conducted in a capillary tube.[15] In early CE instruments, these capillary tubes were adapted directly from capillary gas chromatography, and much of our theoretical understanding of the separation mechanisms in CE has been similarly adapted from chromatographic theory. However, there are some important distinctions. In traditional slab techniques, we referred to the gel as the *matrix,* but in CE the "matrix" is a buffered liquid that has more in common with the mobile phase of a liquid chromatography experiment than the gels of a slab electrophoresis experiment. In fact, the buffer does move through the capillary in a manner similar to a liquid chromatography experiment, so thinking of the buffer as a "mobile phase" has some merit. As a result, the detection systems for CE are adapted directly from liquid chromatography and have been applied to CE instruments with great success. Also, to take full advantage of the mobile phase flow, the capillary tubes in CE must have a much narrower inner diameter than those in GC. We will explain why later in this chapter.

CE offers the analyst a technique that requires much less sample, has much higher resolution, and has a much shorter analysis time than slab gel electrophoresis. These are all characteristics needed for large-scale projects, such as genome analysis.

The Instrument

The CE instrument is conceptually simple. The capillary is a fused silica capillary similar to those used in capillary gas chromatography (see Chapter 16). The capillary is filled with either buffer or gel. The two ends of the capillary are placed in separate reservoirs containing buffer and the electrodes from a high voltage (30 kV) power supply are placed separately at each end. A schematic of a CE instrument utilizing a UV-vis spectrometer

> **Slab Electrophoresis**
>
> With the advent of capillary electrophoresis, traditional techniques that incorporate a gel matrix have become known collectively as slab electrophoresis.

---

[15] Schmitt-Kopplin, P. *Capillary Electrophoresis: Methods and Protocols (Methods in Molecular Biology).* Humana Press: Totowa, NJ, **2008;** Ruttinger, H-H. *Affinity Capillary Electrophoresis in Pharmaceutics and Biopharmaceutics.* Informa Healthcare, **2003;** Altria, K. D. *Capillary Electrophoresis Guidebook: Principles, Operation, and Applications (Methods in Molecular Biology).* Humana Press: Totowa, NJ, **2010.**

In 1953, Watson and Crick published a paper on the helical structure of DNA,[*] and within a few years, the full implications of DNA translation into protein were understood. It was clear that the sequencing of DNA would lead to a better understanding of evolution and genetic diseases. Although classical methods of DNA sequencing were laborious, researchers steadily added new sequences to the database. In 1980, the first automated DNA sequencers became available, and by the end of that year, a public repository of DNA sequences was established (GenBank). Figure 17.18 shows the number of DNA sequences deposited into GenBank as a function of the year. With the advent of automated DNA sequencers, the dream of mapping the entire three-billion base pair human genome seemed within reach, and in 1988, the US Congress funded a Human Genome Project (HGP) coordinated by the National Institutes of Health and the Department of Defense. A large part of the early work of the HGP was devoted to the development of improved technologies for accelerating the sequencing of the genome. One of the fruits of this endeavor was the rapid advancement of capillary electrophoresis (CE). The impact of CE on the HGP is clearly evident in Figure 17.18, where you can see the slope of the trend line increase with the introduction of commercially available CE instruments. The rate of GenBank deposits has quadrupled since the introduction of commercially available CE instruments. ■

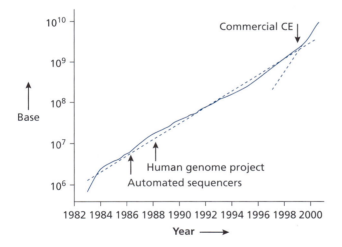

Figure 17.18 DNA sequences deposited into GenBank. The dashed lines represent trend lines for the indicated years. Deposits doubled every 2 years until 1999. Since 1999, deposits have doubled every 6 months.

[*]Watson, J. D.; Crick, F. H. C. *Genetical Implications of the Structure of Deoxyribonucleic Acid. Nature,* **1953**, *171,* 964–967.

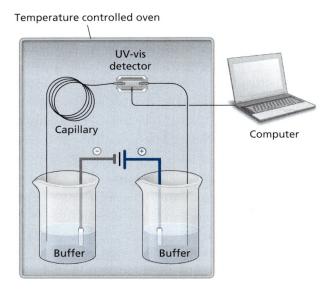

Figure 17.19 Schematic of a CE instrument.

as the detector is depicted in Figure 17.19. In addition to UV-vis detection, CE is quite versatile and easily adaptable to many other detectors. For example, CE can be readily adapted to electrochemical detectors, mass spectrometry detectors, laser-induced fluorescence detectors, radioisotope-counting detectors, and so on.

Separation Efficiency

Just as in the slab techniques, charged particles in CE are separated based on their mobility in an electric field. Similarly, mobility is related to both the charge on the particle and the size/shape of the particle. A simplistic interpretation relates the ionic charge to hydrated volume in solution. The higher the charge-to-volume ratio, the faster the ions will travel through the capillary. In addition, the surface of the capillary walls can be modified by adding coatings to create a combined electrophoretic-chromatographic mode of separation. As you might expect, separation efficiency is modeled using the van Deemter equation. We first introduced the van Deemter equation in Chapter 15, and it is repeated here for convenience and comparison.

$$H = A + \frac{B}{v} + C \cdot v$$

where

H = height equivalent of a theoretical plate
A = eddy diffusion coefficient
B = longitudinal diffusion term
v = linear velocity of the mobile phase
C = resistance to mass transport term

Recall that H represents the distance between theoretical plates,[16] and the ability to separate two similar ions is a function of how many theoretical plates you can stack in a single column.[17] So as H decreases, the efficiency of the column increases. However,

---

[16] The idea of a "plate" is borrowed from a distillation column in which the efficiency is governed by how closely you can stack the condensation plates.

[17] The number of theoretical plates is denoted with an upper case N.

there is a significant difference between gas chromatography efficiency and CE efficiency. In CE, the analyte is carried in a uniformly flowing liquid, so two of the factors that contribute to theoretical plate height are essentially eliminated for CE. These are eddy diffusion (A) and mass transport (C). This greatly simplifies our model, and in practice the van Deemter equation for CE can be simplified to:

$$H = \frac{B}{v}$$ **Eq. 17.8**

where

B = longitudinal diffusion term
v = linear velocity of the mobile phase

Ideally, the only significant contribution to band broadening in CE is longitudinal diffusion, and the result is extremely high separation efficiencies. In a typical CE experiment, the analyst can expect efficiencies in excess of half a million theoretical plates, far surpassing the number of that found in a gas chromatography or high-performance liquid chromatography experiment by an order of magnitude or more.

CE also benefits from an ability to dissipate heat more easily than slab electrophoresis. The Joule heating in CE is estimated as:

$$Q = E^2 \Lambda c$$ **Eq. 17.9**

where

Q = Joule heat (Ω/cm^3)
E = voltage gradient (V/cm)
Λ = molar conductivity of the electrolyte ($cm^2mol^{-1}W^{-1}$)
c = molar concentration of the electrolyte (mol/L)

In CE, the surface area of the capillary walls is large with respect to the volume contained within, and so the walls can dissipate the Joule heating much more efficiently than the slab gels. As a result, CE experiments can be run at much higher voltages than traditional slab techniques. The end result for the user is a much shorter run time.

Electro-osmotic Flow

In a manner similar to slab electrophoresis, the separation of ions in CE is the result of different migration rates of ions in an electric field. However, in a CE experiment, unlike in a slab electrophoresis technique, the matrix (buffer) also flows through the capillary. The surface of a fused silica capillary tube is covered with silanol (Si-OH) groups, which remain protonated below a pH of about 2. Above a pH of 2, the silanol groups are ionized to SiO^-. Solvated cations from the buffer solution serve to counterbalance the negatively charged surface of the capillary tube. Under the influence of the applied electric field, these cations migrate toward the cathode, and the entrained solvent ions are likewise transported toward the cathode. Because the capillaries used in CE are so narrow, the effect extends all the way to the center of the capillary tube. The result is a net flow of solvent toward the cathode (see Figure 17.20). This net flow of solvent through the capillary is called *electro-osmotic flow* (EOF). So, in addition to the migration of ions through an electric field, the conditions of CE include mass transport of the buffered solvent, and therefore all of the components of the analysis (regardless of charge, or even neutrality) typically elute at one end of the capillary.

In CE, the velocity of an ion as a function of electric field strength is called the *electro-osmotic velocity* (v_{eo}) and is a function of the electric field strength and the *electro-osmotic mobility* (μ_{eo}). The derivation is similar to the electrophoretic mobility we derived in Equation 17.3.

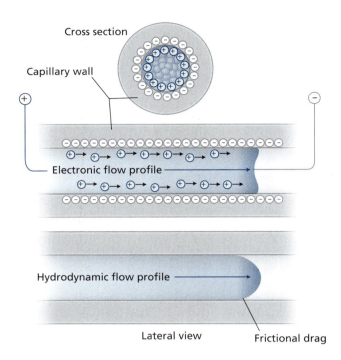

Figure 17.20 Schematic demonstrating EOF. Panel (A) is a cross-section of a capillary showing hydrated ions and the ionic double layer at the capillary surface. Panel (B) is a lateral section of a capillary showing the migration of cations toward the cathode and the EOF of entrained solvent. Panel (C) is a lateral section of a capillary showing the hydrodynamic profile of a solvent being pushed through a capillary by pumping action.

$$V_{eo} = \mu_{eo} E \qquad \qquad \textbf{Eq. 17.10}$$

where

V_{eo} = electro-osmotic velocity of an ion in an electric field $\left(\dfrac{m}{s}\right)$

μ_{eo} = electro-osmotic mobility $\left(\dfrac{m \cdot C}{N \cdot s} = \dfrac{C \cdot s}{Kg}\right)$

E = electric field strength $\left(\dfrac{Newton}{Coulomb} = \dfrac{Volt}{meter}\right)$

In thinking about Equation 17.10, we must consider the fact that the electro-osmotic mobility of an ion is influenced by two variables: EOF and electrophoretic migration. As we discussed earlier, electrophoretic migration of an ion in an electric field is inhibited by the viscous drag (η) of the matrix on the ion. On the other hand, EOF of the solvent is the result of the flow of ions at the ionic bilayer. The density of the ionic bilayer is influenced by the surface potential at the capillary wall. This surface potential is called the *zeta potential* (ζ). Likewise, the reach of the zeta potential into the buffer is quantified by the dielectric constant (ε) of the buffer solution. Combining all of these effects, we can derive an expression for the electro-osmotic mobility known as the *Henry equation*:

$$\mu_{eo} = \frac{\varepsilon \zeta}{4 \pi \eta} \qquad \qquad \textbf{Eq. 17.11}$$

where

μ_{eo} = electro-osmotic mobility $\left(\dfrac{m \cdot C}{N \cdot s} = \dfrac{C \cdot s}{Kg}\right)$

ε = relative permittivity (or the dielectric constant) of the buffer solution unitless[18]

ζ = zeta potential $\left(\dfrac{m}{C \cdot s} \right)$

η = viscosity $\left(\dfrac{N \cdot s}{m^2} = \dfrac{Kg}{m \cdot s} \right)$

Substitution of Equation 17.11 into Equation 17.10 yields:

$$v_{eo} = \left(\frac{\varepsilon \zeta}{4 \pi \eta} \right) E \qquad\qquad \textbf{Eq. 17.12}$$

Equation 17.12 shows us that the analyst has several variables that can be manipulated in order to control the rate of EOF. These are the zeta potential, the viscosity of the buffer, and the applied potential. The zeta potential is a function of the electrolyte concentration (C_i), the temperature (T), and the charge density of the capillary wall (σ_{Si}).[19] All three variables can be manipulated to optimize separation and resolution.

Adjusting C_i affects the thickness of the ionic bilayer. As the ionic strength increases, the depth of the ionic bilayer at the capillary wall decreases and at high ionic strength, the bilayer collapses, essentially shutting off EOF. To adjust σ_{Si}, the analyst can protonate or partially protonate the capillary surface by adjusting the pH. At pH values much less than 2, the capillary walls are essentially fully protonated. Elimination of the SiO^- groups at the surface will effectively shut off EOF. Alternatively, the analyst can chemically modify the capillary surface. For instance, PA or methylcellulose is often used to suppress the ionization of the Si-OH groups at the capillary surface.

Apparent Mobility We have described two ways in which an ion can migrate in a CE instrument. The first is electrophoretic migration and the second is electro-osmosis. It is important to recognize that these two forces are not necessarily working together. Take another look at Figure 17.20. In that diagram, the cations are migrating toward the cathode under the influence of an electric field. Likewise, EOF is also traversing toward the cathode. In this case, the two influences are additive and work to decrease the elution time (increase the rate of migration) of your cationic analytes from the column. However, anions in this same system would experience an electrophoretic attraction toward the anode while the EOF would be dragging the anions toward the cathode. The two influences are working against each other. When the pH is above pH of about 3, the effects of EOF are typically stronger than electrophoretic migration and anions also elute at the cathode end. However, as the pH approaches 2, the EOF becomes weak and will even stop altogether. At low pH, anions will elute at the anode end of the column. In order to detect anions using CE at low pH, you must reverse the polarity of the power supply in order to elute your analyte at the detector end of the column. Neutral analyte species do not experience any electrophoretic migration and move under the influence of EOF only. As a result, there is no electrophoretic separation of neutral species in a CE column. However, if the column has been coated so as to provide a chromatographic mode of separation, it is possible to resolve neutral species in a CE system.

In general, the resolution of neutral species is poor in CE (see the profile "Micellar Electrokinetic Capillary Chromatography"). The vector sum of the electrophoretic mobility and the electro-osmotic mobility is termed the *apparent mobility*:

$$\mu_{app} = \mu_{ep} + \mu_{eo} \qquad\qquad \textbf{Eq. 17.13}$$

where

μ_{app} = apparent mobility
μ_{ep} = electrophoretic mobility
μ_{eo} = electro-osmotic mobility

---

[18] The relative permittivity is the ratio of the permittivity of a substance relative to a vacuum.

[19] Hayes, M. A.; Kheterpal, I.; Ewing, A. G. *Anal. Chem.*, **1993**, *65*, 27.

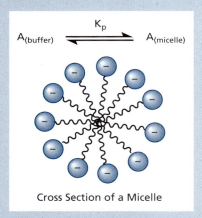

Cross Section of a Micelle

Micellar electrokinetic capillary chromatography is an amalgamated technique that accomplishes the separation of neutral analyte species by taking advantage of both electrophoretic and chromatographic separation modes. Surfactants (detergent and soap) will form micelles if the concentration exceeds a critical value. Micelles are a colloidal aggregation of surfactant molecules suspended in a polar solvent. The orientation of the surfactant molecules aligns all of the hydrophobic ends of the surfactant on the interior of a sphere. In micellar electrokinetic capillary chromatography, surfactant is added to the buffer above the critical concentration, and the polar heads of the micelles undergo electrophoretic migration like any other charged particle. However, the nonpolar interior of the micelle represents a phase boundary between the relatively polar buffer and the interior of the micelle. Polar-organic analyte molecules can partition between the two phases with a mechanism analogous to reverse phase chromatography. The distinction is that when the analyte enters the micelle, it will travel under the influence of electrophoresis. The separation is based on the differential partitioning of an analyte between the polar solvent and the nonpolar interior of the micelle. ■

Micelle

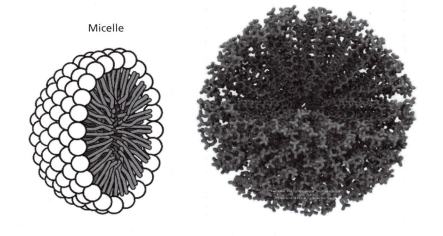

The rate at which an analyte migrates through the capillary column is the product of the apparent mobility and the applied electric field:

$$v_{app} = \mu_{app} E$$ Eq. 17.14

Problem 17.17: The units for E and μ were given in Equations 17.2 and 17.3, respectively. By analyzing the units, show that the product of $\mu_{app}E$ in Equation 17.14 does indeed yield a velocity.

Problem 17.18: You are performing a CE experiment with a field strength of 20 V/m and you are separating two analytes (A and B) with μ_{app} values of 0.001 and 0.0015, respectively. Assume a base line width of 2 seconds. How long must your column be in order to ensure baseline resolution? (See Equation 15.10 for a review of this concept.) Is this a reasonable length? Defend your answer.

Sample Loading and Throughput

Samples are typically loaded onto a CE column by one of three methods: *siphoning, electrokinetic injection,* and *hydrodynamic injection.*

Siphoning is the simplest method. As one end of the capillary is held in a lower position, the higher end is placed into a sample reservoir. It is very difficult to quantify the amount of sample loaded using siphoning, and as a result, siphoning is best used for qualitative work only.

Electrokinetic injection involves using an electric field to draw sample into the column. In electrokinetic injection, the buffer reservoir at the appropriate end of the capillary is replaced with a sample solution. The power is turned on briefly to allow the sample to be pulled into the capillary with the current.

If you are performing a quantitative analysis on a sample, you need to ensure that the relative concentrations of analyte injected onto the capillary is the same as in the actual sample. When using electrokinetic injection, the number of moles of each analyte loaded onto the column is estimated by:

$$n_i = v_{app_i} \left(\frac{k_b}{k_s} \right) t \pi r^2 M_i \qquad \textbf{Eq. 17.15}$$

where

n_i = moles of the ith analyte injected

v_{app_i} = apparent velocity of the ith analyte injected

r = radius of the capillary

t = injection time

M_i = molarity of the ith analyte in the injection solution

$\left(\dfrac{k_b}{k_s} \right)$ = ratio of the buffer conductivity and the sample conductivity

From Equations 17.14 and 17.15, we can see that one of the drawbacks of electrokinetic injections is the fact that the concentration of a specific analyte loaded onto the capillary is a factor of each individual analyte's apparent velocity and the analyte's conductivity. This is referred to in the electrophoresis literature as *loading bias.* The result is that the concentrations of the various components of the sample loaded onto the capillary will be different than the concentrations of the analytes in the original sample. This is not an ideal situation for quantitative work. For this reason, many commercial instruments use hydrodynamic injection.

Hydrodynamic injection involves driving the sample onto the column using a pressure differential between the anode and cathode solvent reservoirs. Once the sample is loaded, the buffer is replaced and the electric field gradient is resumed for analysis. When using hydrodynamic injection, the injection volume is calculated using the *Poiseuille equation:*[20]

$$V_{inj} = \frac{\Delta P \pi d^4 t}{128 \eta L} \qquad \textbf{Eq. 17.16}$$

where

V_{inj} = injection volume (nL)

ΔP = pressure difference between the two ends of the capillary (Pa)

d = inside diameter of the capillary (m)

t = injection time (sec)

η = viscosity of the injected sample (Pa·sec)

L = the total length of the column (m)

---

[20] The Poiseuille equation is taken from classical fluid dynamics and describes the volume of a fluid moving in a pipe when subjected to a pressure differential. Also, see Sutera, S. P.; Skalak, R. The History of Poiseuille's Law. *Annual Review of Fluid Mechanics,* **1993**, *25*, 1–19.

Problem 17.19: For a hydrodynamic injected sample, determine the injection volume given, $\Delta P = 70$ kPa, $d = 0.25$ mm, $t = 2$ sec, $h = 1.3 \times 10^{-3}$ Pa·sec, and $L = 3$ m. If the analyte species were each 3 mM, how many moles of sample were injected?

In the crime scene investigation shows on television, we often see technicians swabbing a piece of evidence for DNA. Even with polymerase chain reaction amplification, DNA, RNA, and even protein samples are typically very small. So, regardless of the injection method, a very desirable advantage of CE is the very small sample injection volumes. A typical injection volume will range from picoliters to nanoliters. This process can be easily automated. The total time from analysis to analysis is approximately 8 minutes, and one can expect to run as many as seven CE experiments in an hour.

Dynamic Coating

At biological pH values (~6.0–8.0), the walls of a fused silica column have a net negative charge, the result being an EOF that sweeps all of the analyte toward the cathode. However, at biological pH values, many important biological analytes are anionic.[21] It had been long known that high ionic strength buffers suppressed EOF. However, in 1983, researchers discovered that addition of the cationic surfactant cetyltrimethylammonium bromide (CTAB)[22] to the buffer solution reversed the direction of EOF rather than the expected suppression of EOF.[23]

If one desires a reversed EOF, it is now common to add 0.5 mM CTAB to a buffer (see Figure 17.21). This practice has become known as *dynamic coating*. Although the practice of dynamic coating has been in use for some time, we would wait until 1996 for a thorough explanation of the process.[24] The CTAB adheres tightly to the fused silica wall, and the hydrophobic tails form a hemimicelle, effectively creating a cationic surface at the capillary wall (see Figure 17.22). The associated counteranions migrate toward the anode and the entrained solvent is carried toward the anode in the resulting EOF. Today, many different commercially available coatings allow for fine-tuning the EOF for specific analyte separations.[25]

Problem 17.20: Read the opening profile on Arne Tiselius ("The Father of Electrophoresis"). Discuss how CE is similar to free boundary electrophoresis. Also, discuss how CE is different than free boundary electrophoresis. Be sure to incorporate aspects of the separation mechanisms and resolution in your discussion.

Figure 17.21 Structure of CTAB.

---

[21] For example, DNA is polyanionic and many glycoproteins are anionic in the pH range of electrophoretic separations.

[22] Also known as cetrimonium bromide.

[23] Reijenga, J. C.; Aben, G. U. A.; Verheggen, T. P. E. M.; Everaerts, F. M. *J. Chromatogr.*, **1983**, *260*, 241.

[24] Lucy, C. A.; Underhill R. S., *Anal. Chem.*, **1996**, *68*, 300–305.

[25] Some commercial vendors include MicroSolve, Waters Inc., Bio-Rad, and Life Technologies.

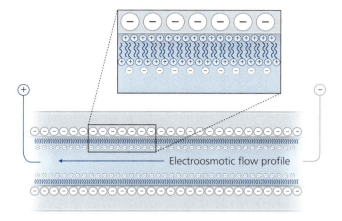

Figure 17.22 Expanded view of the surface of a CTAB coated, fused silica capillary.

Detection

For most CE instruments, detection is done using the capillary as the detection cell. This type of system is referred to as on-capillary detection.

Ultraviolet-visible Detection   Figure 17.19 depicts an example of on-capillary detection, and Figure 17.23 is an expanded view of just the detector portion of a diode array UV-vis CE instrument. In this example, the capillary tube represents a flow cell that passes perpendicularly through the optical train of the UV-vis spectrometer. The advantage of on-capillary detection is in the simplicity of the instrument design. You do not need to provide any additional components to pump the sample into the detection cell. However, from a design perspective, there are some challenges to be overcome. Most capillary tubes are cylindrical (rectangular tubes are now available), and the curvature of the liquid-filled tube creates a convex lens whose refractive index changes with the buffer solution. Collecting the maximum amount of usable light from the sample cell requires a more complex set of lenses. On-capillary detection also suffers from low sensitivity. Beer's law dictates that the absorbance is directly proportional to the path length ($A = \varepsilon bc$). The curvature of the capillary also reduces the effective path length of an already short path length by approximately 40%.[26] A typical capillary has a diameter of approximately 50 μm, leaving you with an effective optical path length of only approximately 30 μm. Absorbance values for UV-vis detection in CE are typically very low, and an absorbance of 0.001 is considered an acceptable value. The small diameter of the capillary also adds another contributing factor to low sensitivity for UV-vis detection. Because the volume of sample is low relative to the total volume of the capillary, a significant percentage of the sample cell is fused silica and not the sample matrix. A proportionately large quantity of light reaches the detector without having passed through the sample matrix. This is essentially stray light (see Equation 6.10), which reduces the absorbance values even further. A design approach for increasing path length in CE is the use of Z-cells. We first introduced Z-cells in Chapter 15 (see Figure 15.30). A Z-cell creates two 90-degree turns in the capillary, and the optical train of the spectrometer is then directed down the linear portion between the two turns. A properly designed Z-cell can also eliminate much of the stray light issues associated with passing the light at a right angle through the silica capillary and improve detection limits by an

---

[26] Petersen, J.; Mohammad, A. A. (editors). *Clinical and Forensic Applications of Capillary Electrophoresis.* Springer: New York, NY, **2001**.

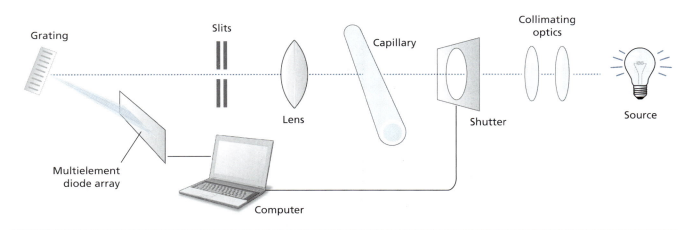

Figure 17.23 Schematic of a diode array UV-vis detector for a CE instrument.

order of magnitude.[27] However, the use of a Z-cell adds the challenge of focusing the incident light down to a diameter of around 30 μm for the full length of the cell.

UV-vis detection is primarily limited to organic analytes that contain at least one π-system and preferably a conjugated π-system or to organic analytes that are easily derivatized. A reasonable lower detection limit of 10^{-7} M can be achieved with UV-vis detection.[28]

Fluorescence Detection   If your analytes have a strong fluorescent chromophore, or can be derivatized to include one, then fluorescence spectroscopy can prove to be a suitable detection system in CE. Recall from Chapter 8 that in order to measure a fluorescence, you must first excite the sample using an external light source. The light source can be an ultraviolet lamp or a laser system. Figure 17.24 shows a schematic of a CE fluorescence detection system. It might also be advisable to review Figure 8.2. You might recall from Chapter 8 that a properly designed fluorescence spectrometer benefits from very high signal-to-noise ratios. This can improve the sensitivity of a CE experiment by as much as three orders of magnitude[29] and detection limits as low as 10^{-13} M.[30] The drawback, of course, is that not all analytes are fluorescent.

Electrochemical Detection   Electrochemical detection is broadly divided into potentiometric and amperometric modes. Potentiometric detection is not very common because the application requires derivatization of the electrode, followed by very careful alignment of the electrode in the capillary body. The process is tedious and does not lend itself to automation. In addition, the response is governed by the Nernst equation and the electrodes require regular calibration. Lastly, the lifetime of an electrode can be as short as 2 or 3 days. However, for specific applications, potentiometric detection has some clear advantages. For example, potentiometric detection has proven to be quite suitable for the detection of main group inorganic analytes such as Ca^{2+}, which are not suitable for UV-vis or fluorescence detection. The use of Ca^{2+}-selective probes[31] with CE has shown detection limits as low as 10^{-8} M.[32] The fabrication of a great number

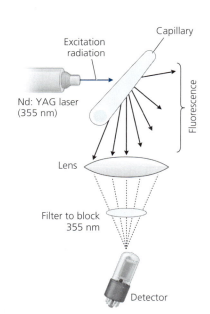

Figure 17.24 Schematic of a fluorescence detector for a CE instrument.

[27] Want, T.; Aiken, J. H.; Huie, C. W.; Hartwick, R. A. *Anal Chem.,* **1991**, *63*, 1372–1376.

[28] Mainka, A.; Bachmann, K. *J. Chromatogr. A*, **1997**, *767*, 241–247.

[29] Wu, S.; Dovichi, N. J. *Talanta,* **1992**, *173*, 39.

[30] Landers, J. P. (editor). *Handbook of Capillary Electrophoresis.* CRC Press, New York, **1997**.

[31] See Chapter 18 for details on the construction of ion-selective probes.

[32] Haber, C.; Silverstri, I.; Roosli, S.; Simon, W. *Chimia,* **1991**, *45*, 117–121.

of ion-selective electrodes are described in the literature,[33] and some are featured in Chapter 18.

Amperometric detectors require the analyte to be redox active. In an amperometric detection system, the electrode is biased with a direct current voltage and as the analyte comes in contact with the electrode surface, the analyte is either oxidized or reduced. The measured signal is the current that passes in the circuitry as a result of the electron exchange between the electrode and the analyte. Amperometric detectors are more robust than potentiometric detectors, but they are also less selective, making them more appropriate as a universal detector and less appropriate if you wish to study a single analyte in a complex mixture. Typical detection limits for amperometric detectors is in the μM range.[34]

Mass Spectrometry Detection   The instrumental design in CE–mass spectrometry shares much in common with that of liquid chromatography–mass spectrometry; it usually uses an electrospray interface to the mass spectrometer. Electrospray ionization was first introduced in Chapter 13 when we introduced tandem mass spectrometry techniques (see Figure 13.8) and again in Chapter 15 when we discussed liquid chromatography–mass spectrometry. The interface of a CE instrument to an electrospray–mass spectrometer is essentially the same as the interface of LC instrument to an electrospray–mass spectrometer. There is one important modification. Because the eluent volume of a CE instrument is so small compared to a liquid chromatography instrument, the eluent volume in CE is increased by a take-up solution provided by an axial needle that sheaths the capillary. The use of an electrospray interface also has some experimental considerations. Electrospray involves the volatilization of the buffer solvent; so it is best to use buffer salts that are somewhat volatile (e.g., ammonium formate). This will reduce the volume of buffer salts that accumulate inside the electrospray interface. In any case, routine cleaning of the electrospray is necessary for CE–mass spectrometry.

One of the key advantages of CE–mass spectrometry is the truly tandem nature of the instrumental design. Because the electrospray interface is connected to the eluent end of the capillary, the mass spectrometer can be located after other common detector types. In fact, it is more common to find a UV-vis/CE–mass spectrometer or a fluorescence/CE–mass spectrometer than it is to find a stand-alone CE–mass spectrometer.

Recent Developments in Capillary Electrophoresis

The rapid development of CE technology is predicated on the high efficiency and rapid analysis time of the technique. Yet, traditional CE is still performed on a bench-top instrument. Recent developments in CE have focused on adapting CE technology to miniaturized microfluidic platforms,[35] with the goal of producing portable, low-cost, "labs on a chip" technology. Soft lithography has been successfully used to etch channels onto chips with dimensions in the 10- to 100-μm range,[36] which has allowed for the rapid and inexpensive fabrication of a wide array of bioanalytical devices. For example, in a 2002 paper in the journal *Analytical Chemistry*,[37] researchers described the fabrication of a planar CE system on a poly(dimethylsiloxane)-glass chip that utilized hydrodynamic injection with analyte detection performed using microdisk electrodes

---

[33] Bakker, E.; Bühlmann, P.; Pretsch, E. Carrier-Based Ion-Selective Electrodes and Bulk Optodes. 1. General Characteristics. *Chem. Rev.*, **1997**, *97*, 3083–3132; Timerbaev, A. R. Element Speciation Analysis Using Capillary Electrophoresis: Twenty Years of Development and Applications. *Chem. Rev.*, **2013**, *113(1)*, 778–812.

[34] Swinney, K., Bornhop, D. J., *Electrophoresis*, **2000**, *21*, 1239–1250.

[35] Microfluidics is the study of the manipulation of very small volumes of liquid and crosses the boundaries of analytical chemistry, biochemistry physics, and nanotechnology.

[36] McDonald, J. C.; et al. Fabrication of Microfluidic Systems in Poly(dimethylsiloxane). (Review.) *Electrophoresis*, **2000**, *21*, 27–40.

[37] Backofen, U.; Matysik, F.; Lunte, C. E. *Anal. Chem.*, **2002**, *74*, 4054–4059.

COMPARE AND CONTRAST

A Look Back at Five Different Separation Techniques

TECHNIQUE	ANALYTE RANGE	SEPARATION MODE	APPARATUS
Liquid chromatography	In a liquid solvent Can be ionic, polar, or nonpolar Can be easily adapted for preparatory scale separations Easily automated	Motive force is a mobile solvent. Separation is the result of partitioning between the mobile (solvent) and stationary (packing material) phases.	Can be as simple as a burette packed with a stationary phase or very complex and expensive in the case of high-performance analysis, where you need a very stable pumping system, a very precisely packed column, and an expensive detector (e.g., ultraviolet-visible or fluorescence spectrometer).
Gas chromatography	Analyte dissolved in a gas Analyte typically volatile organic molecule Not typically used for preparatory scale separations Easily automated	Motive force is a mobile gas. Separation is the result of partitioning between the mobile (gas) and stationary (capillary wall or packing material) phases.	A carrier gas under pressure is passed through a column, which resides in a temperature-controlled oven. The oven is the most expensive part of the instrument. Detectors for GC are relatively inexpensive.
Paper electrophoresis	Used to separate a large variety of ionic analytes; not generally used for large synthetic or natural polymers or for preparatory scale separation Not easily automated	Motive force is an electric field. Separation is the result in differences in mobility within the electric field as the analyte passes over the surface of the paper. In the case of underivatized cellulose, a partitioning mechanism also exists.	Involves two electrodes, a buffer solution, a power supply, and a piece of paper—a very inexpensive and simple device Two-dimensional experiments involving changing the orientation of the electric field are simple to perform and greatly improve resolution.
Gel electrophoresis	Used for the separation of large biopolymers such as proteins and DNA; however, has been adapted for separation of large ionic polymers Can be easily adapted for preparatory scale separations Not easily automated	Motive force is an electric field. Separation is the result in differences in mobility within the electric field as the analyte passes through the gel. In the case of a protein analyte, one can use pH gradients to control migration.	Two electrodes, a buffer solution, a power supply and a gel—a very inexpensive and simple device. The nature of the gel has a large effect on the types of analytes that can be studied and on the resolution of the separation. Two-dimensional experiments involving changing the orientation of the electric field are simple to perform and greatly improve resolution.
Capillary electrophoresis	Typically used for the separation of large biopolymers such as DNA; however, has been adapted for the separation of a great many ionic analytes Not suitable for preparatory scale separations Easily automated	Motive force is an electric field. However, EOF also produces a net flow of buffer. Separation is the result in differences in mobility within the electric field as the analyte passes through the column. Use of surface coatings and/or micelles also allows for chromatographic separation modes.	Two electrodes, a buffer solution, a power supply, and a long capillary column The narrow diameter of the column creates EOF. Surface coatings can be used to control EOF and/or to impart a chromatographic mode of separation. Micelles can also be incorporated into the experiment, creating a chromatographic mode of separations for neutral analytes.

EOF, electro-osmotic field.

Figure 17.25 The Mitos capillary electrophoresis chip by Dolomite Inc. Total dimensions are 15 × 45 × 2 mm. The channel is 20 mm deep and 30 mm long.

in an amperometric detection mode. They demonstrated lower detection limits for ascorbic acid of approximately 5 µM. The technology has advanced rapidly since then, with commercially available devices already on the market[38] (see Figure 17.25).

As the technology continues to develop, chip-based electrophoresis will certainly become an integral component of self-contained portable field "labs" capable of performing sophisticated chemical analysis without the need to bring samples back to the laboratory.

Problem 17.21: Imagine adding thin-layer chromatography to the Compare and Contrast feature "A Look Back at Five Different Separation Techniques" seen directly above. Develop text for the three column headings.

17.10 Further Reading

BOOKS

Garcia C. D.; Chumbimuni-Torres, K. Y.; Carrilho, E. *Capillary Electrophoresis and Microchip Capillary Electrophoresis: Principles, Applications, and Limitations.* John Wiley & Sons: Hoboken, NJ, **2013**.

Martin, R. *Gel Electrophoresis: Nucleic Acids (Introduction to Biotechniques).* Taylor & Francis: New York, **1996**.

Mikkelsen, S. R.; Corton, E., *Bioanalytical Chemistry.* John Wiley & Sons: Hoboken, NJ, **2004**.

Voet. D.; Voet, J. *Biochemistry.* John Wiley & Sons: Hoboken, NJ, **2010**.

Westermeier, R. *Electrophoresis in Practice: A Guide to Methods and Applications of DNA and Protein Separations.* John Wiley & Sons: Hoboken, NJ, **2005.**

JOURNALS

Ewing, A. G.; Wallingford, R. A.; Olefirowicz, T. M. Capillary Electrophoresis. *Anal. Chem.* **1989,** *61(4),* 292A–303A.

Leube, J.; Roeckel, O. Quantification in Capillary Zone Electrophoresis for Samples Differing in Composition from the Electrophoresis Buffer. *Anal. Chem.* **1994,** *66,* 1090.

Melanson, J. E.; Baryla, N. E.; Lucy, C. A. Dynamic Capillary Coatings for Electroosmotic Flow Control in Capillary Electrophoresis. *TrAC Trends in Analytical Chemistry,* **2001,** *20(6–7),* 365–374.

---

[38] Micralyne Inc., ProteinSimple Inc., Dolomite Inc.

17.11 Exercises

EXERCISE 17.1: In the gel electrophoresis of a protein sample:

(a) Explain why some proteins migrate farther in a given amount of time than do other proteins?

(b) How does pH affect the migration rates of proteins in a gel electrophoresis experiment?

EXERCISE 17.2: Sketch an agarose gel for use in separating DNA. Make sure you label:

(a) The location of the wells

(b) The positive electrode

(c) The negative electrode

(d) The direction of migration of the DNA in the gel

EXERCISE 17.3: Toward which pole would a negatively charged protein migrate (anode or cathode)?

EXERCISE 17.4: Relative to a protein's isoelectric point, how would one adjust the pH in order to make the protein migrate toward the cathode?

EXERCISE 17.5: Relative to a protein's isoelectric point, how would one adjust the pH in order to make the protein migrate toward the anode?

EXERCISE 17.6: Why is control of the pH a more important variable in protein electrophoresis than it is in DNA electrophoresis?

EXERCISE 17.7: Describe the composition of the following polyacrylamide gels:

(a) $T_{17}C_5$ (b) T_7C_2 (c) $T_{15}C_4$

EXERCISE 17.8: The structures of two common tracking dyes are given below. Under identical conditions, which dye will migrate farther in a gel electropherogram? Defend your answer.

Pararosalin   Methyl violet

EXERCISE 17.9: The chemical structure of basic yellow dye #2 is given here. The pK_{a1} = 9.8 and the pK_{a2} = 10.7. Describe the relative migration rates for an electrophoresis experiment of yellow dye #2 conducted at pH = 8, pH = 10, and pH = 12. Explain.

EXERCISE 17.10: Draw the structure of ethidium bromide.

EXERCISE 17.11: What is the main function of ethidium bromide in a gel electrophoresis experiment?

EXERCISE 17.12: Research and describe three common visualization techniques used in electrophoresis. Cite your references.

EXERCISE 17.13: What are the theoretical reasons CE enjoys such a high separation efficiency compared to either high-performance liquid chromatography or gas chromatography techniques?

EXERCISE 17.14: Recommend an appropriate gel formulation (if appropriate) and an appropriate staining reagent for each of the following analyses:

(a) A paper electrophoresis analysis of proteins with a molecular weight range of 15,000 to 30,000

(b) A Western blot gel electrophoresis analysis of proteins with a molecular weight range of 15,000 to 30,000

(c) Proteins with a molecular weight range of 100,000 to 130,000

(d) A mixture of glycine, leucine, alanine, and proline

(e) Electrophoretic analysis of a mixture of antibody proteins

(f) Genomic analysis of fruit fly DNA subjected to restriction enzymatic digestion

EXERCISE 17.15: List and describe the three common methods of sample injection used in CE.

EXERCISE 17.16: Compare and contrast the typical injection volumes needed in slab versus CE.

EXERCISE 17.17: Sometimes one will observe curved bands in a gel electrophoresis experiment. What experimental parameter is responsible for the curvature of the bands?

EXERCISE 17.18: A composite gel is made of both PA and agarose. How do you think the combination of these materials would affect pore size? Mechanical stability? Melting point? Do a literature search to find an application of a composite gel. Does the application meet your expectations?

Chapter 18

Potentiometry and Probes

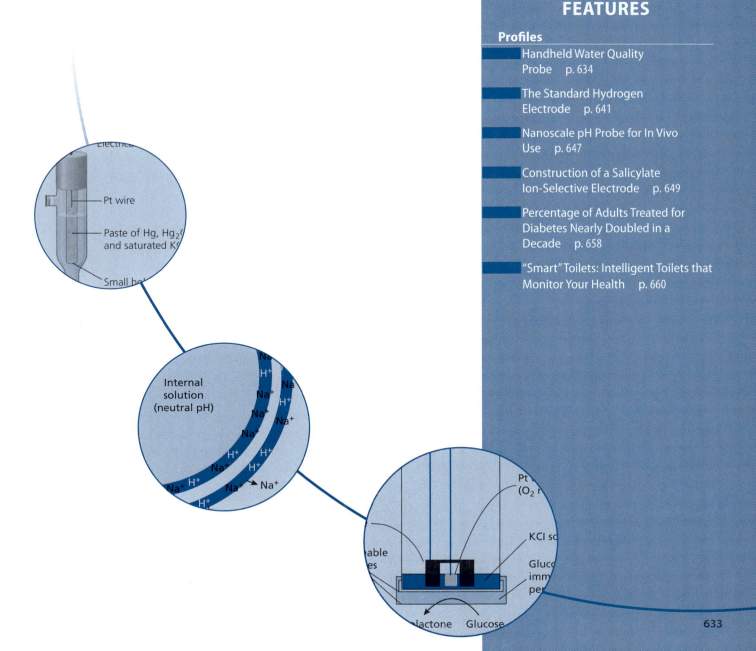

Electrice

Pt wire

Paste of Hg, Hg₂ and saturated K

Small ho

Internal solution (neutral pH)

Na
H⁺
Na
Na⁺
Na⁺ H⁺
Na⁺ Na⁺
Na⁺
H⁺ H⁺
Na⁺ H⁺
Na⁺ Na⁺ → Na⁺
H⁺

Pt (O₂ r

KCl sc

Gluc imm per

able es

lactone   Glucose

Clean water is rapidly becoming one of the most precious resources on our planet, and ecologists struggle to keep pace with the need to monitor important chemical qualities of watershed regions, industrial and municipal wastewater effluent, and aquifers, all of which have been impacted by large-scale agriculture. Historically, this work involved field sampling bodies of water and conducting most of the actual chemical measurements in fixed-location laboratories. However, the past few decades have seen a dramatic shift toward the use of multiparameter water quality probes, such as the one pictured here by Geo Scientific. This sensor includes electrochemical probes for the simultaneous field measurement of dissolved oxygen, pH, nitrate, ammonium, and conductivity. It also has the ability to conduct spectroscopic measurements of chlorophyll, algae, and rhodamine, which is a fluorescent dye sometimes added to mobile bodies of water (e.g., rivers, streams) to monitor flow dynamics. ■■

18.1 Basic Principles: Probes and Biosensors

Instrumental electrochemistry is generally categorized into specific methods, including voltammetry (Chapter 19), amperometry, coulometry, and potentiometry. In a broader sense, however, instrumental electrochemistry can be fundamentally broken down into two general areas—passive and active electrochemistry. Historically, most electrochemical *probes* have been passive in nature; that is, they used static electrochemical instrumentation to simply measure a characteristic of the sample. Active methods, on the other hand, use electricity as an energy source to control a chemical reaction. Many of the more interesting and useful modern probes involve the use of active electrochemical methods, including biosensors. Biosensors are probes that utilize biological substances (organisms or molecules) to indicate the presence of specific molecules, and are often used to provide in situ measurements in biological systems. Although some biosensors are spectroscopic in nature, the advent of nanotechnology has allowed the development of very small active electrochemical probes for biological monitoring. In this chapter, we will discuss some of the passive *potentiometric* probes that have been around for many years but which are still very important tools in chemical analysis. Furthermore, we will survey some important probes that have been developed more recently and which, in many cases, depend on active electrochemical methods to conduct a measurement. The electrochemical processes

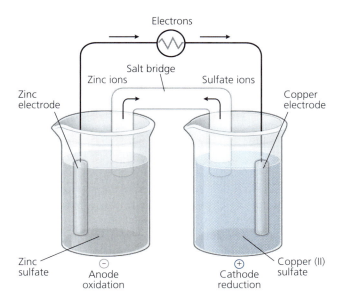

Figure 18.1 A Daniell cell is a galvanic cell in which the voltage of the spontaneous reaction is measured as current flows from the anode to the cathode.

involved in such probes will only be briefly described here; they are considered in more detail in Chapter 19.

Before we describe and discuss any particular probes, we need to give an overview of some basic principles that apply to all probes. In your general chemistry course, and possibly a previous analytical chemistry course, you learned that an electrochemical cell is a device that allows a user to utilize the free energy of a reduction-oxidation (redox) reaction in the form of electricity. A galvanic (also called voltaic) cell captures the electrons that transfer from the oxidized species to the reduced species in a spontaneous reaction. A battery is an example of the use of a galvanic cell reaction. An electrolytic cell uses electrical energy to force a nonspontaneous redox reaction to occur. Sodium metal and chlorine gas are produced from molten sodium chloride through electrolysis.

In *passive probes*, current is neither generated nor imparted to the system; such probes simply monitor the difference in electrical potential (voltage) that results from differences in the chemical nature of two solutions connected through a conductive barrier (e.g., a salt bridge). An active probe utilizes a redox reaction to study a chemical system. With an active probe, you measure the current (or voltage) generated by the redox reaction. Active probes can be either electrolytic or galvanic. A review of basic electrochemistry (cells, half-cells, potentials) will be important in understanding how passive probes work.[1]

All electrochemical cells, whether they are galvanic or electrolytic, have some features in common. You will have discussed these features in previous classes, and so we will only briefly review them here. An electrochemical (redox) reaction involves two half-reactions—the oxidation reaction and the reduction reaction—and to understand the reaction process we consider the half-reactions separately.[2] The reduction half-reaction takes place in the cathode half-cell, and the oxidation half-reaction takes place

---

[1] For additional review of basic electrochemistry, see general chemistry and/or quantitative analysis textbooks.

[2] The half-reactions might be separated by a barrier to allow charge transport but minimize solution mixing, thus allowing us to capture electron flow through circuitry. However, even if we mix the reactants in a single vessel, we can consider the reaction as a sum of the separate half-reactions in order to better understand the electrochemistry of the reaction.

in the anode half-cell. The classical example of this is the Daniell cell (Figure 18.1), which harvests the electrons transferred from zinc metal to copper(II) ions, a spontaneous process. The reaction in the Daniell cell can be represented as:

Anode reaction: $Zn(s) \rightleftharpoons Zn^{2+}(aq) + 2e^-$ $E^0 = +0.76\,V$ **Eq. 18.1a**

Cathode reaction: $Cu^{2+} + (aq) + 2e^- \rightleftharpoons Cu(s)$ $E^0 = +0.34\,V$ **Eq. 18.1b**

Cell reaction: $Zn(s) + Cu^{2+}(aq) \rightleftharpoons Zn^{2+}(aq) + Cu(s)$ $E^0_{cell} = +1.10\,V$ **Eq. 18.1c**

In the Daniell cell, the half-cells are separated by a salt bridge to prevent intimate mixing of the solutions, forcing electron transfer to occur through the circuitry, providing us with a power source. The reaction would, of course, take place without the physical half-cells; that is, if zinc powder $(Zn\,(s))$ were sprinkled into an aqueous solution of copper sulfate $(Cu^{2+}\,(aq))$, the same reaction would take place and copper metal would form as a precipitate in the solution. It is important to understand that the only *real* reaction that takes place is the cell reaction; neither of the half-reactions could happen in isolation. In every case, when an oxidation reaction occurs, a balancing reduction reaction also occurs.

In Equations 18.1a to c, the symbol E^0 represents the *standard potential*, which is the potential measured with respect to the normal hydrogen electrode under standard thermodynamic conditions (25°C, 1 M concentration for aqueous species, 1 atm pressure for gaseous species). Because we can add the chemical equations in 18.1a and 18.1b to obtain the full equation in 18.1c, we can add the standard potentials associated with the half-reactions to get the E^0 of the full reaction. This is simply an extension of Hess's law. Standard half-cell potentials can be found in reference tables, where all half-reactions will be presented as reduction reactions. To obtain the standard potential for an oxidation reaction from such a table, simply reverse the half-reaction (to make it an oxidation reaction) and reverse the sign on E^0. In a reference table (e.g., Table 18.1), the zinc reaction in Eq. 18.1a would have been written as the reduction, and the potential would have been given as $E^0 = -0.76\,V$. If we changed the zinc electrode to a different metal, we would change the voltage and in essence, we would have created a probe, albeit a cumbersome one, that indicates what type of metal is connected to the anode. In addition, we would see changes in the voltage by simply changing the concentration of metal ion in either of the half cells. This cell arrangement could be used to measure metal ion concentrations as well.

Note that in some cases, reference tables will list a value called the *formal potential*, $E^{0'}$, which is the potential measured under conditions that are not necessarily standard, but which are specified. For instance, if the concentrations of all species were not 1 $\underline{M}$, the actual concentrations would be indicated. (See Table 18.1.)

Problem 18.1: What would the standard electrolytic cell potential be if the reaction in Example 18.1 took place in a neutral solution?

Problem 18.2: What would the standard electrolytic cell potential be if the reaction in Example 18.1 took place in an alkaline solution?

Problem 18.3: Automobile batteries produce electricity by converting PbO_2 to $PbSO_4$ in the presence of 30% sulfuric acid at one electrode and converting elemental lead to $PbSO_4$ at the other electrode. Use data from the half-reactions and electrochemical potentials given in Table 18.1 potentials to propose the most likely scheme for this galvanic cell.

Example 18.1

$H_2(g)$ and $O_2(g)$ can be produced by the electrolysis of water, in which gaseous H_2 and O_2 are produced by applying electricity to an aqueous solution of an acid. Estimate the potential needed to produce hydrogen and oxygen from a dilute sulfuric acid solution.

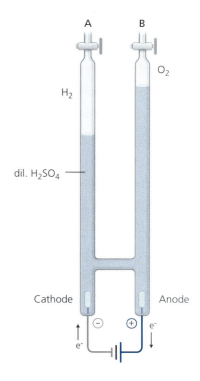

ANALYSIS –

In a molecule of water, hydrogen exhibits a +1 oxidation state and oxygen exhibits a –2 oxidation state. To produce elemental gases, then, hydrogen must undergo a reduction (+1 to 0) at the cathode and oxygen must undergo an oxidation (–2 to 0) at the anode. We peruse our standard reduction half-cell table (Table 18.1) to find half-reactions that might work in this scenario. Note that regardless of whether a cell is galvanic or electrolytic, the definitions of cathode and anode remain the same. At the cathode, two possible reactions could take place:

$$2\,H_2O\,(l) + 2e^- \rightleftharpoons H_2\,(g) + 2\,OH^-\,(aq) \qquad E^0 = -0.83\;V$$

$$2\,H^+\,(aq) + 2e^- \rightleftharpoons H_2\,(g) \qquad E^0 = 0.00\;V$$

The second reaction is thermodynamically favored (more positive E^0), but under conditions of *neutral* pH, the concentration of H^+ is low, so the first reaction would play a significant role. However, to simplify matters, we have specified *acidic* conditions, which provides ample aqueous proton. For our conditions, therefore, the second reaction will predominate at the cathode. There is only one viable candidate for the anode reaction, listed in the table as:

$$O_2\,(g) + 4\,H^+\,(aq) + 4\,e^- \rightleftharpoons 2\,H_2O\,(l) \qquad E^0 = 1.23\;V$$

However, we know that the anode reaction is an oxidation, so the actual half-reaction is the reverse of this. So, the two reactions occurring in the cell are:

Anode reaction: $2\,H_2O\,(l) \rightleftharpoons O_2\,(g) + 4\,H^+\,(aq) + 4\,e^-$ $E^0 = -1.23\;V$

Cathode reaction: $2\,H^+\,(aq) + 2e^- \rightleftharpoons H_2\,(g)$ $\qquad\qquad E^0 = 0.00\;V$

The anode reaction produces four electrons but the cathode reaction only consumes two, so we must correct this imbalance by doubling the cathode reaction. Unlike with K_{eq}, ΔH, and ΔG, such changes in the stoichiometry of half-reactions do not change the value of E^0. E^0 is an intrinsic property of the material.

SOLUTION –

Anode reaction: $2\,H_2O\,(l) \rightleftharpoons O_2\,(g) + 4\,H^+\,(aq) + 4\,e^-$ $E^0 = -1.23\;V$

Cathode reaction: $4\,H^+\,(aq) + 4\,e^- \rightleftharpoons 2\,H_2\,(g)$ $\qquad\qquad E^0 = 0.00\;V$

Cell reaction: $2\,H_2O\,(l) \rightleftharpoons O_2\,(g) + 2\,H_2\,(g)$ $\qquad\qquad E^0 = -1.23\;V$

We sum the half-reactions to get the overall reaction and sum their respective E^0 values to get the overall cell potential of –1.23 V. Note that the negative sign on the cell potential indicates that the reaction is not spontaneous, so 1.23 V must be *applied* in order to make it take place.[3]

---

[3] Recall that a negative ΔG value indicates a spontaneous reaction, but as we see in the Flashback on page 639, $\Delta G = -nFE$.

TABLE 18.1:   Potentials for Standard Reduction Half-Reactions at 25°C with Respect to the Standard Hydrogen Electrode

REACTION	E°, V	REACTION	E°, V
$F_2 + 2H^+ + 2e \rightleftharpoons HF$	3.053	$2H^+ + 2e \rightleftharpoons H_2$	0.0000
$F_2 + 2e \rightleftharpoons 2F^-$	2.866	$Fe^{3+} + 3e \rightleftharpoons Fe$	−0.037
$O(g) + 2H^+ + 2e \rightleftharpoons H_2O$	2.421	$Pb^{2+} + 2e \rightleftharpoons Pb$	−0.1262
$O_3 + 2H^+ + 2e \rightleftharpoons H_2O + O_2$	2.076	$PbSO_4(s) + 2e^- \rightleftharpoons Pb_{(s)} + SO_4^{-2}(aq)$	−0.310
$OH + e \rightleftharpoons OH^-$	2.02	$Sn^{2+} + 2e \rightleftharpoons Sn$	−0.1375
$Au^+ + e \rightleftharpoons Au$	1.692	$In^+ + e \rightleftharpoons In$	−0.14
$PbO_2(s) + HSO_4^- + 2H^+ + 2e^- \rightleftharpoons PbSO_4(s) + 2H_2O$	1.690	$O_2 + 2H_2O + 2e \rightleftharpoons H_2O_2 + 2OH^-$	−0.146
$Mn^{3+} + e \rightleftharpoons Mn^{2+}$	1.5415	$Mo^{3+} + 3e \rightleftharpoons Mo$	−2.00
$Au^{3+} + 3e \rightleftharpoons Au$	1.498	$Cu(OH)_2 + 2e \rightleftharpoons Cu + 2OH^-$	−0.222
$PbO_2 + 4H^+ + 2e \rightleftharpoons Pb^{2+} + 2H_2O$	1.455	$V^{3+} + e \rightleftharpoons V^{2+}$	−0.255
$Au^{3+} + 2e \rightleftharpoons Au^{2+}$	1.401	$Ni^{2+} + 2e \rightleftharpoons Ni$	−0.257
$Cl_2(g) + 2e \rightleftharpoons 2Cl^-$	1.358	$PbCl_2 + 2e \rightleftharpoons Pb + 2Cl^-$	−0.2675
$Tl^{3+} + 2e \rightleftharpoons Tl^+$	1.252	$Co^{2+} + 2e \rightleftharpoons Co$	−0.28
$O_3 + H_2O + 2e \rightleftharpoons O_2 + 2OH^-$	1.24	$Tl^+ + e \rightleftharpoons Tl$	−0.336
$O_2 + 4H^+ + 4e \rightleftharpoons 2H_2O$	1.23	$In^{3+} + 3e \rightleftharpoons In$	−0.338
$MnO_2 + 4H^+ + 2e \rightleftharpoons Mn^{2+} + 2H_2O$	1.224	$Cd^{2+} + 2e \rightleftharpoons Cd$	−0.447
$Ir^{3+} + 3e \rightleftharpoons Ir$	1.156	$Fe^{2+} + 2e \rightleftharpoons Fe$	−0.447
$RuO_2 + 4H^+ + 2e \rightleftharpoons Ru^{2+} + 2H_2O$	1.120	$S + 2e \rightleftharpoons S^{2-}$	−0.4763
$Pt^{2+} + 2e \rightleftharpoons Pt$	1.118	$Ag_2S + 2e \rightleftharpoons 2Ag + S^{2-}$	−0.691
$Br_2(l) + 2e \rightleftharpoons 2Br^-$	1.066	$Cr^{3+} + 3e \rightleftharpoons Cr$	−0.744
$Pd^{2+} + 2e \rightleftharpoons Pd$	0.951	$Zn^{2+} + 2e \rightleftharpoons Zn$	−0.762
$2Hg^{2+} + 2e \rightleftharpoons Hg_2^{2+}$	0.920	$Cr^{2+} + 2e \rightleftharpoons Cr$	−0.913
$Ag^+ + e \rightleftharpoons Ag$	0.800	$Te + 2e \rightleftharpoons Te^{2-}$	−1.143
$Fe^{3+} + e \rightleftharpoons Fe^{2+}$	0.771	$V^{2+} + 2e \rightleftharpoons V$	−1.175
$Rh^{3+} + 3e \rightleftharpoons Rh$	0.758	$Mn^{2+} + 2e \rightleftharpoons Mn$	−1.185
$Ag_2SO_4 + 2e \rightleftharpoons 2Ag + SO_4^-$	0.654	$Al^{3+} + 3e \rightleftharpoons Al$	−1.662
$Rh^+ + e \rightleftharpoons Rh$	0.600	$Be^{2+} + 2e \rightleftharpoons Be$	−1.847
$Rh^{2+} + 2e \rightleftharpoons Rh$	0.600	$H_2 + 2e \rightleftharpoons 2H^-$	−2.23
$MnO_4^- + 2H_2O + 3e \rightleftharpoons MnO_2 + 4OH^-$	0.595	$Mg^{2+} + 2e \rightleftharpoons Mg$	−2.372
$Te^{4+} + 4e \rightleftharpoons Te$	0.568	$Eu^{3+} + 3e \rightleftharpoons Eu$	−2.407
$AgNO_2 + e \rightleftharpoons Ag + NO_2^-$	0.564	$Nd^{3+} + 3e \rightleftharpoons Nd$	−2.431
$MnO_4^- + e \rightleftharpoons MnO_4^{2-}$	0.558	$Mg^+ + e \rightleftharpoons Mg$	−2.70
$I_2 + 2e \rightleftharpoons 2I^-$	0.5355	$Na^+ + e \rightleftharpoons Na$	−2.71
$Cu^+ + e \rightleftharpoons Cu$	0.521	$Ca^{2+} + 2e \rightleftharpoons Ca$	−2.868
$Ru^{2+} + 2e \rightleftharpoons Ru$	0.455	$Sr^{2+} + 2e \rightleftharpoons Sr$	−2.89
$[Ferricinium]^+ + e \rightleftharpoons [Ferrocene]$	0.400	$Ba^{2+} + 2e \rightleftharpoons Ba$	−2.912
$[Fe(CN)_6]^{3-} + e \rightleftharpoons [Fe(CN)_6]^{4-}$	0.358	$Cs^+ + e \rightleftharpoons Cs$	−2.92
$Cu^{2+} + 2e \rightleftharpoons Cu$	0.3419	$K^+ + e \rightleftharpoons K$	−2.931
$Re^{3+} + 3e \rightleftharpoons Re$	0.300	$Rb^+ + e \rightleftharpoons Rb$	−2.98
$Ru^{3+} + e \rightleftharpoons Ru^{2+}$	0.2487	$Li^+ + e \rightleftharpoons Li$	−3.040
$Cu^{2+} + e \rightleftharpoons Cu^+$	0.153	$3N_2 + 2H^+ + 2e \rightleftharpoons 2NH_3$	−3.09
$Sn^{4+} + 2e \rightleftharpoons Sn^{2+}$	0.151	$Ca^+ + e \rightleftharpoons Ca$	−3.80
$2H^+ + 2e \rightleftharpoons H_2$	0.0000	$Sr^+ + e \rightleftharpoons Sr$	−4.10

It is often unproductive, inconvenient, or even impossible to undertake experiments under standard conditions. Consider that the function of a potentiometric probe is to measure the concentration of an analyte under varying conditions. Therefore, a probe will be operating under nonstandard conditions the majority of the time. As you have learned in previous courses, we have the Nernst equation to help us relate the observed potential under nonstandard conditions (E_{obs}) to the theoretical standard potential, E^0.

$$E_{obs} = E^0 - \frac{RT}{nF} \ln(Q_{eq})$$ **Eq. 18.2**

where

R = gas law constant $(J/mol\cdot K)$
T = temperature (K)
n = number of electrons transferred in the reaction
F = Faraday's constant, 96,485 C/mol
Q_{eq} = reaction quotient[4]

We can also relate the potential to the Gibbs free energy (ΔG) and standard Gibbs free energy (ΔG^0) of a reaction through:

$$\Delta G = -nFE$$ **Eq. 18.3**

$$\Delta G^0 = -nFE^0$$ **Eq. 18.4**

From Equation 18.3, we can see that a negative ΔG would result in a positive value for E. Thus, positive potentials are associated with spontaneous reactions, and negative potentials relate to nonspontaneous reactions. We can apply Equations 18.2 to 18.4 to half-reactions, but the extension to the concept of spontaneity can *only* be applied to cell potentials (the full net redox reaction).

Flashback

You will remember from general chemistry these relationships between ΔG^0, $E_{cell}{}^0$, and K for a given redox couple.

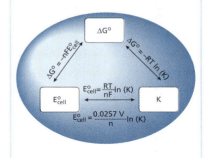

Example 18.2

Determine the cell voltage for $Zn(s)|Zn^{+2}(1\,M)||Cu^{+2}(0.1\,M)|Cu(s)$ at 0.1°C.

STRATEGY –

We must first use the standard reduction tables to find E°. Then recognize that we are not at standard state. Standard state is 25°C and all aqueous reagents at 1 M. Use Equation 18.2 and E° to find E_{cell}.

SOLUTION –

The two half-reactions and the net reaction are:

$$Zn(s) \rightarrow Zn^{+2}(aq) + 2e^- \qquad\qquad E^{\circ}_{1/2} = +0.76V$$

$$Cu^{+2}(aq) + 2e^- \rightarrow Cu(s) \qquad\qquad E^{\circ}_{1/2} = +0.34V$$

$$Zn(s) + Cu^{+2}(aq) \rightarrow Zn^{+2}(aq) + Cu(s). \qquad E^{\circ}_{cell} = 1.10V$$

From the net equation, we see that n = 2e⁻ and $Q = \dfrac{\left[Zn^{+2}\right]}{\left[Cu^{+2}\right]} = \dfrac{1}{0.1} = 10$.

Substitution into the Nernst equation gives:

$$E_{cell} = 1.10V - \frac{\left(8.314\dfrac{J}{K\cdot mol}\right)(273.1K)}{(2)\left(96,500\dfrac{J}{V\cdot mol}\right)} \ln\frac{1}{0.1} = 1.07V$$

---

[4] The reaction quotient has the same mathematical structure as an equilibrium constant. However, it is understood that you are not necessarily at equilibrium. Recall from general chemistry that for the general reaction

$aA + bB \rightleftharpoons dD + eE,\ Q_{eq} = \dfrac{[D]^d[E]^e}{[A]^a[B]^b}.$

Problem 18.4:

(a) What potential would you expect to measure in the Daniell cell at a temperature of 0.0°C if $[Cu^{2+}] = 0.50 \underline{M}$ and $[Zn^{2+}] = 1.25 \underline{M}$?

(b) What would be the Gibbs free energy of the reaction under those conditions?

(c) Mathematically demonstrate how this cell could be used as a probe of copper ion concentration.

(d) Mathematically demonstrate how this cell could be used as a temperature probe.

Problem 18.5: Are there any conditions of temperature or partial pressures of relevant gases that would make the conversion of liquid water into elemental oxygen and hydrogen spontaneous? Explain.

18.2 Potentiometric Probes

Recall that a spectrometer does not measure a given analytical signal (e.g., absorbance) directly. Rather, what the instrument does fundamentally is provide a measure of the current generated when photons strike a detector and cause electrons to flow in the electrical system. Analogously, a potentiometric (*potentio* = voltage, *metric* = measure) probe does not measure an analyte directly. For instance, a pH meter neither measures pH nor specifically $[H_3O^+]$. Rather, it passively measures the voltage difference between two electrodes, and our understanding of the chemistry of the system allows us to deduce that a change in that potential results from a change in the hydronium ion concentration. All potentiometric probes work basically the same way, and so it is incumbent to ensure that the chemical system with which we are working acts in a sufficiently predictable way that we can make a reliable correlation between the signal (potential) and the desired analyte concentration.

The signal obtained from a potentiometric probe is the proportional change in potential (measured in volts) as a function of the analyte concentration. To measure a potential, we must have two electrodes: an *indicator electrode*, which responds to changes in the concentration of the analyte, and a *reference electrode*. The reference electrode (see below) involves a known and stable redox equilibrium to provide a constant potential against which those changes can be compared. The probe is sometimes constructed as the combination of both of these electrodes, as is the case with most pH probes (Figure 18.2[A]). Other probes require a separate reference electrode to be used in conjunction with the probe in order to obtain a measurement, as is the case with a fluoride ion electrode (Figure 18.2[B]). The indicator's voltage varies with analyte concentration and the measured signal is the potential that arises between the indicator and reference electrodes:

$$E_{cell} = E_{probe} = E_{indicator} - E_{reference} \qquad \textbf{Eq. 18.5}$$

Although half-reaction potentials are often reported in reference to the standard hydrogen electrode, in practice the most commonly used reference electrodes are the saturated calomel electrode (SCE) and the silver/silver chloride electrode. An SCE (Figure 18.3[A]) consists of calomel (Hg_2Cl_2) in contact with elemental mercury (usually mixed together with KCl as a paste), with an inert electrode used to make an electrical connection. The paste is enclosed in a tube with a frit that separates it from a solution containing chloride ion (Equation 18.6). The Ag/AgCl

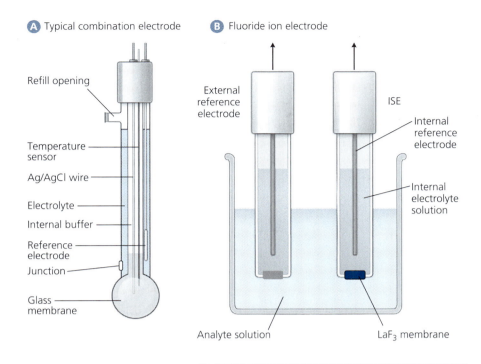

A Typical combination electrode

Refill opening

Temperature sensor

Ag/AgCl wire

Electrolyte

Internal buffer

Reference electrode

Junction

Glass membrane

B Fluoride ion electrode

External reference electrode

ISE

Internal reference electrode

Internal electrolyte solution

Analyte solution

LaF₃ membrane

Figure 18.2 Examples of potentiometric probes: (A) a typical pH electrode in which both electrodes are combined into a single probe and (B) a fluoride ion electrode, in which the indicator electrode is separate from an external (user-supplied) reference electrode.

Under standard state conditions of 25°C, 1 atmosphere, and a concentration of $[H^+] = 1$ M, the reduction of hydrogen ions to hydrogen gas is defined as *exactly* zero volts:

$$2H^+(aq)(1\ M) + 2e^- \rightarrow H_2(1\ atm) \qquad E° = 0\ V$$

and represents the reduction that occurs at a *standard hydrogen electrode (SHE)*, where **E°** is the **standard reduction potential**. The SHE is sometimes referred to as the normal hydrogen electrode. If one were to measure the voltage of a copper electrode at standard conditions against a SHE, you would find that copper has a voltage of +0.34V versus SHE (see Table 18.1). In this case, the copper is being reduced by H_2 gas. Because the potential of the SHE is for the reduction of hydrogen, and we have set hydrogen's reduction potential at 0.00 V, the +0.34 we see for copper means that it is +0.34 volts easier to reduce copper than it is to reduce hydrogen. If we think of the voltage in a wire as being analogous to the pressure in a water pipe, we can think of the reduction potential of copper like this: *it takes 0.34 volts less electron pressure[5] to put electrons onto copper ions than it does to put electrons onto H^+.* ■

PROFILE

The Standard Hydrogen Electrode

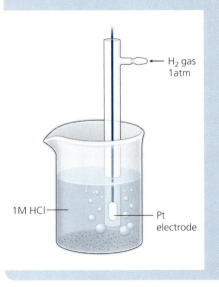

H₂ gas 1atm

1M HCl

Pt electrode

---

[5] *Remember! Electrons are negative so a negative voltage equals more electron density.*

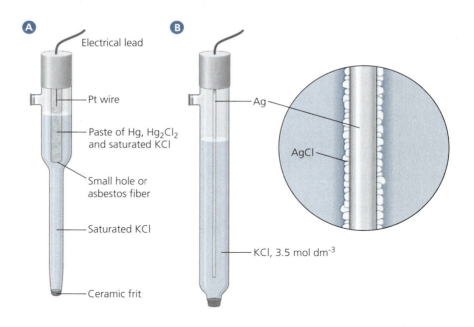

Figure 18.3 Common reference electrodes: (A) a saturated calomel electrode and (B) a silver/silver chloride electrode.

electrode (Figure 18.3[B]) has AgCl deposited directly onto a silver electrode, which is then placed in a solution containing chloride ion (Equation 18.7). Recall that condensed phases are not included in equilibrium expressions, so in both cases, as long as the chloride concentration is maintained at a relatively constant concentration, the potential of the reference electrode half-cell remains the same (see Equation 18.2).

$$Hg_2Cl_2\,(s) + 2e^- \rightleftharpoons 2\,Hg\,(l) + 2\,Cl^- \qquad \textbf{Eq. 18.6}$$

$$AgCl(s) + e^- \rightleftharpoons Ag\,(s) + Cl^-(aq) \qquad \textbf{Eq. 18.7}$$

Problem 18.6: Use Equations 18.2, 18.6, and 18.7 to write the Nernst equation for the SCE and Ag/AgCl half-reactions.

Problem 18.7: Use Table 18.1 and your answer from Problem 18.6 to estimate the potential of the Ag/AgCl reference electrode presented in Figure 18.3(B).

In most cases, probes are calibrated using standard solutions, and so the constant reference potential does not contribute to the final measurement. Hence, the analytical measurement is theoretically related only to the change in concentration of the analyte (A), and the potentiometric response can be related to [A] by Equation 18.8:

$$E_{obs} = \frac{RT}{F}\ln[A] + k \qquad \textbf{Eq. 18.8}$$

Here, E_{obs} is the analytical potential measurement and $[A]$ is the molarity[6] of the analyte of interest. The constant k is a combination of any potential differences (e.g., junction potentials) due to the various interfaces associated with the construction of the probe. For instance, in the pH electrode in Figure 18.2(A), the net E^0 is zero because both the reference and sensor electrodes use the Ag/AgCl redox couple. However, there would be a net k value expected, although small, that would arise from the junction between the reference cell (light blue) and the bulk solution.

Of course, when possible, it is most desirable that the indicator electrode be *specific*—that is, be responsive only to a single analyte. Many times, we must settle for *selective* electrodes that respond to a small number of analytes or a general class of analytes. In these circumstances, we must rely on our understanding of the chemistry of the system to allow us to minimize interferences among similar analytes. For instance, it would be ideal if a pH electrode were specific for hydronium ion. That is, it would respond only to hydronium ion in solution. However, in reality, the pH electrode responds to all +1 cations in solution, with a strong preferential response for hydronium ion.

Generally, indicator electrodes in potentiometric probes are classified as either solid conductor (metal or carbon) or membrane electrodes.

Solid conductor electrodes can be further subclassified.

1. Metallic electrodes of the ***first kind*** serve as indicators of their own ions. An example would be silver metal as an indicator electrode for silver ion in waste effluent from a factory that produces photographic film.

2. Metallic electrodes of the ***second kind*** consist of a metal in contact with one of its insoluble salts. This type of electrode responds to changes in concentration of the anion in the salt. The Ag/AgCl and saturated calomel reference electrodes are both electrodes of the second kind, but the anion concentration (Cl^- in both cases) is kept constant so the potential of the reference electrode remains constant.

3. There are some electrodes known as ***third kind***, but they are generally not widely used because of their reliance on complex, and often sluggish, multiequilibria systems. The equilibrium potential is still a function of the concentration of a metal cation. However, the electrode is not constructed of the same metal as the analyte cation. Construction of the electrode involves a metal in contact with a mixture of two sparingly soluble salts. The first salt contains the ion of the metal contact, and the second salt contains the analyte ion. Both salts share the same anion (e.g., a copper wire in contact with a mixture of copper oxalate and calcium oxalate).

Inert metallic electrodes (e.g., Pt or Au) often simply serve as an electron sink to measure total redox potential versus a reference electrode immersed in the same solution.

Membrane electrodes also are subclassified as either crystalline or noncrystalline. The pH electrode described in the next section is a noncrystalline membrane electrode, and the fluoride ion sensitive electrode seen in Figure 18.2(B) uses a high-purity lanthanum(III) fluoride (LaF_3) crystal as its sensory membrane.

The pH Probe

Potentiometric pH probes utilize a very thin silicate (glass) bulb as the membrane separating the indicator electrode from the reference electrode.[7] Glass is an amorphous solid, so a pH probe is categorized as a noncrystalline membrane electrode. Figure 18.2(B)

---

[6] Strictly speaking, the *activity* of the analyte would be used, but in most cases the concentration is relatively low so using the molarity yields a good approximation.

[7] This glass membrane is quite fragile and can be easily broken. Care should be taken to avoid any impact between it and a hard surface; never use the probe to stir your solution!

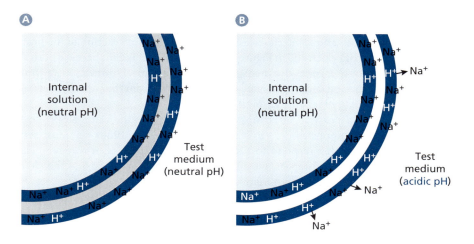

Figure 18.4 The pH membrane electrode. In the absence of H⁺, the inner and outer surfaces (dark blue) are composed of sodium silicate. In the presence of H⁺, some of the Na⁺ within the surface layer of the glass is replaced by H⁺. In (A) both internal and external solutions are neutral, with equivalent displacement of Na⁺ by H⁺, and in (B) there is an acidic external solution, with excess displacement of Na⁺ by H⁺ (more H⁺ in the glass membrane surface layer, after having displaced sodium ions).

shows the bulk construction of the pH probe, consisting of two Ag/AgCl electrodes, one of which is in direct contact with the solution being tested. The other electrode is enclosed in a chamber terminating in a thin-walled glass bulb that is also in contact with the test solution. Within that chamber is also a solution of KCl maintained at constant concentration, so the internal Ag/AgCl electrode stays at a constant reference potential.

Figure 18.4 gives a clearer picture of how the electrode works. The wall of the silicate bulb has exposed Si-O⁻ terminal groups that are charge-balanced by cations, usually sodium or lithium in the absence of any protons. However, H⁺ displaces Na⁺ on the external wall (the one exposed to the test solution), and a potential difference develops because of the change in the membrane junction. The greater the concentration of H⁺, the greater the displacement of Na⁺, which results in a change in the potential between the indicator and reference electrode. This change in potential is our analytical signal. Alternately, the excess hydroxide ions in an alkaline solution will extract protons from the external wall, resulting in the opposite potential shift. In this case, the equation giving the potentiometric response can be written as:

$$E_{obs} = \frac{RT}{nF} \ln\left[H^+\right] + k \qquad\qquad \textbf{Eq. 18.9}$$

where $\left[H^+\right]$ is the concentration of the protons in the external test medium and the constant k encompasses the potential difference due to the solid glass membrane, the potential drop due to solution resistance between the two electrodes, and other similar chemical and physical factors. Of course, pH = ⁻log[H⁺], so:

$$E_{obs} = -\frac{2.303\,RT}{nF} pH + k \qquad\qquad \textbf{Eq. 18.10}$$

and at a given temperature, all values in the expression $\left(\dfrac{2.303\,RT}{nF}\right)$ are constant and E_{obs}, becomes directly proportional to the pH as:

$$pH = -k_1 E_{obs} + k_0 \qquad\qquad \textbf{Eq. 18.11}$$

In practice, then, we calibrate the electrode with standard pH solutions and use the expected linear relationship between pH and measured potential to determine the pH of unknown test solutions. At 25°C, we expect the slope of E versus pH (Equation 18.11) to be 59.1 mV. When a pH meter is calibrated using two buffer standards at a given temperature, the meter will usually report a percent linearity based on the expected slope of Equation 18.11. Note that Equation 18.11 is simplified from Equation 18.10, and so temperature is a factor of the observed signal. When calibrating a pH meter, it is best to do so at or very near the temperature at which you intend to make your analytical measurements.

$$\% \, linearity = \frac{measured \, slope}{expected \, slope} \times 100 = \frac{(measured \, slope)(nF)}{2.303 \, RT} \quad \textbf{Eq. 18.12}$$

Problem 18.8: Demonstrate how the slope of 59.1 mV mentioned in the discussion of Equation 18.11 is determined.

Example 18.3

Here is the derivation of Equation 18.11 from Equations 18.9 and 18.10.

STRATEGY –

We simply need to apply logarithmic rules for converting between base systems and make assumptions regarding values that we expect to remain constant in the equations.

SOLUTION –

$$E_{obs} = \frac{RT}{nF} \ln(H^+) + k \qquad \textbf{Equation 18.9}$$

To convert from ln (natural log, base e) to $\log_{10}$, we multiply by $\ln(10)$, or 2.303:

$$E_{obs} = (2.303) \times \frac{RT}{nF} \log[H^+] + k$$

Because pH = –log[H$^+$], can replace **log[H$^+$]** with **–pH**, giving us

$$E_{obs} = -\frac{2.303 \, RT}{nF} pH + k \qquad \textbf{Equation 18.10}$$

We know that R and F are constants. Also, for any given half-reaction, **n** is constant, having a value of **1** in this case for the half-reaction $H^+ + e^- \rightleftharpoons \frac{1}{2} H_2$. Finally, we will design our experiments such that we calibrate and make analytical measurements at the same temperature, so for a given experiment, we can set T to a constant value as well. Thus, for a given experiment, $\dfrac{2.303 \, RT}{nF}$ will have a constant value, k′, leaving us with Equation 18.11.

Problem 18.9:   How would the slope in Problem 18.8 change if you were conducting experiments at body temperature, 37°C?

As noted above, it is important to understand and, as much as possible, control the chemistry of the system to avoid errors in measurement. Some noteworthy sources of error include the:

1. Dependence of E_{obs} on temperature

2. Increase in junction potential (one of the factors embedded in k) as the difference in ionic strength between the internal buffer and external test solution increases

3. Saturation of the glass membrane with H^+ when $[H^+]$ is high (called the acid error)

4. Response of the electrode to high concentrations of metal ions when $[H^+]$ is low, as in solutions of NaOH or other strong bases (the alkaline error)

The fact that the electrode can respond to any cation might lead to problems any time the activity of a metal ions (particularly small monovalent ions such as Li^+ and Na^+) is appreciable. In such situations, the voltage signal, E_{obs}, includes a measurable response to the metal ions normalized by the relative selectivity of the glass membrane for each species, as seen in Equation 18.13.

$$E_{obs} = \frac{2.303\,RT}{F} \log\left[\left[H^+\right] + \sum \left(f_{\frac{H}{M}} \right)\left[M^+\right] \right] + k \qquad \textbf{Eq. 18.13}$$

The *selectivity coefficient*, $f_{H/M}$, is the normalizing factor that takes into account the equilibrium-based response a given glass electrode has toward a certain metal ion, M^+, with respect to hydronium ion. If multiple interfering ions are present, we would sum their respective effects, as seen in Equation 18.13. Selectivity coefficients are usually quite small (e.g., < 1 ppm for the pH electrode), so corrections for interference by a metal ion are only necessary when the concentration of that ion is relatively large.

Example 18.4

A solution has a nominal pH of 7.00. Given that the selectivity coefficient of a glass electrode for H^+ over Na^+ ($f_{H/Na}$) is 2.5×10^{-8}, what concentration of sodium ion would cause a 1% decrease in the measured pH at 25°C, assuming that $k = 0$?

STRATEGY –

We can use Equation 18.13 to approach this problem. Without the Na^+ interference, the pH would be 7.00, meaning the $[H^+] = 1.0 \times 10^{-7}$ M. A 1% decrease due to Na^+ interference would result in an *observed* pH of 6.99. That is, $pH_{obs} = -\log([H^+] + f_{H/Na} \cdot [Na^+])$.

SOLUTION –

$$6.99 = -\log\left([H^+] + f_{H/Na} \cdot [Na^+]\right)$$

$$6.99 = -\log\left([1.0 \times 10^{-7}] + (2.5 \times 10^{-8}) \cdot [Na^+]\right)$$

$$[Na^+] = 0.093 \text{ M } Na^+ \text{ ion}[8]$$

This would cause a 1% decrease in pH.

---

[8] Sea water is 0.47 M Na^+, so as the sidebar indicates, pH probes are not generally useful to measure ocean water.

Researchers at the University of Southern California and the National University of Singapore recently reported on the development of a nanoscale pH electrode. Antimony demonstrates a potentiometric response to $[H^+]$ without the need for the standard glass membrane, and electrode wires made of pure antimony could be produced with diameters as small as 80 nm, which is around 500 times smaller than a human cell. The electrodes exhibited a high degree of linearity, and when coated with the sulfonated polymer Nafion showed excellent stability over several weeks. ■

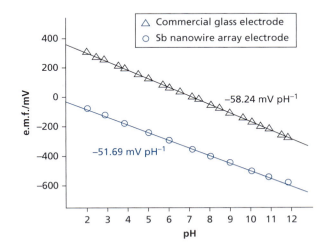

Chang, P-C., Chen, H-Y., Ye, J-S., Sheu, F-S., Lu, J. G. ChemPhysChem, 2007, 8(1), 57–61.

PROFILE

Nanoscale pH Probe for In Vivo Use

Problem 18.10: Why does the glass membrane electrode have a higher affinity for protons than for other metal cations?

Problem 18.11: A pH meter was calibrated using two standard buffer solutions, measuring −234 mV for the pH 4.00 buffer and −403 mV for the pH 7.00 buffer. In an unknown solution, the voltage measured was −478 mV. What would the meter report as the pH of the unknown solution? If this experiment were to be repeated, can you suggest an improvement in methodology?

Problem 18.12: What would the pH meter used in Problem 18.11 have reported as the percent linearity for the calibration step?

Problem 18.13: If the pH_{obs} of a sample of sea water is measured to be 8.05, what is the actual pH if the concentration of sodium ion is 0.47 M at 25°, given that $f_{H/Na}$ = 2.5×10^{-8}? What is the percent error in the measurement?

Problem 18.14: Repeat Problem 18.13, but include interferences by K^+ (0.015 $\underline{M}$, $f_{H/K}$ = 2.0×10^{-10}) and Ca^{2+} (0.015 $\underline{M}$, $f_{H/Ca}$ = 1.6×10^{-8}).

Problem 18.15: What is the estimated potential measured in a pH 4.00 buffer at 25°C, assuming nonanalytical factors (i.e., k) are negligible? What would the expected error be if the laboratory temperature increased by 5°C during the day?

Figure 18.5 Eutrophication leading to an algal bloom.

The Nitrate Probe

Many selective potentiometric probes fall into the category of *ion-selective electrodes* (ISEs). Like pH probes, ISEs are selective for a specific species, or a very small class of species, but will experience interference from other ions of similar size and charge. Many ISEs have the same general construction as glass pH electrodes, with the thin glass membrane replaced by a membrane that is selective for the given ion. In fact, some manufacturers sell ISE assemblies in which only the membrane is swapped out in order to measure the concentration of one ion or another. We will use the nitrate ISE to demonstrate this class of membrane electrode.

The nitrate ISE is an important tool for environmental chemists and ecologists. Nitrate is a form of fixed nitrogen that serves as a fertilizer for all plant life, including algae. Although in nature, nitrate concentrations are generally less than 1 ppm, anthropogenic sources can increase that by a factor of 10 to 100, which can lead to widespread eutrophication that can clog and smother bodies of water (see Figure 18.5). Elevated levels of nitrate in drinking water can lead to a sometimes fatal disease in infants, the so-called blue baby syndrome. The nitrate ISE was developed as a quick and accurate method for monitoring nitrate concentrations in natural waters.

The specific method for construction of a nitrate ISE varies widely. Figure 18.6 shows the general form of a membrane ISE. The membrane is typically a polymer (often polyvinylchloride [PVC]) doped with a substance that provides the selectivity for the ion of interest. For nitrate, the selective dopant is usually a quaternary ammonium nitrate salt, such as tetraoctylammonium nitrate. The ISE might require a separate reference electrode, or it can be constructed as a single unit similar to the pH electrode shown in Figure 18.2(A). The functionality of the nitrate ISE also is similar to the pH electrode. The part of the PVC membrane in contact with the standard internal solution develops a fixed equilibrium involving an exchange between the nitrate in $R_4N^+NO_3^-$ and the anion(s) in the solution. When the electrode is exposed to a test solution, a potential develops across the membrane as a function of the concentration of nitrate in the solution. As with the pH probe, the nitrate ISE must be calibrated with standard nitrate solutions before use, but is capable of measuring nitrate at the ppb level. Also, like the pH probe, the nitrate ISE is susceptible to interference by other anions. In most cases, the interference is small enough as to be easily handled, but even small concentrations of perchlorate and chlorate must be avoided or mathematically corrected. The operational mathematical relationship is given in Equation 18.14, where I^- represents an interfering anion.

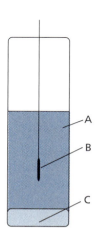

Figure 18.6 The nitrate ISE: (A) internal standard filling solution— usually KCl and/or $(NH_4)_2SO_4$, (B) Ag/AgCl electrode, and (C) PVC doped with $R_4N^+NO_3^-$.

$$E_{obs} = \frac{2.303\,RT}{nF} \log\left[[NO_3^-] + \sum \left(f_{\frac{NO_3}{I}} \right)[I^-] \right] + k \qquad \textbf{Eq. 18.14}$$

Creager and colleagues described a simple method for construction of a salicylate ion-selective electrode. Polyvinylchloride (PVC) is codissolved with tetraoctylammonium salicylate in tetrahydrofuran in a standard 15 mL beaker. After the solvent is allowed to evaporate, the membrane was removed with forceps and 1 cm diameter disks were cut out of it with a cork borer. Using open-top caps, the PVC membrane disks were sealed onto electrode bodies made from 3-dram vials with the bottoms removed. The electrode body was filled with aqueous 0.1 M NaCl/0.1 M Na-salicylate and capped with a rubber septum with a Ag/AgCl electrode inserted through it. A standard Ag/AgCl reference electrode completed the cell. Although simple in design and construction, the electrode yielded a detection limit as low as 10^{-4} M salicylate. However, benzoate was found to exhibit a relatively significant interference, with a $f_{salicylate/benzoate}$ of 3×10^{-3}. ■

Creager, S. E.; Lawrence, K. D.; Tibbets, C. R. J. Chem. Ed. **1995**, *72*, 274–276.

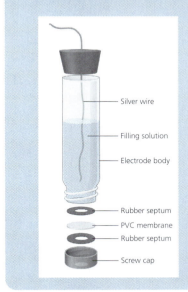
The Oxygen Probe

The development of the oxygen sensor has made it possible for you to have a quality of life far superior than might otherwise been possible. Some models of automobiles had oxygen sensors prior to 1986, although in that year their use, as well as that of the catalytic converter, became mandatory in the manufacture of all new automobiles in the United States and Australia. Since that time, oxygen sensors (also called lambda sensors) have been implemented in factories and power plants that use fossil fuel combustion as an energy source. Potentiometric feedback from the lambda sensor allows for subtle adjustments to be made to the fuel:air ratio, maximizing combustion and thus minimizing emission of volatile organic compounds and carbon monoxide.

The sensor consists of a yttrium-stabilized zirconia (YSZ) membrane sandwiched between two gas permeable platinum electrodes. One of the electrodes is exposed to a sample gas and the other to a reference gas (Figure 18.7). Any difference in the partial pressure of oxygen on either side of the YSZ results in a Nernstian response as defined by Equation 18.15. In this case, n = 4 because the pertinent redox couple is $O_2/2O^{2-}$. The reference gas is usually just atmospheric air. In order to function properly, the lambda sensor must be heated to around 320°C, and so in practice its feedback is shunted during engine warmup, or an internal heater is included to reach the operating temperature rapidly.

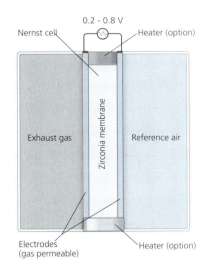

Figure 18.7 Planar oxygen (lambda) sensor.

$$E_{obs} = \frac{RT}{nF} \ln \frac{p(O_2)_{exhaust}}{p(O_2)_{reference}} + k \qquad \textbf{Eq. 18.15}$$

Example 18.5

You are interested in measuring the nitrate ion concentration in a sample of pond water that contains 0.0050 M bicarbonate ion (HCO_3^-) and 1.4×10^{-4} M iodide ion. You expect the nitrate ion concentration to be around 0.020 M. If the selectivity coefficients are $f_{NO3/HCO3} = 0.02$ and $f_{NO3/I} = 20$, what percent error can you expect in your measurement of $[NO_3^-]$?

STRATEGY –

We can use Equation 18.14 to approach this problem. The only part of Equation 18.14 that would be affected by the interferences is the parenthetical term associated with the log function.

SOLUTION –

$$[NO_3^-]_{Obs} = ([NO_3^-] + (f_{NO3/HCO3}) \cdot [HCO_3^-] + (f_{NO3/I}) \cdot [I^-])$$
$$= ([0.020 \underline{M}] + (0.02)(0.005 \underline{M}) + (20)(1.4 \times 10^{-4})]$$
$$= 0.0229 \underline{M}$$

Because our nominal nitrate ion concentration is 0.02, we can calculate the percent error as:

$$\%\,error = \frac{|theoretical - actual|}{theoretical} \times 100 = \frac{|nominal - observed|}{nominal} \times 100$$

$$\%\,error = \frac{|0.020 - 0.0229|}{0.020} \times 100 = 14.5\%\ error$$

Problem 18.16: Compare Equations 18.9 and 18.15. What difference(s) do you observe? Why are they different?

Problem 18.17: For a certain nitrate ISE, the nitrate/nitrite selectivity factor is 0.062, that of nitrate/chloride is 4.0×10^{-3}, and that of nitrate/perchlorate is 990. What is the percent error expected for the measurement of 1.5 mM nitrate in the presence of each interfering anion (individually, not all present at once) at a concentration of 0.1 mM?

Problem 18.18: Given the same conditions as presented in Problem 18.17, what is the expected percent error if all three interferences are present?

Problem 18.19: Assuming that k is negligible, calculate the partial pressure of oxygen in the exhaust gas of an automobile if the lambda sensor is reporting a potential of 317 mV. Assume the reference gas is air.

18.3 Nonpotentiometric Probes

Many modern probes rely on active electrochemical methods; that is, the response of the probe depends on the voltage or current associated with a redox reaction involving the analyte of interest. These probes are often referred to as galvanic or amperometric probes or sensors. Galvanic cells involve a spontaneous redox reaction that results in a

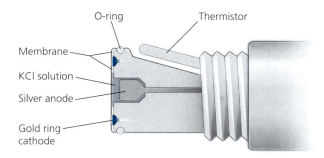

O-ring   Thermistor

Membrane

KCl solution

Silver anode

Gold ring
cathode

Figure 18.8 The Clark cell—a standard dissolved oxygen probe.

measurable voltage, and amperometric cells require an applied voltage to cause a redox reaction to occur, with the subsequent measurement of current as the analytical signal.

The Dissolved Oxygen Probe

The level of dissolved oxygen (DO) is an important measure of the quality of a natural body of water. Insufficient molecular O_2 dissolved in water causes aquatic life to suffer—and eventually to suffocate and die. Furthermore, without adequate levels of DO, bacteria that normally break down, or decompose, dead plants and animals are unable to do so efficiently, and anaerobic processes begin to dominate, often producing the rotten egg smell associated with stagnant swamps. Historically, a modified Winkler method has been used to determine DO, but this requires laboratory analysis, and so it was necessary to sample the water to be tested and rapidly transport it back to the laboratory, because decomposition reactions that consume oxygen would be continuing. The development of the Clark cell with a gas-permeable membrane made possible the on-site determination of DO.

The Clark cell (Figure 18.8) consists of an inert cathode (usually platinum or gold) and an anode made of a reactive metal (usually silver or copper). The electrodes are immersed in a standard (1 M) KCl solution and a voltage of –0.6 to –0.8 V is applied across the electrodes, resulting in a redox reaction that produces a current proportional to the amount of dissolved molecular O_2.

Anode reaction: $Ag\,(s) + Cl^-\,(aq) \rightleftharpoons AgCl\,(s) + e^-$ $E^0 = -0.22$ V   **Eq. 18.16a**

Cathode reaction: $\frac{1}{4}\,O_2\,(aq) + \frac{1}{2}\,H_2O\,(l) + e^- \rightleftharpoons OH^-\,(aq)$ $E^0 = +0.40$ V   **Eq. 18.16b**

Cell reaction: $Ag\,(s) + Cl^-\,(aq) + \frac{1}{4}\,O_2\,(aq) + \frac{1}{2}\,H_2O\,(l) \rightleftharpoons$

$AgCl\,(s) + OH^-\,(aq)$ $E^0_{Cell} = +0.18$ V **Eq. 18.16c**

Note that the E^0_{Cell} is positive, denoting a spontaneous reaction, at least from a purely thermodynamic point of view. However, for the direct reduction of oxygen at a solid electrode, there is an *overpotential*[9] needed to overcome kinetic barriers related to the electron transfer reaction, so the reducing potential is applied to ensure complete and instantaneous reduction of all molecular O_2 that diffuses to the platinum electrode

---

[9] We can estimate thermodynamically the potential at which we expect a given redox reaction to occur; for instance, E^0_{Cell} = +0.18 V in the cell reaction given in Equation 18.16c is a thermodynamically determined potential. In practice, however, it is usually necessary to apply a slightly higher voltage than that thermodynamically predicted due to kinetic barriers or voltage drops across resistances within the cell, such as across a porous membrane. That extra voltage needed to force the reaction to occur is called an *overpotential*.

surface. It is also important to realize that the silver anode is partially consumed during the reaction, resulting in the growth of a layer of AgCl (s) on the electrode surface. The anode in the Clark cell must be periodically polished to allow continued operation. The large applied voltage (0.8–1.0 V more negative than the E_{Cell}) also helps compensate for the *passivation*[10] of the anode during the redox process.

The determination of DO as described above can be accomplished in the laboratory with just bare wires dipped into the aqueous solution, as long as the solution is relatively free of other potential redox species. However, in natural waters, many substances will be codissolved with the oxygen that could cause a false signal. For instance, Fe(III), which is ubiquitous in natural waters, would be reduced to Fe(0) at the applied potential. Therefore, in the typical DO amperometric probe, the reaction cell is separated from the test solution (e.g., pond water) by a thin (~20–25 μm) polyethylene or Teflon membrane that is permeable to molecular oxygen but not to most ionic species that would be in the water (see Figure 18.8). Because of the small volume of the inner chamber, the chloride concentration decreases during the analysis (see Equation 18.16a), and the electrolyte solution must be periodically replenished.

Most commercial probes can measure DO in the range of 1 to 20 ppm with an accuracy of around ± 0.1 ppm. This degree of accuracy is necessary because DO in natural waters typically ranges from 5 to 15 ppm.

The Chloride Probe

Although chlorine and aqueous chlorine reaction products do not occur naturally to any significant extent, in developed countries those species have become virtually omnipresent in all water systems. Chlorine is commonly added to water as a disinfectant. Municipalities add it to wastewater to destroy microorganisms before pumping the water back into natural systems, and they also add it in smaller concentration to drinking water to prevent the growth of microorganism cultures in that system. Industrial plants use chlorine in their water-cooling systems to minimize fungal and algal growths that can clog system pumps, pipes, and valves. It is important to maintain the proper balance to provide adequate disinfection but minimize bad flavor (in drinking water) and hazards to nonharmful biological organisms.

In measuring chlorine, the goal is actually to determine *total available chlorine* (TAC), because molecular Cl_2 reacts with water to form hypochlorite and Cl_2 will only exist at very low pH (see Figure 18.9). Hypochlorite is, of course, the conjugate base of the weak acid hypochlorous acid, and so TAC exists as a pH-dependent complex equilibrium involving all three species.

$$Cl_2 \text{ (aq)} + H_2O \text{ (l)} \rightleftharpoons HOCl \text{ (aq)} + H^+ \text{ (aq)} + Cl^- \text{ (aq)} \quad K_1 = 3.9 \times 10^{-4} \quad \textbf{Eq. 18.17}$$

$$HOCl(aq) \rightleftharpoons H^+(aq) + OCl^-(aq) \qquad\qquad\qquad K_2 = 2.9 \times 10^{-8} \quad \textbf{Eq. 18.18}$$

In most natural water systems, the pH will range from 6.5 to 8.5, depending on the industrial and environmental features of the locale. Unfortunately, this is exactly the pH range over which the concentrations of HOCl and OCl⁻ vary to the greatest extent. In some cases, only the concentration of HOCl is important, because it is the most active disinfecting agent. That would be the case for analysis at a water treatment plant, because the primary concern there is achieving the appropriate level of chlorine to kill microbes.

Usually we are interested in the TAC, which entails the measurement of both HOCl and OCl⁻. Because Cl_2 only becomes significant below a pH of around 2.5, we typically

---

[10] During certain redox processes, the surface of the electrode will become coated with a nonconductive layer, either due to the formation of an oxide or by deposition of an electrochemical reaction product. Because of this layer, the solution species that come into contact with the surface of the electrode experience a voltage lower than that applied; this effect is called *passivation* of the electrode.

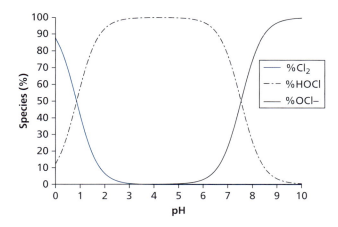

Figure 18.9 Equilibrium distribution of chlorine species in aqueous solution as a function of pH.

consider its contribution negligible. There are three common approaches to measuring TAC with a chlorine probe. A known volume of the water sample can be adjusted to a pH of around 5 prior to measurement, thereby converting all the free chlorine to HOCl, followed by measurement of HOCl. Alternatively, a measurement of HOCl can be made in the original water sample, with a secondary measurement of pH also made, and then the HOCl result can be adjusted mathematically for the pH level. Finally, a two-phase probe can be used that allows for the independent determination of both HOCl and OCl⁻, and TAC is simply the sum of the two.

An amperometric chlorine probe relies on the reduction of HOCl and/or OCl⁻ at the cathode, with the concurrent anode reaction identical to that used in the oxygen probe (Equation 18.16a).

Cathode reaction:

$$HOCl(aq) + H^+(aq) + 2e^- \rightleftharpoons Cl^-(aq) + H_2O(l) \quad E^0 = +1.49\,V \qquad \textbf{Eq. 18.19}$$

and/or

$$OCl^-(aq) + H_2O(l) + 2e^- \rightleftharpoons Cl^-(aq) + OH^-(aq) \quad E^0 = +0.90\,V \qquad \textbf{Eq. 18.20}$$

In either case, the net cell reaction will be spontaneous, and in fact the probe can be operated in the galvanic mode if desired. However, the galvanic potential is greatly dependent on other redox equilibria occurring in solution, as well as on the concentration of Cl⁻, which will change over time. Furthermore, other halogenated species can be severe interferences, particularly organochlorides. The probe is usually operated in the amperometric mode, with a potential of approximately 0.5 V applied across the electrodes in order to allow the desired reactions to occur while preventing simultaneous reduction of organohalides.

The probe itself has the same general construction as the DO probe shown in Figure 18.6, but the membrane is usually polyvinylidene fluoride, which is selectively permeable to HOCl. Most commercial probes can measure chlorine in the range of around 0.01 to 20 ppm, reported as TAC.

Problem 18.20:   What is the E^0_{Cell} for the chloride probe set to measure only HOCl? What is the E^0_{Cell} for the measurement of OCl⁻?

FYI

Polyvinylidene fluoride (PVDF) is structurally similar to Teflon. PVDF is derived from 1,1-difluoroethylene, and Teflon is derived from 1,1,2,2-tetrafluorethylene.

PVDF

Teflon

Problem 18.21: You want to replenish the electrolyte in the DO probe shown in Figure 18.6 when the chloride concentration, originally 1.0 M̲, has dropped by 10%. How long could you operate the probe continuously before refreshing the electrolyte if the inner chamber has a volume of 500 μL, the polytetrafluoroethylene (PTFE) membrane is 25 μm thick, and O_2 diffuses through PTFE at a rate of 1.05×10^{-7} g/s·cm?

Problem 18.22: Consider the equilibria presented in Equations 18.17 and 18.18. Use Excel to generate a plot of $\%Cl_2$, %HOCl, and $\%OCl^-$ versus pH (as in Figure 18.9) for water in an area of industrial runoff where the chloride concentration is 112 ppm and the initial Cl_2 concentration is 4.5 ppm. *Hint: The TAC involves only active chlorine—that is Cl_2, HOCl, and OCl^-; therefore, assume the chloride (Cl^-) concentration remains constant, and use the K_1 expression to find HOCl and the K_2 expression to find OCl^-.*

The Total Salinity Probe

Total salinity refers to the total concentration of ions resulting from all dissolved and dissociated salts in a water sample. To achieve a true measurement of total salinity, we would need to conduct a full chemical analysis of the sample to determine the amount of NaCl, KCl, Na_2SO_4, KNO_3, and so on. This is time and cost prohibitive in most cases. Instead, we measure the conductance of the solution (i.e., that is, its ability to conduct electricity). Of course, the total conductance of a complex mixture such as a natural water sample is a *function* of the total ion concentration, but the actual conductivity associated with a given ion depends directly on its mobility in an electrical field (its migration rate), which in turn is directly related to charge and inversely to its hydrated size.

The salinity probe relies neither on potentiometry nor on redox reactions; rather, it uses a simple conductometric measurement and relates the results to a single species (chloride ion). That is, the meter measures resistance (R, in ohms) across a fixed linear distance between two electrodes, and converts that to *conductance* (G), which is the reciprocal of resistance. A basic schematic of a conductivity cell is shown in Figure 18.10. Previously, the unit of conductance was the mho ("ohm" spelled backwards), but has been replaced by the SI unit, the Siemens. The Siemens, named after the nineteenth-century inventor and industrialist Ernst Werner von Siemens, is equal in magnitude to the mho. The analytical value reported is the *conductivity* (C). The reported value takes into account the area of the electrodes (A) and the distance (d) between the electrodes. The electrodes are constructed of an inert material such as platinum, gold, or carbon.

$$G = \frac{1}{R} \text{(units are Siemens, S)} \qquad \textbf{Eq. 18.21}$$

$$C = G \cdot \frac{d}{A} \text{(units are S/cm)} \qquad \textbf{Eq. 18.22}$$

The determination of salinity (S_{Cl}) used to be based on a conversion factor related to the amount of silver needed to precipitate the chloride in the solution (called the chlorinity). However, the current method uses the measured conductivity and temperature (T) in °C of a sample and an empirically determined set of complex polynomial functions.[11]

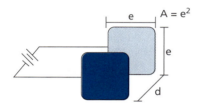

Figure 18.10 Basic schematic of a conductivity cell composed of two square electrodes separated by a distance, d. In a standard conductivity cell, the electrodes are made of platinum and d = 1.0 cm.

---

[11] This approach, described in *Standard Methods for the Examination of Water and Wastewater*, is valid only for samples with salinity greater than 2 PSU and lower than 42 PSU. Sea water has a salinity around 35 PSU at 25°C. Fresh water has a salinity well below the cutoff (< 1 PSU), but total salinity is not usually an important measure for fresh water.

Example 18.6

A conductivity cell uses square electrodes that are 0.50 cm in length and are positioned 0.75 cm apart. When placed in a solution, the resistance is found to be 1400 ohms. Calculate the conductivity of the solution.

SOLUTION –

We can use Equation 18.21 to calculate the conductance as:

$$G = \frac{1}{R} = \frac{1}{1400\,ohms} = 7.143 \times 10^{-4}\,ohms^{-1} = 714.3\,\mu S$$

Then we use Equation 18.22 to calculate the conductivity as

$$C = G \cdot \frac{d}{A} = (714.3\,\mu S) \cdot \frac{0.75\,cm}{(0.50\,cm)^2} = 2142\,\mu S/cm = 2100\,\mu S/cm$$

$$S_{Cl} = a_0 + a_{\frac{1}{2}} R_t^{\frac{1}{2}} + a_1 R_t + a_{\frac{5}{2}} R_t^{\frac{5}{2}} + a_2 R_t^2 + a_{\frac{5}{2}} R_t^{\frac{5}{2}} + \Delta S \qquad \textbf{Eq. 18.23}$$

$$\Delta S_{Cl} = \left[\frac{T-15}{1+0.0162(T-15)} \right] \left[b_0 + b_{\frac{1}{2}} R_t^{\frac{1}{2}} + b_1 R_t + b_{\frac{3}{2}} R_t^{\frac{3}{2}} + b_2 R_t^2 + b_{\frac{5}{2}} R_t^{\frac{5}{2}} \right] \qquad \textbf{Eq. 18.24}$$

The value R_t used in the empirical Equations 18.23 and 18.24 is a ratio of the measured sample conductivity to a known standard, or $R_t = C_{samp}/C_{KCl}$.[12] This standard is known to have a salinity of exactly 35 practical salinity units (PSU) and a conductivity of 53.06 mS/cm. Although PSU has no specific concentration units, in practice the magnitude of salinity measured in PSU is very close to the total ion concentration as measured in parts per thousand. The coefficient values used in Equations 18.23 and 18.24 were found experimentally and are given in Table 18.2.

There are tables available in the literature for finding the conductivity of the KCl standard at various temperatures (the denominator in the R_t term in Equation 18.24). Alternatively, Equation 18.25 can be used to estimate C (KCl standard at temperature T).

$$C(KCl\,Std,\,\mu S/cm) = -0.2672\,T^3 + 4.6637\,T^2 + 861.3\,T + 23035 \qquad \textbf{Eq. 18.25}$$

Problem 18.23: A conductivity cell uses circular electrodes that are 0.50 cm in diameter and are positioned 1.00 cm apart. When placed in a solution, the resistance is found to be 1100 ohms. Calculate the conductivity of the solution.

Problem 18.24: What is the salinity of the water samples described in Example 18.6 and Problem 18.23, assuming the measurements were made at 25°C? It is probably easier to do this calculation using a spreadsheet.

---

[12] The standard is made from 32.4356 g KCl dissolved in water to achieve a final solution mass of 1.000 kg: $R_t = C_{samp}/C_{KCl}$.

TABLE 18.2 : Empirical values of coefficients *a* and *b* for use in Equations 18.23 and 18.24, respectively

X*	a_X	b_X
0	0.0080	0.0005
$\frac{1}{2}$	−0.1692	−0.0056
1	25.3851	−0.0066
$\frac{3}{2}$	14.0941	−0.0375
2	−7.0621	0.0636
$\frac{5}{2}$	2.7081	−0.0144

* The x term refers to the different factors in Equations 18.23 and 18.24. For instance, the second term in Equation 18.23 is $a_{\frac{1}{2}} R_t^{\frac{1}{2}}$, so we would use $a_X = a_{\frac{1}{2}} =$ −0.1692.

18.4 Probes for Measurements in the Human Body

The Glucose Probe: A Biosensor

Electrochemical biosensors are probes that use organisms or molecules of biological origin to detect specific molecules. They are often designed to be highly selective for biological molecules or substances found in biological systems, such as glucose, alcohol, cholesterol, and urea. In function, electrochemical biosensors are very similar to ISEs. A common approach in developing biosensors has been to use a modified Clark cell (Figure 18.8), with the outer oxygen-permeable membrane still in place, but with an additional polymeric membrane in which is immobilized an enzyme that catalyzes the oxidation of the target molecule (e.g., glucose oxidase, alcohol oxidase, cholesterol oxidase). The oxidation reaction consumes oxygen, which is observed as a decrease in the amperometric signal due to reduction of oxygen at the indicator electrode. We will use the first-generation glucose amperometric sensor (which is a Clark cell analog) to describe this in more detail, and then we will discuss some more recent improvements in the sensor design.

People with diabetes must regularly check their glucose levels as a preventive measure against hyperglycemia. Most diabetics monitor their glucose levels using the fingerprick method, drawing a single drop of blood that is transferred to an instrument that reports back their glucose level. Some such instruments are spectroscopic in nature, but many are electrochemically based. A significant amount of research has gone into the development of continuous glucose monitors (CGMs)—microscale-to-nanoscale devices that can be inserted subcutaneously and left to work continuously for an extended period. Although many different forms of CGMs have been commercialized, none have yet received widespread acceptance. We will focus on a few forms of non-CGM amperometric glucose sensors.

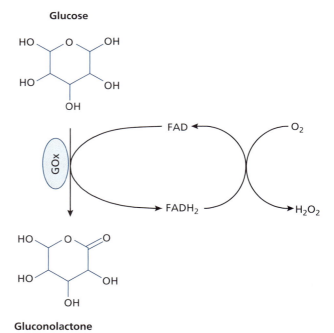

Figure 18.11 Scheme for the glucose oxidase catalyzed oxidation of glucose, resulting in the reduction of O_2 to H_2O_2.

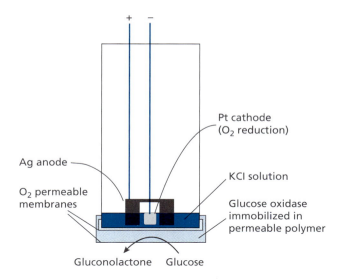

Figure 18.12 Modified Clark cell for the indirect detection of glucose by monitoring decreased oxygen levels.

All glucose biosensors utilize the catalytic oxidation of glucose by glucose oxidase (GOx), as seen in Figure 18.11. Glucose oxidation is catalyzed by GOx with the help of the cofactor flavin adenine dinucleotide (FAD), which is reduced to $FADH_2$ during the process. The $FADH_2$ is then reoxidized by ambient O_2, which is itself reduced to H_2O_2 in the process. *First-generation sensors* initially used a modified Clark cell as depicted in Figure 18.12.

The biosensor measures a baseline current due to the constant level of oxygen diffusing to the platinum electrode in the absence of glucose. When glucose is introduced into the system, it diffuses into the polymer layer containing GOx. The subsequent reaction results in a decrease in the oxygen reaching the platinum cathode and hence a decrease in the signal compared to the baseline current. Another variation of the first-generation sensor measures the level of H_2O_2 produced during the final step of the glucose oxidation scheme. This is favored over the Clark cell because it does not depend on the diffusion of oxygen through two PTFE membranes *and* the GOx immobilizing polymer layer. However, both versions of the first-generation sensors depend on continuous and constant availability of oxygen and the application of relatively high potentials (around 0.6 V) at which other common species (e.g., ascorbic acid, acetaminophen) can be electrochemically oxidized.

The problem of interferences can be at least partially addressed using specialized membranes that permit only a small number of molecules (glucose, of course, included) through to the GOx immobilizing layer. Although various schemes have been proposed for resolving the irregularity of oxygen supply, none are wholly satisfactory. *Second-generation glucose sensors* have dealt with the problem by eliminating oxygen as the electron sink in the final step, replacing it with a reversible redox species such as ferrocene, $Fe(Cp)_2$. The indicator electrode essentially has three layers:

1. Ferrocene (or a derivative) is deposited directly on a platinum or carbon electrode surface, creating a modified electrode.

2. GOx is codeposited with FAD in a permeable polymer, such as Nafion.

3. A thin outer membrane, such as polycarbonate or PVC, is added as a protective coating.

A positive potential (around 400 mV for ferrocene) is applied to the indicator electrode, causing the oxidation of the ferrocene to ferrocenium ion at the electrode surface.

$$Fe(Cp)_2 \text{ (Pt anode)} \rightarrow Fe(Cp)_2^+ + e^- \qquad \textbf{Eq. 18.26}$$

When glucose is introduced to the system, it diffuses into the Nafion layer containing the immobilized GOx and is oxidized.

$$Glucose + FAD(GOdx) \rightarrow gluconolactone + [FAD(GOx)]_{reduced} \qquad \textbf{Eq. 18.27}$$

$$[FAD(GOx)]_{reduced} + 2\,Fe(Cp)_2^+ \rightarrow FAD(GOx) + 2\,Fe(Cp)_2 \qquad \textbf{Eq. 18.28}$$

The ferrocenium ion serves as an electron sink for the reduced FAD(GOx) (Equation 18.28). It is then is immediately reoxidized to ferrocenium by the electrode, resulting in a current signal proportional to the amount of glucose present.

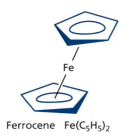

Ferrocene $Fe(C_5H_5)_2$

PROFILE

Percentage of Adults Treated for Diabetes Nearly Doubled in a Decade

According to the Agency for Healthcare Research and Quality, the percentage of US adults treated for diabetes (both type 1 and 2) nearly doubled from 4.6% in 1996 to 8.5% million in 2007. Because the population has increased in that time, the number of adults treated has more than doubled, from 9 million in 1996 to 19 million in 2007. Medical spending to treat diabetes has also more than doubled from $18.5 billion in 1996 (scaled to 2007 dollars) to $40.8 billion in 2007. The trend is continuing, according to the National Institutes of Health, with almost two million new cases reported in 2010. In 2007, diabetes was the seventh leading cause of death in the United States; noted as the cause of more than 71,000 deaths that year, it was listed as a contributing cause of death in an additional 160,000 deaths. The risk of death among people with diabetes is about twice that of people without the disease, making clear the need for continued research into simple and accurate methods for monitoring glucose on a continuous basis. ■

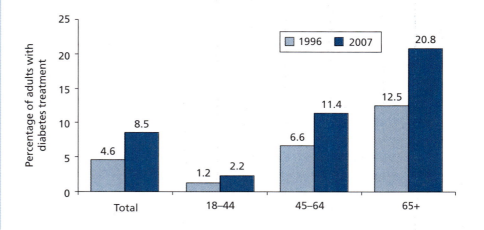

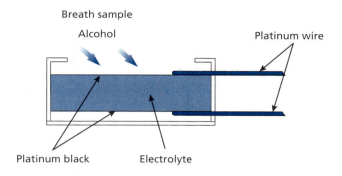

Figure 18.13 Alcohol fuel cell biosensor design.

The Alcohol Fuel Cell Probe

Fuel cells use a different form of active electrochemistry than we have so far discussed. Rather than applying a voltage or current, a fuel cell depends on the occurrence of a spontaneous redox reaction that provides a measurable electrical signal. One version of the modern alcohol breath analyzer, the Alco-Sensor (manufactured by Intoximeters) uses this type of electrochemical method.

Although some details of the alcohol fuel cell sensor are proprietary, the basic design is known and the chemistry is well understood. The cell (Figure 18.13) consists of a thin (2–5 mm) porous polymer coated with finely divided platinum (called platinum black). The porous polymer is filled with an acidic electrolyte solution and is sandwiched between two platinum electrodes. The breath sample is directed onto the platinum black surface, and a spontaneous reaction occurs that results in the oxidation of ethanol to acetic acid.

$$C_2H_5OH + H_2O \rightarrow CH_3COOH + 4\,H^+ + 4\,e^- \qquad E^0 = +0.75\text{ V} \qquad \textbf{Eq. 18.29}$$

The electrons are collected by the top platinum electrode, which by definition is the anode, and flow through the circuitry, providing the signal. The protons produced in Equation 18.29 diffuse through the porous polymer electrolyte and combine with oxygen (from external air) which is reduced at the other platinum electrode by the electron flow from the ethanol oxidation.

$$O_2 + 4\,H^+ + 4\,e^- \rightarrow 2\,H_2O \qquad E^0 = +1.23\text{ V} \qquad \textbf{Eq. 18.30}$$

The net reaction is

$$C_2H_5OH + O_2 \rightarrow CH_3COOH + H_2O \qquad E^0_{net} = +1.98\text{ V} \qquad \textbf{Eq. 18.31}$$

Problem 18.25: Why do you think the porous polymer electrolyte is coated with platinum black rather than just having it in contact with a standard platinum electrode?

Problem 18.26: Calculate the Gibbs free energy (kJ/mol) associated with the net reaction in the alcohol fuel cell sensor. What does that ΔG_{rxn} tell you about the overall reaction?

The alcohol fuel cell biosensor has been tested for potential interferences against a wide range of substances. Unlike its progenitor, the Breathalyzer, which relied on a

TOTO, the leading Japanese manufacturer of toilets and other bathroom appliances, announced in 2005 its "Intelligent Toilet II." The high-end (more than $5,000) toilet monitors certain health-related parameters and includes a temperature sensor, a biosensor for urine sugar levels, a conveniently located blood pressure cuff, and a scale and body mass index (BMI) measurement station located by the toilet. Results are compiled on a per-visit basis and can be automatically downloaded to a computer or other electronic device through the toilet's Internet connection. ■

colorimetric chemical reaction, the fuel cell sensor does not respond to the acetone that might be exhaled adventitiously by a diabetic person. It does respond to other alcohols, but at a much lower efficiency than for ethanol. For instance, it exhibits a 22% efficiency toward methanol and a 10% efficiency toward isopropanol (rubbing alcohol). Most states and provinces have accepted the alcohol fuel cell sensor as a reliable field sobriety test.

Figure 18.14

18.5 Further Reading

BOOKS

Minear, R. A. *Water Analysis: Inorganic Species, Part 2.* Academic Press: Orlando, FL, **1984**.

Namiesnik, J.; Szefer, P. (editors). *Analytical Measurements in Aquatic Environments.* CRC Press Analytical Series: Boca Raton, FL, **2009**.

Pico, Y., ed. *Chemical Analysis of Food: Techniques and Applications.* Academic Press: Orlando, FL, **2012**.

Rice, E. W.; Baird, R. B.; Eaton, A. D.; Clesceri, L. S., eds. *Standard Methods for the Examination of Water and Wastewater.* American Water Works Association: Washington, D.C., **2012**.

Wagner, R. J.; Boulger, R. W.; Oblinger, C. J.; Smith, B. A. *Guidelines and Standard Procedures for Continuous Water-Quality Monitors: Station Operation, Record Computation,* *and Data Reporting.* US Department of the Interior/US Geological Survey: Washington, D.C., **2006**.

JOURNALS

Creager, S. E.; Lawrence, K. D.; Tibbets, C. R. An Easily Constructed Salicylate-Ion-Selective Electrode for Use in the Instructional Laboratory. *J. Chem. Ed.,* **1995**, 72, 274–276.

Wang, J. Electrochemical Glucose Sensors. *Chemical Reviews.* **2008**, 108, 814–825.

ONLINE RESOURCES

Toxicological Profile for Chlorine. US Department of Health and Human Services, **2010**. http://www.atsdr.cdc.gov/toxprofiles/tp172.pdf.

18.6 Exercises

EXERCISE 18.1: Write the anode and cathode half-reactions for a galvanic (spontaneous) cell constructed using tin and silver electrodes. Also, write the overall cell reaction and find the standard cell potential.

EXERCISE 18.2: Calculate the Gibbs free energy of the redox reaction described in Problem 18.20.

EXERCISE 18.3: Write the expected anode and cathode reactions for the electrolysis of water conducted in a neutral solution using NaI as the electrolyte.

EXERCISE 18.4: What is the minimum potential you would need to apply to the system described in Problem 18.3 in order to produce either H_2 or O_2 gas?

EXERCISE 18.5: Why are ISEs usually selective, but rarely specific, for the analyte of interest?

EXERCISE 18.6: What is the estimated potential measured for a glass membrane electrode in a pH = 10.00 buffer at 21°C, assuming nonanalytical factors (i.e., k) are negligible? What would the expected error be if the laboratory temperature rose to 25°C by the end of the day with no additional calibrations?

EXERCISE 18.7: Why is the Clark cell usually operated in the amperometric mode, even though the overall reaction is spontaneous?

EXERCISE 18.8: Estimate the actual cell potential for a Clark cell at 22°C if $[Cl^-] = 0.91$ M, the pH = 10.12, and dissolved oxygen is present at 7.8 ppm.

EXERCISE 18.9: If the pH_{obs} of a sample of water from the Great Salt Lake in Utah is measured to be 7.65, what is the actual pH if the concentration of sodium ion is 1.56 M and the Mg^{2+} concentration is 1.89 M 25°C, given that $f_{H/Na} = 2.5 \times 10^{-8}$ and $f_{H/Mg} = 1.7 \times 10^{-8}$? What is the percent error in the measurement, assuming only sodium and magnesium ions are significant enough to affect the measurement?

EXERCISE 18.10: The concentration of salts in seawater is approximately 0.6 M. By what percent would this affect your pH reading if you failed to correct for salinity? Assume the total salt content can be represented by NaCl.

Analytical Voltammetry

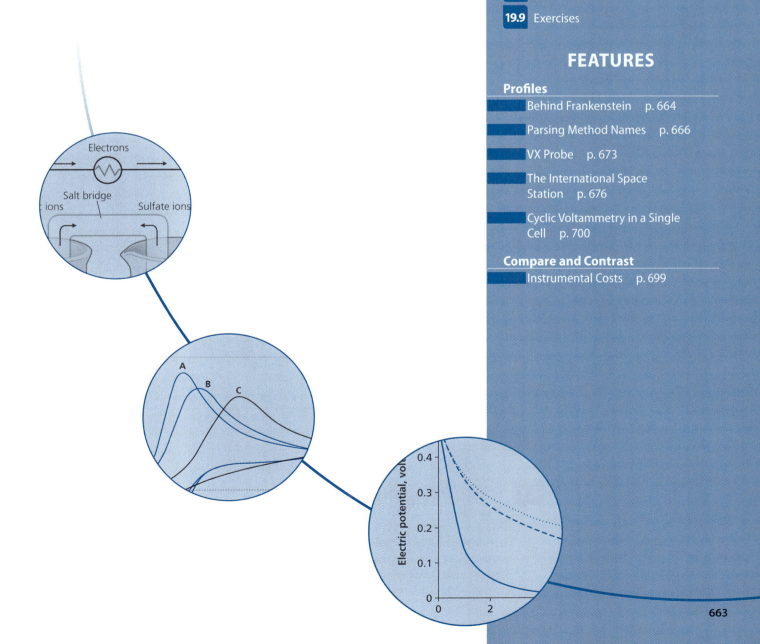

Mary Shelley was inspired to pen her novel *Frankenstein; or, The Modern Prometheus* by a scientific controversy begun by one of the earliest electrochemists, Luigi Galvani, who was a Professor of Anatomy and Surgery at the University of Bologna in the late eighteenth century. While dissecting a frog, Galvani was surprised to observe that when he touched both a nerve and a leg muscle with a conducting tool, such as scissors or a scalpel, the leg muscle twitched! He conducted a series of additional experiments and in 1791, he published a 70-page treatise describing his studies and noting his conclusion that electricity is a vital, innate force of life, which he called "animal electricity."

Galvani was ultimately shown to be wrong—he had not discovered a new form of electricity. However, he was a well-respected scientist. His name lives on in terms such as "biological galvanism," which is the stimulation of muscles with electric current, and "galvanic," which describes an electrochemical cell that produces electricity via a spontaneous reaction. ■■

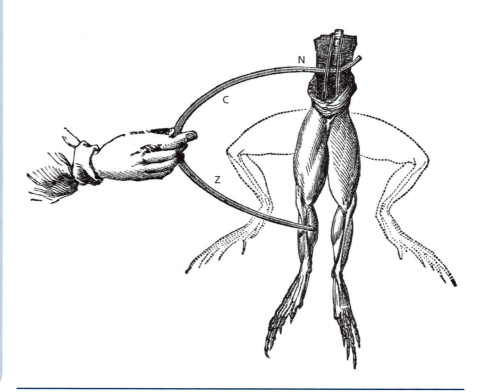

19.1   Basic Principles

Electroanalytical chemistry refers to any technique in which electrochemistry is used in the analysis of a sample. One of the earliest techniques that can be considered dynamic electroanalysis is *chronopotentiometry*, in which a constant current is applied across a pair of electrodes and the resulting potential measured over time. The measured potential provided qualitative information regarding which metal ion(s) might be present in the solution, and quantitative information is related to the amount of time associated

TABLE 19.1: Some Common Dynamic Electroanalytical Chemistry Methods

TECHNIQUE	PARAMETER MEASURED	MEASURED AS A FUNCTION OF	CONTROLLED PARAMETER
Chronoamperometry	Current (amp)	Time	Potential
Chronopotentiometry	Potential	Time	Current
Chronocoulometry	Charge	Time	Current or potential
Chronoabsorptometry	Absorbance	Time	Current or potential
Voltammetry	Current	Potential	Potential

with a given redox event.[1] Chronopotentiometry enjoyed decades of being the primary electroanalytical method because the control of current and measurement of potential both require fairly unsophisticated electronic equipment. It is still used occasionally today, but with the advent of inexpensive solid state transistors and semiconductor circuitry, it has been largely replaced by controlled potential methods such as amperometry and voltammetry. Some dynamic electroanalytical methods are listed in Table 19.1.

As noted in Chapter 18, electroanalytical chemistry can be divided into two general regimes: *passive* and *dynamic*. The techniques discussed in Chapter 18 are, for the most part, *passive* methods.[2] The dynamic methods in the current chapter involve controlling one electrochemical parameter (typically potential or current) and observing the effect on a different parameter or on the chemical system. Although the probes we discussed in Chapter 18 are vital for quantitative analysis, dynamic methods often find greater use and application for qualitative analysis of chemical systems. A diverse menu of methods is available to the analyst in the area of dynamic electrochemistry. The focus of this chapter is on voltammetric methods—and in particular, on cyclic voltammetry and square wave voltammetry. On a day-to-day basis, the most commonly used methods are cyclic and square wave voltammetry, and we will endeavor to gain a firm understanding of those methods, with the anticipation that this basic knowledge will enable a learner to extend his or her expertise to more advanced applications of dynamic electrochemistry as the need arises. Because the analytical signals observed in voltammetry are based on what happens at an electrode when a potential is applied, we will first take a look at chronoamperometry in order to explore the fundamental concepts we will need to understand voltammetry.

It is essential to understand the basics of electricity in order to understand electroanalytical chemistry. In Chapter 4 we presented a more thorough treatment of the importance of electricity and electrical circuits and in Chapter 5 we discussed current and voltage as analytical signals.

However, because they are important to understanding electroanalytical methods, we will define here five fundamental concepts as they relate to those methods. *Current* is the rate of electron flow and is measured in amperes (amps, where an amp = coulombs/second); it is given the symbol i or I. *Potential* or voltage is an energy term that relates to the electrochemical force being applied or generated and is measured in volts (V = joules/coulomb); when discussing electricity in general, its symbol is V, but in electrochemistry it is often given the symbol E. In all electrical systems, there exists a natural *resistance* to the flow of electrons through matter or to the flow of charged particles through a solution; resistance is given the symbol R and is measured in ohms.

---

[1] When current is applied to a system containing a substance that can be oxidized or reduced, the potential will shift to that at which the most easily oxidized (or reduced) species (A) reacts. The potential will remain constant until all of species A is reacted and then will shift to the potential at which the next most easily oxidized species will react, and so on. The time during which the potential is constant is called the transition time and is represented by the Greek letter tau, τ.

[2] Passive methods are those used to measure a steady state parameter (usually potential) that is related to a chemical attribute of the system under study.

The names used to identify the techniques used in dynamic electrochemistry are actually descriptions of the techniques themselves. Familiarity with common Greek and Latin roots (e.g., *chrono* for time) can help you identify the parameters of the experiment, and the names are parsed in reverse, starting with the root at the end of the word. For instance, when the word *chronoamperometry* is parsed in reverse, *metry* + *ampere* + *chrono* can be interpreted as "measure amps as a function of time." We would expect the data we obtain to be a plot of current (*y* axis) versus time (*x* axis). Similarly, a plot prepared from an experiment in voltammetry ("measure amps as a function of volts") should be a graph of current versus potential. The parameter that is actually controlled in these experiments is often obvious, but is not necessarily a part of the name (see Table 19.1). ∎

Charge is current integrated over time and has the units of coulombs, C; charge usually is given the symbol q or Q. Finally, *time* (seconds) is always a factor in electroanalytical studies. Resistance, charge, and time are not always explicitly considered during typical experiments, but they are fundamental attributes that we must always keep in mind.

Problem 19.1:   Parse the experimental method *voltabsorptometry*, giving the measured parameter, what that parameter is measured as a function of, and what you believe the controlled parameter would be.

Problem 19.2:   Describe a possible chemical system in which an experimenter might find voltabsorbtometry to be helpful. *Hint:* Consider Chapters 6 and 11.

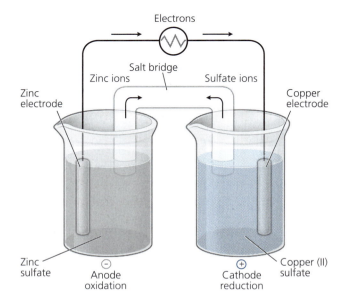

Figure 19.1 The Daniell cell is a galvanic cell that produces a voltage of +1.1 V.

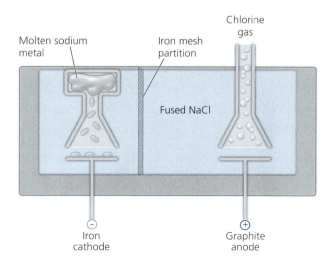

Chlorine gas

Molten sodium metal

Iron mesh partition

Fused NaCl

⊖
Iron cathode

⊕
Graphite anode

Figure 19.2 The Downs cell is an electrolytic cell. From molten (fused) NaCl, sodium ion (Na^+) is reduced to sodium metal at the iron cathode, and chloride ion (Cl^-) is oxidized to Cl_2 gas at the graphite anode.

19.2 The Three-Electrode Cell

We learned in general chemistry that (1) an oxidation-reduction (redox) reaction can be used to generate electricity and electricity can be used to do work (your basic battery) and (2) electrical energy can be applied to a chemical system to drive an otherwise nonspontaneous reaction. In both cases, the process occurs in an electrochemical *cell* that contains at least two electrodes and the chemical system. You are already familiar with two electrode cells. A *galvanic* cell is one in which a spontaneous chemical redox reaction is used to generate electricity. For example, the Daniell cell, consisting of zinc metal and Zn(II) ion in one half-cell and copper metal and Cu(II) ion in the other half-cell, will generate a voltage of approximately 1.1 V (see Figure 19.1). An *electrolysis* cell is one in which electricity is used to drive a reaction such as the recharging of a battery or the production of chlorine gas and sodium metal from liquid sodium chloride (see Figure 19.2). The two electrodes are given names that define the type of reaction occurring at each. The *cathode* is the electrode at which reduction occurs, and the *anode* is the electrode at which oxidation occurs.

> **Problem 19.3:** Using skills gained in Chapter 18, write the two half-reactions and the overall redox reaction associated with the Daniell cell (Fig. 19.1). Calculate the Gibbs free energy of the cell and confirm that it is galvanic (i.e., that the reaction is spontaneous).

> **Problem 19.4:** Using skills gained in Chapter 18, write the two half-reactions and overall redox reactions associated with the Downs cell (Figure 19.2). Calculate the Gibbs free energy of the cell and confirm that it is electrolytic (i.e., that the reaction is nonspontaneous).

Although it is possible to perform electroanalytical experiments using only two electrodes (as with chronopotentiometry), the greatest analytical results are obtained from a three-electrode cell (see Figure 19.3). We do not usually use the terms *cathode* and *anode* when describing an electroanalytical cell, preferring instead terms such as *reference electrode*, *working electrode*, and *counter electrode*, which provide more information about the

Keeping It Straight

Oxidation occurs at the anode and **o**xidation and **a**node both start with a vowel.

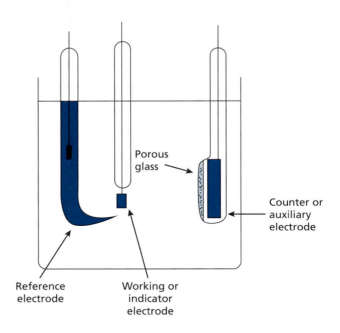

Porous
glass

Counter or
auxiliary
electrode

Reference
electrode

Working or
indicator
electrode

Figure 19.3 Schematic of a three-electrode electroanalytical cell.

function of each electrode in the cell. In using the cell, we apply a controlled waveform[3] of either current or potential to the working electrode[4] (often referred to as the indicator electrode) and the response waveform (the analytical signal, potential or current, respectively) is measured. During the experiment, current flows between the working and the counter electrode (also known as the auxiliary electrode). The reduction or oxidation process of interest occurs at the working electrode, but reduction cannot occur without oxidation and vice versa. The opposite process must occur at the counter electrode. Because we are not specifically controlling the process at the counter electrode, unknown and possibly interfering species might be generated there, so we usually separate the counter electrode from the bulk solution using a porous glass separator that allows for charge transport but hinders the mobility of the unknown species.

The Downs cell (Figure 19.2) is not a dynamic electroanalytical cell, because no signal is being measured as a result of an applied potential or current. However, we can it to demonstrate the concepts discussed so far. A specific voltage is applied across the two electrodes, resulting in the production of liquid sodium at the cathode and chlorine gas at the anode. If the two products were to come into contact, they would spontaneously react to form NaCl, resulting in a decrease in the efficiency of the process. To prevent the recombination of the molten sodium and chlorine gas, the two electrodes are separated by a porous iron mesh.

As noted in Section 19.1, the solution in which the chemical system of interest exists will exhibit some electrical resistance. We will discuss later how we increase solution conductivity, but even after we take such measures, there will be resistance between the working and counter electrodes. Kirchhoff's voltage law (Chapter 4) tells us that we should expect a voltage drop across this resistance. Thus, when only those two electrodes are used, we cannot be certain that the intended potential is actually that experienced by

---

[3] A waveform is a pattern of values used in an experiment. For instance, a controlled potential waveform might be a linear progression of applied voltages from –0.5 V to +0.5 V in steps of 1 mV. The two most common waveforms—those used for cyclic and square wave voltammetries—will be demonstrated later in the chapter.

[4] Note that the instrument does not operate this way functionally, but this is how we think about the process in terms of interpreting the analytical results.

the chemical system at the working electrode. This is where the third electrode comes into play. To partially correct for the voltage drop between the two primary electrodes, a reference electrode is added to the cell[5] to serve as a passive sensor of the working electrode potential. The reference electrode provides feedback to the instrument so that adjustments can be made to the working electrode potential to ensure its accuracy.

Problem 19.5: Name and briefly describe three common reference electrodes. *Hint:* See Chapter 18.

Problem 19.6: Consider Figure 19.2. What can you infer about the density of molten Na compared to that of molten NaCl?

Problem 19.7: Why is it important that the divider that separates the working and counter electrodes (or the two electrodes in Figure 19.2) be porous?

19.3 Chronoamperometry

The Experiment

As the name implies, *chronoamperometry* (CA), the most fundamental of the dynamic electroanalytical techniques, is an experimental methods that measures current as a function of time. In CA, a steady-state voltage is applied to the working electrode (see Figure 19.4[A]), and the current generated by any redox reaction(s) at the working and counter electrodes is measured as a function of time (see Figure 19.4[B]). In the figure, a negative potential is applied to the working electrode (see the y axis in Figure 19.4[A]); therefore, the analyte must be initially in an oxidized form (Ox) and is being converted to a reduced form (Red). A negative potential indicates that electrons are being pushed into the system, and the reaction looks something like Eq. 19.1:

$$Ox\,(aq) + e^- \rightleftharpoons Red\,(aq) \qquad\qquad \textbf{Eq. 19.1}$$

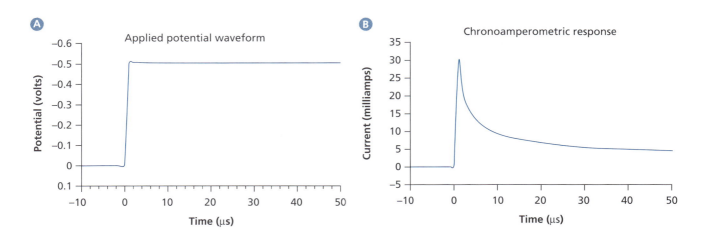

Figure 19.4 A typical CA experiment. (A) Applied potential waveform. At zero seconds, the voltage at the working electrode was changed from 0.0 V to −0.5 V. (B) Measured current waveform. The current measured between the working and counter electrode initially spiked to 30 mA and decayed over time to a steady-state current of 7 mA over time. Experimental parameters: 3 mm diameter electrode, 1 electron reduction reaction, 5.0 mM analyte concentration, and diffusion coefficient of 7.6×10^{-6} cm²/s.

---

[5] Reference electrodes are discussed in Chapter 18.

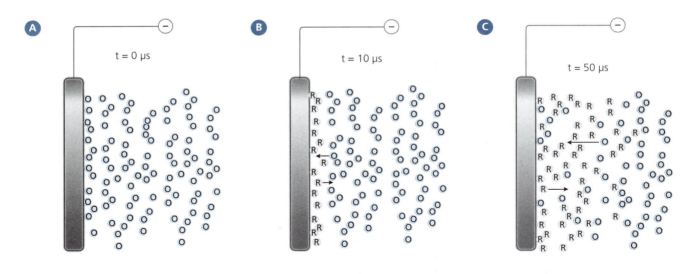

Figure 19.5 Schematic showing diffusion layers of oxidized and reduced species at an electrode surface during a CA experiment. The arrows represent diffusion to and from the electrode surface.

As electrons flow from the working electrode into the system, a current is generated between the working and counter electrodes. Note that the current generated varies over time (see Figure 19.4[B]). We can understand the shape of the current response curve by considering what is going on in the system. Initially, analyte species, Ox, are homogeneously distributed in solution, so it is expected that there are some analyte species very near the working electrode. In fact, some analyte molecules or ions are expected to be adsorbed to the electrode surface (see Figure 19.5[A]); we will address that phenomenon in more detail later. At time = 0, the voltage is instantly stepped to a value of –0.5 volts. The great surge in current observed initially after that voltage step is due to the immediate reduction of those Ox analyte particles[6] adsorbed to the electrode and in the immediate vicinity of the electrode. The number of analyte particles in the Ox form that are available for reduction decreases as the reaction goes forward, and the Ox analyte particles are converted to Red (see Figure 19.5[B]). The concentration of Ox is diminished to zero near the electrode. Because of diffusion, the Ox particles in the bulk solution move toward the electrode while the newly generated Red particles diffuse away from the electrode into bulk solution. At some point, the *depletion zone* around the electrode becomes large enough that the continued current response becomes entirely dependent on simple diffusion of Ox particles from the area of high concentration (bulk solution) to the electrode (see Figure 19.5[C]). At that point, the reaction is said to be *diffusion controlled* and the current plateaus at a nonzero value that is proportional to the concentration and diffusion coefficient of Ox in the bulk solution. Figure 19.4(B) shows the diffusion controlled current. Note that the final current is not zero.

Problem 19.8:   Sketch the diffusion layers of oxidized and reduced species (similar to that shown in Figure 19.5) for the application of a positive voltage to an electrode at which a reduced substance is oxidized.

---

[6] Note that "particle" here is used as a generic term meaning *analyte molecule* or *analyte ion*, because the species of interest could be in either form.

Noise in Chronoamperometry and Related Methods

As in all analytical methods, we try to maximize the desired signal and minimize all other nonanalytical signals (noise). In the case of electroanalytical chemistry, we attempt to maximize the *faradaic current,* due to the redox processes involving the analyte of interest, and minimize faradaic current from impurities present in solution. At extreme potentials, even the solvent may undergo reduction or oxidation and therefore would contribute to the faradaic current. The goal of the experimentalist is to ensure that only the analyte of interest contributes toward the measured faradaic current. Sources of nonfaradaic current and faradaic current from impurities must be minimized. Sources of noise can be minimized through careful preparation before the execution of the CA experiment. *Nonfaradaic current* arises from a change in distribution of redox-inactive species near the electrode surface. The primary component of nonfaradaic current, and one that cannot be eliminated, is *charging current,* which will be present following any change in the potential applied to an electrode regardless of the presence or absence of electroactive species in solution.

Charging Current

If a negative potential is applied to an electrode in a vacuum, with no opportunity to discharge or diffuse the charge, only a very few electrons could be forced to collect at the electrode surface before the electrode becomes highly polarized and thus resistant to continued current flow. In other words, the static potential on the electrode surface would be equal and opposite to the potential applied to the electrode. This is exactly analogous to the buildup of charge on a capacitor (see Chapter 4), and charging current is often referred to as *capacitive charging*. However, when a potential is applied to an electrode in solution, ions and polar species in solution are drawn to the electrode due to the electric field produced. These species provide an effective charge balance, allowing greater polarization and thus a greater flow of current before resistance to additional current flow ends the process. It is theorized that the electrode then appears as a capacitor, with electrons aligned along the electrode surface and ions of opposite charge aligned in solution along the electrode surface. That layer of ions in turn attracts ions of opposite charge in what is termed an *ionic double layer* or *electrical double layer*, and there follows an ever more diffuse series of layers until the magnitude of the electrical field is in essence reduced to zero as the normal conditions of the bulk solution are attained (see Figure 19.6). At the electrode surface, the layers of ions essentially act as a charged electrical capacitor.[7]

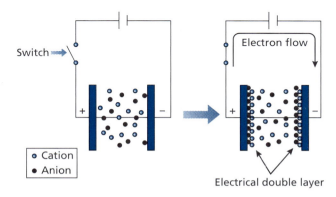

Figure 19.6 Schematic diagram of the electrode surface before and after the application of a potential to the electrode.

---

[7] The theoretical treatment of the double layer describes a much more complicated situation than this, but the simple representation is sufficient for what we need to consider in this chapter.

In CA, the current due to charging of this double layer, the charging current, is measurable and interferes with the acquisition of the desirable signal. It is fortunate that the time required to charge the double layer is short, at least on the time frame of the CA experiment, and so double-layer charging current can be avoided by using only data obtained later in the experiment (e.g., during the diffusion-controlled region). In Figure 19.4 (B), we see this as the plateau region that starts at about 25 μsec. There is, however, an additional problem—*migration*—associated with polarization of the electrode and the nature of the double layer that we must consider. In order to do so, we must address the concept of mass transport in the solution phase.

Mass Transport

Mass transport, or how matter moves from one place to another in solution, occurs via three modes: *diffusion*, *migration*, and *convection*.

Diffusion is the motion of particles from a region of high concentration to one of lower concentration; the flux[8] of particles via diffusion is given by *Fick's first law* (see Equation 19.2):

$$J_k = -D_k \frac{\Delta C_k}{\Delta x}$$

Eq. 19.2

where

J_k = flux density of species k due to diffusion (mol/m^2s)
D_k = diffusion coefficient of species k (m^2/s)
x = distance from the electrode surface (m)
C_k = concentration of k at distance x from the electrode surface (mol/m^3)

The quantity $\frac{\Delta C_k}{\Delta x}$ represents the concentration gradient, and it is the driving force for diffusion of molecular species in solution. The negative sign in front of D_k indicates that species are moving from regions of high concentration to regions of low concentration or from the bulk solution toward the electrode. Instantly after the application of the reducing (or oxidizing) potential, all of the electroactive substance in contact with the electrode surface is reduced, and so $C_k = 0$ at $x = 0$. At some distance from the electrode, the concentration of the substance is at a maximum and is equal to the bulk (initial) concentration. The flux of analyte to the electrode surface then is defined by the inherent mobility of the species in solution (as given by the diffusion coefficient, D_k) and the concentration gradient from the electrode surface to the bulk solution.

Migration is the motion of a charged or asymmetrically polar species within an electrical field. When a potential is applied across two electrodes (the working and counter electrodes in this case), a potential gradient is created. In many cases, analytes in solution are charged species, and so migration can affect their mass transport from the bulk solution to the electrode surface. The flux due to migration is given by Equation 19.3:

$$J_k = -\frac{zF}{RT} D_k \cdot C_k \frac{\partial \phi(x)}{\partial x}$$

Eq. 19.3

where

J_k = flux density of species k due to migration (mol/m^2s)
z = charge on the species of interest (unitless)
$\frac{\partial \phi(x)}{\partial x}$ = electrical field gradient due to the potential applied to the electrode (V/cm)
F = Faraday's constant $(96,485\ C/mol)$
R = gas constant, $8.3145\ (J/mol \cdot K)$

---

[8] Flux (or flux density) is the amount of substance, usually in moles, that travels through a given area in a unit of time (usually $m^2 \cdot s$).

T = temperature (K)

D_k = diffusion coefficient of species k (m²/s)

C_k = concentration of species k at distance x from the electrode surface (mol/m³)

x = distance from the electrode surface (m)

So, the flux due to migration will depend largely on the concentration of the analyte, the charge on the analyte, and the magnitude of the electric field.

Convection is motion due to mechanical mixing, either deliberate (stirring) or from accidental vibrations in the laboratory or from thermal gradients. In the case of mechanical stirring, the flux of material in the solution is proportional to the rate of stirring, $v(x)$, as shown in Equation 19.4.

$$J_k = C_k \cdot v(x)$$

Eq. 19.4

where

J_k = flux density of species k due to convection (mol/m²s)

C_k = bulk solution concentration of species k (mol/m³)

$v(x)$ = rate of mechanical stirring (m/s)

x = distance from the electrode surface (m)

Mass transport occurs via all three modes (diffusion, migration, and convection), and so the total flux is the sum of Equations 19.2 to 19.4. Such a scenario is quite

---

The Geneva Protocol bans the use of "asphyxiating, poisonous or other gases" because such wartime activities are widely believed to be among the most heinous committed by humans against each other. Because nerve gases inspire great fear in a populace, they are a primary tool for potential terrorist groups.

In 2013, a research group in the Czech Republic developed a chronoamperometric sensor that can detect picomolar quantities of well-known nerve gases such as VX and sarin. The sensor depends on the fact that nerve gases inhibit the enzyme acetylcholinesterase (AChE), and so when AChE is immobilized on an electrode, the presence of nerve gases can be detected through simple chronoamperometric detection. ◼

PROFILE
VX Probe

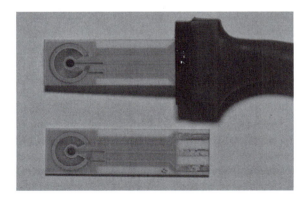

Pohanka, A.; Vojtech, A.; Kizek, R. An Acetylcholinesterase-Based Chronoamperometric Biosensor for Fast and Reliable Assay of Nerve Agents. Sensors, **2013***, 11498–11506.*

complicated and does not support straightforward mathematical modeling of the electrochemical behavior. It would be quite difficult to get analytically useful data from these experiments if all three modes of mass transport had a significant contribution to the total mass transport. Therefore, it is necessary to minimize or eliminate one or more modes of mass transport in order to interpret the data we obtain.

Problem 19.9: For each of Equations 19.2, 19.3, and 19.4, confirm that the units balance out on both sides of the equation.

Problem 19.10: Consider Equation 19.3 and the parameters z, D, T, and C. What is the effect on migration flux density when each is increased? Think about the physical phenomena of flux (motion of particles toward the electrode) and explain the effects you noted with each of the parameters.

Problem 19.11: Define the following terms:

(a) Chronopotentiometry

(b) Voltammetry

(c) Cathode

(d) Diffusion

(e) Capacitive charging

(f) Mass transport

(g) Faradaic current

(h) Galvanic

Controlling Mass Transport

Diffusion is a natural process over which the experimenter has no control; it is inherently controlled by the viscosity of the solvent[9] and the size of the diffusing species. Because we cannot alter the mass transport due to diffusion, in order to simplify the system we must do our best to mitigate the effects of migration and convection.

Migration is also a natural process, but note that the flux in Equation 19.3 is dependent on parameters we can control in the laboratory. We can effectively negate the effect of migration on the mass transport of solute species by the addition of an excess concentration of an electrochemically inactive salt—a *supporting electrolyte* (SE). If the concentration of ions due to the SE is very high compared to that of the analyte species, it will be the SE ions that undergo the greatest migrational flux. Furthermore, a large concentration of SE ions allows the double layer to discharge the electrical field in a very short distance, usually within nanometers of the electrode surface, so very few analyte ions will experience the electric field at all. Figure 19.7 demonstrates the effect of SE concentration on the potential field as a function of distance from a planar disk electrode. Ideally, we select our SE such that the ions are small compared to the particle size of the analyte species, so that the diffusion coefficient of the SE ions will be much larger than that of the analyte, again shifting the migrational flux toward the SE ions in preference to the analyte ions. The addition of SE has further benefits; it significantly decreases the solution resistance, decreasing the ohmic (IR) voltage drop, and the excess concentration of SE makes it possible to fully charge the double layer in

---

[9] In Chapter 17 (gel electrophoresis), we used the gel as a sieving agent *and* to mitigate the effects of diffusion. The very viscous nature of the gel greatly inhibits diffusion of the analyte.

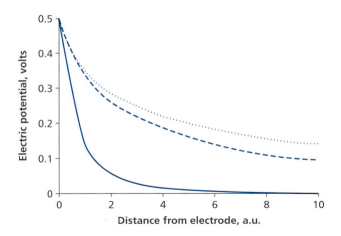

Figure 19.7 Depiction of the electric field gradient on application of a +0.5 V potential to a disk electrode with supporting electrolyte concentrations of 0.5 $\underline{M}$ (solid line), 0.01 $\underline{M}$ (dashed line), and 0.001 $\underline{M}$ (dotted line).

an extraordinarily short time, minimizing the interference by charging current. Ohmic drop, or IR drop, is discussed in somewhat more detail later in Section 19.4.

Mass transport due to convection can easily be minimized by not stirring the analyte solution, avoiding areas where vibrations are common, and maintaining constant temperature during the time frame of the experiment (not usually a problem in modern laboratories). A solution under these conditions is said to be *quiescent*.

Through two simple approaches—addition of a salt and using a quiescent solution—we dramatically minimize mass transport from two of the modes, resulting in transport due primarily to diffusion.

The Cottrell Equation

The *Cottrell equation* (Equation 19.5) describes mathematically the behavior of a CA system and allows us to use CA as a quantitative method by relating signal (measured current) to the concentration of analyte.[10] It is important to note that one of the assumptions underlying the Cottrell equation is that the potential applied is sufficient to overcome any redox equilibria—that is, all particles of the analyte species reaching the electrode are instantly reduced (or oxidized).

$$i(t) = \frac{nFAC_O^* \sqrt{D_O}}{\sqrt{\pi t}}$$

Eq. 19.5

where

$i(t)$ = current measured at time t (s)

n = number of electrons involved in the redox equation (mol e^-/mol reactant)

A = area of the electrode (m^2)

D_O = diffusion coefficient of the analyte (m^2/s)

C_O^* = bulk concentration of the redox active analyte (mol/m^3)

F = Faraday's constant, 96,485 C/mol e^-

[10] The Cottrell equation, as given here, is derived under the assumption that the electrode is planar (i.e., not cylindrical or spherical, for instance), that the potential applied is sufficient to cause all of the given analyte to undergo a redox reaction, that migration and convection are both negligible, and that the reaction is electrochemically reversible (see Section 19.4, *Reversibility*. For a complete discussion of the derivation of the Cottrell equation, see Chapter 4 of Bard, A.; Faulkner L. *Electrochemical Methods*. John Wiley & Sons, New York, **2001**.

PROFILE

The International Space Station

The International Space Station (ISS) has a "taste" for electroanalytical chemistry. One of the primary tools used to monitor water quality on the ISS is the electronic tongue, or E-Tongue. The E-Tongue utilizes a array of electrochemical tools in its mission to help maintain clean water for the ISS astronauts. Included in the array are electrodes for cyclic voltammetry, conductivity, and an assortment of ion-selective electrodes (see Chapter 18). Also included is an array of galvanic micro cells that will be used for corrosion studies. The entire E-tongue is less than 1.5 inches in diameter. ■■

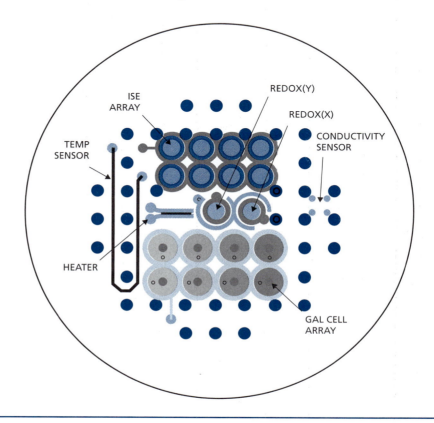

We can see that a plot of $i(t)$ versus $t^{-1/2}$ is expected to be linear, and in practice it is. In most cases, we use only data after the first few microseconds—that is, after the charging current has decayed to zero, when we can be confident that current is limited by diffusion of analyte species to the electrode surface (recall Figure 19.4). Analytically, $i(t)$ is directly proportional to concentration. If we know n, A, and D, we can use the current measured under conditions of diffusion control (the plateau region in Figure 19.4) to directly determine the concentration of analyte in a solution. If any of those three parameters are not known, we can still use CA for quantitative analysis through the use of a calibration curve.

Problem 19.12:   CA with an oxidizing potential was conducted with a 1.13 μM solution of luminol using a 3.0 mm diameter carbon disk electrode. The slope of the $i(t)$ versus $t^{-1/2}$ was 3.7×10^{-8} A·s$^{1/2}$. Luminol has a diffusion coefficient of 4.6×10^{-6} cm^2/s.[11] Estimate the number of electrons transferred in the oxidation of luminol. (Use Example 19.1 as a guide.)

---

[11] Pastore, P.; et al. *Ann Chim.*, **2002**, 92, 271–280.

Example 19.1

CA with a reducing potential was conducted with a 3.05 mM solution of a new $Rh(phen)L_4^{2+}$ complex using a 1.0 mm diameter platinum disk electrode. The slope of the i(t) versus $t^{-1/2}$ was 3.66×10^{-6} A·s$^{\frac{1}{2}}$. A literature search of similar complexes allows us to estimate the diffusion coefficient as approximately 9×10^{-6} cm^2/s. Estimate the number of electrons transferred in the reduction of the $Rh(phen)L_4^{2+}$ complex. (The term *phen* stands for the ligand phenanthroline and L refers to a generic monocoordinating ligand.)

STRATEGY –

First, as a linear equation (y = mx + b), the Cottrell equation describes a plot of i(t) versus $t^{-1/2}$, which will give the slope (m):

$$m = \frac{nFAC_O^* \sqrt{D_O}}{\pi^{1/2}}$$

Solving the slope equation for *n* yields:

$$n = \frac{m\sqrt{\pi}}{FAC_O^* \sqrt{D_O}}$$

Second, we need to make sure all of our values are in the proper units so that they will cancel appropriately throughout the calculation. Then we will plug the values into our equation for *n*.
We use the conversion factor 1 A·s/C to convert coulombs to amp·s.

$$F = 96{,}485 \text{ C/mol} \times 1 \text{ A·s/C} = 96{,}485 \text{ A·s/mol}$$

We use the geometry of the electrode to calculate the Area (A). In this case the area of a circular disk is calculated by πr^2:

$$A = \pi r^2 = \pi(0.50 \text{ mm})^2 \times (1 \text{ m}/10^3 \text{mm})^2 = 7.\underline{8}54 \times 10^{-7} \text{ m}^2$$

The diffusion coefficient that we found in the literature was in units of cm^2/s, and our area term was in meters squared, so we converted the diffusion coefficient to meters squared:

$$D = 9 \times 10^{-6} \text{ cm}^2/\text{s} \times (1 \text{ m}/100 \text{ cm})^2 = 9 \times 10^{-10} \text{ m}^2/\text{s}$$

Finally, the concentration was given in molarity (moles per liter), and in order to match our other volume units, we need to convert L to m^3.

$$C_O = 3.05 \times 10^{-3} \text{ mol/L} \times (10^3 \text{ L}/ 1 \text{ m}^3) = 3.05 \text{ mol/m}^3$$

SOLUTION –

Plugging each of the recalculated values into the Cottrell slope equation yields:

$$n = \frac{m\sqrt{\pi}}{(96{,}485 \frac{\text{A·s}}{\text{mol}})(7.854 \times 10^{-7} \text{ m}^2)(3.05 \frac{\text{mol}}{\text{m}^3})\sqrt{9 \times 10^{-10} \text{ m}^2/\text{s}}}$$

n = 0.93 = 1 electron or 1 mole of electrons per mole of $Rh(phen)^{2+}$

The number of electrons is usually a whole number.

Problem 19.13: CA is often used with a known standard solution to measure the real electrode area, rather than depending on geometry to estimate the area. If $Rh(phen)L_4^{2+}$ were known to have a diffusion coefficient of 9.25×10^{-6} cm^2/s, what is the actual area of the electrode used in Example 19.1?

Problem 19.14: A second metal complex, $M(org)^{2-}$, is discovered that has a ionic diameter approximately 40% smaller than that of $Rh(phen)^{2+}$ described in Example 19.1. If the same electrode and same concentration were used in a chronoampero-metric experiment with $M(org)^{2+}$, would you expect the slope of $i(t)$ versus $t^{-1/2}$ to be higher or lower than that seen with $Rh(phen)L_4^{2+}$? Explain your answer. *Hint:* What part of the Cottrell equation is affected by the size of the particle under study?

19.4 Linear Sweep and Cyclic Voltammetry

Background

In *voltammetry*, the potential (voltage) is controlled by the user, and the current that flows between the working and counter electrodes is measured. In other words, the current is the analytical signal. In voltammetry, the potential is changed or "swept" over time, whereas in CA, the potential is held constant. Also, in voltammetry, current is plotted as a function of the applied potential, whereas in CA, current is measured as a function of time.

One of the challenges in voltammetry is that the electroanalytical cell has an inherent resistance associated with it, resulting from the electrical resistance of the solution and the physical barrier between the working and counter electrodes. Ohm's law (Equation 19.6)[12] shows us that the potential drop across a resistance is directly proportional to the current and resistance.

$$E = I \cdot R \qquad \text{Eq. 19.6}$$

where

E = potential (volts)
I = current (amps)
R = resistance (ohms)

When a voltage is applied across a resistor, the voltage decreases by the amount calculated using Ohm's law. An analogous phenomenon occurs in an electrochemical cell. The inherent resistance in the cell causes the voltage to decrease, and so the potential *actually* applied to the working electrode is something less than what was intended. This is often referred to as *IR drop* (see Figure 19.8). In voltammetry, identifying the potentials at which reduction and/or oxidation processes take place is often of key interest, so IR drop is far more problematic in voltammetry than it is in CA.

As noted in Section 19.2, the reference electrode provides a way of at least partially correcting for the voltage drop (see Figure 19.3). It is placed in the solution as close to the working electrode as is practical without interfering with mass transport to the electrode. Because no current flows between the reference and working electrodes, there is no IR drop between them, and so the reference electrode provides essentially a true measure of the potential at the working electrode. In the circuit, it feeds back a continuous measure of the potential at the working electrode, allowing for automatic and instantaneous adjustment of that potential to the desired value as the potential is swept over a range of values. Recall from Section 19.3, *Controlling Mass Transport*, that we

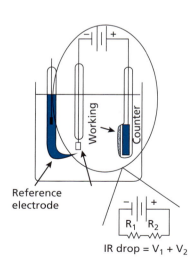

Figure 19.8 Depiction of the IR drop in a three-electrode cell. R_1 is the solution resistance, and R_2 is the resistance imparted by the porous barrier between the working and counter electrodes. The total IR drop is the sum of the voltages lost over both resistances.

[12] Also see Equations 17.7 and 4.1.

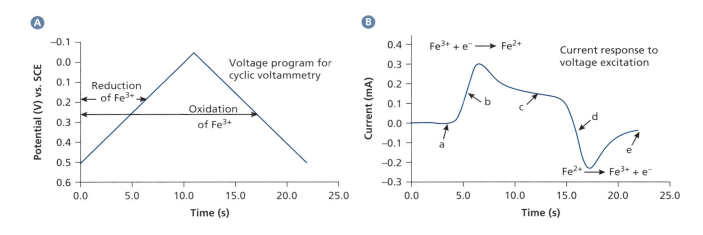

Figure 19.9 Applied potential waveform and current response for CV.

discussed the role of the SE in decreasing the effect of solution resistance; other methods for reducing the effect of IR drop are discussed in Section 19.6, *Ultramicroelectrodes and Nanoelectrodes*.

The Experiment

Probably the most common electroanalytical method used today is *cyclic voltammetry* (CV). In this technique, a linearly varying potential is applied to the electrode and the response current is plotted as a function of the applied potential. Typically, the initial potential is selected as one at which no reduction or oxidation occurs, and then the potential is swept (1) in a negative direction to force reduction of the analyte or (2) in a positive direction to force oxidation of the analyte. If we end the experiment after having only swept the potential in one direction, it is considered to be *linear sweep voltammetry*. In CV, the initial sweep is followed by a second sweep in the reverse direction, usually until the applied voltage returns to the initial potential. For CV, Figure 19.9(A) shows the applied potential waveform as a function of time, and Figure 19.9(B) shows a typical resulting current response as a function of time.

Although the panels in Figure 19.9 show nicely how the experiment is carried out across time, this is not typically how the results are plotted in the end. We usually plot current versus potential (rather than time). The current response of the reverse sweep is plotted along the same potential scale (the *x* axis), resulting in the familiar "duck" shape of a classical CV (seen in Figure 19.10).[13] The entire experiment, conducted at typical sweep rates, takes no more than a few minutes—and often less than a minute.

As we did with CA, we can explain the shape of the CV curve using the chemical reactions that are happening in the system. We begin at a potential where no redox behavior is observed (+0.5 V), so the measured current is zero. Find this starting position in Figure 19.9. As we sweep the potential (in this case in a negative direction from +0.5 V to –0.1 V), a very slight slope is observed in the baseline (see position "a" in Figures 19.9 and 19.10), even before we approach a potential where the analyte species (Fe^{3+}) will be reduced. This initial slight slope is due to charging current. Recall that any change in potential causes charging current, so each successively more negative voltage

---

[13] Note that the CV in Figure 19.9 is presented in the "American style," where negative voltage is plotted to the right and cathodic (reducing) current is plotted up and positive. In many other countries, negative voltage is plotted to the left and reducing current is plotted down and negative.

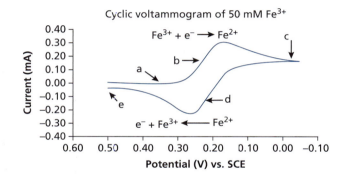

Figure 19.10 Typical CV plot of current versus potential.

is essentially (from the perspective of the working electrode) a new CA experiment with a new potential pulse applied.

As we approach the standard potential for the reduction of Fe^{3+} in this solution, the current begins to rise because the energy states of the different Fe^{3+} ions in solution exhibit a distribution (some will be slightly more easily and some slightly less easily reduced). At potentials beyond the standard potential, E^0, for the Fe^{3+}/Fe^{2+} couple, the distribution of energy states is at its maximum, but this occurs at a position about two-thirds up the CV peak[14] (see position "b" in Figures 19.9 and 19.10). That is because, although we have reached the potential at which most Fe^{3+} ions will be reduced when they reach the electrode, the rate at which we are changing the applied potential outruns the development of the depletion zone around the electrode (as described in Figure 19.5 and the associated text). In other words, we have not yet reached diffusion-controlled conditions, but the potential has moved on well past E^0. The current is observed to peak and then taper off as diffusion control is reached (see position "c" in Figures 19.9 and 19.10). (If we continued sweeping the potential in a negative direction, we would achieve a steady-state current.) For the sake of comparison, if we conducted a CA experiment stepping the potential from +0.5 V to –0.1 V, the diffusion-limited current we would observe would be approximately equal to the current seen at about 12 seconds into the CV curve (Figures 19.9 and 19.10, region labeled "c").

Looking at Figure 19.9, we see that from about 6 seconds (the peak of the Fe^{3+} reduction current) until we reverse the potential at 12 seconds, the current is tapering off. This is due to the completion of the depletion zone around the electrode. That region is devoid of Fe^{3+} ions except for those ions diffusing in from the bulk solution, but it becomes ever more highly concentrated in the reduced form, Fe^{2+} (of course, the Fe^{2+} is simultaneously diffusing away from the electrode).

Finally, when we reverse the direction of the potential sweep, and after a few seconds, we reach the potential at which the accumulated Fe^{2+} becomes oxidized (see position "d" in Figures 19.9 and 19.10). Then we observe a current peak in the opposite direction.

Problem 19.15: What is the analytical signal in CA and voltammetry? What are the implications from Ohm's law (Equation 19.6) with respect to the IR drop?

---

[14] The actual position of E^0 is midway between the upward reduction peak and the downward reoxidation peak, which occurs at approximately two-thirds up the reduction peak rising slope in a reversible system.

Reversibility

When using dynamic electroanalytical methods, we must consider the reversibility of the system from two standpoints: *chemical* reversibility and *electrochemical* reversibility.

Consider the generalized reduction process we used at the beginning of the Section 19.3, in which the oxidized form of a species (Ox) is converted to the reduced form (Red) when a sufficiently negative potential is applied:

$$Ox\,(aq) + e^- \rightleftharpoons Red\,(aq) \qquad \textbf{Eq. 19.7}$$

If the species Red is stable, then a sufficiently positive potential will cause the reverse reaction to occur:

$$Red\,(aq) \rightleftharpoons Ox\,(aq) + e^- \qquad \textbf{Eq. 19.8}$$

In such a case, the system exhibits *chemical reversibility*. You have been dealing with chemical reversibility since general chemistry, when you solved Hess's law problems.

If the product of the original reaction (Red, in Equation 19.7) is unstable and thus undergoes a secondary chemical reaction to a more stable form (P), then the system might not be chemically reversible, as depicted in Equation 19.9:

$$Ox\big(aq\big) + e^- \rightleftharpoons Red\big(aq\big) \xrightarrow{k_c} P \qquad \textbf{Eq. 19.9}$$

where

k_c = rate constant of the following chemical reaction (M/s)
P = a more stable product to which Red is converted

In such a case, the appearance of reversibility will depend on the rate at which the following chemical reaction takes place relative to the speed at which the experiment is conducted. A system having a small k_c might appear chemically reversible if the scan rate of a CV experiment is very high, whereas a system with a large k_c might appear irreversible even at high scan rates.

Figure 19.11 demonstrates the effect of a chemical reaction that follows an electrochemical one where the electrochemical product (Red) is converted to a more stable

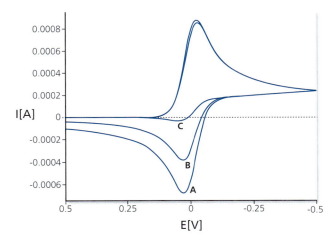

Figure 19.11 CV plots simulated with Digisim$_{TM}$ software for a system exhibiting a chemical reaction that followed an electrochemical one. The rate of potential sweep was the same in all cases, 1 V/s. The rate constant, k_c for the following chemical reaction was (A) 0 s^{-1} (chemically reversible), (B) 0.01 s^{-1}, and (C) 10 s^{-1} (chemically irreversible).

product. When k_c is large (19.11[C]), little or no reoxidation (downward) peak is observed because there is no longer Red available to be oxidized. Note that in Figure 19.11, the rate of potential sweep[15] was kept constant at 1 V/s but the value of k_c was varied. A nearly identical figure would have been obtained if k_c were kept constant at $10 \, s^{-1}$, but the sweep rate was changed from 1 V/s (Figure 19.11[C]), to 10 V/s (Figure 19.11[B]), and to 100 V/s (Figure 19.11[A]). In the latter case, running the experiment very quickly would allow us to reach the oxidation potential for Red before the following chemical reaction could significantly decrease its concentration near the electrode. Unfortunately, there are disadvantages to using such high scan rates, as we will discuss later in the chapter.

Electrochemical reversibility deals with the ease with which electrons are transported from the electrode to the analyte species and vice versa. Electron transport at an electrode in a solution is inherently a heterogeneous process, and as such exhibits rate constants that depend on factors such as the magnitude of the applied potential, the concentration of the analyte species, and the conductivity of the electrode material. Such processes exhibit first-order kinetics. As the potential varies, the rate constants of the forward (reduction) and reverse (oxidation) reactions vary. However, at the standard potential, E^0, the system is at equilibrium, and the forward and reverse rate constants are equal. The rate constant under equilibrium at standard potential is defined as the *standard heterogeneous rate constant*, k_s, and is used to define electrochemical reversibility.

$$Ox \, (aq) + e^- \overset{k_s}{\rightleftharpoons} Red \, (aq) \qquad \textbf{Eq. 19.10}$$

The units of k_s are usually expressed as cm/s (or the SI units of m/s). These might appear to be uncommon units for first-order rate constants, but they are derived from the first-order rate law for heterogeneous (or interfacial) processes, where the area of the electrode (in cm^2 or m^2) is a factor in the overall rate:

$$Rate = k_s C \qquad \textbf{Eq. 19.11}$$

where

$\quad k_s$ = standard heterogeneous rate constant (cm/s [or m/s])
$\quad C$ = concentration of the reacting species (mol/cm^3 [or mol/m^3])

The rate for an interfacial process has the units mol/(s·cm^2) [or mol/(s·m^2)]. If the value of k_s is relatively high (> 0.02 cm/s), the kinetics of electron transport are sufficiently rapid that the reaction will appear reversible (Figure 19.12[A]). If k_s falls below around 10^{-5} cm/s, the system will appear irreversible, and a significant potential will be required to overcome that in order to force the reverse reaction to take place. In a CV, this would appear as a shift in the position of peak maxima (as seen clearly with the oxidation peak in our example of Figures 19.9 and 19.10). In other words, the reduction peaks and oxidation peaks would be farther apart in the CV. The case where k_s lies between 0.02 and 10^{-5} cm/s is called *quasi-reversible* (Figures 19.12[B] and 19.12[C]).

Problem 19.16:   Use Equation 19.11 to show that the units of k_s are cm/s.

---

[15] The potential sweep rate, ν, is simply the rate at which we shift the voltage from one value to another, and usually has the units V/s.

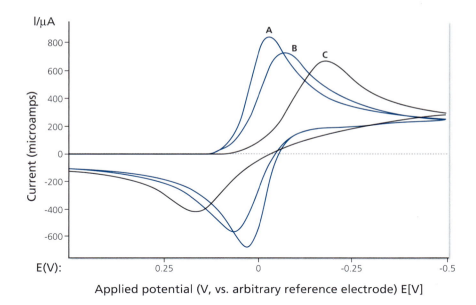

Figure 19.12 CV plots simulated with Digisim$_{TM}$ software for a system exhibiting a various levels of electrochemical reversibility. The rate of potential sweep was the same in all cases, 1 V/s. The standard heterogeneous rate constant, k_s for the electron transfer was (A) 1 cm/s (electrochemically reversible), (B) 0.01 cm/s, and (C) 0.001 cm/s (quasi-reversible).

Quantitative Analysis with Cyclic Voltammetry: The Randles-Sevcik Equation

The signal in a CV experiment is the current measured as a function of the applied potential. CV can be used as an appropriate quantitative analysis method because the current measured at the peak potential (the peak current) is proportional to the concentration of the analyte. This relationship is described by the *Randles-Sevcik equation:*[16]

$$i_p = 0.4463 \, A \, C_O^* (nF)^{3/2} \left(\frac{D_O}{RT}\right)^{1/2} v^{1/2} \qquad \textbf{Eq. 19.12}$$

where

i_p = current measured at CV peak (amp)
v = potential sweep rate (scan rate) (volt/s)
n = number of electrons involved in the redox equation (mol e$^-$/mol reactant)
A = area of the electrode (m^2)
D_O = diffusion coefficient of the analyte (m^2/s)
C_O^* = bulk concentration of the redox active analyte (mol/m^3)
F = Faraday's constant (96,485 C/mol e$^-$)
R = gas constant (8.314 J/mol·K)
T = temperature (K).

---

[16] As with the Cottrell equation, the Randles-Sevcik equation was derived with the assumptions that the electrode is planar and that migration and convection are both negligible. For a complete discussion of the derivation of the Randles-Sevcik equation, see Chapter 5 of Bard, A.; Faulkner L. *Electrochemical Methods.* John Wiley & Sons, New York, **2001**.

If the experiment is conducted at standard thermodynamic temperature of 25°C, Equation 19.12 resolves to Equation 19.13:

$$i_p = (2.687 \times 10^5) A C_O^* n^{3/2} D_O^{1/2} v^{1/2}$$ **Eq. 19.13**

where

i_p = current measured at CV peak (amp)
v = potential sweep rate (scan rate) (volt/s)
n = number of electrons involved in the redox equation (mol e^-/mol reactant)
A = area of the electrode (m^2)
D_O = diffusion coefficient of the analyte (m^2/s)
C_O^* = bulk concentration of the redox active analyte (mol/m^3)

The area of the electrode (A) is typically determined experimentally using a CA experiment with a well-characterized standard redox agent but can be estimated geometrically if its dimensions are known. Here we can see also that i_p is proportional to the $n^{3/2}$, the square root of the sweep rate, as well as the square root of the diffusion coefficient of the redox active species in the solution. So a very simple plot of i_p versus $v^{1/2}$ will yield a slope (m) of:

$$m = (2.687 \times 10^5) A \, n^{3/2} D_O^{1/2}$$ **Eq. 19.14**

If we know two of the three variables (C_O, n, or D_O) then we can use CV experiment and a plot of i_p versus $v^{1/2}$ to find the third variable.

Problem 19.17: Confirm that the units in Equation 19.12 are correct.

Problem 19.18: What are the units of the constant (2.687×10^5) in Equations 19.13 and 19.14?

Problem 19.19: The CVs in this figure[17] were obtained on a 1.0 mm diameter platinum disk electrode at scan rates of 20, 50, 100, 200, 500, and 1,000 mV/s in a 2.0 mM solution of potassium ferricyanide in aqueous 1 M KNO_3 at 25°C. Estimate the peak currents and calculate the diffusion coefficient of ferricyanide ion in this solution. Assume that ferricyanide ion undergoes a one-electron reduction to ferrocyanide ion. (Use Example 19.2 as a guide.)

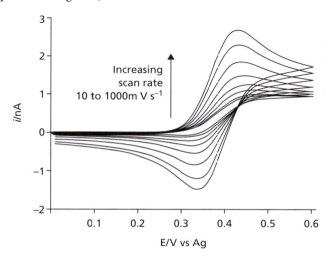

[17] From Rogers E. I.; et al. *J. Phys. Chem. C*, **2008**, *112*, 2729–2735.

Example 19.2 Using the Randles-Sevcik Equation

CV was conducted at 22.3°C with 58 mM ferrocene in a nonaqueous solvent in which the diffusion coefficient of ferrocene is 3.8×10^{-7} cm^2/s. The working electrode was a 1.5 mm diameter platinum disk. The data below were collected from the CVs obtained at different scan rates (each point is the average of three scans). Estimate the number of electrons transferred in the redox reaction.

Sweep Rate (mV/s)	Peak Current (i_p, μA)
20.0	28.9
50.0	40.9
100.	55.5
200.	77.4

STRATEGY –

Because the experimental temperature is not 25°C, we must use Equation 19.12. As a linear equation (y = mx + b), the Randles-Sevcik equation describes a plot of i_p versus $v^{1/2}$, which will give the slope (m):

$$m = 0.4463\,A\,C_O^* \, (nF)^{3/2} \left(\frac{D_O}{RT}\right)^{1/2}$$

Solving the slope equation for n yields:

$$n = \left(\frac{m\sqrt{RT}}{0.4463\,A C_O^* F^{3/2} \sqrt{D_O}} \right)^{2/3}$$

Second, we need to make sure all of our values are in the proper units so that they will cancel appropriately throughout the calculation, and then we will plug the values into our equation for n.

$$F = 96{,}485 \text{ C/mol} \times 1 \text{ A·s / C} = 96{,}485 \text{ A·s/mol}$$
$$A = \pi r^2 = \pi (0.75 \text{ mm})^2 \times (1 \text{ m}/10^3 \text{ mm})^2 = 1.\underline{77} \times 10^{-6} \text{ m}^2$$
$$D = 3.8 \times 10^{-7} \text{ cm}^2/\text{s} \times (1 \text{ m}/100 \text{ cm})^2 = 3.8 \times 10^{-11} \text{ m}^2/\text{s}$$
$$C_O = 58 \times 10^{-3} \text{ mol/L} \times (10^3 \text{ L}/1 \text{ m}^3) = 58 \text{ mol/m}^3$$
$$T = 22.3°C + 273.15 = 295.\underline{45}\text{K}$$

Finally, we need to find the slope of the i_p versus $v^{1/2}$ plot using the proper units of V/s for sweep rate and amps for current. Using Excel, we find the slope to be 1.767×10^{-4} A·s/V.

SOLUTION –

Plugging each of the recalculated values into the slope equation yields:

$$n = \left(\frac{(1.767 \times 10^{-4}\,\text{A·s}^{1/2}/\text{V}^{1/2})\sqrt{(8.3145\,\frac{\text{V·C}}{\text{mol·K}})(295.45\text{K})}}{0.4463(1.77 \times 10^{-6}\,\text{m}^2)(58\,\frac{\text{mol}}{\text{m}^3})(96{,}485\,\text{A·s/mol})^{3/2}\sqrt{3.8 \times 10^{-11}\,\frac{\text{m}^2}{\text{s}}}} \right)^{2/3}$$

$n = 1.02 = 1$ electron or 1 mole of electrons per mole of ferrocene

Problem 19.20: In a CV experiment, a solution of water from Lake Mary Nell in Elon, North Carolina, was analyzed for Cr^{3+}. A 25.00 mL sample showed an i_p of 31.6 μA. A 25.00 mL sample of standard Cr^{3+} (4.56×10^{-7} M) exhibited an i_p of 52.4 μA. What is the concentration of Cr^{3+} in Lake Mary Nell?

Problem 19.21: For important measurements taken in the medical field, the experimental parameters needed to consider the elevated temperature of the human body (compared to standard 25°C). Derive a simplified form (as in Equation 19.13) of Equation 19.12 for the scientific medical community, taking into account standard human body temperature of 37°C.

Problem 19.22: At 25°C the Rh(III) complex, $Rh(bpca)_2^+$, exhibits a reversible two-electron redox peak at –0.82 V versus Ag/AgCl and a diffusion coefficient of 5.8×10^{-6} cm^2/s. A cyclic voltammogram was obtained at a potential sweep rate of 100.0 mV/s using a 3.00 mm diameter glassy carbon disk electrode and the peak height, i_p, was found to be 67.3 μA. What was the millimolar concentration of $Rh(bpca)_2^+$? (The ligand *bpca* is the bis(2-pyridylcarbonyl) amide anion, which exhibits a –1 ionic charge.)

Qualitative Analysis with Cyclic Voltammetry

Many researchers find that the greatest power of CV lies not in using it for *quantitative* analysis but rather in *qualitative* analysis. Full treatment of this topic can be a course in and of itself, and several excellent monographs have been produced on the topic.[18] We will consider an overview of some approaches to the use of CV in a qualitative sense.

CV data can be used, in conjunction with other methods, to determine fundamental characteristics of new substances. For reversible systems, using CV data in a plot of i_p versus $v^{1/2}$ yields a linear relationship with a slope of $2.687\,A\,C\,n^{3/2}D_o^{1/2}$ at 25°C. Usually the area of the electrode is known or can be estimated with some accuracy, and the bulk concentration of the analyte is known as well. Unfortunately, for a new substance, the values for n and D_O are usually not known, leaving us with an equation having two unknown quantities. Sometimes the value of n can be found explicitly in other ways (e.g., through electrolysis[19]), but we can often achieve a reasonable estimate of both n and D quickly from CA and CV, as demonstrated in Example 19.3.

Problem 19.23: Would you expect to obtain a value for n that is reasonably close to a whole number if the values of both the CA and CV slopes in Example 19.3 were halved? Explain.

Problem 19.24: Confirm your expectation from Problem 19.23 by following through with the calculations following Example 19.3.

Problem 19.25: At 25°C, 2.35 mM solution of potassium ferricyanide, $K_3Fe(CN)_6$, was used to collect series of CVs at varied sweep rates and a 3.00 mm diameter glassy carbon electrode. A plot of i_p versus $v^{1/2}$ yielded a slope of 1.59×10^{-4} amp·$s^{1/2}/V^{1/2}$. Using the same system, a plot of i versus $t^{-1/2}$ from a CA experiment gave a slope of 3.3×10^{-5} amp·$s^{1/2}$. Determine the number of electrons (**n**) transferred in the redox reaction and the diffusion coefficient (**D**) of the ferricyanide in this solution.

---

[18] See Section 19.7, Further Reading.

[19] Other methods for finding n can be time consuming, and we can achieve reasonable estimates of n and D using only CA and CV.

Example 19.3 Estimating n and D from Chronoamperometry and Cyclic Voltammetry

The cobalt(III) complex, $Co(acac)_4^{3+}$, exhibits a reversible redox peak at -0.41 V versus Ag/AgCl. A series of CVs were obtained at varied sweep rates using a 5.00×10^{-3} M solution and a 3.00 mm diameter glassy carbon electrode. A plot of i_p versus $v^{1/2}$ yielded a slope of 5.1×10^{-4} amp·s$^{1/2}$/V$^{1/2}$. Using the same system, a plot of i versus $t^{-1/2}$ from a CA experiment gave a slope of 7.1×10^{-5} amp·s$^{1/2}$. In all cases the temperature was 25°C.

Determine the number of electrons (**n**) transferred in the reduction wave and the diffusion coefficient (**D**) of the Co(III) complex. (The ligand *acac* refers to the neutral acetylacetone.)

STRATEGY –

From the Cottrell equation, we know that the slope of the $i(t)$ versus $t^{-1/2}$ plot gives:

$$slope = \frac{nFAC^* \sqrt{D_O}}{\pi^{1/2}}$$ and so we can solve for the quantity $k_{CA} = (nD^{1/2})$

From the Randles-Sevcik equation, we know that the slope of the i_p versus $v^{1/2}$ plot gives:

$$slope = (2.687 \times 10^5) A C n^{\frac{3}{2}} D_O^{\frac{1}{2}}$$, so we can solve for the quantity $k_{CV} = (n^{3/2} D^{1/2})$

Using the ratio of k_{CV} / k_{CA}, we can solve directly for n, and then use that value to find D.

SOLUTION –

$$k_{CA} = \left(nD^{\frac{1}{2}} \right) = \frac{(slope)(\pi^{\frac{1}{2}})}{FAC} = 3.69 \times 10^{-5} \, m/s^{1/2}$$

$$k_{CA} = \left(n^{\frac{3}{2}} D^{\frac{1}{2}} \right) = \frac{slope}{(2.687 \times 10^5) AC} = 5.37 \times 10^{-5} \, m/s^{1/2}$$

$$\frac{k_{CV}}{k_{CA}} = \frac{n^{3/2} \cdot D^{1/2}}{n \cdot D^{1/2}} = n^{1/2} \frac{5.37 \times 10^{-3}}{3.69 \times 10^{-3}} = 1.455$$

$$n = 2.11$$

We know that n must be a whole number, so n = 2. Using that value we can calculate D from:

$$k_{CA} = \left(nD^{\frac{1}{2}} \right) = 3.69 \times 10^{-3} \, m/s^{1/2}$$

$$D = 3.40 \times 10^{-10} \, m^2/s$$

Electrochemists have some simple tools for determining if a system is reversible or not as well. Equation 19.15 predicts that we should expect the potentials of the reduction and oxidation peaks for a one-electron redox couple to be separated by approximately 59 mV at 25°C if the redox system is fully reversible (both chemically and electrochemically).[20] In practice, however, we seldom see a 59 mV peak-to-peak separation due to several factors, such as nonideal mass transport or electron transfer from the electrode, adsorption of either the reactant or product of the redox reaction onto the electrode, or relatively slow chemical reactions that succeed the redox reaction. We usually consider anything between 59 and 100 mV peak-to-peak separation (at 25°C) to be an acceptable indicator of a reversible system.

$$|E_{p,a} - E_{p,c}| = \frac{2.303\, RT}{nF}$$ **Eq. 19.15**

where

$E_{p,a}$ = potential of the anodic (oxidation) CV peak (V)
$E_{p,c}$ = potential of the cathodic (reduction) CV peak (V)
F = Faraday's constant (96,485 C/mol)
R = gas constant (8.314 J/mol·K)
T = temperature (K)
n = number of electrons involved in the redox equation (mol e -/mol reactant)

Also, for a reversible system, we can estimate the standard electrode potential,[21] E^0, for a redox system as the mid-point between the oxidation and reduction peaks. Because the value of E^0 determined this way can be affected by experimental parameters related to the potential sweep rate, many electrochemists prefer to consider the mid-point the $E_{1/2}$, or the half-wave potential (Equation 19.16).

$$E^0 \approx E_{1/2} = \frac{(E_{p,c} + E_{p,a})}{2}$$ **Eq. 19.16**

where

E^0 = standard potential for the redox couple (Ox/Red) (V)
$E_{1/2}$ = half-wave potential for the redox couple (Ox/Red) (V)
$E_{p,a}$ = potential of the anodic (oxidation) CV peak (V)
$E_{p,c}$ = potential of the cathodic (reduction) CV peak (V)

Problem 19.26:   Solve Equation 19.15 to demonstrate that ΔE_p is expected to be around 59 mV for a reversible system at 25°C.

Problem 19.27:   For a given electrochemical system, at a temperature of 25°C, you measured a ΔE_p of 0.0298 V. What logical explanation(s) can you give for this result?

Problem 19.28:   If you conducted a CV of an analyte dissolved in molten NaCl at 900°C, what ΔE_p would you expect to measure, given that it is a one-electron process?

Problem 19.29:   Estimate the ΔE_p for the CV shown in Figure 19.10. Would you consider that CV to represent a reversible system? Explain.

Problem 19.30:   Estimate E^0 for the system depicted in Figure 19.10. Do you think that your estimate is reliable? Why or why not?

---

[20] You might recognize the right-hand side of Equation 19.15 from the Nernst equation you learned about in general chemistry. In fact, Equation 19.15 is fundamentally derived from the Nernst equation, but the actual derivation is not as straightforward as you might expect. For a full explanation of the 59 mV peak-to-peak separation limit, see Chapter 6 of Bard, A.; Faulkner L. *Electrochemical Methods.* John Wiley & Sons, New York, **2001**.

[21] Strictly speaking, what we get from Equation 19.16 is the *formal* potential, $E^{0'}$, which differs from the standard potential when the concentrations of all relevant species are not set to 1.0 molar.

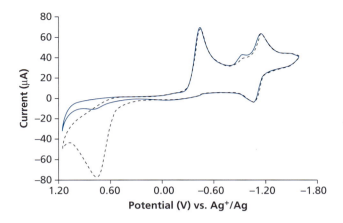

Figure 19.13 CVs of Pt(dpk)Cl$_4$ in CH$_3$CN containing 0.2 M TBAP as supporting electrolyte.

It is not uncommon for species under investigation to exhibit multiple reduction and/or oxidation peaks, and it is rare that CVs have the idealized appearance of Figure 19.10 or Figure 19.11(A). Figure 19.13 shows the CV of di-2-pyridylketone platinum(IV) chloride in acetonitrile. In the dashed-line curve, a reduction sweep was started at 0.00 V and swept negative to –0.80 V, then the voltage sweep was reversed to +1.18 V and reversed one more time back to 0.00 V. In the solid-line curve of Figure 19.13, the potential was initially swept in a positive direction. When we initially sweep positive, we do not see the oxidation peak at +0.8 V. Comparison of the two curves demonstrates that the large oxidation peak at +0.8 V seen in the dashed-line curve is due to reoxidation of the reduced species generated at around –0.5 V during the reduction sweep. The reduction and associated oxidation peaks are separated by more than 1 V, far above the ideal value of 59 mV.

If we push the initial reduction sweep further and make it more negative, it is clear that there are at least two additional reduction processes that follow the initial one. Only the last of those exhibits clear reversibility (based on the ΔE_p between the reduction peak at –1.14 V and its associated oxidation peak at –1.07 V).

Similarly, Figure 19.14 shows a CV obtained from a solution of a bimetallic osmium and rhodium complex in acetonitrile with tetrabutylammonium hexafluorophosphate (TBAP) as the SE. The CV sweep starts at 0.00 V and initially proceeds in a positive direction to +1.5 V, at which point the sweep is revered and proceeds in a negative

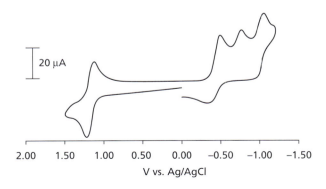

Figure 19.14 Cyclic voltammogram of [(bpy)$_2$Os(dpp)RhCl$_2$(phen)](PF$_6$)$_3$ in CH$_3$CN with 0.1 M TBAP. The analyte exhibits both oxidation (+1.2 V vs. Ag/Ag$^+$) and reduction (–0.5, –0.75, –1.2 V) of the initial species.

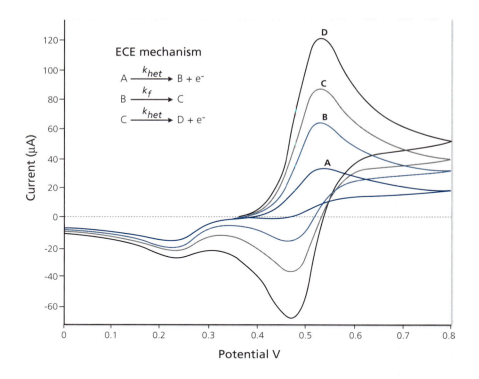

Figure 19.15 Demonstration of nonideal CV response and the utility of electrochemical analysis software to determine the mechanism(s) involved, using DigiSim software. Scan rates were (A) 0.1 V/s, (B) 0.5 V/s, (C) 1.0 V/s, and (D) 2.0 V/s.

direction to -1.25 V. Four separate redox processes are obvious: one oxidation of the original species (seen as the downward peak at ~ 1.25 V) followed by reduction back to the original oxidation state (seen as the upward peak at ~ 1.19 V) and three additional reduction events as the sweep proceeds negative to -1.25 V. Only the oxidation appears to demonstrate near ideal reversibility as indicated by the small potential separation (~ 60 mV) between the oxidation peak (1.25 V) and the reduction peak (1.19 V).

With appropriate software (e.g., DigiSim™), complex electrochemical behavior can be resolved into a clear understanding of the electrochemical and chemical processes that occur during the application of a CV experiment. Such knowledge leads to further understanding of the fundamental chemical nature and reactivity of the species in question. For example, Figure 19.15 shows a set of CVs obtained at different scan rates.[22] It is clear that this system is not reversible at low sweep rates. At a sweep rate of 0.1 V/s (Figure 19.15[A]), a single reduction peak (~ 0.53 V) and a single oxidation peak (~ 0.22 V) are observed, but the potential separation of the two is around 300 mV, about five times the ideal value of 59 mV. As the potential sweep rate is increased, an additional oxidation peak becomes evident (at ~ 0.48 V), with a peak separation from the reduction wave (at ~ 0.53 V), which is very close to the ideal 60 mV value when a sweep rate of 2 V/s is used (Figure 19.15[D]).

Without using any additional methods, we could surmise from the low sweep rate experiment (Figure 19.15[A]) that the initial species is reduced (A → B) but then undergoes a further reaction to generate some species (P) that is not easily oxidized. As the scan rate is increased, the potential changes at a rate that is eventually

---

[22] Figure 19.15 shows simulated CVs prepared using the DigiSim software. The figure is taken from the DigiSim manual.

(Figure 19.15[D]) faster than the rate of the following chemical reaction (B → P). Therefore, we are able to reoxidize the initial reduced substance (B) back to A.

Without specifically designed electrochemical analysis software, it would be difficult to discern the third step in the reduction process. However, with the software, it can be determined with reasonable certainty that the mechanism follows an "ECE" process—that is, an electrochemical process followed by a chemical process and then a second electrochemical process. Using a technique that only requires a few hours of time and an instrument that costs less than $5,000, electrochemistry allows us to learn a tremendous amount about the mechanisms of the reduction process for this species.

Solvents, Electrolytes, and the Electrochemical Window

In real solutions, we are unable to sweep the potential to simply any voltage. Both the solvent and the SE are present in concentrations far higher than our analytes, and so we are bound by the potentials at which the solvent and SE become oxidized or reduced.

Figure 19.16 shows a CV in which the voltage sweep was taken to the practical limits imposed by the solution. The large increase in current on the right is due to the reduction of the solvent (acetonitrile) itself, and the large decrease in current on the left is due to the oxidation of the solvent. The set of potential limits imposed by the system is termed the *electrochemical window* (i.e., the range of potentials over which a solvent is electrochemically stable). Table 19.2 gives the electrochemical windows for several common electrochemical solvents. If we attempt to sweep the potentiostat beyond this range, we will observe significant current due to the redox chemistry of the solvent, and that will overwhelm any current peaks we might see that are related to analytes. The use of a nonaqueous solvent such as acetonitrile or a room temperature ionic liquid can significantly increase the electrochemical window over that seen in water. Because water undergoes reduction and oxidation within the electrochemical windows of most nonaqueous solvents, removal of water from the solvent prior to use in electrochemical studies is usually desirable. For scrupulous solvent drying, distillation followed by storage over a water trapping substance (such as sodium) is often a preferred method. For moderate drying, simply storing over a good drying agent is sufficient. Although most salts and metals have redox potentials outside the range of typical electrochemical solvents, we still must consider any limits that might be imposed by the redox potential of the SE as well as the electrode material selected.

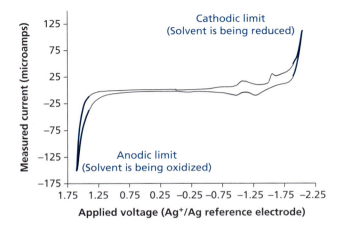

Figure 19.16 CV of Pd(bpca)$_2$ in acetonitrile exhibiting the potential limits imposed by the electrochemical window of the solvent with 0.1 M TBAPF$_6$ added as the SE.

TABLE 19.2: A List of Common Electrochemical Solvents and the Potential Limits of the Respective Electrochemical Windows[*]

SOLVENT	POTENTIAL WINDOW −E/+E V vs. Ag /Ag	TYPICAL SUPPORTING	TYPICAL ELECTRODE
BMIM-PF$_6$	−1.5/+3.5	None[†]	Carbon, solid metals
Tetrahydrofuran	−3.5/+1.5	Alkali metal salts of PF$_6^-$, BF$_4^-$, organic anions	Carbon, solid metals
Dimethylsulfoxide	−4.0/+1.0	Organic salts	Carbon, solid metals
Dimethylformamide	−3.5/+1.5	Alkali metal salts of PF$_6^-$, BF$_4^-$, organic anions	Carbon, solid metals
Acetonitrile	−3.0/+2.5	Alkali metal salts of PF$_6^-$, BF$_4^-$, organic anions	Carbon, solid metals
Water	−2.5/+1.5[‡]	Alkali metal halides and nitrates	Carbon, solid metals, mercury

[*]**To take full advantage of a solvent's electrochemical window, the solvent must be anhydrous. See Table 19.7 for a list of common drying systems.**

[†]**Butylmethylimidazolium hexafluorophosphate (BMIM-PF$_6$) is a room temperature ionic liquid. It is essentially a molten salt and so needs no supporting electrolyte.**

[‡]**For water in particular, the electrochemical window is highly dependent on the electrode material and the presence of acid or base.**

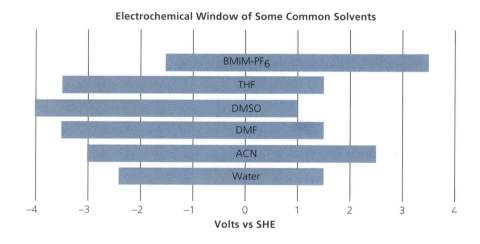

Electrochemical Window of Some Common Solvents

19.5 Square Wave Voltammetry

Square wave voltammetry[23] (SWV) can be used to yield quantitative results superior to CV. SWV also can provide important qualitative analysis about the nature of reactants and reaction mechanisms, but it has not been used as extensively as CV for the latter purpose. We will focus on understanding the method and its application as a quantitative technique.

Recall that the peak shape observed in the CV signal is due to the fact that the potential sweep outruns the development of the depletion zone—that E^0 is found at a position about two-thirds of the peak height. Also, as we will discuss shortly, the charging current becomes more evident at faster scan rates. However, if we collect a voltammogram at very low sweep rates (e.g., 1 mV/s), we would observe a sigmoidal current

---

[23] Sometimes this is referred to as Osteryoung voltammetry or Osteryoung SWV in honor of the authors who reported the initial studies of the quantitative and qualitative applications of the method.

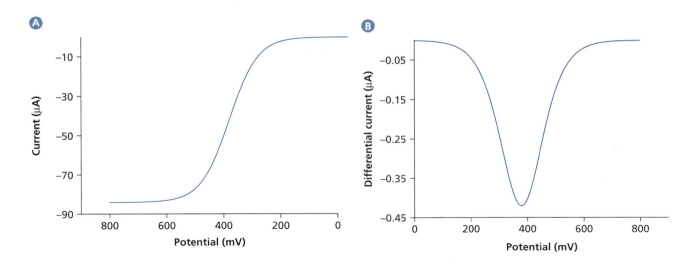

Figure 19.17 A simulated linear sweep voltammogram (only one-half of a CV scan) of ferrocene. When taken at a very slow sweep rate (1 mV/s), the shape is sigmoidal (A) and can be differentiated to produce a peak-shaped curve (B).

signal, as seen in Figure 19.17(A), with the point of inflection occurring at E^0 for the redox couple in question. If we were to mathematically differentiate the sigmoidal curve, the result would be a peak-shaped curve with an essentially flat baseline and the peak occurring at E^0 and with the peak height proportional to the concentration of the analyte, as shown in Figure 19.17(B). Of course, sweeping the potential from, for instance, 0.5 V to –0.1 V at a rate of 1 mV/s would take nearly 10 minutes. SWV is a differential method that yields nearly the same results as a differentiated slow-scan CV but in a small fraction of the time.

Figure 19.18(A) shows the applied potential waveform as a function of time for the SWV experiment. The potential waveform is basically a linear sweep (as with CV) with a square wave superimposed (pulses). The current is sampled near the end of each forward ($i_{forward}$) and reverse ($i_{reverse}$) pulse (indicated by the "sample periods" in Figure 19.18(B). Like the differential plot in Fig. 19.17(B), the net current is the difference of the two sampled currents ($i_f - i_r$) and that difference is plotted versus the mean potential of the two pulses. Figure 19.18(C) shows the SWV voltammogram with the plot of calculated current versus average potential. Sampling the current at the end of a pulse greatly reduces the effect of charging current, and reporting the difference of the two currents enhances the magnitude of the signal. That is, $i_{forward}$ is negative (for an oxidation in this case) and $i_{reverse}$ is positive, so $i_f - i_r$ has a higher magnitude than either. The peak occurs very close to the standard potential, at a value of $(E^0 - \Delta E)$, where ΔE is the height of the square wave pulse (typically only a few millivolts), rather than two-thirds of the way up the peak as in CV. This increased precision is the result of minimizing the charging current. SWV is also faster than CV. An entire SWV sweep typically requires less than a minute—and often less than 10 seconds.

The mathematical model describing the current response for SWV is significantly more complicated than that of CV. Equation 19.17 gives the generalized form of the SWV equation.

$$i_p = \Delta i = \frac{nFAC^* D_O^{\frac{1}{2}}}{(\pi\tau)^{\frac{1}{2}}} \cdot \Delta\psi \qquad\qquad \textbf{Eq. 19.17}$$

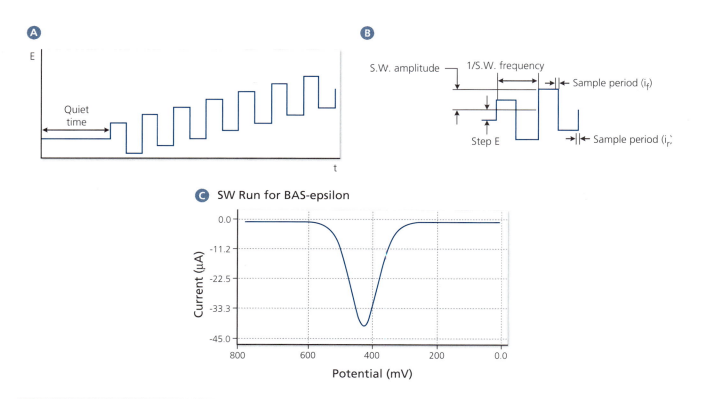

Figure 19.18 Simulated applied potential waveform and current response for SWV.

where

i_p = current measured at SWV peak (amp; the i_p is actually $\Delta i = i_{forward} - i_{reverse}$)

n = number of electrons involved in the redox equation (mol e^-/mol reactant)

A = area of the electrode (m²)

D_O = diffusion coefficient of the analyte (m²/s)

C_O^* = bulk concentration of the redox active analyte (mol/m³)

F = Faraday's constant (96,485 C/mol e^-)

τ = total time for a given square wave (both forward and reverse pulses; s)

$\Delta\psi$ = a dimensionless parameter that is a function of the mean potential, the height of the potential pulse, the frequency of the square wave, the length of the pulse, and the delay time after the pulse before which the current is sampled[24]

We can see from Equation 19.17 that the peak height is directly proportional to the concentration of the analyte. So SWV can be used very efficiently for quantitative analysis. In Example 19.4 and Problem 19.32, we will see how SWV can be used to determine the concentration of unknown metal ions in solution, first using an internal spike and second using a set of standards with a calibration curve.

Problem 19.31: Compare the Cottrell equation (Equation 19.5) with the SWV equation (Equation 19.17). How are they similar? Thinking about what you now know about SWV, explain why you would expect such similarities.

---

[24] Most electrochemists agree that the greatest power for SWV lies in its utility in determining mechanisms and other qualitative parameters. The theoretical analysis of SWV for such purposes can be fairly complex. Many research articles have been devoted to the theoretical treatment of a wide variety of mechanistic situations by SWV, but we recommend, as a solid general compendium, the following: Mirceski, V.; Komorsky-Lovric, S.; Lovric, M. *Square Wave Voltammetry*, Springer, **2007**.

Example 19.4 Using the Square Wave Voltammetry Equation

A sample of drinking water from a house with old pipes was collected, and supporting electrolyte (SE) was added to the bulk sample. A 20.00 mL aliquot of deionized water was added to 250.0 mL of the sample with SE. A SWV conducted with the resulting solution showed a peak at −0.41 V versus the AgCl/Ag reference electrode (this is the voltage at which Pb^{2+} is expected to be reduced). The height of the peak was found to be 4.85 μA. Next, to 250.0 mL of the sample (with SE already dissolved) were added 20.00 mL of 1.32×10^{-5} M $Pb(NO_3)_2$. In the SWV of the resulting solution, the peak at −0.41 V had a height of 10.57 μA. What was the concentration of the Pb^{2+} in the original sample?

STRATEGY –

This is a classical single standard addition problem. Because we diluted both samples by the same amount (20 mL spike added to 250 mL of sample), we can take the difference in peak height as being due only to the Pb^{2+} from the spike of $Pb(NO_3)_2$ solution. From there we can use a proportionality to find the concentration of Pb^{2+} in the sample without the spike and then use the dilution equation to determine the Pb^{2+} concentration in the original sample.

SOLUTION –

Concentration of Pb^{2+} from the spike in the second sample:

$$(M_{added})(V_{added}) = (M_{final})(V_{final})$$

$$(1.32 \times 10^{-5} \text{ M}) (20.00 \text{ mL}) = (M_{final}) (250.0 + 20.00 \text{ mL})$$

$$M_{final} = 9.7\underline{7}8 \times 10^{-7} \text{ M } Pb^{2+} \text{ from the spike}$$

Current due to the spike = 10.57 − 4.85 μA = 5.72 μA

$$\frac{i_{spike}}{M_{spike}} = \frac{i_{unspiked}}{M_{unspiked}} \qquad \text{so} \qquad M_{unspiked} = \frac{i_{unspiked} \cdot M_{spike}}{i_{spike}}$$

$$M_{unspiked} = \frac{(4.85 \mu A)(9.778 \times 10^{-7} \text{ M})}{(5.72 \mu A)} = 8.291 \times 10^{-7} \text{ M}$$

Concentration of Pb^{2+} in the original sample:

$$(M_{original})(V_{original}) = (M_{diluted})(V_{diluted})$$

$$(M_{original})(250.0 \text{ mL}) = (8.291 \times 10^{-7} \text{ M})(270.0 \text{ mL})$$

$$M_{original} = 8.95 \times 10^{-7} \underline{\text{M}} Pb^{2+}$$

Problem 19.32: To determine the amount of cadmium in a sample of drinking water, a set of calibration solutions were prepared and analyzed with SWV.

Cd^{2+} (PPB)	SWV PEAK HEIGHT (nA)
10	0.59
20	1.15
40	2.28

The SWV peak height of the drinking water sample was measured to be 1.36 nA. What was the concentration of Cd^{2+} in the drinking water?

Electrical lead

Electrode body

Solid disk electrode

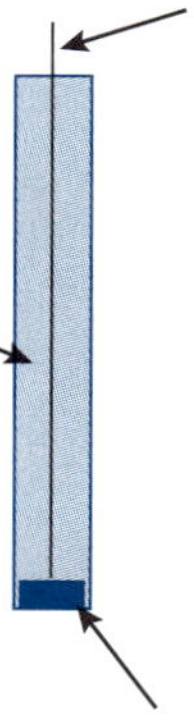

Figure 19.19 Schematic diagram depicting typical construction of disk electrodes made using metal wire and glassy carbon rod or metal disk.

19.6 Working Electrodes

Common Working Electrodes

One of the most important decisions an electrochemist faces in research is the material from which the working electrode is made. Any conducting material can be used as an electrode, but not all materials are well suited to electrochemical analysis. Table 19.3 lists the desired characteristics of the ideal working electrode material. Although virtually all metals have been used as electrodes in one application or another, there are a few materials that best fit the ideal characteristics given in Table 19.3 and are used on a regular basis. Gold and platinum are the most common metals used in constructing working electrodes, and carbon is the most common nonmetal material. Graphite is still used to make carbon electrodes, but glassy carbon is by far more common in modern applications.

The simplest design for an electrode is to immerse a wire or foil of the electrode material directly into the solution. However, in such a case, the area of the electrode cannot be estimated with much accuracy, and cleaning/polishing of the electrode is often ineffective. Most commonly, disk electrodes with dimensions on the millimeter scale are used for normal electrochemical methods. Figure 19.19 includes a general schematic depicting the construction of metal and carbon electrodes. Working electrodes can be made in-house by sealing metal wire in glass, using a carbide saw to expose the disk and polishing the disk with a series of increasingly fine particulates suspended in water. However, the time and cost invested in such endeavors usually outweighs the cost of simply purchasing appropriate electrodes from companies that make them in bulk.

Ultramicroelectrodes and Nanoelectrodes

We have discussed in Section 19.3 how to minimize charging currents under typical experimental conditions. Charging currents are not a serious problem at normal scan rates of 0.02 to around 1 V/s; note the very slight slope in the baselines seen in Figures 19.10, 19.13, and 19.14. However, mechanistic analysis of cyclic voltammograms requires that we obtain CVs over as wide a range of scan rates as possible—preferably at least a two-decade[25] span. The wider the range used, the greater the confidence we have in the mechanism determined. The Randles-Sevcik equation (Equation 19.12) showed us that the faradaic current (i_p) increases linearly with $v^{1/2}$. Unfortunately, charging currents increase linearly with v. The result is that as v increases, the charging current increases at a faster rate than the faradaic current. In other words, the noise increases faster than the signal. The charging current typically become a significant source of noise as the scan rate is increased above values of around 5 to 10 V/s. At high scan rates, the background noise due to charging current essentially overwhelms the analytical signal (see Figure 19.20). To get past this intrinsic limitation, analysts have developed

TABLE 19.3:   The Ideal Working Electrode

DESIRED CHARACTERISTIC	RATIONALE FOR DESIRABILITY
Highly conductive	Low resistance to flow of electrons; minimal potential drop within the electrode
Chemically inert	Can be used in a wide variety of solvents and pH ranges; does not react with analytes
High oxidation potential	Is not damaged during use and does not limit the electrochemical window
Achieves high polish	Area of the electrode can be estimated with accuracy from simple geometric calculations (e.g., $A = \pi r^2$ for a disk electrode)

---

[25] The term "decade" is often used to refer to an order of magnitude. For example, a range of 0.01 to 10 would span three orders of magnitude, or three decades.

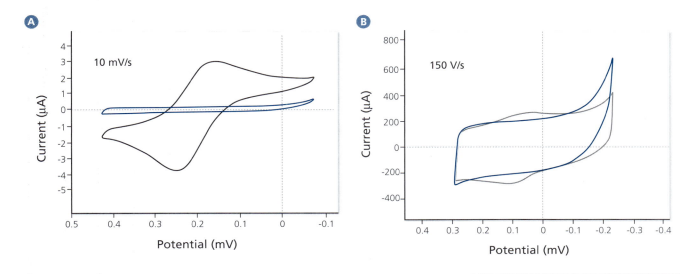

Figure 19.20 Comparison of the blank and solute CVs at scan rates of 10 mV/s and 150 V/s. At 10 mV/s (A), the charging current seen in the blank is significantly smaller than the analyte scan. However, at a scan rate of 150 mV/s (B), the charging current is the dominant source of current. The analyte signal is barely discernible above the "background" current.

equipment that allows for the use of ultramicroelectrodes (UMEs) that minimize charging current noise because of their small size.

As implied by examples used previously in this chapter (for instance, Examples 19.1 and 19.2), electrodes used in common electrochemical studies have dimensions on the millimeter scale. For electrodes of that size (and larger) the primary mode of mass transport to the electrode is by planar diffusion—that is, diffusion of species to the electrode occurs in a path perpendicular to the electrode surface. Of course, some particles approach the electrode at the edges in a nonplanar fashion, but the circumference of the millimeter-sized disk is very small with respect to the area of the electrode. The amount of material reaching the electrode by a nonplanar path is negligible compared to the bulk material approaching the face of the electrode (see Figure 19.21[A]). In fact, one of the boundary conditions used in deriving most electrochemical equations, such as Equations 19.5 and 19.12, is that only planar diffusion is important.

As the size of the electrode decreases, the relative importance of nonplanar diffusion increases, because the ratio of the circumference to the area increases. UMEs, also sometimes referred to simply as microelectrodes, are those that have at least one dimension on the micron scale, and at such small scales, nonplanar diffusion (hemispherical in the case of a disk electrode) provides a significant amount of analyte material to the electrode. Figure 19.21(B) demonstrates the difference between planar and nonplanar diffusion to an electrode surface.

Recall that in using CV to determine a reaction mechanism, it is desirable to use a wide range of scan rates. However, we are limited in the magnitude of scan rates

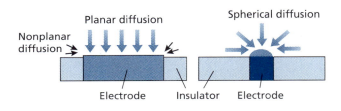

Figure 19.21 Comparison of planar diffusion at a millimeter-scale electrode (A) and nonplanar (spherical) diffusion at a UME (B).

available because charging current increases more rapidly than faradaic current as scan rates increase. Both charging current and faradaic current decrease directly with the area of the electrode, but because nonlinear diffusion only affects faradaic current (in a good way, from the point of view of the experimentalist), the smaller the area of the electrode, the less effect charging current has relative to our faradaic signal. The effect becomes even more dramatic as the dimension of the electrode approaches the nanometer scale (nanoelectrodes [NEs]).

TABLE 19.4: Comparison of Electrodes of Decreasing Diameter

DISK DIAMETER	AREA (cm²)	CIRCUMFERENCE (cm)	CIRCUMFERENCE: AREA RATIO
1 cm	7.8×10^{-1}	3.14	4
1 mm	7.8×10^{-3}	3.14×10^{-1}	40
100 µm	7.8×10^{-5}	3.14×10^{-2}	400
10 µm	7.8×10^{-7}	3.14×10^{-3}	4,000
1 µm	7.8×10^{-9}	3.14×10^{-4}	40,000
100 nm	7.8×10^{-11}	3.14×10^{-5}	400,000
10 nm	7.8×10^{-13}	3.14×10^{-6}	4,000,000

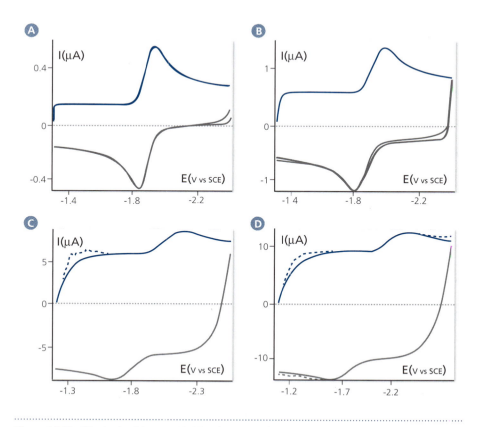

Figure 19.22 CVs obtained from a 10 mM anthracene solution at a UME at scan rates of (A) 22 kV/s, (B) 56 kV/s, (C) 320 kV/s, and (D) 1.7MV/s. The working electrode was a 5 µm gold disk electrode.

COMPARE AND CONTRAST

Instrumental Costs

Consider other instruments that are used to conduct mechanistic studies through reaction rates. A high-quality stop-flow system with high-speed optics will cost on the order of $30,000 to $60,000, with annual maintenance costs of $1,000 to $2,000. A moderate-quality superconducting nuclear magnetic resonance spectrometer will cost more than $200,000, with annual maintenance costs of $10,000 to $20,000. However, a high-speed potentiostat with the ability to measure nanoamp currents will cost $20,000 to $25,000, with nearly zero maintenance costs.

Table 19.4 provides some comparison data for disk electrodes of decreasing diameter. We can take the ratio of circumference to area as a measure of the relative increase in nonplanar diffusion compared to planar diffusion. In switching from a 1 mm diameter electrode to a 10 μm diameter electrode, we see a 1,000-fold increase in that ratio, whereas both the charging current and faradaic current due to planar diffusion have decreased by a factor of 100,000 due to the decrease in area. On a relative scale, then, the charging current has decreased 1,000 times more than the faradaic current, meaning we gain a 1,000-fold increase in the signal to noise ratio. Figures 19.22 and 19.23 show the practical benefit of using UMEs. Even at 1.7 *million* V/s (Figure 19.22[D]), the faradaic peak shape on the 5 μm electrode is superior to that seen on the 1 mm electrode at 150 V/s in Figure 19.20(B). By accessing extremely fast scan rates, we can make kinetic measurements of reaction rates far beyond what was possible with electrodes above the microscale.

Figure 19.23 shows CVs of anthracene taken at two different scan rates. Even at 100 V/s, the anthracene oxidation appears to be entirely irreversible, implying that an extremely fast chemical reaction follows the electrochemical oxidation, producing a very stable end product. A scan rate of 100 V/s would be beyond the limits of normal (mm scale) electrodes, but the charging current is barely noticeable on the 5 μm platinum disk. When the scan rate is increased 100-fold, the reduction wave associated with the couple is visible. This means that in using a 10,000 V/s scan rate, we have outrun the rate of the following chemical reaction and so can observe the reverse oxidation peak. From the CV, we can estimate kinetic parameters. In the 10 KV/s scan, the oxidation (upward) peak occurs at around 1.0 V, and then the sweep continues to +1.3 V before reversing. The reduction peak occurs at around 0.9 V. So, between the oxidation peak and the reduction peak, the instrument has swept a total of 0.7 V at 10,000 V/s. From this we can surmise that the chemical reaction that follows the oxidation requires more than 70 μs to take place, so its rate constant is lower than $1.4 \times 10^4 \text{ s}^{-1}$.

An associated advantage to using very small area electrodes relates back to Ohm's law. Recall that the problem of IR drop comes about because we want high signal (current), but the inherent resistance in the solution causes significant potential loss. However, with the very small currents measured using UMEs and NEs, the IR drop is negligible. Similarly, mass transport due to migration is diminished greatly at small electrodes. In fact, it is possible to conduct electrochemistry in solutions with no supporting electrolyte whatsoever, and even in the gas phase.[26]

Why, then, do we not use UMEs and NEs on a regular basis? No advantage comes without trade-off, and the answer to that question was foreshadowed in the previous paragraph and in Figure 19.23. The very small currents observed at these tiny electrodes are far below the noise level exhibited by most electrochemical instruments. Even instruments that are specifically designed to detect nanoamp currents suffer from

---

[26] Many such examples exist, but we recommend the special issue of *Sensors* titled Ultramicroelectrode Electrochemistry—Theory and Applications, Lyons, M. (editor), *Sensors*, **2013**, *13*.

problems with signal-to-noise ratio at such low levels. The cell and much of the wiring must be encased in a Faraday cage, which shields against electromagnetic interference, and all wiring between the cell and the instrument must be shielded as well. In other words, there is a cost—both in money and time—associated with conducting electro-chemistry at electrodes of this scale.

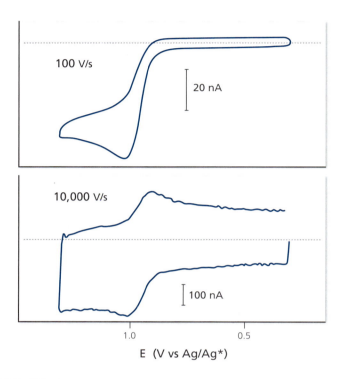

Figure 19.23 CVs showing the oxidation of 2.4 mM anthracene in acetonitrile (0.6 M tetraethyl ammonium phosphate as the SE) taken at 100 V/s and 10,000 V/s at a 5 μm platinum disk electrode.

PROFILE

Cyclic Voltammetry in a Single Cell

Wightman and colleagues used 5 μm carbon fiber electrodes to conduct cyclic voltammetry in live rats. The CV shows the voltammetric response to the reduction and oxidation of dopamine in a rat brain. Because an ultramicroelectrode was used, no effort was needed to enhance solvent conductivity with supporting electrolyte. ■

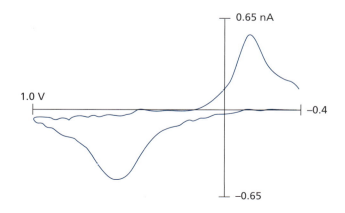

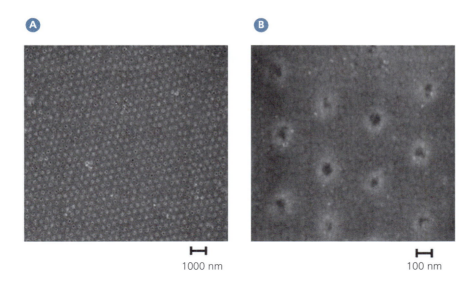

A

B

1000 nm

100 nm

Figure 19.24 An example of an ultramicroarray electrode. The full array (A) has more than 5 million UMEs (100 nm diameter each) in a 1 mm² grid. A close-up view of the electrodes is seen in (B).

One way to balance those costs with the advantages attained using UMEs and NEs is to use array electrodes (Figure 19.24). Because the electrodes are of such small size, multiple electrodes can be arrayed together with separations on the order of 10 to 50 times the diameter of the electrode. If we array a set of 1 μm electrodes separated from each other by 25 μm in a 50-by-50 grid, the total electrode size is only around 1 cm diameter and the summed currents yields a 2500-fold increase in signal, shifting from the nanoamp region to the microamp region that is easily accessible by normal instrumentation.

Problem 19.33:   You want to create a microarray electrode that contains electrodes having a diameter of 1 μm. To avoid having the electrodes interfere with each other, each should be separated from all adjacent electrodes by a distance of at least 10 diameters (10 μm). Given that you have a 1.5 × 1.5 mm square support for the electrodes, how many electrodes can you fit in your array?

Problem 19.34:   Consider Table 19.4 and think about the array electrode you designed in Problem 19.33. By what factor do you expect your signal-to-noise ratio to increase compared to a 1 mm diameter electrode? By what factor do you expect the current given by one of your electrodes to decrease compared to the 1 mm electrode? What is that current factor when you use your entire array rather than just one electrode?

19.7   Useful Data

Tables 19.5 and 19.6 contain information that may be helpful in understanding the effect of temperature on voltammetric measurements. Table 19.7 describes some methods for drying common nonaqueous solvents used in electrochemistry.

TABLE 19.5:   Temperature Dependence of Reference Electrodes*

T(°C)	CALOMEL 0.1 M KCl	CALOMEL SATURATED	Ag/AgCl 1.0 M KCl
0	336.7	259.18	236.55
5	—	—	234.13
10	336.2	253.87	231.42
15	336.1	251.1	228.57
20	335.8	247.75	225.57
25	335.6	244.53	222.34
30	335.4	241.18	219.04
35	335.1	237.6	215.65
40	334.5	234.49	212.08
45	—	—	208.35
50	331.5	227.37	204.49
55	—	—	200.56
60	324.8	223.5	196.49
70	—	—	187.82
80	—	208.3	178.7
90	—	—	169.5

* E°_{mV} **versus standard hydrogen electrode.**

From Shugar, G. J.; Dean, J. A. *The Chemist's Ready Reference Handbook.* **McGraw-Hill: New York, NY, 1990.**

TABLE 19.6:   Temperature dependence of $\dfrac{2.3026RT}{F}$ (mV)

T(°C)	$\dfrac{2.3026RT}{F}$	T(°C)	$\dfrac{2.3026RT}{F}$	T(°C)	$\dfrac{2.3026RT}{F}$
0	54.199	35	61.144	70	68.088
5	55.191	40	62.136	75	69.080
10	56.183	45	63.128	80	70.073
15	57.175	50	64.120	85	71.065
20	58.167	55	65.122	90	72.057
25	59.159	60	66.104	95	73.049
30	60.152	65	67.096	100	74.041

From: Shugar, G. J. and Dean, J. A. *The Chemists Ready Reference Handbook,* **McGraw-Hill, New York, 1990.**

TABLE 19.7:   Solvent Drying Techniques*

THF	Tetrahydrofuran. The THF should be predried over 3Å sieves or CaH_2. Solid sodium and benzophenone are added to a pot of THF and refluxed under N_2. When the THF is rigorously dry, the free radical from the reduction of benzophenone will produce a deep blue color in the THF pot. The dry THF is then distilled under a nitrogen atmosphere. When the pot turns orange, it is time to recharge the still with fresh sodium and benzophenone.
DMSO	Dimethyl sulfoxide. The most convenient way to dry DMSO is over 4Å sieves. However, this method reports water content in the 6–10 ppm range.[†] DMSO has a very high boiling point, so distillation methods must be performed under vacuum in order to achieve good results. Reported distillation methods include distillation over CaH_2, $MgSO_4$, or $CaSO_4$ with storage of the dry DMSO over 4Å sieves.
DMF	Dimethyl formamide. DMF decomposes at high temperatures or in the presence of acids or bases. Allow the DMF to stand over 4Å sieves for at least 12 hours, and then decant and distill under vacuum to prevent overheating the DMF. The dry DMF can be stored over 4Å sieves.
ACN	Acetonitrile should be predried over 4Å sieves, $CaSO_4$, or CaH_2. Calcium hydride is added to a pot of ACN and distilled over N_2. The dried ACN can be stored over 4Å sieves.

*For additional information on common solvent drying methods, see pages 445–447 of *The Chemist's Companion* **by A. J. Gordon and R. A. Ford, John Wiley & Sons, 1972.**

†Gaylord Chemicals—Technical Notes.

19.8 Further Reading

BOOKS

Bard, A. J.; Faulkner, L. R. *Electrochemical Methods: Fundamentals and Applications*, 2nd ed. John Wiley & Sons: New York, **2000**.

Gosser, D. K. *Cyclic Voltammetry: Simulation and Analysis of Reaction Mechanisms*. John Wiley & Sons: New York, **1993**.

Monk, P. M. S. *Fundamentals of Electroanalytical Chemistry*. Wiley: Chichester, UK, **2001**.

Wang, J. *Analytical Electrochemistry*. Wiley-VCH: Hoboken, NJ, **2006**.

JOURNALS

Dickinson, E. J. F.; Limon-Peterson, J. G.; Rees, N. V.; Compton, R. G. How Much Supporting Electrolyte Is Required to Make a Cyclic Voltammetry Experiment Quantitatively "Diffusional"? A Theoretical and Experimental Investigation. *J. Phys. Chem. C*, **2009**, *113*, 11157–11171.

Jo, A., Do, H., Jhon, G.-J., Suh, M., Lee, Y. Electrochemical Nanosensor for Real-Time Direct Imaging of Nitric Oxide in Living Brain. *Anal. Chem.*, **2011**, *83*, 8314–8319.

Wang, H., Pilon, L. Accurate Simulations of Electric Double Layer Capacitance of Ultramicroelectrodes. *J. Phys. Chem. C*, **2011**, *115*, 16711–16719.

19.9 Exercises

EXERCISE 19.1: What values of the standard heterogeneous rate constant, k_s, in units of m/s, would designate electrochemically reversible, quasi-reversible, and irreversible reactions?

EXERCISE 19.2: What purpose does each electrode in the three-electrode cell serve?

EXERCISE 19.3: What is the purpose of the porous separator between the counter electrode and the bulk solution in the three-electrode cell?

EXERCISE 19.4: Explain why the current rises then drops rapidly in a CA experiment. Why does it not drop to zero?

EXERCISE 19.5: Define the following terms:

(a) Chronoamperometry

(b) Anode

(c) Migration

(d) Convection

(e) Sweep rate

(f) Mechanism

(g) Ultramicro

(h) Electrolytic

EXERCISE 19.6: A 1.75×10^{-3} M solution of a metal ion was used to conduct a CV experiment on a square electrode with an edge length of 0.5 cm (only one face of the electrode was exposed to the solution). Using a sweep rate of 50 mV/s, a peak current of 62.9 μA was measured. What is the diffusion coefficient of the metal ion?

EXERCISE 19.7: A disk electrode of unknown size was used in a CV experiment of a known organic cation with a diffusion coefficient of 7.21×10^{-10} m^2/s. The cation is known to undergo a two-electron reduction. A 2.0 mM solution of the cation was used with a sweep rate of 100 mV/s and a peak current of 78 mV was observed. What was the area of the electrode?

EXERCISE 19.8: Chronoamperometry with a reducing potential was conducted with a 3.05×10^{-3} $\underline{M}$ solution of a new Rh(phen)L$_4^{2+}$ complex using a 1.0 mm diameter platinum disk electrode. The slope of the i$_{cat}$ versus t$^{-1/2}$ was 3.66×10^{-6} A·s$^{1/2}$. A literature search reveals that the diffusion coefficient of Ru(phen)L$_4^{2+}$ is known to be 1.1×10^{-5} cm^2/s and the diffusion coefficient of Pd(phen)L$_4^{2+}$ is 7.2×10^{-6} cm^2/s. (*Hint:* Look at a periodic table.) Give a reasonable estimate of the number of electrons transferred (ν) in the chronoamperometric reduction of Rh(phen)L$_4^{2+}$. (The term *phen* refers to the neutral phenanthroline ligand.)

EXERCISE 19.9: For a given electrochemical system, you measured a ΔE_p of 0.203 V. What logical explanation(s) can you give for this result? *Hint:* Consider Figure 19.12 and the accompanying text.

EXERCISE 19.10: If you conducted a CV of a species in vivo (in the body) at 37°C, what ΔE_p would you expect to measure, assuming that it is a two-electron process?

EXERCISE 19.11: Estimate the ΔE_p for each of the CVs shown in Figure 19.12. In each case, would you consider that CV to represent a reversible system? Explain.

EXERCISE 19.12: Estimate E^0 from each CV depicted in Figure 19.12. What do your results tell you about the reliability of using Equation 19.16 for systems that are not reversible?

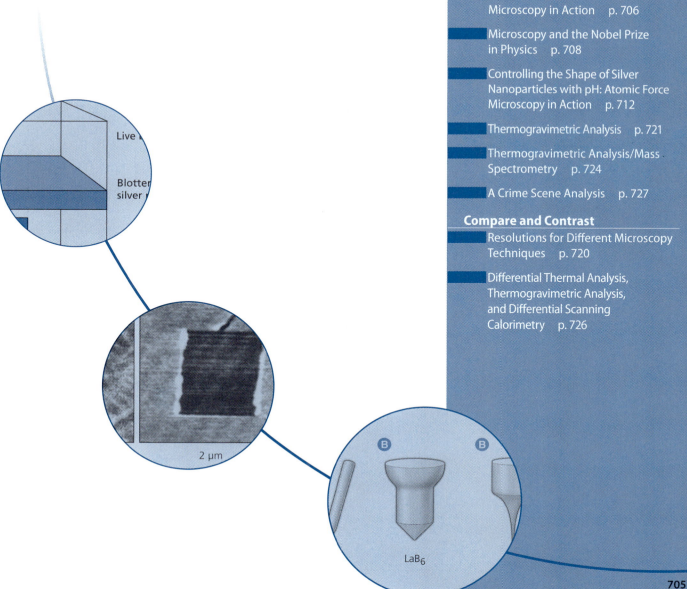

Chapter 20

Material and Surface Analysis Techniques

Live

Blotter
silver

2 μm

B   B

LaB_6

PROFILE

Characterizing Metal Nanoparticles for Water Purification: Electron Microscopy in Action

It is estimated that more than 1 billion people do not have access to clean drinking water. Many researchers are working toward solving this problem. Ideally, they will develop solutions that are inexpensive, safe, and require low to no power consumption. For example, Theresa Dankovich and Derek Gray have been researching the use of silver nanoparticle–impregnated filter paper. Silver nanoparticles have been shown to have antimicrobial action. The authors report promising antibacterial effects while keeping the silver nanoparticles safely in the paper. A sketch of their filter design in shown in figure (A). In order to characterize the sizes of the silver nanoparticles, they imaged the impregnated filter paper using transmission electron microscopy, an instrument which will be discussed in this Chapter. The magnified view of the filter surface showing the silver nanoparticles (inset gray bar is 20 nm) is presented in (B). In addition to the image, the authors report the size distribution of their nanoparticles. The authors state in their paper that the smaller nanoparticles have the larger antibacterial effect. Two other images, taken with the scanning electron microscope (another instrument discussed in this chapter), are shown: (C) at 50,000× magnification (size of horizontal is ~1 μm) and (D) at 100,000× magnification (size of horizontal is ~500 nm). The larger structures in the images are the pulp fibers, and the small light structures are the silver nanoparticles. ■

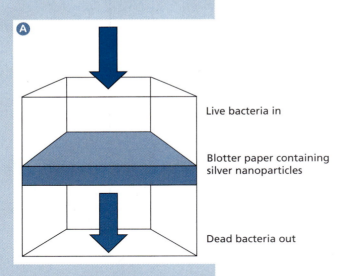

Live bacteria in

Blotter paper containing silver nanoparticles

Dead bacteria out

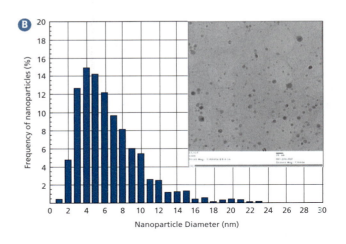

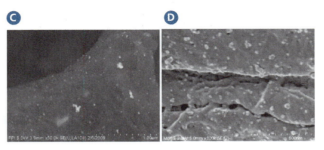

*Dankovich, T. A.; Gray, D. A. Bactericidal Paper Impregnated with Silver Nanoparticles for Point-of-Use Water Treatment. Environmental Science & Technology, **2011,** 45 (5), 1992–1998.*

20.1 Introduction

What we now recognize as modern material science is a relatively new field that exists on the border of analytical chemistry, material engineering, and applied physics. At one time, the instruments and techniques surveyed in this chapter would have been thought of as belonging to material scientists instead of to analytical chemists, who would use them in instrumental analysis. However, this distinction has become more obscure, and it is now common to find analytical chemists working alongside engineers and physicists. Material analysis is the investigation of how the structure of atoms and molecules and their relative phases in a material affect the bulk properties of a material. The analysts strive to characterize the surface topography and/or the relationship between internal phase changes and temperature (or stress). Using thermodynamic principles and an understanding of microstructure, they relate these characteristics to the performance of a bulk material under specific environmental conditions. In addition to designing and selecting specific materials during the design phase, material scientists are also found performing forensic analysis as they try to determine why a particular material failed to perform at the designed specifications.[1]

Even though modern materials science has blurred the boundaries between well-established disciplines of science and engineering, material analysis is a *very old* form of applied science. For example, anthropologists have bracketed human history using the technological materials of the ages with phrases such as the *Stone Age*, the *Bronze Age,* and the *Iron Age*. It is not difficult to imagine that future anthropologists will refer to our time in terms of the materials of the day—perhaps the plastic age, semiconductor age, or composite material age.

Although material science has been around as long as recorded history, a modern analytical approach to material science did not begin until the mid-nineteenth century when Josiah Gibbs (as in Gibbs free energy) demonstrated that phase changes and the thermal properties of materials could be related to the atomic and molecular structures composing those materials (thermodynamics was born). As you will see later in this chapter, the study of thermodynamics remains a fundamental component of material science, but the study of materials has expanded well beyond simple correlation of heat to bulk properties of a material. It now incorporates a wide array of analytical techniques. In this chapter, we will survey three broad instrumental categories used in material science analysis: microscopy, thermal analytics, and mechanical stress analysis.

20.2 Microscopy

Our ability to build and characterize atom-sized structures has increased rapidly since physicist and Nobel Laureate Richard Feynman stated in 1959 that "...there's plenty of room at the bottom...," referring to a field that did not yet have a name—nanoscience. The instruments that help guide scientists in nanoscience are now an important part of the array of instruments available to the chemist. *Microscopy*, a term used to describe techniques that allow us to view objects that cannot be seen with the unaided eye, is a critical tool in nanoscience. In this section, we will survey four general instrumental techniques used in microscopy: *atomic force microscopy (AFM)*, *scanning tunneling microscopy (STM)*, *scanning electron microscopy (SEM)*, and *transmission electron microscopy (TEM)*.

The general field of microscopy can be divided into two specific instrumental types. In the first type, a tip interacts with (probes) the sample surface; AFM and STM are both examples of this type of microscopy. In the second type, an electron beam images below the surface in a fashion similar to optical microscopes; SEM and TEM are both examples of this type of microscopy.

---

[1] Zambrano, O. A.; Coronado, J. J.; Rodriguez, S. A. Failure analysis of a Bridge Crane Shaft. *Case Studies in Eng. Failure Analysis, 2,* **2014;** Bibel, G. *Beyond the Black Box: The Forensics of Airplane Crashes.* Johns Hopkins University Press: Baltimore, MD, **2008**.

PROFILE

Microscopy and the Nobel Prize in Physics

The 1986 Nobel Prize in Physics was awarded for the invention and design of the scanning tunneling microscope (STM) by Gerd Binnig (left) and Heinrich Rohrer (middle) and for the work on electron optics and the design of the electron microscope by Ernst Ruska (right). Both Binnig and Rohrer did their work on the STM, culminating in 1982 at the IBM Research Laboratory in Zurich, Switerland. Much of Ruska's work on electron optics took place in the 1920s and 1930s in Germany. The Nobel Prize summary from 1986 notes that the electron microscope is "one of the most important inventions of this century." Interestingly, Binnig, one of the inventors of the STM, also invented, along with Gerber and Quate, the AFM in 1986, the same year he won the Nobel Prize for his role in the development of the STM. ∎

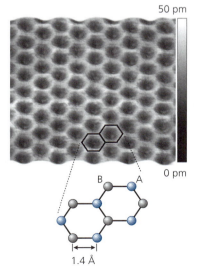

Each of the microscopy techniques can provide better resolution than a conventional optical microscope and allow scientists to probe surface topography, in other words, yielding information on the structure or arrangement of atoms at or below the boundary between the bulk material and its local environment. As an example of the analytical importance of microscopy, a STM image of graphene is shown in Figure 20.1. The STM image of graphene shows the spatial arrangement of atoms on the surface along with atomic distances between atoms.

In this section, we will address the basics of how each type of microscopy works, the sample types used, the resolution possible under specific experimental conditions, as well as limitations of each technique.

AFM, STM, TEM, and SEM differ from surface spectroscopies such as Auger spectroscopy, X-ray photoelectron spectroscopy, or X-ray fluorescence (XRF) spectroscopy,[2] where the information gathered is related to the types of atoms that constitute the surface or below the surface (but not their arrangement). Combining microscopy techniques with spectroscopic techniques can yield information about both atomic composition *and* structure and can be a particularly powerful approach when investigating nanoscale systems.

Figure 20.1 Atomic scale STM image of graphene.

[2] See Chapter 10 for a review of these techniques.

Atomic Force Microscopy

The AFM provides a three-dimensional atomic view of a surface. The essential components of the AFM are shown in Figure 20.2. A cantilever with a small tip as shown in Figure 20.3, is moved across the surface (x and y axes) by a *piezoelectric stage*. The piezoelectric stage allows for fine control of sample position with respect to the small tip. *Piezoelectric materials* are a class of materials where *applied electric potential alters the size of the piezoelectric material*. The opposite also occurs in piezoelectric materials—an applied force on the material can create a voltage. Figure 20.4 shows how the size of a piezoelectric material changes with a positive and negative applied potential.

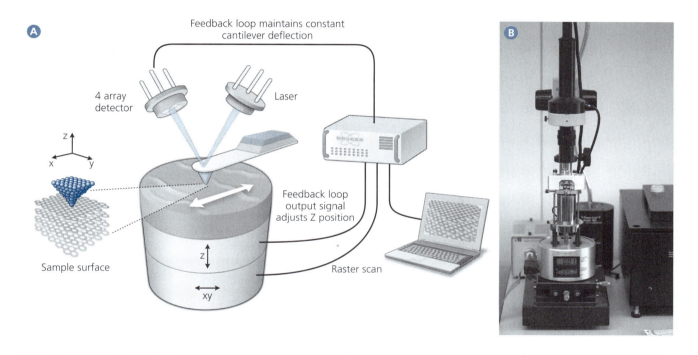

Figure 20.2 Panel (A): Basic components of an AFM. Panel (B): Photograph of an AFM.

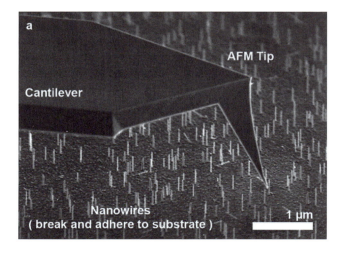

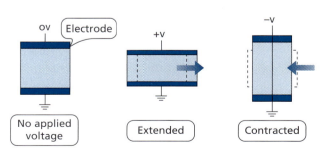

Figure 20.3 Image of AFM cantilever (taken by SEM) on silicon nanowire surface.

Figure 20.4 Piezoelectric material changing size with applied potential.

The laser light is reflected off the end of the cantilever and is detected on a photo-diode with four unique detector faces called a *four-zone photodiode*. As the tip, in contact with the surface, moves up and down in the z direction while moving in the x–y plane, the laser spot moves on the photodiode array. Because the tip is in physical contact with the surface, this is known as *contact mode* operation. AFM tips are generally either silicon or silicon nitride.

There are two versions of contact mode operation: *constant force mode* and *constant height mode*. In constant force mode, a feedback system controls the piezoelectric sample stage to move up and down (z axis) so that the laser beam does not move on the photodiode array. The signal for the image is related to the potential applied to the piezoelectric stage required to keep the laser beam stationary on the photodiode. When the reflected light from the AFM tip is stationary on the photodiode, the force between tip and surface is constant. In constant height mode, the tip scans the surface, moving up and down, and the signal used is related to the deflection of the laser beam on the photodiode array as the tip scans over the atomic "hills" and "valleys."

The AFM can also operate in *tapping mode*. In this mode, a sinusoidal force drives the cantilever so that the tip taps on the surface. Figure 20.5 shows an AFM in tapping mode along with the moving laser spot on the photodiode. Because the tip spends less time in physical contact with the surface in tapping mode, tapping mode is less likely to damage the sample than in contact mode. An example of a silicon AFM scan in contact mode and the associated sample change compared to tapping mode is shown in Figure 20.6. The user should use tapping mode for soft materials.

Fundamentally, the interaction force between the tip and surface is probed with AFM. The force between tip and sample is strongly dependent on the distance

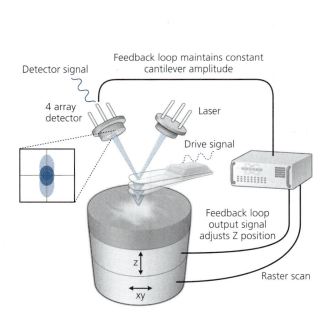

First scan   Second scan

1 µm   2 µm

Contact

First scan   Second scan

1 µm   2 µm

Tapping

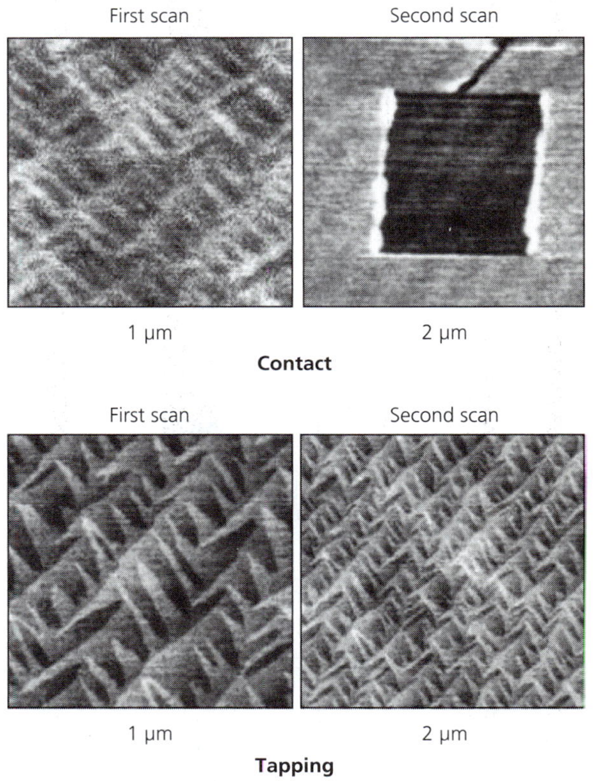

Figure 20.6 Comparison of sample change using AFM in contact and tapping mode—Silicon (100).

Figure 20.5 Tapping mode AFM.

Detector signal

Feedback loop maintains constant cantilever amplitude

4 array detector

Laser

Drive signal

Feedback loop output signal adjusts Z position

z

xy

Raster scan

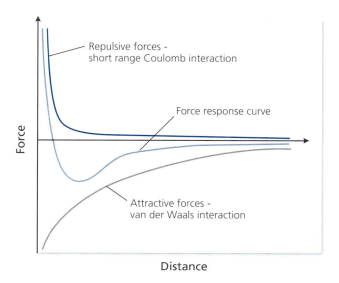

Figure 20.7 Force between tip and sample versus tip–sample distance.

between atoms in the tip and the atoms in the sample. There are two key forces involved, the repulsive electrostatic force between the tip and surface, which is significant for very small tip sample distances (less than roughly 2 Å), and the attractive van der Waals force, which is significant for larger tip sample distances. A plot of the force between tip and sample versus tip sample distance is shown in Figure 20.7. The repulsive force is shown in blue and the attractive forces are shown in gray. The total force between the tip and sample is shown in light blue. In contact mode, the tip sample distance is very small and the dominant force is repulsive. In noncontact mode, the tip sample distances are larger and attractive van der Waals forces dominate. In tapping mode, the tip moves such that both the attractive and repulsive regimes are sampled.

Samples can be explored with AFM in air, in vacuum, and in fluid. Vacuum AFM allows for the highest resolution. Although it depends on a variety of factors, AFM can be capable of resolution of less than 1 nm. One benefit of using AFM is that sample preparation is fairly easy. The sample is simply placed on a small puck with adhesive. The sample for AFM is not restricted by conductivity; samples can be conductors, semiconductors, or insulators. AFM can also be used to study biological systems.

There are several advanced techniques that use the AFM principle. These include options to investigate chemical (chemical force microscopy), electrical (conductive AFM), magnetic (magnetic force microscopy), mechanical (force modulation microscopy), and thermal properties (scanning thermal microscopy).

Scanning Tunneling Microscopy

In STM, the analytical signal is the current that flows from the STM tip through the sample. STM shares several features with AFM, including the use of a piezoelectric stage to control tip sample distances. Like AFM, STM uses an atomically small tip and generates a three-dimensional image of the surface. Because STM measures conduction through the tip, STM tips are usually made of tungsten, instead of Si or Si_3N_4, as is the case with AFM. A key difference between AFM and STM is that STM requires the sample to be conducting, so the range of materials that can be studied with STM is more restrictive than with AFM. Given the similarities between the two techniques, there is a trend for microscope manufacturers to combine AFM and STM approaches into one instrument.

Metal particles with triangular shape, known as nanoprisms, have received a large amount of attention from researchers. Under certain conditions, they possess unique interactions with light. The basis for this particular interaction between light and the surface electrons in a metal is referred to as plasmonics. Plasmonic effects have strong shape and size dependency, which can lead to a wide array of useful behaviors. Although the topic of plasmonics is beyond the scope of this text, the instruments used to characterize some of the nanostructured systems are not. There are a vast number of publications in this area. For example, a recent study by Ying Chen and colleagues demonstrated silver nanoprisms shape and color can be changed by changing the pH of the solution. Because the shape of the particles can have substantial effects on the interaction with light, an easy technique to change shape can be quite useful. Atomic force microscopy has been used to study the pH dependency on nanoprism shape and is shown here. In (A) the authors show the silver nanoprisms on a quartz substrate, in (B) the nanoprisms were dipped in pH 4.0 solution and the triangles begin to lose their points, and in (C) the nanoprisms were dipped in pH 2.2 solution and the silver resembles nanodisks. The average edge length in (A) is 48 nm. The authors also report that the peak absorption of the silver nanoparticles moves to shorter wavelengths with a more acidic dipping. ■■

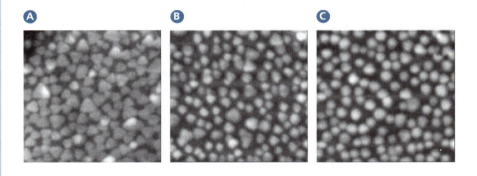

Chen, Y.; Wang, C.; Ma, Z.; Su, Z. *Controllable Colours and Shapes of Silver Nanostructures Based on pH: Application to Surface-Enhanced Raman Scattering. Nanotechnology,* **2007,** *18, 1–5.*

The basic components of STM are shown in Figure 20.8. A potential difference (voltage) between the microscope tip and sample allows for a very small current to quantum mechanically tunnel across the air gap potential barrier. The ability of STM to resolve small distances is due to the fact that the tunneling current is strongly dependent on the sample–tip distance. The tunneling current, I is related to the tip–surface distance d by:

$$I = AVe^{-\frac{\sqrt{2m\Phi}}{h}d}$$

Eq. 20.1

where Φ is the work function of the surface, m is the mass of an electron, h is Planck's constant, V is the tip bias potential, and A is a constant. Tunneling currents in the

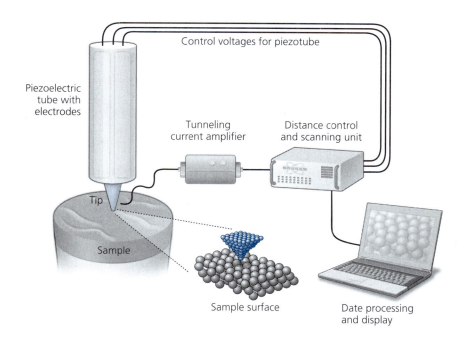

Figure 20.8 Basic components of STM. On deflection of the tip, the piezoelectric tube generates a voltage signal in the electrodes attached to the tip.

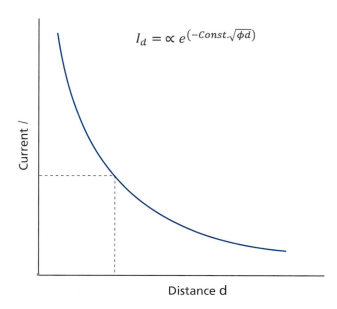

$$I_d = \propto e^{\left(-Const.\sqrt{\phi d}\right)}$$

Figure 20.9 STM: tunneling current versus distance between the tip and the sample surface.

pA range are possible. Figure 20.9 shows the tunneling current, in pA, as a function of the tunneling distance, in angstroms. Note the logarithmic scale for the tunneling current.

The STM can work in a mode where the stage is moved to keep the tunneling current constant. This constant current mode is similar to the constant force mode with the AFM. Similarly, the tip can simply scan across the sample, and the current can be measured and converted to an image.

Although the requirement that the sample be conducting is limiting in STM, sample preparation is similarly easy to AFM. The advantage to using STM is the fact that STM, in air or solution, can provide better resolution than the AFM. STM and non-contact AFM at ultra-high vacuum (UHV) can obtain atomic resolution.

Problem 20.1: Using a tungsten tip (work function, 4.3 eV) to study a surface, what is the ratio of currents at two distances: d = 3.5 Å and d = 4.0 Å?

Transmission Electron Microscopy and Scanning Electron Microscopy

Comparing and contrasting SEM, TEM, and light microscopy instruments will provide insight into their operation. With any form of microscopy, it is important to understand what parameters limit the resolution. The spatial resolution of the light microscope is limited by the wavelength of light as well as the quality of the optics. The theoretical limit for spatial resolution of a light microscope is roughly the wavelength of the light used. The SEM and TEM both use an electron beam as the illumination source and take advantage of the de Broglie wavelength for high-energy electrons instead of the wavelength of light. Similarly, the resolution for the SEM and TEM is related to the wavelength of the electrons directed onto the sample. Strictly speaking, the resolution of a TEM or SEM is always larger than the wavelength of the electrons because of lens aberrations. See Chapter 3 for a description of lens aberrations.

The nonrelativistic de Broglie wavelength λ for the electron that has been accelerated through a potential V (assuming that this energy is converted to kinetic energy) is:

$$\lambda = \frac{h}{p} = \frac{h}{\sqrt{2meV}} = \frac{1.23 \times 10^{-9} \left(m \cdot volt^{1/2} \right)}{\sqrt{V}}$$ **Eq. 20.2**

where h is Planck's constant, p is the electron's momentum, m is the mass of an electron, e is the charge on an electron, and V is the accelerating potential (volts).

Commercial TEM instruments may have accelerating potentials up to around 300 kV. SEM accelerating potentials are lower, typically around 30 kV.

Problem 20.2: What accelerating voltage is needed to give an electron a de Broglie wavelength of 4 pm?

Problem 20.3: A voltage of 5 kV is used to accelerate electrons in an SEM. Is this wavelength longer or shorter than a 300 nm ultraviolet photon?

Example 20.1

What is the wavelength for an electron which has been accelerated through a voltage of 200 kV?

SOLUTION –

Using Equation 20.2, we find:

$$\lambda = \frac{h}{p} = \frac{h}{\sqrt{2meV}} \frac{6.63 \times 10^{-34} \text{ Js}}{\sqrt{2(9.1 \times 10^{-31} \text{ kg})(1.6 \times 10^{-19} \text{ C})(200 \times 10^{3} \text{ V})}}$$

$$= 2.74 \times 10^{-12} \text{ m} = 2.74 \text{ pm}$$

The TEM and SEM are very similar instruments, as we will see with a side-by-side comparison. Figure 20.10 shows an SEM and TEM in action. The key difference between TEM and SEM is that in TEM the image is formed by electrons transmitted through the sample while in SEM the image is formed by scanning the electron beam at a particular location on the sample surface. Another difference between the TEM and SEM is that in the TEM, the image is projected on a fluorescent viewing screen (and is captured with a digital camera). The TEM instrument does not utilize scanning of the electron beam (rastering) while the SEM has the electron beam scan across the sample and electrons are detected directly without the use of a viewing screen. The optics that guide electrons in the SEM and TEM operate using a fundamentally different approach compared to the optics used in optical microscopes. Both the SEM and TEM use an electromagnet to exert a force on the electrons in order to collimate and focus the electron beam. The electron optics used to focus an electron beam consists of a current carrying coil (which creates the magnetic field which exerts the force on the electrons), as shown in Figure 20.11. The electron beam passes through the opening in the coil.

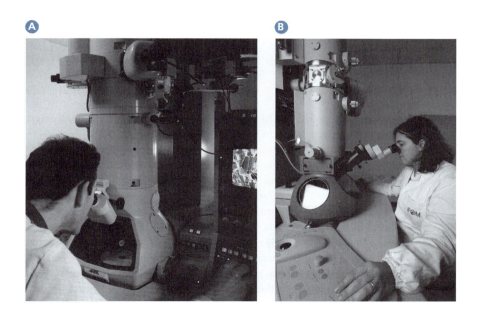

Figure 20.10 Panel (A): SEM. Panel (B): TEM.

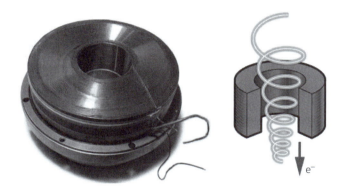

Figure 20.11 An electromagnetic lens taken from an electron microscope.

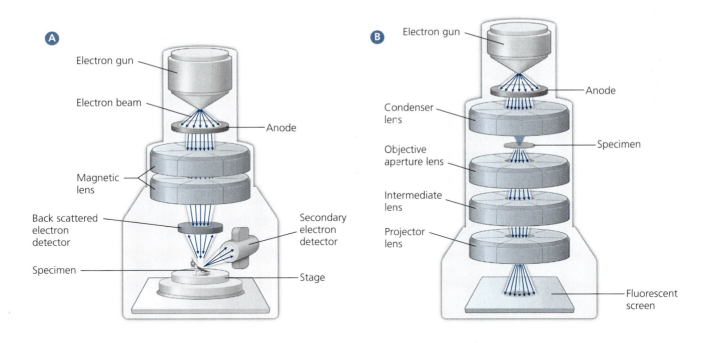

Figure 20.12 Basic components of electron microscopes, panel (A) showing an SEM instrument, and panel (B) showing an TEM instrument.

The basic components of the SEM and TEM are shown in Figure 20.12. There is some variation in instrument design, so one will find deviations from the general figure. The components are all housed in a vacuum chamber and are kept at low pressure to reduce interactions between the electrons and air, which would alter the electron speed and beam shape. In both the SEM and TEM, an electron gun provides electrons to an anode and then to a set of condenser lenses.

In the TEM, the beam passes through the condenser lenses and is focused onto the sample, and then a set of lenses is used to project an image of the sample onto a fluorescent screen (e.g., made of ZnS) where a digital camera can store the image. For most TEM instruments, the entire beam is transmitted through the sample. The electrons detected in TEM are the primary electrons associated with the electron gun. In an optical microscope, the light is reflected and/or transmitted through a sample, and some parts of the sample absorb more light than others, so we see features as contrasts between light and dark areas. In its simplest form, at low magnification (still high relative to optical instruments), TEM works the same way, depending on absorption or scattering of electrons by the sample. At higher magnifications, the ability to observe picometer-scale features depends on the interactions between the electron wave functions and those of the materials through which it is being transmitted.

In the SEM, the electron beam passes through a set of condenser lenses and then through a computer-controlled scanning coil before striking the sample. The scanning coil controls the rastering of the beam so that the beam can scan different parts of the sample. Instead of recording the electrons that are transmitted through the sample, the SEM collects electrons scattered off the surface. Adjusting the electron energy will induce different electron–surface interactions and allow the user to adjust the depth at which the sample is probed within the first few microns of the material's surface. Electrons coming off the sample come in several categories depending on the type of incident electron–sample interaction. These include back-scattered electrons, secondary electrons, and Auger electrons.[3]

---

[3] See Chapter 10 for a review of Auger and XRF spectroscopy.

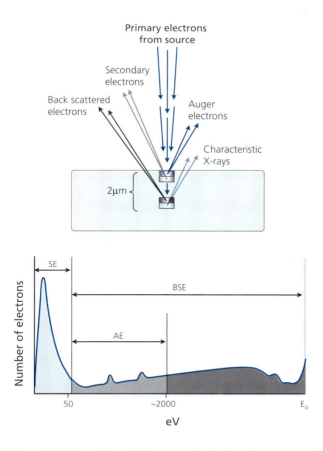

Figure 20.13 Various electron–surface interactions and general electron spectrum for secondary electrons (SE), back-scattered electrons (BSE), and Auger electrons (AE).

In addition, under certain conditions, SEM can produce characteristic X-rays and function in a manner similar to an XRF spectrometer.[4] A sketch of the electron beam sample processes is shown in Figure 20.13. A characteristic spectrum comparing number of electrons versus energy is shown in Figure 20.13. E_0 is the initial energy of the electron beam. We will now briefly discuss each category of scattered electrons.

Back-scattered electrons are the result of elastic scattering with the surface. The peak in back-scattered electrons occurs close in energy to the incident electron beam. The scattering probability is related to the atomic number of the species under investigation. The back-scattered electrons are emitted from roughly 10 to 100 nm into the surface. Back-scattered electrons can be used for chemical composition determination.

Secondary electrons are the result of inelastic scattering with the incident electrons and the inner shell electrons of the sample. We first encountered secondary electrons in Chapter 10 when we studied Auger and XRF spectroscopy. These secondary electrons are much lower in energy compared to the incident electron beam. Secondary electrons generally have energies less than 50 eV. This is shown in Figure 20.13 in the electron spectrum. These electrons are emitted up to a few nm from inside the material.

Secondary electron detection (SED) is the most common technique used in SEM imaging. The most common detection scheme uses the Everhart-Thornly detector. The Everhart-Thornly detector uses a scintillator similar to the one seen in Figure 6.27 for UV-Vis spectroscopy. When the secondary electrons strike the scintillator, electrons cause fluorescence in the scintillator and the fluoresced photons travel down a waveguide where the intensity is measured using a photomultiplier tube (PMT). The primary design difference between a traditional scintillating PMT and an Everhart-Thornly

SED is the application of the voltage drop to draw electrons to the surface and the utilization of a waveguide (also known as a light pipe) between the scintillator and the PMT. This light intensity is used to construct the image of the sample.

The process of secondary electron emission also results in the emission of characteristic X-rays and/or Auger electrons. The topics of XRF and Auger electron spectroscopy were first introduced in Chapter 10. When the secondary electron is ejected, an outer shell electron moves into an inner orbital. During this process, an X-ray can be emitted. These X-rays can originate as far as several microns into the sample, making the X-rays the deepest probe discussed. The excess energy from the electron moving to an inner orbital can alternatively be transferred nonradiatively to an outer shell electron, causing this outer shell electron to be ejected from the sample. These are *Auger electrons*, and their energy is in the range of 50 eV to a few thousand eV, much less than the energy of the incident electron beam. They originate from as far as about 75 Å into the sample. The spectrum of Auger electrons can be used for determining chemical composition. Some SEM instruments have Auger electron and X-ray detection capabilities. Figure 20.14 reviews the underlying quantum mechanics of these processes.

The electron gun, a common component for both SEM and TEM, comes in two general types: *thermionic emission* and *field emission*. The type of source used ultimately affects

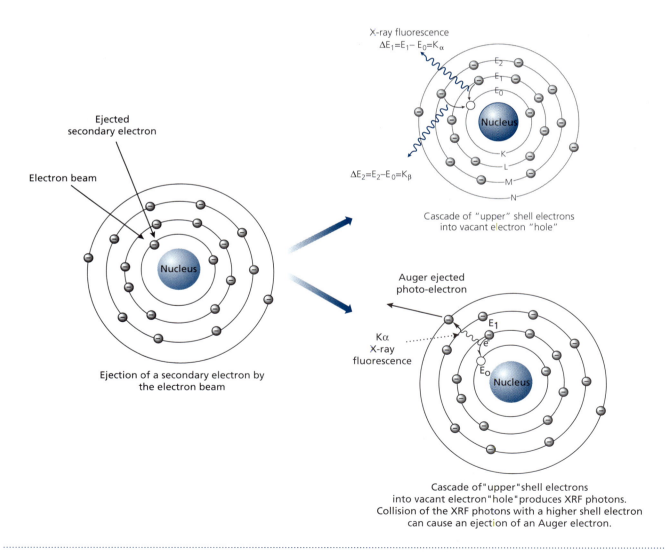

Figure 20.14 Schematic demonstrating the process of producing secondary electrons (left), characteristic X-rays (top right), and Auger electrons (bottom right).

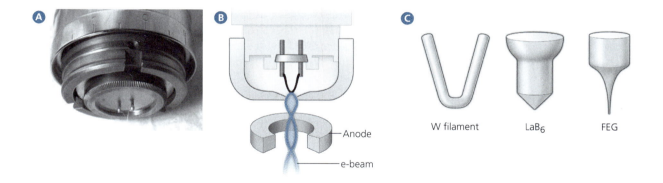

Figure 20.15 Panel (A): Filament (tungsten) for electron gun for SEM and TEM. Panel (B): Sketch of filament, electrode, and anode. Panel (C): Tungsten, LabB₆, and FE filaments.

the spatial resolution of the instrument. In the thermionic electron gun, a heating current runs through a filament (tungsten or LaB_6). The filament is heated until enough thermal energy is provided to eject electrons (2,700K for tunsgten and 1,800K for LaB_6). An accelerating voltage is then applied to the electrons. A photograph and sketch of the filament are shown in panels (A) and (B) of Figure 20.15. In a field emission (FE) source, an electric field is set up at the tip of the filament. This electric field is large enough to strip electrons from the filament. A circular anode provides an accelerating potential to the ejected electrons. The thermionic tungsten and LaB_6 filaments as well as the FE filament are shown in panel (C) of Figure 20.15. FE sources differ from thermionic sources as they have smaller size tips. Two particular types of FE sources are common: the cold field emitter and Schottky field emitter. The cold field emitter has a tungsten single crystal tip, and the filament is not heated. The Schottky field emitter has a zirconium oxide–coated tip, and the filament is heated (~1,800K). The tip is coated with a zirconium oxide to lower the work function.

We can rank the sources according to the magnifications they provide. The field emitter sources (cold emitter and Schottky) provide the largest magnifications possible, with LaB_6 thermionic sources next, followed by tungsten thermionic sources. The FE filaments are more costly than the thermionic filaments. The type of filament will also alter the minimum pressure required in the vacuum chamber. For example, LaB_6 thermionic filaments require lower pressure (~10^{-10} atm) compared to tungsten thermionic sources (~10^{-9} atm). Single-crystal tungsten Schottky FE sources require pressures around ~10^{-12} atm and cold FE sources require ~10^{-13} atm. The lower the pressure required, the more costly the vacuum system. Tungsten filaments also suffer from shorter lifetimes compared to the other filament options.

TEM and SEM each require different sample preparation processes. SEM samples must be solid and can be essentially any thickness. Samples for the SEM can be as large as 5 to 10 cm on edge. Most SEM instruments require the sample to be conducting, so nonconducting samples are sputter coated with gold, gold/palladium, or some other metal. In addition, there exists a more specialized instrument, the environmental SEM (ESEM), which allows for nonconducting samples and liquid phase samples. This is accomplished by splitting the instrument chamber in two, where the electron optics (top part of the instrument) are under vacuum and the sample area (bottom part) is at a higher pressure and usually filled with water vapor. ESEM is particularly useful when the metal sputter coating might alter your sample.

Sample preparation is more challenging for the TEM.[4] Samples need to be very thin, roughly 100 nm thick, and overall sample size is generally a few mm². A variety of

---

[4] Sridhara Rao, D. V.; Muraleedharan, K.; Humphreys, C. J. TEM Specimen Preparation Techniques. In *Microscopy: Science, Technology, Applications and Education*, Méndez-Vilas, A.; Díaz, J. (editors); Vol. 2, **2010**, 1232.

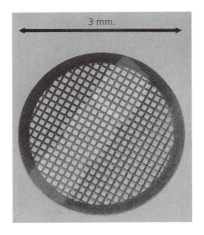

Figure 20.16 TEM copper grid, generally 3 mm in diameter.

processes are used to prepare thin samples, including ion beam milling and mechanical milling. However, one advantage of TEM is the fact that, the sample need not be conducting. Only a small area (on the scale of a few microns by a few microns and smaller) of the sample is imaged in TEM. It is common, when imaging particles in solution, to dip a carbon-coated copper grid into the solution containing the particles. The particles adsorb onto the grid (see Figure 20.16). The grid and sample are then placed on the TEM sample holder.

Problem 20.4:   Your laboratory has an excellent SEM equipped with secondary electron detection, Auger electron detection, and X-ray detection. You have been investigating the chemical composition of a polymer film that is up to 5 μ thick. The required thin metal film has been sputtered onto the sample. Which detection approach is most likely to help tell you of the chemical composition several microns into the film?

COMPARE AND CONTRAST

Resolutions for Different Microscopy Techniques

In choosing a microscopy instrument, a variety of options must be considered—for example, the necessity and relative ease of sample preparation, cost and availability of the instrument, and the x–y and/or z resolution needed. A set of images from atomic force microscopy and scanning electron microscopy for a similar system are shown in Figure 20.17. Vertical resolution (z axis) is seen in panel (A) of Figure 20.17 and resolution in the x–y plane is seen panel (B) and (C) of Figure 20.17.

In the table in this feature, we show the maximum x–y and z resolutions for several microscopy techniques. The values assume optimal instrument and sample conditions and are approximate. The table also shows the scan range size—the length that the instrument can image in one scan. The scanning electron microscope has by far the largest scan size at 1 mm.

A General Resolution Comparison Between Scanning Probe, Electron, and Light Microscopies

	RESOLUTION x–y	RESOLUTION z	SCAN SIZE RANGE
Atomic force microscopy	0.2 nm	0.05 nm	100 μm
Scanning electron microscopy	3 nm	N/A	1 mm
Transmission electron microscopy	0.2 nm	N/A	100 nm
Optical microscopy	200 nm	N/A	1 μm

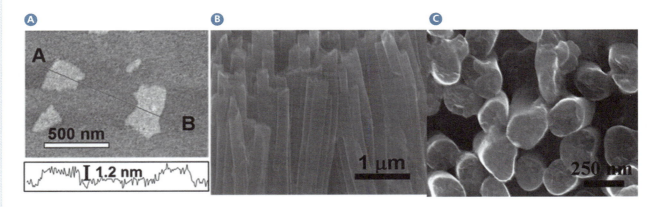

Figure 20.17 Images of graphene (A) AFM image of graphene oxide (B) side view SEM of polypyrrole/graphene oxide nanowires (C) top view SEM of polypyrrole/graphene oxide nanowires.

Problem 20.5:   You are investigating a sample with embedded nanoprisms which are around 5 nm tall and are a maximum of 10 nm in size in the x and y dimensions. You want to effectively measure the x–y and z dimensions of the nanoprisms. Using the table in the previous Compare and Contrast feature, which technique would be most effective given these dimensions?

20.3 Thermoanalytic Techniques

Thermoanalytic techniques are used to study physical properties of materials. The types of properties studied include the temperature dependence of the solid–solid and solid–liquid phase transitions, glass transitions, and decomposition of materials. There are quite a few well-defined thermoanalytic techniques. However, we will survey only the most commonly encountered, which include differential thermal analysis, thermogravimetric analysis, simultaneous thermal analysis, and differential scanning calorimetry.

Differential Thermal Analysis

Differential thermal analysis (DTA) is used to measure the enthalpy (ΔH) changes associated with specific phase transitions such as crystallization, glass transitions, melting/freezing, and sublimation. We use DTA to identify the stable temperature range of a material and the temperature at which a material will undergo a phase change. For instance, when selecting a material for an engine gasket, the engineer needs to know if the material is stable at all of the temperatures experienced by the engine. The head gaskets of most modern engines consist of thin sheet of steel coated with a rubber-like plastic called Viton. The rubber-like plastic must be able to withstand the high temperatures of an internal combustion engine as well as the cold temperatures of a northern winter. Viton is an elastomeric copolymer derived from vinylidene fluoride and hexafluoropropylene. Recall that Teflon is derived from tetrafluoroethylene. So Viton can be thought of as a cousin of Teflon. It is the presence of the fluorine groups in Viton that allow it to withstand the high temperatures of the engine block. Other common materials for head gaskets include asbestos[5] or copper.

DTA curves also serve as a useful fingerprint for quality-control purposes in manufacturing and find wide use in the polymer and ceramic industry. In addition, DTA analysis is used to construct phase diagrams and decomposition rates in various atmospheres. Lastly, DTA can be used to reverse engineer a product by identifying the composition of an unknown material.[6]

In a study of gold nanoparticles (AuNPs), researchers[*] were able to show that the AuNPs could be made soluble in organic solvents by modifying the AuNPs with alkyl groups. The nature of the alkyl group strongly affected the absorption properties of the AuNPs. Thermogravimetric measurements indicated ligand desorption evidenced by a one-step weight-loss process for the AuNPs. In addition, the inclusion of π-conjugation in an alkyl chain greatly increased the thermal stability of the AuNPs. ▬

[*] *Xin Su. Chem. Lett.,* **2012,** *41,* 708–710.

PROFILE
Thermogravimetric Analysis

[5] Less common because of health concerns.

[6] For example, identifying the difference between low-density polyethylene (LDPE), high-density polyethylene (HDPE), and polyvinyl chloride (PVC).

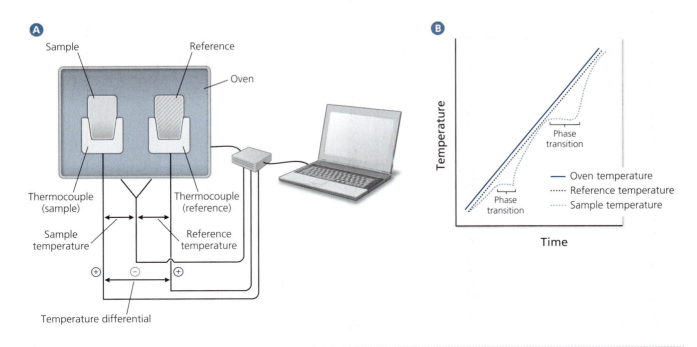

Figure 20.18 Panel (A): Schematic of a DTA instrument. Panel (B): Representation of DTA data.

The basic design of a DTA instrument involves two thermocouples,[7] a voltmeter, two sample containers, an oven, a temperature controller, and a recording mechanism. A sketch of a DTA instrument is shown in Figure 20.18. One thermocouple is placed in a reference sample and the other into the analyte sample. The reference material must be stable, with no phase transitions over the temperature range of the thermal cycle. In general chemistry, you learned that materials will absorb or lose heat during a phase change but they will not change temperature during a phase change. Phase changes in the analyte sample are induced as the oven cycles through a programmed temperature range. The latent heat of the phase change is registered as a change in potential on a multimeter, which is then translated into temperature and displayed on a computer screen.

The technique is referred to as *differential* because you are measuring the difference in the temperatures of two materials when subjected to identical thermal environments. Both materials are placed into the same oven and experience identical thermal cycles. Therefore, the heat capacity of the instrument itself is automatically subtracted out of the analysis. Any temperature difference between the analyte sample and the reference is recorded. The data is collected as the difference in temperature of the two materials versus time. Alternatively, one can also plot the differential temperature versus programmed temperature. A plot of the differential temperature versus time will result in a peak for any exothermic or endothermic processes in the analyte sample. Figure 20.19 shows some typical DTA data. The area under the peak is proportional to the enthalpy of the process.

In panel (A) of Figure 20.19, we see a representation of a DTA curve for a polymer sample. Five different physical changes can be identified as a function of temperature: a glass transition, crystallization, melting, oxidation, and decomposition. A curve of this type allows an analyst to determine the suitability of a material for a given application. In addition, one can use DTA to identify an unknown material. In panel (b) of Figure 20.19, we see an overlay of butter and margarine. It is quite

---

[7] See Section 11.5 on *Thermal Detectors* for a review of thermocouples.

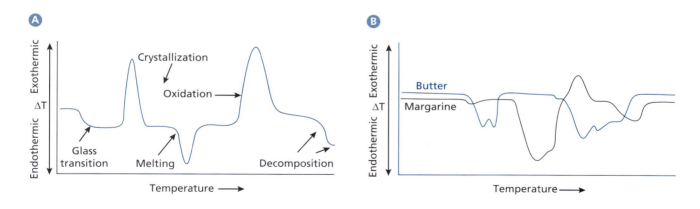

Figure 20.19 Sketch of typical DTA data. Panel (A): A polymer sample showing a glass transition, crystallization, melting, oxidation, and decomposition. Panel (B): A comparison of butter and margarine.

easy to determine the difference between the two substances. Figure 20.18 shows a basic schematic of a DTA instrument along with some representative raw data. In the schematic, the sample is placed in the left hand side of the compartment and a well-characterized reference is placed in the right-hand side. The sample and the reference are each connected to a separate thermocouples. The oven moves through a programmed heating cycle, and the analytical signal is the difference between the potentials of the two thermocouples.

Thermogravimetric Analysis

Thermogravimetric analysis (TGA) is used to measure the mass changes of a material as a function of temperature. Most manufacturers no longer make a stand-alone DTA instruments but instead have created hybrid instruments that simultaneously perform both DTA and TGA analysis.

TGA is a thermal analysis technique that measures the changes in mass of a sample during a programmed thermal cycle. Sample sizes are approximately a gram with precisions as low as 0.001%.[8] TGA is useful for identifying decomposition events, evaporation events, and oxidation events as a function of temperature and environment. For example, TGA can be used to determine the organic content of a sample (or conversely the inorganic content) and finds wide use in the ceramic and concrete industry. TGA is suitable for determining the thermally stable range for petroleum products such as motor oil blends, resins, stains, adhesives, and paints. TGA is also used extensively in the polymer industry for identifying the useful temperature ranges for thermoplastics, thermosets, elastomers, composites, plastic films, fibers, coatings, and paints. The atmosphere in the sample chamber may be purged with an inert gas to prevent oxidation or the atmosphere may be specifically altered in order to examine the reactivity of a material under specific atmospheric conditions.

Conceptually, a TGA instrument is very similar to a DTA instrument. In the case of TGA, the sample and reference thermocouples have been replaced by microbalances. In addition to stand-alone TGA instruments, many manufacturers are now producing TGA-FTIR and TGA-MS instruments. In these instruments, the volatile components lost during the thermal cycle are swept into either the FTIR or MS for further analysis. These tandem instruments fall into a class of instrumentation collectively referred to as *simultaneous thermal analysis* (STA).

---

[8] *Thermogravimetric Analysis: A Beginner's Guide*, PerkinElmer Inc., **2010**.

Kaolinite is a white clay used to make porcelain. Kaolinite has the formula $Al_2[Si_2O_5](OH)_4$. It is a layered inorganic polymer consisting of siloxane and the aluminum hydroxide ore gibbsite. The oxygen atoms forming the base of the tetrahedra of the siloxane layer (SiO_4) are linked to a gibbsite-type layer containing aluminum ions and OH groups. A closely related mineral called halloysite has the formula $Al_2[Si_2O_5](OH)_4 \cdot H_2O$. In a 2010 paper,[*] researchers used TGA-MS to characterize samples of kaolinite and halloysite from China. They found that below 100°C, the only physical change was the desorption of water from halloysite and the irreversible conversion to kaolinite (drying). At temperatures above 330°C, the kaoline converts to metakaoline by dihydroxylation.

$$Al_2[Si_2O_5](OH)_4 \rightarrow Al_2O_3 \cdot SiO_2 + 2H_2O$$

In addition, one can observe dehydration of halloysite impurities within the kaolinite at temperatures starting at 225°C as well as evolution of CO_2 from carbonate impurities. The tandem use of TGA-MS allowed researchers to determine what physical changes had occurred during each phase transition. ■

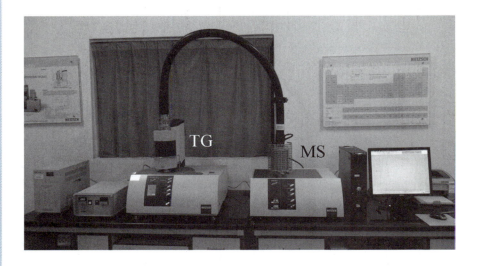

*[*] Hongfei, C.; Yang, J.; Liu, Q.; He, J.; Frost, R. L. Thermogravimetric Analysis–Mass Spectrometry (TG–MS) of Selected Chinese Kaolinites. Thermochimica Acta, **2010**, 507, 106–114.*

Differential Scanning Calorimetry

In *differential scanning calorimetry* (DSC), an analyte and a reference material are subjected to identical heating cycles and the amount of heat needed to maintain a constant temperature between an analyte sample and a reference material is measured as a function of time. The information obtained from DSC is very similar to the information obtained from DTA and like DTA, DSC is routinely used to identify materials and/or in the quality control of manufactured products. Figure 20.20 shows a DSC of a polymer resin. The graph shows the flow of heat in watts/gram as a function of the applied temperature. The manufacturer's specifications call for a 5:1 resin-to-hardener mix, and a DSC of a 5:1 mix is shown as the black line in Figure 20.20. An improper

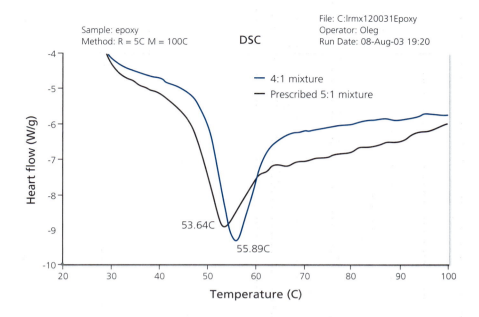

File: C:lrmx120031Epoxy
Operator: Oleg
Run Date: 08-Aug-03 19:20

Figure 20.20 A DSC plot of an epoxy blend. The specifications call for a 5:1 blend of resin to hardener and the blue line in the graph demonstrates how DSC can be used to identify improper blending.

4:1 resin-to-hardener mix is shown as the blue line in Figure 20.20 and is easily discernible from the 5:1 mix. In DTA, we measured the difference in temperature between a sample and a reference material as a function of time during a thermal cycle. In DSC, we also subject a sample and reference through a thermal cycle, but we keep the two materials at the same temperature throughout the cycle and measure the amount of heat that flowed into or out of the material. In short, DSC measures the difference in the amount of *heat* required to increase or decrease the temperature of a sample versus a reference material; both the sample and reference are maintained at nearly the same temperature throughout the experiment.

The reference sample should have a well-characterized heat capacity over the range of temperature cycle to be scanned. In a manner similar to DTA, DSC is used to identify phase transitions within a material. When a sample undergoes a phase transition, more or less heat will need to flow into the analyte sample relative to the reference sample in order to maintain both samples at the same temperature. By observing the difference in heat flow between the analyte sample and the reference sample, DSC can measure the amount of heat absorbed or released during each transitions. DSC may also be used to observe more subtle physical changes, such as glass transitions.

A schematic of a DSC instrument is depicted in Figure 20.21. Note the schematic similarities and differences between a DTA (Figure 20.19) and a DSC. In the DSC, the sample cell and reference cell are independently heated by a resistive heating element. The user programs a thermal cycle and the controller adjusts the amount of current that must pass through the two heating elements so that each cell's temperature profile remains the same over time. The measured signal is the difference in current (Δi) that passes through the two heating elements as a function of time. With proper calibration, the current is proportional to total heat. DSC finds wide use as a quality-control measure in industry, especially in evaluating sample purity and for studying polymer curing.

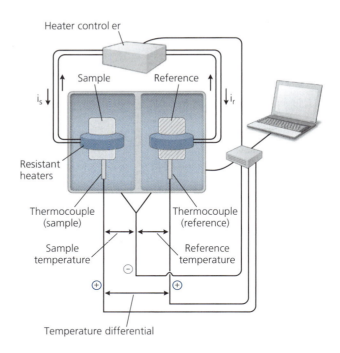

Figure 20.21 Schematic of a DSC. The current in the sample and reference heaters are adjusted so that the two chambers have the same temperature over time. With proper calibration, the current is directly proportional to heat. Therefore, Δi is proportional to ΔH.

COMPARE AND CONTRAST

Differential Thermal Analysis, Thermogravimetric Analysis, and Differential Scanning Calorimetry

	Dependent Variable	Independent Variable	Objective
DTA	ΔT between Reference and Sample	$\dfrac{dT_{oven}}{dt}$	Identification of Phase Transitions
TG	Δm between Reference and Sample	$\dfrac{dT_{oven}}{dt}$	Identify decomposition, evaporation, and redox events as a function of temp. and/or environment
DSC	Δi between Reference and Sample heating elements	$\dfrac{dT_{ref}}{dt} \approx \dfrac{dT_{sample}}{dt}$	Δi is converted into ΔH. Determination of the enthalpy associated with specific phase transitions.

ΔT = Difference in Temperature

Δm = Difference in Mass

Δi = Difference in the amount of current going to the heating elements.

$\dfrac{dT_{(oven)}}{dt}$ = The change in oven temperature as a function of time.

French criminologist Edmond Locard was quoted as saying "Wherever he steps, whatever he touches, whatever he leaves, even unconsciously, will serve as a silent witness against him." Scientists would soon develop the equipment to allow Locard to identify certain kinds of trace evidence.

At the turn of the twentieth century, the chemical industry was called on to meet the domestic demand for synthetic fibers for use in textile manufacturing. As a result, the variety and complexity of different fibers proliferated as manufacturers sought out optimal properties, such as mechanical strength, dye fastness, crease resistance, and perspiration wicking. Professors Fiona Gray, Michael Smith, and Magda Silva published a paper demonstrating the use of differential scanning calorimetry (DSC) and thermogravimetric analysis (TGA) to identify different textile fibers. The left side of the figure shows DSC thermograms of five different fiber types. The data demonstrates how one might use DSC to identify an unknown fiber. The right side of the figure shows TGA scans of five different fiber types. Again, it is clear that TGA can be used to distinguish one fiber type from another. ■

PROFILE

A Crime Scene Analysis

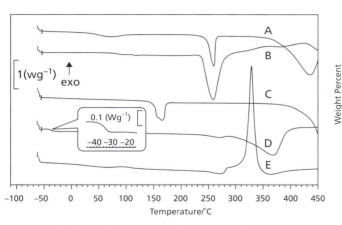

Normalized DSC thermograms of selected synthetic fibers: (A) nylon-6,6; (B) vinyon; (C) olefin; (D) spandex; and (E) modacrylic.

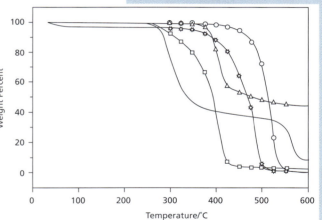

TGA of selected synthetic fibers: olefin (O); modacrylic (Δ); nylon-6,6 (★); spandex (□); and vinyon (–). Symbols are intended as a guide to the eye.

Gray, F. M.; Smith, M. J.; Silva, M. B. *Identification and Characterization of Textile Fibers by Thermal Analysis* J. Chem. Ed., **2011**, 88, 476–479.

Problem 20.6: You are investigating a homicide, and the victim had a piece of fabric clutched in her hands. A DSC analysis produced the thermogram seen here. Using the data from the profile "A Crime Scene Analysis," what type of fabric was in the victim's hand?

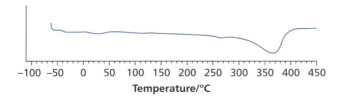

Problem 20.7: You are investigating a burglary, and several fabric pieces were found on the soles of the alleged perpetrator's shoes. A TGA produced the plot seen below. Using the data from the profile "A Crime Scene Analysis," what type of fabric was on the suspect's shoes?

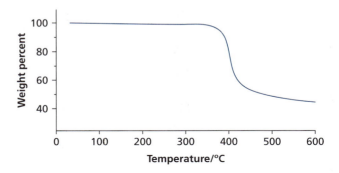

Problem 20.8: Based on the data given in the profile "A Crime Scene Analysis," what would be the material of choice for the construction of an oven mitten (hot mitt)? Justify your answer.

Problem 20.9: Based on the data given in the profile "A Crime Scene Analysis," what material would be the least appropriate for the construction of a cryogenic mitten? Justify your answer.

Problem 20.10: Imagine you are talking to a nonscientist. In your own words, write a three-to-five sentence description of:

(a) DTA   (b) TGA   (c) DSC

Problem 20.11: Imagine you are the quality-control technician in a candy factory. Use the data in Figure 20.22 and describe how you would ensure that your supplier is delivering pure sucrose.

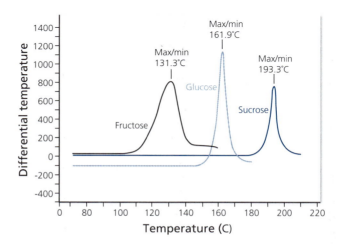

Figure 20.22 DSC curves for the melting of fructose, sucrose, and glucose. Even though fructose and glucose have the same molecular weight, a DSC curve can quickly identify the difference between the two substances.

20.4 Mechanical Stress Analysis

Many materials must meet requirements associated with their response to applied forces. Therefore, the analyst needs to experimentally determine a material's (1) tendency to flow or deform as a function of temperature and stress and (2) ability to withstand repeated deformations. In addition, the analyst needs to understand the stiffness of a material (modulus) and/or the elasticity (ability to recover from a deformation). These properties are conveniently studied by *mechanical stress analysis*.[9] A tensile strength measurement is a type of mechanical stress analysis, which involves taking a wire or fiber of known diameter and hanging weights from the fiber until a mass is reached that causes the wire to break. The tensile strength is reported as the ratio of the mass that breaks the sample to the cross-sectional area of the wire at the breaking point. Some materials (such as most metals) will deform gradually until they reach a maximum strain and then break. Others will reach a point where their stiffness (modulus) suddenly changes and then become more elastic. Plastics often exhibit this behavior under increasing strain.

When performing mechanical stress analysis, the deformation (γ) of a material is measured against the applied stress (σ) at a specified temperature. The stress is simply a measure of force divided by the area over which the force is applied. The *strain* (also known as *engineered strain*) is defined as the change in length (or volume) of a material as a function of the original length (or volume).

$$\gamma = \frac{\Delta L}{L} \qquad\qquad \textbf{Eq. 20.3}$$

where
  γ = engineered strain or deformation
  ΔL = change in length of the material
  L = initial length of the material

A typical stress–strain curve is seen in Figure 20.23. The initial slope of Figure 20.23 is called the *Young's modulus* and is a measure of the material's initial stiffness. The modulus along any point in the curve can be taken as the instantaneous slope at that point (the tangent to the curve). As seen in Figure 20.23, the modulus is often a function of the applied stress and changes as the stress (load) increases. In addition, the modulus is dependent on the temperature, so stress–strain curves are usually obtained over a range of applicable temperatures. This helps the analyst determine a material's usefulness for a specific application. Will the material hold up under extreme heat or cold? Will it retain its strength under repeated loads, vibrations, and so on? For instance, the modulus of a polymer heated beyond its glass transition will often decrease by several orders of magnitude (i.e., it will become rubbery). Conversely, many plastics become brittle in extreme cold. It is also a common property of some metal alloys to become "work hardened" if bent repeatedly.

Problem 20.12: Determine the Young's modulus of the material depicted in Figure 20.23.

Problem 20.13: Determine the instantaneous modulus for the material depicted in Figure 20.23 under a strain of:

  (a) 0.1 N   (b) 0.2 N   (c) 0.3 N

---

[9] In this context, the formal definition of *stress* is the sum of the forces experienced by a particle in a continuous material by all of a neighboring particle in that material.

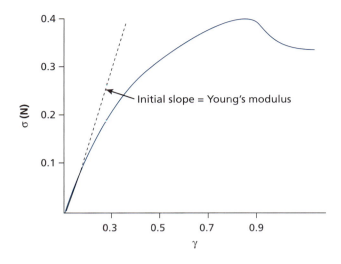

Figure 20.23 The slope at any point along the stress–strain curve is a measurement of the materials stiffness (modulus).

Problem 20.14:   For the material in Figure 20.23, how much strain can the material withstand before its modulus abruptly changes?

Dynamic Mechanical Analysis

Some materials become stiffer (increased modulus) as a result of repeated deformation. This phenomenon is called *work hardening* (also known as *strain hardening* or *cold working*). Many metal alloys exhibit work hardening. Other materials gradually weaken (lose stiffness) under repeated strain. Therefore, it is often desirable to study the stress–strain relationship of a material as it undergoes a repeated change in stress. These sorts of studies are called *dynamic mechanical analysis* (DMA) or *forced resonance analysis*. DMA is the study of a material's response to an applied oscillating force and involves a study of a materials deformation as a function of applied force. Its widest application is in the polymer industry, where it is used to characterize the *viscoelastic* behavior of polymers. *Viscoelastic* materials display both properties associated with viscosity[10] and properties associated with elasticity.[11] In addition to the polymer industry, DMA finds regular use in metallurgy and in the study of wood products (i.e., plywood).

Modern DMA instruments often use a technique called *force-resonance*, in which a drive motor on a piston applies a sinusoidal stress on the analyte material. The measured signal is the time-dependent deformation and recovery of the material (see Figure 20.24). A transducer at the bottom of the sample cavity measures the strain the sample experiences and a position sensor measured the deformation of the material. Modern DMA instruments are capable of performing time-dependent programmed analysis that allows the user to sweep over a designated frequency and/or temperature range. For this reason, some analysts have started referring to these types of analysis as dynamic material thermal analysis (DMTA).

These types of studies allow the analyst to examine a material's suitability under different thermal and vibrational environments. In panel (A) of Figure 20.25, we see

---

[10] The viscosity is a measure of a material's resistance to flow as a function of applied force.

[11] The elasticity is a measure of a material's tendency to return to its original shape after being deformed by a force.

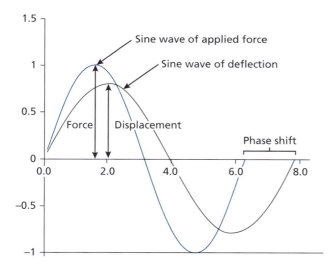

Figure 20.24 The applied force on the analyte and the deflection of the analyte as a function of time to material deflection.

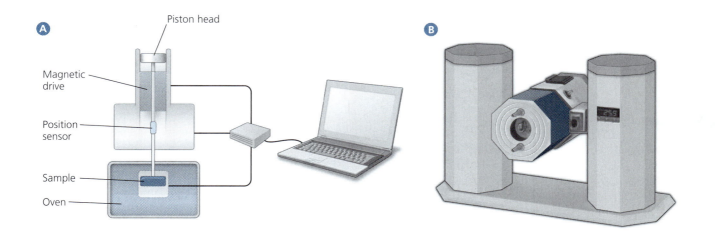

Figure 20.25 Panel (A): Schematic of a dynamic stress analyzer. Panel (B): Drawing of the PerkinElmer DMA 8000.

a schematic of a DMTA. An electromagnet is used to pull down on a piston, the force being a function of the current flowing through the magnet. A piston transfers that force to a sample. Deformations in the sample are indicated by the position of the piston. Panel (B) of Figure 20.25 shows a photograph of a PerkinElmer DMA 8000.

The modulus is determined as described above and is the ratio of the applied force to deformation at any particular point along the sinusoidal signal. Forced DMA also gives us information on damping. Damping is phase and temperature dependent, and so being able to adjust the temperature as well as the frequency of the DMA experiment is important.

20.5 Further Reading

BOOKS

Callister, W.; Rethwisch, D. *Materials Science and Engineering: An Introduction,* 9th ed.; John Wiley & Sons: Hoboken, NJ, **2013**.

Höhne, G. W. H.; Hemminger, W.; Flammersheim, H-J. *Differential Scanning Calorimetry: An Introduction for Practitioners.* Springer, New York, **1996**.

Menard, K. *Dynamic Mechanical Analysis: A Practical Introduction,* 2nd ed.; CRC Press, Boca Raton, FL, **2008**.

Shackleford, J. *Introduction to Materials Science for Engineers,* 8th ed.; Prentice Hall, Upper Saddle River, NJ, **2014**.

Williams, D. B.; Carter, C. B. *Transmission Electron Microscopy—A Textbook for Materials Science,* 2nd ed.; Springer, New York, **2009**.

JOURNALS

Binnig, G.; Quate, C. F.; Gerber, C. Atomic Force Microscope. *Phys. Rev. Lett.,* **1986**, *56*, 930–933.

Binnig, G.; Rohrer, H.; Gerber, C.; Weibel, E. Surface Studies by Scanning Tunneling Microscopy. *Phys. Rev. Lett.,* **1982**, *49*, 57–61.

Frank, A. J.; Cathcart, N.; Maly, M. E.; Kitaev, V. Synthesis of Silver Nanoprisms with Variable Size and Investigation of Their Optical Properties: A First-Year Undergraduate Experiment Exploring Plasmonic Nanoparticles. *J. Chem. Ed.,* **2010**, *87*, 1098–1101.

Lyman, B. M.; Farmer, O. J.; Ramsey, R. D.; Lindsey, S. T.; Stout, S.; Robison, A.; Moore, H. J.; Sanders, W. C. Atomic Force Microscopy Analysis of Nanocrystalline Patterns Fabricated Using Micromolding in Capillaries. *J. Chem. Ed.,* **2012**, *89*, 401–405.

ONLINE RESOURCES

Companies that design and manufacture micrsocopy equipment many times have excellent resources on instrument options. Examples include:

- Bruker, www.bruker.com/ (accessed Aug 2015)
- Hitachi, http://www.hitachi-hta.com/ (accessed Aug 2015)
- JEOL, http://www.jeolusa.com/ (accessed Aug 2015)
- Park, http://www.parkafm.com/ (accessed Aug 2015)
- Veeco, http://www.veeco.com/ (accessed Aug 2015)
- Zeiss, http://microscopy.zeiss.com/ (accessed Aug 2015)

20.6 Exercises

EXERCISE 20.1: Determine the instantaneous modulus at 25°C, 45°C, and 100°C for the material in Figure 20.26 under a strain of:

(a) 0.1 N   (b) 0.2 N   (c) 0.3 N

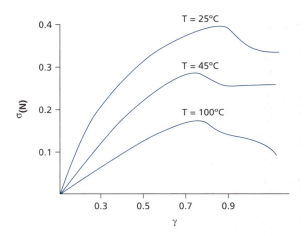

Figure 20.26 Stress–strain curves for an imaginary material as a function of temperature.

EXERCISE 20.2: Based on your answer to Problem 20.14, make a recommendation for temperature and stress tolerances for the material if the design specifications require a modulus of −0.14 or stiffer. For what temperature ranges would the material be suitable? For what stress ranges would the material be suitable?

EXERCISE 20.3: Define the following acronyms:

(a) STM   (c) DSC   (e) TGA   (g) AFM   (i) STA
(b) SEM   (d) DTA   (f) TEM   (h) MTA

EXERCISE 20.4: There are several online simulators for SEM and TEM. These allow for simulated control of the instruments and give a sense for how various parameters affect image quality. Perform an online search for a simulator and, after experimenting with the available options, write a brief paragraph on how various options affect image quality.

EXERCISE 20.5: DSC is a technique useful in determining;

(a) Modulus transition points of amorphous polymers

(b) Modulus transition points of glasses

(c) Glass transition temperatures for polymers and glasses

(d) Melting point of a crystalline materials

(e) All of the above

EXERCISE 20.6: Which technique would be most appropriate for simultaneously measuring the enthalpy of a phase transition—the mass loss during the transition and the composition of the effluent gases?

(a) DTA   (d) SEM   (g) TEM
(b) AMF   (e) STA   (h) DMTA
(c) TGA   (f) TGA-MS   (i) TG-IR

EXERCISE 20.7: If you needed to know the oxidation states of the elements on the surface of a material, which of the following techniques would yield the most complete information? Defend your answer. You may wish to review Chapter 10.

(a) AFM

(b) STM

(c) SEM

(d) XRF spectroscopy

(e) X-ray photoelectron spectroscopy

(f) Auger electron spectroscopy

EXERCISE 20.8: Which of the techniques described in this chapter would be most appropriate for studying the rate of oxidation of an iron alloy?

$$4Fe_{(s)} + 3O_{2(g)} + 2H_2O_{(l)} \rightarrow 2Fe_2O_3{\cdot}H_2O_{(s)}$$

Describe the experiment, the data one would obtain, and how you would manipulate the data in order to obtain a rate constant.

EXERCISE 20.9: You are choosing the electron gun for a TEM instrument. Your goal is to get the best resolution possible. Which type of electron gun should you purchase?

EXERCISE 20.10: The figure shows the DSC/TGA scan for a 28.629 mg sample of $Na_2WO_4 \cdot xH_2O$. The highlighted

area in the DSC plot is associated with the melting of the sample. Using the data given in the plot, calculate the number of water molecules in the crystal structure ("x") and the change in enthalpy of fusion (ΔH_{fus}) for the compound.

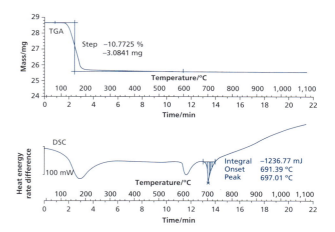

EXERCISE 20.11: Paracetamol (acetaminophen) is sometimes used as a cutting agent by cocaine distributors. DSC can be used to determine the purity of cocaine and to learn something about how a given cocaine sample was prepared, which can help lead investigators to the origins of the sample. The figure shows a series of DSC curves for pure paracetamol and cocaine, followed by those for some paracetamol–cocaine mixtures. The feature at around 95°C represents the melting of cocaine. Assuming that the peak height is proportional to the area in these plots, estimate the percentage of cocaine in the mixed samples.

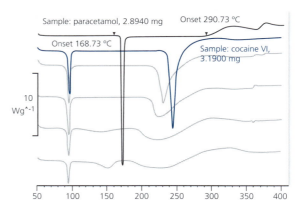

Chapter 21

Understanding the Nuclear Magnetic Resonance Experiment

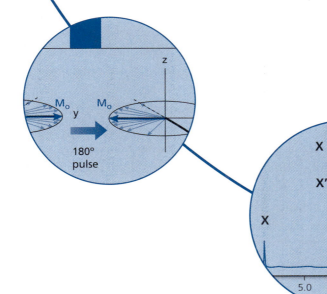

21.1 Introduction

In this chapter, we further develop the theory and concepts presented in Chapter 14 so that you may have a basic understanding of how a nuclear magnetic resonance (NMR) experiment is constructed. This includes an introduction to NMR in the rotating frame (a useful model for understanding magnetic transitions) and the development of pulse sequence strategies, which are essential for understanding multidimensional NMR. In bringing forth these ideas, we also come back to the idea of nuclear coupling—the ways in which near-neighbor nuclei influence one another. These interactions are at the heart of understanding the vast amount of information that can be learned about molecular structure by NMR. The profile on this page of Adriaan Bax demonstrates how NMR can be used for solving important questions about complex molecular structures. If you have not yet studied Chapter 14, you are advised to start there, because the topics in this chapter are more advanced and rely on your good understanding of the fundamentals and vocabulary that were laid down in that chapter.

21.2 Resonance in the Rotating Frame

In this section we will take another look at what happens to the nucleus, or more precisely, to the nuclear magnetic moment during and after the radiofrequency (RF) pulse so that we can further our understanding of the information that those signals provide about molecular structure. The idea of the rotating frame is a convention that is used

PROFILE
Adriaan "Ad" Bax

Adriaan "Ad" Bax is a Distinguished Investigator at the National Institutes of Health (NIH) in Bethesda, Maryland. A naturalized US citizen, he was born in the Netherlands and educated at the Delft University of Technology and Oxford University. As a graduate student, Bax developed two-dimensional nuclear magnetic resonance (NMR) techniques, resulting in a book that was an internationally popular text for many years. He has been a pioneer in the development of standard methods in biomolecular NMR spectroscopy. Bax went on to make contributions in solid-state NMR as a postdoctoral researcher at Colorado State University. After joining the NIH in 1983, he developed a wide variety of multidimensional and multinuclear NMR techniques and spearheaded the introduction of triple resonance NMR spectroscopy, a technique that allows scientists to understand protein backbone dynamics. His work and developments in NMR have made it possible (1) to push the boundaries on determining the three-dimensional structures of biomolecules, including proteins and nucleic acids, and (2) to expand the size of molecules that can be investigated by NMR, including ever larger and more interesting biomolecules and complexes. His most recent work utilizes liquid crystalline solvent to orient molecules of interest in the NMR as a means of obtaining additional coupling information, which can extend and enhance structural information in large molecules. This information typically is averaged out in solution. ∎

in NMR to provide a classical basis for visualizing the origins of signals and the effects of RF pulses.

As we saw in Chapter 14, Section 4, the RF pulse disturbs the equilibrium state of the net magnetization of the nuclear spins, which are under the influence of an externally applied magnetic field (B_o). Although we represent the net magnetic moment with a single arrow, aligned with B_o along the +z axis at equilibrium, it is important for you to remember that the individual nuclei continue to adopt many and various orientations as before. The use of a net magnetization vector is a way to present a simplified model of the system's bulk magnetization. This is a useful model, but it obscures our view of the precessional motion of the spins as they relax toward equilibrium. It is important that you keep both images in mind (review Figures 14.7, 14.10, and 14.11). Because the magnetic moment of the net magnetization (M_o) has a precessional motion and is therefore tilted slightly off of the z axis, there exists a magnetic component in the x-y plane as well as along the z axis. Consider taking a bird's-eye view (looking down z) of the coordinate system in which the precession is happening, depicted in Figure 21.1. From above, looking down the z axis, we can see that there are x contributions and y contributions to both energy and position of the magnetic moment as it moves in its precessive path. If the net magnetic moment is close to alignment along +z, then the x-y component is very small.[1] However, if the net magnetic moment is tipped away from +z, then the x-y component can be substantial.

A useful analogy for this physical behavior is the gyroscope (or a spinning top) and its momentum along the z axis. In this case, we have defined the z axis as the vector representing the force of gravity. With enough spinning energy, the gyroscope will be keenly oriented to the z axis. However, the gyroscope's behavior will begin to change as it begins to lose energy. The precession begins to take on more and more contributions in both the x and y directions. The precession becomes wider as the spinning gyroscope loses energy (see Figure 21.2), and the momentum of the gyroscope has a greater

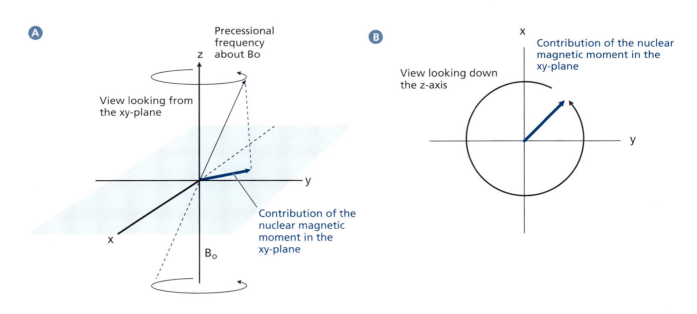

Figure 21.1 Contribution of the nuclear magnetic moment in the x-y plane as the nucleus precesses about B_o (z axis).

---

[1] Use of the rotating frame and magnetic vectors allows us to use simple geometry and trigonometry to describe the properties of the spin states. For example, if we know that the net magnetization vector M_o is tilted off the +z axis by an angle Θ, we can calculate that the energy of the system will be a function of the cosine Θ. See Section 21.8, Further Reading, which provides resources for a more detailed description.

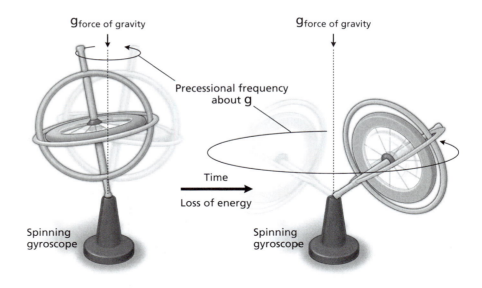

Figure 21.2 The spinning gyroscope's momentum has an increasing contribution in the *x-y* plane as the gyroscope tips further and further from the *z* axis. The gyroscope demonstrates a type of motion called *precession*. We can measure the frequency of the precession around the *z* axis.

contribution in the *x-y* plane. Relaxation of a spinning nucleus in a magnetic field (along the *z* axis) likewise involves a changing contribution in the *x-y* plane. However, in this case it is the magnetic moment, not momentum, that shifts to the *x-y* plane.

We will apply the idea of a spinning gyroscope to our spinning nucleus using the net magnetic moment vector model. Figure 21.3 shows an energized or excited nucleus "flipped" along the $-z$ axis, relative to the externally applied B_o field. This would be the situation after an RF pulse of sufficient energy. This is represented in the first frame of Figure 21.3 as a blue arrow $\left(M_o \right)$ pointing down, oriented along the $-z$ axis $\left(M_{o-z} \right)$. The excited state, only transiently stable, will begin to undergo the process of relaxation. This is the tendency of the system to return to equilibrium.

In the second frame of Figure 21.3, we see that the net magnetization along $-z$ has decreased slightly (M_{o-z} is a little shorter) and there is a corresponding increase in magnetization in the *x-y* plane $\left(M_{xy} \right)$. This is like our spinning gyroscope. As it loses orientation along *z*, it gains *x-y* characteristics. Following the excitation pulse, as the energy dissipates, the contributions in *x* and *y* will increase as the net magnetization fades along $-z$. Moving along in Figure 21.3, see that once the vector has diminished along $-z$, where it has its maximum contribution in the *x-y* plane, the magnetization will continue toward equilibrium as it returns fully to its original $+z$ position. (Our spinning gyroscope analogy breaks down at this point because the gyroscope will succumb to gravity unless additional energy is added.) As the nuclei return to the equilibrium state, the magnetization grows along $+z$ and the contributions in the *x-y* plane will decrease again.

It is important to understand what is being measured by the NMR spectrometer. Instrumentally, we are measuring the strength of the signal of the *magnetic moment in the x-y plane*. See Figure 21.4 (same image as Figure 14.11). Note that both the transmitter coils $\left(B_1 \right)$ and the detector coils are in the *x-y* plane.[2] Let us imagine the signal from this perspective. At equilibrium, there is very little magnetic energy (signal) in the *x-y* plane (most of it is around *z*). The RF pulse is applied, and the magnetization flips

---

[2] In some designs, the transmitter coils and the receiver coils are physically the same. After the coils transmit the RF pulse, the same coils are used to detect the magnetic emission as the nuclei relax.

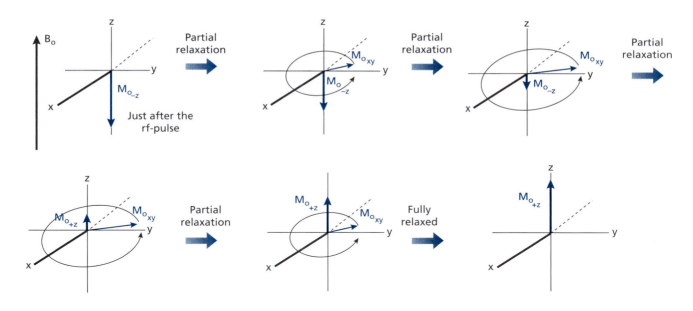

Figure 21.3 Contribution of the *net* nuclear magnetic moment in the *x-y* plane as the nuclei precess about the *x-y*. The top left image represents the net nuclear magnetic moment just moments after the sample has undergone a spin flip. As the excited state nuclei relax to the ground state, the contribution in the *x-y* plane increases to a maximum and then decreases to a minimum as the nuclei reestablish a ground state equilibrium along the *z* axis.

to $-z$. Then as the nuclei relax, the signal builds to a maximum, then it fades away as the magnetic moment returns to the $+z$ axis. The process whereby the net magnetization returns back to its original position on $+z$ is called *longitudinal relaxation* and will be discussed in greater detail in Section 21.3.

What we have endeavored to describe here is a description of the resonance in what is called the *rotating frame,* using the vector model of NMR. By obscuring our view of the precessional motion of the nuclei, the rotating frame allows us to view the three-dimensional coordinate axis as if it were rotating at the Larmor frequency of the nucleus, thus simplifying our view of the changing magnetic moment without trying to superimpose all of the various motions into a single model. A very good analogy for the rotating frame is that of a person watching a carousel (merry-go-round). When you watch a carousel, you see many motions. The horses are moving up and down and also moving forward and backward in a galloping motion. In addition, they are moving in a circular path because of the rotation of the carousel. As an observer, if you were to try to sketch all of these motions, it would be very difficult to do in a single model. However, the task becomes easier if we imagine the viewer traveling around the carousel at the same rate as the rotation of the carousel (in synchronicity). In this rotating frame, the motion of the horse is simplified to the up-and-down and back-and-forth motions and could be easily described in an *x-y-z* coordinate axis. Our model is similar in that *our vector diagram in the three-dimensional axis is rotating (as is our perspective on the diagram) at the Larmor frequency.*

21.3 The Pulse Experiment

Two of the parameters that an NMR spectroscopist can control are the frequency and duration of the RF pulse; these two parameters affect the amount of perturbing energy applied to the system. Despite the simplicity of the idea that the magnetic vector can have one of two possible orientations (up or down with respect to the B_o field), the nuclear magnetic moments can and do adopt a great many orientations in space. (Review Section 14.4 and Figure 14.10.) Recall that the net magnetic vector (M_o) represents the population excess, calculated by the Boltzmann distribution, which is in the slightly more stable state of alignment in the magnetic field. If the RF pulse is of the right energy and

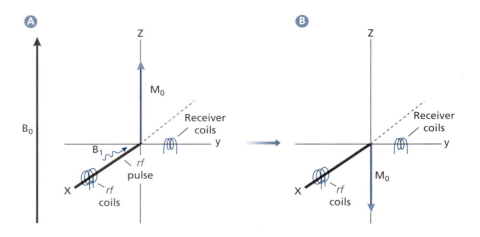

Figure 21.4 (A) Diagram showing the net magnetization aligned with B_o. This represents the equilibrium state of the net magnetization. Note that at equilibrium, the net magnetization lies along $+z$ axis in the coordinate frame. (B) As an RF pulse is applied (B_1) along $+x$ axis, we see reorientation of the net magnetization along $-z$. This is also known as a *180° or π-pulse*.

duration,[3] the net magnetic vector (M_o) will be flipped completely from $+z$ to the $-z$ orientation, as was shown previously in Figures 21.3 and 21.4. A pulse that completely flips M_o is called a *180° pulse* (also known as a *π-pulse*). A 180° pulse provides enough energy to redistribute the population excess so that more spins are flipped than are in alignment to B_o.

However, as we have just shown, the signal strength is strongest when M_o is in the x-y plane. Thus, a pulse that tips M_o only "half-way" over into the x-y plane would be very useful. This is called a *90° pulse* or *π/2 (pi-over-two) pulse*. A 90° pulse provides enough energy to redistribute the population excess so that more spins are perpendicular to B_o than are in alignment (see Figure 21.5). The 90° pulse is useful because it creates the maximum signal in the x-y plane, with the shortest possible relaxation time back to equilibrium.[4] A short relaxation time is desirable if you are signal averaging. We will discuss signal averaging later in this chapter.

NMR experiments are described using *pulse schemes*, also called *pulse sequences* (see Figure 21.6), which describe the manipulation of the net magnetization in the rotating frame and the evolution of signal as equilibrium is reestablished. A wide variety of pulse schemes have been developed that provide different types of information, both simple and complex, about the molecular structures under investigation. You may have heard of or used some of these in your own laboratory work: COSY, NOESY, TOCSY, HSQC, DEPT, INEPT, WALTZ, DANTE, HETCOR, EXSY, MLEV, BIRD, and so on. But before we look into any of these in more depth, first let us consider what causes the nucleus to return to its equilibrium state.

Relaxation of the Excited State

In NMR spectroscopy, the signals that we are trying to resolve are energetically very close together. Fortunately, the line widths of NMR signals are also intrinsically narrow. In fact, they are so narrow that Heisenberg Uncertainty broadening[5] is a dominant contributor to the ultimate resolution of an experiment. The Heisenberg Uncertainty

> For an NMR experiment, the relaxation time is typically between 0.1 and 10 seconds but can be as long as 100 seconds.

---

[3] The *rf* pulse width is the time the RF signal is applied (t_p). It is a function of the frequency of the RF energy (ω_1) and the flip angle (β_1). For a 180° pulse, the flip angle is 180° or π radians. To convert the frequency in radians to hertz, $\omega_1/2\pi$ is ν_1 and $\nu_1 = 1/2t_{180}$ for a 180° pulse.

[4] Using our gyroscope analogy from earlier, a 90° pulse would be like hitting the top of the spinning gyroscope to knock it down into the x-y plane. If it has enough energy, it will recover from being knocked down and pull itself up to realign with the z axis.

[5] We first introduced Heisenberg broadening in Chapter 7 in our discussion of the intrinsic line width of atomic transitions.

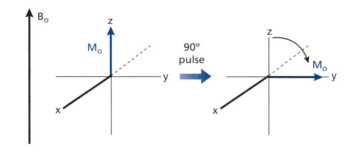

Figure 21.5 A schematic of a 90° pulse, also known as a π/2 pulse.

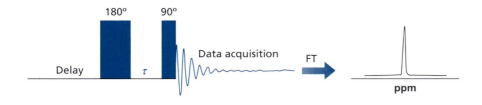

Figure 21.6 Simple NMR pulse scheme. The delay time at the start of the sequence is set to a sufficiently long time so that equilibrium is reestablished. The τ is a delay time that will be manipulated during the experiment. In a typical experiment, the pulse scheme will be repeated and data acquired multiple times for the purposes of signal averaging. Also see Figure 21.8.

principle dictates that ΔE of a quantum transition is inversely proportional to the lifetime, Δt, of the excited state:

$$\Delta E \Delta t \geq \frac{\hbar}{2} \therefore \Delta E \geq \frac{\hbar}{2\Delta t}$$

Eq. 21.1

If the relaxation is very fast, signals become broad and we lose the ability to distinguish interactions between nuclei. In extreme cases, we may not be able to detect the signal at all. An understanding of the relaxation process and how long relaxation takes is important to the NMR spectroscopist and helps us design effective experiments.

Oxygen often quenches the excited state, speeding relaxation and broadening the line widths. Therefore, when high resolution is required, it is common to degas the sample to remove oxygen from the system. It is important to emphasize the trade-off that is being made here. By degassing the sample, you can improve resolution, but you also increase the acquisition time because you now have to increase the delay time between each pulse. This is important to consider if you are using signal averaging in order to achieve a desired signal-to-noise (S/N) ratio. Degassing your sample does not provide any advantage if your resolution is satisfactory for your desired purposes. Example 21.1 demonstrates these principles.

Example 21.1

Determine the intrinsic line width in hertz of an excited state with a lifetime of 0.1 seconds. What would be the intrinsic minimum line width in ppm for a 60 MHz NMR?

STRATEGY –

Use Equation 21.1 to find ΔE for the transition and use the relationship E = hν to find the frequency range for this transition. Then use Equation 14.8 to find the ppm range for a 60 MHz NMR.

(Continued)

ANSWER –

From Equation 21.1:

$$\Delta E \geq \frac{\hbar}{2\Delta t} = \frac{1.0546 \times 10^{-34}\, \text{J} \cdot \text{s}}{2(0.1\text{s})} = 5.273 \times 10^{-34}\, \text{J}$$

Using the relationship E = hv:

$$v = \frac{E}{h} = \frac{5.273 \times 10^{-34}\, \text{J}}{6.626 \times 10^{-34}\, \text{J} \cdot \text{s}} = 0.80\, \text{Hz}$$

From Equation 14.8:

$$\left(\frac{8.0\,\text{Hz}}{60,000,000\,\text{Hz}} \right) \times 10^6 = 0.013\,\text{ppm}$$

Problem 21.1: Using Equation 21.1, determine the intrinsic line width in hertz of an excited state with a lifetime of:

(a) 1 s   (b) 3 s   (c) 7 s   (d) 9 s

Problem 21.2: For the line width found in Problem 21.1(a), determine the minimum ppm separation that could be resolved for the following NMR field strengths:

(a) 60 MHz   (b) 200 MHz   (c) 400 MHz   (d) 900 MHz

Problem 21.3: By degassing your sample, you were able to lengthen the relaxation time of your desired signal from 0.2 to 6 seconds.

(a) By what factor will your line width decrease as a consequence?

(b) If your experiment required a signal-averaged scan of 64 scans in order to achieve the desired S/N ratio, how much longer will your experiment take as a result of degassing the sample?

NMR signals are intrinsically weak, so unless your sample is sufficiently concentrated you will make use of signal averaging to increase the S/N ratio of your Fourier transform–NMR experiments. Signal averaging was first discussed[6] in Chapter 5 and reintroduced in Chapters 11 and 14. When programming the experiment, it is important to understand how long we need to pause after data acquisition ends before we initiate a new RF pulse. Because the energy difference between m = –½ and m = +½ states is very small, the excess population of nuclei in the ground state compared to the excited state at equilibrium is also small. When the delay between experiments is too short, we can *saturate* the transition between states (equalize populations of excited and ground states) and decrease an already weak signal. However, if our RF pulses are too far apart, the experiment will take an unnecessarily long time to complete. So for these reasons, we want to know the *relaxation time* (rate of decay) of the excited state

Boltzmann and Signal Strength

If the populations of the two states were exactly equal, then no net magnetization would be observed and no signal could be obtained. A population imbalance must exist in order for the signal to be generated.

---

[6] See Equation 5.7.

when we program the pulse sequence in a signal-averaged Fourier transform–NMR experiment.

After the RF pulse, there are two ways that the excess energy becomes dissipated and the net magnetization returns to equilibrium. The first is *longitudinal* or T_1 *relaxation,* which is also called *spin-lattice relaxation* because it describes the transfer of magnetic energy from the nuclear spin system to the neighboring molecules (the lattice). The second is *transverse* or T_2 *relaxation,* which describes the loss of phase coherence in the *x-y* plane due either to inhomogeneity in the magnetic field or from transfers of magnetic energy between spinning nuclei. Because of the latter process, T_2 relaxation is sometimes called *spin-spin relaxation*. The mathematics that describes the various contributions in the *x-y-z* coordinate systems for the net magnetic moment describes both T_1 and T_2 as first-order processes. This is an important point because it means that the rate of the decay of the excited state is an intrinsic property of the nucleus under discussion. Thus, the terms *longitudinal* and *transverse* are more descriptively correct when applied to relaxation (as compared to the descriptors spin-lattice and spin-spin relaxation, which by name tend to imply a second-order process, a common misconception).

As we have shown, an understanding of T_1 and T_2 relaxation is important in designing signal-averaged experiments. However, more importantly, knowledge of the T_1 values is important in the programming of multidimensional experiments. Multidimensional experiments (e.g., two- and three-dimensional NMR techniques) exploit the difference in the relaxation delay between different nuclei, and in some cases, the T_1 values are used directly in the characterization of a variety of compounds. We will come back to two-dimensional experiments again later in this chapter. Furthermore, as was discussed in Chapter 14, the T_2 relaxation event is highly dependent on the mobility (viscosity) of the sample and is a major part of magnetic resonance imaging, or MRI, where T_2 rates are correlated to differing tissue densities.

Longitudinal Relaxation (Spin-Lattice): T_1

We have already described longitudinal relaxation of the nuclear magnetic moment as the reestablishment of the equilibrium position of M_o along the *z* axis in alignment to the B_o field.[7] Figure 21.7 shows that, immediately after a 180° pulse, the magnitude of the net magnetization vector (M_o) along the *z* axis is negative and thereafter it slowly increases as a function of time until it reaches the equilibrium position. The time at which the net magnetization along *z* is 0 corresponds to the time at which the magnetization is fully in the *x-y* plane and is the point of maximum signal as measured in the *x-y* plane.

The time constant for the process is called T_1 and is derived from the *Bloch equations*[8]:

$$M_o(t) = M_o(eq) - [M_o(eq) - M_o(0)]e^{-t/T_1} \qquad \textbf{Eq. 21.2}$$

According to Equation 21.2, the net magnetization (M_o) as a function of time t is an exponential relaxation, e^{-t/T_1}, multiplied by a factor that describes the magnitude of the vector along *z*, where M_o(t = eq) is the equilibrium net magnetization and M_o(t = 0) is the net magnetization at time 0 (time of the pulse). Equation 21.2 shows the integrated equation for the change in net magnetization along *z* as a function of time, and Figure 21.7 shows the changing magnetization as a function of time (as the excited state relaxes back to equilibrium). Figure 21.7 also indicates the time when the net magnetization along the *z* axis is zero (t = null). This is the point at which we have maximum signal in the *x-y* plane.

---

[7] The longitudinal direction is along the *z* axis.

[8] The Bloch equations, which describe the change in magnetization over time, are sometimes referred to as the "equations of motion."

As a rule of thumb for quantitative work, the time delay between pulses needs to be at least five multiplied by T_1. However, for fast acquisition of qualitative data, a time interval of slightly longer than one T_1 will suffice.

Problem 21.4: In general chemistry, you learned that the integrated first-order rate equation has the form $\ln(A)_t = -kt + \ln(A)_o$. Using Equation 21.2 as a starting point, derive an analogous equation for M_o.

Problem 21.5: Rework Problem 21.3, assuming a delay between successive scans of $5 \times T_1$.

For an isolated nucleus, longitudinal relaxation is a slow process. T_1 times are long. In this context, a long time is several seconds. However, the relaxation process can be influenced by dissipating energy to the environment (the lattice), which is a typically slow process. As a result, the rate of decay for the excited state M_o vector is strongly influenced by the local environment. T_1 relaxation is also a function of the gyromagnetic ratio of the nucleus and dependent on B_o. As a result, T_1 values are not transferable from instrument to instrument and therefore must be measured in situ.

As the mobility of the lattice increases, the vibrational and rotational frequencies of all of the molecules in the lattice increase, and this increases the probability that one of these frequencies can couple with the excited state nuclear spins ($m = -\frac{1}{2}$) and stimulate the transition from high- to low-energy states. As a result, T_1 times decrease with increasing temperature. Likewise, paramagnetic species in solution (such as O_2) can also couple with the excited state and reduce T_1 relaxation times. Most relaxation times observed in routine NMR are between 0.1 and 10 seconds unless the sample has been degassed of oxygen or if the sample is run at a cold temperature. Variable temperature NMR is discussed in more detail in Section 21.6.

Measuring T_1: Inversion Recovery

It is important to note that T_1 relaxation must be measured on the spectrometer you are using. The pulse sequence for measuring T_1 relaxation is the inversion recovery scheme, which is shown in Figure 21.6 and is diagrammed again using the rotating frame model in Figure 21.8. The experiment begins with a delay time. This time (D_1) will be set to a relatively long time, at least five times longer than the expected T_1 time. The *delay time* is a constant in this experiment and provides sufficient time (after data acquisition is complete) for the system to reestablish equilibrium before the sequence repeats. The delay (D_1) allows the sample to come to equilibrium, establishing the net magnetization along +z, aligned with B_o. The first pulse, a 180° pulse, inverts the M_o vector, flipping the net magnetization. The nuclei are then allowed to relax for a short time, tau (τ), and *during that time* the net magnetization begins to relax. This relaxation is interrupted by a second pulse, a 90° pulse, which tips the net magnetization into the *x-y* plane where the signal is then measured. The experiment is carried out multiple times as τ (the delay between pulses) is steadily increased. Figure 21.9 shows the results from several different perspectives.

To understand the results, consider the case in which the delay between the 180° and 90° pulses is very short—so short, in fact, that the net magnetization (M_o) will be mostly inverted (very little relaxation has occurred) at the time when the 90° pulse is applied. This will produce a relatively large negative peak in the transformed spectrum (shown in Figure 21.9 as the largest negative). As τ is increased in subsequent experiments, the negative signal gets smaller and smaller (as seen in Figure 21.9).

Why does the signal diminish? As the time increases, the net magnetization, flipped to –z by 180° pulse, is returning to equilibrium, along +z because of longitudinal relaxation (T_1). Because the 90° pulse interrupts this process (like a freeze-frame) by dropping the magnetization into the *x-y* plane (for maximum signal strength), by varying τ between the 180° and 90° pulse we are observing the relaxation of M_o along the z axis. Notice that as τ becomes very long, the net magnetization

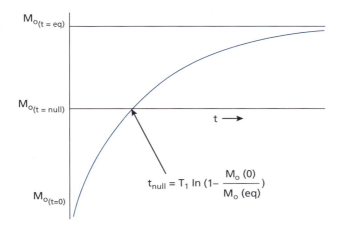

Figure 21.7 Net magnetization (M_o) along the *z* axis as a function of time after the application of a 180° pulse. The net magnetization along *z* becomes zero when the magnetization is completely in the *x-y* plane (point of maximum signal).

completely returns to the equilibrium position and the Fourier transformed spectrum is a positive peak.

Over the course of the change in τ, we can identify the time at which exactly half of M_o has returned to the ground state (the half-life of the decay). The null time ($τ_{null}$), or the time at which the signal along *z* goes to 0, is the half-life for T_1. When exactly half of the nuclei in the excited state have relaxed, the 90° pulse will produce no net change in the magnetic vector (M_o) and therefore we will get no signal (the signal is saturated). This value for τ is termed the *null time* and is depicted in Figure 21.9 as $τ_{null}$.

Panels (A) and (B) of Figure 21.9 demonstrate how one can find $τ_{null}$ experimentally by finding the delay time at which the 90° pulse creates no net signal. Panel (C) of Figure 21.9 shows a plot of peak intensity versus τ, which has been fit to a first-order decay. One can find T_1 by substitution into a traditional first-order half-life equation $kt_{1/2} = \ln(2)$, as shown in Equation 21.3a:

$$\frac{1}{T_1}\,τ_{null} = \ln(2) \qquad\qquad \textbf{Eq. 21.3a}$$

Solving for T_1:

$$T_1 = \frac{τ_{null}}{0.693} \qquad\qquad \textbf{Eq. 21.3b}$$

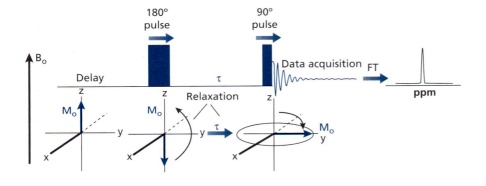

Figure 21.8 Pulse scheme for the inversion recovery experiment, an experiment used to measure T_1. The time τ is varied. Starting with a very short τ, the time is lengthened until the signal inverts.

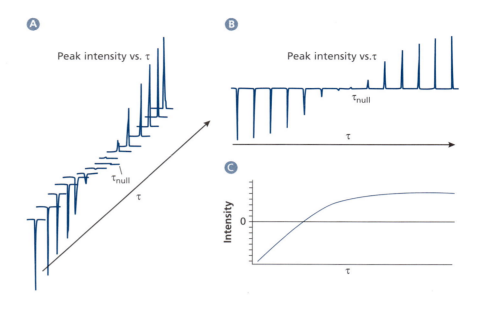

Figure 21.9 Inversion recovery response: peak intensity as a function of time τ. Panels (A) and (B) show peak intensity and orientation as a function of τ. Panel (C) shows a fit of peak intensity versus τ starting at τ_{null} and proceeding forward in time.

Or if a more precise value of T_1 is needed, one can substitute into an integrated first-order rate law, as shown in Equation 21.4:

$$ln(I)_\tau = -\frac{1}{T_1}\tau + ln(I)_o \qquad \textbf{Eq. 21.4}$$

where
    I_τ = peak intensity at τ
    I_o = peak intensity when t is zero
    τ = time delay between pulses
    T_1 = time constant for the longitudinal relaxation process

Using Equation 21.4, one can plot $ln(I)_\tau$ versus τ and obtain a slope of $-\dfrac{1}{T_1}$.

Problem 21.6: The following data were collected for an inversion recovery experiment. I_τ is the peak intensity with a delay between the 180° and 90° pulse of τ. Determine the value of T_1.

$\tau_{(s)}$	0.1	0.5	1	2	4	8	16	32	64
$I_{(\tau)}$	19,643	18,279	16,705	13,945	9,735	4,739	1,123	63	0.20

Problem 21.7: The following data were collected for an inversion recovery experiment. I_τ is the peak intensity with a delay between the 180° and 90° pulse of τ. Determine the value of T_1.

$\tau_{(s)}$	0.1	0.5	1	2	4	8	10	13	16
$I_{(\tau)}$	474.14	383.41	294.01	172.88	59.78	7.15	2.47	0.50	0.10

Example 21.2

The following data were collected in an inversion recovery experiment. I_τ is the peak intensity with a delay between the 180° and 90° pulse of τ. Determine the value of T_1.

$\tau_{(s)}$	$I_{(t)}$	$Ln[I_{(t)}]$
0	7,817	8.96
1	7,125	8.87
2	6,345	8.76
3	5,032	8.52
4	3,165	8.06
8	1,252	7.13
16	196	5.28
32	5	1.57
50	0	−2.60

SOLUTION –

This problem is best solved using a spreadsheet. Input the data into a spreadsheet and fit the data to Equation 21.4.

ANSWER –

A plot of $\ln(I)$ versus τ gives us a straight line with a slope of $-\dfrac{1}{T_1}$ and a y-intercept of $\ln(I)_o$. The value of T_1 is 4.3 s.

First order fit of ln(I) vs τ
y = −0.2318x + 8.9872

If we know that T_1 relaxation is the rate-limiting event in our system, we can estimate T_1 directly from the line width at half height ($v_{1/2}$) of an NMR peak (Equation 21.5). For many applications, a quick estimate of T_1 is sufficient to designing a good pulse scheme.

$$v_{1/2} = \frac{1}{\pi T_1}$$

Eq. 21.5

Problem 21.8: Show the derivation of Equation 21.5 from Equation 21.4.

Problem 21.9: You have an NMR spectrum taken on a 400 MHz NMR with a nominal peak width at half-height of approximately 1.8 ppm. Use Equation 21.5 to estimate the value of T_1. *Hint*: You will need to use Equation 14.6 to convert ppm into hertz.

Transverse Relaxation (Spin-Spin): T_2

Transverse relaxation is in the x-y plane, and the rate constant is called T_2. In our rotating frame, we can think of T_2 relaxation as a fanning out of the individual spins away from the net magnetization vector over time. For an ideal system, when we tip the net magnetization (M_o) into the x-y plane with a 90° pulse, we expect a single frequency to be registered by the receiver coils. However, because of field inhomogeneity (B_o)

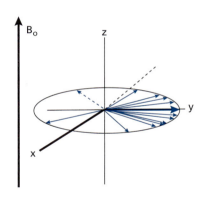

Figure 21.10 A representation of T_2 relaxation in the *x-y* plane.

and interactions with the local environment, some frequencies become a little higher than the Larmor frequency (move ahead of the rotating frame) and some frequencies become a little lower than the Larmor frequency (fall behind the rotating frame). Thus, what starts off as a tightly bunched net magnetization (M_o), after the pulse ends will begin to relax in terms of precessional frequency as measured in the *x-y* plane. This leads to peak broadening and eventually signal loss.

Because the detector coils are in the *x-y* plane, we can view this dispersion of Larmor frequencies as magnetic vectors that are "spread out" in the *x-y* plane (see Figure 21.10). Whereas T_1 accounts for an emission of energy (an enthalpy effect), T_2 is related to a dissipation of energy (an entropy effect in terms of phase). What starts off as a coherent (in phase) net magnetic vector slowly becomes more disorganized in terms of phase as T_2 relaxation takes effect. Although the rate of transverse relaxation is also affected by field inhomogeneity in B_o, for most modern instruments the entropy effect is the predominate contribution to T_2 relaxation.

It may be useful for you to think of T_2 relaxation in terms of the carousel analogy introduced earlier. In the rotating frame, our perspective remains at pace with one of the horses (at the Larmor frequency) on the carousel. One can imagine, however, that keeping all viewers at equal pace with the horse is difficult; some viewers might fall a little behind, and other viewers might run a little ahead. Therefore, the net view is blurred by the entropy effect of all viewers not being exactly in phase with the target frequency.

Measuring T_2: Spin-Echo

A *spin-echo experiment* (see Figure 21.11) starts with a 90° pulse that drops M_o into the *x-y* plane followed by a short time delay, τ. During time τ, transverse relaxation of M_o occurs according to the intrinsic time constant T_2 and the influences of any field inhomogeneity.

As shown in Figure 21.11, at the completion of the time τ, the sample is subjected to a 180° pulse that inverts the net magnetization[9] from the +*y* axis to the –*y* axis. During another equal time, τ, the magnetization refocuses along –*y*. For example, if a nucleus experienced a 5° phase shift in its Larmor frequency during the time τ, it will experience a –5° phase shift during the same time τ increment when flipped to –*y*. The T_2 relaxation that is present in the first time increment τ is present during the second time τ, and the nuclear spins that were inverted by the 180° pulse are refocused during the second τ. Data are acquired after the second time τ. Note that the spin-echo refocuses all of the spins, regardless of their off-set value.[10] The spin-echo is an important building-block sequence for other more elaborate pulse schemes. Therefore, it is important to understand conceptually.

The *Carr-Purcell-Meiboom-Gill* (CPMG) pulse sequence is another type of spin-echo experiment and is shown in Figure 21.12. The CPMG sequence is derived from the Hahn spin-echo but autocorrects for any error in the pulse width of 180° pulse applied after the first τ. By using a train of time delays (τ), if there is any inaccuracy in the 180° pulse, it will not degrade the signal (as in the Hahn spin-echo) but will be self-corrected

---

[9] It should be noted here that a 90° pulse, if applied along +*x*, would by convention and using the right-hand rule, be flipped 90° onto the –*y* axis as a "positive" rotation. In order to flip the net magnetization vector to +*y* axis from +*z*, the B_1 field would be applied along the –*x* axis. Pulses may be applied along any axis in order to shift the net magnetic vector accordingly. For our purposes, to introduce the concept of pulse schemes, and to focus attention on the signal evolution, we are leaving the excitation axis as intentionally vague so as not to confuse the issue. However, the references provided in Further Reading (Section 21.8) give more information about spin dynamics and may be of interest to the reader.

[10] *Off-set* is a term used to describe the difference between the frequency of the resonance and the frequency of the rotating frame. If the rotating frame is exactly at the frequency of the NMR, the off-set will be zero. However, we know that each nucleus will have a slightly different frequency due to its local environment; these differences will be squelched in a spin-echo experiment. Therefore, the only measured signal will be due to T_2 relaxation (entropy effects).

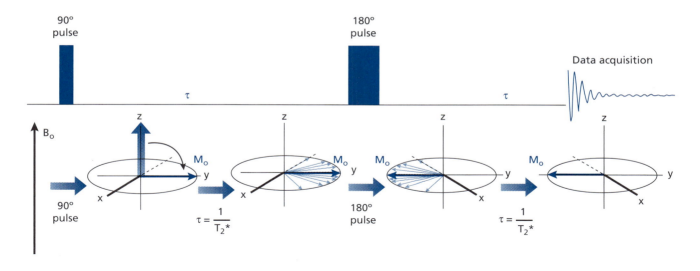

Figure 21.11 The Hahn spin-echo pulse scheme for a T_2 experiment, showing the refocusing of M_o in the *x-y* plane.

by following with the same delay and a subsequent 180° pulse. A pulse scheme for a CPMG sequence showing the double focusing of M_o is seen in Figure 21.12.

$$\tau = \frac{1}{T_2} \qquad\qquad \textbf{Eq. 21.6}$$

In the spin-echo experiment, the T_2 time is measured as a function of the signal intensity at intervals of τ per the relationship:

$$\ln I_\tau = \ln I_o - \frac{\tau}{T_2} \qquad\qquad \textbf{Eq. 21.7a}$$

Equation 21.7a can be rearranged into a linear $(y = mx + b)$ format to yield Equation 21.7b.

$$\ln I_\tau = -\frac{1}{T_2}\tau + \ln I_o \qquad\qquad \textbf{Eq. 21.7b}$$

A set of spectra are acquired at varying values for τ (e.g., $\tau = 2\,\tau,\,4\,\tau$) and a plot of $\ln I_{(\tau)}$ versus $-\tau$ will yield a line with a slope of $1/T_2$.

Problem 21.10: The following data were collected for a Hahn spin-echo experiment. I_τ is the peak intensity with a delay between the 90° and 180° pulse of τ. Determine the value of T_2.

$\tau_{(s)}$	0.05	0.5	1	2	3	4	5	6	10	15
$I_{(\tau)}$	2,908	2,195	1,606	859.5	460.1	246.3	131.8	70.55	5.29	0.25

Problem 21.11: The following data were collected for a Hahn spin-echo experiment. I_τ is the peak intensity with a delay between the 90° and 180° pulse of τ. Determine the value of T_2.

$\tau_{(s)}$	0.05	0.5	1	2	3	4	5	6	10
$I_{(\tau)}$	1,443	1,021	695.1	322.1	149.2	69.15	32.04	14.85	0.684

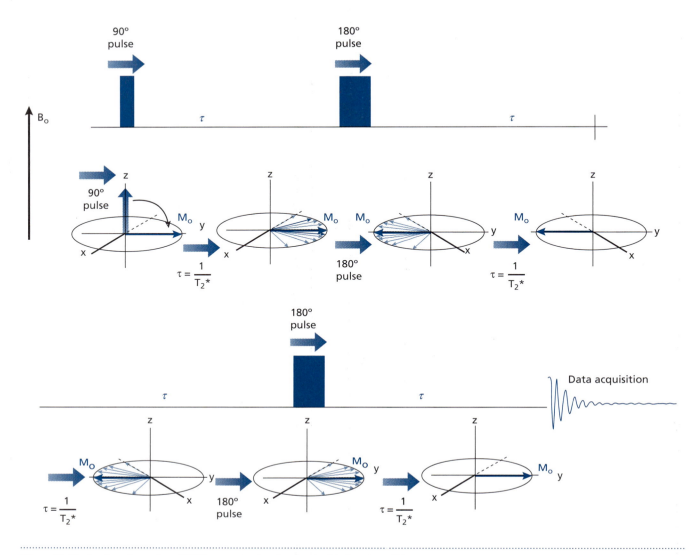

Figure 21.12 A pulse scheme for a CPMG sequence showing the double focusing of M_o.

21.4   The Influence of Nuclear Neighbors

Now that you have a good understanding of the behavior of a nucleus when under the influence of an externally applied magnetic field (B_o) and how that behavior results in a measurable signal when the nucleus is subjected to an RF pulse, let us expand our discussion to consider how the nuclear behavior is affected by neighboring nuclei. Up to this point, we have been considering and evaluating a single nucleus (or a set of identical or equivalent nuclei with the same resonant energy). As we discussed in Chapter 14, the resonance energy will be near the Larmor frequency of the nucleus type under study (e.g., ^1H, ^{13}C) but will be shielded to some extent by the electrons (bonding and nonbonding) that surround it and influenced by other nuclei nearby. This results in the production of a unique chemical shift (the value, typically in parts per million [ppm], along the x axis of the NMR spectrum) for each nuclear environment in a molecule. These chemical shift data are our first clue about molecular structure. Chemical shift is an indication of shielding; peaks that are more "upfield," closer to 0 ppm, are more shielded (may be associated with electron-donating groups), whereas peaks that are more "downfield," further from 0 ppm, are less shielded (may be associated with electron withdrawing groups).

J-Coupling

NMR is also affected by the relationships that the spinning magnetic nuclei have to each other—through bonds and through space. The nuclear magnetic moments (like tiny bar magnets) will have small effects on each other and push or pull the resonance frequency away from its specific chemical shift. We call this *spin-spin coupling*.[11] Here we will consider *indirect spin-spin coupling* or *scalar coupling*, also known as *J-coupling*. These are the *influences nuclei have on each other that are mediated through the bonds connecting them*.

You are already familiar with the application of this signal information through the n + 1 rule, which we reviewed for propane in Chapter 14. For the purposes of this chapter, we will use a more conventional and introductory approach to understanding J-coupling through the n + 1 rule and the use of splitting diagrams. However, you should be cognizant of the fact that J-coupling is essentially an energy term in the Hamiltonian that describes the total energy of the system; J-coupling may contribute to the total energy in a positive or in a negative sense. Likewise, the value of the J-coupling contribution will depend on the nuclei involved and their relative positions to one another. Thus, the entirety of what is described in this section could be more rigorously approached through a quantum mechanical model. We will come back to a more energy-based consideration a bit later in the chapter when we introduce the nuclear Overhauser effect in the context of dipole–dipole coupling.

J-coupling arises because of the numbers of ways in which the spins can couple or associate with each other. Let us revisit the NMR spectrum of propane and analyze how the *a*-hydrogens are affected by their proximity to the *b*-hydrogens[12] (see Example 21.3).

> **Problem 21.12:** Repeat the analysis of J-coupling from Example 21.3 for a *b*-hydrogen in propane.

Example 21.3 shows us that the number of possible interactions a unique proton can experience is related to the number of coupled neighboring protons by the relationship:

$$M = n + 1 \qquad \textbf{Eq. 21.8}$$

where
   M = total number of unique dipole orientations (multiplicity)
   n = number of J-coupled protons (neighbors)

Furthermore, the relative intensity (height of the peak) of the J-coupled multiplet is a result of the total number of microstates that are possible for a specific orientation. This is also demonstrated by the figure in Example 21.3; frame II shows that there are two possible ways to achieve a combination in which the two *b*-protons are antiparallel relative to each other, and frames I and III show that there is only one possible way to achieve a microstate in which the two *b*-protons are parallel relative to each other. The result is a 1:2:1 triplet splitting of the *a*-hydrogens. So when we see a triplet with relative intensities of 1:2:1 in an ^{1}H-NMR spectrum, it indicates that the hydrogen producing that signal is coupled to two neighboring protons (*a*-CH_2-).

---

[11] Not to be confused with spin-spin relaxation (T_2).

[12] Recall from Chapter 14 and as demonstrated in Example 21.3, it is *very* common for NMR spectroscopists to interchange the words "protons" with "hydrogen nuclei" as synonymous terms.

Example 21.3 An analysis of J-coupling for the hydrogen nuclei in propane.

We are asking the question, "how might the dipole moments of the *b*-protons affect the magnetic environment of the *a*-protons?" In this example analysis, we consider one *a*-proton, represented by the blue arrow.

STRATEGY –

Using arrows to represent nuclear dipole moments, draw all of the possible orientations of the *b*-hydrogens relative to the *a*-hydrogen's magnetic moment.

SOLUTION –

What we discover from this analysis is that the *b*-proton dipole moments can align in three possible combinations relative to the *a*-proton's dipole moment. The three combinations are:

1. Both of the *b*-hydrogens' dipole moments are parallel to the *a*-hydrogen's dipole moment (Frame I).

2. Both of the *b*-hydrogens' dipole moments are anti-parallel to the *a*-hydrogen's dipole moment (Frame III).

3. One of the *b*-hydrogens has a dipole moment that is aligned parallel with the *a*-hydrogen and the other *b*-hydrogen has a dipole moment that is aligned anti-parallel to the *a*-hydrogen's dipole moment. There are two possible ways this orientation can be formed (Frame II).

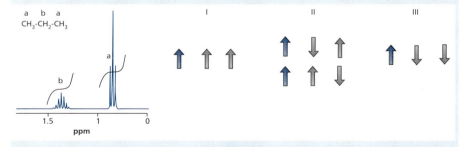

The result is that the *a*-proton's chemical shift gets split into a *triplet* with the relative peak heights proportional to the number of microstates that exist for that particular dipole-dipole orientation; in this case 1:2:1.

A useful mnemonic[13] for obtaining the splitting pattern and relative intensities is a coupling tree diagram (Figure 21.13), which shows the splitting and relative intensities of each peak. Consider, for example, a coupled system where nucleus A is coupled to two equivalent neighbors, X; this is an AX_2 spin system.[14] Our n + 1 rule tells us that A will be split into a triplet, whereas our X will be split into a doublet. Can you explain these results in terms of a tree diagram?

---

[13] The useful memory device here is Pascal's triangle. It starts with 1 at the top. Carry down the 1 to the second row, then sum the numbers as you bring them down to create the third and lower rows. This chart can be reproduced using the summing trick without thinking through all of the possible combinations of spins.

[14] Spin system nomenclature follows the Pople convention, where equivalent nuclei are given the same letter of the alphabet and nonequivalent nuclei are given letters based on their differences in chemical shift. This naming convention bears the name of mathematician and theoretical chemist John Pople, who won the Nobel Prize in Chemistry in 1998. Letters used tend to be A, B, M, N, X, and Y. An AX_2 system has two equivalent nuclei (both X) and a nonequivalent nucleus (A) that has a chemical shift very different from the X nuclei.

Example 21.4

Use Figure 21.13 to predict the splitting pattern of an *a*-hydrogen in glutaric acid, shown below.

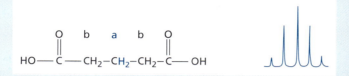

STRATEGY –

Looking at the chemical structure and connectivity, we see that an *a*-proton has four equivalent neighboring protons, labeled b. A proton coupled to four equivalent protons will get split into a multiplet, where n + 1 = 4 + 1 = 5. We can work through the number of possibilities of each using arrow diagrams (as demonstrated in Example 21.3) or consult Figure 21.13 to determine the ratios.

SOLUTION –

You should determine that multiplet will be a quintet and that the relative intensities of peaks across the quintet will be 1:4:6:4:1.

Problem 21.13: Using the J-coupling tree diagram (Figure 21.13), sketch the antici-pated splitting pattern for the *b*-protons from Example 21.4.

J-coupling constants have values in ppm (or hertz) and split the peaks around the resonance at the chemical shift. Thus, if the A nucleus had a chemical shift of 4.3 ppm and it was coupled with four identical X nuclei that influenced a 0.1 ppm coupling con-stant, the peak would split into a quintet centered on 4.3 ppm, and each peak of the quintet would be separated by 0.1 ppm. (See Figure 21.14.) Likewise, and helpfully, the J-coupling constant will continue to be 0.1 ppm for the other nucleus (X) which, when coupled to the neighboring A nucleus, would result in a doublet centered on the

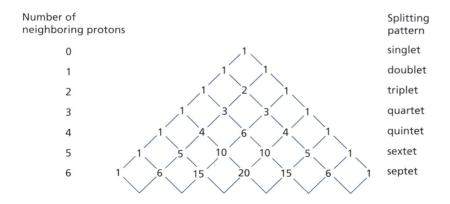

Figure 21.13 J-coupling tree showing the splitting pattern and relative intensities of J-coupled protons. On the left, find the number of neighboring protons for the nucleus of interest. Then go across to the right to find the splitting pattern; as you cross the coupling tree, you will identify the relative intensities of the peaks.

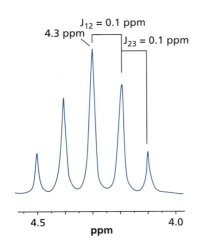

Figure 21.14 A quintet centered at 4.3 ppm showing the J-coupling to each peak, where J_{AX} is 0.1.

chemical shift of nucleus X and having a 0.1 ppm difference between the two peaks of the doublet. Finding the matching J-coupling constants in an NMR spectrum is another clue that helps spectroscopists identify structures.

Before we leave this topic, let us look at one last system, a system that is a bit more complicated: the AMX spin system. In this system, there are three nonequivalent nuclei coupled to each other and all three nuclei have very different chemical shifts. In this case, there are three coupling constants to be considered: J_{AM}, J_{AX}, and J_{MX}. J_{AM} is the coupling between A and M, J_{AX} is the coupling between A and X, and J_{MX} is the coupling between M and X.

There are many ways in which an AMX system can be connected and coupled. For the purposes of this discussion, let us say that X and M have very weak coupling (J_{MX} is very small) and that J_{AM} is greater than J_{AX}. In considering the splitting for the A nucleus, the coupling of A to M is considered first because it is the strongest. The coupling of A to M will produce a doublet with a coupling constant of J_{AM} (see Figure 21.15). The A nucleus will also couple with X, producing a doublet of each of the original doublets with a coupling constant of J_{AX}. The final result is a doublet of doublets centered about the chemical shift of A.

It is not the intention of this chapter to cover the complexities of NMR spectral analysis. Much of that discussion occurred in your organic chemistry course, and there are entire chapters and books devoted to spin-spin coupling in complex molecular structures and their spectral analysis. We will leave it to you at this point to follow up on particular systems of interest, and we have provided some recommended reference texts in the further reading section at the end of this chapter.

Problem 21.14: Using Figure 21.15 as a reference and model, sketch the expected splitting pattern for the M nucleus. Assume that J_{MX} is 0.05 ppm and that the peak is centered at 3.5 ppm.

Problem 21.15: Label each peak in the ^{1}H-NMR spectrum of ethanol (see Figure 21.16). Explain your rationale for the peak assignments using expected J-coupled splitting patterns. What are the chemical shift values (in hertz) for the J-coupling constants? Recall that for a 200 MHz NMR, 1 ppm is 20 Hz.

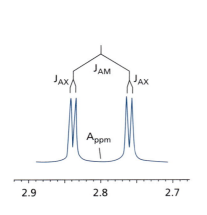

Figure 21.15 A doublet of doublets produced by a J_{AMX}-coupled system.

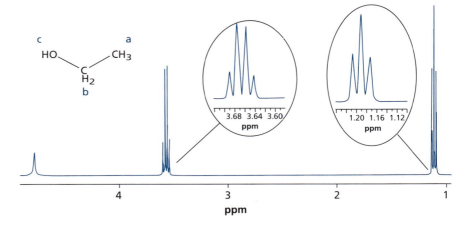

Figure 21.16 ^{1}H-NMR of ethanol taken on a 200 MHz NMR.

Dipolar Coupling and the Nuclear Overhauser Effect

The *nuclear Overhauser effect* (NOE; also known as a *nuclear Overhauser enhancement*) also results from spin-spin coupling but is not mediated by bonds. In this case, the spin-spin coupling effect is direct (nucleus to nucleus) *through space* in the form of *dipolar coupling.* The coupling arises from the effect that one nucleus has on another because of their proximity in space. Thus, the nuclei under consideration have to be close together but are not necessarily bonded. For this discussion, let us simplify things by considering J to be zero—there is no scalar coupling. This does not have to be the case, but it will make the explanation more straightforward.

Let us think again about our AX spin system (but now $J_{AX} = 0$, so each peak in the AX spectrum would be a singlet). In such a system, there are four possible ways we can spin-spin couple (four energy levels): up-up, up-down, down-up, and down-down (see Figure 21.17).[15]

Figure 21.17 shows two possible pathways for dipole-coupled nuclei to transition from a state where both nuclei are spin aligned with B_o to a state where both nuclei are spin flipped with respect to B_o. The pathway on the left side of Figure 21.17 shows the A nucleus flipping first followed by the X nucleus; and the pathway on the right side of Figure 21.17 shows a pathway in which the X nucleus has flipped first, followed by the A nucleus flip.

In addition to the transitions shown in Figure 21.17, there are two more possible transitions: the double quantum transition (where $\alpha\alpha \rightarrow \beta\beta$) and the zero quantum transition (where $\alpha\beta \rightarrow \beta\alpha$). See Figure 21.18. In the case of the double quantum transition, $\Delta m = 2$ and is spectroscopically forbidden[16] in terms of excitation. However, the relaxation pathway is allowed (possibly even preferred). And because NMR is based on resonance, and the signal is observed as the system relaxes back to equilibrium, so long as the population of the two states are different, a signal can be observed.

An NOE is induced by eliminating the population difference for one of the nuclei in a system (for example, A in an AX system). We do this by *saturating* one nucleus with energy at its Larmor frequency so that the population of the higher state equals the population of the lower state for that nucleus.[17] Thus, the signal (both the excitation and the relaxation) for that nuclear transition is eliminated. However, in a coupled system, if we equalize the population of the A nuclei in the ground and excited states, the X nuclei populations are also affected. We will use an energy-level description. (Refer to Figures 21.18 and 21.19 as we work through this explanation.) If we transmit a continuous signal through the sample of $E = \Delta E_{A_{\alpha\rightarrow\beta}}$, we will equalize the population of A_α and A_β. This is represented in Figure 21.19 by the blue lines. This will in turn enhance the population of the two A_β states (circled in Figure 21.19). Because the analytical signal is the emission of RF energy as the excited state relaxes to the ground state, the $A_\beta X_\beta$ state can relax by the double quantum transition seen in panel

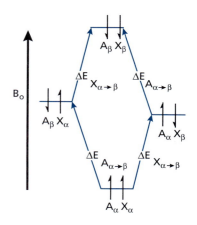

Figure 21.17 Possible spin states relative to an external magnetic field (B_o). At the bottom of the figure is a representation of spin up (+½) for each nucleus (AX). At the top of the figure is a representation of spin down (−½) for each nucleus (AX). ΔE for each level represents the Larmor frequency ($\Delta E = h\nu$) needed to obtain the indicated spin flip.

---

[15] Although Figure 21.17 may resemble a molecular orbital diagram, the resemblance is only superficial. The large arrow on the ordinate represents the external magnetic field and the small arrows represent the relative spin orientation of two magnetically coupled nuclei. The implication is that the relative energy of the coupled spin states increases as you move up the ordinate. In other words, $A_\beta X_\beta$ is your highest energy spin state. The Pauli Exclusion principle does not apply in this case.

[16] Δm is the change in spin quantum number. The change +½ to −½ is a Δm of −1 and the change −½ to +½ is a Δm of +1. Spectroscopically allowed transitions have Δm values of +1 or −1. The double quantum transition is +½ and +½ going to −½ and −½; therefore Δm is 2, and this transition is not allowed.

[17] On-resonance selective pulses, also known as "soft" pulses, are applied at a very specific frequency, at low intensity, and over a longer period of time. They have the effect of stimulating only the very particular nuclei that resonates at that very particular frequency. This is in contrast to the pulses we have looked at earlier in this chapter, which are nonselective or "hard" pulses that are broader in frequency, applied at high intensity, and are very short in duration. Hard pulses have the effect of stimulating all of the resonances in a region of frequencies. Using our wind chime analogy, a strong burst of wind, which sets all the chimes in motion at once, is like a hard pulse. A gentle tapping at the top of the wind chime at a specific resonance could get one of the chimes to begin to ring in resonance.

Albert Warner Overhauser (1925–2011) was a US physicist best known for his development of spin-polarization theory (the Overhauser effect). He obtained his doctorate in physics in 1951 from the University of California, Berkeley, and developed his theories of nuclear spin polarization while a postdoctoral student at the University of Illinois (1951–1953). Overhauser was on the faculty at Cornell University from 1953–1958 and left to work for the Ford Motor Company until 1973, when he took a faculty position at Purdue University. He remained at Purdue University for the remainder of his career. He was elected to the National Academy of Sciences in 1976 and received the National Medal of Science from President Clinton in 1994. ■

(A) of Figure 21.18 and will proceed at a rate determined by k_1. By increasing the population of the $A_\beta X_\beta$ state, we will see an enhancement of the X nucleus signal as the X nucleus relaxes from $X_{\beta \to \alpha}$. Alternatively, the $A_\beta X_\alpha$ state can relax by the zero quantum transition seen in panel (B) of Figure 21.18; with respect to the X nucleus, this is an *absorption* event ($X_{\alpha \to \beta}$) and results in a decrease of the X-nucleus signal. During an NOE experiment, the population of X is affected by either the double quantum or zero quantum pathways (or both).

Depending on which pathway is affected the most, the effect can be positive (an enhancement) or negative (the signal is decreased). However, in either case, the signal for X has been affected by the saturation of A. This effect is dependent on the extent to which the two nuclei are dipole coupled, which *can be measured as a distance effect through space*. Thus, the magnitude of the NOE is a function of the distance between the two nuclei[18] and can be used to elucidate three-dimensional structure of complex molecules in situ.

NOE information is obtained through a *difference spectrum*. The experiment is conducted twice, once without saturating one of the nuclei and again with saturation. The

---

[18] Saturation of one nucleus (A) causes a change in the intensity of the signal of the dipolar-coupled nucleus (X), which is related to the inverse sixth power ($1/r^6$) of the distance between A and X. However, this relationship is rather complex, and a routine conversion cannot be applied across all systems. In general, a qualitative comparison is the best use of the results.

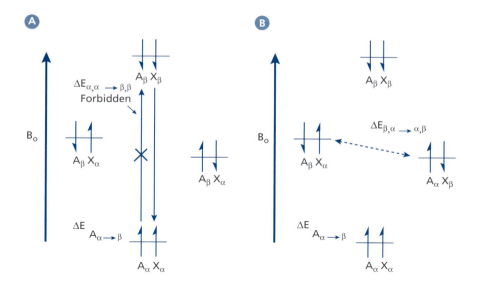

Figure 21.18 Panel (A): Double quantum excitation is quantum mechanically forbidden. However, the relaxation event is allowed. Panel (B): The zero quantum transition is likewise forbidden ($\Delta m = 0$) but may be used as a route of relaxation.

free induction decay (FID) data for the two experiments are then subtracted one from the other and then transformed via Fourier transform analysis to create the difference spectrum. In the difference spectrum, the signal from the saturated nucleus will be inverted, signals from nuclei that are not affected by the NOE will disappear, and signals that are affected by the NOE will either appear as positive or negative signals depending on which mechanism in Figure 21.18 dominated the relaxation process. There will also typically be some signals that appear out of phase, which are remnant of less-than-perfect spectra subtraction. NOEs are reported as percentage enhancements.

Within a single molecule, signals with larger percentage enhancements can be identified as closer in space to the saturated nucleus. The following example demonstrates. An example of a one-dimensional NOE is seen in Figure 21.20. The top panel shows the normal ^{1}H-NMR spectrum for ethyl methacrylate taken on a 200 MHz NMR. By saturating the A protons (the methyl group), the investigators were trying to unambiguously assign the two proton signals X and X′ to the peaks at 5.54 and 6.1 ppm, respectively. The bottom panel shows the difference spectrum for the one-dimensional NOE experiment with the methyl group saturated. Note the complete inversion of the A signal at 1.95 ppm. The inset for the NOE spectrum zooms in on the 5.0 to 6.5 ppm region (olefinic region). A 9% NOE is shown for the proton at 5.54 ppm, which must therefore be the H in the *cis* position (closer to A in space), whereas the NOE for the *trans* proton, X′, at 6.1 ppm is indeed much smaller, confirming its further distance from A.

The NOE is a dipolar interaction and is described as a dipolar coupling. The NOE is measured as a direct link or force of influence from one nucleus to another nucleus without interference or contributions from the electrons. The NOE does *not* require the two nuclei to be connected through bonds, only that they be close enough in space to have influence over one another.

These dipolar interactions diminish quickly over distance; thus NOEs are limited to distances that are less than 6 Å. Also, theory predicts that for small molecules, the maximum NOE effect that can be seen for a homonuclear system is 50%, whereas heteronuclear NOE effects can be much greater and are dependent on the magnetogyric ratios (γ) of the nuclei involved.

Interestingly, for very large molecules having very large molecular weights, such as proteins, the NOEs are negative and can be large (up to ~100% enhancement). This

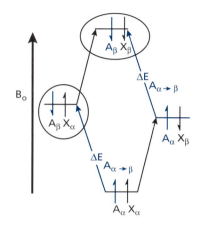

Figure 21.19 Saturation of the A nucleus.

Jean Jeener (b. 1931) is credited with inventing two-dimensional nuclear magnetic resonance in 1971, when he first proposed the pulse scheme and double Fourier technique we now refer to as COSY (see Section 21.5). Professor Jeener retired from the Université Libre de Bruxelle (Belgium) in 1996, where he had been on the faculty for 36 years. ■

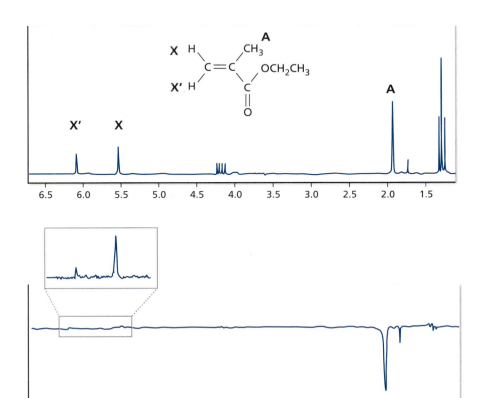

Figure 21.20 ¹H-NMR of ethyl methacrylate (top) and one-dimensional NOE with the methyl group (A) saturated (bottom).

is useful for structural determination of proteins and other large biological molecules, especially when combined with two-dimensional and multinuclear pulse schemes. We will touch on this again later in the chapter.

We have limited this discussion to spin ½ nuclei and, because both ^{1}H and ^{13}C are spin ½ nuclei, this encompasses the vast majority of NMR analysis. Spin ½ nuclei are spherical in terms of their charge distribution around the nucleus, and as a result there is no quadrupolar coupling in these systems. However, you should be aware that, for the many NMR active nuclei with spins greater than ½, the unequal distribution of charge around the nucleus can have coupling effects with the electric field gradients in the instrument. These quadrupolar coupling effects can cause spectral broadening. Because of the magnitude of the effects, the information about chemical shift coupling, J-coupling, and dipolar coupling can be lost.

21.5 Introduction to Two-Dimensional Nuclear Magnetic Resonance

Simple, traditional, or *one-dimensional*, NMR experiments are sufficiently well resolved to obtain useful information for small molecules. However, as molecules of interest become increasingly more complex, their NMR spectra become overcrowded and difficult to interpret in a simple one-dimensional axis. Resolution can be improved by moving to higher and higher fields, but there is a practical limit in terms of cost and access.

Two-dimensional, three-dimensional, and higher dimensional NMR experiments are instrumental approaches to solving the resolution problem for larger molecules. In these *multidimensional techniques*, the NMR signals are spread out from the single one-dimensional axis into two or more dimensions. In this section, we will focus on a few examples of the major experimental two-dimensional techniques and help you develop a general sense of the multidimensional aspect of the experiment.

Two-dimensional NMR (2D NMR) experiments involve magnetic correlations—between spins both through bonds and through space (i.e., NOE)—detected along a new spectral dimension. The peaks in the two-dimensional spectrum arise from these correlations and allow the analyst to determine the proximity of a nucleus to its neighboring nuclei. Because the signals are spread out in two dimensions, resolution is greatly enhanced, which improves our ability to assign peaks along each of the one-dimensional axes. Experiments are either *homonuclear*, involving correlations between nuclei of the same type (e.g., proton–proton) or *heteronuclear*, involving correlations between nuclei of different types (e.g., proton–carbon).

What we call a 2D NMR spectrum actually has three dimensions (see Figure 21.21). These spectra have one dimension in ppm along the axis of one nucleus and one dimension in ppm along the axis of other nuclei, and they also have a height dimension (signal strength). What we see as a typical two-dimensional spectrum is a topographical map of that three-dimensional space (a three-dimensional *contour map*). Like viewing a range of mountains from above, peak elevation is represented by a series of concentric circles—or sometimes a color-coded height scale.

The analyst can adjust the scale of the view to move the viewing range higher or lower, as needed. The view can also be adjusted by zooming in on particular regions of the entire spectral space. Typically, you would want to adjust your view so that the signals of interest are resolved, but not so low (on the topographical plane) that too much noise starts to clutter the spectrum. With computer technology, these perspectives are as easy to adjust as a mouse click.

2D NMR experiments are characterized by pulse sequences that collect data as a function of two different time domains: often a delay time between pulses and the amount of time you collect your FID. The Fourier transform converts both time dimensions into frequency dimensions. Thus, two-dimensional experiments include pulses that establish a time domain for the evolution of the signal and a time domain for the FID. The vocabulary here is a little confusing, because it is common to use the terms

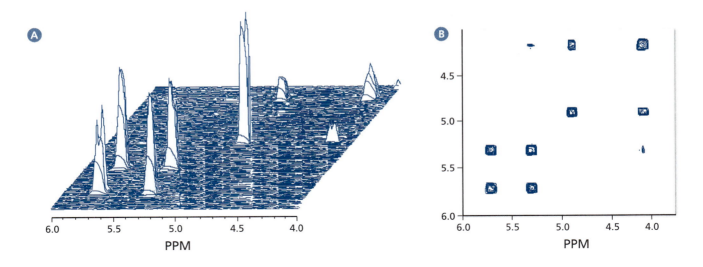

Figure 21.21 Example of two-dimensional correlation spectroscopy. Panel (A) shows a stacked plot of a 2D NMR COSY spectrum in which the frequency domains are plotted along the *x-y* axes and the intensity is plotted along the *z* axis. Panel (B) shows the same data as a contour plot and the intensity of each peak is indicated topographically.

t_1 and t_2 for these two time domains despite the use of the same terms for T_1 and T_2 relaxation. However, t_1 and t_2 are conceptually different from T_1 and T_2 relaxation and should not be confused with them. The two time domains can be labeled as the *evolution period* and the *acquisition period*, respectively.

This might be easier to think of as a series of experiments in which t_1 is varied sequentially. Figure 21.22 shows a pulse sequence for a two-dimensional experiment. By varying t_1 prior to the acquisition time, we create a series of FID data as a function of t_1. This particular pulse sequence is for an inversion recovery experiment. As particular nuclei come into and out of resonance, the J-coupling to the nearest neighbors will likewise vary with t_1. There are hundreds of two-dimensional pulse sequences, and this is just one example.

Together, the data set is a matrix[19] in both t_1 and t_2. When transformed in the first dimension (t_1), an interferogram results in the form of a function that has both a frequency domain and a time domain.[20] Next the function is transformed again such that the spectrum is a function of two frequency domains[21] and the signals identified as coordinates associated with both frequency 1 and frequency 2 (see Figure 21.23).

Correlation Spectroscopy and Total Correlation Spectroscopy

The 2D NMR technique called *correlation spectroscopy* (*COSY*; pronounced "cozy") is a relatively simple two-dimensional experiment in terms of data acquisition. A related technique, *total correlation spectroscopy* (*TOCSY*; pronounced "tocksey" and rhymes with "oxy"), takes advantage of a similar correlation. Therefore, we will discuss them together. Both COSY and TOCSY[22] experiments are designed to exploit magnetic correlations between nuclei that are connected through bonds. In COSY, correlations that connect nuclei can be measured for one to three bonds in terms of distance. In TOCSY, the correlation distance can be increased through as many as five or six bonds.

---

[19] A matrix in the form of $s(t_1, t_2)$.

[20] A matrix in the form of $s(t_2, \omega_1)$.

[21] A matrix in the form of $s(\omega_1, \omega_2)$.

[22] TOCSY is also known as a HOHAHA experiment, which comes from homonuclear Hartmann Hahn.

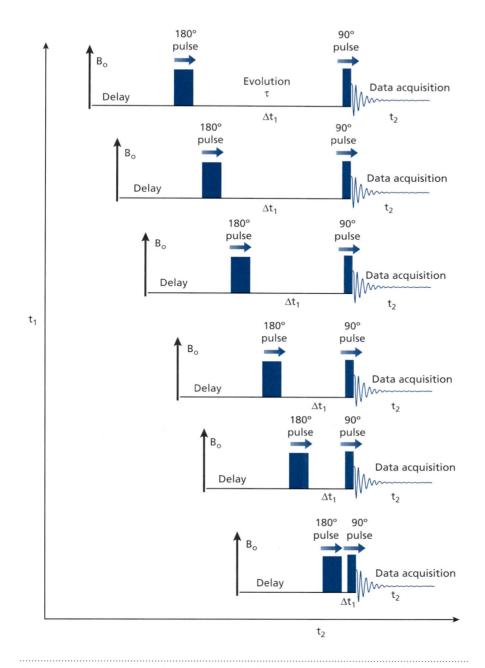

Figure 21.22 A pulse sequence for a two-dimensional experiment. Note that t_1 is varied as a change in the evolution time and t_2, the acquisition time, encodes all of the frequency data as an FID. Together, these two dimensions are used to develop the two-dimensional contour plot of the spectrum.

The pulse scheme for a COSY experiment (Figure 21.24) should look somewhat familiar. (You may want to review the information on pulse schemes from Section 21.2 and/or explore this topic in more depth by consulting with some of the resources listed in Further Reading (Section 21.8) at the end of this chapter.)

In this sequence, the first 90° pulse flips the magnetic vector into the *x-y* plane, where it continues its precession as it decays. The evolution time, τ, in this experiment is varied (creating the second dimension of the experiment) to allow for correlations of different types to "evolve" in the data set during that relaxation time. A second 90° pulse moves the evolved signals back into the *x-y* plane for maximum signal to be measured as the FID for each of the excited nuclei.

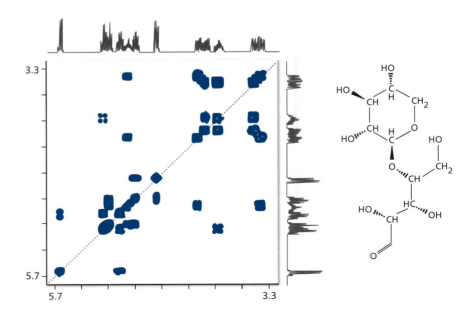

Figure 21.23 A two-dimensional ^{1}H-^{1}H COSY NMR of disaccharide xylobiose with its structure shown. The diagonal peaks are emphasized with the dotted line.

In a homonuclear experiment, the resulting three-dimensional contour map will include a set of peaks that lie on the diagonal of the x-y plane. These diagonal peaks represent the one-dimensional NMR spectrum and correspond to the peak positions where the chemical shift along x equals the chemical shift along y. Typically, these are of little use for analysis and are ignored. The off-diagonal peaks (also known as cross peaks) are the signals produced by the magnetic correlations, and those are identified and analyzed in terms of position and volume. They are assigned to the molecular structure, helping to identify nuclei that are connected through bonds. Spectral symmetry is expected; the part of the spectrum above the diagonal should be identical to (by symmetry) the section of the spectrum below the diagonal. You can use a processing function on the NMR computer to enhance the spectral symmetry so that, from an analysis perspective, you obtain the best of both sections.

The signal strength of the correlations provides additional information about the correlation. The closer the two nuclei are, the stronger the signal will be. Thus, in addition to being able to identify the connectivity of the nuclei, we can also use the information to determine how closely they are connected. Figure 21.25 shows a ^{1}H–^{1}H COSY of 2-butanone. The ^{1}H spectra are displayed on the x and y axes, and the two-dimensional contour spectrum is plotted in the middle of the graphic. You can clearly see the one-dimensional contour spectrum running along the x-y diagonal, and the cross peaks between the b-protons and the c-protons have been emphasized using gray lines. The spacing of the contours represents the peak intensity. The cross peaks, seen between the b-protons and the c-protons, tell us that these protons are J-coupled. In other words, spectral interpretation allows us to deduce that these protons reside on adjacent carbon atoms. We do not see any cross peaks for the a-protons because they are isolated by the ketone and are not J-coupled to any neighboring protons. TOCSY experiments provide similar information and allow the spectroscopist to examine longer distance correlations.

In processing a 2D NMR data set, the analyst measures peak areas using an integration function, a process known as "peak picking," which is supported by the computer software. In practice, the spectroscopist isolates the peak of interest in the spectrum by zooming in, places a set of cursors around the peak (boxing it in), and presses

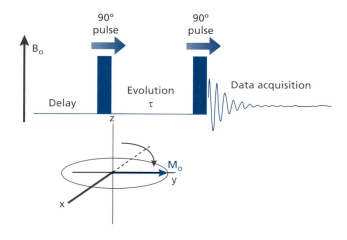

Figure 21.24 Pulse sequence for a COSY experiment.

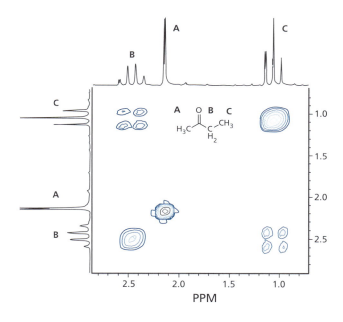

Figure 21.25 ^{1}H-^{1}H COSY of 2-butanone. The cross peaks between the *b* and *c* signals allow you to deduce that these two signals were produced by protons on adjacent carbon atoms.

a command function to determine the *peak volume*. Note that in a two-dimensional experiment, peaks are integrated in terms of volume (three-dimensional space). Peak *areas* are determined in a simple, one-dimensional spectrum. Computer-generated data tables list the chemical shift coordinates (ppm in x and ppm in y for the center of the peak) and the measured volume.

Nuclear Overhauser Effect Spectroscopy

For complex molecules, the COSY/TOCSY information is not sufficient to fully understand three-dimensional molecular structure. In particular, biomolecules such as proteins, carbohydrates, lipid assemblies, DNA, and RNA have additional levels of three-dimensional structure due to their specific folding patterns and structural details that are important for binding and recognition.

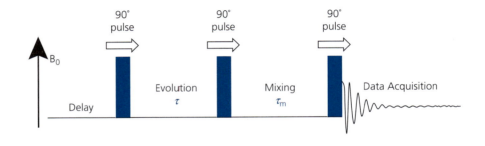

Figure 21.26 Pulse sequence for a NOESY experiment.

Thus, to obtain a complete picture of the molecular structure, we have to understand how the molecule is organized in three dimensions. We need to understand which nuclei are close together—not only in terms of bonds but also in terms of space. These through-space correlations are provided by the NOE. You may want to refer to Section 21.4 for an overview of NOE. The 2D NMR experiment that provides NOE correlations between nuclei is called *nuclear Overhauser effect spectroscopy* (NOESY; pronounced "nosey" and rhymes with "cozy").

A pulse sequence for a NOESY experiment is seen in Figure 21.26. However, this is not the entire RF story. The NOESY experiment is more complicated because we have to add in an RF pulse that will saturate one of the nuclei (eliminating the population difference between the high-energy and low-energy spin states) in order to stimulate the NOE at the coupled nucleus. A series of 90° pulses generates the desired correlations.

It is beyond the scope of this book to delve into the spin dynamics that result from the NOESY pulse sequence. However, by now you should be able to understand that the net magnetization is being flipped from the z axis to the x-y plane, back again to the z axis, and finally again into the x-y plane. The time delays are providing the correlations; the first time, τ, is performing the same function as in the COSY experiment. It provides an "evolution" of the differences among the spins and is varied to provide the two-dimensional component to the experiment.

The second delay in the NOESY experiment, τ_m, is called the *mixing time* and can be varied from experiment to experiment as one of the experimental parameters. Longer mixing times allow for more signals to "report," but the relationship between signal intensity and distance is diminished. In contrast, shorter mixing times limit the data in terms of the number of signals, but each one has a stronger distance-to-volume relationship. Mixing time is also related to the size of the molecule and can be optimized experimentally. The spectroscopist decides how to set up the experiment for the type of analysis and information needed about the structure.

In the homonuclear NOESY experiment, we see that (1) the diagonal peaks again represent the one-dimensional experiment and (2) the off-diagonal peaks result from the NOE-coupled nuclei. The signals are processed as described earlier to generate a table of NOE volumes, which correspond to each set of coupled nuclei. The NOE volumes correspond to distances, with large volumes corresponding to close distances and small volumes corresponding to longer distances.

Together, the COSY/TOCSY experimental data, along with the NOESY data, can be combined into an information-rich data set that can be interpreted to reveal the three-dimensional structures of some very complicated molecules. Spectroscopists, and their colleagues in computational chemistry, import the NMR data into computational programs that use the data to determine the lowest energy conformation of the molecular structure. By providing a reasonable starting structure and information about the atomic parameters (e.g., type of nucleus, van der Waals radii) and connectivity, the NMR data provide additional structural information and limitations (constraints) that are used to calculate the structure of a best fit. These NMR solutions to structure can be compared to alternative descriptions, such as those derived from crystal structures through X-ray analysis, to provide a more complete picture of the structure of the molecule under the solution conditions of the sample in the NMR tube. These solution conditions can be made to resemble those in vivo, thus providing a solution-state structure for the molecule. Furthermore, the NMR timescale can also provide information about motion within the structure under the solution conditions vis-à-vis areas of the structure that are less well resolved from the data and computational results. This information is very useful for understanding how biomolecules move, which is important for understanding their functions and function–structure relationships.

However, before we leave this topic, let us examine the scope of information that can be obtained in a 2D NMR experiment by reviewing a couple of heteronuclear 2D NMR experiments. *Heteronuclear single quantum coherence* (HSQC) spectroscopy is a very powerful heteronuclear technique that relies on J-coupling between ^1H and a heteronucleus, such as ^{13}C. Along one dimension is the ^1H spectrum and along the second dimension is the ^{13}C spectrum. Cross peaks are observed at resonances where the ^1H and ^{13}C nuclei are coupled through bonds. Note that there is no diagonal spectral line in a two-dimensional heteronuclear NMR spectrum. A related technique, used especially for the determination of protein structures, is *heteronuclear multiple quantum coherence* (HMQC). For protein analysis, HMQC spectroscopy is used to correlate ^{15}N with ^1H to obtain bond connectivity information. Again, because there is no diagonal, all the data are used in terms of correlations. The pulse sequence transfers magnetization from ^1H to ^{15}N nucleus to boost the signal from the ^{15}N nuclei through a process called *cross polarization*.[23] The ^{15}N nucleus has a low gyromagnetic constant and therefore weak signals, but ^1H is much stronger. By enhancing the correlation between a coupled H–N pair, we can measure the N signal. Another pulse scheme, *heteronuclear multiple bond coherence* (HMBC), is designed for long-range heteronuclear couplings. This process is started by a pulse sequence called *INEPT*, insensitive nuclei enhancement by polarization transfer, which enhances the signal of the ^{15}N nuclei through the transfer of the energy of the ^1H. If you are interested in studying the pulse schemes that provide this type of energy transfer in NMR, there are a number of good references in Further Reading (Section 21.8) at the end of this chapter. We can see from the two-dimensional HSQC spectrum (Figure 21.27) that the detail of information provided would be very useful for study in a complex system, especially for proteins where the amino acid backbone is composed of a repeating pattern of amide (N–H) linkages.[24]

The use of multidimensional NMR allows us to gain detailed information about the connectivity and spatial arrangement of the atoms in our analyte molecule, and this in turn leads to better models of the in situ structures of our analyte. Many multidimensional NMR experiments can be used to obtain additional, useful information about structure. For example, a two-dimensional technique known as *diffusion ordered NMR spectroscopy* (DOSY) helps researchers learn more about the transport properties of molecules and ions, contributing information on size, mobility, and diffusion rates. Diffusion coefficients can yield valuable information about translational motion of particles in a mixture. The technique has found use in pharmaceuticals, polymers, and in biological fluid analysis. Unfortunately, an overview of more useful and interesting multidimensional NMR experiments is best left to a course dedicated to NMR spectroscopy.

---

[23] To take advantage of the difference in population between the two nuclei, the net magnetization of the more populated state (the ^1H in this example) can be transferred to the less populated state (the ^{13}C) in order to observe the ^{13}C nuclei more easily. This cross polarization, or CP, is carried out electronically through the pulse scheme of the RF emitter.

[24] Phi (Φ) and psi (Ψ) values are dihedral angles that define the protein backbone conformation. Repeating patterns of Φ and Ψ define major secondary structures, such as helices and β-sheet conformations, and also deviations from those, such as turns or irregular structures.

PROFILE

G. Marius Clore

G. Marius Clore received the 2011 Centenary Prize from the Royal Society of Chemistry for his work pioneering computational chemistry coupled with multidimensional nuclear magnetic resonance (NMR) data. His work has greatly expanded our ability to characterize biological macromolecules using NMR. Clore is a distinguished research fellow at the US National Institutes of Health and holds dual British and US citizenship. His laboratory studies the dynamics and structure of proteins, protein–protein complexes, and protein–nucleic acid complexes. Clore's work has led to an expanded understanding of genetic signal transduction and transcriptional regulation as well as AIDS and AIDS-related proteins. ◼

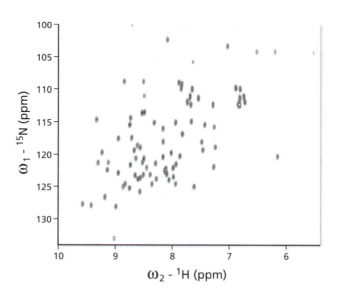

Figure 21.27 2-D NMR, ^{1}H-^{15}N HSQC spectrum of the protein ubitquitin.

Kurt Wuthrich was awarded the 2002 Nobel Prize in Chemistry for his work in "developing multidimensional NMR techniques based upon the nuclear Overhauser effect (NOE) for the determination of the three-dimensional structure of biological macromolecules in solution." He received his PhD from the University of Basel, Switzerland, in 1964. He presently maintains a research laboratory at The Scripps Research Institute in La Jolla, California, and a laboratory at ETH Zürich, the Swiss Federal Institute of Technology. ▬

http://en.wikipedia.org/wiki/File:Kurt_wuethrich.jpg

21.6 Special Topics in Nuclear Magnetic Resonance

Variable Temperature Nuclear Magnetic Resonance

Variable temperature NMR (VT-NMR) is a special application in which the spectrum is collected at nonambient temperature. VT-NMR experiments can be conducted at temperatures higher than room temperature (heated experiments) or lower than room temperature (cooled experiments). In either case, the spectroscopist must have special training due to the potential of causing damage to the NMR if the VT experiment is not properly set up. Mistakes can cause needed repairs exceeding $50,000 and significant instrument down time that affects many other people and their work.

In planning a VT-NMR experiment, the analyst needs to consider the temperature range, solvent properties, sample properties, NMR tube suitability, the spinner qualities, and NMR probe limitations. Among these considerations, solvent boiling point is a key limitation for high-temperature experiments. Furthermore, because the change in the temperature is imprecise, instruments tend to overshoot and then fall back from the target temperature. Thus, you never want to be working in a temperature range that is too close to a phase transition of the solvent or a critical failure point of any of the hardware (NMR tubes and probes).

VT experiments can be time consuming compared to routine analyses. It is not uncommon for the change in temperature to necessitate the need to retune the probe, a process that is typically carried out by a facilities manager. And it will always be the case that the temperature change necessitates a reshimming[25] of the magnetic field. In some VT situations, sample spinning may not be possible or advisable, which

---

[25] Shimming was introduced in Chapter 14.

can lead to additional time needed to enhance field homogeneity with more refined shimming.

The VT equipment capabilities vary by instrument manufacturer. A typical situation involves a temperature-controlled gas that comes into the spectrometer near the probe and surrounds the sample. For low-temperature experiments, the VT gas is passed through a gas line that runs through a heat exchanger (ice water, dry ice/isopropanol, or liquid nitrogen); once cooled, it then enters into the sample chamber. In low-temperature experiments, the compressed air has to be switched to dry nitrogen gas to avoid condensation. For high-temperature experiments, the gas is heated as it enters into the sample chamber. There is a thermocouple near both the probe and the sample that measures the temperature; however, it is only an approximation of the actual temperature. Gas flow rate is a critical consideration, and the standard operating procedure for each instrument should be explicitly followed. Temperature is adjusted slowly, and the sample equilibrium is adjusted in a stepwise fashion. Temperature gradients (top to bottom and outside to inside) can be minimized with sufficient airflow and time for equilibration. Shorter sample heights (using a modified NMR tube) can offer less chance for a temperature gradient within the sample tube thus leading to a loss of resolution.

Because the temperature control is imprecise, it may be necessary for you to do a temperature calibration on your instrument to determine the actual temperature of the sample. Methanol (for low-temperature experiments) and ethylene glycol (for high-temperature experiments) have been well studied and can be used to calibrate the temperature for the experiment.

VT-NMR is widely used to determine activation energies or kinetics related to conformational changes, such as (1) *cis*-to-*trans* isomerism or chair-to-boat interconversions or (2) dynamic processes such as proton exchange, keto-to-enol tautomerization, protonation, or complex formation. The change in temperature allows the analyst to speed up an *exchange process* so that you do not see it (making it faster than the NMR time domain) or to slow the exchange process down, so that the two nuclei involved in exchange can be resolved. By moving into and out of the NMR time domain, the analyst can control what type of information is obtained. In a classic example of the effect of VT on NMR spectra, cyclohexane-d11 (which has one H and eleven D on its six-membered carbon ring) is analyzed at several temperatures (see Figure 21.28). The spectrum shows two well-resolved peaks at low temperature: one peak corresponding to the H in the axial position on the cyclohexane ring and a second peak for the H in the equatorial position on the ring. As temperature increases, the two peaks broaden and then merge, coalescing into a single peak, when the axial to equatorial exchange is in rapid equilibrium and too fast for the NMR time domain. Note that the single peak has a resonance at the average chemical shift between the two isomers. More information on interpreting VT-NMR spectra and deriving kinetic constants from the VT-NMR results is provided in Section 21.8, Further Reading.

VT-NMR demonstrates how powerful NMR spectroscopy is for understanding not only molecular structure but also the various dynamic dimensions of structure. This information can be extended to understanding intermolecular dynamics. It has helped NMR become one of the most powerful tools in the chemist's toolbox for probing biophysical properties and the intermolecular interactions and motions associated with biomolecular structure and assembly.

Solid-State Nuclear Magnetic Resonance

In nearly every context in this chapter up to this point, we have been discussing samples in the liquid phase and most especially samples that are dissolved in a solvent in a solution state. You may have mistakenly thought that NMR is limited to liquid samples, and if so, let us correct that impression here. The technique of NMR spectroscopy can be applied to solid-state samples using a technique called *solid-state NMR* (*SSNMR*).

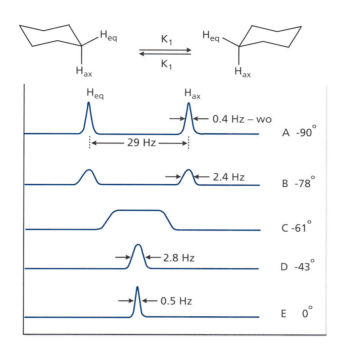

Figure 21.28 VT-NMR spectrum of cylcohexane-*d11*. (A) The low-temperature limit at –90° shows two distinct signals for each of the two H positions. (B) At –78°, slow exchange is evident. (C) At –61°, coalescence begins as the two protons start to become indistinguishable. (D) In the fast exchange regime at –43°, the two signals are fully merged. (E) At the high-temperature limit for the exchange of these two signals, note how the line width also changes as a function of temperature.

SSNMR is used, in the broadest sense, to characterize and analyze materials. It has found applications in diverse fields such as archaeology and pharmacology. Not only is it an important technique for discovering structural information, but it reveals information at the fundamental levels of atomic and molecular assembly. SSNMR provides complementary information to X-ray crystal analysis. It can be used alongside this analysis, or as an alternative technique, to solve ever more complicated structural problems such as those presented by the proteins embedded in lipid bilayers or the malformed proteins (e.g., beta-amyloid fibrils) that characterize Alzheimer's disease.

The basics are the same. The nuclear spin, under the influence of an external (B_o) magnetic field, will undergo a nuclear magnetic transition at the Larmor frequency of the nucleus as influenced by its local environment. For our previous considerations, we limited our discussion to spin–1/2 (the nuclear spin quantum number, $I = \frac{1}{2}$); these are systems that have two possible states (characterized by quantum numbers $m = +\frac{1}{2}$ and $m = -\frac{1}{2}$). In SSNMR, we will also consider some nuclei with higher spin order $(I > \frac{1}{2})$. These "quadrupolar" nuclei have four possible spin states (e.g., $2I + 1 = 4$ when $I = 3/2$).

In solution, the molecules are moving around, and therefore the slight differences in signal average out. However, in SSNMR, line width is going to be broad because the molecules in the sample are not moving and differences in signal do not average out. SSNMR uses a technique called *magic angle spinning* (MAS) to improve line width in combination with heteronuclear and homonuclear spin decoupling (to isolate spins) in multidimensional experiments. Isotopic labeling can also be used to improve spectral information. Together these methods, along with advances in magnetic field technology and computational power, have helped SSNMR move into the forefront of instrumental techniques used for the analysis of noncrystalline materials.

Chemical shift anisotropy[26] means that the chemical shift is going to vary with the angle of orientation between the crystal and the B_o field. Chemical shift anisotropy leads to line broadening, because each signal is slightly shifted relative to the others and overlapped with all of the other signals. In a solid material (or powder), these positions are fixed and they are all on top of each other, resulting in a broad line spectrum. In solution, the molecules tumble and do not maintain a fixed orientation, so the anisotropy is averaged away.

As a simple example, let us consider a single crystal of a molecule that has a carbonyl (or other) functional group. Recall that the electron distribution in the molecule causes shielding and deshielding of the nucleus under the influence of the external magnetic field (B_o). Therefore, we can think of the molecule as a kind of ellipsoid, and the relative position of that ellipsoid in the magnetic field matters. The intensity of the peak at any particular chemical shift will be proportional to the probability of the molecule having a particular orientation (or energy), relative to the B_o field. And the spectra will be the sum of the various orientations. Figure 21.29 shows three possible orientations of a carbonyl group with respect to an external magnetic field.

Because of the chemical shift anisotropy, which comes from the distribution of electrons, symmetry, and resulting shielding and deshielding, the overall NMR signal is spread out across the range of frequencies, which makes the signals in SSNMR low. Thus, the technique suffers from poor sensitivity. Chemical shift anisotropy is a function of $(3\cos^2\Theta - 1)$, where Θ as shown in Figure 21.30 is the angle between the long axis of the molecule's electron distribution and the B_o field. It was discovered in 1958 by E. R. Andrew and colleagues[27] that if the sample spins rapidly at a specific angle to B_o, called MAS, then the chemical shift anisotropy can be eliminated. The magic angle is 54.74 degrees. This spin axis is called the z_R axis. Chemical shift anisotropy is averaged out with MAS, which increases sensitivity and resolution (see Figure 21.30). However, even with the improved resolution imparted by MAS, J-coupling is too small to be seen in SSNMR.

MAS also improves SSNMR spectra by minimizing dipolar coupling. You may recall from earlier in this chapter that dipolar coupling is a nuclear phenomenon related to the interactions of nuclei in close proximity. If you think of a nucleus as a small

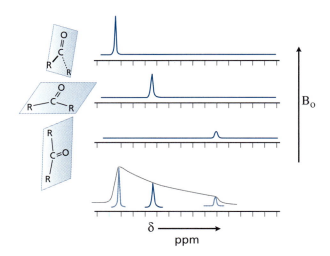

Figure 21.29 Chemical shift anisotropy in an external magnetic field.

---

[26] *Anisotropy* is a term that indicates a substance properties have a directional component.

[27] Andrew, E. R.; Bradbury, A.; Eades, R. G. *Nature,* **1958,** *182,* 1659.

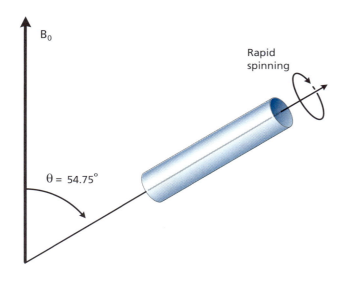

Figure 21.30 A schematic of MAS relative to B_o.

magnet, then you know that two small magnets in close proximity to each other will begin to exert greater effects on one another the closer they approach.[28] As discussed before, effects that depend on orientation are anisotropic and will lead to line broadening in a sample that is static, like a crystal. Again, the magic angle of the z_R axis will reduce these effects. The rate of spin about the z_R axis should be equal to or greater than the coupling constant; however, in practice this is somewhat difficult to achieve. When the spin rate is not fast enough, the spectrum may include what are known as spinning sidebands. These peaks appear to the left and right of the isotropic[29] chemical shift at values that are multiples of the spin rate. Spinning sidebands can be seen in Figure 21.31, a spectrum from Sigma-Aldrich, which shows a ^{1}H MAS SSNMR spectrum of an aluminum hydride.

Dipole–dipole coupling between ^{1}H can be a significant factor in SSNMR because, in a given molecular structure and in a given solid sample, the number of ^{1}H nuclei is substantial and the nuclei can be close together. In contrast, dipole–dipole coupling between less abundant nuclei, such as ^{13}C–^{13}C coupling, is much less significant, not only because of their lower abundance but also because of their lower gyromagnetic constants (γ). For heteronuclear dipole–dipole coupling (e.g., ^{1}H nuclei that are dipole coupled to ^{13}C nuclei), SSNMR adapts a magnetization transfer technique, cross polarization, which we considered in our discussion of heteronuclear multidimensional NMR in Section 21.5.

Other Spin-Active Nuclei

NMR discourse tends to be heavily focused on ^{1}H NMR—and for good reason. ^{1}H is the most abundant isotope of hydrogen, and that element is found in many molecular structures. In addition to having a high natural abundance, ^{1}H also has a large magnetogyric ratio (see Table 21.1). This makes ^{1}H NMR easy to observe with widespread applicability. ^{13}C NMR, which also finds widespread use, is a less abundant isotope of carbon. However, it is important enough from a structural perspective for spectroscopists to have devised technologies and methodologies of making its

---

[28] This magnetic dipole–dipole coupling is a function of the distance between the nuclei to the third power (r^3) and is a function of their relative orientation, which is measured as the angle between the two nuclei in relation to the B_o field, and is a function of Θ ($3\cos^2\Theta - 1$).

[29] *Isotropic* means uniform in all directions; isotropic chemical shifts do not depend on orientation in space.

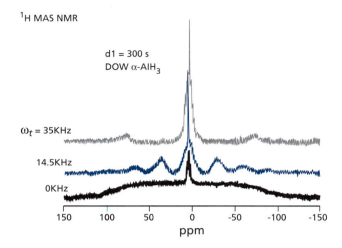

Figure 21.31 ^{1}H MAS SSNMR of an aluminum hydride material being studied for its potential as a hydrogen storage system for automotive applications by scientists at the California Institute of Technology. These materials are studied by SSNMR because the instrumental technique provides valuable information about dynamics, equilibrium, reversibility, and mechanism. In these spectra, ω indicates the spin rate. The bottom spectrum, 0 kHz, was collected under "static" or nonspinning conditions and the anisotropic broadening is evident. At the 14.5 kHz spin rate, the spinning sidebands (in multiples of 14.5) can be seen both to the left and right of the central peak (the isotropic chemical shift, at 0 ppm in this figure). As the spin rate is increased to 35 kHz, the spinning sidebands are relocated to multiples of 35, indicating that even at this high rate of spin, the dipole–dipole coupling of the protons is very strong and not completely eliminated. The sharper peak, evident in the static spectrum, is a signal that is attributed to residual gaseous hydrogen trapped in the material.

TABLE 21.1:   **Magnetic Data for Some Common Nuclides**

NUCLIDE	NATURAL ABUNDANCE (%)	SPIN (I)	MAGNETIC MOMENT (μ)*	MAGNETOGYRIC RATIO (γ)†
^{1}H	99.984	½	2.793	26.75
^{13}C	1.108	½	0.7022	6.728
^{31}P	100	½	1.131	10.84
^{15}N	0.368	½	−0.283	−2.713
^{195}Pt	33.8	½	1.056	5.839
^{19}F	100	½	2.627	25.18

˙μ has units of nuclear magnetons = $5.0508 \times 10^{27} JT^{-1}$.

†γ has units of 10^7 rad $T^{-1}s^{-1}$.

lower abundance a tolerable circumstance. It is used quite extensively especially in multidimensional applications. Hence, our discussions to this point in NMR have been focused on these elements. In this final section, you will get a slightly broader view, albeit not at all extensive, of the variety of applications related to other nuclei in NMR spectroscopy. As we have mentioned, the ability to "see" a nucleus is related to its natural abundance. It is also important to remember that the magnetogyric ratio of a nucleus also influences how strong an NMR signal will be. Table 21.1 shows the natural abundance and magnetogyric ratios of some common nuclei with a spin of ½.

Phosphorus-31 ^{31}P-NMR finds wide-ranging applications in nearly all areas of chemistry. In biochemistry, the phosphorus atom plays an essential role in energy transduction, structure, and function across all major biomolecular classes. For example, in a 2010 *Journal of Organic Chemistry* paper by scientists from the Johns Hopkins University Biomolecular NMR Center, the authors report on a two-dimensional ^{1}H–^{31}P–^{31}P pulse sequence for elucidating structures of small di- and triphosphorylated compounds, such as ATP, molecules that are important intermediates in metabolism. ^{31}P is also used to characterize DNA structures, protein–DNA complexes, and membrane assembly. In food science, ^{31}P has been used to analyze foods such as olive oils and peanut oils for phospholipid component composition. Organophosphorus compounds, such as those used in chemical weapons, are sometimes a focus of NMR analysis. Environmental samples such as soil and water can be effectively screened for suspect organophosphorus compounds against a complicated background matrix using ^{31}P-NMR, under situations where ^{1}H or ^{13}C-NMR alone would be overcome due to the number of background signals. ^{31}P-NMR also finds a use in materials science: for characterization and development of novel materials such as superconducting polymers and ceramics, glasses with various transmission properties, and catalysts with increased efficiencies.[30]

Nitrogen-15 ^{15}N-NMR, like ^{31}P-NMR, also finds widespread use across various chemical disciplines, typically under "enriched" conditions. That is, the samples have been intentionally enriched with the ^{15}N isotope because its natural abundance is very low (< 1%). In biochemistry, ^{15}N is used to characterize proteins and protein structures as well as organizational assemblies and to probe conformational flexibility and motion. ^{15}N-NMR also finds uses in carbohydrate analysis as well as metabolite profiling. In a 2013 paper in *Analytical Chemistry*, Daniel Raftery from Purdue University and his collaborators report on the use of ^{15}N-NMR in combination with mass spectrometry to profile metabolites in biological samples (Figure 21.32).

^{15}N NMR also finds use in environmental chemistry for quantitative and qualitative analysis of analytes (such as pesticides and explosives) in soils, water, and a variety of biological materials, as well as in agricultural sciences for better characterization of metabolic processes. For example, researchers at the Tokyo Metropolitan University used ^{15}N-NMR to study large proteins that had been created from isotopically labeled ^{15}N. Studying very large protein structure using ^{1}H-NMR is often difficult due to the crowded nature of the spectrum. Using ^{15}N-NMR, Masatsune Kainosho and coworkers were able to elucidate the structure of a 41 kDa protein.[31] In Chapter 14, there is a profile of Angela Gronenborn. In a 2014 publication, Gronenborn and colleagues reported on the dynamics and structures associated with binding domains of HIV-1 to the human homolog of a yeast DNA repair protein. It is hoped that an understanding of these protein structures may eventually allow researchers to produce a vaccine. She used several different ^{1}H-^{15}N 2D NMR experiments as well as several ^{13}C-^{15}N 2D NMR experiments in order to determine the backbone connectivity of the protein.[32]

Platinum-195 ^{195}Pt-NMR is well known for a variety of applications related to pharmaceuticals and pharmacology. Platinum-based anticancer drugs have been a target of study for several decades, and ^{195}Pt-NMR has helped to elucidate structures, activities, and the design of ever more selective ways of delivering these potent compounds directly to their intended target. So-called "magic bullet" effects can be studied and used to manipulate delivery systems. Researchers at Purdue University used ^{195}Pt-NMR in conjunction with ^{15}N-NMR to study the interaction of $[Pt(Cl)_4]^{-2}$ with ^{15}N-enriched peptides. They were able to show that coordination of the platinum to the peptide proceed by amine coordination followed by sequential deprotonation and coordination of

[30] Gorenstein, D. G. *Phosphorus-31 NMR: Principles and Applications*, Academic Press: New York, NY, **1984.**

[31] Kainosho, M.; et al. *Nature,* **2006**, *440*, 52–57.

[32] Gronenborn, A. M.; et al. *J Biol Chem.,* **2014**, *289*, 2577–2588.

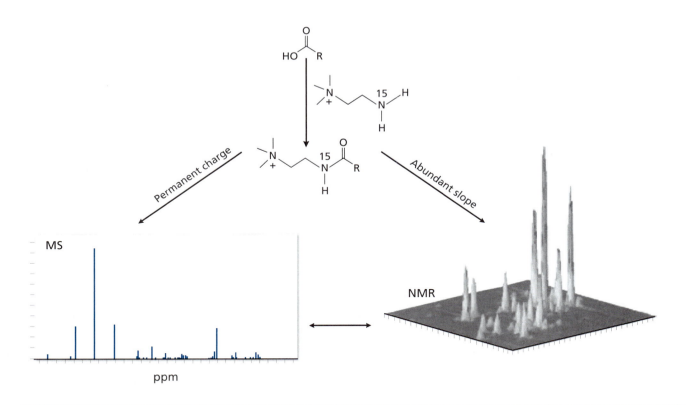

Figure 21.32 Image demonstrating the tagging of a biological intermediate in metabolism (containing a carbonyl group) with a ^{15}N-cholamine functional group for the simultaneous analysis of these components by both mass spectrometry and NMR.

available peptide nitrogens.[33] Also, ^{195}Pt-NMR is used for the study and characterization of platinum-containing catalysts, electrochemical applications, reaction kinetics, and equilibria. Many of these applications have the potential for greater solar energy conversion as well as future applications in space science and development of habitable environments on other planets.

Fluorine-19 ^{19}F-NMR is mostly utilized by pharmaceutical industry to probe the structure of organofluorine compounds. One of the concerns of pharmaceutical chemistry is the diastereomeric purity of starting materials and the resulting products. In a 2009 paper, researchers from the University of Bath reported on a derivatization process in conjunction with the use of proton decoupled ^{19}F-NMR for the determination of the diastereomeric purity of diol esters.[34] In addition, the polymer industry makes wide use of ^{19}F-NMR in studying fluorosubstituted polyethylene compounds.[35] Because fluorine has a high magnetogyric ratio and a high natural abundance, fluorine is relatively easy to study using NMR spectroscopy.

Problem 21.16:   Other NMR-active nuclei include ^{19}F, ^{29}Si, ^{57}Fe, ^{77}Se, ^{89}Y, ^{103}Rh, ^{107}Ag, ^{109}Ag, ^{111}Cd, ^{113}Cd, ^{117}Sn, ^{119}Sn, ^{125}Te, ^{129}Xe, ^{169}Tm, ^{183}W, ^{189}Os, ^{199}Hg, ^{207}Pb, ^{203}Tl, and ^{205}Tl. Do a library search using the American Chemical Society database (pubs.acs.org) to find an application of NMR using one of these other nuclei. Write a short summary of your findings or give a brief presentation to your class.

[33] Schwederski, B. E.; et al. *Inorg. Chem.* **1990**, *29*, 3569–3578.

[34] Yeste, S. L.; et al. *J. Org. Chem.* **2009**, *74*, 427–430.

[35] Teflon is polytetrafluoroethylene, or PTFE.

21.7 Useful Data

Tables 21.2, 21.3, and 21.4 present useful information for NMR experimentation. The tables all come from the same source.[36]

TABLE 21.2: Characteristics of Nuclear Magnetic Resonance (NMR) Nuclei

ISOTOPE	ABUNDANCE (%)	SPIN (I)	γ	NMR FREQUENCY AT INDICATED FIELD STRENGTH (KG)*				
				10.000	14.092	21.139	23.487	51.568
^1H	99.985	½	5.5856	42.5759	60.0000	90.0000	100.0000	220.0000
^{13}C	1.108	½	1.4044	10.7054	15.0866	22.6298	25.1443	55.3174
^{14}N	99.635	1	0.4035	3.0756	4.3343	6.5014	7.2238	15.924
^{15}N	0.365	½	−0.5660	4.3142	6.0798	9.1197	10.1330	22.2925
^{17}O	0.037	⁵⁄₂	−0.7572	5.772	8.134	12.201	21.557	29.825
^{19}F	100	½	5.2545	40.0541	42.3537	63.5305	94.0769	206.9692
^{29}Si	4.70	½	−1.1095	8.4578	11.9191	17.8787	19.8562	43.7035
^{31}P	100	½	2.2610	17.235	24.288	36.433	40.481	89.057
^{35}Cl	75.53	³⁄₂	0.5473	4.17187	5.8790	8.8184	9.7983	21.5562
^{79}Br	50.54	³⁄₂	1.3993	10.667	15.032	22.594	25.054	55.119
^{81}Br	49.46	³⁄₂	1.5084	11.498	16.204	24.305	27.006	59.413

*1 kG = 10⁻¹⁰ T.

TABLE 21.3: ^1H and ^{13}C Chemical Shifts for Useful Nuclear Magnetic Resonance Solvents

SOLVENT	CHEMICAL SHIFTS (PPM)		LIQUID RANGE (°C)	DIELECTRIC CONSTANT
	δ^1H	$\delta^{13}C$		
Acetone-d₆	2.17	29.2, 204.1	−95 to 56	20.7
Acetonitrile-d₃	2.00	1.3, 117.7	−44 to 82	37.5
Benzene-d₆	7.27	128.4	6 to 80	2.28
Carbon tetrachloride	NA	96.0	−23 to 77	2.24
Chloroform-d₃	7.25	76.9	−64 to 61	4.81
Cyclohexane-d₁₂	1.43	27.5	6 to 81	2.02
Dichloromethane-d₂	5.33	53.6	−95 to 40	9.08
Dimethylformamide-d⁷	2.9, 3.0, 8.0	31, 36, 132.4	−60 to 153	36.7
Dimethylsulfoxide-d₆	2.62	39.6	19 to 189	46.7
1,4-Dioxane-d₈	3.7	67.4	12 to 101	2.21
Methanol-d₄	3.4, 4.8*	49.3	−98 to 65	32.63
Nitrobenzene-d₅	7.5, 7.6, 8.2	124, 129, 134,149	6 to 211	34.8
Pyridine-d₅	7.0, 7.6, 8.6	124, 136, 150	−42 to 115	123
Silicon tetrachloride	NA	NA	−70 to 57.6	2.25
Tetrahydrofuran-d₈	1.9, 3.8	25.8, 67.9	−108 to 66	7.54
Water-d₂	4.7	NA	0 to 100	78.5

*Variable with concentration.

[36] Bruno, T. J.; Svoronos, P. D. N. *Handbook of Basic Tables for Chemical Analysis,* 2nd ed. CRC Press: Boca Raton, FL, **2003**.

TABLE 21.4:   **Chemical Shift Range and Reference Standards of Nuclear Magnetic Resonance Nuclei**

ISOTOPE	CHEMICAL SHIFT RANGE (ppm)	REFERENCE STANDARD	REFERENCE FORMULA
^{1}H	15	Tetramethylsilane	$(CH_3)_4Si$
^{13}C	250	Tetramethylsilane	$(CH_3)_4Si$
^{15}N	930	Ammonium nitrate	NH_4NO_3
^{17}O	700	Water	H_2O
^{19}F	800	Trichlorofluoromethane [Freon 11, R-11]	CCl_3F
^{29}Si	400	Tetramethylsilane	$(CH_3)_4Si$
^{31}P	700	Tetramethylphosphite	$(CH_3O)_3P$
		Phosphoric acid (85%)	H_3PO_4
^{35}Cl	820	Sodium chloride	$NaCl$

21.8 Further Reading

BOOKS

Alberty, R. A.; Silbey, R. J. *Physical Chemistry*, 2nd ed., John Wiley & Sons: New York, **1997.** (for information on the theory of exchange rates and calculating k values)

Apperley, D.; Harris, R.; Hodgkinson, P. *Solid State NMR: Basic Principles and Practice,* Momentum Press; New York, **2012.** (for information on SSNMR)

Bruno, T. J.; Svoronos, P. D. N. *Handbook of Basic Tables for Chemical Analysis*, 2nd ed. (Chapter 9), CRC Press: Boca Raton, FL, **2003.**

JOURNALS

Gorenstein, D. G. Conformation and Dynamics of DNA and Protein-DNA Complexes by ^{31}P-NMR. *Chem. Rev.,* **1994,** *94,* 1315–1338.

Johnson, C. S. Jr. Diffusion Ordered Nuclear Magnetic Resonance Spectroscopy: Principles and Applications. *Progress in Nuclear Magnetic Resonance Spectroscopy,* **1999,** *34,* 203–256. (contains information on DOSY)

Majumdar, A.; Sun, Y.; Shah, M.; Meyers, C. L. F. Versatile ^{1}H–^{31}P–^{31}P COSY 2D-NMR Techniques for the Characterization of Polyphosphorylated Small Molecules. *J. Org. Chem.,* **2010,** *75(10),* 3214–3223.

Meier, U. C. Application of Nonselective 1D ^{1}H–^{31}P Inverse NMR Spectroscopy to the Screening of Solutions for the Presence of Organophosphorus Compounds Related to the Chemical Weapons Convention. *Anal. Chem.,* **2004,** *76(2),* 392–398.

Tayyari, F.; Gowda, G. A. N.; Gu, H.; Raftery, D. ^{15}N-Cholamine—A Smart Isotope Tag for Combining NMR- and MS-Based Metabolite Profiling. *Anal. Chem.,* **2013,** *85(18),* 8715–8721.

Van Geet, A. L. Calibration of the Methanol and Glycol Nuclear Magnetic Resonance Thermometers with a Static Thermistor Probe. *Anal Chem.,* **1968,** 40, 2227. (for information on calibrating VT-NMR temperatures)

21.9 Exercises

EXERCISE 21.1: By degassing your sample, you were able to increase the relaxation time of your desired signal from 0.1 to 4 seconds.

(a) By what factor will your line width decrease as a consequence?

(b) If your experiment required a signal averaged scan of 500 scans in order to achieve the desired S/N ratio, how much longer will your experiment take as a result of degassing the sample?

EXERCISE 21.2: Using your answer from Exercise 21.1, determine the percent improvement in resolution in ppm for the following NMR field strengths:

(a) 60 MHz   (b) 200 MHz   (c) 400 MHz   (d) 900 MHz

EXERCISE 21.3: Is T_1 relaxation a first-order or second-order process? Write out the integrated rate equation that describes a T_1 relaxation process.

EXERCISE 21.4: What are the two names often given for a T_1 relaxation process? Which is the descriptively more accurate term?

EXERCISE 21.5: Is T_2 relaxation a first-order or second-order process? Write out the integrated rate equation that describes a T_2 relaxation process.

EXERCISE 21.6: What are the two names often given for a T_2 relaxation process? Which is the descriptively more accurate term?

EXERCISE 21.7: Why is it sometimes desirable to degas an NMR sample before collecting the spectrum?

EXERCISE 21.8: Discuss two ways an NMR spectroscopist can manipulate a sample in order to increase the spectral resolution.

Chapter 22

Statistical Data Analysis

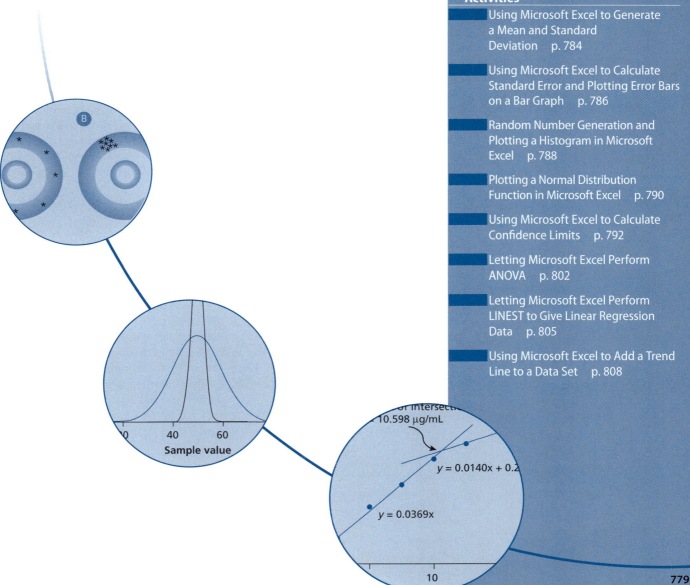

Sample value

of intersection
10.598 µg/mL

$y = 0.0140x + 0.2$

$y = 0.0369x$

10

22.1 Introduction

When we make an instrumental measurement, we want the measurement to be correct. So it makes sense for us to start this discussion with a look at what the word *correct* means to a scientist. When we make a measurement, there is a fundamental limit to how well we can *know and be confident of* the answer.[1] Therefore, a real measurement cannot have a single *true* value. To be complete, a statement of the uncertainty in the number must accompany a real measurement. For example, you might see a reported mass as 2.15 ± 0.01 grams. In order for a scientific measurement to be "correct," it must represent the best estimate of the mean of a set of replicate measurements and be accompanied by an estimate of the uncertainty in the mean (i.e., scientific error). The typical interpretation of the reported value in this example would be that our best estimate of the "true" value is 2.15 grams and the standard deviation of the mean lies in the range of 0.01 above and below the value of 2.15 (range, 2.14–2.16). Unfortunately, the interpretation of the ±0.01 is not consistent throughout all disciplines of science. Although we have stated here that the ±0.01 typically represents standard deviation, it is possible that the ±0.01 represents the standard error or the confidence interval.[2] It is a *best practice* to always include a statement describing how you are reporting error in all of your scientific reports.

> It is the analyst's objective to minimize and quantify error.

22.2 Types of Error

"Nature does not give up her secrets lightly,"[3] and in the pursuit of nature's secrets, it is accepted that the first measurement will yield a false representation of the truth. In other words, any single data point will inherently contain error.[4] The word *error* comes from Latin and loosely translates as "wandering," and it is used in science to represent the inherent uncertainty in a measurement. For our purposes, we define error as the difference between the experimentally obtained value and the true value. Ironically, if we knew the true value, we would have no need to conduct the experiment in the first place. This leads us to a philosophically important conclusion: the goal of an experiment is to obtain a "true" measured value, but because all measured data points contain error, we can never know with absolute certainty the true value of an experimentally obtained result. All experimentally obtained results contain uncertainty. Therefore, it is the analyst's objective to minimize and quantify error.

It is generally recognized that there are three broad categories of error: *gross error*,[5] *systematic error*,[6] and *random error*. These are discussed in the following paragraphs.

Gross Error

Gross error occurs when the analyst makes a mistake. For example, the analyst might misread a balance or strike the wrong button on his or her calculator. These gross errors are often obvious. However, not all mistakes are equivalent. For example, if you were to make replicate measurements of the volume of your favorite coffee mug and you obtained a set of volumes such as 298 mL, 302 mL, 299 mL, *80.53 liters*, 297 mL, 301 mL, 299 mL, 295 mL, 301 mL, and 270 mL, you would immediately recognize that the 80.53-liter measurement was *completely incorrect*. You obviously made a mistake! The purist might say that you must keep the 80.53-liter data point until you can statistically justify the exclusion of that point. However, in practice, few analysts will keep a data point if it is completely obvious that a gross mistake was made. But be careful. Casually throwing out data points that you do not like is against *best practices*. There are good reasons why the purist will always justify an exclusion using statistical tools. Taking another look at the

---

[1] This idea is implied by the Heisenberg Uncertainty principle: Heisenberg, W. *Z. Phys.*, **1927**, *43*, 172–198.

[2] We will define *standard error* and *confidence limits* later in this chapter.

[3] Greene, B. *The Fabric of the Cosmos: Space, Time, and the Texture of Reality.* First Vintage Books: New York, NY, **2004**, 470.

[4] We will use the terms *uncertainty* and *error* interchangeably.

[5] Also known as *human* error, *operator* error, or *illegitimate* error.

[6] Also known as *bias*.

data, you might also wonder about the 270 mL data point. If you exclude the 80.53-liter data point *and* the 270 mL data point, you obtain an average value of 299 mL. It would appear that the 270 mL data point is ≈ 30 mL "too low." You might be tempted to ignore that data point, but again, this would be a violation of *best practices*—and in this case, it is not so obvious that the answer is incorrect. Within the precision of your technique, the 270 mL data point might be legitimate. For example, if you keep the 270 mL data point, you obtain an average of 295.7 mL and your original data set had a data point of 295 mL. By casually throwing out the 270 mL data point, you may have artificially raised the mean of your data set. You would first need to statistically justify the exclusion of the 270 mL data point before you could legitimately ignore it.

Data points that statistically fall outside the range of a data set are called *outliers*. We will explore the notion of outliers further in Section 22.4 when we discuss *Q-tests* and *Grubbs' tests*.

Systematic Error

Systematic error can be described as a measurement that is always too high or always too low, and the magnitude of the deviation from the "true" value is constant. Systematic error is often difficult to identify. The origin of systematic error is typically either chemical and/or instrumental. Instrumental systematic errors can result from drift noise,[7] external interference, or improper calibration of the instrument. For instance, an improper ground wire may result in a bias on the detector that artificially raises or lowers the instrument response to your measurement. Likewise, if your instrument's critical components are not properly shielded, an external magnetic or radiofrequency signal can cause your instrument's response to shift from its original calibrated value. Instrumental systematic errors are identified by analyzing carefully constructed standards on a regular basis. For example, baseline drift is a common problem when conducting atomic absorption spectrometry analysis. For this reason, it is common for these spectrometry methods to incorporate a blank and a known standard in the analysis after every five or ten samples.

Chemical systematic error occurs in many ways. For instance, any error in the construction of standards used to calibrate an instrument will necessarily impart a systematic error to the instrumental response. Or a chemical systematic error might result from chemical steps used in preparing the sample for analysis. For example, it is common to esterify carboxylic acids prior to gas chromatography/mass spectrometry analysis. If the derivatization step had a yield of 85%, the analyst would need to correct for the 15% loss; otherwise, there would be a negative systematic error of 15% in the final results. Likewise, you can imagine a similar loss of sample if there was an inefficient extraction step in the sample preparation.

Random Error

Random error is an unpredictable high or low fluctuation in the measurement of a physical property. This fluctuation can arise from an environmental change, such as a moment-to-moment fluctuation in pressure or temperature, or result from slight variation in procedural steps. Fortunately, random error can be quantified using statistical tools. In the absence of any gross or systematic error, if one repeats an experiment several times, the mean value of a normally distributed data set will appear close to the true value, and the scatter about the mean can be used to quantify the confidence we have in that mean. We will discuss each of these ideas in more detail later in this chapter.

22.3 Precision versus Accuracy

In the simplest case, *accuracy* is used to quantify the correctness of an analysis—how close the measured value is to the "true" value. *Precision* is used to quantify the reproducibility of our technique—how close to the previous measurements will our next measurement be? A common analogy used when discussing the terms *accuracy* and *precision* is that of hitting a target. In Figure 22.1(A), we have a situation in which the

The **precision** of an experiment is influenced most by our ability to control **random error**. The **accuracy** of an experiment is influenced most by our ability to control **systematic error**.

---

[7] See Chapter 5 for a review of noise sources.

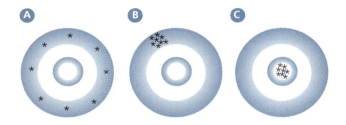

Figure 22.1 Three targets. Target (A) has relatively high accuracy but relatively low precision. Target (B) has relatively low accuracy but relatively high precision. Target (C) has relatively high accuracy and relatively high precision.

reproducibility of each attempt is poor, but if we average the distance of each attempt from the bullseye, we obtain an average value very close to a perfect bullseye. We would say that the precision is poor, but the accuracy can be made acceptable if enough data are collected and the results averaged. Conversely, in Figure 22.1(B), we have a scenario in which the reproducibility (the precision) is relatively high, but the shooter has consistently failed to hit the bullseye. We would say that the precision is high but the accuracy is low. Averaging these shots will *not* yield a result close to the bullseye. Relating these results to the previous section, we would conclude that this shooter has a systematic error of shooting high and to the left in addition to the random error one normally sees with target shooting. Finally, in Figure 22.1(C), we have a scenario in which the precision and accuracy are both relatively high.

Tying these ideas together, we recognize that the *precision* of an experiment is related to our ability to minimize random error. In targets (B) and (C) of Figure 22.1, we see relatively small random error. They are both precise, but only target (C) is accurate. The accuracy of an experiment is related to our ability to minimize systematic error. For example, target (B) shows a systematic error resulting in a high and left pattern, yielding an inaccurate result.

22.4 Statistical Tools

Population versus Sample

Before we delve too deeply into specific statistical tools, we need to define some terms. The term *population* is used when an infinite sampling occurred or all possible subjects were analyzed. Obviously, we cannot repeat a measurement an infinite number of times, so quite often the idea of a population is theoretical; in such cases, we take a representative *sample* of the entire population. For example, if you wanted to know the average height of the human race, you would have to take a representative sample of people and measure their heights. Your result would be an estimate, and you would necessarily report the uncertainty of your estimate. However, if the parameters of an experiment are specifically defined, one can analyze an entire population. For example, if you want to know the average height of your immediate family, then your population has been well defined, and it is now possible to measure the height of the entire population. Despite your ability to collect data on the entire population, you still have random error associated with each measurement.

Be careful to distinguish the statistical use of the word *sample* from the way a chemist often uses the word *sample* in the context of laboratory or field work. For example, if we were analyzing the soil in a field for arsenic concentration, we might go out to the field and collect 20 representative soil "samples" and bring them back into the laboratory. The 20 soil "samples" would give us 20 data points. The statistician would consider that the "sample" is the entire set of 20 data points because the 20 data points are being used to represent the entire population (all of the soil in the field). It can be a confusing tangle of words, so take a moment to think through it.

Mean

The term *mean* is synonymous with the term *average* and is obtained by summing all of the results from an analysis and dividing by the total number of individual results (N). The symbol for a *sample mean* is $\bar{x}$. If we have a large number of individual measurements (N is very large), then we use a different terminology, the *population mean,* which uses the symbol μ.

$$\mu = \frac{\sum_{i=1}^{i=N} \mu_i}{N} \qquad \text{Eq. 22.1}$$

$$\bar{x} = \frac{\sum_{i=1}^{i=N} x_i}{N} \qquad \text{Eq. 22.2}$$

In either case the individual measurement is represented by μ_i or x_i respectively and the mean of the set of measurements is represented by μ or $\bar{x}$ respectively.

As N approaches infinity, $\bar{x}$ approaches μ. How quickly $\bar{x}$ approaches μ is dependent on the relative amount of random error (precision) associated with each individual measurement, x_i. We quantify the random error using two statistical tools—the standard deviation and the variance.

Standard Deviation and Variance

A *standard deviation* is a mathematical representation of the overall range of values within a set of numbers. A small standard deviation indicates that all of the values in the data set are close or consistent. A large standard deviation indicates that the values are dispersed or spread apart. The equations for calculating a standard deviation of a population (N is large) and the standard deviation of a sample (N is small) are given in Equations 22.3 and 22.4. The symbol for a *population standard deviation* is σ and the symbol for a *sample standard deviation* is s.

$$\sigma = \sqrt{\frac{\sum_{i=1}^{i=N}\left(x_i - \mu\right)^2}{N}} \qquad \text{Eq. 22.3}$$

$$s = \sqrt{\frac{\sum_{i=1}^{i=N}\left(x_i - \bar{x}\right)^2}{N-1}} \qquad \text{Eq. 22.4}$$

If we take a close look at Equation 22.3, we see that the term $(x_i - \mu)$ is nothing more than the deviation of an individual data point from the population mean. We then square the deviation values for each data point to eliminate the negative sign. By summing all of the squares, dividing by N, and taking the square root, we are left with an average absolute deviation. So for a population, the standard deviation is simply the absolute value of the average deviation from the mean. However, when determining the standard deviation of a sample, we have a slight modification to the equation necessitated because of the small value for N. In Equation 22.4, you can see that we use $(N - 1)$ in the denominator instead of N. The term $(N - 1)$ is defined as the *degrees of freedom* for a sample set. Degrees of freedom represents the number of repeated measurements (also known as replicates) that are free to vary. The mean of a sample set is constrained by the mean of the population; therefore, the last data point is not "free to vary" because the average of all data points must represent the mean of the population. Degrees of freedom show up in several other statistical tools, so it is important that you take a moment to learn this term.

On many calculators, the buttons for calculating standard deviation are labeled σ and σ_{n-1}, where σ_{n-1} is the sample standard deviation that we have represented here with the symbol "s" as defined in Equation 22.4. One rarely samples an entire population in a laboratory experiment, so in most situations you will want to use Equation 22.4 or your σ_{n-1} button on your calculator to calculate "s."

ACTIVITY

Using Microsoft Excel to Generate a Mean and Standard Deviation

Recreate the spreadsheet seen in Figure 22.2 in Microsoft Excel. Select cell B13 and click on the f_x button to open the Insert Function dialog box (see Figure 22.3). From the drop-down window in the Insert Function dialog box, select Statistical. In the Select Function window, select Average. The Function Arguments dialog box will open (see Figure 22.4). The Average function will use Equation 22.2 to calculate the average of the data set. In the Number 1 field, enter the range of addresses for the numbers you wish to average. In this example, the range of addresses is B3:B12. Or you can click the grid button (circled in gray in Figure 22.4) and drag-and-drop the range of values to be averaged. Click OK and the average of cells B3 → B12 will be returned in cell B13. Now select cell B14 and repeat the above sequence of steps, but this time select the STDEV.S function instead of the Average function. STDEV.S uses Equation 22.4 to calculate the standard deviation of a sample. The Function Arguments box will open again, and you will need to enter the range of values for the data set (B3:B12), or you can use the drag-and-drop function. Your final spreadsheet should resemble the one shown in Figure 22.2. We will revisit this data set when we discuss standard error and confidence limits, so take a moment to save your spreadsheet as "Fish".

	A	B	C
1	Analysis of Mercury in Fish		
2	Trial	ppb(Hg)	
3	1	5.4	
4	2	2.9	
5	3	5.1	
6	4	4.2	
7	5	5.6	
8	6	4.7	
9	7	7.9	
10	8	4.8	
11	9	7.6	
12	10	3.2	
13	Avg.	5.1	
14	StDev	1.6	
15	S.E.		
16	95% C.L.		

Figure 22.2 Spreadsheet demonstrating the use of Microsoft Excel to calculate a mean and a standard deviation.

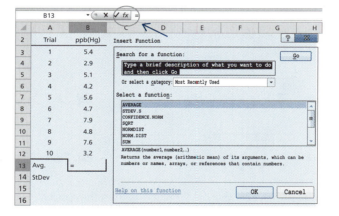

Figure 22.3 Insert Function dialog box.

Figure 22.4 Function Arguments dialog box.

Most scientific calculators can perform the statistics function we have outlined in the Activity "Using Microsoft Excel to Generate a Mean and Standard Deviation," although the keystrokes differ from calculator to calculator, the steps typically involve entering the data points into a data array (often symbolized with a $\Sigma+$ button). As you enter each data point, the total number of points in the array will be displayed as N = #. Once you have entered your data array, you can press the $\bar{x}$ button to display the average or the σ or $\sigma_{(n-1)}$ buttons to display the appropriate standard deviation.

Problem 22.1: Using the same data set we examined in the activity "Using Microsoft Excel to Generate a Mean and Standard Deviation," use the statistical functions on your calculator to determine the mean and the standard deviation of the data set. You may need to review the owner's manual or visit the manufacturer's website for instructions on statistical functions on your calculator.

Problem 22.2: Use Excel or a similar spreadsheet program to determine the mean and standard deviation of the following data sample. Repeat the analysis using your calculator's statistical functions.

Lead in Drinking Water

Replicate	1	2	3	4	5	6	7	8	9	10
Value (ppm)	2.002	1.996	2.000	1.995	1.999	1.987	2.010	2.014	2.007	2.004

Problem 22.3: Use Excel or a similar spreadsheet program to determine the mean and standard deviation of the following data sample. Repeat the analysis using your calculator's statistical functions.

Lead in a Paint Chip

Replicate	1	2	3	4	5	6	7	8	9	10
Value (ppm)	1001.9	989.0	1020.4	996.1	1002.4	990.0	1019.4	991.3	999.2	1002.4

Standard Error and Error Bars

In the introduction to this chapter, we reported a mass as 2.15 ± 0.01 grams and mentioned that the ±0.01 indicated one standard deviation unit above and below the mean. In the activity "Using Microsoft Excel to Generate a Mean and Standard Deviation," we reported the concentration of mercury in fish as 5.1 ± 1.6 ppb. The conventional way to report error graphically is to include "error bars." Chemists typically report error using standard deviation; however, not all disciplines of science share the same conventions. Another very common way to represent error is to report a value called the standard error. The standard error is a function of the standard deviation as seen in Equation 22.5:

$$\text{S.E.} = \frac{s}{\sqrt{N}} \qquad\qquad \textbf{Eq. 22.5}$$

Note that for a given set of measurements, the standard error will always be less than the standard deviation.

Microsoft Excel allows the user to report error bars on a graph as either the standard deviation, standard error, or as a percentage of the mean. In addition, Excel allows you to add a customized value for the error bars. It is important that you specify how you are reporting your uncertainty in your numbers. This is appropriately done in the figure caption.

ACTIVITY

Using Microsoft Excel to Calculate Standard Error and Plotting Error Bars on a Bar Graph

Open the spreadsheet Fish that you created in the previous Activity ("Using Microsoft Excel to Generate a Mean and Standard Deviation"). We are going to program Equation 22.5 for the standard error into cell B15. First, we need to calculate the square root of N. Select cell B17 and type "=sqrt(10)." Excel will return a value of 3.16. Next select cell B15 and type "=B14/B17." Excel will return a standard error of 0.51.

To display the standard error as error bars on a graph, first create a graph of your data. In this activity, we have created a column graph. Next, place your cursor in the graph and "left click."[8] This will display the Chart Tools group. From the Chart Tools group, select the Layout tab. Next select Error Bars from the Analysis Group and fill in the correct parameters. Your spreadsheet should now resemble the one in Figure 22.5. We will return to this spreadsheet when we discuss confidence limits, so be sure to save your work.

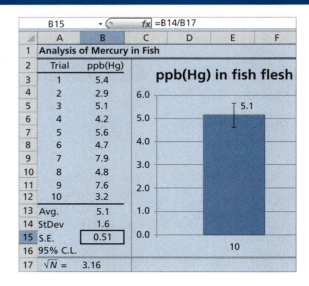

Figure 22.5 Determining standard error and displaying it on a graph.

Normal Distributions

For data in which the error is truly random, the probability of obtaining a specified value for an individual data point (x_i) is a function of the population mean (μ), and the standard deviation of the analytical method being used (σ). Equation 22.6 shows a normal probability distribution function:

$$f_{(x)} = \frac{1}{\sigma\sqrt{2\pi}}e^{-\frac{(x-\mu)^2}{2\sigma^2}}$$

Eq. 22.6

where x is the value of a particular data point, σ is the standard deviation, μ is the mean of the population, and f_x is the probability of obtaining a particular value of x. Expressing Equation 22.6 in plain English, the probability of obtaining a particular value of x when sampling a population is a function of the true value for that population (μ) and the precision of the technique used (σ). Equation 22.6 is referred to as a *normal probability function* (npf), also known as a *Gaussian distribution*; it is colloquially referred to as a "bell curve."

---

[8] If you are using an Apple computer, the "left click" commands can be obtained by holding down the Apple command key while clicking.

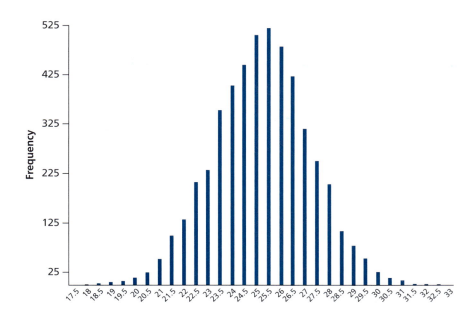

Figure 22.6 Histogram demonstrating a normal distribution of points about a mean: $\bar{x} = 25$, $n = 5,000$, $\sigma = 2$.

In modern instruments, data are collected digitally and therefore data are discrete.[9] You do not get a true bell curve but instead a histogram of points that fall within the digital resolution of the processor. For an npf, the histogram will resemble a bell-shaped distribution about the mean. Figure 22.6 shows a histogram for a measurement in which the error followed an npf and the "true" value was 25. Random error in the analysis returned a range of values with a mean value approximately centered at 25. If you trace a line through the top of each bar in the graph, the histogram approximately conforms to a normal distribution function.

An npf represents the way data is scattered about a mean when the error in the sampling is the result of random error. Figure 22.9 shows a normal probability distribution with the area under the curve integrated as a function of standard deviation. We see that 68.2% of all data points fall within a range of $\pm\sigma$ from the mean, 95.4% of all data points fall within $\pm2\sigma$ of the mean, and by the time we get to $\pm3\sigma$ from the mean, we have incorporated 99.7% of all data points. If we repeated an analysis 1,000 times, we could reasonably expect that only three data points would fall outside the $\pm3\sigma$ range. Knowing the standard deviation allows us to predict the likelihood of the next sampled data point residing within a specified range from the mean.

---

[9] See Chapters 4 and 5 for a review of analog-to-digital conversion.

ACTIVITY

Random Number Generation and Plotting a Histogram in Microsoft Excel

The point of this activity is to help you visualize how N, $\bar{x}$, and σ affect the distribution of data points within a sample set.

Many of the advanced statistical tools available in Excel are found in the Analysis ToolPak. The Analysis ToolPak is not included in the default installation of Excel, so you may need to "turn it on" if you have never used advanced statistical tools in your copy of Excel. Each version of Excel has different steps for activating the Analysis ToolPak. Activate the Help screen on your copy of Excel and select Analysis ToolPak and then follow the instructions for your particular version of Excel.

First we will use Excel's random number generator. Select Random Number Generation from the (data) Analysis ToolPak. The Random Number Generation dialog box will open (see Figure 22.7). Fill in the fields as shown. The random number generator will return a string of numbers with a mean of 25 and a standard deviation of 1.

Next select Histogram from the (data) Analysis ToolPak. The Histogram dialog box will open (see Figure 22.8). Fill in the Histogram dialog box as shown and select OK. Excel will generate a data table similar to the one shown bottom right. To generate a histogram, plot the Bin number versus Frequency as a Column Graph. Your graph should resemble Figure 22.6. You should notice that the histogram has the beginnings of a bell curve, but the existence of random error is

visibly evident. Now repeat this activity with a much larger N values, such as 1,000 or 2,000. Observe how the shape of the histogram has changed. Repeat the exercise again, and this time decrease the standard deviation. What effect do N and σ have on the shape of the histogram?

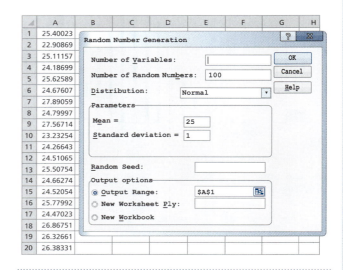

Figure 22.7 Random Number Generator dialog box.

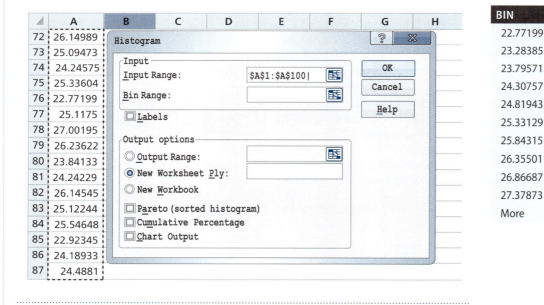

Figure 22.8 Histogram dialog box.

BIN	FREQUENCY
22.77199	1
23.28385	5
23.79571	4
24.30757	17
24.81943	17
25.33129	18
25.84315	15
26.35501	10
26.86687	6
27.37873	4
More	3

Problem 22.4: In your own words, explain how changing N and changing σ affect the histogram generated in the Activity "Random Number Generation and Plotting a Histogram in Microsoft Excel."

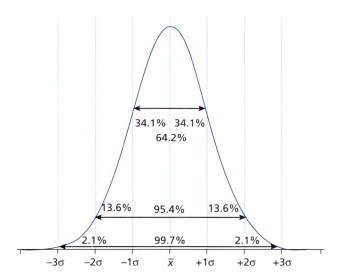

Figure 22.9 The npf. The ranges indicate the percentage of all data points as a function of the distance from the mean. The *x* axis is in standard deviations from the mean.

Example 22.1 The Bell Curve's Shape as a Function of the Standard Deviation

Figure 22.10 shows two different npf curves. Imagine that these two curves represent the analysis of a chemical sample under different experimental conditions. Each experiment produced a sample mean of 50. However, one technique produced a data set with a standard deviation of 5, whereas the other data set had a standard deviation of 10. In the case where s = 5, nearly 99.7% of all data points fell within the range of 40 to 60. In the case where s = 10, we have to expand the range to 20 to 80 in order to capture 99.7% of all data points. If we could only afford to repeat the analysis a few times (time is money), we would have a lot more confidence that our sample mean is close to the population mean for the technique where s = 5 than we would for the technique where s = 10.

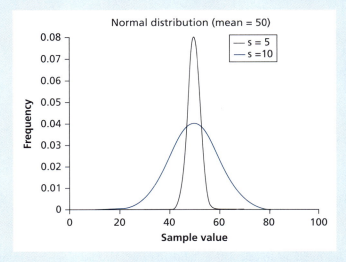

Figure 22.10 Two npf curves. The narrow curve has a standard deviation of 5. The wide curve has a standard deviation of 10.

ACTIVITY

Plotting a Normal Distribution Function in Microsoft Excel

Create the following worksheet in Excel (see Figure 22.11). Create a column of numbers from 2 to 600 in intervals of two in cells A2 to A601. Place a mean value of 301 in cell D1 and your standard deviation of 50 in cell D2. Then select cell B2 and click the Insert Function link (f_x) and choose NORMDIST. The Function Arguments box will open. For the x argument, choose cell A2. For the Mean argument box, type $D$1 and for the Standard Deviation argument box, type $D$2. The dollar signs in the cell addresses lock the addresses and prevent them from scrolling. In the Cumulative argument field, type the word FALSE. The NORMDIST function will use Equation 22.5 to return a probability value for obtaining a value of 2 in cell B2. Select cell B2 again and drag-and-drop it to cell B601. The "B" column now contains the probability of obtaining the values listed in the "A" column. Plot an x–y scatter plot of cells A2:A601 versus B2:B601 and insert the graph in your worksheet. You should see a classic bell curve. Now play with your mean and standard deviation values and observe how the shape of the Gaussian distribution changes as a function of each variable.

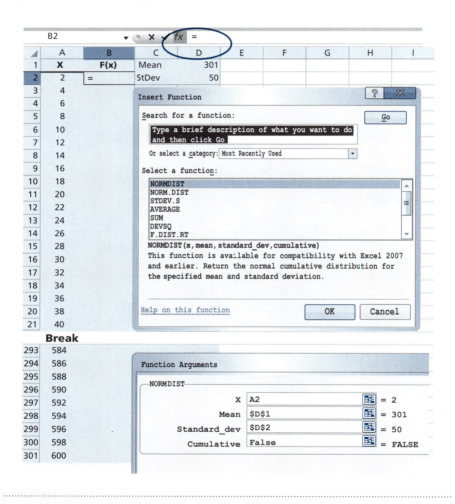

Figure 22.11 Sample spreadsheet for programming a Gaussian curve.

Confidence Limits

Earlier we learned how to calculate a standard error. Another common statistical tool for reporting the uncertainty (precision) of a measurement is the *confidence limit* (CL), which tells us roughly where under the bell curve our data sits. For example, we might report the percent alcohol in a solution as 13% ± 2, where the ±2 represents the uncertainty at the 95% CL (within the ±2σ portion of the bell curve).

Unless otherwise stated, the reported CL is assumed to be at the 95% CL and represents the range in which we are 95% certain the "true" answer lies. The reason the 95% CL is the accepted norm is because 95.4% of all data points in a normal distribution are encompassed by a range of approximately ±2σ. It is reported at 95% instead of 95.4% for purposes of simplicity. However, as you will soon see, it is possible to calculate CL values other than the 95% CL.

We define CL using σ. Recall that σ is the standard deviation of the entire population. If we know the standard deviation for the entire population, then the 95% CL[10] is simply:

$$95\% \text{ CL} = \pm 2\sigma \qquad \textbf{Eq. 22.7}$$

and we report the mean as shown before: $\mu \pm 2\sigma$.

However, we seldom know the mean or the standard deviation of an entire population. *All chemical analyses* deal with sampled populations. When we do not know σ for the entire population, we use "s" with an added statistical term. The CL for a sample is given in Equation 22.8:

$$\text{Confidence limit} = \pm t \frac{s}{\sqrt{N}} \qquad \textbf{Eq. 22.8}$$

and we report the mean[11] as: $\bar{x} \pm t \dfrac{s}{\sqrt{N}}$

where $\bar{x}$ is the mean of the sample, s is the standard deviation of the sample, N is the number of data points in the sample, and t is a statistical factor taken from a table of t-values, such as in Table 22.1. The value of t is determined based on degrees of freedom in a normal distribution. The term $(N - 1)$ in Table 22.1 is the *degrees of freedom* for the sample set.

Using Spreadsheets to Determine Confidence Limits As we have seen, modern spreadsheets such as Excel are capable of very sophisticated statistical analysis. The following Activity will walk you through the steps of calculating the CL for a sample mean.

TABLE 22.1: Confidence Limits t-Values as a Function of (N – 1)*

N – 1	90%	95%	99%	99.5%
2	2.920	4.303	9.925	14.089
3	2.353	3.182	5.841	7.453
4	2.132	2.776	4.604	5.598
5	2.015	2.571	4.032	4.773
6	1.943	2.447	3.707	4.317
7	1.895	2.365	3.500	4.029
8	1.860	2.306	3.355	3.832
9	1.833	2.262	3.205	3.690
10	1.812	2.228	3.169	3.581

*The term (N–1) is the degrees of freedom for the data set.

Confidence Limits versus t-tests

When comparing two different measurements of similar systems (i.e., the alcohol content of two bottles of the same brand of beer), it is desirable to know if the differences in the measured means are statistically significant. An older statistical tool called a t-test is often used and is still in common use. However, confidence limits are a more rigorous statistical tool. Students are encouraged to compare CL intervals rather than performing t-tests. If the confidence intervals of the two means *do not* overlap, then the two means are statistically different.

---

[10] To be completely accurate, the 95% confidence limit is actually the 95.4% confidence limit because it represents ±2σ from the mean (see Figure 22.5).

[11] Recall that we defined $\dfrac{s}{\sqrt{N}}$ as the standard error in Equation 22.5.

ACTIVITY

Using Microsoft Excel to Calculate Confidence Limits

Open the spreadsheet Fish that you created in the previous Activities. We are going to use Excel to determine the 95% CL of our data set.

Select cell B16 and click on the f_x button once again. From the Insert Function dialog box, select CONFIDENCE.NORM. The CONFIDENCE.NORM dialog box will appear. To calculate a 95% CL you need to input the uncertainty in the Alpha field as 1.00 – CL. You need to input the CL as a decimal (1.00 – 0.95 = 0.05). Enter the standard deviation (or the address of the standard deviation) into the Standard_dev field. In the Size field, enter the value of N, the total number of data points (10), and click OK. At this point, your spreadsheet should resemble the one shown in Figure 22.12. You can calculate other CLs by changing the value of alpha.

In this Activity, we calculated the 95% CL for the analysis of mercury in fish. If we were to report the answer to the hundredth place, we would say that the average concentration of mercury in fish is 5.10 with a 95% CL of ±1.00. The implication of the CL is that we are 95% certain that the "true" value lies between 4.1 and 6.1 ppb. Or to state this another way, if we repeated the experiment one more time, we are 95% confident that the next data point will lie within this range.

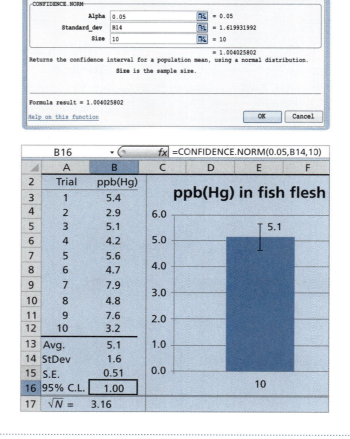

Figure 22.12 Data for the analysis of mercury in fish. The data includes the mean, standard deviation, standard error, 95% CL, and a plot of the data showing the standard error as error bars.

Problem 22.5: For the data set used in the activity "Using Microsoft Excel to Calculate Confidence Limits," determine the 90% and the 99% CL.

Problem 22.6: For the data set used in the activity "Using Microsoft Excel to Calculate Confidence Limits," determine the 95% CL using Equation 22.8 and compare your answer to your answer from the activity.

Propagation of Error

Reporting the standard deviation, the standard error, or the CL for measured data points is an acceptable way of portraying the precision of a measurement. But what do you do if you use two or more measured values in a computation? How do you report the confidence in the computed value? For example, imagine you determined the density of an object by independently measuring the mass and the volume. Each of those measurements contains error. In other words, you have an error associated with both the volume measurement and the mass measurement, and when we divide the mass by the volume to get density we want to be able to report the composite error of the resulting density. We need to know how to propagate the standard deviations through various mathematical manipulations. Table 22.2 outlines this process.[12] The standard deviation of a computed result is given as S_R where R is the computed result.

Once you have propagated the standard deviation through the mathematical manipulations, the 95% CL can be approximated as ±2s. Similarly, the 99.7% CL can be approximated as ±3s. However, if you wish to calculate a CL other than the 95% CL or the 99.7% CL, you will need to determine the degrees of freedom (df) for the calculated value using Equation 22.9 and then use Equation 22.8 or Microsoft Excel to find the CL:

TABLE 22.2: Propagation of Error*

CALCULATION	EXAMPLE	STANDARD DEVIATION OF RESULT
Multiplication/division	$R = \dfrac{(\alpha \times \beta)}{\gamma \times \delta}$	$S_R = R\sqrt{\left(\dfrac{s_\alpha}{\alpha}\right)^2 + \left(\dfrac{s_\beta}{\beta}\right)^2 + \left(\dfrac{s_\gamma}{\gamma}\right)^2 + \left(\dfrac{s_\delta}{\delta}\right)^2}$
Addition/subtraction	$R = \alpha - \beta + \gamma + \delta$	$S_R = \sqrt{s_\alpha + s_\beta + s_\gamma + s_\delta}$
Logarithm	$R = \log(\alpha)$	$S_R = 0.434\dfrac{s_\alpha}{\alpha}$
Inverse log	$R = \text{inv-log}(\alpha)$	$S_R = R(2.303 S_\alpha)$
Exponents	$R = \alpha^x$	$S_R = RX\left(\dfrac{s_\alpha}{\alpha}\right)$

*α, β, γ, and δ are experimentally derived data with standard deviations of s_α, s_β, s_γ, and s_δ, respectively; R = computed result; S_R = standard deviation of result.

---

[12] For more on propagation of error, see Bevington, P. R.; Robinson, K. *Data Reduction and Error Analysis for the Physical Sciences*, 3rd ed., McGraw Hill: New York, **2002**, or Monk, P.; Munro, J. *Math for Chemistry*, 2nd ed., Oxford University Press: Oxford, UK, **2010**.

$$df = \frac{\left(\frac{s_\alpha^2}{N_\alpha} + \frac{s_\beta^2}{N_\beta} + \frac{s_\gamma^2}{N_\gamma} + \frac{s_\delta^2}{N_\delta}\right)}{\frac{s_\alpha^4}{N_\alpha(N_\alpha-1)} + \frac{s_\beta^4}{N_\beta(N_\beta-1)} + \frac{s_\gamma^4}{N_\gamma(N_\gamma-1)} + \frac{s_\delta^4}{N_\delta(N_\delta-1)}}$$ Eq. 22.9

where N_α, N_β, N_γ, and N_δ are the number of replicate data points for the experimentally derived data sets α, β, γ, and δ, with standard deviation of s_α, s_β, s_γ, and s_δ, respectively.

Example 22.2

Imagine that we were determining the volume of an unknown solid by displacement of water in a graduated cylinder ($\Delta V = V_f - V_i$). The initial volume was 23.40 mL, the final volume was 24.95 mL, and $\Delta V = 24.9 - 23.2 = 1.7$ mL. You might be tempted to conclude that the uncertainty is ±0.1 mL. However, if you were to be rigorous in your propagation of error, you would recognize there was an implied ±0.1 mL uncertainty in both the initial and final volume readings. Table 22.2 showed us that the proper way to estimate error when subtracting two numbers is:

$$S_{\Delta V} = S_{\Delta V}\sqrt{S_{V_f} + S_{V_i}} = \sqrt{0.2} = 0.44$$

Now let us imagine we determined the mass on a digital balance and obtained a value of 3.003 grams. If you recall what you were taught about significant figures, the implication is that the uncertainty is in the thousandth place and a reasonable estimate of the standard deviation would be ±0.001 g. What is the uncertainty in the density?

$$d = \frac{mass}{volume} = \frac{3.003g}{1.7mL} = 1.77\,g\,/\,mL$$

If you simply applied the rules for reporting significant figures, you might assume the uncertainty in this number were ±0.01. However, because we have a calculated data point resulting in measurements made with different precisions, a more rigorous application of propagation of error is required. Take another look at Table 22.2. The equation for propagation of errors for multiplication and division is:

$$S_d = RS_d = R\sqrt{\left(\frac{s_V}{V}\right)^2 + \left(\frac{s_m}{m}\right)^2} = 1.77\sqrt{\left(\frac{0.14}{1.77}\right)^2 + \left(\frac{.001}{3.003}\right)^2} = 0.08$$

We would want to report our final density as:

$$d = 1.77\,g\,/\,mL \pm 0.08$$

Problem 22.7:   Assume that you measured the mass of 1.0014 grams of potassium oxalate ($K_2C_2O_4$) on a digital balance and placed it in a one-liter volumetric flask with a rated precision of 0.001 L. Calculate the molarity of the final solution and report the molarity with a 95% CL using the appropriate propagation of error equation.

Analyzing Data Sets

In addition to reporting the error associated with an individual data set, the analytical chemist often needs to compare and analyze the variance in data gathered under different circumstances. The different circumstances can be as benign as collecting data on different days or potentially more significant such as collecting data using different instruments or data collected by different technicians. For example, imagine you are perfecting a C-18 reverse phase high-performance liquid chromatography method for the purification of a pharmaceutical product. In the final protocols, how important is it that you purchase your C-18 columns from the same manufacturer each time you replace the column? Are the changes you see in the data when you change suppliers statistically significant? We could ask the same question of the solvent. Is it statistically important that we use the same supplier of solvent every time we run the procedure? We can investigate these types of questions by using several different statistical tools.

Because of random error anytime you repeat an analysis, you expect to obtain different results. But are the observed differences within the expected variance of the technique? This is a fundamental question in an analytical laboratory. You may have a data point that seems significantly different than the other replicates in the data set, and you would like a statistical basis for keeping or rejecting that data point. Or you may want to know the effect of a particular experimental parameter on the overall variance of a method. For instance, when comparing the means of data taken by two different laboratory technicians, are the observed differences in the means statistically significant? Or you may want to compare the results of an analysis using two different instruments (i.e., two different ultraviolet-visible spectrometers) or two different techniques. Again, are the observed differences statistically significant? In the next few sections, we will introduce tools that you can use to help answer these types of questions.

Identifying Outliers: Q-Test

Although the International Organization of Standardization now recommends that we use the Grubbs' test for identifying outliers, the Q-test still remains a very commonly used method and we introduce it here because you are likely to encounter it in your careers. We will examine the Grubbs' test in the next section.

Sometimes you obtain a set of replicate data, and there is one (or more) data point that just intuitively "seems wrong." For example, Table 22.3 shows the results for the $N = 10$ replicate analysis of caffeine in coffee. The data points tend to cluster around 80 ppm with the exception of cup 5, which has a lower reading of 72 ppm. The sloppy analyst might be tempted to throw out the cup 5 data based solely on intuition; however it is quite possible that 72 ppm falls within the 95% CL for this distribution of points.

TABLE 22.3: Analysis of Caffeine in Coffee

CUP	CAFFEINE (PPM)
1	78
2	82
3	81
4	77
5	72
6	79
7	82
8	81
9	78
10	83
Average	79.3
Standard deviation	3.3
95% confidence limit	2.0

It is unethical to simply assume data is inaccurate and discard that data point. You should include all data in a report, even outliers, and if you decide to reject a point in your final analysis, you must have a statistical justification for that decision.

A *Q-test* is a statistical tool used to identify an outlier within a data set.[13] To perform a Q-test, you must first arrange your data in a progressive order (low to high or high to low) and then using Equation 22.10, you calculate an experimental Q-value (Q_{exp}). If Q_{exp} is greater than the critical Q value (Q_{crit}) found in Table 22.4, then you are statistically justified in removing your suspected outlier from further consideration.[14] You then recalculate the mean, standard deviation, and the 95% CL with the outlier removed from the calculations.

$$Q_{exp} = \frac{\left| x_q - x_{n+1} \right|}{w}$$
Eq. 22.10

where

X_q = suspected outlier
X_{n+1} = next nearest data point
w = range (largest – smallest data point in the set)

Example 22.3

Perform a Q-test on the data set from Table 22.3 and determine if you can statistically designate data point 5 as an outlier within a 95% CL. If so, recalculate the mean, standard deviation, and the 95% CL.

STRATEGY –

Organize the data from highest to lowest value and use Equation 22.10 to calculate Q_{exp}.

SOLUTION –

Ordering the data from Table 22.3 from highest to lowest results in

Cup	10	7	2	8	3	6	9	1	4	5
ppm caf	83	82	82	81	81	79	78	78	77	72
Range = 83–72=	11									

Substitution into Equation 22.10 yields:

$$Q_{exp} = \frac{\left| x_q - x_{n+1} \right|}{w} = \frac{|72-77|}{11} = 0.455$$

Using the Q_{crit} table, we see that $Q_{crit} = 0.466$. Because Q_{exp} is less than Q_{crit}, you must keep the data point.

---

[13] Dean, R. B.; Dixon, W. J. *Anal. Chem.*, **1951**, *23*, 636–638.

[14] Rorabacher, D. B. *Anal. Chem.*, **1991**, *63(2)*, 139–146.

TABLE 22.4: Critical Rejection Values for Identifying an Outlier: Q-test

	Q_{CRIT}		
N	90% CL	95% CL	99% CL
3	0.941	0.970	0.994
4	0.765	0.829	0.926
5	0.642	0.710	0.821
6	0.560	0.625	0.740
7	0.507	0.568	0.680
8	0.468	0.526	0.634
9	0.437	0.493	0.598
10	0.412	0.466	0.568

Problem 22.8: Use the data in Table 22.3 and determine the value (in ppm) cup 5 would have to be before Equation 22.10 would identify it as an outlier. Show your work.

Problem 22.9: Imagine that this set of five replicate data points were collected for the analysis of lead in drinking water.

Trial	1	2	3	4	5
Pb (ppm)	1.3	1.4	1.0	1.3	1.4

(a) Calculate a mean, standard deviation, and 95% CL using this data set (you may want to use a spreadsheet).

(b) Perform a Q-test on the data set. How does the performance of a Q-test alter your answer in part (a)?

Identifying Outliers: Grubbs' Test

The recommended way of identifying outliers is to use the Grubbs' test. A *Grubbs' test* is similar to a Q-test; however, G_{exp} is based on the mean and standard deviation of the distribution instead of the next-nearest neighbor and range (see Equation 22.11).

$$G_{exp} = \frac{|x_q - \bar{x}|}{s}$$

Eq. 22.11

If G_{exp} is greater than the critical G value (G_{crit}) in Table 22.5, then you are statistically justified in removing your suspected outlier from further consideration. You then

TABLE 22.5: Critical Rejection Values for Identifying an Outlier: Grubbs' test

	G_{crit}		
N	90% CL	95% CL	99% CL
3	1.153	1.154	1.155
4	1.463	1.481	1.496
5	1.671	1.715	1.764
6	1.822	1.887	1.973
7	1.938	2.020	2.139
8	2.032	2.127	2.274
9	2.110	2.215	2.387
10	2.176	2.290	2.482

recalculate the mean, standard deviation, and the 95% CL with the outlier removed from the calculations.

> **Problem 22.10:** Perform a Grubbs' test using the data set in Problem 22.9. Report the mean, standard deviation, and the 95% CL based on the results of the test.

Analyzing Variance: F-Tests

The *F-test* allows for the comparison of the variance of two different data sets in order to determine if there is a statistically significant difference. The test is named after Ronald Fisher, who first developed it in the 1920s.[15]

In a working laboratory or field situation, it is common to have data sets that were obtained under different circumstances. For instance, data may have been collected on different days, or you may have two different analysts conducting the same measurements. When the final results vary, you need a way to determine if the difference is statistically significant. In a manner similar to the Grubbs' test and the Q-test, you perform an F-test by calculating an experimental F value (F_{exp}) and comparing that to a critical F value (F_{crit}). If F_{exp} is greater than F_{crit}, then the variance of the two data sets used to calculate F_{exp} is statistically different. F_{exp} is determined by the ratio of the sample variances (square of the standard deviations). The larger variance value goes in the numerator so that F_{exp} is always greater than one.

$$F_{exp} = \frac{s_1^2}{s_2^2}$$

Eq. 22.12

In this case, the null hypothesis is that the two variances represent the same population. To reject (or accept) the null hypothesis, we compare F_{exp} to F_{crit}. The tables for critical F values are tabulated as a function of CLs and degrees of freedom for s_1^2 and s_2^2. As a result, a full set of F tables can be extensive. Table 22.6 is an example of critical F values at the 95% CL for degrees of freedom up to 10.

TABLE 22.6:   F-Test Critical Values (95% confidence limit)*

DENOMINATOR DEGREES OF FREEDOM	NUMERATOR DEGREES OF FREEDOM						
	1	2	3	4	5	7	10
1	161.5	199.5	215.71	224.6	230.2	236.8	241.9
2	18.51	19.00	19.164	19.25	19.30	19.35	19.40
3	10.13	9.552	9.2766	9.117	9.014	8.887	8.786
4	7.709	6.944	6.5915	6.388	6.256	6.094	5.964
5	6.608	5.786	5.4095	5.192	5.05	4.876	4.735
7	5.591	4.738	4.3469	4.12	3.972	3.787	3.637
10	4.965	4.103	3.7082	3.478	3.326	3.135	2.978

* The degrees of freedom used to calculate s_1^2 and s_2^2 represent the column and row headings, respectively.

---

[15] Fisher, R. A. *Statistical Methods, Experimental Design, and Scientific Inference.* Oxford University Press: New York, NY, **1999**.

Fortunately, we do not need a complete set of F tables on hand. Microsoft Excel can be used to perform F-tests. Example 22.4 shows an example data set collected from the high-performance liquid chromatography analysis of residual acrylamide from a batch of polyacrylamide.[16] In this study, two different analysts performed ten replicate studies. The results showed a mean value of 10.1 ppb for analyst 1 and 10.5 ppb for analyst 2, with standard deviations of 0.9 and 1.5, respectively. The mean values of 10.1 and 10.5 may seem similar enough with a gross deviation between the two means of only 0.4, but what you really want to determine is if a gross deviation of 0.4 is within a 95% confidence limit for the standard deviations of the data sets. Example 22.4 walks you through the performance of an F-test.

Example 22.4

The spreadsheet seen below is a right-tailed F-test* comparing the results of two different analysts for the measurement of residual acrylamide monomer in a batch of polyacrylamide (ppb). To perform an F-test using Microsoft Excel, you need to enter your data as shown in the spreadsheet and determine the standard deviation for each data set (e.g., see cells B13 and C13). Then determine the experimental F value using Equation 22.12 and put that value into one of the cells (here we used B14). Next you need to click on the Insert Function button (f_x) and choose F.DIST.RT. The Function Argument dialog box will open as shown.

Enter your experimentally determined F value for x and the numerator degrees of freedom[17] for Deg_fredom1 and the denominator degrees of freedom for Deg_fredom2 and click OK.

In the spreadsheet, the F-test returned a value of 0.077. If we round that to 0.08, then what this test tells us is the two sets of data can be considered the same if we also accept a CL of 92% $(1.00 - 0.08 = 0.92)$. If we need an 95% CL, we need the F-test to return a value of 0.05 or less.

_____

*Data set with largest variance in the numerator.

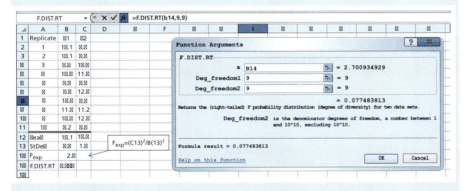

Spreadsheet 22.1 Right-tailed F-test comparing the results of two different analysts for the measurement of residual acrylamide monomer in a batch of polyacrylamide (ppb).

_____

[16] Polyacrylamide is a water-absorbent polymer used in diapers. The monomer is a neurotoxin, so it is critical that each batch be tested for residual monomer concentration before it is sent to market.

[17] See Equation 22.4 for a review of *degrees of freedom*.

Problem 22.11: You have just measured the pH of the water sampled from a local lake. You have ten replicate measurements with two different pH probes. The data are presented below. Conduct an F-test on the data set and comment on the results.

pH OF LOCAL LAKE WATER										
Replicate	1	2	3	4	5	6	7	8	9	10	Average pH
Probe 1	6.74	6.49	6.71	6.62	6.76	6.67	6.99	6.68	6.96	6.52	6.71
Probe 2	6.93	6.83	6.90	6.79	6.88	6.64	7.10	7.18	7.04	6.97	6.93

Problem 22.12: In 2006, the Anne Arundel County Maryland Department of Health tested local wells for elevated levels of arsenic. They found that 35 out of 71 wells showed elevated levels. Atomic absorption spectroscopy is a very convenient way to measure arsenic in water. Imagine you are a laboratory manager and you have given identical arsenic samples to two different technicians. Conduct an F-test on the two sets of data and comment on the results.

Replicate	1	2	3	4	5	6	7	8	9	10	Average Arsenic Levels (ppb)
Tech 1	0.304	0.306	0.301	0.320	0.324	0.276	0.302	0.329	0.304	0.297	0.306
Tech 2	0.331	0.285	0.317	0.298	0.346	0.239	0.307	0.258	0.308	0.326	0.302

ANOVA: *A Two-Dimensional F-Test Analysis of variance* (ANOVA) is another statistical test to help us determine the dispersion in a data set. It is very similar in concept to an F-test, and, in fact, we actually calculate an F value in the analysis. For example, in Example 22.4, we imagined a scenario where two different analysts performed the same test on the same batch of polyacrylamide. Let us imagine next that we sent that same batch of polyacrylamide out to five different laboratories and on receiving the data, we wished to statistically compare the results. We could conduct an F-test on each possible pairing of laboratories, but that would be tedious and the results would be difficult to interpret. A more sophisticated approach would be to compare the average variance that occurs as a result of changing laboratories to the average variance that occurs as a result of performing replicate samplings. Spreadsheet 22.2 shows the raw data along with an ANOVA analysis with inputs conducted by hand for the purposes of demonstration. Fortunately for us, Microsoft Excel will do an ANOVA automatically, and you will not need to program each cell manually (see Spreadsheet 22.3).

When you combine the data from all five laboratories, there are a total of 50 replicate data points. The average result of all 50 points is called the *grand mean*. In this case, we obtained a value of 10.31 ppb. For each data point the deviation from the grand mean was calculated (columns: D, G, J, M, P). This value is termed the *mean corrected value* (d_{ij}). Next, we squared the mean corrected values (d_{ij}^2) to generate positive numbers (columns: E, H, K, N, Q). Next we summed all of the d_{ij} values (SS_c) and then divided SS_c by the degrees of freedom[18] to yield $\overline{SS}_C$. Compare the derivation of $\overline{SS}_C$ to

---

[18] See Equation 22.4 for a review of *degrees of freedom*.

the derivation of a standard deviation (Equation 22.4). The value $\overline{SS}_C$ is essentially the *grand standard deviation* of replicates among laboratories. Similarly, we also calculated a $\overline{SS}_r$ value, which can be thought of as the grand standard deviation of laboratories among replicates. The F value is then determined by dividing $\overline{SS}_C$ by $\overline{SS}_r$. How one uses an F value is demonstrated in the next Activity.

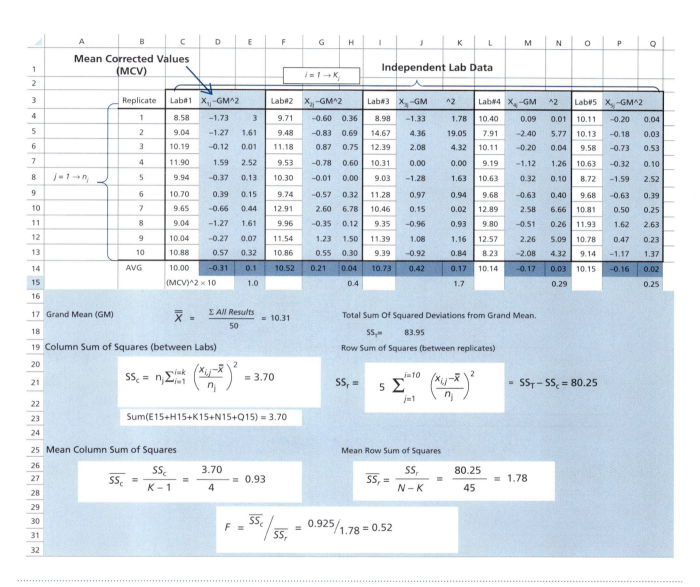

Spreadsheet 22.2 Analysis residual monomer (ppb) found in a batch of polyacrylamide conducted at five different independent laboratories—ANOVA.

ACTIVITY

Letting Microsoft Excel Perform ANOVA

Using the same data we examined in Spreadsheet 22.2, we will perform an ANOVA using the ANOVA statistics function in Excel. From the (data) Analysis ToolPak, select ANOVA: Single Factor. The ANOVA Single Factor dialog box will open (see Spreadsheet 22.3). The input range is the total 50 data points obtained between all five laboratories. In this example, we have the independent laboratories arranged in columns so make sure the Grouped By Columns radial button is selected. Notice that we have also selected the Labels in First Row check box. The default value for α is 0.05, which will calculate a 95% confidence interval for

your ANOVA. You have several options for the output. If you choose to keep your ANOVA output with your raw data, then you have to tell Excel where you want the data table to start. In this case we began our data table at cell W16.

The ANOVA table in Spreadsheet 22.3 has a calculated F value of 0.51897 (the same value we calculated by hand). The p value shown is called the *value of probability*. Because we selected an alpha value of 0.05, we want our p value to be above 0.05 in order for the null hypothesis to hold. In other words, this ANOVA study did *not* find any statistically significant variance between the five laboratories.

Independent Lab Data

$i = 1 \rightarrow K_j$

	Replicate	Lab#1	Lab#2	Lab#3	Lab#4	Lab#5
	1	8.58	9.71	8.98	10.40	10.11
	2	9.04	9.48	14.67	7.91	10.13
	3	10.19	11.18	12.39	10.11	9.58
	4	11.90	9.53	10.31	9.19	10.63
$j = 1 \rightarrow n_j$	5	9.94	10.30	9.03	10.63	8.72
	6	10.70	9.74	11.28	9.68	9.68
	7	9.65	12.91	10.46	12.89	10.81
	8	9.04	9.96	9.35	9.80	11.93
	9	10.04	11.54	11.39	12.57	10.78
	10	10.88	10.86	9.39	8.23	9.14

Data Analysis

Analysis Tools

Anova: Single Factor
Anova: Two-Factor With Replication
Anova: Two-Factor Without Replication
Correlation
Covariance
Descriptive Statistics
Exponential Smoothing
F-Test Two-Sample for Variances
Fourier Analysis
Histogram

OK   Cancel   Help

Anova: Single Factor

Anova: Single Factor

SUMMARY

Input

Input Range: $Y$4:$AC$13

Grouped By: ● Columns ○ Rows

☑ Labels in first row

Alpha: 0.05

Output options
● Output Range: $W$16
○ New Worksheet Ply:
○ New Worksheet

OK   Cancel   Help

Groups	Count	Sum	Average	Variance
Lab#1	10	99.9537	9.99537	0.98611
Lab#2	10	105.215	10.5215	1.21963
Lab#3	10	107.255	10.7255	3.21661
Lab#4	10	101.4	10.14	2.62418
Lab#5	10	101.52	10.152	0.87011

ANOVA

Source of Variation	SS	df	MS	F	P-Value	F crit
Between Groups	3.70201	4	0.9255	0.51897	0.72217	2.57874
Within Groups	80.2498	45	1.78333			
Total	83.9518	49				

Spreadsheet 22.3 A single-factor ANOVA in Microsoft Excel.

22.5 Linear Regression Analysis

The preceding section provided tools useful to the experimenter when working with repetitive data (i.e., measurements that are expected to have essentially the same value every time). When conducting instrumental analysis, however, it is often the case that we do not know in advance the actual magnitude of the measurement but only an estimate of a range in which the measurement might fall. In such cases, we must prepare and measure standard samples[19] that fall in the expected range in order to calibrate the instrumental response for known concentrations. The fundamental signal that is obtained from an instrument is either a voltage or a current, neither of which directly gives us useful information about our sample. In practice, we use standard calibration curves to relate that fundamental signal to one that is more meaningful, such as pH or absorbance. We then plot this signal as a function of known concentrations to yield a calibration curve so that the signal from an unknown sample can be used to determine the analyte concentration. The basic statistical tools outlined above must be further developed for application to measurements made using a calibration curve.

For an instrument response that is linear with analyte concentration, we would expect to obtain a series of data points that fall on a straight line as the concentration is varied. However, we also expect there to be error in the measured values, so the points will have some degree of variance from the anticipated straight line. If we were to graph those values on paper, we could use a straight edge to estimate a best fit line for the points. In the modern electronic age, however, it is more common (and more accurate) to use a method called *linear regression analysis* to determine the best linear approximation from the measured points.

Much of what we need for our analyses can be obtained quickly and easily from Microsoft Excel or another spreadsheet software package. Using the LINEST function in Excel, we can obtain the slope and intercept for the regression, as well as the standard deviations associated with those values. Furthermore, we can extract the *coefficient of determination*, R^2, also known as the *R-squared* value and the standard error for the y-estimate (essentially the standard deviation for the regression), s_y. The R^2 has a value between zero and 1 and is often referred to as the *goodness of fit* or a *correlation coefficient*. An R^2 value of 1 indicates a perfect fit between the actual y values and those calculated using the linear equation. The more the R^2 value deviates from 1, the greater the deviations between the actual and calculated y values. The s_y value is used in calculating the standard deviation of results for measurements of unknown samples obtained using the calibration curve.

It is helpful to understand how the software goes about calculating an equation for the linear regression. To find a best fit line (Equation 22.13), the software is programmed to minimize the sum of the squared differences (sometimes called *residuals*) between the actual y-values and those calculated by the linear equation for each x–y pair. The reason we minimize the residuals in y-values rather than x-values stems from the assumption that the most significant error occurs in the measurement of the signal (which is plotted on the *y*-axis) and not the result of error resulting from poor construction of your standard solutions (which is plotted on the *x*-axis). Therefore it is critically important that signal response be plotted on the *y*-axis. Figure 22.13 provides a visual depiction of what we mean by these residuals. The residuals are squared to eliminate any negative values, and then the slope of the line is adjusted until the sum of the residuals reaches a minimum value. If we call the summed residual values value SS_{y-y}, the software seeks to minimize it in the form of Equation 22.14.

$$y_{i,calc} = mx_i + b \qquad\qquad \textbf{Eq. 22.13}$$

$$SS_{y-y} = \sum_{i=1}^{i=N} \left(y_{i,act} - y_{i,calc} \right)^2 \qquad\qquad \textbf{Eq. 22.14}$$

where
  SS_{y-y} = sum of the squared residuals
  $y_{i,act}$ = actual (measured) value for y in a given (i) of an x–y pair
  $y_{i,calc}$ = y value calculated from the linear equation Equation 22.13

---

[19] Recall that a standard sample is one in which the concentration of analyte is known.

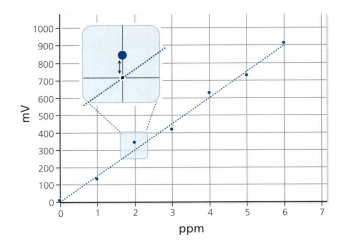

Figure 22.13 Demonstration of the y-residuals. The region around the third point is expanded in the inset. The small black dot represents $y_{i, calc}$, and the large blue point represents $y_{i, act}$. The difference between the two points is $(y_{i, act} - y_{i, calc})$ for $y_i = 3$.

Most of what we need for sample analysis can be obtained fairly directly through Excel (see the Activity "Letting Microsoft Excel Perform LINEST to Give Linear Regression Data" and Example 22.5), but in order to accomplish full statistical analysis, we need to define one additional quantity, S_{x-x}, given in Equation 22.15. With the information obtained from the Excel LINEST function and S_{x-x}, we will be able to calculate a standard deviation for any y value calculated for a sample of unknown concentration using the linear regression of the calibration plot (Equation 22.16):

$$S_{x-x} = \sum_{i=1}^{i=N} (x_i - \overline{x})^2$$

Eq. 22.15

where

  x_i = value for x in a given (i) of an x-y pair
  $\overline{x}$ = the mean of all of the x-values

$$S_c = \frac{S_y}{m} \sqrt{\frac{(\overline{y}_s - \overline{y}_{cal})^2}{m^2 (s_{x-x})} + \frac{1}{N_C} + \frac{1}{N_s}}$$

Eq. 22.16

where

  S_c = standard deviation of a calculated y-value for an unknown sample
  S_y = standard error in the y-estimate (from Excel *linest* function)
  m = slope of the regression line (from Excel *linest* function)
  $\overline{y}_s$ = mean of all y-values for N_s replicates of the unknown sample
  $\overline{y}_{cal}$ = mean of all y-values for N_C samples used in the calibration
  N_c = number of samples used in the calibration
  N_s = number of replicates of the unknown sample

ACTIVITY

Letting Microscoft Excel Perform LINEST to Give Linear Regression Data

Set up a calibration data set as given in Spreadsheet 22.4. To have Excel calculate the pertinent analysis data from the calibration information, we use the LINEST function in the following procedure:

1. Highlight an area encompassing two columns and three rows* (the highlighted area in Spreadsheet 22.4 is in columns F and G, rows 7 through 9).

2. *With that area still highlighted,* place your cursor in the command prompt field and start typing the function = *linest*

3. After the open parentheses, highlight all the y values in the calibration data, then enter a comma.

4. Next, highlight all the x values in the calibration data, then enter a comma.[†]

5. For the next parameter, you need to make a choice.

   (a) If you expect that the calibration data should pass through zero (intercept of zero), then enter a zero followed by a comma.

   (b) If you want the function to calculate an intercept value, enter a 1, followed by a comma.

6. Now enter a 1, telling Excel to calculate statistics beyond just the slope and intercept, close the parentheses, but *do not simply press Enter.*

7. To complete the calculation, *press Ctrl + Shift + Enter* (while holding down the Ctrl key, press the Shift key, and while still holding down both of those, press the Enter key).

In the example given in Spreadsheet 22.4, the completed function looked like this, wherein we allowed *LINEST* to calculate an intercept value:

$$=LINEST(C7:C11,B7:B11,1,1)$$

The information that Excel yields from the *LINEST* function includes the slope (m, in Cell F7), the intercept (b, G7), the standard deviation in the slope (s_m, F8), the standard deviation in the intercept (s_b, G8), the coefficient of determination (R^2, F9), and the standard deviation in the y-estimate (s_y, G9).

Note that if you need to edit your *LINEST* function, you will need to highlight the full 2 × 3 block again, make your edits, then press *Ctrl + Shift + Enter.*

	A	B	C	D	E	F	G	H
1								
2		Fluorescence as a Function of Quinine Concentration						
3								
4		Quinine	Measured					
5		Concentration	Fluorescence					
6		(ppm)	(mV)					
7		1.0	157.3		m =	140.4	5.32	= b
8		2.0	301.1		s_m =	13.59	45.1	= s_b
9		3.0	363.4		R^2 =	0.973	43.0	= s_y
10		4.0	601.5					
11		5.0	709.0					

Spreadsheet 22.4 The LINEST function in Microsoft Excel.

* *Actually, Excel will provide additional statistics. If we highlight an area that is two columns by five rows, however, the additional two rows of statistics are not generally as useful as the first 3.*

[†] *You can accomplish the same thing using the LINEST function in the function dialog box and following the prompts.*

Problem 22.13: Use Equation 22.14 to calculate SS_{y-y} for the example given in the Activity "Letting Excel Perform LINEST to Give Linear Regression Data" (Spreadsheet 22.4).

Problem 22.14: Repeat Problem 22.13, but use a slope that is 1% lower and an intercept that is 1% higher than that seen in Spreadsheet 22.4. Compare the SS_{y-y} you calculate with that found in 22.13. Is the result as expected? Explain.

Example 22.5

Following the calibration represented in Spreadsheet 22.4, three replicates of a quinine sample of unknown concentration were prepared and the fluorescence measured, yielding the values 406.6, 414.6, and 408.2. Calculate the quinine concentration in the sample and the standard deviation in the calculated value.

STRATEGY –

Use Equation 22.13 and the LINEST data in Spreadsheet 22.4 to calculate the quinine concentration. Then use Equation 22.16 to calculate the standard deviation in the calculation.

SOLUTION –

The average measured value (The average measured value ($\bar{y}_s$) for the unknown sample is 409.8, so we can calculate the concentration as:

$$\bar{y}_s = mx + b$$

$$409.8 = 140.4x + 5.32$$

$$x = 2.8813 = 2.9 \text{ ppm}$$

For Equation 22.16, we will use the following values:

$$s_y = 43.0 \qquad m = 140.4 \qquad S_{x-x} = 10$$

$$\bar{y}_s = 409.8 \qquad \bar{y}_{cal} = 426.5$$

$$N_C = 5 \qquad N_S = 3$$

$$S_c = \frac{s_y}{m}\sqrt{\frac{\left(\bar{y}_s - \bar{y}_{cal}\right)^2}{m^2\left(s_{x-x}\right)} + \frac{1}{N_C} + \frac{1}{N_S}}$$

$$S_c = \frac{43.0}{140.4}\sqrt{\frac{\left(409.8 - 426.5\right)^2}{140.4^2\left(10\right)} + \frac{1}{5} + \frac{1}{3}} = 0.224 = 0.22 \text{ ppm}$$

Problem 22.15: The following data were obtained for a set of calibration solutions of p-nitroaniline, measured by absorbance using ultraviolet-visible spectrophotometry.

CONCENTRATION (ppm)	ABSORBANCE (AU)
19.5	0.980
9.74	0.440
4.87	0.255
0.974	0.101

A p-nitroaniline solution of unknown concentration exhibited an average absorbance of 0.181 for five replicate samples. Assuming that the y-intercept is zero for the calibration, calculate the concentration of the unknown solution and the standard deviation in the calculation.

Problem 22.16: Repeat Problem 22.15, but do not assume that the intercept is zero for the calibration. Which set of results do you believe are more accurate? Explain. What additional information would you need in order to make a more definitive judgment?

Problem 22.17: The following data were obtained for the calibration of a flame atomic absorption instrument in the measurement of calcium:

CONCENTRATION OF Ca (ppm)	ABSORBANCE (AU)
0.100	0.010
0.250	0.024
0.500	0.069
1.000	0.093
2.500	0.225
5.000	0.427
7.500	0.628
10.00	0.804

A urine sample was treated to remove interferences, resulting in a dilution factor of 5:2 of the original urine. The mean absorbance for calcium of three replicates of the diluted urine sample was found to be 0.325. Assuming that the intercept is zero for the calcium calibration, calculate the concentration of calcium in the original (undiluted) urine solution. Include a calculation of the standard deviation of your analysis.

If you do not need a full statistical analysis of your calibration curve and simply want the y = mx + b form of the equation and the R^2 value, Excel also offers a feature called *Add Trendline.* The Add Trendline feature is accessed by "right clicking" on the x–y scatter plot of the calibration data. The next activity demonstrates the Add Trendline feature.

22.6 Limit of Detection, Limit of Quantitation, and Linear Dynamic Range

As noted in Section 22.2 we must expect the presence of random error (noise) in every measurement. Sometimes that noise is clearly visible, but other times it is not obvious. This realization makes it necessary to ask the question: at what point can I trust that my measurement is real—and not just noise? Fortunately, statistically sound tools have been developed to help us make that judgment.

The *limit of detection* (LOD) is the lowest value measurable above the background noise. At the LOD, we can be confident that we are measuring *some* analyte, but we cannot be confident about the actual amount. The *limit of quantitation* (LOQ) is the minimum value at which we can be confident in the quantitative value of the measurement. The IUPAC[20] has demonstrated the following for any given analytical method:

$$LOD_y = \overline{y}_{blnk} + k_D \cdot s_{blnk}$$ **Eq. 22.17**

where

   LOD_y = limit of detection of the measurement (y value)

   $\overline{y}_{blnk}$ = mean y values of a set of blank or baseline measurements

   s_{blnk} = standard deviation of a set of blank or baseline measurements

   k_D = multiplicative factor $(k_D = 3$ at 95% CL)

---

[20] Long, G. L.; Winefordner, J. D. *Anal. Chem.,* **55**, 1983, 712–724.

ACTIVITY

Using Microsoft Excel to Add a Trend Line to a Data Set

Using the same data set from our previous activity, create an x–y scatter plot as shown here. Highlight the x–y points in the scatter plot and "right click." If using an Apple operating system, hold down the Apple key and click. A dialog box will open. Select Add Trendline. The Format Trendline dialog box will open (see figures). If you expect your instrument response to be linear, then select the linear radial button. Then select Display Equation on Chart and Display R-squared value on chart. Hit Return. Your graph should now resemble the one seen on the bottom.

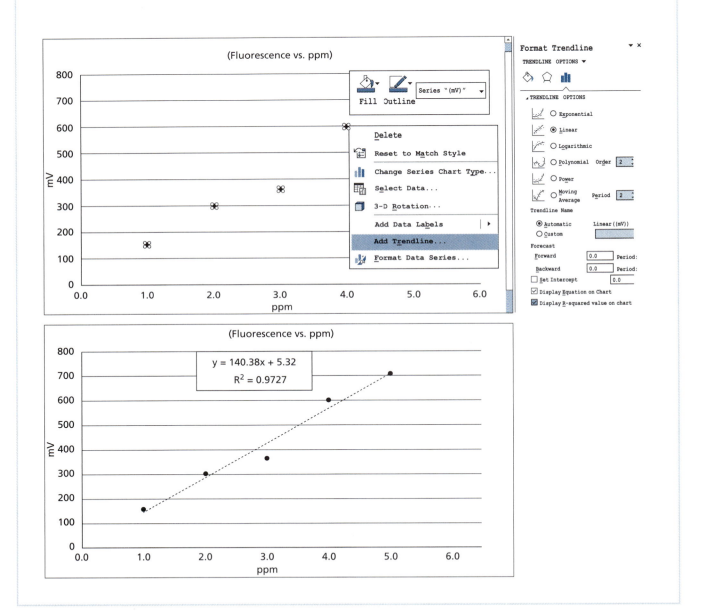

$$LOQ_y = \overline{y}_{blnk} + k_Q \cdot s_{blnk}$$ **Eq. 22.18**

where

   LOQ_y = limit of quantitation of the measurement (y value)

   $\overline{y}_{blnk}$ and s_{blnk} are as above

   k_Q = multiplicative factor (10 for 10% relative standard deviation [RSD] and 20 for 5% RSD)

Note that Equations 22.17 and 22.18 will be relevant for any type of measurement, even if a calibration plot is not used, so long as we can obtain a reasonable estimate of the mean blank signal and its standard deviation.

   In many cases involving calibration plots, it is more desirable to think about LOD and LOQ in terms of actual concentrations (x value) rather than the measured quantity (y value). Because we know the relationship between x and y for a linear relationship (y = mx + b), we can derive expressions that give us LOD and LOQ in terms of concentration. Note that in most instrumental methods, the instrument will be set to a measurement of zero using a blank solution, so we can assume that in the absence of significant drift, the intercept (b) is equal to the average blank measurement, $\overline{y}_{blnk}$, or

$$y = mx + \overline{y}_{blnk}$$ **Eq. 22.19**

The y in an LOD (or LOQ) calculation is the LOD_y (or LOQ_y) from Equation 22.17 (or 22.18), and the x would be the LOD_x (or LOQ_x), which is the LOD in terms of concentration:

$$LOD_y = m \cdot LOD_x + \overline{y}_{blnk}$$ **Eq. 22.20**

   If we combine Equations 22.17 and 22.20 (or 22.18 and the LOQ equivalent of 22.20), we can find our equations for LOD and LOQ in terms of concentration (Equations 22.21 and 22.22):

$$LOD_x = \frac{k_D \cdot s_{blnk}}{m}$$ **Eq. 22.21**

where

   LOD_x = LOD of the concentration
   s_b = standard deviation of a set of blank or baseline measurements
   k_D = multiplicative factor (3 at the 95% CL)

$$LOQ_x = \frac{k_Q \cdot s_{blnk}}{m}$$ **Eq. 22.22**

where

   LOQ_x = LOD of the concentration
   s_b = standard deviation of a set of blank or baseline measurements
   k_Q = multiplicative factor (10 for 10% RSD and 20 for 5% RSD)

Problem 22.18: Demonstrate that Equations 22.21 and 22.22 can be derived from Equations 22.17 and 22.18.

Problem 22.19: Demonstrate that the units of LOD_x (and thus LOQ_x) are concentration units. Assume that the concentration is in units of molarity and that the measurements are made in milliamps from an arbitrary detector.

Problem 22.20: In the experiment represented in Problem 22.17, eight blank measurements (AU) were made: 0.001, 0.000, 0.000, 0.001, 0.002, –0.001, 0.000, and –0.001. Calculate the LOD_y, LOQ_y, LOD_x, and LOQ_x.

Problem 22.21: Consider your results from Problem 22.19 and the data presented in Problem 22.17. If you were presenting these data for publication, would you need to redo the calculations you did in Problem 22.17? Explain.

It is often the case that we do not have available multiple blank measurements for a method involving a calibration plot. In such a case, two alternatives have been proffered. If, in doing the linear regression, an intercept is calculated, then we can substitute s_b (standard deviation of the intercept) for s_{blnk} in Eq. 22.21 and Eq. 22.22. If we set the intercept to zero in the linear regression, we can use the s_y (standard deviation of the y-estimate) in place of s_{blnk}. In both cases, we must expect that our estimates for LOD and LOQ will be higher than they would have been using the formal approach, but it is better to err on the side of caution.

Problem 22.22: Estimate LOD_x and LOQ_x for the data given in Spreadsheet 22.4.

Problem 22.23: Estimate LOD_x and LOQ_x for the data given in Problem 22.15 (zero intercept).

Although we are usually most concerned with knowing the lowest concentration our analytical method will allow us to measure, it is also important to keep in mind that a calibration can fail if the concentration is too high. This can happen for a variety of reasons, many of which are discussed in Chapter 6 when we looked at the upper concentration limits for UV-vis spectroscopy. The *upper limit of quantitation* ($ULOQ$) is indicated by a downward trend of the calibration curve as the concentration increases, as seen in the final three points in Figure 22.14. To estimate the $ULOQ$, we need to determine the intersection of the two lines through the linear portion at lower concentrations and the points past the point when linearity fails (Figure 22.15). The *linear dynamic range* (LDR) is the range of concentrations over which a method exhibits continuous linearity. The LDR is essentially the range of concentrations from the LOQ_x to $ULOQ_x$.

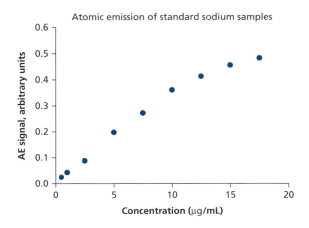

Figure 22.14 Demonstration of a calibration plot with an upper linearity limit.

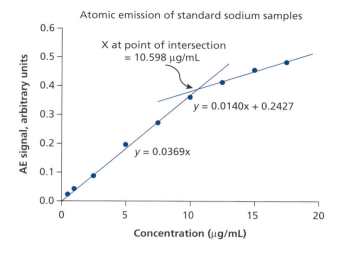

Figure 22.15 Determination of the ULOQ$_x$ for a calibration plot that loses linearity at higher concentrations.

Problem 22.24: The calibration depicted in Figures 22.14 and 22.15 was created from the following data. Calculate the LOQ$_x$ and estimate the LDR.

Concentration (µg/mL)	0.50	1.00	2.50	5.0	7.5	10.0	12.5	15.0	17.5
Atomic Emission Signal (arbitrary units)	0.026	0.045	0.090	0.199	0.274	0.363	0.415	0.458	0.485

22.7 Further Reading

BOOKS

Bevington, P.; Robinson, D. K. *Data Reduction and Error Analysis for the Physical Sciences,* 3rd ed. McGraw-Hill Higher Education: New York, 2003.

Bilo, E. J. *Excel for Chemists: A Comprehensive Guide.* John-Wiley & Sons: Hoboken NJ, 2011.

Caria, M. *Measurement Analysis: An Introduction to the Statistical Analysis of Laboratory Data in Physics, Chemistry and the Life Sciences.* Imperial College Press: London, 2000.

de Levie, R. *Advanced Excel for Scientific Data Analysis,* 3rd ed. Atlantic Academic: Orr's Island, ME, 2012.

Hibbert, D. B.; Gooding, J. J. *Data Analysis for Chemistry: An Introductory Guide for Students and Laboratory Scientists.* Oxford University Press: New York, 2006.

Taylor, J. R. *An Introduction to Error Analysis: The Studies of Uncertainties in Data Analysis,* 2nd ed. University Science Books: Sausalito, CA, 1982.

22.8 Exercises

EXERCISE 22.1: In your own words, define the statistical terms *population* and *sample.*

EXERCISE 22.2: In your own words, define the statistical terms *gross error, systematic error,* and *random error.*

EXERCISE 22.3: In your own words, define and give equations for the statistical terms *mean, standard deviation, variance, standard error,* and *CL.*

EXERCISE 22.4: Input the following data set into a spreadsheet and use the spreadsheet to determine the mean, standard deviation, and the 95% CL of the data set.

Replicate	1	2	3	4	5	6	7	8	9
Pb (ppm)	5.4	2.9	5.1	4.2	5.6	4.7	7.9	4.8	7.6

EXERCISE 22.5: Plot the results of Problem 22.15 as a column graph with standard error bars. Repeat the exercise using standard deviation error bars.

EXERCISE 22.6: Using the statistical functions on your calculator, input the data set from Problem 22.17 and determine the mean, standard deviation, and the 95% CL of the data set.

EXERCISE 22.7: Input the following data set into a spreadsheet and perform a Q-test and a Grubbs' test on the data set. Report the "best" mean, standard deviation, and the 95% CL of the data set.

Replicate	1	2	3	4	5	6	7	8	9
Cu (ppm)	6.54	6.80	6.99	6.54	6.75	6.56	6.75	6.72	6.80

EXERCISE 22.8: Plot the results of Problem 22.7 as a column graph with standard error bars. Repeat the exercise using standard deviation error bars.

EXERCISE 22.9: Input the following data set into a spreadsheet and perform a Q-test and a Grubbs' test on the data set. Report the "best" mean, standard deviation, and the 95% CL of the data set.

Replicate	1	2	3	4	5	6	7	8	9
Fe (ppm)	3.3	3.6	3.8	3.8	3.4	2.1	3.7	4.3	3.6

EXERCISE 22.10: Plot the results of Problem 22.9 as a column graph with standard error bars. Repeat the exercise using standard deviation error bars.

EXERCISE 22.11: Water samples were sent to two different laboratories and analyzed for ppb mercury. Each laboratory conducted 10 replicate analyses. Perform an F-test on the data and comment on the results.

Replicate	1	2	3	4	5	6	7	8	9	10
Hg (ppb) in Lab 1	12.4	12.2	12.2	12.4	12.1	12.4	12.4	12.4	12.2	12.4
Hg (ppb) in Lab 2	12.6	11.6	9.1	13.9	9.8	13.4	14.0	11.3	10.0	11.5

EXERCISE 22.12: Describe the relationships between accuracy and precision with respect to the concepts of systematic error and random error.

EXERCISE 22.13: Use Equation 22.14 to calculate SS_{y-y} for the example given in Problem 22.15.

EXERCISE 22.14: Estimate LOD_x and LOQ_x for the data given in Problem 22.15 with a nonzero intercept.

EXERCISE 22.15: Estimate LOD_y and LOD_x for the data given in Problem 22.24 (zero intercept).

Advanced Exercises

EXERCISE 22.16: Open a spreadsheet and create a number string from 1 to 5,000. Program the spreadsheet to generate and graph npf with a mean of 2,500 and a standard deviation of 500. Properly label your graph and insert it into the spreadsheet. Print the spreadsheet and turn it in.

EXERCISE 22.17: Using the membrane filtration method, a sample of water from a local lake was sent to two different laboratories for a coliform bacteria analysis. Each laboratory tested five replicates and reported total coliform colony counts. The data are presented. Perform an analysis of the data, including a Grubbs' test and an F-test. Comment on the data and report the results of the analysis.

REPLICATE	LAB 1	LAB 2
1	37	35
2	35	37
3	41	35
4	34	35
5	39	27

EXERCISE 22.18: You have just standardized a batch of NaOH by titration against phthalic acid (MW 166.14 g/mol). The phthalic acid standard was made by dissolving 2.0312 ± 0.0001 g of phthalic acid into a 500 ± 0.2 mL volumetric flask. A burette was used to deliver 27.51 ± 0.02 ml of phthalic acid into 250 ± 0.2 mL of NaOH solution. Being mindful to propagate error through your calculations, determine the molarity and $\pm\sigma$ of the NaOH solution.

EXERCISE 22.19: Using the membrane filtration method, a sample of water from a local lake was sent to five different laboratories for a coliform bacteria analysis. Each laboratory tested five replicates and reported total coliform colony counts. The data are presented. Perform an analysis of the data, including a Grubbs' test and an ANOVA. Comment on the data and report the results of the analysis.

REPLICATE	LAB 1	LAB 2	LAB 3	LAB 4
1	12.4	12.6	13.3	11.9
2	12.2	11.6	7.9	11.8
3	12.2	9.1	11.4	11.6
4	12.4	13.9	17.0	11.8
5	12.1	9.8	10.6	12.0
6	12.4	13.4	7.7	11.6
7	12.4	14.0	15.5	11.9
8	12.4	11.3	13.6	11.9
9	12.2	10.0	14.7	11.8
10	12.4	11.5	6.7	12.1

Appendix

Table of Abbreviations and Acronyms

ABBREVIATION	DEFINITION
A	Absorbance
a	Selectivity factor
AAS	Atomic absorption spectroscopy
AC	Alternating current
ADC	Analog-to-digital conversion
AE	Auger electron
AED	Atomic emission detector
AES	Atomic emission spectroscopy
AFM	Atomic force microscopy
Ag/AgCl	Silver/silver chloride (reference electrode)
AMF	Atomic force microscopy
AMS	Accelerator mass spectrometry
ANOVA	Analysis of variance
APCI	Atmospheric pressure chemical ionization
API	Atmospheric pressure ionization
AS	Auger electron spectroscopy
ASTM	American Society for Testing and Materials
ATR	Attenuated total reflection
AU	Absorbance units (sometimes arbitrary units)
b	Intercept in an linear equation
b	Diffraction angle
bis	N,N'-methylenebisacrylamide
BJT	Bipolar junction–transistor
B_o or B	External magnetic field
^{13}C NMR	Carbon-13 nuclear magnetic resonance (spectroscopy)
C	Coulombs
C	Conductivity
CA	Chronoamperometry
CCD	Charge coupled device
CD	Compact disk
CE	Capillary electrophoresis
CE-MS	Capillary electrophoresis–mass spectrometry
CFP	Cyan fluorescent protein

ABBREVIATION	DEFINITION
CGM	Continuous glucose monitors
CI	Chemical ionization
CI	Confidence interval
CL	Confidence limit
cm^{-1}	Wave numbers or reciprocal centimeters
CSP	Chiral stationary phase
CTAB	Cetyltrimethylammonium bromide
CV	Cyclic voltammetry
d	Distance between excitation plates in an FTICR
D	Dispersion (linear)
DAC	Digital-to-analog conversion
D_b	Dispersion (angular)
DC	Direct current
DCP	Direct current plasma
df	Degrees of freedom
D_m	Diffusion coefficient in the mobile phase
DMA	Dynamic mechanical analysis
DMF	Dimethylformamide
DMSO	Dimethylsulfoxide
DMTA	Dynamic material thermal analysis
DNA	Deoxyribonucleic acid
DO	Dissolved oxygen
DPK	Di-2-pyridyl ketone
DSC	Differential scanning calorimetry
DTA	Differential thermal analysis
DTGS	Deuterated trigylcine sulfate
e	Fundamental charge of an electron
E	Potential (voltage)
E	Electric field strength
e/h	Electron-hole pair
E^0	Standard potential
$E^{0\prime}$	Formal potential
$E_{1/2}$	Half-wave potential in CV
ECD	Electron capture detector
E_{cell}	Cell potential
EDL	Electrodeless discharge lamp
EDXRF	Energy dispersive X-ray fluorescence spectrophotometer
EI	Electron ionization or electron impact ionization
$E_{indicator}$	Potential of the indicator electrode
ELCD	Electrochemical detector
EMR	Electromagnetic radiation
EMS	Electromagnetic spectrum
E_{obs}	Observed potential
EOF	Electro-osmotic flow
$E_{p,a}$	Potential of the anodic (oxidation) peak in CV
$E_{p,c}$	Potential of the cathodic (reduction) peak in CV
E_{probe}	Probe potential

ABBREVIATION	DEFINITION
$E_{reference}$	Potential of the reference electrode
ESEM	Environmental scanning electron microscopy
ESI	Electrospray ionization
F	Faraday's constant
F	Farad (base unit)
f	Friction coefficient (electrophoresis)
$f_{(x)}$	Normal probability distribution function
FAAS	Flame atomic absorption spectroscopy
FAD	Flavin adenine dinucleotide
$FADH_2$	Reduced form of flavin adenine dinucleotide
FAMEs	Fatty acid methyl esters
F_B	Force experienced by an ion in a magnetic field
F_c	Average carrier gas velocity (gas chromatography)
F_C	Centripetal force (mass spectrometry)
f_c	Precession frequency of ions in an FTICR
F_{crit}	Critical F value used in a statistical F-test
FDA	Food and Drug Administration
F_E	Force experienced by an ion in an electrostatic field
$Fe(Cp)_2$	Ferrocene
FET	Field effect transistor
F_{exp}	Calculated F value, determined from experimental data; used in a statistical F-test
$f_{H/M}$	Selectivity coefficient of proton over a given metal ion (M)
FID	Free induction decay
FID	Flame ionization detector
fMRI	Functional magnetic resonance imaging
FOV	Field of view
FPD	Flame photometric detector
FRET	Fluorescence (or Forster) resonance energy transfer spectroscopy
FSOT	Fused silica open tubular (column)
FT	Fourier transform
FTICR	Fourier transform ion cyclotron resonance mass analyzer
FTIR	Fourier transform infrared (spectroscopy)
FT-NMR	Fourier transform nuclear magnetic resonance (spectroscopy)
FT-Raman	Fourier transform Raman (spectroscopy)
FZP	Fresnel zone plate
G	Conductance (potentiometry and probes)
γ	Gyromagnetic ratio
GC	Gas chromatography
GC-FTIR	Gas chromatography–Fourier transform infrared spectroscopy
GC-MS	Gas chromatography–mass spectrometry
G_{crit}	Critical G value used in a statistical Grubbs' test
G_{exp}	Calculated G value, determined from experimental data, used in a statistical Grubbs' test
GFAAS	Graphite furnace atomic absorption spectroscopy
GFC	Gel filtration chromatography

ABBREVIATION	DEFINITION
GLC	Gas-liquid chromatography
GO_x	Glucose oxidase
GPC	Gel permeation chromatography
GSC	Gas-solid chromatography
^{1}H NMR	Proton nuclear magnetic resonance (spectroscopy)
H (also HETP)	Height equivalent to a theoretical plate
HCL	Hollow-cathode lamp
HIC	Hydrophobic interaction chromatography
HOMO	Highest occupied molecular orbital
HPLC	High-performance liquid chromatography (also high-pressure liquid chromatography)
Hz	Hertz
hv	Planck's constant times frequency (also a common way to indicate a photon)
I	Direct current (electronics)
$i_{(t)}$	Current measured at time (t) in a chronoamperometry experiment
ICP	Inductively coupled plasma
ICP-AES	Inductively coupled plasma–atomic emission spectroscopy
ICP-MS	Inductively coupled plasma– mass spectrometry
ICR	Ion cyclotron resonance (mass spectrometry)
ICs	Integrated circuits
IEC	Ion exchange chromatography
$i_{forward}$	Current measured on the forward pulse in SWV
i_p	Current measured at CV peak
$i_{reverse}$	Current measured on the reverse pulse in SWV
IRMS	Isotope ratio mass spectrometry
ISE	Ion-selective electrode
J	Joules
j	Gas compressibility correction factor
J_k	Flux density of species k to an electrode
k	Boltzmann constant
K	Equilibrium constant for a reaction
K_α	X-ray emission that results from an electron that fell from one shell above the K-shell
K_β	X-ray emission that results from an electron that fell from two shells above the K-shell
K_{cf}	Capacity factor
K_D	Distribution Coefficient
KE	Kinetic energy
K_{ow}	Octanol:water distribution constant
K_p	Partition coefficient
k_R	Retention factor
k_s	Standard heterogeneous rate constant for a redox reaction at an electrode
L	Mean free path (mass spectrometry)
L	Length
Λ	Molar conductivity (capillary electrophoresis)

ABBREVIATION	DEFINITION
LASER	Light amplification by stimulated emission of radiation
LC	Liquid chromatography
LC-MS	Liquid chromatography—mass spectrometry
LDR	Linear dynamic range
LED	Light-emitting diode
LIBS	Laser-induced breakdown spectroscopy
LOD	Limit of detection
LOQ	Limit of quantitation
LUMO	Lowest unoccupied molecular orbital
M	Analyte molecule (mass spectrometry)
M	Molarity
m	Mass
m	Slope in a linear equation
MO	Molecular orbital
m/z	Mass-to-charge ratio
MALDI	Matrix-assisted laser desorption ionization (mass spectrometry)
MCT	Mercury cadmium and telluride
MDGC	Multidimensional gas chromatography
MECK	Micellar electrokinetic capillary chromatography
μ_{eo}	Electro-osmotic mobility
MHz	Megahertz (1×10^6 s^{-1})
ML_6	Transition metal coordinated to six ligands
M_o	Magnetic moment
MP	Microwave plasma
MP-AES	Microwave plasma–atomic emission spectroscopy
MRI	Magnetic resonance imaging
MS	Mass spectrometry
MS-MS	Dual mass spectrometric tandem method
MS-MS-MS	Triple mass spectrometric tandem method
n	Number of electrons transferred per mole of reactant (in a redox reaction)
n	Principal quantum number
N	Number of measurements in a sample or population
BK7	Borosilicate crown glass
NCL (also NLCL)	Negative ion chemical ionization
Nd-YAG	Neodymium-doped yttrium aluminum garnet; Nd:$Y_3Al_5O_{12}$
NE	Nanoelectrode
NHE	Normal hydrogen electrode
NICI (also NCI)	Negative ion chemical ionization
nm	Nanometer
NMR	Nuclear magnetic resonance (spectroscopy)
NPN	A BJT constructed of an n-type/p-type/n-type semiconductor
N_t	Number of theoretical plates
OD	Optical density
Ox	Oxidized form of a reactant in a redox reaction
^{31}P NMR	Phosphorus-31 nuclear magnetic resonance (spectroscopy)
P	Pressure
p	Electron momentum (SEM)

ABBREVIATION	DEFINITION
PA	Polyacrylamide
PAGE	Polyacrylamide gel electrophoresis
PFGE	Pulsed field gel electrophoresis
P_i	Inlet gas pressure (gas chromatography)
pI	pH at which a protein becomes neutral
PID	Photoionization detector
PLOT	Porous lined open tubular (column)
PMT	Photomultiplier tube
PNP	A BJT constructed of a p-type/n-type/p-type semiconductor
ppb	Parts per billion
ppm	Parts per million
PSU	Practical salinity units
PTFE	Polytetrafluoroethylene (Teflon)
PVC	Polyvinyl chloride
PVDF	Polyvinylidene fluoride
PZT	Lead zirconium titanate
q	Charge on an ion (electrophoresis)
Q or Q_{eq}	Reaction quotient for a reaction not at equilibrium
Q_{crit}	Critical Q value used in a statistical Q-test
Q_{exp}	Calculated Q value, determined from experimental data, used in a statistical Q-test
QIT	Quadrupole ion trap
QSPR	Quantitative structure property relationship
R	Radius of an arc (mass spectrometry)
R	Gas law constant
R	Resistor (electronics)
r	Hydrodynamic volume (electrophoresis)
R^2	Coefficient of determination
RC	Product of a resistor's value times a capacitor's
r_c	Orbital radius for ions in an FTICR
Red	Reduced form of a reactant in a redox reaction
RET	Resonance energy transfer spectroscopy
RF or rf	Radiofrequency
RNA	Ribonucleic acid
RPC	Reversed phase chromatography
R_s	Resolution
S	Total salinity
s	Standard deviation of the sample
SE	Standard error
S/N	Signal-to-noise ratio
S_0	Singlet ground state
S_1	First excited singlet state
SAX	Strong anion exchange chromatography
s_{blnk}	Standard deviation of the blank measurements
s_c	Standard deviation of an x value calculated from a y measurement using a linear regression equation
SCE	Saturated (standard) calomel reference electrode

ABBREVIATION	DEFINITION
SCOT	Solid coated open tubular (column)
SCX	Strong cation exchange chromatography
SDD	Silicon drift detector
SDS PAGE	Sodium dodecyl sulfate polyacrylamide gel electrophoresis
SE	Supporting electrolyte
SEC	Size exclusion chromatography
SED	Secondary electron detection
SEM	Scanning electron microscopy
SHE	Standard hydrogen electrode
SIMS	Secondary ion mass spectrometry
SS_{y-y}	Sum of the squared residuals
STA	Simultaneous thermal analysis
STM	Scanning tunneling microscopy
SWV	Square wave voltammetry
S_{x-x}	Sum of the squared difference from the mean of x-values
s_y	Standard error in the y-estimate (Excel LINEST function)
T	Temperature (usually in Kelvin)
t	t-distribution value for normal distribution for a given value of degrees of freedom
T	Transmittance (optical spectroscopy)
T_1	Longitudinal (or spin-lattice) relaxation (NMR)
T_1	First excited triplet state (electronic excitation)
T_2	Transverse relaxation (NMR)
T_2	Second excited triplet state (electronic excitation)
TAC	Total available chlorine
TCD	Thermal conductivity detector
TEM	Transmission electron microscopy
t_{excite}	Excitation pulse time in an FTICR
TGA	Thermogravimetric analysis
TGA-FTIR	Thermogravimetric analysis–Fourier transform infrared spectroscopy
TGA-MS	Thermogravimetric analysis–mass spectrometry
THF	Tetrahydrofuran
TID	Thermionic ionization detector
TIMS	Thermal ionization mass spectrometry
TLC	Thin-layer chromatography
t_m	Void time
TMS	Tetramethylsilane
TOF	Time of flight
t_r	Retention time
TXRF	Total reflection X-ray fluorescence spectrometer
UME	Ultramicroelectrode
UPLC or UHPLC	Ultra-high performance liquid chromatography
UV-Vis	Ultraviolet-visible (refers to molecular spectroscopic methods and instruments)
V	Voltage (mass spectrometry)
V	Tip bias potential (STM)

ABBREVIATION	DEFINITION
V	Volt (base unit)
VBT	Valence bond theory
V_N	Net adjusted retention volume (gas chromatography)
V_{p-p}	Amplitude of the rf signal in an FTICR
V_r	Phase volume ratio (liquid chromatography)
V_r	Retention volume (gas chromatography)
V_R°	Adjusted retention volume (gas chromatography)
W	Watt
ω	Larmor frequency
WAX	Weak anion exchange chromatography
W_b	Peak width at baseline
WCOT	Wall coated open tubular (columns)
WCX	Weak cation exchange chromatography
WDXRF	Wavelength dispersive X-ray fluorescence spectrophotometer
W_h	Peak width at half height
XPS	X-ray induced photoelectron spectroscopy
x_q	Questioned data value
XRF	X-ray fluorescence spectroscopy
YFP	Yellow fluorescent protein
$y_{i,act}$	Actual y value from experimental data set
$y_{i,calc}$	y value calculated from the linear regression equation
YSZ	Yttrium-stabilized zirconia
z	Charge on an ion (mass spectrometry)
Z	Atomic number
Z	Impedance
ζ	Zeta potential
γ	Deformation or strain (mechanical stress analysis)
ΔG	Gibbs' free energy change for a reaction
ΔG^0	Standard change in free energy for a reaction
ΔH	Enthalpy change
ε	Molar absorptivity (optical spectroscopy)
ε	Dielectric constant or relative permittivity (electrophoresis)
η	Viscosity of a liquid
λ	Wavelength
μ	Mean of the population
μ_{app}	Apparent mobility
μ_{ep}	Electrophoretic mobility
v	Linear flow velocity (chromatography)
v	Sweep scan rate (electrochemistry)
v_{app}	Apparent velocity
v_{eo}	Electro-osmotic velocity
σ	Standard deviation of the population
σ	Applied stress (mechanical stress analysis)
σ_{Si}	Charge density of the capillary wall (capillary electrophoresis)
ϕ	Work function of the surface (STM)
$\bar{X}$	Mean of the sample

Credits

Chapter 1

1.3: Reprinted from Amanda L. Clark, Kathryn Mansfield Matera. "Effect of unsaturation in fatty acids on the binding and oxidation by myeloperoxidase: Ramifications for the initiation of atherosclerosis," Bioorg. Med. Chem. (2010) 20 (19) Pages 5643–5648, with permission from Elsevier.; **1.13:** Reprinted (adapted) with permission from Huerta-Fontela, M., Galceran, M. T., Ventura, F., "Ultraperformance Liquid Chromatography–Tandem Mass Spectrometry Analysis of Stimulatory Drugs of Abuse in Wastewater and Surface Waters," Anal. Chem., 2007, 79 (10), 3821–3829. Copyright (YEAR) American Chemical Society.; **1.15:** DeFrancesco, J. V., et al., "GHB Free Acid: I. Solution Formation Studies and Spectroscopic Characterization by 1HNMR and FT-IR." J. Forensic Sci., 2006, 51(2), pp. 321–329; **1.16:** DeFrancesco, J. V., et al., "GHB Free Acid: I. Solution Formation Studies and Spectroscopic Characterization by 1HNMR and FT-IR." J. Forensic Sci., 2006, 51(2), pp. 321–329.

Chapter 3

3.1A: GIPhotoStock/PhotoResearchers; **3.1B:** E.R. Degginger/PhotoResearchers; **3.1c:** GIPhotoStock/Getty Images; **3.3:** Friedrich Saurer / Science Source; **3.8B:** Benjamin Crowell (Fullerton College). University of California, Davis material licensed under a Creative Commons Attribution-Noncommercial-Share Alike 3.0 United States License.; **3.8C:** Wikimedia Commons/Dicklyon; **3.9A:** GIPhotoStock/Photo Researchers; **3.9C:** GIPhotoStock/Photo Researchers; **3.10A:** GIPhotoStock/Photo Researchers; **3.10B:** Charles Winters/PhotoResearchers; **3.12A:** Sundance Solar; **3.13:** GIPhotoStock/PhotoResearchers; **3.26A:** T.J. Nelson; **3.26B:** Data collected by Tianqi Zhang in the research group of Martin Zanni at the University of Wisconsin-Madison.; **Profile, Optics that Operate by Diffraction:** Scimat/PhotoResearchers; Getty/GIPhotoStock; **Exercise 3.13:** SPL/Photoresearchers.

Chapter 4

Profile, Alessandro Volta: Science Source/Jessica Wilson; **4.1A:** GIPhotoStock/PhotoResearchers; **4.1B:** PhotoResearchers/Science Source; **4.1C:** GIPhotoStock/PhotoResearchers; **4.2:** X-RAY pictures/Shutterstock.com; Krasowit/Shutterstock.com; WachiraS/Shutterstock.com; James Hoenstine/Shutterstock.com; Mihancea Petru/Shutterstock.com; NaMaKuKi/Shutterstock.com; Timothy Hodgkinson/Shutterstock.com; Antoine Beyeler/Shutterstock.com; Feng Yu/Shutterstock.com; Mrs_ya/Shutterstock.com; airobody/Shutterstock.com; ekapotfoto-thai/Shutterstock.com; Mr. Klein/Shutterstock.com

Chapter 5

5.1A: Moerner, W.E. and Kador, L., "Optical detection and spectroscopy of single molecules in a solid," Phys. Rev. Lett., (1989) 62 (21), 2535–2538.; **5.1B:** Kartalov, E.P., "Single Molecule Detection and DNA Sequencing-by-Synthesis", Ph.D. Dissertation, California Institute of Technology, 2004, 31.

Chapter 6

Profile, James Clerk Maxwell: PhotoResearchers; **Profile, HACH DR3900:** BDR3900 Benchtop Spectrophotometer © Hach Comapany, 2016; **6.21:** Elavazhagan T, Arunachalam KD, 2011. "Memecylon edule leaf extract mediated green synthesis of silver and gold nanoparticles." Int J Nanomed 6 p. 1265–1278; **6.26** CC-BY-A 2.5; **6.27B:** GIPhotoStock/PhotoResearchers; **Profile, Walter Hermann Schottky:** Deutsches Museum, Munich.

Chapter 7

Profile, Lightning Over Salty Waters: NOAA, National Oceanic and Atmospheric Administration; **7.16:** Environmental Express HotBlock™; **7.17:** CC-BY-SA 2.0 Jeff Dahl.

Chapter 8

8.1: S Moser et al, Angew. Chem. Int. Ed., 2008, DOI: 10.1002/anie.200803189; Steven Haddock • MBARI; Abronsteinwiki/Wikimedia Commons; **Profile, Is Your $100 Bill Real?:** Chia, T.; Levene, M., Detection of counterfeit U.S. paper money using intrinsic fluorescence lifetime, Opt. Express 2009, 17, 22054 –22061.: http://www.yaleseas.com/levenelab/publications/Chia%20and%20Levene%20Optics%20Express%202009.pdf; **8.7** Benjamin T. Wigton, Balwant S. Chohan, Cole McDonald, Matt Johnson, Doug

Chem. Educator 7(1) 15–18.; **12.30:** Reprinted with permission from Bernhard Zachhuber, Christoph Gasser, Engelene t.H. Chrysostom, and Bernhard Lendl, "Stand-Off Spatial Offset Raman Spectroscopy for the Detection of Concealed Content in Distant Objects," Anal. Chem. 2011, 83, 9438–9442. Copyright 2015 American Chemical Society."

Chapter 13

13.1: Getty / GIPhotoStock; **13.2:** NIST Chemistry Webbook, http://webbook.nist.gov/chemistry; **Profile, Joseph John Thomson:** Portrait of Joseph Thomson from the book 'From Immigrant to Inventor' by Michael Pupin, New York : Scribner's sons, 1927, c1923.; **13.5:** NIST Chemistry Webbook, http://webbook.nist.gov/chemistry; **13.7:** NIST Chemistry Webbook, http://webbook.nist.gov/chemistry; **13.10:** QB3/Chemistry Mass Spectrometry Facility at UC Berkeley; **13.11:** University of Bristol, Chemistry Department, Paul Gates; **13.15:** ©2014 PerkinElmer, Inc. All rights reserved. Printed with permission; **Profile, Graham Cooks:** Photo Credit: Xin Yan; **Profile, Eugene Goldstein:** AIP Emilio Segre Visual Archives, Lande Collection; **13.21:** Image from http://what-when-how.com/proteomics/ft-icr-proteomics/; **13.22:** Figure 3 on http://what-when-how.com/proteomics/ft-icr-proteomics/; **13.24:** Image courtesy of Shimadzu Corporation; Main figure on http://www.shimadzu.com/an/structure_electron.html; **13.25:** Laboratoire Suisse d'Analyse du Dopage; **13.29:** Figure courtesy of Mike Christie; **Profile, Human Scent Fingerprinting:** CC-BY-A / Curran, AM, Prada, PA, Furton, KG., "The Differentiation of the Volatile Organic Signatures of Individuals Through SPME-GC/MS of Characteristic Human Scent Compounds," J. Forensic Sci., (2010) 55(1), 50–57. doi: 10.1111/j.1556-4029.2009.01236.

Chapter 14

14.4: Noboru Ishiyama; **Profile, Richard R. Ernst:** University of California, Berkeley, courtesy AIP Emilio Segre Visual Archives; **14.16:** Dr. Richard Shoemaker, Dept. of Chemistry and Biochemistry, University of Colorado. http://chemnmr.colorado.edu/moreinfo/instruments/; **Profile, Angela Gronenborn:** Courtesy of Angela Gronenborn; **14.20:** Courtesy of FONAR Corporation.

Chapter 15

Activity, Chromatography at Home: Andy Crawford and Tim Ridley / Dorling Kindersley / Science Source; **15.10:** Based on Christopher J Fenk, Nicole M. Hickman, Melissa A. Fincke, Douglas H. Motry, and Barry Levine, "Identification and Quatitative Analysis of Acetaminophen, Acetylsalicylic Acid, and Caffeine in Commercial Analgesic Tablets by LC-MS," 2010 J. Chem Ed, 87(8), 838–841. Copyright (YEAR) American Chemical Society.; **15.11:** Based on Ciardullo, et al "Bioaccumulation Potential of Dietary Arsenic, Cadmium, Lead, Mercury, and Selenium

in Organs and Tissues of Rainbow Trout as a Function of Fish Growth," J. Agric. Food Chem. (2008) 56(7) 2442, 2451. Copyright 2015 American Chemical Society; **Profile, Analysis of Wine:** Mohamed Chamkha, Bernard Cathala, Veronique Cheynier, and Roger Douillard, "Phenolic Composition of Champagnes from Chardonnay and Pinot Noir Vintages," Journal of Agricultural and Food Chemistry, Vol 51, 2003, pp 3179–3184.; **Table 15.1:** J. Sangster. J. Phys, Chem. Ref. Data, Vol. 18, No. 3, P. 1111, 1989.; **15.13:** Haralabos C. Karantonis, Smaragdi Antonopoulou, and Constantinos A. Demopoulos, "Antithrombotic Lipid Constituents from Vegetable Oils. Comparison between Olive Oils and others," Journal of Agricultural and Food Chemistry, 2002, Vol. 50, pp 1150–1160.; **15.18:** Waters Coorporation, 2011; **15.23:** Xu Q, van Wezel GP, Chiu H-J, Jaroszewski L, Klock HE, Knuth MW, et al. (2012) Structure of an MmyB-Like Regulator from C. aurantiacus, Member of a New Transcription Factor Family Linked to Antibiotic Metabolism in Actinomycetes. PLoS ONE 7(7): e41359. doi:10.1371/journal.pone.0041359. Copyright: © Xu et al. This is an open-access article distributed under the terms of the Creative Commons Attribution License, which permits unrestricted use, distribution, and reproduction in any medium, provided the original author and source are credited.; **15.30:** Picture by courtesy of FIAlab Instruments, Inc. (www.flowinjection.com); **Exercise 15.3:** Riffat Naseem, Alice Sturdy, David Finch, Thomas Jowitt, Michelle Webb. "Mapping and conformational characterization of the DNA-binding region of the breast cancer susceptibility protein BRCA1." Biochemical Journal (May 2006) 395 (3), 529–535.

Chapter 16

Profile, Gas Chromatography on Mars: public domain via NASA; **16.2:** Sanders, T.H. "Non-detectable Levels of trans-Fatty Acids in Peanut Butter." J. Agric. Food Chem., 2001, 49 (5), pp 2349–2351; **16.4:** Sanders, T.H. "Non-detectable levels of trans-fatty acids in peanut butter." J. Agric. Food Chem. 2001, 49, 2349-2351; **Profile, NIST 14 Gas Chromatography Library with Search Software:** Reproduced by permission of Scientific Instrument Services, Inc, etrieved from: http://www.sisweb.com/software/nist-gc-library.htm; **16.9:** Image courtesy of CTC Analytics AG.; **16.14:** Sanders, T.H. "Non-detectable Levels of trans-Fatty Acids in Peanut Butter." J. Agric. Food Chem., 2001, 49 (5), pp 2349–2351; **16.19B:** Reproduced by permission of Scientific Instrument Services, Inc, retrieved from http://www.sisweb.com/ms/sis/jetrepar.htm; **16.22:** © Agilent Technologies, Inc. 2007.

Chapter 17

Profile, Father of Electrophoresis: Uppsala University Library; **17.5:** Martin Shields/PhotoResearchers; ethylalkohol/Shutterstock.com; **Profile, DNA Markers:** Photocredits: New England BioLabs Inc, https://www.neb.com/products/n0340-lambda-ladder-pfg-marker;

Chapter 18

Chapter 19

Chapter 20

Chapter 21

Index

References to tables and figures are denoted by an italicized *t* and *f*